51st Conference on Engineering Mechanical Design and Stress Analysis (AIAS2022)

AA003690

IOP Conference Series: Materials Science and Engineering
Volume 1275

Padova, Italy
7 – 10 September 2022

ISBN: 978-1-7138-6982-5
ISSN: 1757-8981

Printed from e-media with permission by:

Curran Associates, Inc.
57 Morehouse Lane
Red Hook, NY 12571

Some format issues inherent in the e-media version may also appear in this print version.

This work is licensed under a Creative Commons Attribution 3.0 International Licence.
Licence details: http://creativecommons.org/licenses/by/3.0/.

No changes have been made to the content of these proceedings. There may be changes to pagination and minor adjustments for aesthetics.

Printed with permission by Curran Associates, Inc. (2023)

For permission requests, please contact the Institute of Physics at the address below.

Institute of Physics
Dirac House, Temple Back
Bristol BS1 6BE UK

Phone: 44 1 17 929 7481
Fax: 44 1 17 920 0979

techtracking@iop.org

Additional copies of this publication are available from:

Curran Associates, Inc.
57 Morehouse Lane
Red Hook, NY 12571 USA
Phone: 845-758-0400
Fax: 845-758-2633
Email: curran@proceedings.com
Web: www.proceedings.com

TABLE OF CONTENTS

IOP Conf. Series: Materials Science and Engineering 1275 (2023) 011001 doi:10.1088/1757-899X/1275/1/011001

Editorial

F. Furgiuele[1], N Bonora[2], M. Quaresimin[3], G. Meneghetti[4], P Livieri[5]
[1] University of Calabria, Rende I-87036, Italy
[2] University of Cassino and Southern Lazio, Cassino I-03043, Italy
[3] University of Padova, Vicenza, I-36100, Italy
[4] University of Padova, Padova, I-35131, Italy
[5] University of Ferrara, Ferrara, I-44121, Italy

giovanni.meneghetti@unipd.it

The Italian Association for Stress Analysis (AIAS) was founded in 1971 by researchers from academia, research centers, and industry. AIAS was intended as a community where to discuss, share, and develop scientific knowledge related to all technical aspects of stress analysis. In the years, from an initial focus on experimental techniques, AIAS contributed considerably to the development of modern numerical methods and computational techniques for mechanical engineering design. In 2015, AIAS turned in the Italian Scientific Society of Mechanical Engineering Design.

Today, AIAS is an institutional partner that supports the instances from academia in the subject area of mechanical engineering design. Every year, AIAS organizes a technical conference offering the possibility to present research updates, share new ideas, and foster collaborations. The AIAS conference has become a fundamental event for all those interested in current developments in mechanical engineering design and stress analysis, where to meet researchers, testing equipment, and software developers.

The 51st AIAS Conference edition was held in Padova, Italy again in presence after two years of online-only events due to the COVID-19 pandemic.

The response of researchers and students has been outstanding: over 200 oral contributions have been presented during the three days of the conference, with four parallel sessions. In addition to the thematic sessions on AIAS traditional subjects, special sessions on *additive manufacturing*, *energy methods for structural analysis, circular design in mechanical engineering*, and *mechanical behavior under extreme conditions*, have been successfully organized with the contribution of the AIAS technical committees.

Among all contributions presented at the conference, 48 have been selected to be published after peer review, in this volume. This was made possible thanks to the active participation of all AIAS members, to the work of the AIAS Scientific Committee and Conference Papers Review panel (Profs Giovanni Meneghetti, AIAS Scientific Coordinator, Luciano Afferrante, Francesco Bucchi, Filippo Cianetti, Enrico Armentani, Marco Sasso). Their outstanding contribution is gratefully acknowledged.

Disclaimer

AIAS2022 Conference was held in presence on Sept. 7-10, 2022. Presentations were arranged in four parallel sessions with a time slot of 15 min assigned to each presenter. Approximately 250 participants attended the conference during the three days event. All sessions have been continuously monitored by

Content from this work may be used under the terms of the Creative Commons Attribution 3.0 licence. Any further distribution of this work must maintain attribution to the author(s) and the title of the work, journal citation and DOI.
Published under licence by IOP Publishing Ltd

the organizers to provide technical support. The conference ran smoothly, and the participants' feedback was very positive.

IOP Conf. Series: Materials Science and Engineering 1275 (2023) 011002 doi:10.1088/1757-899X/1275/1/011002

Peer Review Statement

All papers published in this volume have been reviewed through processes administered by the Editors. Reviews were conducted by expert referees to the professional and scientific standards expected of a proceedings journal published by IOP Publishing Publishing.

- **Type of peer review:** Single Anonymous
- **Conference submission management system:** AIAS proprietary abstract management system
- **Number of submissions received:** 194
- **Number of submissions sent for review:** 62
- **Number of submissions accepted:** 47
- **Acceptance Rate (Submissions Accepted / Submissions Received × 100):** 24.2
- **Average number of reviews per paper:** 2
- **Total number of reviewers involved:** 15
- **Contact person for queries:**
  Name: Nicola Bonora
  Email: bonora@unicas.it
  Affiliation: University of Cassino and Southern Lazio - Civil and Mechanical Engineerign

Content from this work may be used under the terms of the Creative Commons Attribution 3.0 licence. Any further distribution of this work must maintain attribution to the author(s) and the title of the work, journal citation and DOI.
Published under licence by IOP Publishing Ltd

IOP Conf. Series: Materials Science and Engineering 1275 (2023) 012001 doi:10.1088/1757-899X/1275/1/012001

Additive manufacturing of AISI 316L specimens with distributed inner bone-type cavities: processability and characterization

C Bregoli[1,2], F Buccino[2], F Picca[2], S Bagherifard[2], C A Biffi[1], A Tuissi[1] and L M Vergani[2]

[1] National Research Council (CNR), ICMATE, Lecco, Italy
[2] Department of Mechanical Engineering, Politecnico di Milano, Milano, Italy

Email: chiara.bregoli@icmate.cnr.it (corresponding author)
Email: federica.buccino@polimi.it
Email: sara.bagherifard@polimi.it
Email: francesca.picca@mail.polimi.it
Email: carloalberto.biffi@cnr.it
Email: ausonio.tuissi@cnr.it
Email: laura.vergani@polimi.it

Abstract. Bone lacunae are cavities the morphology of which strongly affects the damage propagation inside bone. Nevertheless, the role of eventual variations in their morphological features is not clear yet. In this scenario, the work aims at isolating the effects of lacunar-like pores on the mechanical response of 3D printed samples. The research presents a detailed study on the processability of those cavities by means of laser powder bed fusion process, carefully considers both drawbacks of the process, such as the need of heat treatment to minimize the residual stresses, and the limitation of design constraints, such as the presence of metallic powders trapped inside closed cavities. The identification of the optimized heat treatment is permitted both by X-ray diffractometer analysis and morphological examinations by means of optical and micro-CT investigations of cavities. The selected heat treatment is performed on tensile test samples with lacunar-like cavities to progress with a preliminary mechanical static characterization. Future developments will investigate the fracture modality, both under static and fatigue loadings to comprehend how cavities with different morphology influence the damage propagation.

1. Introduction

A detailed comprehension of the bone structure functionality has always been a crucial topic both for medicine and biomedical engineering. Recently, technology has permitted to focus on bone structures at microscale, the role of which is still under debate [1–3].

Bone Lacunae (BL) and canaliculi, which make the lacunar-canalicular network (LCN), are the main bone structures at the microscale [4–7]. BL are closed micro cavities presented in bone inside of which osteocytes are located: osteocytes are the bone cells responsible for the sensing mechanisms of bones

Content from this work may be used under the terms of the Creative Commons Attribution 3.0 licence. Any further distribution of this work must maintain attribution to the author(s) and the title of the work, journal citation and DOI.
Published under licence by IOP Publishing Ltd

with a relevant impact on bone matrix formation [7,8]. BL have double role: indeed, they act both as stress concentrators and propagation deviators affecting both the microscale toughening and weakening mechanisms [1,6,8,9]. Gauthier [10] and Qiu et al. [11] observed the crucial role of BL in crack propagation: it was stated that micro damage and bone lacunae deficiency take place in the same region of bone.

BL morphology, density and alignment change according to age, pathology, anatomical region, cortical and trabecular site [1,12–14]. In fact, van Hoove et al. [14] observed that osteopenic osteocytes are large and round, osteopetrotic osteocytes are small and discoid shape and osteoarthritic osteocytes are large and elongated. Buccino et al. [1] reported that osteoporotic bones present more roundish and oblate lacunae than healthy bones.

Recent researches [6,15] are moving to deepen the BL morphology role in order to isolate the effects of cavities morphology on the mechanical performance of bone, in particular on damage propagation. D. Burr [6] observed that more roundish lacunae are less inclined to stress concentrating effects than spindle shaped lacunae.

The majority of researches are computational-based due to the difficulties related to test human samples [16,17]. The necessity to identify an alternative tool to experimentally study this structures, lead to the use of Additive Manufacturing (AM) technology. Indeed, AM permits the faithful replication of parts with complex shapes which can be experimentally tested.

The work aims at isolating the effects of lacunar-like pores on the mechanical response of 3D printed samples. The realization of such particular closed cavities is possible only by means of AM process. The closed cavities are replicated inside AISI 316L cube 8x8x8mm^3 realized by using Laser Powder Bed Fusion (LPBF) process. The use of LPBF process permits the correct reproducibility of samples and the isolation of the morphological contribute of closed cavities removing material effects.

The use of LPBF process involves some challenges that the study will address in detail, such as the need to analyse the state of metallic powders entrapped inside the closed cavities and the need to remove residual stresses due to the process. Indeed, the presence of residual stresses in additive manufactured (AMed) specimen may negatively affect the failure during testing making the final output unreliable.

For these reasons, the study aims at the identification of the best heat treatment (HT) in terms of residual stress removal without affecting cavities morphology and the powders state inside them. The processability of cavities in AMed samples and the identification of the best HT are the main goals of the research in order to proceed with an adequate mechanical characterization, both under static and fatigue conditions.

Preliminary static mechanical tensile tests are carrying out, and their first results are reported in the paper: these first mechanical tests pave the way to further studies focused on fatigue mechanical tests.

2. Materials and Methods

2.1 Design

To address the preliminary investigation regarding the processability of cavities inside AMed samples, cubes with three cavities which morphology replicated BL morphology are designed in solid modelling CAD software (Solidworks; Dassault Systèmes Solidworks Corp).

It is considered the lacuna shape typical of osteopenic bone referring to the dimensional characterization conducted by van Hove et al. [14], as reported in figure 1.

Three replicas of cavities in figure 1 are inserted in the cube 8mmx8mmx8mm (figure 2). Four replicas of the cubes are realized.

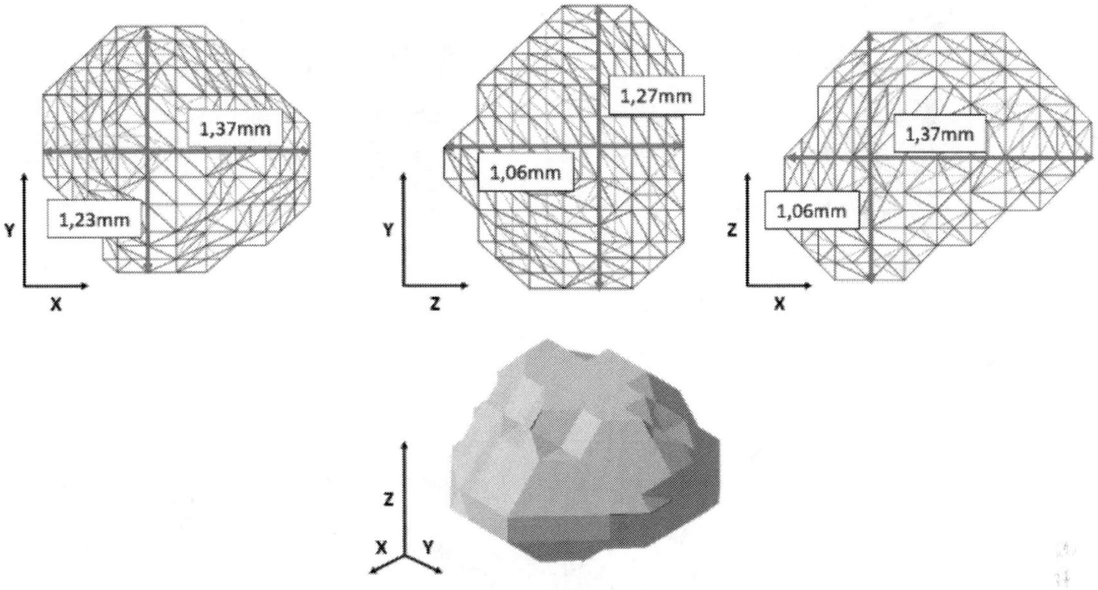

Figure 1. Representation of the cavity in 2D planes and 3D representation of the cavity. The dimension of the cavity refers to the dimensional characterization conducted in [14].

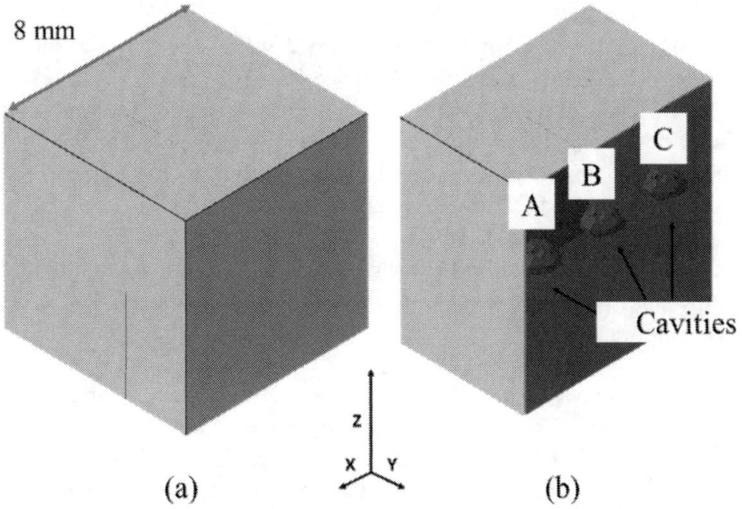

Figure 2. 3D representation of the cube 8x8x8mm^3 (a). Section of the cube and view of the three cavities (cavity "A"," B" and "C"). The cavities have same morphology; indeed, they are replicas of the cavity reported in figure 1 (b).

Once the results on the preliminary analysis of the cube specimens are obtained, tensile test samples including different cavities morphologies and different orientation of cavities are realized to proceed with their mechanical characterization (figure 3). The tensile test specimens were machined: a light milling was made on the surface of the samples to obtain the nominal desired geometry.

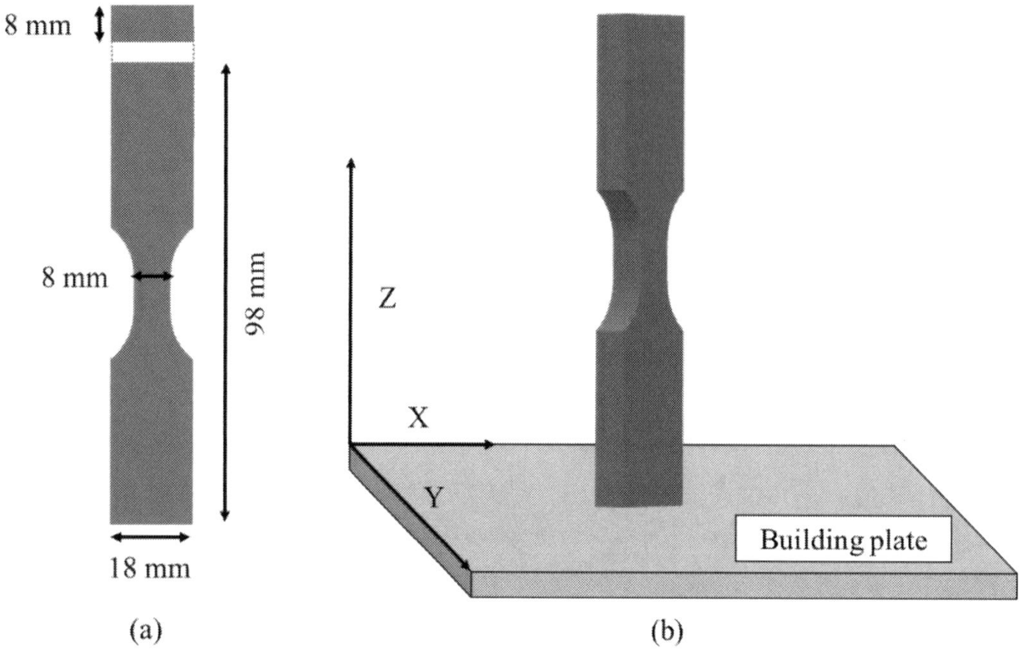

Figure 3. Geometry of the tensile test specimens (a); vertical orientation of the tensile specimen on the building plate (b).

The reference tension samples are designed without any cavities. Two cavities morphologies and two different orientations are considered, as reported in the table 1.

In particular, it is considered the lacuna shape typical of osteopenic (OP) and osteopetrosis (PET) bone referring to the dimensional characterization conducted by van Hove et al. [14]. The distribution of cavities was selected by taking inspiration from the 3D spatial positioning of real bone lacunae in pathological bone conditions such as OP and PET, as indicated in van Hove et al. [14]. PET1 and PET2 differ in the number of cavities, 20 and 13 respectively. PET1_a and PET2_a present all cavities with the same orientation, while PET1_b and PET2_b cavities do not present any alignment.

A transparent rendering of the tension samples was described in figure 4 in order to visualize all CAD cavities in each samples.

Table 1. Names and features of the tensile test samples. OP=tensile test sample with cavity with morphology similar to osteopenic BL. PET= tensile test sample with cavity with morphology similar to osteopetrosis BL. PET1 and PET2 differ by the number of cavities as reported in the second column of the table. "_a" and "_b" characterize aligned disposal and not aligned disposal, respectively.

Name of samples	Number of lacunae	Dimension of lacunae	Orientation
No_cavities	-	-	-
OP	20	Large	Aligned
PET1_a	20	Medium	Aligned
PET1_b	20	Medium	Not aligned
PET2_a	13	Medium	Aligned
PET2_b	13	Medium	Not aligned

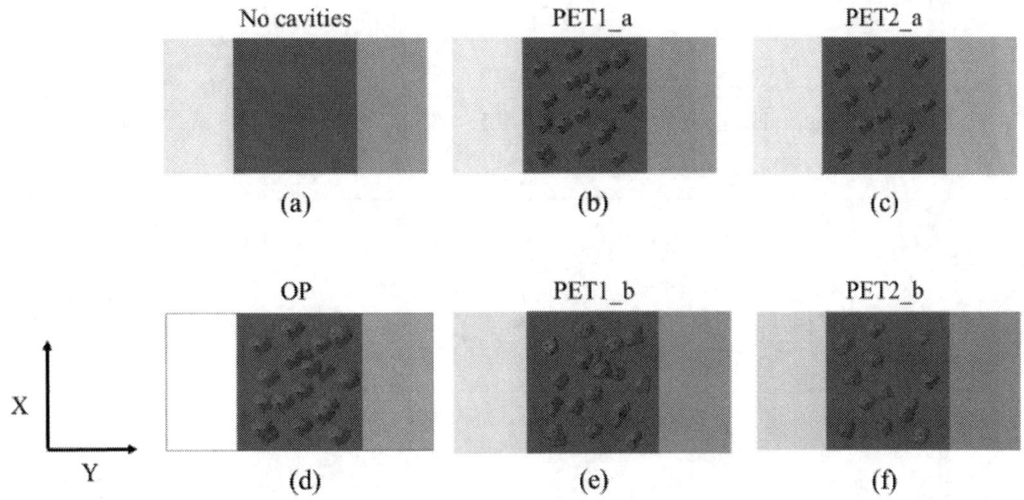

Figure 4. Transparent rendering of tensile test samples: no cavities (a); PET1_a (b); PET2_a (c); OP (d); PET1_b (e); PET2_b (f). The labels of the six tensile test specimens refer to the identification reported in table 1.

2.2 Processability and heat treatments of the samples

A LPBF system (mod. AM400 Renishaw) equipped with a pulsed wave laser was used to produce AISI 316L samples. Optimized processing parameters were used (table 2): a density equal to 99,6% of ideal density was achieved. Four replicas of the aforementioned cube were realized.

Table 2. Process parameters used for the printing of fixtures using LPBF (mod AM400 Renishaw).

Parameters	Values
Scanning strategy	Meander
Laser spot size [μm]	30
Hatch distance [μm]	90
Point distance [μm]	50
Thickness layer [μm]	50
Power [W]	200
Scanning time [μs]	70

The four cubes are thereafter subjected to different HT in temperature range of 700÷900°C lasted 1 hour followed by air cooling (table 3). All samples are treated under argon atmosphere. Tensile test samples are realized with the same LPFB process employed for cubes (table 2) and then they are treated using the HT selected after the preliminary characterization on the previous four cubes. No machining process have been used to refine the geometry of the tensile test samples.

Homologous temperature, T_h, is material dependent and is determined by material melting temperature, T_M, and HT temperature (1): homologous temperatures for each HT were indicated in Table 3. Sintering threshold is the value above which partial sintering process may be visible. For metals, sintering threshold is about 0,50.

$$T_h = \frac{T_{HT}}{T_M} \tag{1}$$

Table 3. Heat treatments: temperature, homologous temperature, duration, cooling. As built condition is the reference condition. Homologous temperature is calculated by means of the equation (1). AISI 316L melting temperature is about 1450°C.

Name of sample	Temperature treatment	Homologous temperature	Duration of treatment	Cooling
As built	No heat treated	-	-	-
CT 700	700°C	0,48	1 h	Air
CT 800	800°C	0,55	1 h	
CT 900	900°C	0,63	1 h	

2.3 Residual stress analysis

X-ray diffraction (XRD) analysis is performed on the specimens by means of an AST X-Stress 3000 X-ray diffractometer and is executed on close to the barycentre of vertical side of the cube (figure 5).

Operating parameters of the conducted measurements are: Chromium tube with a voltage of 30kV, diffraction angle of 152.3°, and collimator diameter of 3 mm. Measurements were fulfilled in depth step by step removing a layer of material with a thickness of 25 μm. The thin layer is removed by using an electro-polishing device in order to obtain the in-depth trend of residual stresses. Beside residual stresses measurements, FWHM (full width half maximum) parameter is also obtained by XRD. FWHM is an indicator associated to material hardening. The measurements of both residual stress and FWHM were carried out at 0 degrees at the barycentre of vertical side of the cube (figure 5b).

(a)

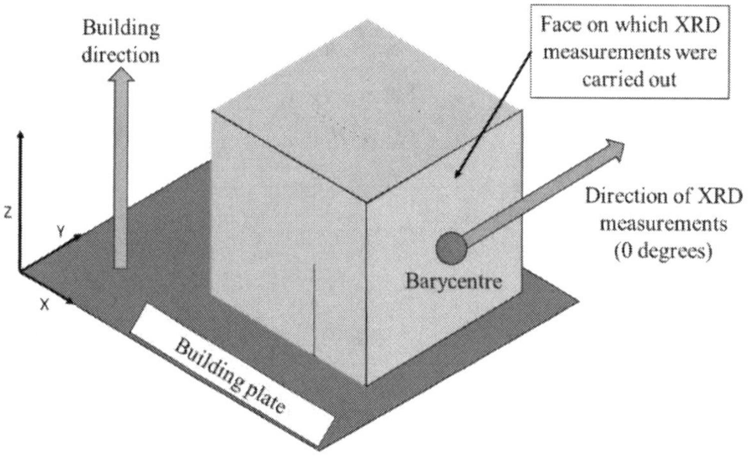

(b)

Figure 5. Representation of XRD measurements method. In particular, the system implemented to perform the XRD measurements (a) and the schematic enlargement of the tested cube (b). XRD measurements were carried out on a vertical side of the cube at 0 degrees.

IOP Conf. Series: Materials Science and Engineering 1275 (2023) 012001 doi:10.1088/1757-899X/1275/1/012001

2.4 Morphological characterization: cavities shape and powder state

All AISI 316L cubes are cut and polished in correspondence of the cavities. Morphological investigations of the cavities on polished sections are carried out by using optical microscope (Leitz Aristomet microscope, 50× magnification).

Micro-CT analysis are fulfilled using Phoenix v|tome|M300 with a voxel resolution of 9x9x9μm to quantify porosities and detect both the integrity of the specimens and the 3D quality of internal cavities. Micro-CT analysis are carried out on as Built and CT 800 cubes.

The powders state inside the cavities are observed by means of optical microscope investigations. Beside the qualitative observation, the identification of homologous temperatures permits the comprehension of the results about the powders state. Qualitative morphological analysis on fracture section of tension specimens are carried out using scanning electron microscope (SEM Leo 430 by Zeiss).

2.5 Mechanical characterization

Designed specimens are tested under tensile loading conditions using the MTS Synergie 200 testing device, equipped by a 1kN load cell. Tests are conducted under displacement control with an imposed strain rate of 2 mm/min. An extensometer is applied along the effective length on the specimen. Force-displacement curves are obtained.

3. Results

3.1 Residual stress characterization

The measurements of residual stresses and FWHM at 0 degrees are reported in figure 6.

The maximum residual stresses are detected on as built cube, while the minimum residual stresses are observed on CT 900 cube, with average values of 300 MPa and 50 MPa, respectively. Residual stresses values on as built cube are comparable with the value reported in the literature [18]. State of the art explored and assessed the promising effect of stress relieving heat treatments on tensile properties and residual stresses in metallic materials produced by LPBF technology [19,20,21].

FWHM represents full width of the diffraction peak at half of the maximum intensity and it is assumed as an index of material hardening. The results show a decreasing trend in FWHM values after the applied heat treatments, indicating that by increasing the temperature higher material softening occurred. This is expected due to the relaxation of residual stresses and micro-strains beside possible grain growth.

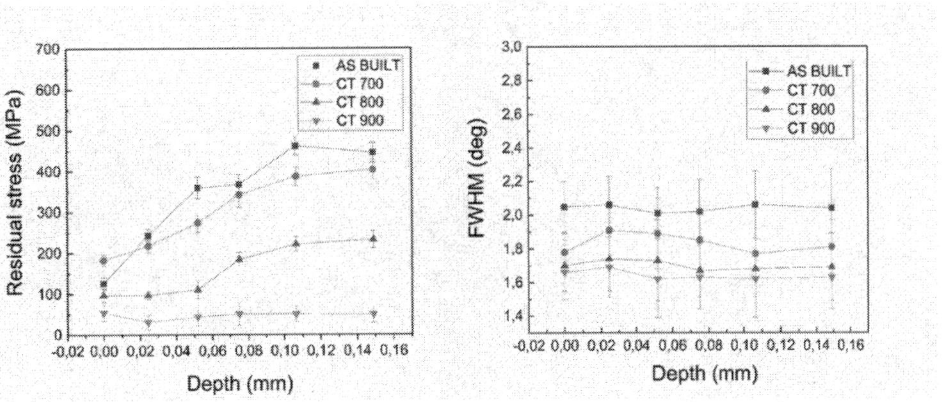

(a) (b)

Figure 6. In-plane residual stresses and FWHM measurements on 0° degrees' plan.

Peak values of residual stresses, as depicted in figure 7, confirm a decreasing trend by increasing the temperature of the HT. CT 700 residual stresses is equal to twice the CT 800 residual stresses.

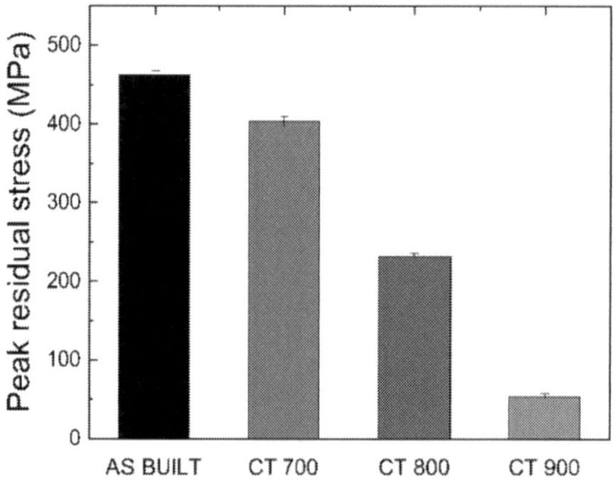

Figure 7. Peak values of residual stresses.

3.2 Cavity morphology and powder state characterization

The HT may induce changes in powders state inside the cavities: indeed, the higher the temperature heat treatment, the higher the homologous temperature which may lead to partial powders sintering.

2D optical images of one cavity for each condition are reported in figure 8. Cavity morphologies are irregular as expected from the CAD shape of cavities (figure 1). Small round shape pores are observed around each cavity.

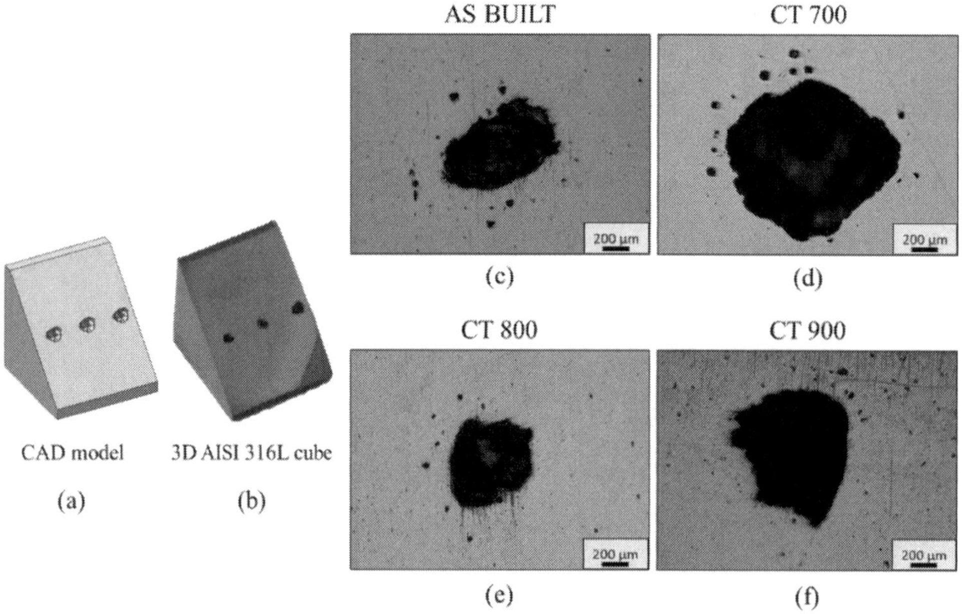

Figure 8. CAD cube (a) and AMed sample sectioned in correspondence of cavities (b). Microscope optical images of one lacuna for each samples with small round shape pores were observed around each cavity circled in red. As built condition (c); CT 700 (d); CT 800 (e); CT 900 (f).

Partial powders sintering is clearly visible in CT 900 (figure 9). As built and CT 700 cubes do not show any powders sintering, while CT 800 presents a very first initial sintering. Nevertheless, sintering necks are more visible in CT 900: the absence of sintering necks in CT 800 optical images is due to the polishing process during which the few sintered powders are probably detached. A more detailed analysis is carried out by means of micro-CT.

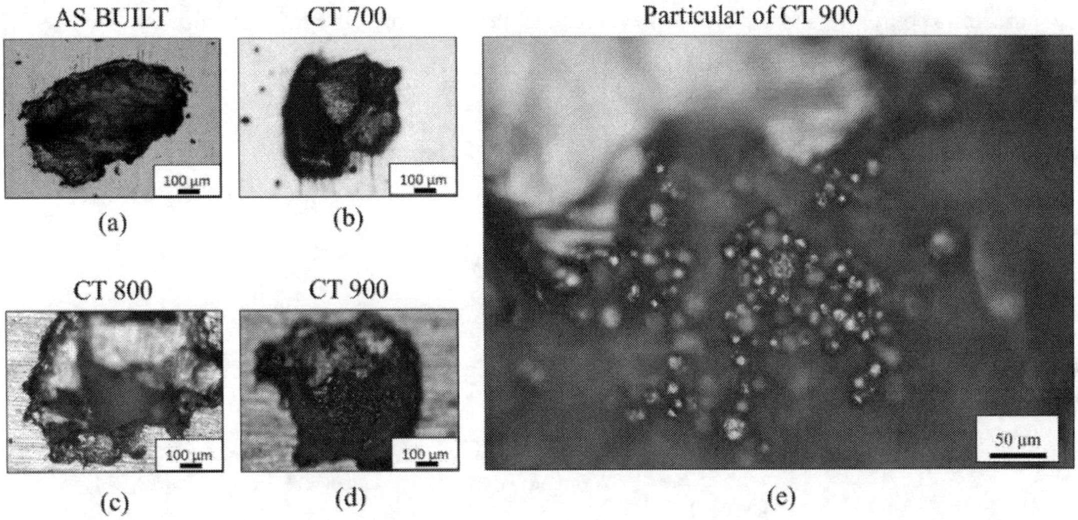

Figure 9. Optical images of cavities: as built condition (a); CT 700 (b), CT 800 (c); CT 900 (d). Particular on CT 900 where sintering necks are visible (e).

MicroCT analysis performed on as Built cube and CT 800 cube permit a 3D morphological analysis of the defects and the three cavities presented in the cubes. The diameter of each single porosity is calculated considering the diameter of the circumscribed sphere to the defect. Porosities are similar in diameter, about 30 µm, while the defects shapes change after the HT. In particular, defects in CT 800 (figure 10b) present more elongated shapes than in as built condition (figure 10a).

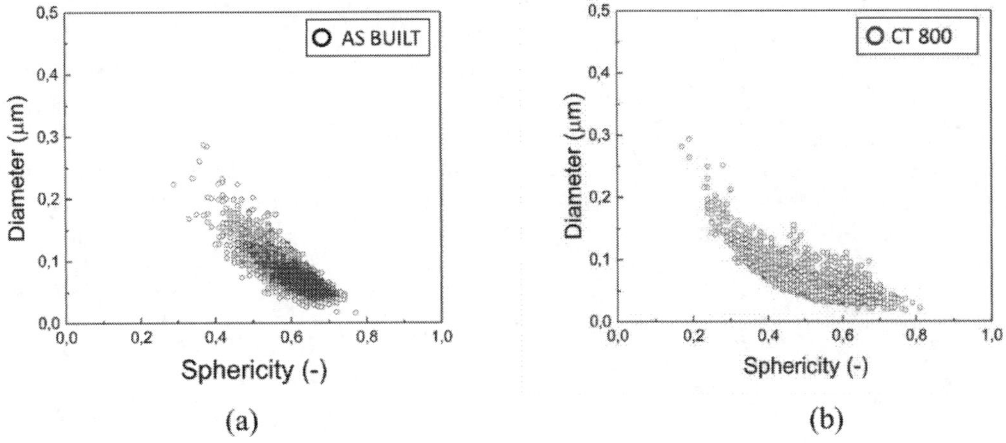

Figure 10. Defects inside cubes: as built condition (a); CT 800 (b). Diameters remained constant while the cavities shape become more spherical.

IOP Conf. Series: Materials Science and Engineering 1275 (2023) 012001 doi:10.1088/1757-899X/1275/1/012001

The cavities, which can be intended as big "defects", present a more elongated shape in CT 800 instead of as built condition, following the same trend of defects distribution. The designed three cavities in as built condition are the same in geometry and morphology A decrease in diameter for two cavities on three is observed in the CT 800 cube (figure 11).

The blue points (figure 11) present a lower sphericity than the black ones: the result is consequent to the HT which cause a partial sintering process. One blue cavity keeps the same diameter as the black ones: this blue point refers to the central cavity of the cube on which the HT influence is lower.

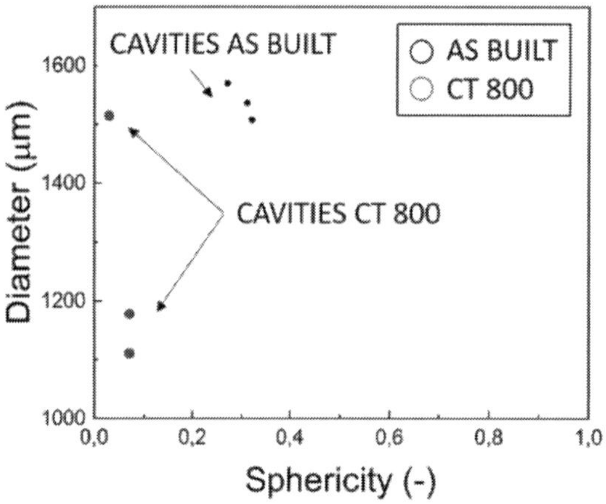

Figure 11. The three black and blue points refer to three cavities in as built and CT 800 condition respectively. The black points are similar in diameter and sphericity.

The micro-CT images of analysed cubes are post-processed in order to remove the presence of powder inside cavities making the cavities easier detectable. By observing 3D micro-CT reconstruction of CT 800, the central cavity (cavity "B" in figure 12) does not show any relevant morphological changes compared to the corresponding cavity in as built condition (figure 12). Indeed, the central cavity (cavity "B" in figure 12) is the cavity on which the HT is less influent because it is the furthest from the external surface of the cube. This cavity is graphically represented by the blue point with the higher diameter in figure 11.

The presence of partial sintered powders in CT 800 is visible in the micro-CT image by observing the "green" particles visible inside the two lateral cavities (cavity "A" and "C" in figure 12). In CT 800 cavity "B", which is less affected by the HT, less "green" particles are evident inside it and it is emptier compared to the two aside cavities, A and C respectively. This qualitative statement was supported by the aforementioned quantitative differences in diameters and sphericity (figure 11). The presence of very initial partial powders sintering in CT 800 is linked to the homologous temperature threshold above which powder particles begin to sinter. Indeed, in CT 800 a partial initial powders sintering is visible and it was justified by the homologous temperature of 0,55.

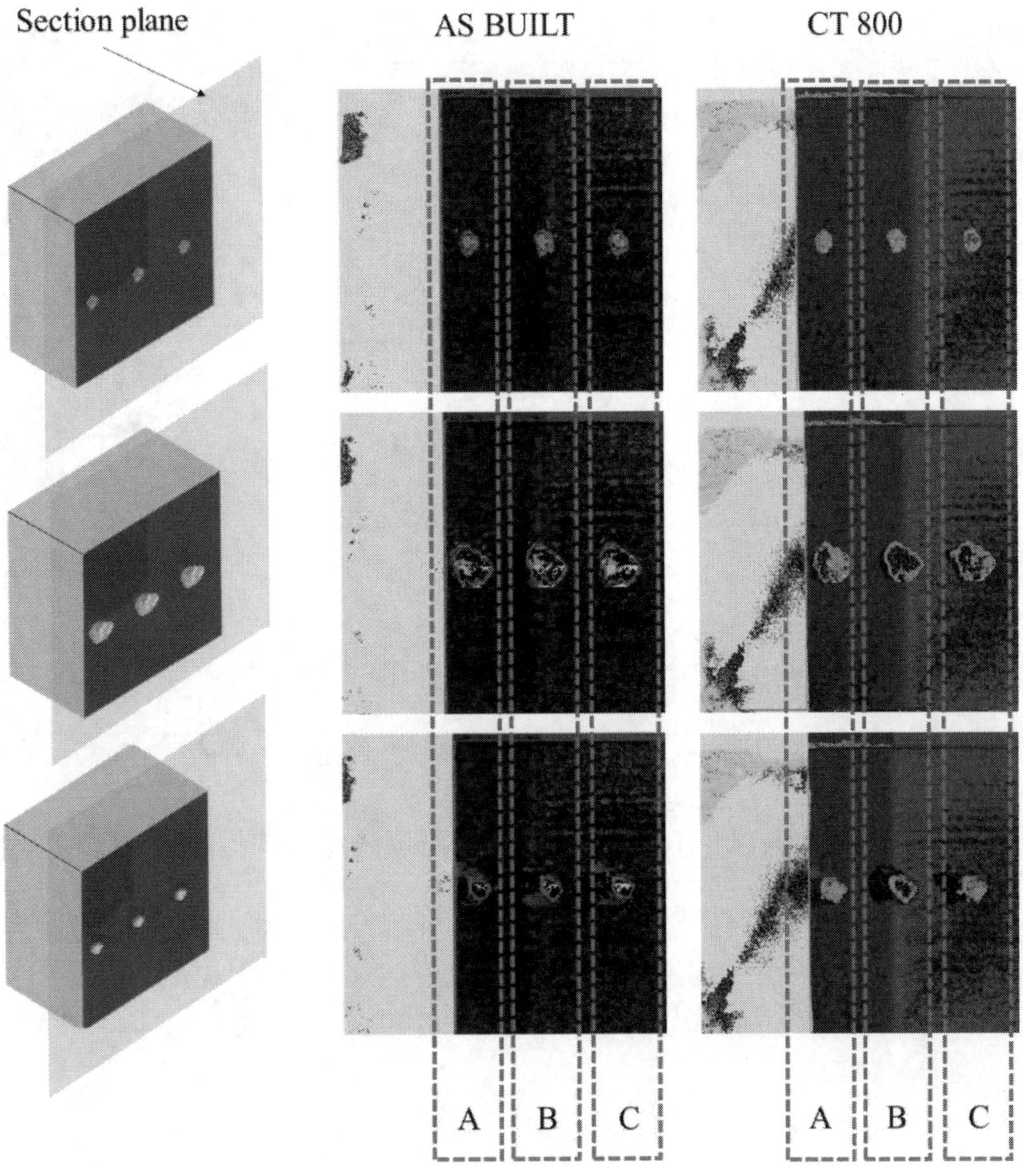

Figure 12. Overlap of micro-CT in as built and CT 800 conditions and designed cubes with cavities. Entrapped powders inside cavities have been removed from the images in order to easier detect the cavities morphologies.

3.3 Mechanical characterization

Mechanical static tensile tests permit a preliminary analysis of the impact of cavity morphology on specimen mechanical response (figure 13). The maximum stress obtained by means of the tensile tests was 600 MPa and the minimum stress is 570 MPa for as built and OP, respectively.

Ductile fracture characterizes all specimens: relevant plastic deformation is reached in each condition and the necking phenomena is evident in the image of the tensile test specimens pre and post mechanical test (figure 13).

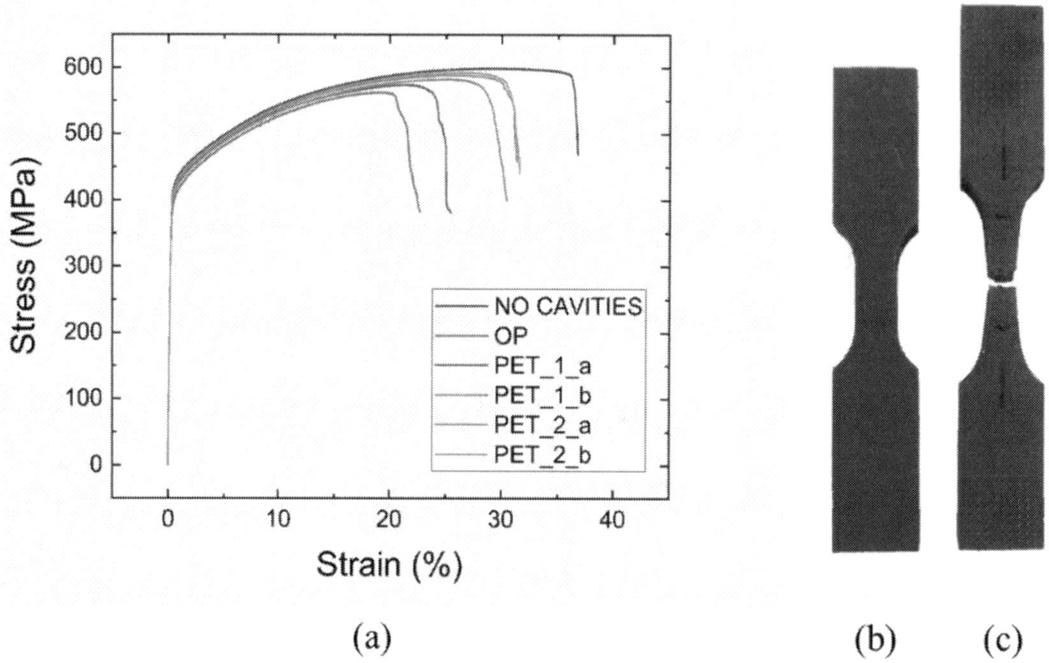

Figure 13. Stress-strain curves obtained by tensile tests (a). Image of samples pre (b) and post (c) mechanical test.

SEM magnification indicates the presence of cavities inside the specimens and confirmed the fracture typical for AMed specimens, where roughness and smooth areas coexist on fracture surface (figure 14) [22]. Dimples, micro cavities typical in ductile fracture, are visible on fracture surface of all fractured samples.

Metallic powders remain partially adhered on the internal side of the cavities (figure 15). The presence of these particles may negatively affect the 3D contour of cavities.

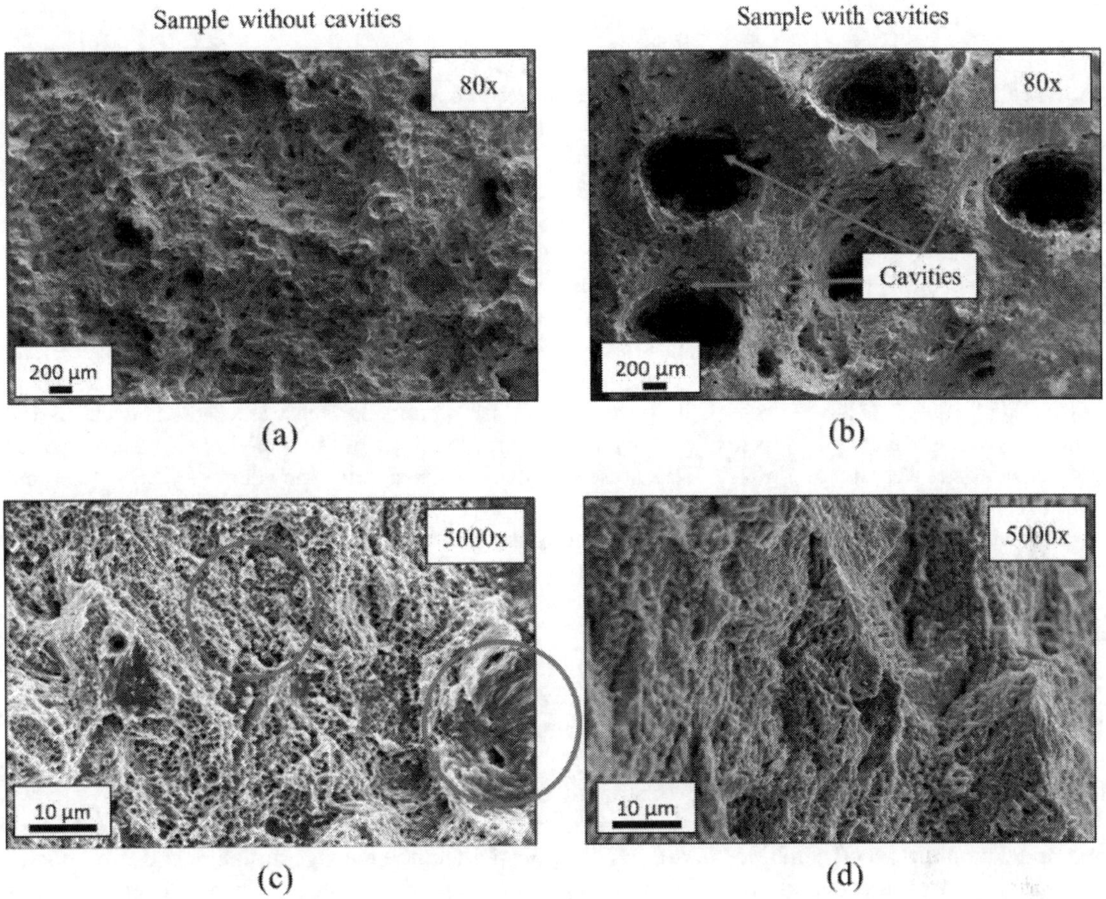

Figure 14. SEM images of representative fracture surfaces of samples without cavities (a) (c) and with cavities (b) (d). Red circles highlighted the rough and smooth areas on fracture surface, typical of LPBF fracture.

Figure 15. SEM magnification of cavities. Focus on particles partially adhered to the cavities.

4. Discussions and conclusions

The performed experimental analysis defines the best HT to preserve cavity shapes in order to proceed with mechanical characterization of the additive tensile test samples including lacunar cavities. The idea to employ 3D printing technology to study parts with complex shape with bone-like morphology has been already carried out in different research groups [23,24]. In particular, the studies reported in [23,24] were focused on replication of mesoscale bone structure, such as trabeculae typical of cancellous bone.

According to the literature [18], HT temperature below 600°C do not minimize the residual stresses: for this reason, the present work focuses on temperature above 600°C.

The higher the HT temperature, the lower the residual stresses. Minimum residual stresses are reached by means of HT 900°C: nevertheless, this temperature induces notable sintering of metallic powders in the cavities. CT 700, which has no sintering effect, is characterized by high residual stresses, close to the yield stress of AISI 316L (≈600 MPa [25]). Such finding negatively affects the mechanical tests. HT of 800°C (HT 800) shows the best compromise among the analyzed HTs. Despite partial sintering process affects powders inside the cavities, the morphology of CAD cavities is well replicated and residual stresses are minimized. The presence of partially sintered powders can be solved and optimized by means of selection of lower temperature HT for a longer duration. Further research will check lower temperature for the HT to improve their influence on such additive samples with cavities.

The optimized HT so far, HT 800, is performed on tensile test specimens. OP with the largest cavities, fails first. OP cavities are closer to the surface: this it can be concluded that cavities close to the surface are more harmful than embedded ones.

The dimensions of cavities strongly affect the mechanical strength of the specimens. PET 1_a and PET 1_b, despite the same number and shape of cavities, show different mechanical behavior: in this first case the orientation of cavities affects the failure of the specimens. PET 2_a and PET 2_b are designed with the same number and morphology of cavities and differ only in cavities orientation. In this second case orientation do not seem to strongly affect the failure.

These finding demonstrated the strong correlation among different variables, such as morphology and orientation that when combined in different way, can affect the damage initiation and evolution in metallic specimens. It is hypothesized that orientation acquires more relevance when number of cavities is considerable (e.g. PET1_a and PET2_b), while its impact is reduced in case the number of cavities decreases (e.g. PET2_a and PET_2b).

The main limitation of this study concerns the impossibility to perform compression tests: indeed, the presence of metallic powders inside the cavities makes the compression test unreliable. Compression state is the most frequent load state applied to the bone and, for this reason, it would be interesting to deepen this aspect. A second drawback concerns the difficulties to print the exact morphology of cavities: indeed, as suggested by the round shape pores around the cavities, the laser may find difficulties in faithful replication of the designed cavities. Nevertheless, the choice of printing parameters and the analysis by means of micro-CT are useful tools to improve the faithful reproducibility of the designed cavities: to date in the current work, an acceptable replication of the designed cavities has been reached.

The current study permits the identification of the best HT to realize additive manufactured samples with lacunae-like morphology cavities. Further analysis on mechanical fatigue performance will be carried out to analyze the crack propagation along lacunae-shaped cavities inside the additive manufactured samples.

Acknowledgements
The author would like to thank Nicola Bennato from CNR-ICMATE in Lecco, Lorenzo Giudici and Giuseppe Ghilardi from Politecnico di Milano, for their assistance in the experiments.

5. References

[1] Buccino F, Bagherifard S, D'Amico L, Zagra L, Banfi G, Tromba G and Vergani L.M. 2022

Assessing the intimate mechanobiological link between human bone micro-scale trabecular architecture and micro-damages, *Eng. Fract. Mech.* **270**

[2] Abdel-Wahab A A, Maligno A R and Silberschmidt V 2012 Micro-scale modelling of bovine cortical bone fracture: Analysis of crack propagation and microstructure using X-FEM, *Comput. Mater. Sci.* **52** 128–135

[3] Yadav R N, Uniyal P, Sihota P, Kumar S, Dhiman V, Goni V G, Sahni D, Bhadada S K and Kumar N 2021 Effect of ageing on microstructure and fracture behavior of cortical bone as determined by experiment and Extended Finite Element Method (XFEM), *Med. Eng. Phys.* **93** 100–112

[4] Yu B, Pacureanu A, Olivier C, Cloetens P and Peyrin F 2020 Assessment of the human bone lacuno canalicular network at the nanoscale and impact of spatial resolution, *Sci. Rep.* **10** 4567

[5] Goff E, Buccino F, Bregoli C, McKinley J P, Aeppli B, Recker R R, Shane E, Cohen A, Kuhn G and Müller R 2021 Large-scale quantification of human osteocyte lacunar morphological biomarkers as assessed by ultra-high-resolution desktop micro-computed tomography *Bone* **152** 116094

[6] Burr D B 2019 Stress concentrations and bone microdamage: John Currey's contributions to understanding the initiation and arrest of cracks in bone, *Bone*. 127 517–525

[7] Hemmatian H, Bakker A D, Klein-Nulend J and van Lenthe G H 2017 Aging, Osteocytes, and Mechanotransduction, *Curr. Osteoporos. Rep.* **15** 401–411

[8] Nicolella D P, Moravits D E, Gale A M, Bonewald L F and Lankford J 2006 Osteocyte lacunae tissue strain in cortical bone, *J. Biomech.* **39** 1735–1743

[9] Reilly G C 2000 Observations of microdamage around osteocyte lacunae in bone *J. Biomech.* **33** 1131–1134

[10] Gauthier R 2017 E.D. N 2017 Crack propagation mechanisms in human cortical bone on different paired anatomical locations : biomechanical , tomographic and biochemical approaches *Universitè de Lyon*

[11] Qiu S, Rao D S, Fyhrie D P, Palnitkar S and Parfitt A M 2005 The morphological association between microcracks and osteocyte lacunae in human cortical bone *Bone* **37** 10–15

[12] Rolvien T et al. 2018 Inter-site variability of the osteocyte lacunar network in the cortical bone underpins fracture susceptibility of the superolateral femoral neck *Bone* **112** 187–193

[13] Akhter M P, Kimmel D B, Lappe J M and Recker R R 2017 Effect of macroanatomic bone type and estrogen loss on osteocyte lacunar properties in healthy adult women *Calcif. Tissue Int.* **100** 619–630

[14] van Hove R P, Nolte P A, Vatsa A, Semeins C M, Salmon P L, Smit T H and Klein-Nulend J 2009 Osteocyte morphology in human tibiae of different bone pathologies with different bone mineral density - Is there a role for mechanosensing? *Bone* **45** 321–329

[15] Mulargia S, Dooley C, Cristofolini L and Taylor D 2014 Fracture and fatigue in osteocytes *J. Mech.Behav. Biomed. Mater.* **39** 231–237

[16] Gustafsson A, Khayyeri H, Wallin M and Isaksson H 2019 An interface damage model that captures crack propagation at the microscale in cortical bone using XFEM *J. Mech. Behav. Biomed.Mater.* **90** 556–565

[17] Donaldson F, Ruffoni D, Schneider P, Levchuk A, Zwahlen A, Pankaj P and Müller R 2014 Modeling microdamage behavior of cortical bone, *Biomech. Model. Mechanobiol.* **13** 1227-1242

[18] Chao Q, Thomas S, Birbilis N, Cizek P, Hodgson P D and Fabijanic D 2021 The effect of post processing heat treatment on the microstructure, residual stress and mechanical properties of selective laser melted 316L stainless steel, *Mater. Sci. Eng. A.* **821** 141611

[19] Weber S, Montero J, Petroll C, Schäfer T, Bleckmann M and Paetzold K 2020 The fracture behavior and mechanical properties of a support structure for additive manufacturing of Ti-6Al-4V, *Crystals.* **10**

[20] Fiocchi J, Biffi C A, Colombo C, Vergani L M and Tuissi A 2020 Ad Hoc Heat Treatments for

Selective Laser Melted Alsi10mg Alloy Aimed at Stress-Relieving and Enhancing Mechanical Performances, *JOM* **72**, 1118–1127

[21] Fiocchi J, Colombo C, Vergani LM, Fabrizi A, Timelli G, Tuissi, A. and Biffi C A 2021 Heat Treatments for Stress Relieving AlSi9Cu3 Alloy Produced by Laser Powder Bed Fusion, *Materials* **14** 4184

[22] Colombo C, Biffi C A, Fiocchi J, Tuissi A and Vergani L M 2019 Effect of Optimized Heat Treatments on the Tensile Behavior and Residual Stresses of Selective Laser Melted AlSi10Mg Samples, *Key Engineering Materials* **813** 364-369

[23] Barak M M and Black M A 2018 A novel use of 3D printing model demonstrates the effects of deteriorated trabecular bone structure on bone stiffness and strength, *J. Mech. Behav.Biomed. Mater.* **78** 455–464

[24] Zach Wooda M M B, Lynna L, Nguyena J T, Blacka M A and Patela M 2019 Are we crying Wolff? 3D printed replicas of trabecular bone structure demonstrate higher stiffness and strength during off- axis loading *Bone* **127** 635–645

[25] Pacheco J T, Meura V H, Bloemer P R A, Veiga M T, de Moura Filho O C, Cunha A and Teixeira M F 2022 Laser directed energy deposition of AISI 316L stainless steel: The effect of build direction on mechanical properties in as-built and heat-treated conditions, *Advances in Industrial and Manufacturing Engineering* **4** 100079

IOP Conf. Series: Materials Science and Engineering 1275 (2023) 012002 doi:10.1088/1757-899X/1275/1/012002

Fatigue assessment of 3D-printed porous PLA-based scaffold structures by Thermographic Methods

R Ahmadi*, D D'Andrea and D Santonocito

University of Messina, Department of Engineering, Contrada di Dio (S. Agata), 98166 Messina, Italy

*email: reza.ahmadi@studenti.unime.it

Abstract Additive manufacturing with Fused Deposition Modelling (FDM) provides new opportunities for biomedical applications by manufacturing more complex structure for tissue scaffolds. In contrast to traditionally manufacturing methods such as injection moulding, FDM-made structures are produced with layer-by-layer adhesion mechanism. Moreover, setting different parameters is critically important to obtain reliable components, consequently fatigue analysis of these structures gets complicated. Therefore, a scaffold must be designed to replace human bone and withstand to static and fatigue loads. Although PLA scaffolds fabricated by FDM have been widely investigated, the fatigue behaviors research on them have been narrowly published; therefore, it is necessary to analyze the fatigue behaviour of PLA scaffolds because of their wide application in medical science. In this study, fatigue analysis of scaffolds-like specimens made by PLA is carried out by monitoring the energy release of such structures. Thermographic Methods are applied to obtain, in a short time and with few materials' consumption, fatigue life information of scaffold-like structures produced with different printing parameters.

1. Introduction

Additive Manufacturing (AM) is a cutting-edge technology that allows for the rapid fabrication of physical models directly from three-dimensional computer-aided design (CAD) data without the use of traditional tooling. Moreover, it is a group of fabrication techniques through which materials are joined, usually layer upon layer, to make objects from three-dimensional virtual models [1]. Fused Deposition Modelling (FDM) is one of the most common additive manufacturing technologies for numerous engineering applications and the determination of proper process parameters considerably affect the quality of FDM fabricated parts, playing an important role to ensure quality of products [2].

Polylactic acid (PLA) is a biodegradable, absorbable and biocompatible thermoplastic aliphatic polyester that is widely used for rapid prototyping, to manufacture tools, jigs, and fixtures designed to maximize the production efficiency and to make biomedical components with complex shape .Recent developments in manufacturing techniques, particularly 3D printing, have empowered the fabrication of more complicated scaffolds and architectures [3]. The flexibility of this technology has attracted several industries to print complex geometrical structures. In contrast to traditionally fabricated polymers, FDM-made polymers have layer by layer adhesion and the influence of printing variables make fatigue analysis complicated. Because

Content from this work may be used under the terms of the Creative Commons Attribution 3.0 licence. Any further distribution of this work must maintain attribution to the author(s) and the title of the work, journal citation and DOI.
Published under licence by IOP Publishing Ltd

of manufacturing of complex structures, additive manufacturing technique is applied in a wide variety of engineering applications such as mechanical manufacturing [4], civil engineering [5], biomedical [6], aerospace[7], and food industries [8,9].

Several studies have been performed to investigate the fatigue behaviour and process optimizations for the parts processed by the original Stratasys FDM rapid prototyping technology, but none of them investigated the strength assessment of PLA Scaffold by Energy Methods. Afrose et al. [10] investigates the fatigue behaviour of PLA parts processed by FDM additive manufacturing process and they have shown that the parts in 45° build orientation showed higher fatigue life than the parts in X and Y build orientations for the same percentage of applied static loads.

In this paper, an experimental work is presented to investigate the fatigue behaviour of FDM-processed PLA scaffold through using optimized specimens under fatigue cyclic loading. For the first time, the Static Thermographic Method (STM) has been applied to scaffold structures to rapidly assess the first damage initiation within the structure.

2. Theoretical background

The main idea of the Static Thermographic Method is to assess the end of the thermoelastic behaviour of a material during a static tensile test.

Risitano and Risitano [3], in 2013, observed how the temperature signal during a static tensile test, show three different phases. Phase I is characterized by a linear decrement of the temperature, due to the thermoelastic law of Lord Kelvin. Phase II is characterized by a temperature decrement but with a different trend compared to the Phase I. At the end of Phase II, the temperature reaches a plateau region, which correspond to the yielding stress of the material. In Phase III, the plastic damage is more predominant compared to the elastic behaviour, hence the temperature begins to rise up to the specimen failure.

The transition point between Phase I and Phase II correspond to the onset of irreversible damage within the material. If we are able to assess this point in the temperature signal, we can correlate it to a macroscopic stress, defined as the "limit stress", σ_{lim}, which produce an irreversible energy release due to the increment of plastic damage. For more information about the STM, we remand to [3–5].

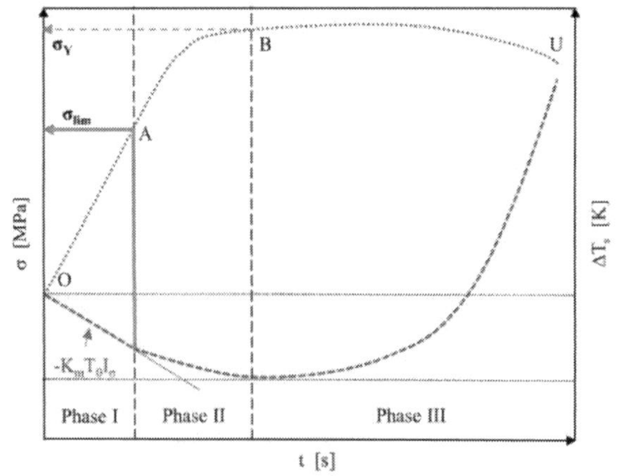

Figure 1. Temperature trend during a static tensile test. Phase I is characterized by a linear decrement due to the thermoelastic constant K_m and the first stress invariant I_σ.

3. Materials and methods

3.1. Preparation of test specimens

PLA is one of the most popular materials used in bio medical 3D printing. It is safe, affordable, and has outstanding material properties compared to its ease of use. It is one of the most popular biomaterials approved by the Food and Drug Administration (FDA), which has excellent biocompatibility and biodegradability [13].

3D-printing is a rapidly developing method for producing bone tissue engineering scaffolds, which allows minimizing time of material processing at high temperatures to avoid thermal degradation of PLA [14]. Two kinds of specimens were 3D-printed in PLA. The first kind of specimens presents the geometry of Figure 2a, while the second kind has the same geometry but with a scaffold region in the middle of the section (Figure 2b). Totally, 19 full and 19 scaffold specimens were printed for the test. All of them were printed with the same parameters as indicated in Table 1.

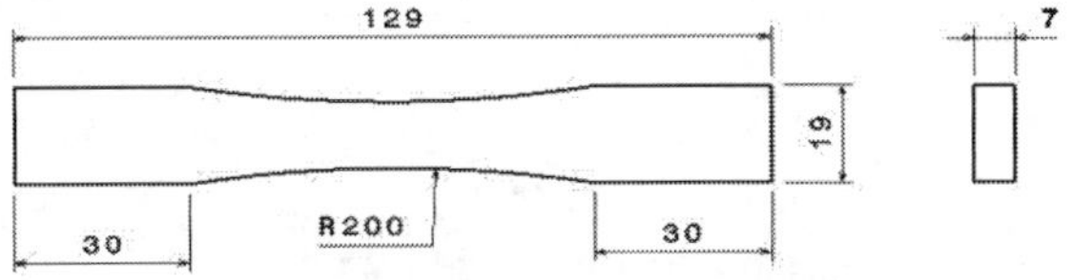

(a)

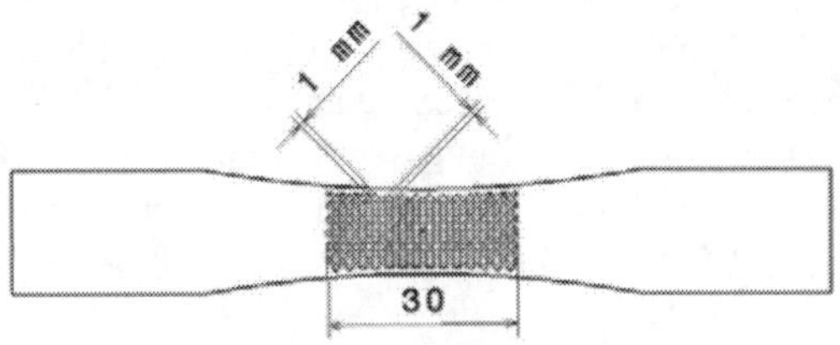

(b)

Figure 2. Dimensions of the adopted specimens: a) Full; b) Scaffold.

Table 1. 3D Printing parameters for specimens.

Nozzle Diameter	0.4 mm
Nozzle Temperature	215 °C
Infill density	100%
Layer height	0.1 mm
Shell thickness	0.4 mm
First layer speed	30 mm/s
Print speed	100 mm/s

The structure of a 3D scaffold was modelled using CATIA® software. PRUSA® slicer software was used for analysing and slicing the generated STL-files. The modelled pore size was 1x1 mm. All pores were interconnected (Figure 2b). 3D-printing of porous scaffolds was performed using PRUSA Printer Machine

(Figure 3) at a nozzle temperature of 215°C. The optimization of process parameters is a major challenge to obtain adequate scaffold. The process parameters involved in the fabrication of scaffolds, such as, deposition velocity, layer thickness, nozzle tip size, filament distance, deposition pattern, and speed movement, were considered. After several printing test, the printing parameter shows in Table 1 were adopted.

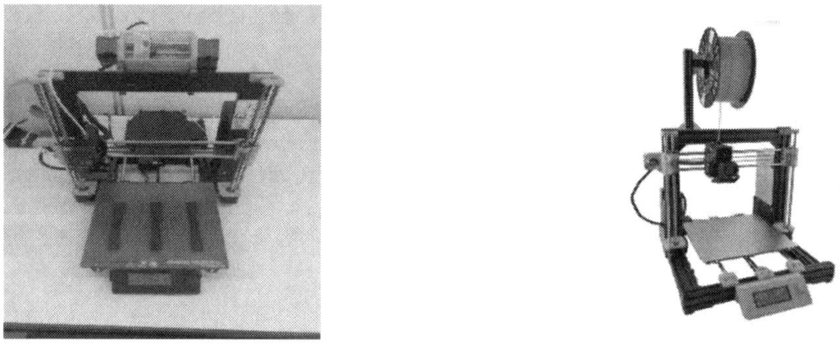

(a) (b)

Figure 3. PRUSA 3D Printer I3MKS

In this study, the geometry of the specimen has been optimized to reduce the printing time and the stress concentration. Figure 2 shows the dimensions of the specimen adopted, which could be printed 6 specimens simultaneously in every print on the PRUSA Printer bed plate (250x210x210 mm, printing volume).

The STL file was imported into the PRUSA software and processed into a 3MF file which contains all the instructions to generate the tool path of the deposition tip for PRUSA machine. The software offers different print modes, methods and settings. Building a part using different print modes and different build directions will affect the part strength and mechanical behaviour so that's why choosing the best option and setting for printing is a challenge because we don't know how they interact with each other.

3.2. Experimental setup

Mechanical testing was performed with an ITALSIGMA-25kN servo hydraulic testing machine using dog bone specimens, both scaffolds and full, with the cross-sectional area of 13 mm x 7mm. The specimens were tested under static tensile and constant amplitude (CA) loading conditions. Static tensile tests have been performed on three scaffold and three full specimens, adopting a crosshead speed of 2.5 mm/min. Constant amplitude fatigue test were performed adopting a stress ratio of R= 0.1 and frequency f= 3Hz. An infrared camera FLIR® System model A40 was adopted, with an acquisition rate of 5 Hz (Figure 4).

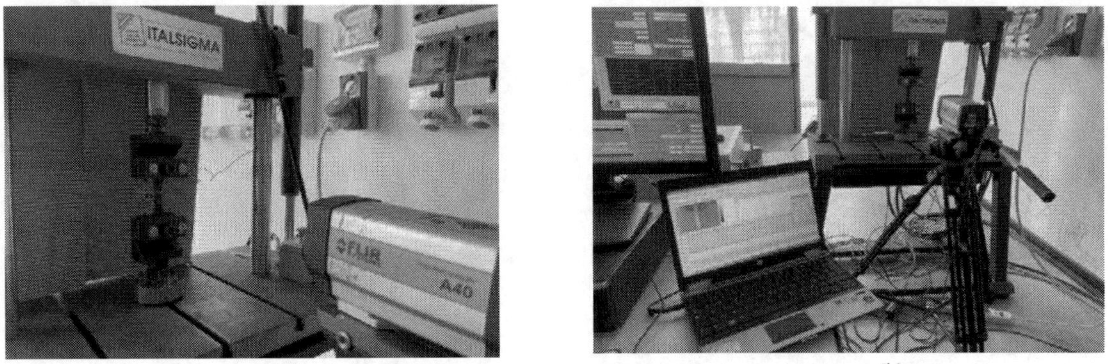

(a) (b)

Figure 4. ITALSIGMA-25kN fatigue test machine

4. Results and discussion

Static tensile tests were performed on full and scaffold specimens adopting a constant crosshead speed, while the superficial temperature evolution was monitored adopting an infrared camera.

Figure 5 shows the trend of temperature for the full specimens. It is clearly visible that there is an initial phase in which the temperature variation has an almost constant slope, which identifies the thermoelastic phase and then tends to decrease, following another trend, during a second phase. The first deviation from linearity, which corresponds to the end of the first phase, can be correlated to the limit stress. It can be identified when a deviation from the first linear thermoelastic trend of the temperature signal is noticed.

By making the intersection between two regression lines (the first for phase I, the latter for phase II), it is possible to assess the limit stress with the STM. The average limit stress is σ_{lim}= 36.1 MPa.

Scaffold specimens (Figure 6) show the same thermal behaviour. An average value of the limit stress can be assessed equal to σ_{lim}= 20.5 MPa. The scaffold specimens have shown that the temperature experiences a high increment when the stress level reaches the failure point (Figure 7).

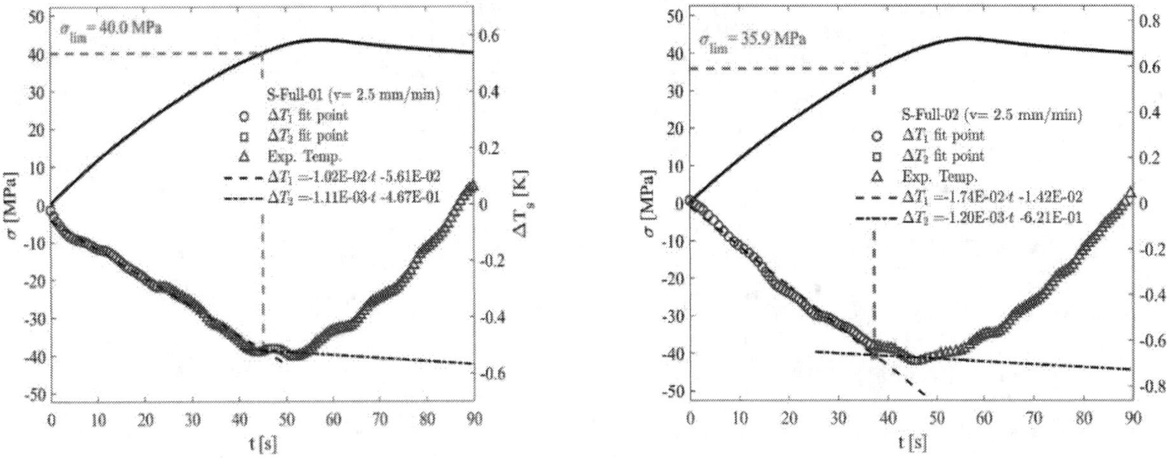

(a) (b)

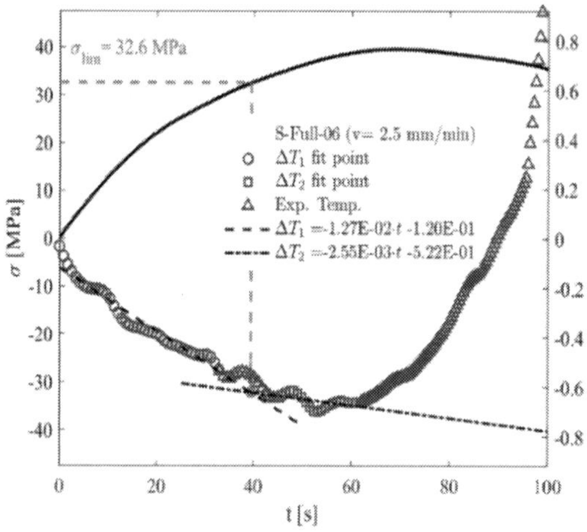

(c)

Figure 5. Temperature evolution during a static tensile test on full PLA specimens.

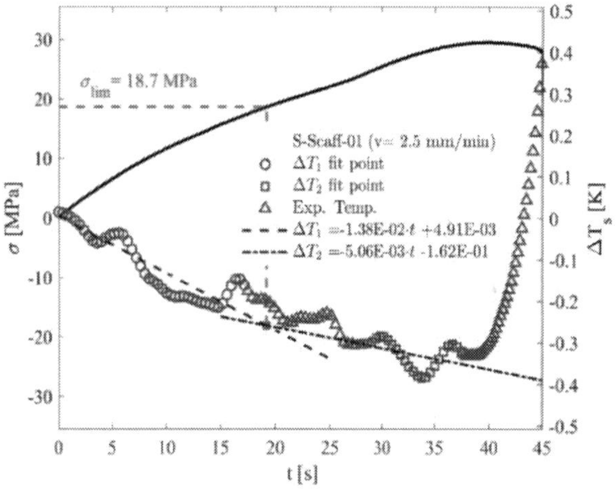

(a)

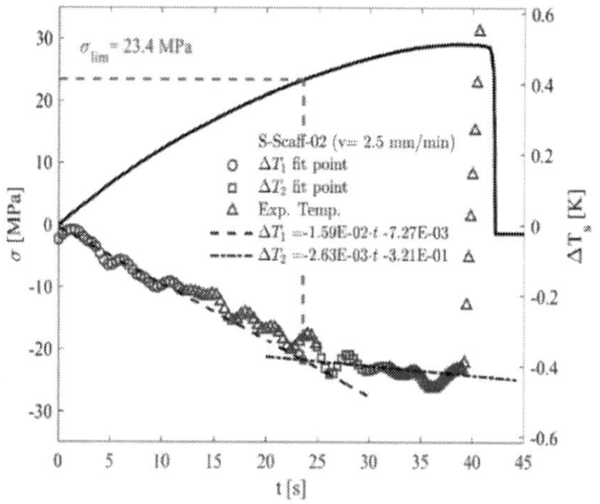

(b)

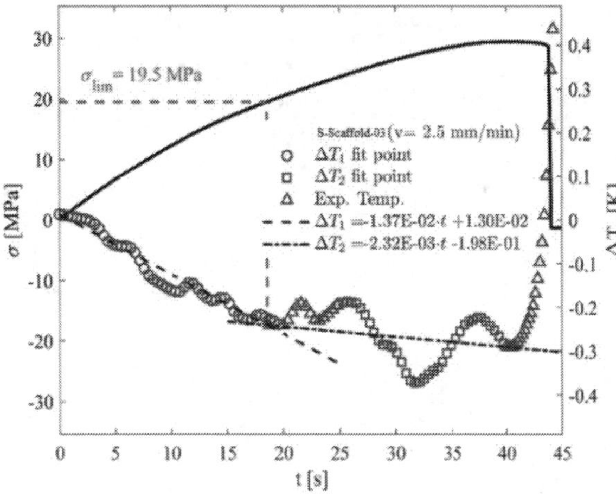

(c)

Figure 6. Temperature evolution during a static tensile test on scaffold PLA specimens.

Figure 7 reports the thermographic images of a static tensile test performed on a scaffold specimen. As it is shown, the temperature increased suddenly and, at the same time, the specimen failed.

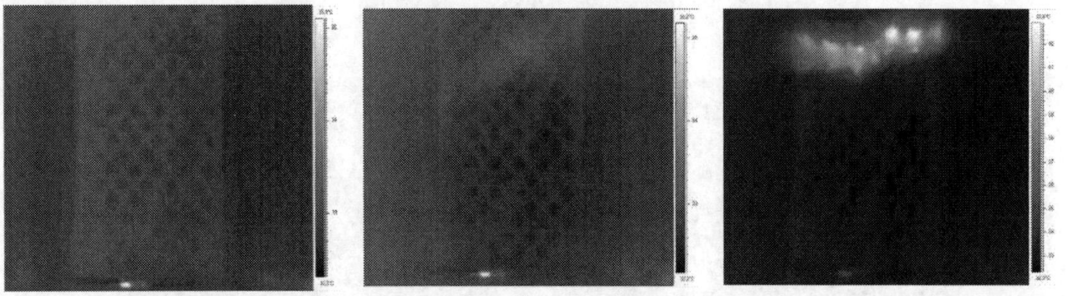

(a) (b) (c)

Figure 7. Thermographic images of a scaffold specimen: a) first phase; b) second phase; c) third phase.

4.1. Constant Amplitude fatigue tests

A total of 38 of full and scaffold were used to perform fatigue life curve (Figure 8). As shown in the diagram and according to the microscopy (Figure 9), fatigue failure of specimens is inherently due to the accumulated damage or due to the growth of the defect to a critical dimension. No specimens, both full and scaffold, reached the un-out condition ($1x10^6$ cycles).

The fatigue behaviour of 3D printed polymeric materials is substantially influenced by the printing parameters, physical properties, environmental conditions, and mechanical loading factors that's why we see a random behaviour in the full specimen curve. On the other hand, the scaffold specimens show an almost linear trend, as the stress level is reduced.

In micrography of failed specimen, there are accumulation of defects, such as micro-delamination at the interfaces between layers, micro cracks in the layers, and an accumulation of local plastic deformation was occurred during the cyclic loading of the material. More investigation must be performed to obtain better printing parameter for PLA scaffold structures.

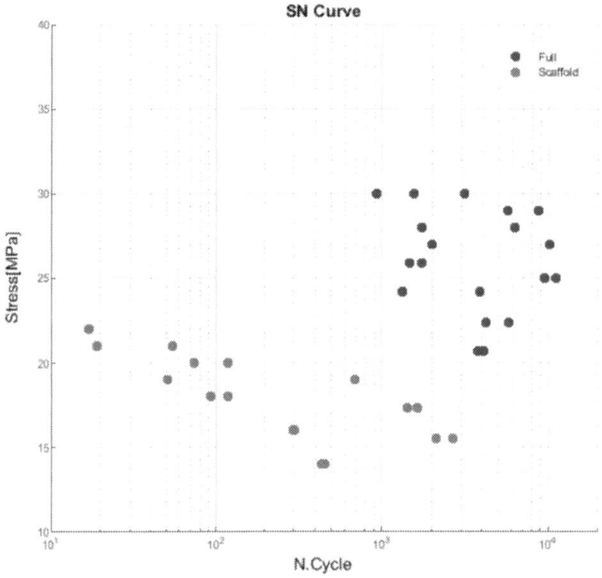

Figure 8.SN Curve of full and scaffold specimens

(a) (b)

Figure 9. Micrography of a specimen

5.Conclusion

In this work, for the first time, the Static Thermographic Method has been applied on porous PLA-based scaffolds obtained by FDM. Two types of specimens were printed and tested under static tensile test. By monitoring the superficial temperature evolution with an infrared camera, it has been possible to assess the limit stress as the first damage within the structure. Constant amplitude fatigue tests have been performed on bith type of specimen. The full specimens exhibit an high scatter of the SN data, without following a trend. On the other hand, the scaffold specimens, even if exhibit an high scatter of the SN data, show a linear trend.

More investigation about the printing parameters must be performed in order to obtain reliable structures that can withstand to fatigue load services.
Energy methods, such as the STM, can severally reduce the testing time to obtain reliable fatigue data for mechanical design.

References

[1] Ezeh OH, Susmel L. On the fatigue strength of 3D-printed polylactide (PLA). Procedia Struct Integr 2018;9:29–36. doi:10.1016/j.prostr.2018.06.007.

[2] Mohamed OA, Masood SH, Bhowmik JL. Optimization of fused deposition modeling process parameters: a review of current research and future prospects. Adv Manuf 2015;3:42–53. doi:10.1007/s40436-014-0097-7.

[3] Miller AT, Safranski DL, Wood C, Guldberg RE, Gall K. Deformation and fatigue of tough 3D printed elastomer scaffolds processed by fused deposition modeling and continuous liquid interface production. J Mech Behav Biomed Mater 2017;75:1–13. doi:10.1016/j.jmbbm.2017.06.038.

[4] Henriques B, Pinto P, Silva FS, Fredel MC, Fabris D, Souza JCM, et al. On the mechanical properties of monolithic and laminated nano-ceramic resin structures obtained by laser printing. Compos Part B Eng 2018;141:76–83. doi:10.1016/j.compositesb.2017.12.044.

[5] Ma G, Wang L. A critical review of preparation design and workability measurement of concrete material for largescale 3D printing. Front Struct Civ Eng 2018;12:382–400. doi:10.1007/s11709-017-0430-x.

[6] Levesque JN, Shah A, Ekhtiari S, Yan JR, Thornley P, Williams DS. Three-dimensional printing in orthopaedic surgery: A scoping review. EFORT Open Rev 2020;5:430–41. doi:10.1302/2058-5241.5.190024.

[7] Ahmed NA, Page JR. Manufacture of an unmanned aerial vehicle (UAV) for advanced project design using 3D printing technology. Appl Mech Mater 2013;397–400:970–80. doi:10.4028/www.scientific.net/AMM.397-400.970.

[8] Sun J, Peng Z, Zhou W, Fuh JYH, Hong GS, Chiu A. A Review on 3D Printing for Customized Food Fabrication. Procedia Manuf 2015;1:308–19. doi:10.1016/j.promfg.2015.09.057.

[9] Shanmugam V, Das O, Babu K, Marimuthu U, Veerasimman A, Johnson DJ, et al. Fatigue behaviour of FDM-3D printed polymers, polymeric composites and architected cellular materials. Int J Fatigue 2021;143:106007. doi:10.1016/j.ijfatigue.2020.106007.

[10] Afrose MF, Masood SH, Iovenitti P, Nikzad M, Sbarski I. Effects of part build orientations on fatigue behaviour of FDM-processed PLA material. Prog Addit Manuf 2016;1:21–8. doi:10.1007/s40964-015-0002-3.

[11] Risitano A, Risitano G. Determining fatigue limits with thermal analysis of static traction tests. Fatigue Fract Eng Mater Struct 2013;36:631–9. doi:10.1111/ffe.12030.

[12] Risitano A, Risitano G, Clienti C. Fatigue limit by thermal analysis of specimen surface in mono axial traction test. EPJ Web Conf 2010;6. doi:10.1051/epjconf/20100638010.

[13] Zhang B, Wang L, Song P, Pei X, Sun H, Wu L, et al. 3D printed bone tissue regenerative PLA/HA scaffolds with comprehensive performance optimizations. Mater Des 2021;201:109490. doi:10.1016/J.MATDES.2021.109490.

[14] Senatov FS, Niaza K V., Stepashkin AA, Kaloshkin SD. Low-cycle fatigue behavior of 3d-printed PLA-based porous scaffolds. Compos Part B Eng 2016;97:193–200. doi:10.1016/j.compositesb.2016.04.067.

Design of hierarchical lattice structures attainable by additive manufacturing techniques

L Musenich, A Stagni and F Libonati

University of Genoa, Department of Mechanical, Energy, Management and Transportation Engineering, Via all'Opera Pia 15/A, 16145, Genova, Italy

flavia.libonati@unige.it

Abstract. Readiness of new materials that are simultaneously lightweight, damage-resistant, multifunctional, and sustainable is a primary need for many technology sectors. Thanks to additive manufacturing, lattice materials appear to be ideal candidates to meet this challenge. By designing their unit cells and structural organization, multiscale materials with unique combinations of properties can be obtained. Nevertheless, many gaps remain to be filled for their effective and efficient design. Nature, exploiting hierarchical architectures on a material scale, actually amplifies the properties of biological materials and combines them in ways we cannot achieve yet in synthetic materials. In materials design, we are still far from such a level of perfection. To narrow this gap and expand the current knowledge on the effects of hierarchy on the mechanical behaviour of materials, we numerically studied the mechanical response of 3D hierarchical lattice specimens under a four-point bending loading scenario. For this, we selected two types of unit cells with different structural behaviour and combined them together into different specimen topologies. The results show that, through hierarchy, it is possible to tailor lattice material performances, achieving benefits in terms of both specific mechanical properties and multifunctionality. The evidence found opens new horizons for applications such as heat exchangers, mechanical filters, scaffolds, energy storage, and packaging.

1. Introduction

Sustainable development requires the search for new technological solutions that can preserve and improve the quality of life by promoting a circular economy aimed at protecting natural resources and minimizing the ecological footprint. In the field of materials design, this translates into the research of methods and tools to fabricate high-performance materials (i.e., lightweight, damage-tolerant, and multifunctional) by using a limited number of resources, thus reducing the environmental impact and increasing the recycling likelihood. Nature is a master in this field. Indeed, natural composite materials such as bone, mother-of-pearl, or wood teach us that such performances can be achieved through complex bottom-up multiscale design and building logic [1–4]. Although the recent advances in manufacturing it is not possible to achieve such details at the material scale yet. Lattice materials, thanks to the development of additive manufacturing techniques, are offering a very promising opportunity to narrow this technological gap that exists between Nature and engineering [5–12].

Lattice materials are made of a combination of elementary building blocks (known as unit cells, u.c.s) that are repeated in space. This peculiarity makes lattice materials to behave as structures at the u.c.

Content from this work may be used under the terms of the Creative Commons Attribution 3.0 licence. Any further distribution of this work must maintain attribution to the author(s) and the title of the work, journal citation and DOI.
Published under licence by IOP Publishing Ltd

scale (meso scale) and as materials at the macro scale [13–15]. Thanks to their structure, lattice materials can be architected, including structural heterogeneities and multiscale substructures, to achieve combinations of properties not found in conventional engineering materials. The multiple design degrees of freedom that lattice materials offer make the characterization of the property-process-structure-performance relationship (PSPP relationship) very complicated. To date, there are many challenges to overcome for their optimal implementation in engineering applications [16–18]. Indeed, from literature, it is known that the apparent mechanical behavior of lattice materials at the macro scale can be described as a function of: (i) the base material properties, (ii) the u.c. connectivity, and (iii) the relative density [14]. Depending on u.c. topology, lattice u.c. have also been classified into two macro-families characterized by distinct mechanical behaviors: bending- and stretching-dominated, respectively [14,19]. However, it is not entirely clear how the combination of different lattice u.c. at different length scales can change the material performance.

Optimizing the structural organization of fractal-like hierarchical 2D lattice materials, elastic properties can be adjusted within a wide range and some of the structure may benefit from the hierarchical architecture [20–22]. In addition, not only the elastic properties, but also crack propagation can be controlled and enhanced through the implementations of a multiscale design [23]. Similar results were also found for three-dimensional ones. Specifically, in terms of stiffness, it has been observed that bending-dominated lattice u.c.s benefit from a hierarchical architecture to a greater extent than stretching-dominated ones [24]. Yet, the presence of more hierarchical levels produces higher stress concentrations that reduce the overall strength of the material. Also, beyond two levels of hierarchy, it is hard to obtain appreciable gains on the performance of lattice materials [25].

Combining bending-dominated and stretching-dominated u.c.s in hybrid-type hierarchical lattice structures unique properties can also be obtained. Such structures have been conventionally identified with names consisting of two parts: "x-y" [26–28], where "x" denotes the type of structure on the largest reference length scale and "y" that on the smallest length scale. We follow this convention in our work. Stiffness and elastic buckling resistance of multiscale bending-bending lattices progressively improve with hierarchy, while for stretching-stretching lattices the increase in buckling resistance is accompanied by a reduction in stiffness. Bending-stretching lattices showed major enhancement in both stiffness and buckling with hierarchy, while stretching-bending lattices are at the same time less stiff and less strong [26]. In determining the mechanical response of lattice materials, the slenderness ratios of u.c. struts at different hierarchical levels play a key role [29].

Literature findings presented above are mainly derived from numerical analyses of the deformation of lattice materials under compressive loading condition and do not provide comprehensive design guidelines. Therefore, to extend the current knowledge on the PSPP relationship of multiscale lattice materials, in this work we analyze the mechanical performance of 3D lattice structures at two levels of hierarchy, subjected to bending. To evaluate the benefits obtainable from exploiting substructures with different behaviors, we chose two different types of "strut-based" elementary u.c., one bending-dominated and one stretching-dominated. Then we combined them into six different structural topologies. Through finite element (FE) simulations, we analyzed the influence of the u.c. struts slenderness ratio on their mechanical behavior. With a view to extending this study, the geometries of the lattice models and their material properties were assigned based on the Fused Filament Fabrication (FFF) 3D printer constraints, for the purpose of future experimental validation.

2. Materials and Methods

2.1. Hierarchical lattice structure modelling

As a case study, we considered a specimen subjected to 4-point bending (4PB) loading. The study of a bending loading scenario allowed us to work on a design space with a simple geometry but, at the same time, characterized by gradient stress states, which opens up the design of lattice structural topologies in view of optimal material utilization. In addition, since in the 4PB configuration a larger portion of the specimen is affected by a constant bending moment, discontinuities, peculiar to lattice structures, are mediated over a larger volume.

We defined our design volume as a 4-by-4 u.c. square section specimen (H= 48 mm) and 32 u.c. along the longitudinal dimension (L = 192 mm). The pin distances were set to 144 mm and 48 mm for the supporting and for the pushing pins respectively.

Then, we voxelized the design space into cubic sub volumes with 12 mm edges with the idea of substituting each voxel with a specific type of lattice unit cell. We defined the dimensions of the voxels, and thus the u.c., on the basis of printing tests carried out with a Prusa i3 MK3S Original kit 3D printer having a 0.4 mm nozzle and using a commercial bio-based filament (polylactic acid, PLA).

To populate the design space, we chose two types of u.c. (i.e., non-hierarchical u.c., NH-u.c.) that are characterized by different mechanical behaviors, good geometric compatibility, and that are easy to be 3D printed: the bending dominated simple cubic cell (SC) and the stretching dominated face-centered cubic cell (FCC). A qualitative representation of their different behavior and a general representation of the followed workflow is visible in figure 1.

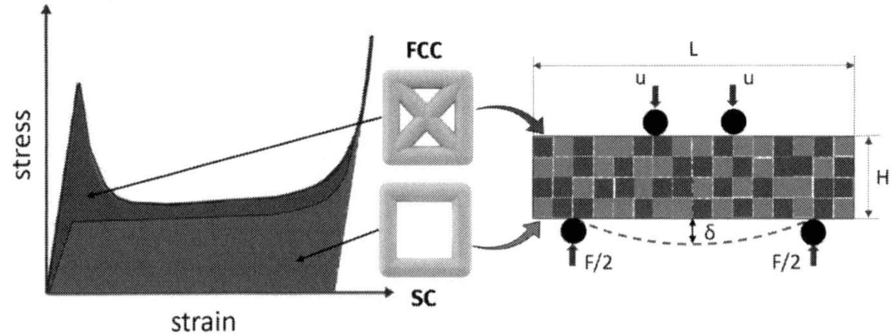

Figure 1. Representation of the study rationale. We first selected the SC and FCC lattice u.c.s for their different mechanical behaviour and good geometric compatibility. Then we populate a beam specimen –voxel by voxel–with the chosen u.c.s. The specimen has square cross section and nominal dimensions L = 192 mm and H = 48 mm and is subjected to a 4PB load. In the figure F represents the reaction force, u the imposed displacement and δ the maximum specimen deflection.

Combining the selected unit cells at different levels, we were able to introduce structural hierarchy in the specimen. To obtain the hierarchical u.c. (H-u.c.), we exploited a nested logic: we emptied the material present in the above-mentioned unit cell struts and substituted it with arrays composed of five SC or FCC unit cells. A representation of the hierarchical unit cell modeling logic adopted is shown in figure 2(a), while figure 2(b) depicts the four types of H-u.c. that we obtained by combining SC and FCC cells on two different length scales. Following the convention used in literature [26–28], the names of the H-u.c. were defined through acronyms consisting of two parts: "x-y", where "x" describes the level 1, and "y" describes the level 2. H-u.c. also can be classified into two families: self-similar H-u.c., i.e., those obtained by combining the same type of NH-u.c. with each other (SC-SC and FCC-FCC models), and hybrid H-u.c., i.e., those obtained by combining different types of NH-u.c. (SC-FCC and FCC-SC models). To comply with the constraints imposed by the 3D printer mentioned above the edge length of the NH-u.c. was set equal to 12 mm while for the H-u.c. edge length was set at 3 mm (figure 2(a)).

With the defined lattice u.c. we created six different architected lattice specimens, two non-hierarchical (NH-specimens) and four hierarchical (H-specimens). The six models were obtained by substituting in the entire design space the SC, FCC, SC-SC, SC-FCC, FCC-SC, FCC-FCC u.c., respectively. The representation of the different specimens is showed in figure 3.

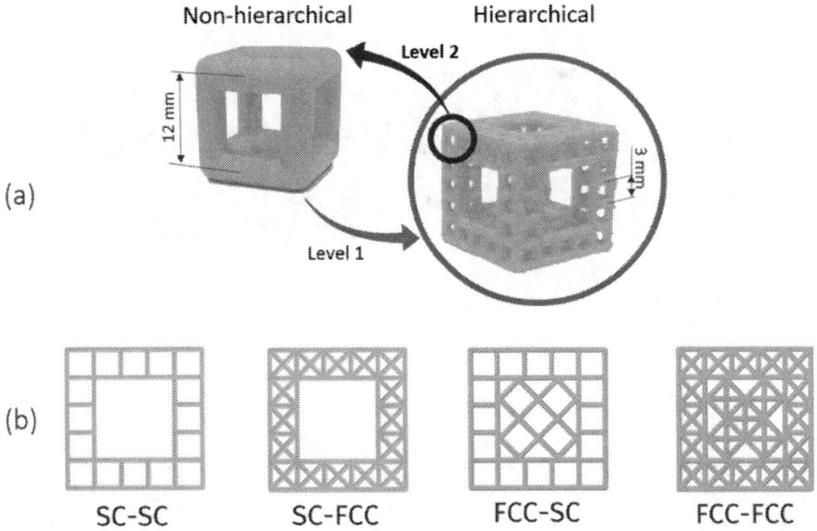

Figure 2. Hierarchical design of lattice unit cells: (a) representation of the modelled structural levels; (b) schematization of the four different hierarchical (or level 2) u.c. used in our work. The edge length dimensions used for this study are equal to 12 mm and to 3 mm for the NH-u.c. and the H-u.c. respectively.

The names of the four lattice specimens' configurations echo those of the unit cells used to populate the design space and are indicative of the behavior of the different structures. For example, the SC-FCC specimen has the SC structure at the macro level and the FCC structure at the meso level. Thus, it is a specimen characterized by hybrid bending-stretching behavior.

2.2. Numerical simulations

To investigate the mechanical behavior of the architected lattice specimens, we set up numerical simulations in ANSYS Mechanical APDL. This software allowed us to efficiently formulate customized calculation routines (i.e., macros) capable of correctly defining the architected u.c. and combine them together to create the specimen geometries in an automated manner. Parametric lattice specimen models were defined using 3D 2 node beam elements with linear shape function (beam 188 with KEYOPT(3)=0 in the Ansys MAPDL element library). An iterative cycle was set up to automatically compute the solution of the different designs. The use of beam elements, despite allowing only a simplified representation of the real phenomena to be studied, made it possible to evaluate about 50 different geometric configurations in a reasonable time frame and with limited computational costs. The exploitation of geometrical symmetry, material properties, and boundary conditions, present in all the designed lattice models, made it possible to further speed up the analyses.

All the simulations were performed in displacement control, assigning a negative deflection $u = 0.3$ mm to the nodes of the finite element models located under the bending pins. A convergence analysis was done to define the maximum size of the FE mesh. Specifically, in the NH-specimens each strut was discretized into ten beam elements, while for the H-specimens each strut of the smallest u.c. was represented by two FEs. A homogeneous and isotropic linear elastic material model was used. Material properties were derived from an experimental campaign of a commercial PLA filament.

IOP Conf. Series: Materials Science and Engineering 1275 (2023) 012003 doi:10.1088/1757-899X/1275/1/012003

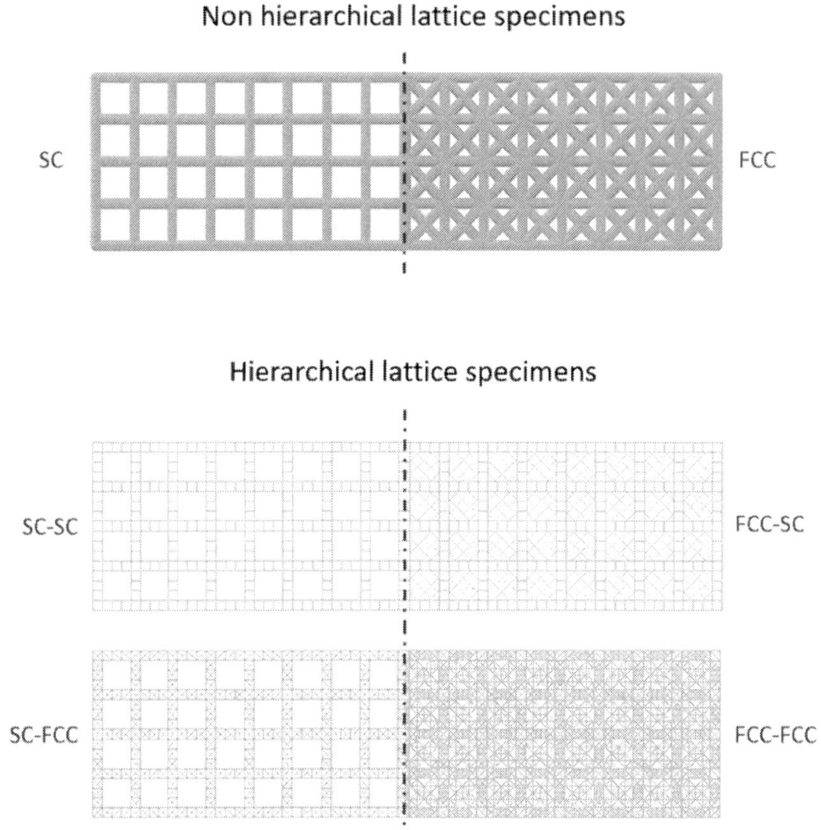

Figure 3. Front view representation of the modelled non-hierarchical (top) and hierarchical (bottom) lattice specimens. The specimens are symmetrical with respect to their midplane, so for each model, we depicted only one half.

2.3. Performance metrics

We analyzed the performance of the lattice specimens designed in terms of stiffness through four different metrics. To characterize our specimens as structures, we used the slenderness ratio (s.r.) of the u.c. lattice struts and the bending stiffness (K_b) of the specimen, defined respectively as:

$$s.r.=d/l \qquad (1)$$

$$K_b = F/\delta \qquad (2)$$

where d and l are the strut diameter and strut length of the smallest u.c. (see figure 4), respectively; F is the force exerted by the bending pins, δ is the maximum deflection experienced by the specimen (see figure 1).

Once the u.c. edge length has been fixed, we set the s.r. range from 0.15 to 0.45, then the strut diameters were calculated accordingly. The same s.r. were used for both the H-u.c. and the NH-u.c.. This strategy for defining the s.r. is consistent with previous studies [29].

To characterize our specimens as materials we employ the relative density (ρ_{rel}) and the relative elastic modulus (E_{rel}) of the lattice material, defined respectively as:

$$\rho_{rel}=\rho^*/\rho_{bulk} \quad (3)$$

$$E_{rel}=E^*/E_{bulk} \quad (4)$$

where ρ^* and E^* are the apparent density and the apparent modulus of the lattice structure, and ρ_{bulk} and E_{bulk} are the density and the Young modulus of the base material, respectively.

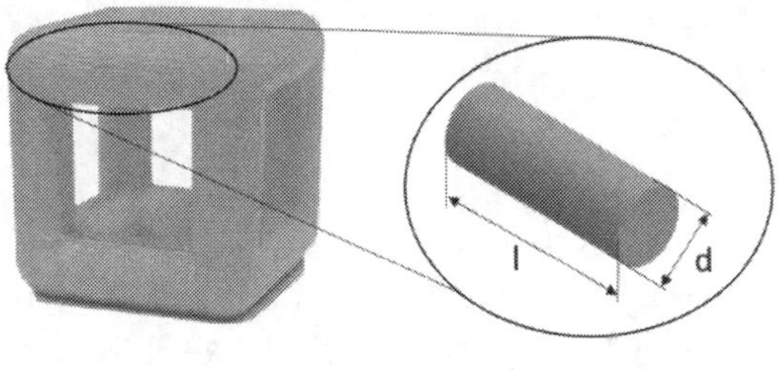

Figure 4. Geometric control parameters of the lattice unit cell: l represents the length of the strut and d the strut diameter.

3. Results and discussion

3.1. Structural hierarchy effect

We first analyzed the effect of structural hierarchy on the stiffness of lattice specimens. figure 5(a) shows the stiffness of the lattice specimens as a function of the slenderness ratio of the u.c. struts, while figure 5(b) shows the same plot but with the stiffness normalized to the weight W of the lattice specimens.

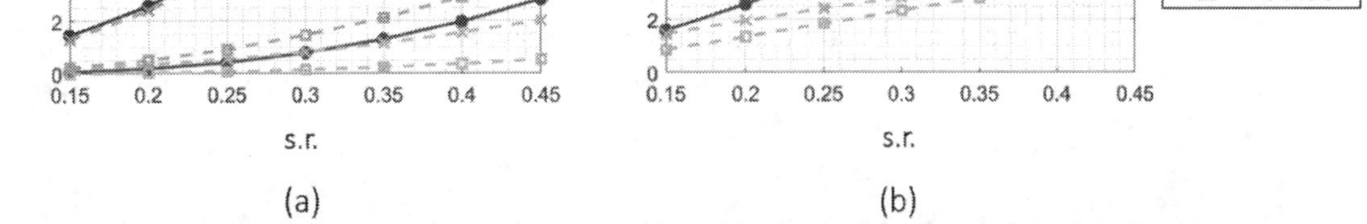

Figure 5. Finite element analysis results of lattice specimen models. (a) Trend of the lattice specimen stiffness as a function of the slenderness ratio of the lattice unit cell struts. (b) Trend of lattice specimen stiffness normalized to their weight as a function of the slenderness ratio of the lattice unit cell struts.

As expected, based on literature results [14], the NH-specimens show significantly different behavior: both show an increase in stiffness as the slenderness ratio of the u.c. struts (and thus their diameter) increases (see figure 5(a)), but the stretching-dominated FCC specimens are much more efficient. Indeed, at the same s.r., they show consistently higher performance than non-hierarchical bending-dominated lattice specimens. Moreover, regardless of their weight, they keep approximately constant structural performance within the considered s.r. range (see figure 5(b)).

Looking at the results for H-specimens, on the other hand, it can be seen that–in terms of stiffness (see figure 5(a))–the introduction of stretching-dominated FCC substructures does not appreciably change the mechanical behavior of lattice specimens, while the introduction of bending-dominated SC substructures is not advantageous, as it leads to worse performance than NH-specimens. Looking at the weight-normalized performances (see figure 5(b)), there is a significant change in slope between hierarchical and non-hierarchical structures, regardless of the type of substructure present at level 2. Specifically, reasoning from a lightening viewpoint, we can say that hierarchy is only beneficial when introduced in bending-dominated NH-specimens since, as the amount of material used decreases, the loss of stiffness they exhibit compared to non-hierarchical models is less. For non-hierarchical stretching-dominated specimens, structural hierarchy is advantageous only if, assigned a specific value of s.r., stretching-dominated lattices are introduced at level 2, as they lead to better performance than the corresponding NH-specimens.

3.2. Scaling laws

Then we investigated the behavior of lattice specimens through homogenized properties and thus as equivalent materials. It is known, from the literature [14], that the scaling law of the relative elastic modulus of lattice materials with their relative density is of the type:

$$E_{rel} \propto \rho_{rel}^{n} \tag{5}$$

where the exponent n takes value 2 for bending-dominated structures and 1 for stretching-dominated structures [14]. We then used these results as a benchmark for the analysis of the models we built. Figure 6 shows the interpolating curves representing the scaling laws of the relative elastic modulus as the relative densities of the lattice specimens we modeled. Table 1 shows the numerical values of the scaling law exponents of the studied lattice materials and the corresponding theoretical values for the bending- and stretching-dominated cases.

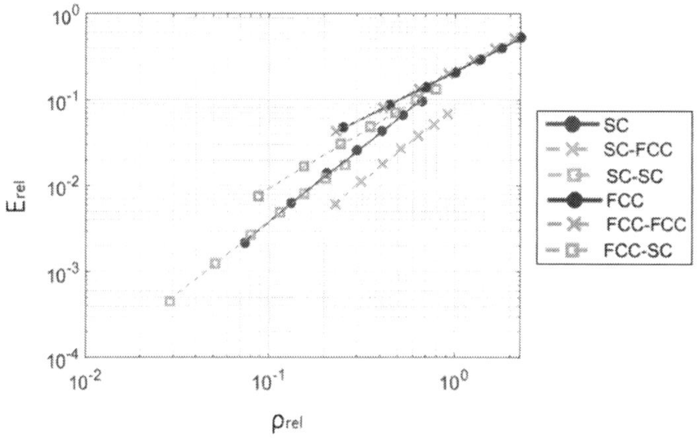

Figure 6. Scaling laws of relative elastic modulus for numerically simulated lattice specimen models.

We can see from these results that the self-similar hierarchical lattice structures, i.e., bending-bending (SC-SC) or stretching-stretching (FCC-FCC) keep the same trends as the corresponding NH-structures (SC and FCC, respectively), allowing the application range of these materials to be extended more or less significantly. On the contrary, the hybrid hierarchical lattice structures, i.e., bending-stretching (SC-FCC) and stretching-bending (FCC-SC), exhibit different trends than the lattice structures from which they were derived. In general, it can be said that hybridization leads to structures that are prone to bending-dominated behaviour.

Table 1. Scaling law exponents: theoretical values and numerical values, obtained from the FE-simulation performed in this study.

Scaling law exponent		
Model	**n**	
Bending dominated	2.00	Theoretical behavior
Stretching dominated	1.00	
SC	1.60	Numerical results
SC-SC	1.66	
SC-FCC	1.73	
FCC	1.10	
FCC-SC	1.30	
FCC-FCC	1.13	

4. Conclusions

In our work we set up a parametric and automated numerical framework to systematically analyse NH- and H- lattice specimens under 4PB loading scenario. In total we look at six different specimen topologies. Two NH-specimens, which we constructed by uniformly filling the design space with bending-dominated SC or stretching-dominated FCC lattice u.c.s, and four H-specimens, which we built by combining the same type of NH-u.c.s (SC-SC and FCC-FCC i.e., self-similar models) and the two different types of NH-u.c.s (SC-FCC and FCC-SC i.e., hybrid models). Changing their geometrical parameters, more than forty configurations were calculated. From the obtained results, the hierarchical self-similar lattice structures showed scaling laws of the relative elastic modulus with respect to the relative density very similar to those of the non-hierarchical lattice structures from which they were derived, whilst in terms of structural stiffness they showed significant changes in the mechanical efficiency of the specimens in bending. This means that self-similar structural hierarchy can be advantageously exploited to lighten structural parts and/or increase their specific mechanical performance (i.e., K_b, E_{rel}). Bending-stretching or stretching-bending structural hybridization, on the other hand, has led to materials with significantly different scaling laws than those of non-hierarchical lattice structures. Also, generalizing according to literature [24,26], is advantageous in terms of stiffness only if the structural hierarchy is designed by introducing stretching-dominated substructures within bending-dominated lattices. For both classes of the above-mentioned lattice specimens, the mechanical behaviour seems to be predominantly determined by the structural organization on the macro scale and thus by level 1 or non-hierarchical.

Looking beyond mechanical functions, the structural hierarchy opens the way for applications where lattice materials can perform multiple functions simultaneously. Specifically, one feature of lattice structures that undergoes significant changes with hierarchy is the surface-to-volume ratio. The latter greatly influences capabilities such as heat transfer, mechanical filtering, and shielding to electromagnetic waves. Therefore, extending the understanding of the PSPP relationship is of paramount importance not only in the field of mechanical design. This preliminary study lays the foundation for further analytical, experimental, and numerical investigations aimed at extending the characterization of hierarchical lattice materials subjected to bending loads from a multidisciplinary perspective, including the effect of local relative density gradients or material properties resulting from combinations of heterogeneous structures over multiple length scales.

References

[1] Huang W, Restrepo D, Jung J Y, Su F Y, Liu Z, Ritchie R O, McKittrick J, Zavattieri P and Kisailus D 2019 Multiscale Toughening Mechanisms in Biological Materials and Bioinspired Designs *Adv. Mater.* **31** 1–37

[2] Chen P Y, McKittrick J and Meyers M A 2012 Biological materials: Functional adaptations and bioinspired designs *Prog. Mater. Sci.* **57** 1492–1704

[3] Libonati F and Buehler M J 2017 Advanced Structural Materials by Bioinspiration *Adv. Eng. Mater.* **19** 1600787

[4] Musenich L and Libonati F 2022 Damage and Failure Mechanisms of Biological Materials *Ref. Modul. Mater.* Sci. Mater. Eng.

[5] Schaedler T A and Carter W B 2016 Architected Cellular Materials *Annu. Rev. Mater. Res.* **46** 187–210

[6] Tamburrino F, Graziosi S and Bordegoni M 2018 The design process of additively manufactured mesoscale lattice structures: A review *J. Comput. Inf. Sci. Eng.* **18** 1–16

[7] Plocher J and Panesar A 2019 Review on design and structural optimisation in additive manufacturing: Towards next-generation lightweight structures *Mater. Des.* **183** 108164

[8] Pham M S, Liu C, Todd I and Lertthanasarn J 2019 Damage-tolerant architected materials inspired by crystal microstructure *Nature* **565** 305–11

[9] Manno R, Gao W and Benedetti I 2019 Engineering the crack path in lattice cellular materials through bio-inspired micro-structural alterations *Extrem. Mech. Lett.* **26** 8–17

[10] Benedetti M, du Plessis A, Ritchie R O, Dallago M, Razavi S M J and Berto F 2021 Architected cellular materials: A review on their mechanical properties towards fatigue-tolerant design and fabrication *Mater. Sci. Eng. R Reports* **144** 100606

[11] Xia X, Spadaccini C M and Greer J R 2022 Responsive materials architected in space and time *Nat. Rev. Mater.* **7** 683-701

[12] Libonati F, Graziosi S, Ballo F, Mognato M and Sala G 2021 3D-Printed Architected Materials Inspired by Cubic Bravais Lattices *ACS Biomater. Sci. Eng. DOI:10.1021/acsbiomaterials.0c01708*

[13] Gibson, Lorna J., Ashby M F 1997 *Cellular Solids: Structure and Properties* Cambridge University Press

[14] Ashby M F 2006 The properties of foams and lattices *Philos. Trans. R. Soc. A Math. Phys. Eng. Sci.* **364** 15–30

[15] Somnic J and Jo B W 2022 Homogenization Methods of Lattice Materials *Encyclopedia* **2** 1091–1102

[16] Olson G B 1997 Computational design of hierarchically structured materials *Science.* **277** 1237–1242

[17] Perricone V, Santulli C, Rendina F and Langella C 2021 Organismal design and biomimetics: A problem of scale *Biomimetics* **6.4** 56

[18] Fischer S C L, Hillen L and Eberl C 2020 Mechanical metamaterials on the way from laboratory scale to industrial applications: Challenges for characterization and scalability *Materials (Basel).* **13** 1–16

[19] Deshpande V S, Ashby M F and Fleck N A 2001 Foam topology: Bending versus stretching dominated architectures *Acta Mater.* **49** 1035–1040

[20] Ajdari A, Jahromi B H, Papadopoulos J, Nayeb-Hashemi H and Vaziri A 2012 Hierarchical honeycombs with tailorable properties *Int. J. Solids Struct.* **49** 1413–1419

[21] Oftadeh R, Haghpanah B, Vella D, Boudaoud A and Vaziri A 2014 Optimal fractal-like hierarchical honeycombs *Phys. Rev. Lett.* **113** 104301

[22] Oftadeh R, Haghpanah B, Papadopoulos J, Hamouda A M S, Nayeb-Hashemi H and Vaziri A 2014 Mechanics of anisotropic hierarchical honeycombs *Int. J. Mech. Sci.* **81** 126–136

[23] Sen D and Buehler M J 2011 Structural hierarchies define toughness and defect-tolerance despite simple and mechanically inferior brittle building blocks *Sci. Rep.* **1** 1–9

[24] Vigliotti A and Pasini D 2013 Mechanical properties of hierarchical lattices *Mech. Mater.* **62** 32–43
[25] Meza L R, Zelhofer A J, Clarke N, Mateos A J, Kochmann D M and Greer J R 2015 Resilient 3D hierarchical architected metamaterials *Proc. Natl. Acad. Sci. U. S. A.* **112** 11502–11507
[26] Banerjee S 2014 On the mechanical properties of hierarchical lattices *Mech. Mater.* **72** 19–32
[27] Bauer J, Meza L R, Schaedler T A, Schwaiger R, Zheng X and Valdevit L 2017 Nanolattices: An Emerging Class of Mechanical Metamaterials *Adv. Mater.* **29** 1701850
[28] Fleck N A, Deshpande V S and Ashby M F 2010 Micro-architectured materials: Past, present and future *Proc. R. Soc. A Math. Phys. Eng. Sci.* **466** 2495–2516
[29] Sha Y, Jiani L, Haoyu C, Ritchie R O and Jun X 2018 Design and strengthening mechanisms in hierarchical architected materials processed using additive manufacturing *Int. J. Mech. Sci.* **149** 150–163

IOP Conf. Series: Materials Science and Engineering 1275 (2023) 012004 doi:10.1088/1757-899X/1275/1/012004

Smart design of customized hip prostheses in additive manufacturing by combining numerical and experimental methodologies

D Milone[1*], C De Marchis[1], F Longo[1], G Merlino[1], L D'Agati[1], D Catelani[2] and G Risitano[1]

[1] Department of Engineering, University of Messina, Contrada di Dio, 98166 Messina, Italy

[2] MSC Software Corporation,Via Santa Teresa 12,10121, Torino, Italy

*Email: dario.milone@unime.it

Thanks to the development of additive manufacturing techniques, prosthetic surgery has reached increasingly advanced levels, revolutionizing the clinical course of patients with joint arthritis. 3D printing has made it possible to obtain customized prostheses based on patient needs, using high-performance materials. However, wear caused by regular gait activities such as walking, sitting, or running, leads to the deterioration of the material used in the joint. Thus, the use of traditional materials has gradually been replaced with more performing ones which have made it possible to obtain customized devices based on patient needs and, therefore, more effective. Numerical techniques have recently been adopted, such as the Finite Element Method (FEM), to support the experimentation, allowing the calculation of the useful life and the optimization of the prostheses' functionality to accurately evaluate the distribution of the load on the prosthesis. The present work aims to develop an algorithm that optimizes hip replacement mechanically using a machine learning algorithm coupled with multi-body and finite element model simulations.

1. Introduction

Thanks to the development of additive manufacturing techniques, prosthetic surgery has reached increasingly advanced levels, revolutionizing the clinical course of patients with joint arthritis, rheumatoid arthritis, post-traumatic arthrosis, and congenital dysplasia. 3D printing has made it possible to obtain customized prostheses based [1] on the patient needs, using high-performance materials such as 316L stainless [2]steel or Ti-6Al-4V. However, wear caused by regular gait activities such as walking, sitting, or running, leads to the deterioration of the material used in the joint [3–5]. The use of traditional materials has gradually been replaced with more performing ones [6–8] which have made it possible to obtain customized devices based on patient needs and, therefore, more effective. Numerical techniques have recently been adopted, such as the Finite Element Method (FEM), which, to support the experimentation, allows the calculation of the useful life and the optimization of the functionality of the prostheses, allowing the evaluation of the load distribution on the prosthesis accurately[9]. The

Content from this work may be used under the terms of the Creative Commons Attribution 3.0 licence. Any further distribution of this work must maintain attribution to the author(s) and the title of the work, journal citation and DOI.
Published under licence by IOP Publishing Ltd

IOP Conf. Series: Materials Science and Engineering 1275 (2023) 012004 doi:10.1088/1757-899X/1275/1/012004

development of human models has increased with the need to understand the biomechanics of movements and their consequences on human comfort [10–13]. These models are always used to simulate the influence of prosthetic design on the human system [14–19].

In recent years, particular importance has been given to modeling the mandible and its interactions with prostheses. The CT scan characterized the bone structure. In the literature, various models allow extrapolation of the mechanical characteristics, restraints and loads on certain bodies subject to specific actions [20–23]. Researchers are increasingly focusing on creating models that simulate the whole body rather than limiting themselves to small and specific sections [24].

Multibody simulations find an excellent application in sports biomechanics [25]. On this basis, the definition of the dynamic equation of human movement and the acquisition of the dynamic parameters of human action are critical steps in the theoretical analysis of the biomechanics of movement. Another field of dynamic simulation is the analysis of the interaction between human exoskeletons [26].
Finally, motion capture often defines the trajectories of the body's joints in multi-body simulations. With this technique, the movement of an object is recorded through various video capture devices placed in the space around it.[27]

From the recorded images, the coordinates (X, Y, Z) of a series of markers affixed to the subject; allow for quantifying the position, speed, and acceleration of these points and, consequently, the movement produced by the subject during a given action.

To investigate the effect of different propulsion on the stresses acting on mechanical components, Previti et Al. [28] proposed a new agile and light methodology, based on automatic evaluation of the cumulative stress through the interactions of different software (multibody, FEM, etc.) and models (consumption, vehicle dynamics, structural, etc.) in the Matlab environment. This new way of verifying and re-designing mechanical components has been called "Smart Design" by the authors. Indeed, in the modern era, it is impossible to think that structural design does not make use of virtual, digital, rapid and secure prototyping methods. The present work aims to develop an algorithm that optimizes hip replacement mechanically using a machine learning coupled with multi-body and finite element methods simulations (Smart Design Method). The innovative aspect is represented by the use of:

- Artificial intelligence to evaluate the real human movement kinematics.
- A parametric humanoid model in the MSC ADAMS® environment.
- A FEM model, which allows the optimization of the prosthesis, thus suggesting the best geometry using the Smart Design Method.

2. Materials and Methods

Most of the musculoskeletal and nervous systems participate in walking, for humans. The forces and twists exchanged by the lower limbs during walking are transmitted through the joints, ligaments, and muscles to the trunk. The accelerations generated by the movement of the upper limbs and, to the extent permitted by the vertebral joints and the trunk, generate forces of inertia to balance the body and combine with those generated by the lower limbs to generate the desired movement [29,30]. Walking was chosen as the reference movement to optimize the prosthesis since it is the most frequent act in daily activities. Once the type of movement to be analyzed was defined, the algorithm to optimize the hip replacement was defined. The algorithm consists of five steps:

1) Motion capture: capturing the subject's pose during movement.
2) Extraction of the joint angles starting from the coordinates of the markers .
3) Creation of humanoid model: the creation of a parametric multibody model having the dimensions and weight distribution of the subject under analysis.
4) FEM analysis of the prosthesis: calculation of the stresses acting on the prosthesis based on the loads to which the hip joint is subjected.
5) Optimization of the results: shape optimization algorithm based on the gradient descent method.

2.1. *First step – Motion Capture*

Human pose estimation is the process of inferring a subject's poses directly from a digital image. Pose estimation requires highly accurate detection and identification of human joints[31–35]. Pose estimation algorithms follow a top-down or bottom-up approach. The execution time of these top-down approaches [36–39] is proportional to the number of people: for each survey, a pose estimator was performed for one person, and the higher the number of people, the higher the computational cost. Conversely, bottom-up approaches are interesting because they offer robustness to early engagement and have the potential to decouple the complexity of the runtime from the number of people in the image. However, bottom-up approaches do not directly use global contextual cues from other parts of the body and other people. In practice, the previous bottom-up methods[40,41] do not preserve the efficiency gains since the final analysis requires a costly global inference. The present paper[42–44] evaluated a model (OpenPose) to define the subject's pose during wolking. OpenPose is a bottom-up, multi-person human pose detection library that detects the human body, foot, hand and face key points on single images. It can detect 135 vital points of the body from a digital image. A single CNN is used for both key point detection and association. A numerical score between 0 and 1 is assigned when key points are found. It is a measure of overall confidence in the estimated key points. OpenPose has been trained to produce three different pose patterns. The difference is represented by the number of points identified, MPI can estimate a total of 15 key points, COCO can estimate a total of 18 points ,BODY_25 can estimate a total of 25 points. The most comprehensive pose model is the third. In addition to the key points estimated by the MPI and COCO models, it contains the feet and pelvic centre descriptors. There is also a new experimental OpenPose model called body_25b (Figure 1).

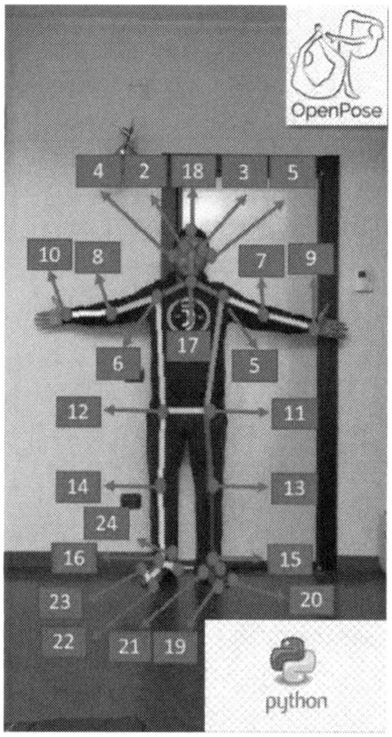

Figure 1. Keypoints from Model body_25b Openpose.

As reported in [45,46], this model has the highest accuracy parameters, and it is, therefore, more accurate than the default body_25 and reduces the number of false positives. The definition of the key point differs from MPI for evaluating the key points of the head and neck and removes the key point of the model body's neck and mid-hip of the model body_25. The first step implemented was classifying all points captured by the algorithm. As in Figure 1, the algorithm can identify 25 points. The subject's movement was captured with an action camera placed in front of the subject.

2.2. Second step – Extraction of joints angles

The calculation of the joint angles characterized the second phase of the experimentation. The subject was positioned in front of the camera in a vertical position. The subject's initial pose was defined with an angle of 180 between the trunk and leg. A video of his movement was recorded. The subject performed the walking exercise. Openpose has extrapolated the pose. As in [47] the knee angle was calculated using the product of the vector point.

$$\theta = \cos^{-1}\left(\frac{\vec{a} * \vec{b}}{|\vec{a}||\vec{b}|}\right) \tag{1}$$

Two vectors were constructed from the hip, knee and ankle coordinates obtained from the pose data. The Figure 2 shows the process of identifying and calculating vectors.[48]

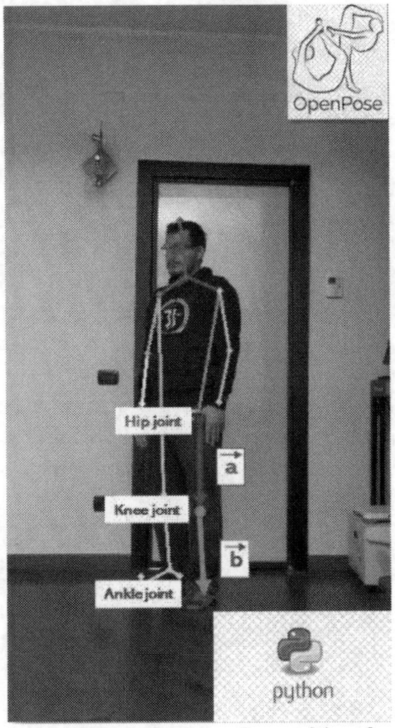

Figure 2. Angle measurements from video frame [46,47]

Once the vectors have been generated, the angles have been defined as in [47]. From the description of angles as functions of time, the laws of motion in MSC Adams were created.

2.3. Third step – Human Model

The development of virtual human models has grown with the need to understand the biomechanical nature of movements to understand the consequences that affect comfort. Simulation techniques such as the multibody study the movement of the whole body as they consider it a set of components connected by joints. Multibody simulations [22] find excellent application in sports biomechanics [25]. On this basis, the definition of the dynamic equation of human movement and the acquisition of the dynamic parameters of human action constitute the critical steps in the theoretical analysis of the biomechanics of movement. The model presented in the paper has been divided into three main macro-areas:

- Upper limbs
- Torso
- Lower limbs.

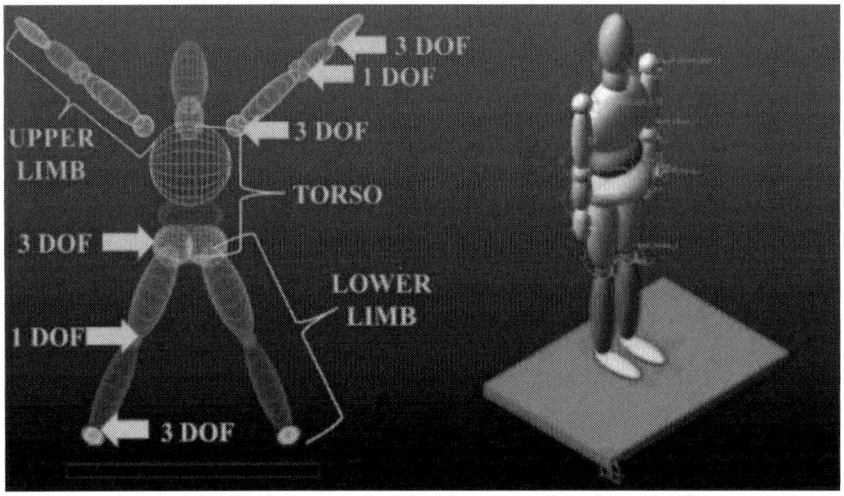

Figure 3. Dummy model schematization from [48]

The manikin was defined as in [48] The dimensions that characterize it are 32 as in [24]. The data reported in Table 1 were considered for evaluating the subject under examination. The dimensions of the manikin were identified as a function of known data, using linear regression,

Table 1. Subject characteristics.

Human characteristics	
Weight	85 kg
Standing Height	1755 mm
Seated Height	918 mm

Once the mannequin model has been defined, as reported previously, the state variables have been created. The set of variables represents the angle variation in the three components of the 25 points identified by the motion capture algorithm. As shown in Figure 4, the system's input is the neural network output. At the same time, the three components relating to the forces along the three axes of rotation of the spherical joint located in the hip are calculated at the output.

IOP Conf. Series: Materials Science and Engineering 1275 (2023) 012004 doi:10.1088/1757-899X/1275/1/012004

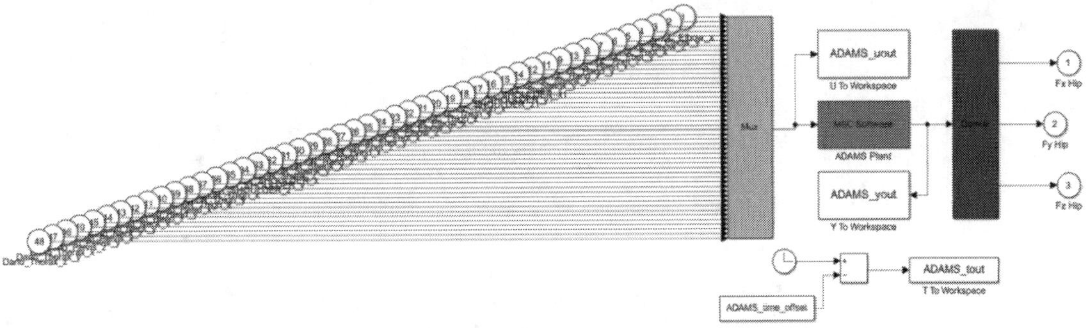

Figure 4. Model for load calculus from the multibody system

The system receives the inputs from Openpose, instant by instant, it transfers the data on the angles reproduced by each joint, and the mannequin reproduces the movement (Figure 5)

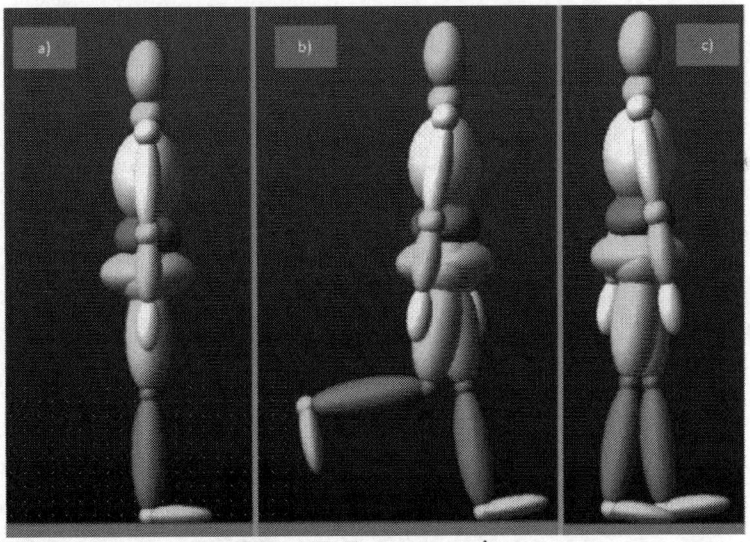

Figure 5. Dummy motion.

The reaction forces due to the contact between the floor and the feet have been calculated. Finally, the forces acting on the hip are calculated (Figure 6) and then transferred to the fem model.

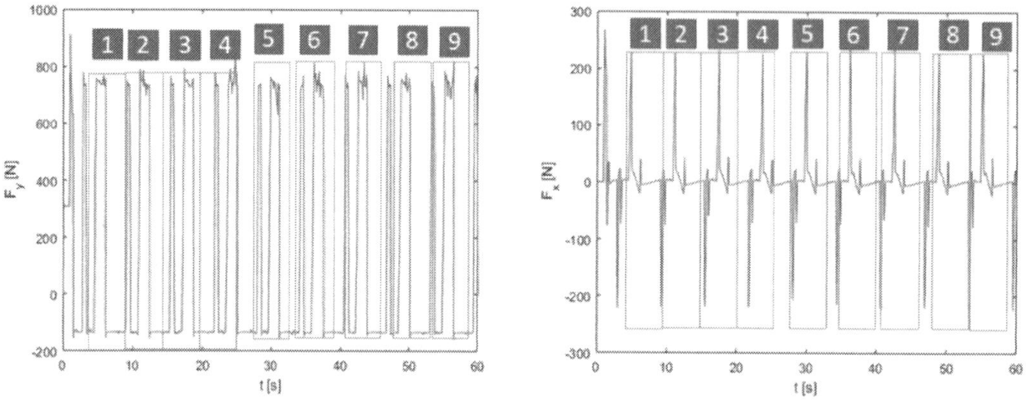

Figure 6. Forces applied on hip joint.

The nine walking cycles within Figure 6 have been highlighted, equivalent to the 18 steps taken during the 60 s simulation.

2.4. Fourth step – FEM Analysis

The fourth phase is characterized by the structural analysis of the "hip prosthesis" component. A sensitivity analysis was performed based on the system displacement values reproduced at the output. For this reason, the mesh was thickened until a displacement value equal or less than 2% compared to the previous simulation was obtained. Figure 7 shows the outputs of this procedure. The mesh used for the geometry consists of second order tetrahedral (SOLID186) elements with a lower limit of 1 mm in size for the prosthesis. The mesh has been generated through the "MultiZone method".

Table 2. Mesh properties

:Mesh properties	
Node	150,000
Elements	105,000

Table *2* shows the average number of nodes and elements for analysis during the optimization process. It is considered an average number because, during each cycle, the geometer changes its size.

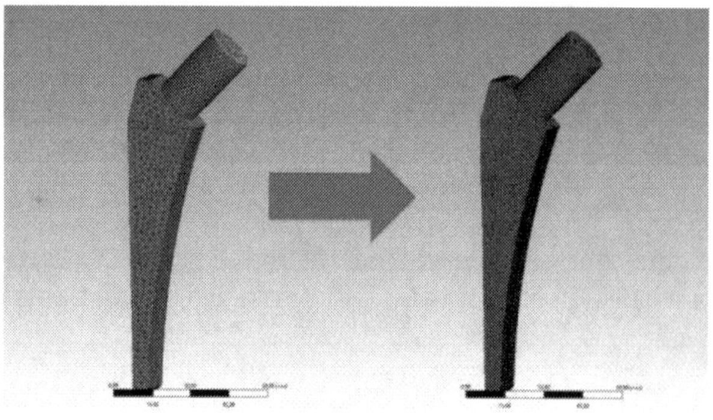

Figure 7. Mesh sensitivity.

The loads calculated in the multi-body system were used as forcing in the model. Schematically, the system was tested as follows:

1) Constraints: the constraint is of the fixed type, applied on the lower faces of the prosthesis.

2) Loads: concentrated load applied on the horizontal face of the prosthesis neck.
The constraints are represented in blue, while the loads are secured according to their lines of action using a different colour for each axis (Figure 8).

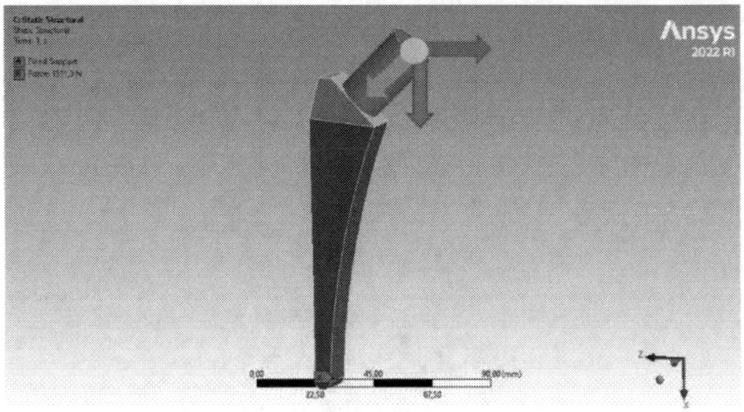

Figure 8. Load and constraint of the system.

For the choice of material, AISI 316 L was selected, a material studied in the literature to produce custom hip prostheses[48]. Specimens were made, and the material was tested to obtain its properties. An analysis was also carried out using the static thermographic method. The characteristics of the material used in the FEM are shown Table 3.

Table 3. AISI 316L properties.

Material Properties	
Density[kg/mm^3]	7954
Young Modulus [GPa]	195
Poisson Ratio	0,25
Bulk Modulus [GPa]	190
Shear Modulus [GPa]	78
Yeld Strength [MPa]	250
Tangent Modulus [MPa]	2091

2.5. Fifth step – Optimization

2.5.1. Shape optimization. The last phase of the analysis is characterized by optimization using the Smart Design Method. The algorithm used is FMINCON, a tool that looks for the minimizer of a scalar function of several variables within a region specified by linear constraints and limits. The algorithm follows the procedure shown in Figure 9.

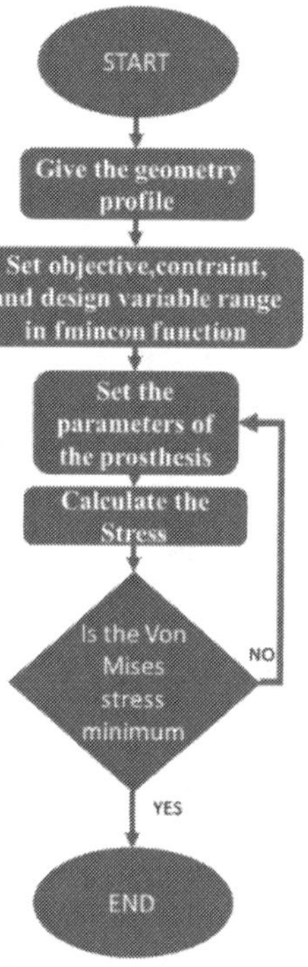

Figure 9. Shape optimization workflow.

The prosthesis dimensions were found in a well-known manufacturer's official catalogue. The starting parameters represent the characteristic measures for the subject under examination. As shown in Figure 10, three independent parameters were identified for the prosthesis study. Each iteration cycle has expected that one / or more of them change at the same time.

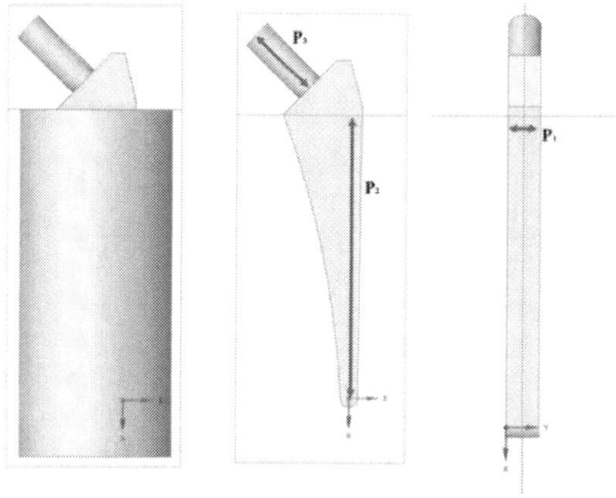

Figure 10. Shape optimization parameters.

Once the parameters were obtained, the geometry was created. An iteration consists in calculating the solution of the FEM model based on the assigned geometry. The first solution with characteristic prosthetic parameters for the chosen subject under examination was identified as the target. Subsequently, the solutions were compared with this goal. When the solution-processed in the "i" loop reported a value lower than the target, it was overwritten on it. The system iterated until it captured the minimum of the objective function.

2.5.2. Lattice optimization. Thanks to the rapid evolution of additive manufacturing today, it is possible to make high-strength open porous scaffolding constructed with beam elements [49–51], which can provide favourable property trade-offs when considering mechanical and biological factors. Parametric approaches can help design lattices for scaffold applications by exploiting parameter coupling and scale relationships, which has been a successful approach for designing complex biomechanical systems [52]. Cubic cells, with cross-rods centred on the body, were used as the basis for the reticular structures. Table 4 shows the characteristics chosen for the unit cell.

Table 4. Lattice cell features.

Cell features	
Minimum density	0,1
Maximum density	0,8
Lattice cell size [mm]	1

Once the properties attributed to the lattice were defined, a subsequent FEM analysis was defined using the same boundary conditions previously used. The following analysis identified the internal areas of the prosthesis in which the stress was lower than the critical one and superimposed them on reticular structures to decrease the mass of the sample while maintaining the same values as in the previous case. Figure 11 shows the obtained output.

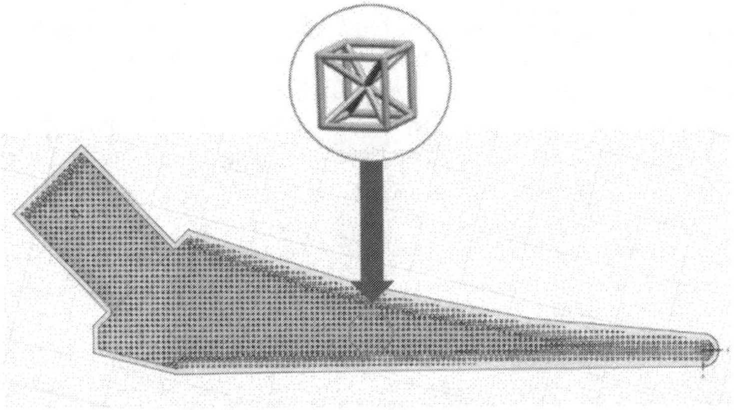

Figure 11. Lattice optimization output.

3. Results and Discussion

The optimizer realized led to a variation in the mass of the sample under examination and the dimensions. In the first phase, focusing on the variation of the parameters, it can be seen that the prosthesis is thicker and longer, but the neck of the same is stockier than in the case initially considered. Table 5 shows the change in size between the initial prosthesis and the one in which the shape has been optimized for loading.

Table 5. Shape optimization parameters.

	Shape Optimization	
	Original Dimensions	Optimized Dimensions
P_1 [mm]	15	18
P_2[mm]	111	114,78
P_3[mm]	26	25

Focusing on the geometry of the prosthesis, Table 6 shows the dimensional variation in terms of volume, it also shows the variation in mass and finally, the von mises stress difference calculated on the system.

Table 6. Volume and mass variation.

	Shape Optimization		
	Original Dimensions	Optimized Dimensions	Variation
Volume [mm^3]	28012	35066	20%
Mass [kg]	0,22	0,28	21%
Stress [MPa]	120	72	39%

It can be seen that:

- There was an increase of around 20% in both the volume and mass of the prosthesis.
- The maximum Von Mises stress calculated by the system was reduced by 39%.

The data reported in the table above suggested using a topological optimization to contain the masses, keeping the stress unchanged. After the evaluation of the masses, after topological optimization, a mass reduction was obtained compared to the initial case of 18%. If the choice had been made to perform shape optimization, the result would have been the opposite - a 21% mass increase. In this case, comparing the two optimizations, there is a 35% decrease in mass in favour of topological optimization compared to simple shape optimization. *Table 7* shows the results obtained.

Table 7. mass properties of the system before optimization and after optimization.

	Shape Optimization		
	Original Dimensions	Shape Optimization	Lattice Optimization
Mass [kg]	0,22	0,28	0,18

The main objective of the activity was to apply a new approach to define a custom geometry for a hip replacement.

The approach using Deep Learning algorithms, allowed us to demonstrate that it is possible to identify the pose of a subject faster and at a lower cost than traditional techniques.

Using a parametric human model has broadened the horizons of subsequent evaluation since it allows to identify the dynamics of several subjects simultaneously. It also allows to define of a specific subject to be analysed.

The combined use of the multibody model with Openpose made it possible to reduce the evaluation times for gait analysis, thus allowing the identification of the loads acting on the hip during this cycle.

The shape algorithm allowed the optimization of the prosthesis's shape for the load to which it is subjected. Unfortunately, the shape optimization alone increased mass and volume, which required a second procedure. The topological optimization made it possible to reduce the masses by about 20% compared to keeping the stress on the system unchanged. Finally, it can be deduced that the AISI 316 L material is confirmed to be correct for the realization of such prostheses since, from the analyses conducted, the Von Mises stress is lower than the yield strength of the material and also the fatigue limit. Future developments of the present work foresee the optimization of the prosthesis under the action of a dynamic load varying over time and to overcome problems related to sudden load variations due to transients.

References

[1] Cucinotta F, Guglielmino E, Longo G, Risitano G, Santonocito D and Sfravara F 2019 Topology optimization additive manufacturing-oriented for a biomedical application *Lecture Notes in Mechanical Engineering* vol 1 (Springer International Publishing) pp 184–93

[2] Milone D, Risitano G, Pistone A, Crisafulli D and Alberti F 2022 A New Approach for the Tribological and Mechanical Characterization of a Hip Prosthesis Trough a Numerical Model Based on Artificial Intelligence Algorithms and Humanoid Multibody Model *Lubricants* **10** 160

[3] Scappaticci L, Risitano G, Santonocito D, D'Andrea D and Milone D 2021 An Approach to the Definition of the Aerodynamic Comfort of Motorcycle Helmets *Veh. 2021, Vol. 3, Pages 545-556* **3** 545–56

[4] Fellah M, Labaïz M, Assala O, Dekhil L, Zerniz N and Iost A 2014 Tribological behavior of biomaterial for total hip prosthesis *Mater. Tech.* **102**

[5] Banchet V, Fridrici V, Abry J C and Kapsa P 2007 Wear and friction characterization of materials for hip prosthesis *Wear* **263** 1066–71

[6] Kourra N, Warnett J M, Attridge A, Dibling G, McLoughlin J, Muirhead-Allwood S, King R and Williams M A 2018 Computed tomography metrological examination of additive manufactured acetabular hip prosthesis cups *Addit. Manuf.* **22** 146–52

[7] Murr L E 2018 Additive manufacturing of biomedical devices: an overview *Mater. Technol.* **33** 57–70

[8] Cortis G, Mileti I, Nalli F, Palermo E and Cortese L 2022 Additive manufacturing structural redesign of hip prostheses for stress-shielding reduction and improved functionality and safety *Mech. Mater.* **165** 104173

[9] Milone D, Fiorillo L, Alberti F, Cervino G, Filardi V, Pistone A, Cicciù M and Risitano G 2022 Stress distribution and failure analysis comparison between Zirconia and Titanium dental implants *Procedia Struct. Integr.* **41** 680–91

[10] Ackland D C, Moskaljuk A, Hart C, Vee Sin Lee P and Dimitroulis G 2015 Prosthesis Loading after Temporomandibular Joint Replacement Surgery: A Musculoskeletal Modeling Study *J. Biomech. Eng.* **137**

[11] Gröning F, Jones M E H, Curtis N, Herrel A, O'Higgins P, Evans S E and Fagan M J 2013 The importance of accurate muscle modelling for biomechanical analyses: a case study with a lizard skull *J. R. Soc. Interface* **10**

[12] Langenbach G E J and Hannam A G 1999 The role of passive muscle tensions in a three-dimensional dynamic model of the human jaw *Arch. Oral Biol.* **44** 557–73

[13] Shi J, Curtis N, Fitton L C, O'Higgins P and Fagan M J 2012 Developing a musculoskeletal model of the primate skull: Predicting muscle activations, bite force, and joint reaction forces using multibody dynamics analysis and advanced optimisation methods *J. Theor. Biol.* **310** 21–30

[14] Huston R L 1996 Multibody Dynamics Since 1990 *Appl. Mech. Rev.* **49** S35–40

[15] Müller A 2009 Generic mobility of rigid body mechanisms *Mech. Mach. Theory* **44** 1240–55

[16] D'Andrea D, Milone D, Nicita F, Risitano G and Santonocito D 2021 Qualitative and Quantitative Evaluation of Different Types of Orthodontic Brackets and Archwires by Optical Microscopy and X-ray Fluorescence Spectroscopy *Prosthesis* **3** 342–60
[17] Milone D, Nicita F, Cervino G, Santonocito D and Risitano G 2021 Finite Element Analysis of OT Bridge fixed prosthesis system *Procedia Struct. Integr.* **33** 734–47
[18] May B, Saha S and Saltzman M 2001 A three-dimensional mathematical model of temporomandibular joint loading *Clin. Biomech.* **16** 489–95
[19] Tamer A, Zanoni A, Cocco A and Masarati P 2021 A numerical study of vibration-induced instrument reading capability degradation in helicopter pilots *CEAS Aeronaut. J.*
[20] Cervino G, Cicciù M, Fedi S, Milone D and Fiorillo L 2020 FEM Analysis Applied to OT Bridge Abutment with Seeger Retention System
[21] Paschold H 2009 Whole-body vibration *ASSE Professional Development Conference and Exposition 2009*
[22] Young S N, VanWye W R and Wallmann H W 2018 Sport simulation as a form of implicit motor training in a geriatric athlete after stroke: a case report *https://doi.org/10.1080/09593985.2018.1488322* **36** 524–32
[23] Wu Y and Liu J 2021 Discover Internet of Things Research on college gymnastics teaching model based on multimedia image and image texture feature analysis **1** 15
[24] Cheng H, Obergefell L and Rizer A 1994 Generator of Body Data (GEBOD) , Manual.
[25] CAMERON P W, SOLTERO N C and BYERS J 2018 Effects of a 60 Minute on Ice Game Simulation on the Balance Error Scoring System *Int. J. Exerc. Sci.* **11** 462
[26] Chung H-J 2021 Optimization Based Dynamic Human Motion Prediction with Modular Exoskeleton Robots as Interactive Forces: The Case of Weight Lifting Motion *Collab. Humanoid Robot. [Working Title]*
[27] Mündermann L, Corazza S and Andriacchi T P 2006 The evolution of methods for the capture of human movement leading to markerless motion capture for biomechanical applications *J. Neuroeng. Rehabil.* **3** 1–11
[28] Previti U, Galvagno A, Risitano G and Alberti F 2022 Smart Design: Application of an Automatic New Methodology for the Energy Assessment and Redesign of Hybrid Electric Vehicle Mechanical Components *Veh. 2022, Vol. 4, Pages 586-607* **4** 586–607
[29] Winter D A 2009 Biomechanics and Motor Control of Human Movement: Fourth Edition *Biomech. Mot. Control Hum. Mov. Fourth Ed.* 1–370
[30] Whittle M W 2007 Gait Analysis *Gait Anal.*
[31] Ramakrishna V, Munoz D, Hebert M, Andrew Bagnell J and Sheikh Y 2014 Pose machines: Articulated pose estimation via inference machines *Lect. Notes Comput. Sci. (including Subser. Lect. Notes Artif. Intell. Lect. Notes Bioinformatics)* **8690 LNCS** 33–47
[32] Bulat A and Tzimiropoulos G 2016 Human pose estimation via Convolutional Part Heatmap Regression *Lect. Notes Comput. Sci. (including Subser. Lect. Notes Artif. Intell. Lect. Notes Bioinformatics)* **9911 LNCS** 717–32
[33] Ramanan D, Forsyth D A and Zisserman A 2005 Strike a pose: Tracking people by finding stylized poses *Proc. - 2005 IEEE Comput. Soc. Conf. Comput. Vis. Pattern Recognition, CVPR 2005* **I** 271–8
[34] Johnson S and Everingham M 2010 Clustered pose and nonlinear appearance models for human pose estimation *Br. Mach. Vis. Conf. BMVC 2010 - Proc.*
[35] Yang Y and Ramanan D 2013 Articulated human detection with flexible mixtures of parts *IEEE Trans. Pattern Anal. Mach. Intell.* **35** 2878–90
[36] Andriluka M, Roth S and Schiele B 2010 Pictorial structures revisited: People detection and articulated pose estimation 1014–21
[37] Wei S E, Ramakrishna V, Kanade T and Sheikh Y 2016 Convolutional Pose Machines *Proc. IEEE Comput. Soc. Conf. Comput. Vis. Pattern Recognit.* **2016-Decem** 4724–32
[38] Simon T, Joo H, Matthews I and Sheikh Y 2017 Hand Keypoint Detection in Single Images

using Multiview Bootstrapping *Proc. - 30th IEEE Conf. Comput. Vis. Pattern Recognition, CVPR 2017* **2017-January** 4645–53

[39] Pfister T, Charles J and Zisserman A 2015 Flowing ConvNets for Human Pose Estimation in Videos

[40] Insafutdinov E, Pishchulin L, Andres B, Andriluka M and Schiele B 2016 DeeperCut: A Deeper, Stronger, and Faster Multi-Person Pose Estimation Model *Lect. Notes Comput. Sci. (including Subser. Lect. Notes Artif. Intell. Lect. Notes Bioinformatics)* **9910 LNCS** 34–50

[41] Pishchulin L, Insafutdinov E, Tang S, Andres B, Andriluka M, Gehler P V. and Schiele B 2016 DeepCut: Joint Subset Partition and Labeling for Multi Person Pose Estimation 4929–37

[42] Cao Z, Hidalgo G, Simon T, Wei S E and Sheikh Y 2018 OpenPose: Realtime Multi-Person 2D Pose Estimation using Part Affinity Fields *IEEE Trans. Pattern Anal. Mach. Intell.* **43** 172–86

[43] Cao Z, Simon T, Wei S-E and Sheikh Y 2017 Realtime Multi-Person 2D Pose Estimation Using Part Affinity Fields 7291–9

[44] Simon T, Joo H, Matthews I and Sheikh Y *Hand Keypoint Detection in Single Images using Multiview Bootstrapping*

[45] Martinez G H, Raaj Y, Idrees H, Xiang D, Joo H, Simon T and Sheikh Y 2019 Single-Network Whole-Body Pose Estimation *Proc. IEEE Int. Conf. Comput. Vis.* **2019-October** 6981–90

[46] Pagnon D, Domalain M and Reveret L 2021 Pose2Sim: An End-to-End Workflow for 3D Markerless Sports Kinematics—Part 1: Robustness *Sensors 2021, Vol. 21, Page 6530* **21** 6530

[47] Viswakumar A, Rajagopalan V, Ray T and Parimi C 2019 Human Gait Analysis Using OpenPose *Proc. IEEE Int. Conf. Image Inf. Process.* **2019-November** 310–4

[48] Milone D, Risitano G, Pistone A, Crisafulli D and Alberti F 2022 A New Approach for the Tribological and Mechanical Characterization of a Hip Prosthesis Trough a Numerical Model Based on Artificial Intelligence Algorithms and Humanoid Multibody Model *Lubr. 2022, Vol. 10, Page 160* **10** 160

[49] Walczak J, Shahgaldi F and Heatley F 1998 In vivo corrosion of 316L stainless-steel hip implants: morphology and elemental compositions of corrosion products *Biomaterials* **19** 229–37

[50] Cucinotta F, D'Aveni A, Guglielmino E, Risitano A, Risitano G and Santonocito D 2021 Thermal Emission analysis to predict damage in specimens of High Strength Concrete *Frat. ed Integrità Strutt.* **15** 258–70

[51] Melvin A D, Lucia A C and Solomos G P 1993 The thermal response to deformation to fracture of a carbon/epoxy composite laminate *Compos. Sci. Technol.* **46** 345–51

[52] Egan P F, Gonella V C, Engensperger M, Ferguson S J and Shea K 2017 Computationally designed lattices with tuned properties for tissue engineering using 3D printing *PLoS One* **12** e0182902

IOP Conf. Series: Materials Science and Engineering 1275 (2023) 012005 doi:10.1088/1757-899X/1275/1/012005

A smart tool for the abridged assessment of potential environmental impacts of automotive solutions

F Del Pero, R Barbieri, M Delogu and M Pierini

Department of Industrial Engineering, University of Florence, Via di S. Marta, 3 – 50139 Italy
Corresponding author: Francesco Del Pero (francesco.delpero@unifi.it)

Abstract. This paper presents a smart tool to support the determination of sustainability of automotive components. Many research and industrial activities have been concentrated in the field of sustainability of passenger vehicles through the development of smart materials, manufacturing technologies and production processes. However, the environmental impacts quantification phase is always the most difficult and challenging in the design process. Literature and technical regulations provide that the Life Cycle Assessment methodology needs a lot of resources (both human and economic resources) to be implemented because of its complexity. The base idea of this work is to design and realize a tool easy-to-use also by non-LCA experts, with a set of input and output GUIs (Graphical User Interfaces) to make the design process easier also from the point of view of impact characterization. As final result, the conceived method is targeted at representing a simple and ready-to-use tool, which allows the application of eco-design methodologies in the application to the automotive sector.

1. Introduction

In the recent years the environmental issues are growing up because of the increase of the human activities and the Global warming is one of the major challenges to be take into account [[1],[2]]. More in details, the industrial processes, residential heating and transportation are the most relevant categories in the global warming increase, and many authors are working on these fields of research [[3],[4],[5]]. As the transportation sector is considered responsible for about 15-20% of the overall CO_2 emissions [[6],[7],8] and the demand for transport has been growing continuously in the last decades, a transition towards a more ecological mobility is a key factor to attain beneficial effects from an environmental point of view [9,10]. Because the automotive sector is one of the most active in the field of research of lightweight design [11,12,13], an approach mandatory to reduce global warming and CO_2 emissions, it can result interesting to widespread this model to other design experiences [14,15,16]. The so called LCA (Life Cycle Assessment) is the automotive sector most used methodology [17,18] to determine how and how much a product or a service could be expensive in terms of environmental impact: the knowledge on "how much" is the very first step to develop a component green re-design [19,20]. However, there are also problems related to LCA model, that could be the motivation that hinders the methodology widespread [21,22]. Firstly, the need for high trained operators with methodology know-how and process know-how, because they have to take decisions and assumptions that could strongly influence the study itself [23,24]. On the other hand, any LCA study is highly influenced by the dataset used to develop the assessment (i.e. in this paper authors provided a study that involved a high number of components [25,26], materials and machining for a single product since its concept design [27,28]),

Content from this work may be used under the terms of the Creative Commons Attribution 3.0 licence. Any further distribution of this work must maintain attribution to the author(s) and the title of the work, journal citation and DOI.
Published under licence by IOP Publishing Ltd

and it is not easy to manage a large set of data. To support the green design, the idea developed in this paper is to design and realize a simpler database that can be used as a gross dimensioning tool. In this database, the designer will face a very simple set of input GUIs (Graphical User Interfaces), masks where some input information are requested, that drive the end user through reports and graphs of final results. Authors expect that the use of simpler and ready-to-use tools will simplify the LCA approach and widespread the model benefits.

2. Development of environmental simulation tool

The conceived environmental tool is based on a database of category-specific indicators that allows the quantification of several impact categories, starting from a limited and pre-defined set of primary inventory data. The model is developed by Microsoft Access, a software dedicated to the management of relational databases (RDBMS - Relational Database Management System), through which environmental indicators are stored in tables composed of a large number of records. These indicators are used to carry out environmental analyses by means of calculation queries, while data entry is performed through specially-developed tables and masks, so that the model is flexible and easy to use even by non-expert users. The developed tool consists of two main sections:

- Manufacturing: this section allows calculating the environmental impacts of the manufacturing phase by taking into account not only the company's internal production but also the extraction and processing of raw materials;
- Use stage: this section is functional to calculate the impacts associated with the use phase of the component, using the Fuel Reduction Value (FRV) approach [29,30].

The following section explains how the tool is built and structured in Microsoft Access for both manufacturing and use phases. For each stage the description is divided into sub-paragraphs which refer to the main elements of the model: tables for data entry, queries for calculations, and masks to make the system user-friendly

2.1. Manufacturing stage

2.1.1. Tables. To calculate the environmental impacts associated with the manufacturing stage (which includes both extraction and processing of raw materials), three main tables are defined: Material indicators, Energy sources indicators and Consumables indicators. All indicators provided in tables above are derived from the GaBi6 software [31] by calculating the impact per unit of elementary flow.

As regards the material indicators table, it contains all data related to the extraction of raw materials, as well as the production of wastes due to the processing of materials. The table is structured by means of the following attributes:

- Material, indicating the type and process of material extraction/processing;
- Quantity, specifying the amount of material used during the production process;
- Scrap Factor, specifying the percentage of material scrap during the production process;
- ID Scrap Scenarios, defining the scenario for the treatment of the scrap from manufacturing. In this regard, three main choices are provided by the tool: Closed Loop recycling (which provides that the waste material is assumed fed back into the production process), Open Loop recycling (which provides that the waste material is recycled), Incineration with energy recovery (which takes into account incineration of the scrap with energy recovery) and Landfilling (which provides landfilling of the scrap material).
- Specific indicators related to the extraction and processing of materials;
- Specific indicators related to the management process of the scrap from manufacturing.

A one-to-many (1:N) relationship is set between the attribute "ID Scenario" and the "ID_Scrap Scenarios" attribute of the "Material Indicators" table, thus enabling the user to choose the most

IOP Conf. Series: Materials Science and Engineering 1275 (2023) 012005 doi:10.1088/1757-899X/1275/1/012005

appropriate scrap management scenario through a combo box. Figure 1 provides both the Materials table and the above relationship between attributes.

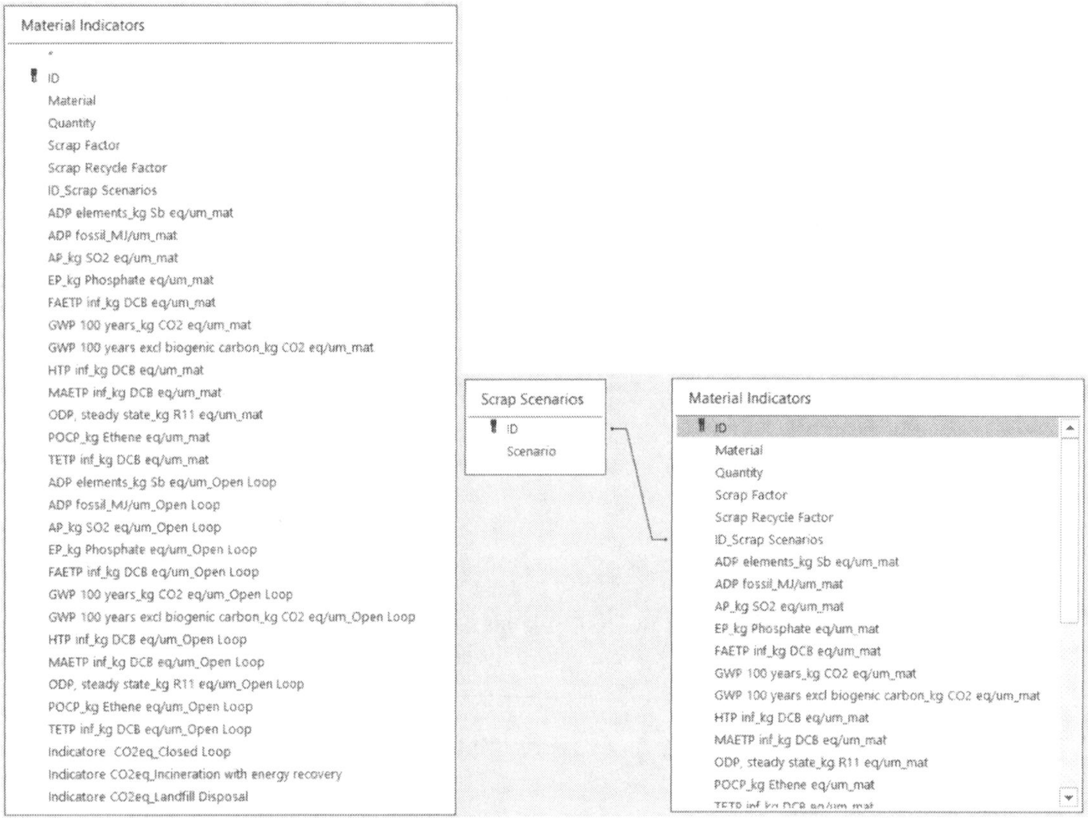

Figure 1. "Material Indicators" table and one-to-many (1:N) relationship between attributes "ID Scenario" and "ID_Scrap Scenarios"

The Material Indicators table is linked to the mask "Mask Data for Manufacturing (Material)" in which the user has the possibility to carry out the following actions:

- Choosing the type of material by which the component is made;
- Entering the amount of material needed to produce the component;
- Entering the waste percentage of the material chosen during the manufacturing process;
- Entering the percentage of recyclable scrap from manufacturing;
- Choosing the scenario for the management of waste from manufacturing.

To this end, back and next buttons are inserted to select the various records in the table, thus giving the user the ability to choose the appropriate materials between those included in the database by simply entering the required amount. Figure 2 provides the mask "Data for Manufacturing (Material)".

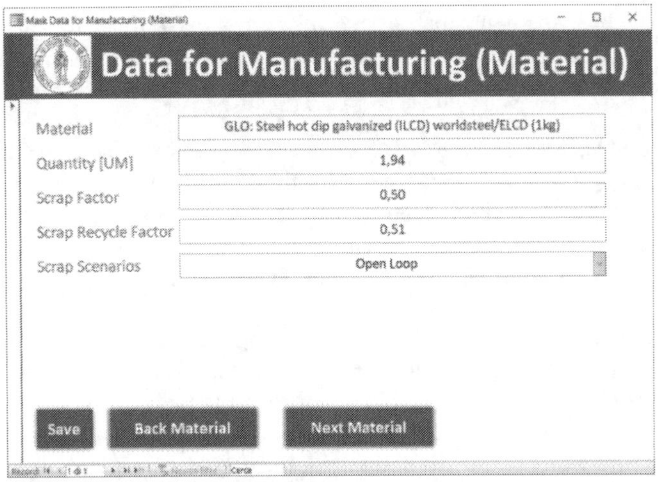

Figure 2. Mask "Data for Manufacturing (Material)"

The second main table relating the manufacturing phase is the "Energy Source Indicators". Such a table contains the resources used for component manufacturing. The attributes of this table are the type of energy resource (Type), the amount of resource used in manufacturing process (Quantity) and the specific impact indicator referred to the different environmental damages taken into account (Indicator). For the selection of energy resources, the "Energy Sources Indicators" table is linked to the mask named "Mask Data for Manufacturing (Energy Sources)". The choice of the resource is made by moving between the records through the Back and Next buttons, and simply entering the amount of the resource in component manufacturing.

The last table refers to consumables that are consumed in production activities. The table, called "Consumables Indicators", consists of the following attributes: "Consumables", indicating the type of consumable material, "Quantity", specifying the quantity used during the company's internal production process and "Specific impact indicators", referred to the different environmental impacts taken into account. Figure 3 provides both "Energy Sources Indicators" and "Consumables indicators" tables.

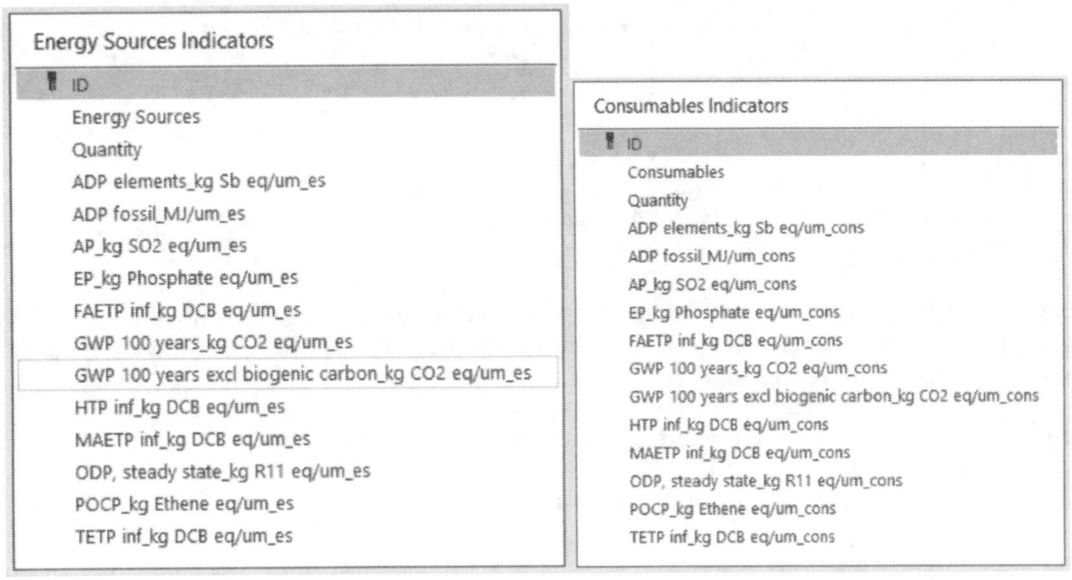

Figure 3. "Energy Sources Indicators" and "Consumables Indicators" tables

IOP Conf. Series: Materials Science and Engineering 1275 (2023) 012005 doi:10.1088/1757-899X/1275/1/012005

Similar as previous tables, the consumables are linked to a mask for data entry. In the mask, named "Mask Data for Manufacturing (Consumables)", the user has the possibility to choose the specific consumables and enter the quantity needed to manufacture the component under consideration.

2.1.2. Calculation model and related queries. Queries are used to calculate the damage for each impact category considered. A query is a tool that is used by the user to query a table or other queries, which are part of a database typically structured according to the relational model. Once a data is queried, it is possible to carry out a series of operations, such as selection, insertion, deletion of data, update and calculations. Microsoft Access provides the following types of queries:

- Selection queries, to select specific data from one or more sources. Such a type of query allows to retrieve only needed data, as well as to combine data from different sources. For a selection query, it is possible to use tables and other selection queries;
- Parameterized queries, to perform frequent variations of a particular query;
- Totalization queries, which allow to group or summarize data (e.g. to display total impacts);
- Cross-field queries, to present data according to an array with a series of headers in both rows and columns;
- Queries for table creation, to create a new table from data stored in other tables;
- Queuing queries, to retrieve data from one or more tables and add it into another;
- Updating queries, to change data in tables and to enter criteria specifying which rows must be updated;
- Deletion queries, to delete data from tables and to enter criteria specifying which rows must be deleted.

The developed tool makes use of selection queries to select data and then to carry out calculations. In addition, grouping queries are also used to obtain the total impacts. Specifically, for the manufacturing phase, three selection queries are used to calculate the following impacts: extraction, processing, and waste management, use of resources in production, and use of consumables in manufacturing. In addition, four other clustering queries are used to sum up, for each category, the impacts related to extraction, processing and waste management. The query related to waste extraction, processing and management, is created by querying the "Material Indicators" table. The following attributes are selected in the creation of the query: "Material", "Quantity", "Scrap Factor" and "ID_Scrap Scenarios". Through these data, the quantity of scrap for each j material used in production is calculated as follows:

$$Scrap_j = Quantity_j * SF_j \tag{1}$$

Where:
$Scrap_j$ = Amount of scrap from manufacturing for material j (kg);
$Quantity_j$ = Amount of input material j to the manufacturing stage (kg);
SR_j = Scrap Rate in manufacturing for material j (-).

The next step provides the calculation of environmental impacts for each impact category i and material j. The impacts depend on the selected scenario for waste management, for which the tool takes into account the following scenarios:

<u>Closed Loop recycling</u>. The waste is fed back into the production process, so that the impact of material j in impact category i is calculated starting from the amount of input material, the scrap rate of material j and the specific indicator of material j in the category i according to the following equation:

$$Impact_{ij} = \left(Quantity_j - Scrap_j\right) * Indicator_{ij} \tag{2}$$

Where:

$Impact_{ij}$ = Impact of material j for impact category i (substance_eq);
$Indicator_{ij}$ = Impact Indicator of raw materials acquisition of material j for impact category i (substance_eq/kg).

Open Loop recycling. It is assumed that the scrap material is open loop recycled. To consider this, the amount of waste material is multiplied by the specific indicator and a Substitution Factor (SF) (which provides the amount of impact effectively avoided thanks to the recycling process) and then subtracted to the impact of primary production, as provided below:

$$Impact_{ij} = (Quantity_j * Indicator_{ij}) - (Scrap_j * Indicator_{ij} * SF) \quad (3)$$

Where:
SF_{ij} = Substitution Factor of material j for impact category i (-).

Incineration with energy recovery. The environmental impact avoided thanks to incineration with energy recovery is subtracted to the impact of the primary production, according to the following equation:

$$Impact_{ij} = (Quantity_j * Indicator_{ij}) - (Scrap_j * Indicator_{\text{EnRec } ij}) \quad (4)$$

Where:
$Indicator_{\text{EnRec } ij}$ = Impact Indicator of incineration with energy recovery of material j for impact category i (substance_eq/kg).

Landfilling. The last scenario is the disposal of waste in landfill: in this case the impact of landfilling is summed up to the impact of the primary production.

$$Impact_{ij} = (Quantity_j * Indicator_{ij}) + (Scrap_j * Indicator_{\text{Land } ij}) \quad (5)$$

Where:
$Indicator_{\text{Land } ij}$ = Impact Indicator of landfilling of material j for impact category i (substance_eq/kg).

Finally, a grouping query is included to sum the impacts of both each material and overall manufacturing impact and related to extraction, raw material processing and waste management, according to the following equation:

$$Impact_{\text{Material } i} = \sum_j Impact_{ij} \quad (6)$$

Where:
$Impact_{\text{Material } i}$ = Total Impact of Materials for impact category i (substance_eq).

Figure 4 provides the queries referred to the manufacturing stage (single materials and total).

Figure 4. Queries related to materials used in manufacturing stage (single materials and total)

The second query for the calculation of manufacturing impacts takes into account the consumption of energy resources and consumables in manufacturing. To set up the query, the table "Energy Sources Indicators" and "Consumables" are queried and the following attributes are taken: "Energy Sources/Consumables" and "Quantity". Through these data and indicators provided by the corresponding table, the impacts of each category *i* and related to each energy resource *j* are calculated. The calculation performed is as follows:

$$Impact_{\mathrm{EnRes}\ ij} = Quantity_{\mathrm{EnRes}\ j} * Indicator_{\mathrm{EnRes}\ ij} \tag{7}$$

$$Impact_{\mathrm{Cons}\ ij} = Quantity_{\mathrm{Cons}\ j} * Indicator_{\mathrm{Cons}\ ij} \tag{8}$$

$$Impact_{\mathrm{Energy}\ i} = \sum_j Impact_{\mathrm{En\ Res}\ ij} \tag{9}$$

$$Impact_{\mathrm{Consumables}\ i} = \sum_j Impact_{\mathrm{Cons}\ ij} \tag{10}$$

Where:
$Impact_{\mathrm{EnRes}\ ij}$ = Impact of energy resources *j* for impact category *i* (substance_eq);
$Quantity_{\mathrm{EnRes}\ j}$ = Amount of energy resource *j* (kWh);
$Indicator_{\mathrm{EnRes}\ ij}$ = Impact Indicator of Energy Resource *j* in impact category *i* (substance_eq/kWh);
$Impact_{\mathrm{Cons}\ ij}$ = Impact of consumable *j* for impact category *i* (substance_eq);
$Quantity_{\mathrm{Cons}\ j}$ = Amount of consumable *j* (kg);
$Indicator_{\mathrm{Cons}\ ij}$ = Impact Indicator of Consumable *j* in impact category *i* (substance_eq/kWh);
$Impact_{\mathrm{Energy}\ i}$ = Impact of energy resources for impact category *i* (substance_eq);
$Impact_{\mathrm{Consumables}\ i}$ = Impact of consumables for impact category *i* (substance_eq).

Finally, the impacts related to materials acquisition, energy resources and consumables are summed through a selection query, according to the following equation:

$$Impact_{\mathrm{Prod}\ i} = Impact_{\mathrm{Material}\ i} + Impact_{\mathrm{Energy}\ i} + Impact_{\mathrm{Consumables}\ i} \tag{11}$$

Where:
$Impact_{Prod\ i}$ = Impact of Production stage for impact category i (substance_eq);
$Impact_{Material\ i}$ = Impact of Materials for impact category i (substance_eq);
$Impact_{Energy\ i}$ = Impact of Energy resources for impact category i (substance_eq);
$Impact_{Consumables\ i}$ = Impact of Consumables for impact category i (substance_eq).

The query is provided in figure 5.

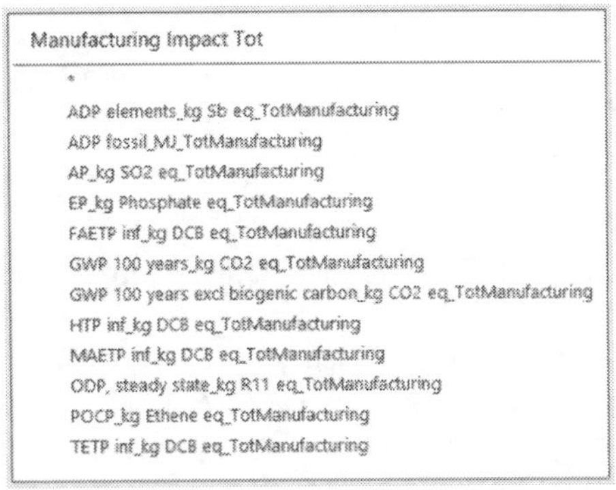

Figure 5. Query related to overall impacts of manufacturing stage

2.2. Use stage

Similar as the manufacturing stage, the model of the use stage consists of tables for data entry, queries for calculations, and masks to make the system user-friendly, which described in the following paragraphs.

2.2.1. Tables. Three tables are created to calculate the impact categories associated with use stage: “Fuels” “Vehicle classes” and “Data for use stage”. The table “Fuels” provides all the fuel data necessary for the evaluation of impacts. Such data are fuel type (gasoline or diesel), fuel density (ρ_{fuel}) (0.741 kg/l for gasoline and 0.820 kg/l for diesel) and Impact indicator of the fuel production (I_{WTT}), obtained through calculations carried out through the Gabi6 software.

The “Vehicle Classes” table takes into account vehicle-related data which are essential for the calculation of fuel consumption: vehicle class, through which it is possible to choose between class A/B, C, D and Fuel Reduction Value (FRV) coefficient.

Finally, the table “Data for use stage” provides use stage boundary conditions. These data are entered directly by the user and are: component mass ($Mass_{Comp}$), mileage over the entire life cycle of the vehicle ($Mileage_{LC}$), fuel type, vehicle class, specific emission of CO_2 ($CO_{2\ km}$) and Sulphur content in the fuel ($ppm_{Sulphur}$). Figure 6 provides “Fuels”, “Vehicle classes” and “Data for use stage” tables.

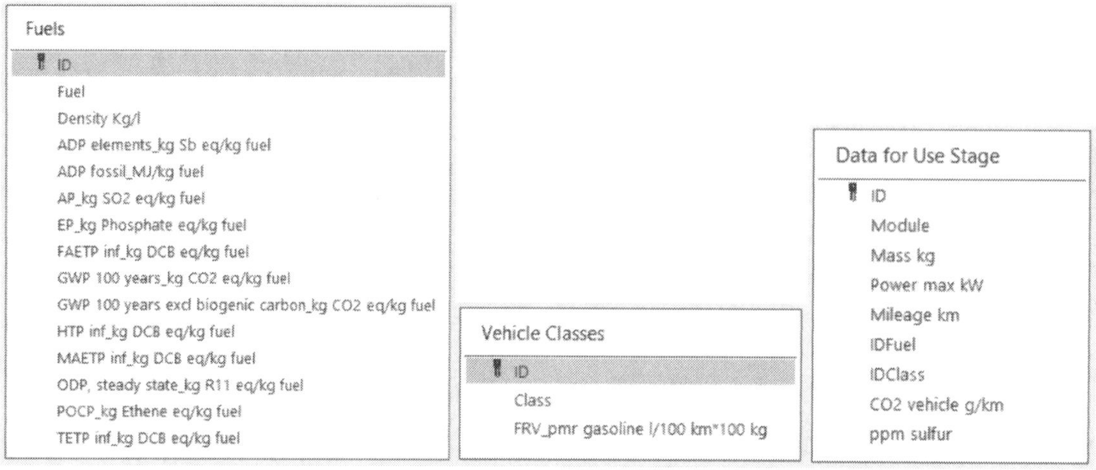

Figure 6. "Fuels", "Vehicle classes" and "Data for use stage" tables

Between the tables above, the model provides 2 one-to-many (1:N) relationships. The first relationship is between the "Fuels" table and the "Data for Use Stage" table, while the second one is between the "Vehicle Classes" and the "Data for Use Stage" table. Such a relationship is shown in figure 7.

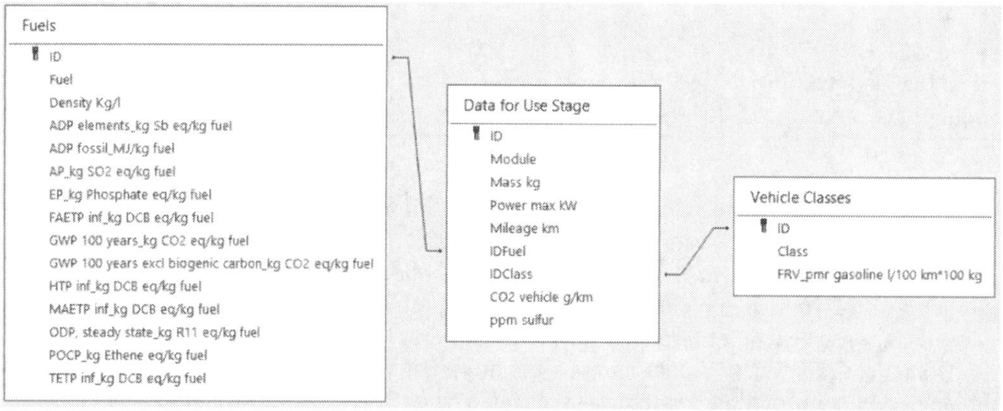

Figure 7. Relationship between tables "Fuels", "Data for Use Stage" and "Vehicle Calsses"

The model provides also a table of impact indicators for CO_2 and SO_2 emissions. The model assesses only these two types of emissions, because all the other pollutants are assumed to be removed by exhaust abatement systems. As regards, SO_2, the indicators refer to impact categories Acidification Potential (AP) [kg SO_2 eq.], Human Toxicity Potential (HTP inf.) [kg DCB eq.] and Photochemical Ozone Creation Potential (POCP) [kg Ethene eq.]. The model provides also a mask attached to the table "Data for Use Stage", to make easier the operation of data entry.

2.2.2. Calculation model and related queries. The determination of use stage impacts is carried out through a calculation model based on the Fuel Reduction Value (FRV) [15,16], according to the analytical framework reported below.

$$FC_{\mathrm{Comp}} = \frac{FRV * Mass_{\mathrm{Comp}} * \rho_{\mathrm{Fuel}}}{10000} * Mileage_{\mathrm{LC}} \tag{12}$$

$$CO_{2\,\mathrm{Comp}} = CO_{2\,\mathrm{km}} * Mileage_{\mathrm{LC}} * \frac{FC_{\mathrm{Comp}}}{FC_{\mathrm{Veh}}} \tag{13}$$

$$SO_{2\,\mathrm{Comp}} = SO_{2\,\mathrm{km}} * Mileage_{\mathrm{LC}} * \frac{FC_{\mathrm{Comp}}}{FC_{\mathrm{Veh}}} \tag{14}$$

$$SO_{2\,\mathrm{km}} = \frac{2 * ppm_{\mathrm{Sulphur}} * FC_{100\mathrm{km}} * \rho_{\mathrm{Fuel}}}{1000000 * 100} \tag{15}$$

$$Impact_{\mathrm{WTT}\,i} = FC_{\mathrm{Comp}} * Ind_{\mathrm{WTT}\,i} \tag{16}$$

$$Impact_{\mathrm{TTW}\,i} = CO_{2\,\mathrm{Comp}} * Ind_{\mathrm{CO2}\,i} + SO_{2\,\mathrm{Comp}} * Ind_{\mathrm{SO2}\,i} \tag{17}$$

$$Impact_{\mathrm{Use}\,i} = Impact_{\mathrm{WTT}\,i} + Impact_{\mathrm{TTW}\,i} \tag{18}$$

Where:
FRV = Fuel Reduction Value (l/100km*100kg);
$Mass_{\mathrm{Comp}}$ = Mass of the Component;
ρ_{Fuel} = Density of the Fuel (kg/l);
$Mileage_{\mathrm{LC}}$ = LC mileage of the vehicle (km);
$CO_{2\,\mathrm{Comp}}$ = CO_2 emissions associated with the Component (g);
$CO_{2\,\mathrm{km}}$ = per kilometer CO_2 emissions of the vehicle (g);
$SO_{2\,\mathrm{Comp}}$ = SO_2 emissions associated with the Component (g);
$SO_{2\,\mathrm{km}}$ = per kilometer SO_2 emissions of the Vehicle (g);
FC_{Comp} = Fuel Consumption associated with the Component (kg);
ppm_{Sulphur} = Sulphur content in the fuel (ppm);
$Impact_{\mathrm{WTT}\,i}$ = Impact of Well-To-Tank stage for impact category i;
$Impact_{\mathrm{TTW}\,i}$ = Impact of Tank-To-Wheel stage for impact category i;
$Ind_{\mathrm{WTT}\,i}$ = impact Indicator of Well-To-Tank stage for impact category i (substance_eq/kg);
$Ind_{\mathrm{CO2}\,i}$ = characterization factor of CO_2 for impact category i (kg substance_eq/kgCO_2);
$Ind_{\mathrm{SO2}\,i}$ = characterization factor of SO_2 for impact category i (kg substance_eq/kgSO_2);
$Impact_{\mathrm{Use}\,i}$ = Impact of Use stage for impact category i.

To calculate the fuel consumption associated with the component (FC_{Comp}), the following tables are queried:

- "Data for Use Stage", from which the following attributes are selected: $Mass_{\mathrm{Comp}}$, ID_{Fuel}, ID_{Class}, $CO_{2\mathrm{Veh}}$;
- "Fuels", from which the attribute ρ_{Fuel} is selected.

The WTT impacts are calculated by selecting the following data from table "Fuels" and the query "FC module": $Mass_{\mathrm{Comp}}$, $Mileage_{\mathrm{LC}}$, ρ_{Fuel}, FRV, $Ind_{\mathrm{WTT}\,i}$. On the other hand, the TTW impacts are calculated by selecting the following data from the table "Data for Use Stage" and the query "FC module": $CO_{2\,\mathrm{km}}$, LC_{Mileage}, ID_{fuel}, ρ_{fuel}, FC_{Veh}, FRV, ppm_{Sulphur}. Finally, the overall use stage impact is provided through a selection query which sums up the impacts related to WTT and TTW stages. The queries for calculation of fuel consumption and impacts in WTT, TTW and total use stage are reported in figure 8.

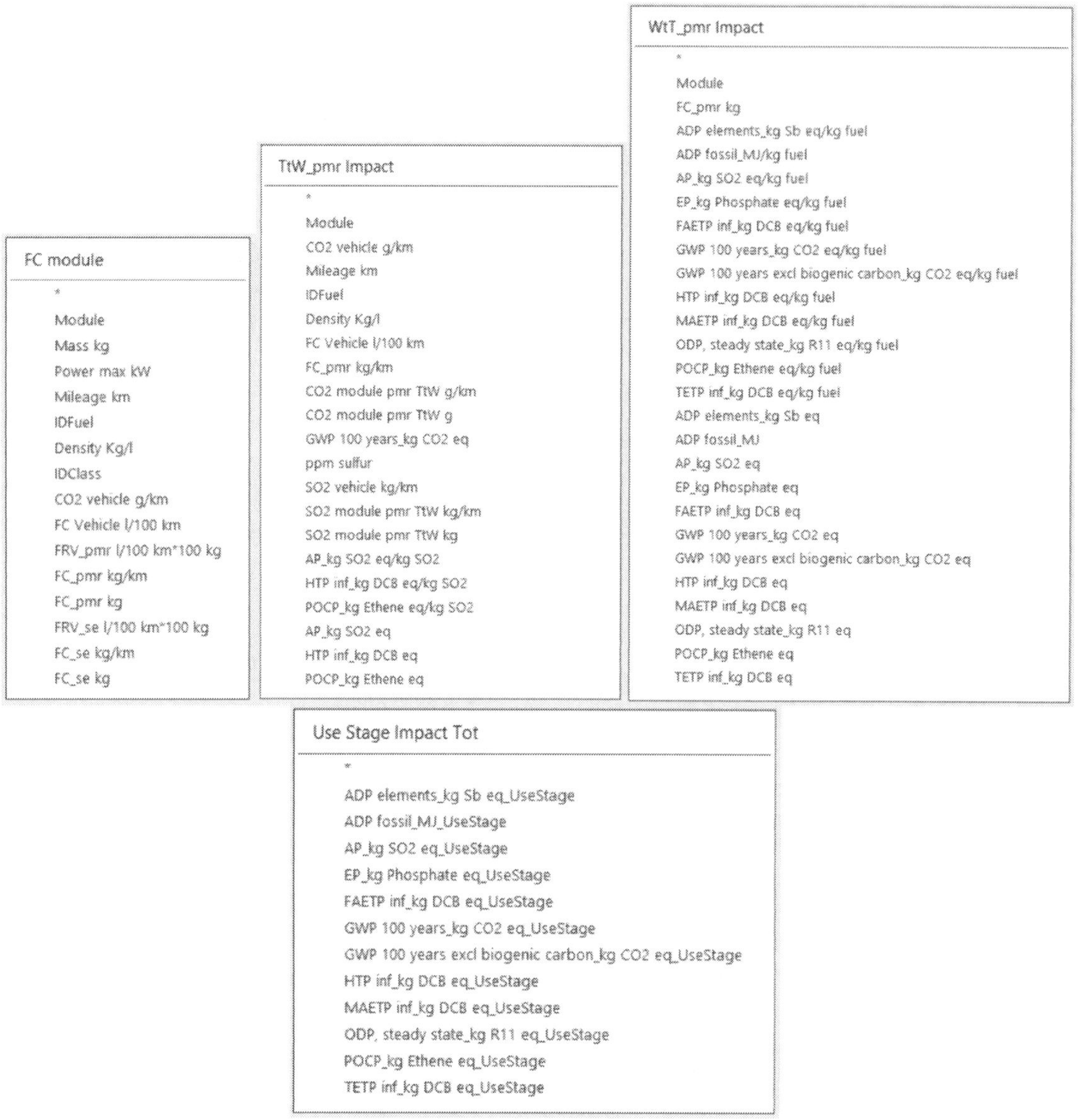

Figure 8. Queries for calculation of fuel consumption and impacts in WTT, TTW and total use stage

The total impacts related to manufacturing and use are calculated through a selection query which performs the summation of the two LC stages, according to the following equation:

$$Impact_{\text{Tot}\,i} = Impact_{\text{Manuf}\,i} + Impact_{\text{Use}\,i} \tag{19}$$

Where:
$Impact_{\text{Tot}\,i}$ = Total Impact (Production + Use stages) for impact category i (substance_eq);
$Impact_{\text{Prod}\,i}$ = Impact of Production stage for impact category i (substance_eq);
$Impact_{\text{Use}\,i}$ = Impact of Use stage for impact category i (substance_eq).

The query related to the calculation of the total impacts is reported in figure 9.

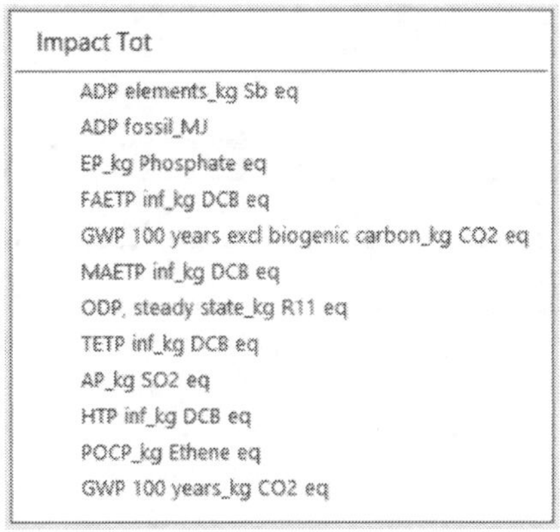

Figure 9. Query for calculation of total impacts (manufacturing and use)

Finally, six reports are created to make the results more understandable:

- Report "Manufacturing Impact", presenting the overall results associated with the manufacturing phase;
- Report "FC module", providing the fuel consumed during the entire life-cycle and the Fuel Reduction Value (FRV) coefficient;
- Report "Tank to Wheel", providing the impacts related to the Tank to Wheel use sub-phase;
- Report "Well to Tank", providing the impacts related to the Well to Tank use sub-phase;
- Report "Use Stage", providing the overall impacts associated with use stage;
- Report "Impact Total", providing the overall impacts of both manufacturing and use stages.

3. Validation through application to a practical case study

The conceived assessment tool is applied to a practical case study, and the results obtained are compared with those obtained from traditional modelling (carried out through the GaBi6 environmental software) for validation purposes.

The chosen case study concerns an automotive component produced through deep drawing process made of 1.94 kg galvanized steel sheet. During production, 0.98 MJ of electrical energy and 0.272 MJ of thermal energy is used. In addition, 50% waste is produced from the process, 51% of which is recycled and the remaining amount disposed of in landfills. The component obtained at the end of the process has a mass of 0.97 kg and it is mounted on a passenger car with the following technical features functional to set the simulation model:

- Sulphur content in the fuel: 10 (ppm);
- Vehicle class D;
- CO_2 emissions: 179 (g/km).

These data are entered into the developed system and the results are reported in table 1 (reporting breakdown of impacts between different LC stages and sub-stages) and the percentage difference with respect to the GaBi6 modelling.

As provided in the table, the specific impacts for each category are compared with the results obtained through the modeling calculated by the GaBi6 software, obtaining a maximum percentage difference lower than 1%. Since the discrepancy is very low, the conceived model is considered validated and ready to be used in the industry for affordable assessment of automotive case studies.

Table 1. Environmental impacts of case study obtained through the developed tool and GaBi6 simulation modelling

	Tool				GaBi6	Variation (%)
	Production	Use		Total LC	Total LC	
		WTT	TTW			
ADP elements [kg Sb eq]	1,11E-04	7,09E-07	-	1,11E-04	1,11E-04	0
ADP fossil [MJ]	5,67+01	1,47+02	-	1,47+02	1,50+02	-0,2
AP [kg SO_2 eq]	2,00-02	7,92E-03	7,06E-05	2,00-02	2,00-02	0
EP [kg Phosphate eq]	1,11E-03	1,40E-03	-	2,51E-03	2,53E-03	-0,1
FAETP inf [kg DCB eq]	1,00-02	6,00-02	-	7,00-02	7,02-02	-0,7
GWP 100 years [kg CO_2 eq]	5,46+00	1,92+00	9,41+00	1,68+01	1,71+01	-0,3
HTP inf [kg DCB eq]	4,53E-01	3,30-01	5,65E-06	7,80-01	7,80-01	0
MAETP inf [kg DCB eq]	1,24+03	1,52+02	-	1,39+03	1,39+03	0
ODP, steady state [kg R11 eq]	5,90E-08	4,03E-13	-	5,90E-08	5,92E-08	-0,2
POCP [kg Ethene eq]	2,41E-03	1,19E-03	2,82E-06	3,60E-03	3,60E-03	0,0
TETP inf [kg DCB eq]	8,56E-03	4,50E-03	-	1,00-02	1,01-02	-0,5

4. Conclusions

The environmental impact assessment system developed is specific to the automotive sector and allows a simplified LCA analysis, positioning itself as a partial replacement tool for commercial LCA software currently on the market. The tool is based on a set of environmental indicators that allow the quantification of impacts according to a from-cradle to use approach which includes raw material extraction and processing, in-house production and use phases. In addition, thanks to the development in the Microsoft Access environment, the users can easily add indicators related to the manufacturing phase, thus being able to evaluate a wide range of materials, energy resources and consumables. The tool is finally validated through the application to a real-life automotive case study, providing similar results to the ones obtained through traditional environmental modelling performed through GaBi6 software. Future developments of this work are: 1. extension of the analysis to electric and hybrid vehicles; 2. expansion of the indicators database to a bigger range of materials and technologies; 3. inclusion of the assessment of end-of-life phase.

References

[1] U.S. Environmental Protection Agency 2018 *DRAFT Inventory of U.S. Greenhouse Gas. Emissions and Sinks: 1990-2016*; U.S. Environmental Protection Agency: Washington, DC, USA, 2018

[2] European Commission 2011 *White Paper: Roadmap to a Single European Transport Area—Towards a Competitive and Resource Efficient Transport System*; European Commission: Brussels, Belgium, 2011; /* COM/2011/0144 final */

[3] Baer P, Mastrandrea M 2006 *High Stakes: Designing Emissions Pathways to Reduce the Risk of Dangerous Climate Change*; Institute for Public Policy Research: London, 2006

[4] NRC 2009 *Special Report 298 - Driving and the Built Environment: The Effects of Compact Development on Motorized Travel, Energy Use, and CO2 Emissions*; Transportation Research, The National Research Council of the National Academies, 2009

[5] Yang C, McCollum D, McCarthy R, Leighty W 2009 Meeting an 80% reduction in greenhouse gas emissions from transportation by 2050: a case study in California Transp. Res., Part D 2009. https://doi.org/10.1016/j.trd.2008.11.010

[6] Witik RA, Payet J, Michaud V, Ludwig C, Manson JAE 2011 Assessing the life cycle costs and environmental performance of lightweight materials in automobile applications *Compos. Part A 2011, 42(11):1694-1709*, https://doi.org/10.1016/j.compositesa.2011.07.024

[7] Ferreira V, Merchán M, Egizabal P, García de Cortázar M, Irazustabarrena A, López-Sabiróna, A M, Ferreira G 2019 Technical and environmental evaluation of a new high-performance material based on magnesium alloy reinforced with submicrometre-sized TiC particles to develop automotive lightweight components and make transport sector more sustainable *J. Mater. Res. Technol. 2019, 8(3)*, 2549–2564. https://doi.org/10.1016/j.jmrt.2019.02.012

[8] Kelly J, Sullivan JL, Burnham A, Elgowainy A 2015 Impacts of Vehicle Weight Reduction via Material Substitution on Life-Cycle Greenhouse Gas Emissions *Environmental Science & Technology, 2015, 49: 12535–12542*, https://doi.org/10.1021/acs.est.5b03192

[9] Milovanoff A, Kim H C, De Kleine R, Wallington TJ, Posen ID, MacLean HL 2019 A Dynamic Fleet Model of U.S Light-Duty Vehicle Lightweighting and Associated Greenhouse Gas Emissions from 2016 to 2050 *Environ. Sci. Technol. 2019, 53, 2199–2208*

[10] Egede P 2016 Sustainable Production, Life Cycle Engineering and Management *Environmental Assessment of Lightweight Electric Vehicles*, Springer International Publishing: Cham, Switzerland, 2016

[11] Tisza M, Czinege I 2018 Comparative study of the application of steels and aluminium in lightweight production of automotive parts *Int. J. Lightweight Mater. Manuf. 2018, 1, 229–238*

[12] Das S, Graziano D, Upadhyayula VKK, Masanet E, Riddle M, Cresko J 2016 Vehicle lightweighting energy use impacts in U.S. light-duty vehicle fleet *Sustain. Mater. Technol. 2016, 8, 5–13*

[13] Baroth A, Karanam S, and McKay R 2012 Life Cycle Assessment of Lightweight Noryl * GTX * Resin Fender and Its Comparison with Steel Fender *2012 SAE World Congress, Detroit, MI, April 24-26, 2012*

[14] Deng Y, Guo Y, Wu P, and Ingarao G 2019 Optimal Design of Flax Fiber Reinforced Polymer Composite as a Lightweight Component for Automobiles from a Life Cycle Assessment Perspective *Journal of Industrial Ecology 23 (2019): 986-997*, https://doi.org/10.1111/jiec.12836

[15] Sun X, Meng F, Liu J, and McKechnie J 2019 Life Cycle Energy Use and Greenhouse Gas Emission of Lightweight Vehicle—A Body-in-White Design *J Clean Prod 220 (2019):1-8*, https://doi.org/10.1016/j.jclepro.2019.01.225

[16] Fassi HF, Ourihi R, and Elahrach K 2020 An Integrated Method Involving Design-Manufacturing-Environment Applied in Structural Optimization *Materials Today: Proceedings 38, no. 1 (2021): 135-138*, https://doi.org/10.1016/j.matpr.2020.06.118

[17] Dhingra R, Das S 2014 Life Cycle energy and environmental evaluation of downsized vs. lightweight material automotive engines *J. Clean. Prod. 2014, 85, 347-358*

[18] ALIVE Project 2019 *Advanced High Volume Affordable Lightweighting for Future Electric Vehicles*, www.projectalive.eu, accessed December 2019

[19] Delogu M, Del Pero F, Romoli F, and Pierini M 2015 Life Cycle Assessment of a Plastic Air Intake Manifold *Int J Life Cycle Assess 20 (2015): 1429-1443*, https://doi.org/10.1007/s11367-015-0946-z

[20] Nehuis F, Kleemann S, Egede P, Vietor T, Herrmann C 2014 Future Trends in the Development of Vehicle Bodies Regarding Lightweight and Cost, *Innovative Design, Analysis and Development Practices in Aerospace and Automotive Engineering*; Springer Science & Business: Berlin/Heidelberg, Germany, 2014; ISBN 978-81-322-1871-5

[21] Visentin C, Da Silva Trentin AW, Braun AB, and Thomé A 2020 Life Cycle Sustainability Assessment: A Systematic Literature Review through the Application Perspective, Indicators, and Methodologies *J Clean Prod 270 (2020): 122509*, https://doi.org/10.1016/j.jclepro.2020.122509

[22] Sullivan JL, Hu J 1995 Life cycle energy analysis for automobiles *Total Life Cycle Conference and Exposition*. Society of Automotive Engineers: Vienna, Austria, October 16−19, 1995

[23] Kim HC, Wallington TJ 2013 Life-cycle energy and greenhouse gas emission benefits of lightweighting in automobiles: Review and harmonization *Environ. Sci. Technol. 2013, 47, 6089−6097*. https://doi.org/10.1021/es3042115

[24] Russo D, and Rizzi C 2014 An ECO-DESIGN Approach Based on Structural Optimization in a CAD Framework *Computer-Aided Design & Application 11, no. 5 (2014): 579-588*, http://doi.org/10.1080/16864360.2014.902691

[25] Aruldoss M, Lakshmi TM, and Venkatesan VP 2013 Survey on Multi Criteria Decision Making Methods and Its Applications *Am J Inf Syst 1 (2013): 31-43*, https://doi.org/10.12691/ajis-1-1-5

[26] Onat NC, Gumus S, Kucukvar M, and Tatari O 2020 Application of the TOPSIS and Intuitionistic Fuzzy Set Approaches for Ranking the Life Cycle Sustainability Performance of Alternative Vehicle Technologies *Sustainable Production and Consumption 6 (2020): 12-25*, http://doi.org/10.1016/j.spc.2015.12.003

[27] Delogu M, Maltese S, Del Pero F, Zanchi L Challenges for Modelling and Integrating Environmental Performances in concept Design: The Case of an Automotive Component Lightweighting *International Journal of Sustainable Engineering 11, no. 2 (2018): 135-148*, http://doi. org/10.1080/19397038.2017.1420110

[28] Costa D, Quinteiro P, Dias, AC 2019 A Systematic Review of Life Cycle Sustainability Assessment: Current State, Methodological Challenges, and Implementation Issues *Science of the Total Environment 686 (2019): 774-787*, https://doi.org/10.1016/j.scitotenv.2019.05.435

[29] Delogu M, Del Pero F, Pierini M 2016 Lightweight design solutions in the automotive field: Environmental modelling based on Fuel Reduction Value applied to diesel turbocharged vehicles *Sustainability 2016, 8, 1167*

[30] Del Pero F, Delogu M, Pierini M 2017 The effect of lightweighting in automotive LCA perspective: Estimation of mass-induced fuel consumption reduction for gasoline turbocharged vehicles, *J. Clean. Prod. 2017, 154, 566–577*

[31] GaBi software. https://gabi.sphera.com/italy/index/. Accessed: August 2022

Assessment of tensile behaviour of plain weave fabric CFRP composites using acoustic emission technique and deep learning

C Barile, C Casavola, G Pappalettera and V Paramsamy Kannan

Dipartimento di Meccanica, Matematica e Management, Politecnico di Bari, Via Orabona 5, 70125 - Bari, Italy

pk.vimalathithan@poliba.it

Abstract. The applications of Carbon Fibre Reinforced Plastic (CFRP) composites in aerospace applications have increased exponentially in the last decade. CFRP composites are now replacing conventional materials used in high temperature applications. Different types and strategies of reinforcements are used in these composites to improve their applicability to high temperature applications. One of the reinforcement strategies that has been used frequently in the recent years is the plain weave fabric configuration of reinforcing fibres in polymer matrix. A comprehensive damage assessment is essential to evaluate the characteristics of these composites at elevated temperatures. The Acoustic Emission (AE) signals generated during the damage evolution stages are used for studying the damage evolution stages when the plain weave fabric composites are tested at temperatures close to their glass transition state. A new information-theoretic parameter, Lempel-Ziv (LZ) complexity and the deep learning neural network is used for understanding the damage evolution stages. Furthermore, the neural network is used for validating the utilization of LZ complexity as a potential AE parameter for damage classification applications. The results are promising with the damage classification strategy using LZ complexity exhibits an accuracy of 85.1% which is validated through the neural network.

1. Introduction

In the last three decades, the Fibre Reinforced Polymer (FRP) composites have found themselves at the centre of constructing complex components in aerospace and automotive applications. Due to their high strength-to-weight ratio, resistance to various chemical attacks, easy manufacturing and repairing capabilities, and the ease of applications, they have gained the interest of the modern world. Various strategies in reinforcement, ranging from altering the fibre orientations to surface modifications by various chemical treatments or addition of nano-fillers have been explored by researchers over the years [1,2]. Among them, woven fabric reinforcement strategies of the FRP composites have shown excellent commercial applications. Their high impact damage resistance, adequate load-carrying and thermal insulation capabilities and excellent drapeability made them suitable for complex shapes and structures in low-load carrying air crafts.

Nonetheless, the woven fabric composites are not without their limitations. The polymeric resin used in these composites are amorphous and thermosets in nature and they begin to decompose when the operating temperature reaches above their glass transition state [2,3]. Besides, the presence of defects such as delaminated strands of fibres, improper curing, resin rich regions, and voids can aggravate the

Content from this work may be used under the terms of the Creative Commons Attribution 3.0 licence. Any further distribution of this work must maintain attribution to the author(s) and the title of the work, journal citation and DOI.
Published under licence by IOP Publishing Ltd

failure of these composites. Thus, it is essential to study the damage characteristics of these composites at the temperatures close to their glass transition state.

Acoustic Emission (AE) technique is one of the formidable non-destructive evaluation techniques, which have gained tremendous success in the damage characterization of FRP composites. Their advantages and applications are well-documented over the last 50 years [4–6]. However, the reliability of the descriptors that define the AE signals are under constant scrutiny. Researchers attempt to improve the capability of improving the available descriptors or by introducing a new descriptor [6–8]. In this study, a new information-theoretic descriptor named Lempel-Ziv (LZ) complexity, which is commonly used in time-series data applications, is used as a potential AE descriptor. This parameter has been introduced in several bio-medical and machine diagnostic applications [8–11]. However, validation is a crucial part in introducing a descriptor in an extraneous field. The machine learning tools, such as Neural Networks, are viewed as predictive and decision-making tool in various fields. In this study, a deep learning neural network tool is used for evaluating the damage stages and validating the application of LZ complexity as an AE descriptor in damage evaluation.

For this study, CFRP composites with the reinforcement in plain weave fabric configuration are tested. Static tensile tests are carried out at a temperature close to the glass transition stage of the composites and the resulting AE signals are recorded and analysed using LZ complexity, other AE descriptors, signal processing techniques, and the Neural Network.

2. Materials and Methods

2.1. Materials

The CFRP composites are prepared by hot melt method for resin impregnation. Epoxy resin, Toray 2510 and high-strength carbon fibres in plain weave fabric configurations are used for preparing the prepreg. The prepregs are stacked over one another and cured at a pressure of 22 inches of Hg at an ambient temperature of 132 °C for about 120 to 150 min. More details about the resin, fibre and prepreg are presented in Table 1. Further details about the preparation of composites can be found in the authors' previous research works [12].

Table 1. General Properties of Resin, Fibre and Prepreg.

Properties	Description
Glass Transition Temperature	141 ± 5°C
Density of the Prepreg	1.267 g/cm^3
Resin Percentage	41.5%
Gel Time	8 min (approx.)
Reinforcement Area Weight	192 g/m^2
Density of Reinforcement	1.8 g/cm^3
Average Fibre Density in Warp	1.78 g/cm^3
Average Fibre Density in Fill	1.8 g/cm^3
Tensile Strength of Fibre	4900 MPa
Tensile Modulus of Fibre	240 GPa

From the cured CFRP slabs, tensile test specimens are cut according to ASTM D3039 standard longitudinally along the fibre warp direction. End tabs are bonded to the specimen using Hysol EA9628 adhesive. The specimens are mounted on to the test vices of an INSTRON servo-hydraulic test machine. The entire setup is enclosed inside an environmental chamber and the temperature is ramped up to 120

°C. The specimens are placed inside the chamber ideally for 5 min before the commencement of the test. The test is carried out at a constant displacement speed of 2 mm/min. A total of 5 specimens are tested in this study, which are named as T1 through T5.

2.2. Acoustic Emission Technique
AE signals generated during the tensile loading of the CFRP composites are recorded using a piezoelectric sensor, which has an operating frequency of 200 kHz to 750 kHz. The sensor is mounted on the gauge length of the specimen and silicone grease is smeared between the transducing unit of the sensor and the specimen surface. More details about the instrumentation can be found in Table 2 and procedures used can be found in the authors' previous research works [7].

Table 2. Acoustic Emission Test Setup.

Test Setup	
Sensor	Pico Sensor, Physical Acoustics
Amplifier	2/4/6 – Differential Amplifier
Amplification	40 dB
Acquisition Threshold	35 dB
Bandpass Filters	100 kHz / 1 MHz
Length of the Waveform	1K
Sampling Frequency	2 MHz

Two conventional AE descriptors, peak amplitude and weighted peak frequecny (W-Freq) along with the novel parameter LZ complexity are used as the primary tools for AE signal processing in damage identification. The definition, advantages and limitations of peak amplitude and W-Freq are well-documented in literature. The details about LZ complexity are presented in the next section.

3. Acoustic Emission Signal Processing
In this work, first, the basic AE descriptors peak amplitude and W-Freq are extracted from the recorded AE waveforms, which are generated during the tensile test. Next, the LZ complexity values of the AE waveforms are calculated. Using a multi-parameter approach, the failure modes during the tensile tests are evaluated using these three parameters. Two modes are used for validating the evaluation strategy used. One, the frequency components of the AE signals are analysed using Continuous Wavelet Transform (CWT) and two, the deep learning neural network. Brief details of each of the technique developed and used are illustrated in this section.

3.1. Lempel Ziv Complexity
Complexity is the measure of the extent to which a sequence resembles a random one. It can also be defined as the measure of information. Lempel and Ziv [13], in 1976, developed a complexity measure related to the number of steps in a self-delimiting production process, by which the given sequence is presumed to be generated. For this, the AE signal data must be first transformed into a finite series with the elements of few symbols. Popularly, this is achieved by using Hilbert transform. This involves converting the signal into a sequence of binaries.

Consider an acoustic signal S of length n.

$$S = \{s_1, s_2, s_3, \dots, s_n\} \quad (1)$$

The signal is converted into its analytical form using Hilbert transform.

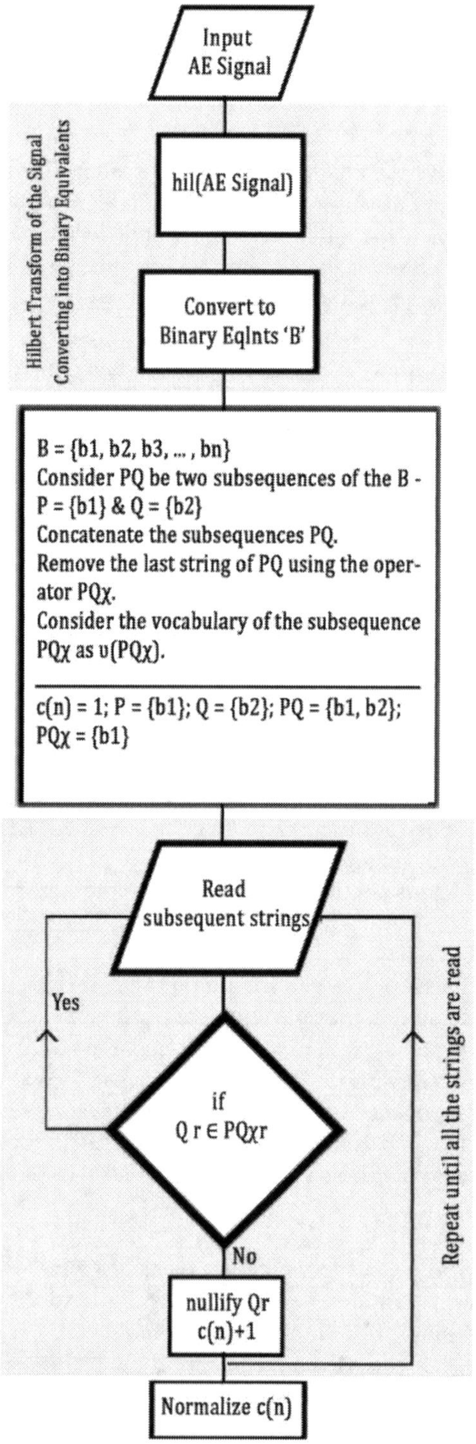

Figure 1. Algorithm for Calculating Lempel-Ziv Complexity.

$$H = \{h_1, h_2, h_3, \ldots, h_n\} \tag{2}$$

H is the Hilbert transform of the signal S in its analytical form. The median of H can be selected as the threshold t_h for converting the signal into binaries. The following conditions are followed.

$$b_i = \begin{cases} 0, if\ h_i < t_h \\ 1, otherwise \end{cases} \quad (3)$$

Thereby, the analytical form obtained by Hilbert Transform H is converted into a binary form B.

$$B = \{b_1, b_2, b_3, \dots, b_n\} \quad (4)$$

Following the conversion, the algorithm showed in Figure 1 is used for calculating the LZ complexity of the signal. For further details, the original research work by Lempel and Ziv or a simplified version by other authors can be referred.

3.2. Continuous Wavelet Transform

The time-frequency characteristics of an AE signal can be characterized using various techniques such as Continuous Wavelet Transform (CWT), Short-Time Fourier Transform (STFT), Discrete Wavelet Transform (DWT), Wavelet Packet Transform (WPT), Hilbert-Huang Transform (HHT) etc [4,8]. Among them, CWT provides a great visual representation of the time-frequency features of an AE signal. This capability of CWT is quite useful for deep learning, since the histograms generated by CWT are used as the training data for the neural network. CWT is a formidable and conventional tool, which has been explained greatly in rich textbooks. In this work, the programming module of MATLAB is used for generating the CWT histograms.

3.3. Deep Learning Neural Network

For this study, a deep Convolutional Neural Network (CNN) based on the popular SqueezeNet is designed and trained for reaching the desired objective. SqueezeNet has an advantage over other CNN due to its smaller architecture, fewer parameters, less memory consumption but equally high accuracy. This is achieved by using a Fire module in the micro-architecture of a CNN. Fire module is a micro-architecture which contains a Squeeze (containing 1x1 convolutional filters) and Expand (containing 1x1 and 3x3 convolutional filters) convolutional layers. This means that the deeper layers are not required, and high accuracy can be achieved with less number of parameters and less number of convolutional layers.

Table 3. SqueezeNet Training Parameters, Training and Validation Data Information.

Training Parameters	
Initial Learning Rate	0.001
Learning Schedule	Piecewise
Drop Rate Factor	0.1
Drop Rate Period	8
Maximum Epochs	20
Minibatch Size	75
Data for Training and Validating SqueezeNet	
Training Data	6000 Images per Class
Test Data	2500 Images per Class

The original SqueezeNet, which is available in MATLAB Deep Network Designer module has 68 layers and 75 connections, which is not suitable for this study. Due to the deeper configuration of the

IOP Conf. Series: Materials Science and Engineering 1275 (2023) 012006 doi:10.1088/1757-899X/1275/1/012006

network, over-fitting of data might occur, since the training data are simple histograms of CWT. Therefore, the network is modified to reduce the number of layers and the connections. Besides, a dropout layer is provided at the end to saturate the output of the final convolutional layer. The modified SqueezeNet has 39 layers and 42 connections.

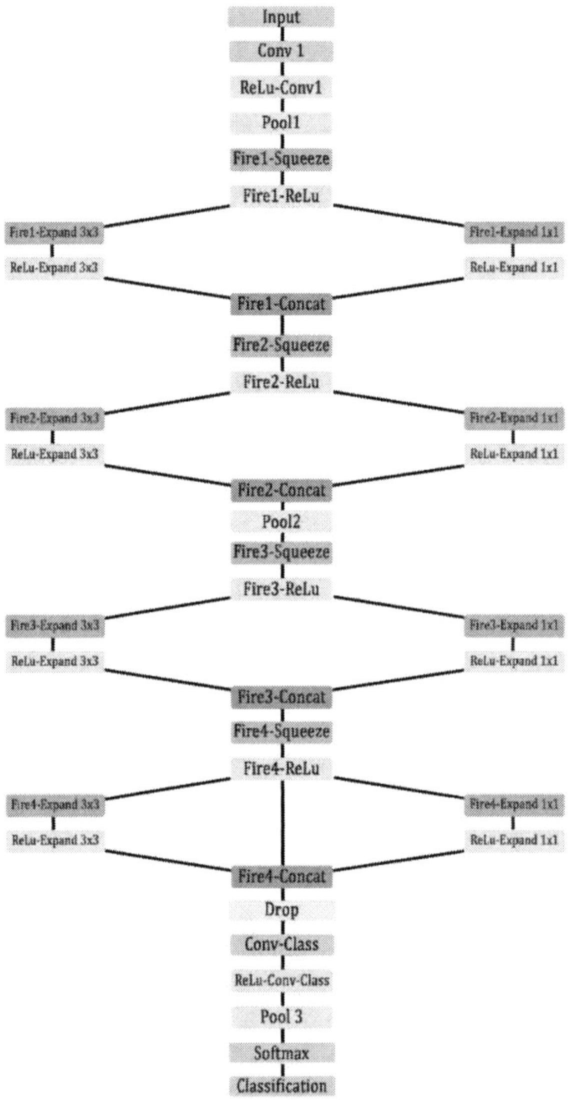

Figure 2. Architecture of the Modified SqueezeNet.

The architecture of the modified SqueezeNet is provided in Figure 2. For further details, refer to the authors' previous research work.

For training and testing the SqueezeNet, the CWT histograms of the AE signal generated from different damage modes are used. The classification methodology using the multi-parameter approach will be discussed in detail in the next section. This section is dedicated to the architecture of the CNN and the training and evaluation parameters, which are illustrated in Table 3.

IOP Conf. Series: Materials Science and Engineering 1275 (2023) 012006 doi:10.1088/1757-899X/1275/1/012006

4. Results and Discussions

The mechanical properties such as Young's modulus and Tensile strength at rupture of the CFRP plain weave fabric composites are measured and calculated in accordance with ASTM D3039 standard. Since the composites are tested at elevated temperatures, it is essential to compare these properties with the CFRP specimens tested at room temperatures. This information is exhaustive for the current research objective, which focuses on two important aspects. One, to assess the mechanical behaviour the composites using AE technique and two, to validate the efficacy of LZ complexity as an efficient AE parameter for the damage classification using deep learning neural networks. Readers, who are interested in the mechanical properties of the plain weave fabric CFRP such as the tensile strength, strain at failure and Young's modulus are directed to the authors' previous works [12].

4.1. Classification of Damage Modes using AE Descriptors

Peak amplitude has been one of the most commonly used AE descriptors for damage classification. Many researchers, after extensive research campaigns, have established standards to use peak amplitude for damage classification in FRP composites. The general consensus is that the AE signals with amplitude less than 45 dB may represent micro-failures in the polymeric matrices or matrix cracking. Delamination and debonding, which are the common failure modes in CFRP composites (FRP Composites, in general) generates AE signals with amplitude between 46 dB and 65 dB. Fibre breaking, due to the high stiffness of the fibres, always generates high amplitude signals, generally greater than 65 dB. The accumulation of micro-failures causes crack to propagate through the thickness of the material, which also generates AE signals with higher amplituide, similar to the fibre breakage events.

To alleviate the confusion in differentiating fibre breaking and through-thickness crack growth, several researchers recommended the utilization of the frequency characteristics of the signal, say peak frequency [14,15]. Besides, the attenuation in peak amplitude due to the propagating distance also comes into the discussion. The general idea is that the low stiffness matrix of the composite generates AE signals with less frequency compared to the high stiffness fibres. Supposedly, failures in matrices are associated with low frequency AE signals and fibres are with high frequency AE signals.

However, in the last decade, this association of AE signals with the peak frequency component is highly scrutinized. Because the through-thickness crack growth through the matrix also generates AE signals with high frequency [16]. To avoid this confusion, W-Freq has been utilized.

Table 4. Classification Strategy for Identifying Failure Modes.

Failure Modes	Transverse Cracking	Delamination	Fibre Rupture	Crack Growth
Peak Amplitude	<45 dB	46 dB – 65 dB	>65 dB	>65 dB
LZ Complexity	>0.6	0.3 – 0.8	<0.6	0.3 – 0.8
W-Freq	-	<300 kHz	-	>300 kHz

Nonetheless, the time-frequency analysis using tools such as CWT are more efficient in associating the generated AE signal with the damage mode. But this is a post-processing technique, which is exhaustive and time consuming. For this reason, several new parameters are proposed over the years to utilize multi-parameter approach to associate the AE signals with their damage sources.

In the pursuit of effective damage classification, the information-theoretic parameter LZ complexity is utilized. The authors have tested this parameter on a large amount of data and proposed a classification strategy in our previous works [4,7]. Using peak amplitude, W-Freq and LZ complexity, four different damage modes can be classified efficiently. The classification methodology is illustrated in Table 4.

Using the classification strategy shown in Table 4, the AE signals generated during the tensile tests of five CFRP plain weave fabric specimens are classified into four classes, each representing a damage mode. The number of AE signals generated from each damage mode is presented in Table 5.

The results show that majority of the AE signals are generated from matrix cracking and delamination/debonding events. In plain weave fabric composites, when the load is applied along the warp direction of the fibres, initially, the fibres will try to align themselves to the loading direction, which squeezes the polymer matrix material. This localized compression of the polymeric material causes matrix cracking [17,18]. The majority of the damage mode occurs due to the delamination and debonding of the fibres, which finally leads to the catastrophic failure. The indication that the majority of the signals are from Class 1 and Class 2 seems to prove the experimental results.

Table 5. Classification of AE Signals from Each Class of Damage Modes during the Tensile Test.

Specimen	Total Signals	Class 1	Class 2	Class 3	Class 4
T1	3701	2042	1256	63	119
T2	2394	1400	764	25	107
T3	3035	1596	1064	80	227
T4	2091	979	715	53	324
T5	2149	1070	854	39	148

Despite the test being conducted at elevated temperature of 120 °C, the mechanical behaviour of the plain weave fabric CFRP and their failure modes is similar to failure modes of specimens tested at room temperature. It has been reported by several researcher that the tensile behaviour of the composites starts to degrade as the material reaches the glass transition state. Since the test is conducted at temperatures less than the glass transition temperature of the CFRP in this study (which is 141 °C), the tensile behaviour did not change drastically. Besides, the increase in testing temperature affects the off-axis properties such as in-plane shear strength or interlaminar shear strength rather than the tensile properties. Because these properties are dependent on the matrix material, which is more susceptible to high temperatures [1–3].

Nonetheless, the classification strategy proposed in Table 4 clearly shows that the CFRP specimens have failed mostly through matrix cracking and debonding failure, which coincides with the experimental results and the results presented in literature. This validates the utilization of LZ complexity as an effective parameter for damage classification. Further validation is given through CWT and neural networks in the subsequent sections.

4.2. Validation of Damage Classification using CWT

The signals from each class of the damage mode are analysed in this time-frequency domain and their results are discussed in this section. Typically, the AE signals from transverse cracking of matrices will generate low amplitude and low frequency AE signal, which have quite wider distribution of its frequency content in the time domain. In Figure 3, it can be observed that the maximum spectral energy of the signal is distributed at low frequency (around 120 kHz) and is distributed for about 75 ms. The maximum magnitude of the signal is also quite low, confirming its origin from a matrix cracking event.

In Figure 4, the frequency of the signal is centered around 350 kHz and the magnitude is quite high compared to the on in Figure 3. There is a second frequency component with a magnitude of about 0.015, which is distributed for a longer duration in the time domain. These are the characteristics of the delamination or debonding events.

Figure 5 shows the time-frequency characteristics of the signal from fibre rupture. The high magnitude of the spectral energy, which is 0.035 (max), which is localized in a narrower time domain compared to the other signal classes. Due to the high stiffness of the fibres, fibre rupture always generates high frequency and high magnitude signals. The reverberations present in the signal is associated with the propagating direction and the location of the fibre failure.

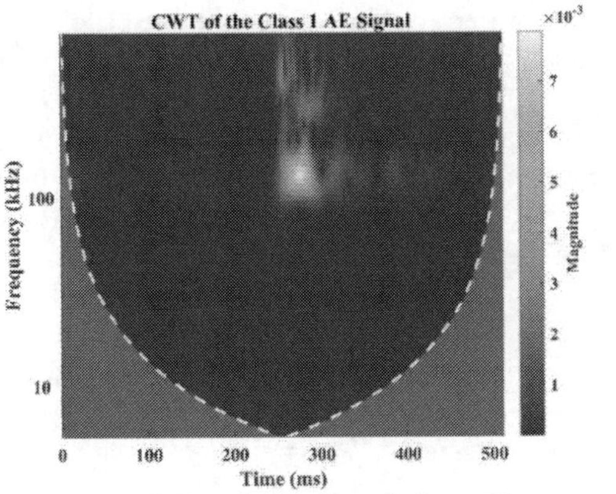

Figure 3. CWT of AE Signal from Transverse Cracking.

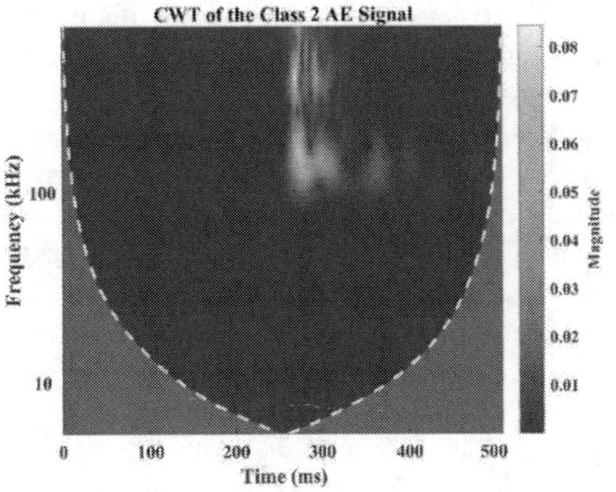

Figure 4. CWT of AE Signal from Delamination or Debonding.

Finally, the time-frequency features of the signal in Figure 6 also have high frequency component localized in a narrower time domain, which is similar to the signal from fibre breakage. This similarity has been addressed by several researchers, who believed that the W-Freq or time-frequency analysis is the most appropriate way to differentiate the signals from fibre failure and through-thickness crack growth. It can be observed that despite the similarity in the frequency characteristics of these signals, the spectral magnitude of the signal in Figure 6 is quite low (0.014 max). Besides, more reverberations can be observed in the high frequency component, which is absent in the signal from fibre rupture in Figure 5. Therefore, this can be identified as the signal from through-thickness crack growth event.

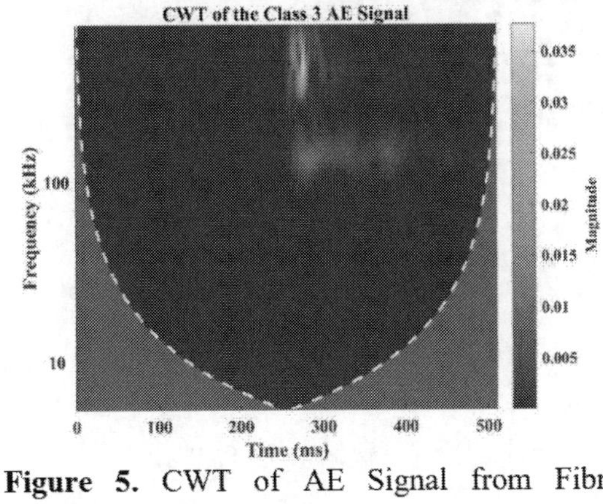

Figure 5. CWT of AE Signal from Fibre Breakage or Fibre Rupture.

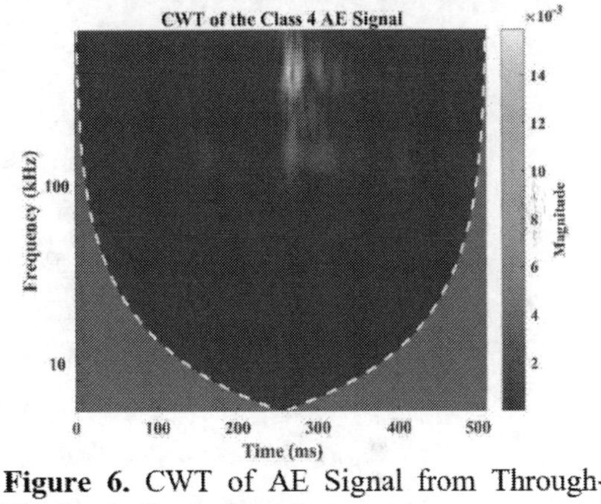

Figure 6. CWT of AE Signal from Through-Thickness Crack Growth.

The classification strategy proposed in Section 4.1 is validated through the CWT analysis. The multi-parameter approach, not the say the least, the LZ complexity, is rather successful in classifying the AE signals based on their damage sources. Nonetheless, only a handful of CWT histograms are provided here as an example to validate the classification strategy using LZ complexity. In order to obtain a comprehensive validation, the modified SqueezeNet introduced in Section 3.3 is used.

4.3. Validation of Damage Classification using SqueezeNet

The image-based classification capability of the CNN, SqueezeNet is used for validating the efficacy of the classification strategy proposed in Section 4.1. For training the SqueezeNet, a set of training data and validation data is required. The classified signals in Table 5 are used for generating the training and validation data. The signals generated during the testing of specimens T1, T2 and T3 are used for training the SqueezeNet, while the data from the remaining two tests T4 and T5 are used for validating. Since the classified data from a same group of specimens is used for training and validating, the efficiency of the classification can be considered as the overall efficiency of the classification strategy proposed using LZ complexity.

For training the SqueezeNet, the input data must be equal for all the classes for obtaining an efficiency network. For this reason, the signals must be augmented by adding to random noise. In this study, a random white noise is added to augment the signal. Class 3 and Class 4 signals comprise only about 9.68\% of the overall signals. Therefore, augmenting them to reach the number of signals in Class 1 and Class 2 will lead to over-fitting of data. Therefore, Class 3 and 4 are merged together. The signals are augmented in the way that each class will have 6000 signals for training and 2500 signals for validation. It is recommended that higher number of training data will increase the classification efficiency of the SqueezeNet. However, the goal of this work is not to improve the efficiency of the SqueezeNet but rather to validate the classification strategy using LZ complexity. Therefore, the augmentation is limited to 6000 signals for training and 2500 signals for testing.

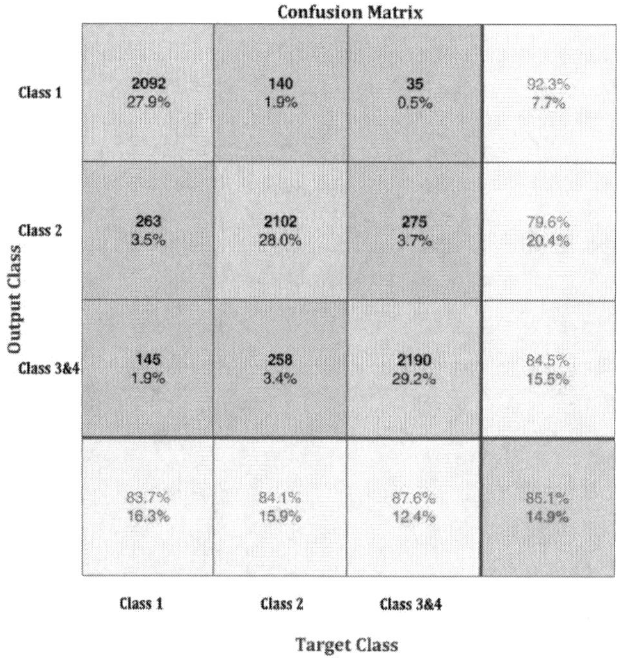

Figure 7. Confusion Matrix of the Classification by SqueezeNet.

After a training hour with the augmented data obtained from the test specimens T1, T2 and T3 of 1.5 hours, 20 epoch and 75 minibatch size for each epoch, the designed SqueezeNet trained for 100%. Validation data from the test specimens T4 and T5 are used for testing the efficiency of the network. The confusion matrix of the classification is presented in Figure 7.

The confusion matrix shows that the SqueezeNet was able to classify the AE signals at an overall efficiency of 85.1%. The signals in Class 1 are classified at an impressive 92.3% efficiency, while the signals in Class 3&4 are at 84.5%. The signals in Class 2 are classified at a significantly lower efficiency of 79.6%. This classification efficiency can be improved quite easily by increased the number of training dataset. In our previous works, SqueezeNet was trained to classify 4 classes of signals at an efficiency close to 100%. However, the goal here is not to improve the efficiency, but rather to identify the true efficiency without augmenting the training data.

From these observations, it is safe to say that the multi-parameter approach using LZ complexity, peak amplitude and W-Freq, the classification strategy has an efficiency around 85.1%, which is quite successful considering the limited data available for classification. Nonetheless, LZ complexity is an efficient parameter, which can be used for damage identification in fibre reinforced composites using Acoustic Emission technique.

5. Conclusions

The tensile behaviour of plain weave fabric CFRP specimens tested at elevated temperatures below the glass transition state is studied. The failure modes of the specimens show that the test temperature did not affect the tensile behaviour of the composites. The AE technique is used for assessing the failure modes and the tensile behaviour of the CFRP composites. A new parameter called Lempel-Ziv complexity is introduced for analysing the AE data. The primary observation is that LZ complexity, when utilized along with the peak amplitude and W-Freq of the AE signals is capable of identifying the damage sources of the AE signals. Using this classification strategy, the failure modes of the plain weave fabric CFRP specimens under tensile loading is assessed. To validate the efficacy of this novel parameter, the classified signals are studied in their time-frequency domain using CWT. The results seem to agree with the experimental results. Finally, to obtain a quantitative efficiency of this new parameter, a modified Computational Neural Network named SqueezeNet is trained and tested using the AE signals generated from the tensile tests. The results show that the LZ complexity has an efficiency of 85.1\% in classifying the damage modes using AE data, when it is used as a multi-parameter module along with peak amplitude and W-Freq. This novel parameter seems promising and has quite wider application, which must be scrutinized in the future.

References

[1] Hawileh R A, Abu-Obeidah A, Abdalla J A and Al-Tamimi A 2015 *Constr. Build. Mater.* **75** 342–8
[2] Bazli M and Abolfazli M 2020 *Polymers* **12** 2600
[3] Cao S, Wang X and Wu Z 2011 *J. Reinf. Plast. Comp.* **30** 799–807
[4] Barile C, Casavola C, Pappalettera G and Kannan V P 2020 *Eng. Fract. Mech.* **235** 107083
[5] Liptai R G, Harris D O and Tatro C A 1972 *An introduction to acoustic emission Acoustic Emission* (Pennsylvania: ASTM) p 505
[6] Hamstad M A 2000 *Acoustic emission beyond the millennium* (Amsterdam: Elsevier) p 77–91
[7] Barile C, Casavola C, Pappalettera G and Vimalathithan P K 2019 *Compos. Part B Eng.* **178** 107469
[8] WenQin H, Ying L, AiJun G and Yuan F-G 2016 *Appl. Compos. Mater.* **23** 155–78
[9] Li Y, Wang X, Liu Z, Liang X and Si S 2018 *IEEE Access* **6** 66723–41
[10] Aboy M, Hornero R, Abásolo D and Álvarez D 2006 IEEE *Trans. Biomed. Eng.* **53** 2282–8
[11] Cui L, Gong X, Zhang J and Wang H 2016 *J. Sound. Vib.* **385** 372–88
[12] Barile C, Paramsamy Kannan V, del Core L and Casavola C 2022 *Polym. Compos.* **43** 6536–47
[13] Lempel A and Ziv J *IEEE Trans. Inf. Theory.* **22** 75–81
[14] de Groot P J, Wijnen P A M and Janssen R B F 1995 *Compos. Sci. Technol.* **55** 405–12
[15] Sause M G R, Müller T, Horoschenkoff A and Horn S 2012 *Compos. Sci. Technol.* **72** 167–74
[16] Oz F E, Ersoy N and Lomov S V 2017 *Compos. Part A Appl. Sci. Manuf.* **103** 230–5
[17] Gao F, Boniface L, Ogin S L, Smith P A and Greaves R P 1999 *Compos. Sci. Technol.* **59** 137–45
[18] Gao F, Boniface L, Ogin S L, Smith P A and Greaves R P 1999 *Compos. Sci. Technol.* **59** 123–36

IOP Conf. Series: Materials Science and Engineering 1275 (2023) 012007 doi:10.1088/1757-899X/1275/1/012007

Increased interaction between Ultra-High Molecular Weight Polyethylene fibres and epoxy matrices for advanced composite materials

E Galvanetto, S Caporali, G Zonfrillo, M S Gulino, L Pugliese and D Vangi

Department of Industrial Engineering of Florence, Università degli Studi di Firenze, Via di Santa Marta 3, 50139, Florence, Italy

dario.vangi@unifi.it

Abstract. Ultra-High Molecular Weight Polyethylene (UHMWPE) fibre is considered as an excellent strengthening component due to its high specific stiffness and strength, good wear resistance, low coefficient of friction and great chemical resistance. However, very weak interfacial adhesion with resin has hampered the development of UHMWPE-based composites with high performance. With the aim of improving the interface adhesion properties, the surface of the fibres has been decorated with PE crystal by a suitable chemical treatment. Adhesion property has been examined by pull-out tests, while tensile tests have been performed to evaluate the strength of the fibre. Preliminary results indicate that, compared to unmodified UHMWPE fibres, the overall mechanical properties of the composite with treated UHMWPE fibres and epoxy resin have been greatly improved. Moreover, the applied treatment does not weaken the strength of the fibre. The study results imply that the modified fibres have a great potential for use as reinforcing elements in high-performance composites.

1. Introduction

The growing demand for lightweight high-performance materials has expanded the use of composite materials, especially polymers reinforced with various kinds of fibers. Among all the reinforcing fibers, Ultra-High Molecular Weight Polyethylene (UHMWPE) has attracted large attention from researchers and industrial engineers, on account of its interesting mechanical and chemical properties. UHMWPE fibers are composed of linear chain macromolecules, grouped in a very compact and oriented configuration. They are characterized by very high mechanical properties: low density (equal to 0.95 g/cm^3), excellent corrosion resistance, chemical inertness, low water absorption, low coefficient of friction and high resistance to abrasion [1-7]. At the same weight, they are about 15 times stronger than steel and 2 times stronger than glass fibers, making them particularly suitable for use in various technological areas, such as the biomedical sector or the production of sails or bulletproof vests.

Considering specific mechanical properties, UHMWPE fibers have higher specific strength than carbon and higher specific stiffness than Kevlar (Figure 1). UHMWPE would therefore represent an ideal reinforcement in composite materials, but industrial outlets are severely limited especially by the reduced adhesion between fiber and matrix. In fact, the polymer surface is strongly hydrophobic and non-reactive, and under load the fiber tends to pull away from the matrix.

Content from this work may be used under the terms of the Creative Commons Attribution 3.0 licence. Any further distribution of this work must maintain attribution to the author(s) and the title of the work, journal citation and DOI.
Published under licence by IOP Publishing Ltd

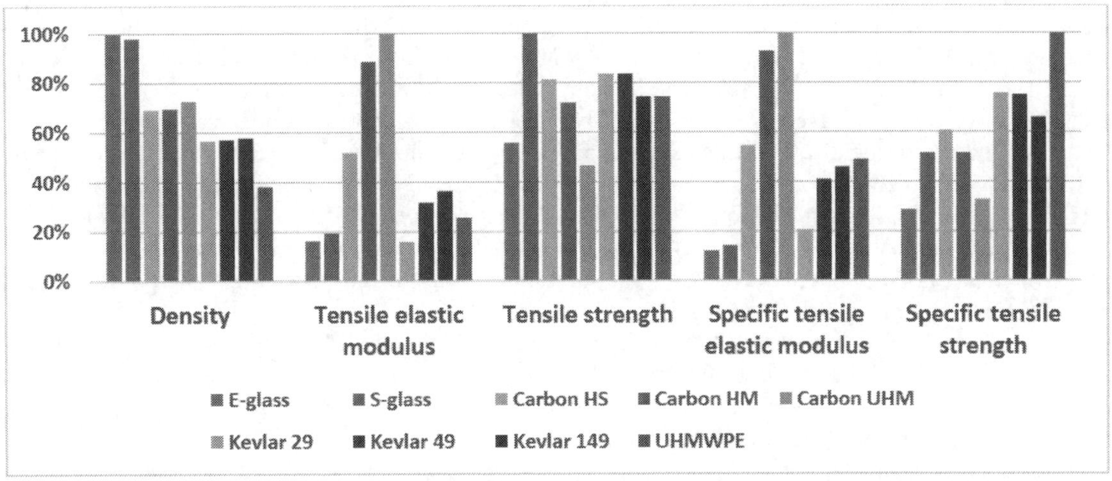

Figure 1. Qualitative comparison between UHMWPE and fibres commonly used in industrial practice.

The purpose of the present work is to study and experimentally investigate methods to increase the adhesion between epoxy matrix and fiber, a necessary step for future use of fiber as reinforcement in industrial settings. The topic has been covered for years and has attracted considerable scientific interest and research activity [8-11].

Several methods to improve fiber-matrix adhesion are described in the literature, which can be grouped into three macro-groups: chemical interaction [12-17], physical [18-23], mixed physical-chemical [24] and mechanical interaction [25-28]. Fiber-matrix chemical interaction can be increased by altering the surface chemical properties of the fibers to create reactive groups, which interact with the resin to form stronger bonds. One of the most common ways to achieve this issue is through plasma techniques [22-24]. There are various chemical treatments aimed at creating reactive groups in addition to the use of plasma. An example, it's by oxidation or sulfation in reagents, such as chromic acid and chlorosulfonic acid (these processes are often carried out after pretreatment cycles in benzene organic solvents) [12, 13, 29].

Mechanical treatments can be used to roughen the fiber surface and thus provide increased interface strength. This is achieved by increasing the adhesion area of the fiber and by the formation of microscopic mechanical blocks, such as knots or twisting of multiple fibers, which restrain sliding between fiber and matrix. Surface roughening and the use of mechanical blocks is not recommended for fibers such as carbon and glass, due to the presence of micro-stress concentration zones not tolerated by brittle materials, while it can be used for UHMWPE fibers [30]. However, there is a weakening of mechanical strength characteristics, which goes hand in hand with the improvement of adhesion properties. This often results in the loss of the specific strength of UHMWPE fibers over glass or Kevlar fibers.

This phenomenon was confirmed by some preliminary tests carried out on fibers with the presence of knots, which showed a degradation in tensile strength of more than 50 percent compared to the smooth fiber. The present work has been therefore directed toward comparing the strength properties between fibers with and without a recent surface treatment, aimed at making the outside of the fiber rougher and consequently improving adhesion with the matrix.

2. Experimental

The Teijin Endumax® UHMWPE fibers used in the experiment consist of a film in the form of a sheet, made from a roll of 2D braided 90° carbon fiber/UHMWPE. The tapes have a thickness of 0.055 mm, a width of 2 mm and a length of 129 mm and exhibit high values of both modulus of elasticity (170 GPa) and breaking strength (2200 MPa). Additional characteristics are available in the manufacturer's technical documentation [31]. The fibers thus obtained were labeled as specimens A.

Some fibers (referred to as B samples) were subjected to the treatment described in [32]. PE-decorated UHMWPE fibers were obtained by solution crystallization. The UHMWPE fibers were cleaned and activated by means of chromic acid treatment. This treatment aims to remove organic contaminants from the surface as well as to activate the surface to get the polymer crystallization. For PE crystallization, toluene was used as solvent and LDPE was dissolved in it. Heating the solution to 110°C allow the complete dissolution of the LDPE up to a 4wt% concentration. The pretreated UHMWPE fibers was then added to the hot LDPE toluene solution and kept at 110°C for 10 min. Then the mixture was cooled to 90°C to promote the PE crystallization on the fiber surface. After 1 hour at 90°C, the treated fibers were removed from the solution, cooled down to room temperature, rinsed with toluene and air dried.

The resin used (IN2 Epoxy Infusion Resin manufactured by Easy Composites Ltd) is an ultra-low viscosity epoxy resin, ideal for infusion processes. The resin has clear UV stable appearance, excellent mechanical properties, and variable hardener speed. For details of the features, please refer to the technical notes provided by the manufacturer [33].

Both fiber-to-resin adhesion tests and tensile tests were performed on both types of fiber.

Adhesion test specimens consisted of tape specimens of varying lengths with the ends immersed in resin cylinders, which in turn were inserted into aluminum cylinders of 12.5 mm inside diameter and 1.5 mm thickness. The lengths of the cylinders (and consequently the portion of fiber in contact with the epoxy resin) were varied (L equal to 20, 40 and 80 mm) to determine the value of the creep load or the corresponding minimum length required for complete load transfer to the fiber. On the samples thus made, post cure treatment was then carried out at a low temperature (about 50°C) for 24 h to ensure complete polymerization of the resin. For tensile strength tests, the fiber was wrapped around discs and constrained to them by adhesive. The two types of specimens are shown in Figure 2.

A servo-hydraulic tensile machine (MTS 810), equipped with a 2 kN full-scale load cell, was used to determine the mechanical properties. Tests were carried out under displacement control, maintaining a crosshead speed of 3 mm/min. Experimental load and displacement data were collected by a computer-controlled data acquisition system. For each condition, three tests were carried out on nominally equal samples to get an indication of the dispersion of the results.

2.1. SEM analysis

SEM microscope images of UHMWPE fibres before and after solution crystallization are reported in Figure 3. After the treatment the surface of UHMWPE fibres results covered with LDPE fine structures. Similar structures have been obtained using a similar approach on micro-UHMWPE fibre by Li et al. in 2016 and 2019 [32, 34]. Using the described treatment, the whole surface of PE-decorated UHMWPE fibres results deeply modified, presenting a rough "net-like" surface morphology. These samples are characterized by the presence of micron-sized growths and bulges, periodically spaced and often mutually interconnected forming a "net-like" appearance.

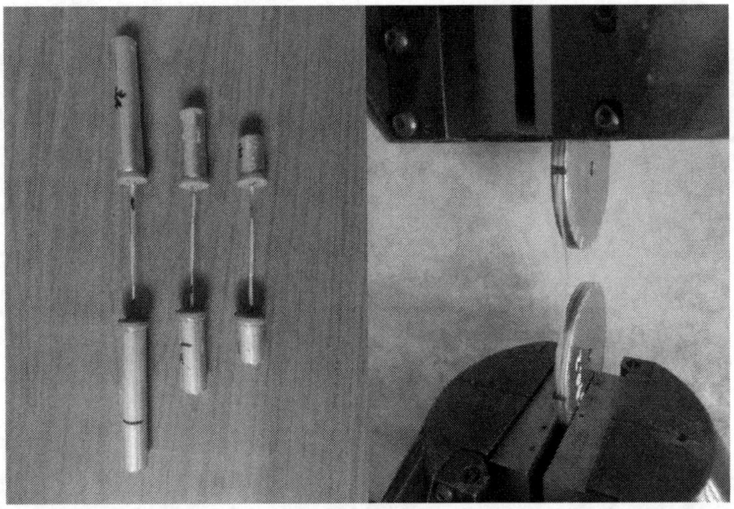

Figure 2. Specimens for pull-out (left) and tension (right) tests.

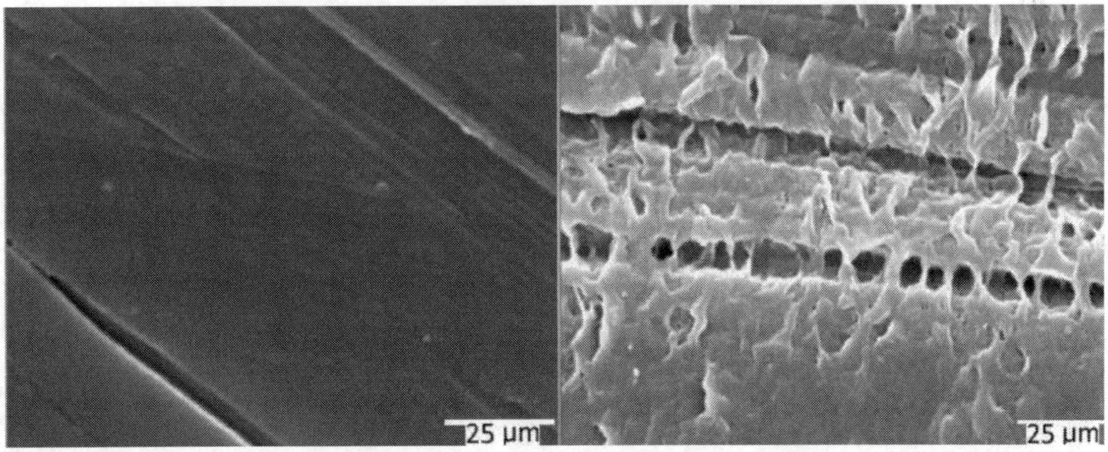

Figure 3. SEM micrographs of the surface of non-treated (left) and treated (right) fibres.

3. Results and discussion

3.1. Adhesion test

Figure 4 shows, in a displacement-load representation, the results of tests conducted on untreated specimens. Only one of the three tests conducted on nominally identical specimens has been shown in the graph for clarity; the "A" in the legend stands for chemically untreated UHMWPE-based samples. As the length of the contact section between fiber and resin varies, different behaviors are observed. In the 20-mm-long specimens, after an initial phase of fiber stretching, there is a constant gradient until the maximum value is reached. At this point the detachment of the fiber from the resin begins, practically over the entire length of contact; therefore, the load has an abrupt decrease and then settles to an almost constant value and, due to the friction, the fiber encounters in sliding through the resin. In specimens with greater length of the bonding section, initially the situation is similar, but after reaching a certain value, a load growth is observed with a sawtooth trend, interpolating with a straight line of lesser slope to the previous one. The most plausible interpretation of this behavior consists of a progressive debonding: when the load at the interface reaches a critical value, a detachment between fiber and resin occurs, but this detachment does not affect the entire length of the contact. The load is transferred to a "further upstream" zone, and the detachment of this new zone occurs for a greater load than the previous one because friction appears in the detached zone. The maximum value of the load corresponds to the situation where the detachment has occurred along the entire contact length.

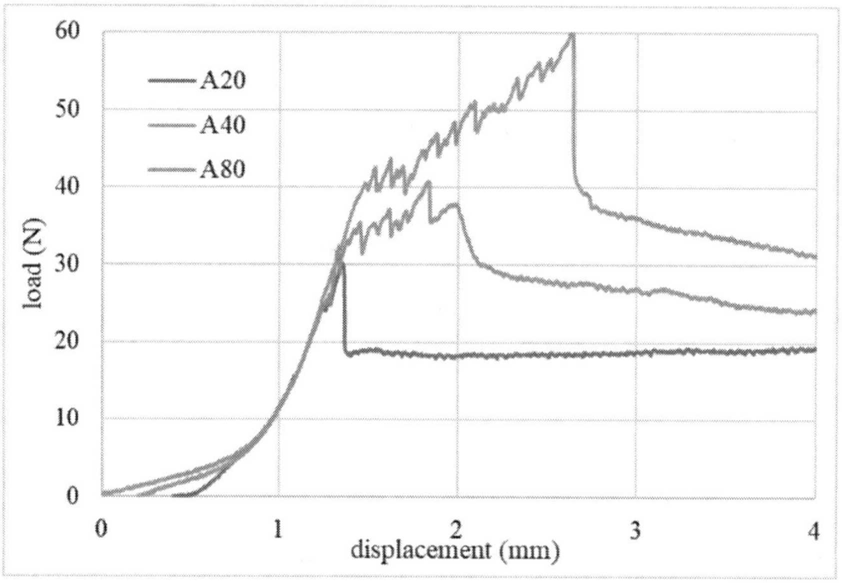

Figure 4. Pull-out test on untreated UHMWPE-based specimens.

A broadly similar situation is observed in the case of the treated specimens, shown in Figure 5 (the "B" in the legend stands for chemically treated UHMWPE-based samples). In the specimen with contact length 20 mm, the first detachment load coincides in two of the three tests with the maximum value, while in the third it is slightly lower. The samples with longer lengths all exhibit the load fluctuations before the peak value is reached.

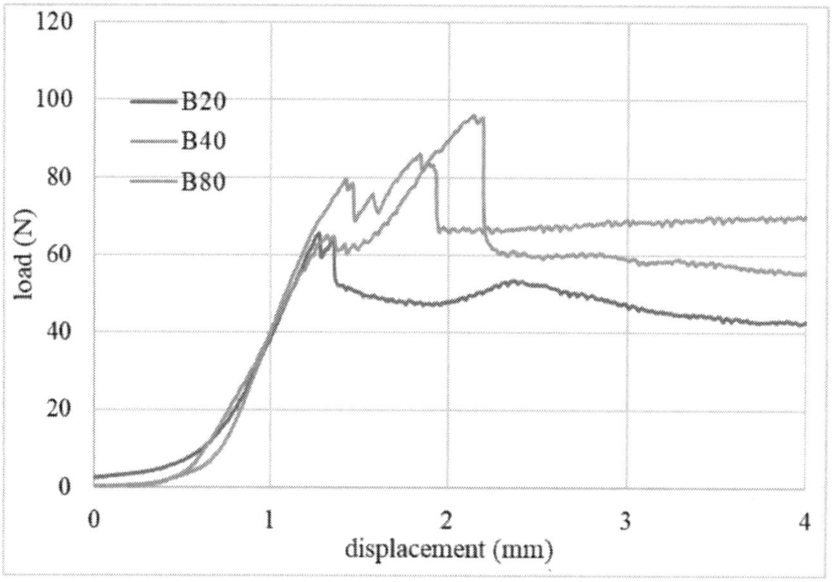

Figure 5. Pull-out test on treated UHMWPE-based specimens.

3.2. FEM Analysis

The test for the contact length of 40 mm was reproduced by FEM modeling, using the HyperWorks® software. A non-linear static analysis was carried out considering 4 contact groups (node-to-surface type), one on each side of the fiber, at the interface with the upper matrix. The lower matrix was omitted from the analysis for the sake of computational efficiency, as were the aluminum cylinders containing the resin. The fiber, of rectangular cross-section, was embedded in the lower portion at the interface zone with the matrix. The upper matrix was constrained through its outer surface both radially and tangentially, with reference to a cylindrical system aligned with the fiber. The applied load was divided equally on each node of the outer surface of the upper matrix.

The numerical values of the elastic constants and friction parameters introduced into the FEM model were taken from the technical literature. Therefore, the simulation results are useful for understanding the situation and qualitative stress trends, but it is not conceivable to have a reliable quantitative comparison with experimental test results. The convergence parameters embrace the software's default proposal (10 increments for the applied load split, 25 iterations maximum for each increment, convergence check on displacements, load and work).

Figure 6 shows, for three different values of the applied force, the stress trend at the interface at one of the four contact surfaces. In the figure, the part of the fiber not in contact with the matrix has been omitted and replaced with an arrow representing the direct downward force acting on it. The results are in line with the expected ones, i.e., stresses at the interface are not uniform, with high values in the area where the fiber exits the array (bottom area in Figure 6), values that gradually decrease as they proceed upward. Beyond a certain contact length, the stress becomes negligible.

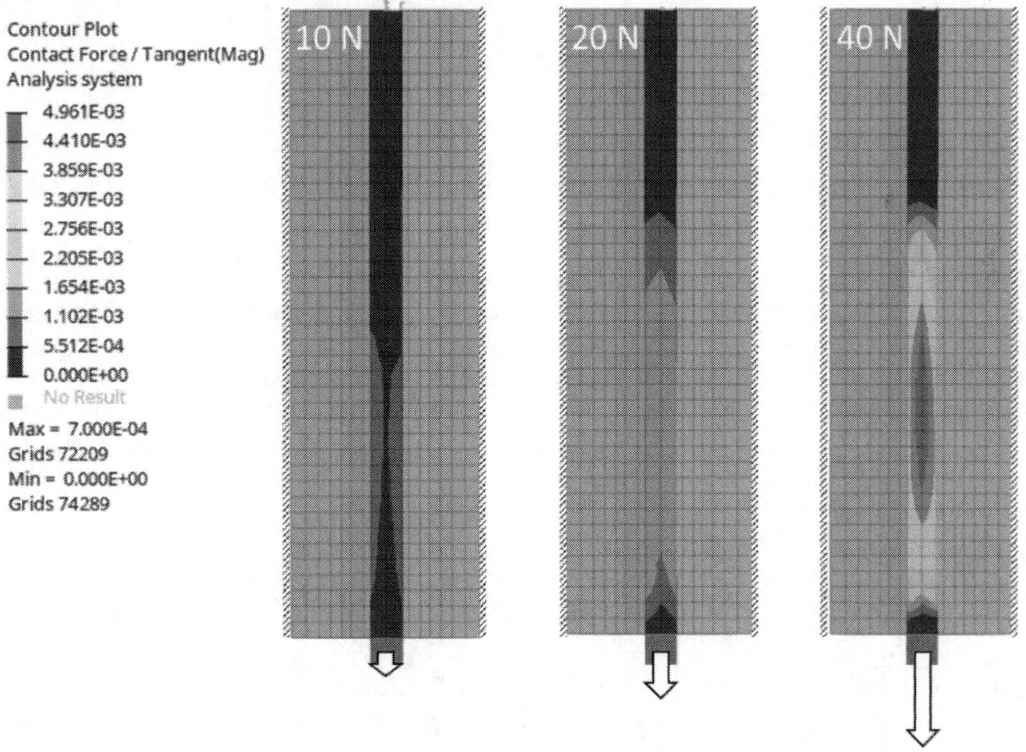

Figure 6. Contact force according to FEM analysis at three different load values

3.3. Comparison

To quantitatively define the behavior and compare treated and untreated samples, three different load values were considered, taken in the various tests in particular situations:

- the load at which debonding begins, F_d
- the maximum value achieved in the test, F_m
- the value following the sharp decrease after the maximum, F_s.

For the latter parameter, the value of the load was chosen at the point where the first derivative of the displacement does not show significant changes (less than 1% of the maximum span, Figure 7). In the tests with 20 mm adhesion length, the value of F_d (if not present) was considered coincident with that of F_m. As mentioned earlier, F_d gives indications of the value of the load at the start of detachment between fiber and resin, F_m on that at which detachment occurs over the entire length of contact, F_s on the friction the fiber encounters in sliding into the matrix.

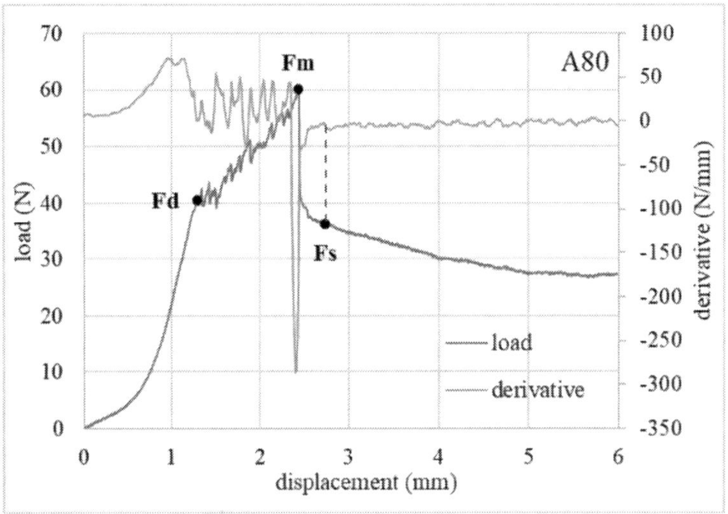

Figure 7. Example of evaluation of the load parameters considered.

The parameters thus defined are shown in Table 1. Each value is the average of the data obtained in the three tests performed on nominally equal samples. The load for which debonding begins is practically independent of the length of the fiber-to-matrix contact section for both treated and untreated specimens. However, treatment results in a large increase in the value of F_d. The treated specimens also show higher values for the maximum load value, with an increase of almost 40 N for each of the three lengths under test. As a function of length, there is an increasing trend in both F_m and F_S. Considering the latter parameter, in the untreated specimens F_s is about 50 percent of the corresponding value found in the treated specimens. Regardless of the quantitative details of the behavior, it is clear that treatment results significantly increase fiber-resin bonding properties.

Table 1. Average values of the load parameters measured in the tests for treated and untreated UHMWPE-based composite specimens.

L (mm)	F_d (N)		F_m (N)		F_s (N)	
	Untreated	Treated	Untreated	Treated	Untreated	Treated
20	32.1	62.4	32.1	70.5	21.9	49.1
40	32.8	60.5	44.2	84.0	29.2	58.9
80	36.0	59.2	56.2	94.4	35.5	69.5

3.4. *Tensile tests*

The results of tensile tests on both treated and untreated specimens are shown in Figure 8. The behavior of the two types of fibers is essentially identical, in terms of both stiffness and strength. The breaking strength for the untreated fiber averages 178 N, while that of treated fiber is 175 N. Therefore, it can be derived that the treatment does not change the mechanical properties of the fiber.

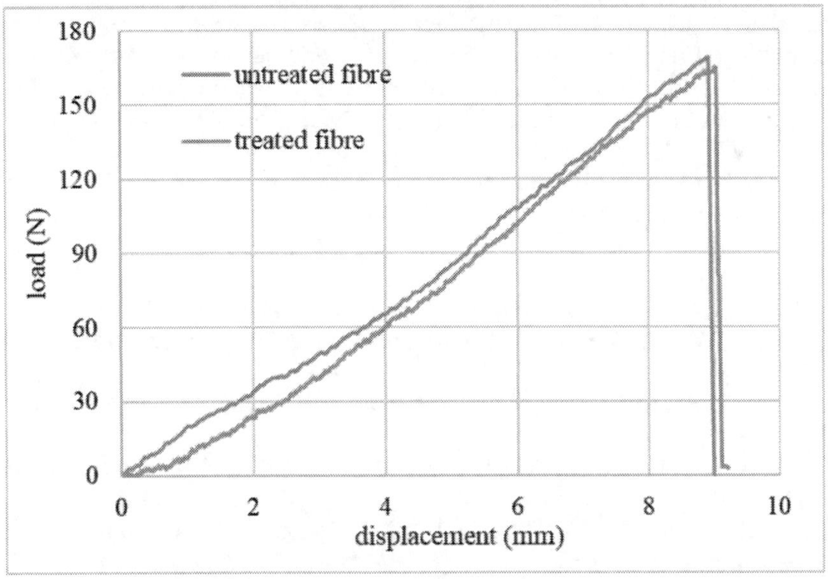

Figure 8. Tensile test on treated and untreated fibres.

4. Conclusions

This work explored the possibility of using Ultra-High Molecular Weight Polyethylene fibers to produce epoxy matrix-based composites. A method to improve particularly the interfacial adhesion between fibers and matrix has been investigated, resulting in the formation of a rougher surface morphology of the fibers. For this purpose, both adhesion tests for composite specimens, varying the length of the contact section, and tensile tests on fibers were conducted.

The results showed that the treatment brought an effective increase in adhesion properties, evaluated both in terms of the value of the debonding initiation load, and that at the complete detachment between fiber and matrix. The treatment also did not change the strength properties of the fiber, as conversely obtained with the introduction of mechanical blocks. The results are useful for understanding the dependence of strength on interface structure and may provide suggestion for the preparation of high-performance composites.

References

[1] Gao S and Zeng Y 1993, *J.Appl. Polym. Sci.* **47** pp 2065–71
[2] Gao S and Zeng Y 1993 *J. Appl. Polym. Sci.* **47** pp 2093–101
[3] Silverstein M S, Breuer O and Dodiuk H 1994 *J. Appl. Polym. Sci.* **52** pp 1785–95
[4] Cohen Y, Rein D M and Vaykhansky L 1997 *Compos. Sci, Technol.* **57** pp 1149-54
[5] Chukov D I, Kharitonov A P, Tcherdyntsev V V, Zherebtsov D D and Maksimkin A V 2018 *J. Compos. Mater.* **52** pp 1689-98
[6] Maksimkin A, Kaloshkin S, Zadorozhnyy M and Tcherdyntsev V 2014 *J. Alloys Compd* **586** S214-17
[7] Sobieraj M C ad Rimnac C M 2009 *Journal of the mechanical behavior of biomedical materials* **2** pp 433-443

[8] Postema R 1968 *Polym. Bull.* **16**
[9] Strobel M 1994 *Plasma Surface Modification of Polymers: Relevance to Adhesion*
[10] Ward M 1983 *A study of the adhesion of drawn polyethylene polymeric resin systems J. Mater. Sci.* **18** pp 533–45
[11] Chhetri S and Bougherara H 2021 *Composites Part A: applied science and manufacturing* **140**
[12] Chanunpanich N, Ulman A, Strzhemechny Y M, Schwarz S A, Janke A, Braun H G and Kraztmuller T 1999 *Surface modification of polyethylene through bromination* **15** pp 2089-94
[13] Zhang Z, Jiang G, Wu Y, Kong F, and Huang J 2018 *Applied Surface Science* **427** pp 410-415
[14] Sa R, Wei Z, Yan Y, Wang L, Wang W, Zhang L, Ning N and Tian M 2015 *Composites Science and Technology* **113** pp 54-62
[15] Holloway J L, Lowman A M, VanLandingham M R and Palmese G R 2013 *Composites science and technology* **85** pp 118-125
[16] Ahmadi M, Zabihi O, Masoomi M and Naebe M 2016 *Composites Science and Technology* **134** pp 1-11
[17] Du G and Wang J J 2017 *Surface and Interface Analysis* **49** 10 pp 940-944
[18] Wang L, Gao S, Wang J, Wang W, Zhang L and Tian M 2018 *The Journal of Adhesion* **94** 1 pp 30-45
[19] Zheng Z, Tang X, Shi M and Zhou G 2004 *Journal of Polymer Science Part B: Polymer Physics* **42** 3 pp 463- 472
[20] Oosterom R, Ahmed T J, Poulis J A and Bersee H E N 2006 *Medical engineering and physics* **28** 4 pp 323-330
[21] Gilman A B, Kuznetsov A A, and Ozerin A N 2017 *Russian Chemical Bulletin* **66** 4 pp 577-586
[22] Piskarev M, Gilman A, Kuznetsov A and Ozerin A 2017 *In J Phys Conf Ser* **927**
[23] Moon S I and Jang J 1999 *Composites science and technology* **59** 4 pp 487-493
[24] Jin X, Wang W, Xiao C, Lin T, Bian L and Hauser P 2016 *Composites Science and Technology* **128** pp 169-175
[25] Tze W T Y, Gardner D J, Tripp C P and O'Neill S C 2006 *J Adhes Sci Technol* **20** pp 1649–68,
[26] Silverstein M S and Breuer O 1993 *J Mat Sci* **28** pp 4153-4158,
[27] Karger-Kocsisa J, Mahmood H and Pegoretti A 2015 *Prog Mat Sci* **73** pp 1–43
[28] Zhou Y, Fan M and Chen L 2016 *Compos Part B* **101** pp 31–45
[29] Potter A 1981 *Directed chlorination in some substituted alkanes (Doctoral dissertation University of St Andrews)*
[30] Zeng J and Netravali A 2004 *Polymer Surface Modification: Relevance to Adhesion* **3** pp 159–182
[31] https://www,teijinaramid,com/en/product-details/endumax-film/ (last accessed 1/9/2022)
[32] Li W, Meng L, Wang L, Mu J and Pan Q 2016 *Surf Interface Anal* **48** pp 1316–1319
[33] https://www,easycomposites,co,uk/in2-epoxy-infusion-resin (last accessed 1/9/2022)
[34] Li W, Liu X, Feng M and Yang J 2019 *Composites Science and Technology* **182**

IOP Conf. Series: Materials Science and Engineering 1275 (2023) 012008 doi:10.1088/1757-899X/1275/1/012008

Vibro-Acoustic Modulation with broadband pump excitation for efficient impact damage detection in composite materials

G Loi[1], G Marongiu[2], M C Porcu[2] and F Aymerich[1]

[1] Department of Mechanical, Chemical and Materials Engineering, University of Cagliari, Piazza d'Armi, 09123 Cagliari, Italy

[2] Department of Civil, Environmental Engineering and Architecture, University of Cagliari, Piazza d'Armi, 09123 Cagliari, Italy

Email address: gabriela.loi@unica.it

Abstract. In the past few decades, the need for efficient and reliable Structural Health Monitoring strategies has led to the development of several approaches for damage detection and characterization purposes. Among them, the Nonlinear Vibro-Acoustic Modulation (VAM) exploits the modulation arising from the interaction of two concurrently applied driving waves, namely the probe and the pump excitations, in the presence of nonlinear scatters such as cracks and defects. Therefore, the VAM provides information on the emergence of internal damage by extracting the nonlinear modulated components of the response of a damaged system. Originally proposed for granular media, the method has shown to be effective in detecting the presence of defects also in metals and composite materials. Nonetheless, its efficacy is highly affected by the excitation frequencies, which are usually chosen among the system resonances. The need for a preliminary modal analysis and, at once, the risk of selecting pump-probe frequency combinations with low sensitivity to damage may make the procedure time-consuming and not fully reliable, preventing the VAM technique from being widely accepted as a robust monitoring tool. To overcome these limitations, a broadband excitation may be used. This study assesses the effectiveness of the VAM technique when a combination of a frequency-swept pump excitation and a mono-harmonic probe wave is applied to drive the sample. Experimental tests were conducted on a composite laminated beam mounted on an electrodynamic shaker and tested in both pristine and damaged conditions. Low-profile surface-bonded piezoceramic transducers were used for both probe excitation and sensing. Barely visible impact damage (BVID) was introduced in the composite beam to examine the potential of the approach for the detection of very small, localized damage. The results show that the use of VAM with a broadband low-frequency excitation may be an effective option for identifying nonlinearities associated with typical damage occurring in composite structures.

1. Introduction

Structural composites are multiphase materials that have been increasingly used in aeronautical, chemical, automotive, and civil engineering, due to their advantageous mechanical properties (e.g., high specific strength and stiffness, excellent resistance to fatigue and corrosion) and flexibility in design. Nonetheless, their susceptibility to impact damage questions the use of composite materials in critical load-bearing applications. For instance, events like bird strikes, hail impacts, and runaway debris during takeoff or landing might trigger the onset of internal damage (e.g., indentation, matrix cracking, delamination, fiber fracture) in aerospace components, severely compromising their structural integrity.

Content from this work may be used under the terms of the Creative Commons Attribution 3.0 licence. Any further distribution of this work must maintain attribution to the author(s) and the title of the work, journal citation and DOI.
Published under licence by IOP Publishing Ltd

The need to detect and monitor damage at its early-stage resulted in the development of several Non-Destructive Testing (NDT) techniques, which include methods based on visual inspection, acoustic emission, X-rays, ultrasonic waves, thermography, and shearography [1-5]. Despite some attempts to automate inspections, current NDT techniques remain, however, generally time-consuming, labor-intensive, and greatly dependent on the operator's ability.

For these reasons, there is a growing interest in the development and application of Structural Health Monitoring (SHM) techniques [6-9] that may provide a real-time evaluation of the integrity of the structure. Several SHM approaches have been proposed and applied to detect the initiation and growth of damage and defects in structural systems. Guided ultrasonic waves [10, 11] and nonlinear acoustic methods [12] have been widely exploited for damage detection in metal and composite structures due to their ability to inspect large structures with a relatively small number of transducers. Although the pioneering work in the field dates to the 1970s [13, 14], nonlinear acoustic methods, which rely upon the extraction and the analysis of specific features of the system response, have gained increasing interest over the past few decades because of their higher sensitivity to small, localized damage in comparison to linear techniques. Application examples include the Scaling Subtraction Method (SSM) [15, 16], higher harmonic generation [17, 18], resonance spectra [19], frequency mixing [20], modulation analysis [21-25]. In particular, the damage detection approach based on the *Vibro-Acoustic Modulation* (VAM), also referred to as *Nonlinear Elastic Wave Spectroscopy* (NEWS) [20] exploits the vibro-acoustic interaction between two impinging waves (namely, the pump and the probe waves) concurrently applied to the inspected system. The interaction of the two waves at a damaged region generates nonlinear modulating effects, which may be observed as additional frequency components (sidebands) in the spectral response of the system and provide an indication of the presence of damage in the material. Introduced in the 1990s [26, 27] for the detection of contact-type defects (e.g., cracks), VAM was subsequently applied to a variety of materials and structural defects, such as fatigued steel beams [21], fatigued aluminum beams and plates [28, 29], loose bolt connections [30], and kissing bonds [31], showing high sensitivity to damage initiation and advantageous low-cost hardware implementation [24, 32].

The effectiveness of VAM in detecting impact-induced damage in composite structures was first assessed by Meo and Zumpano [33], who analyzed the nonlinear features of the vibro-acoustic response of a sandwich composite previously impacted by a sharp object. Later, Aymerich and Staszewski [34] applied the VAM approach to monitor the progression of internal damage in a multidirectional composite plate subjected to multiple low-velocity impacts and highlighted the role of the boundary conditions on the performance of the method. Klepka et al. [23] investigated the effect of the pump and probe frequency, usually chosen among the system resonances, on the modulation intensity. The study shows that higher nonlinear features effects are achieved by combining a probe frequency matching the local damage resonance with a pump frequency that excites the damage interfaces in the out-of-plane direction. The need for a preliminary modal analysis to determine the resonance frequencies of the system and the lack of an established procedure of general validity to select the most suitable excitation frequencies are the main current limitations of the VAM approach. In order to circumvent these limitations, broadband excitations, covering a wide range of probe or pump frequencies in a single test, may be used.

Many attempts to apply a wide-band probe signal to identify the occurrence of internal damage have already been reported in the literature [29, 35-37]. In contrast, only a minor effort has been dedicated to investigating the use of a broadband pump excitation. Donskoy et al. [38] examined the response of fatigued steel samples subjected to an impulsive pump excitation consisting of the superposition of multiple bending modes. They showed that the perturbation introduced in the cracked region and, thus, the modulation intensity, greatly depends on the applied vibration mode. More recently, the VAM technique was used by the authors with a multi-modal pump excitation to detect barely visible damage in a composite beam [39]. The method was found to be effective, though affected by the position of the sensor with respect to the bending shape of the selected pump frequency. The response of the system was acquired using single-axis accelerometers, which measure along the direction perpendicular to the mounting surface and thus along the direction of the bending deflections of the examined beam. Low-profile PZT sensors are however much more commonly used than accelerometers in SHM applications, because of their low cost, small size, and good frequency response. Moreover, they do not require dedicated signal conditioners or amplifiers. These sensors exhibit their highest sensitivity to in-plane deformations, in contrast to the transverse sensitivity of accelerometers. In this context, the present study aims to assess the effectiveness of a VAM testing approach based on broadband pump excitation for impact damage detection when low-profile PZT transducers bonded to the surface are used to sense the response of the beam. The VAM tests were conducted on a composite beam simultaneously vibrated through a shaker and a surface bonded PZT actuator. Low-profile surface-bonded piezoceramic transducers were used for sensing. The beam was tested first in the pristine and then in a damaged condition, obtained by subjecting the beam to a low-velocity impact, to assess the potential of the VAM method for detecting barely visible impact damage in composite structures.

2. Nonlinear Vibro-Acoustic Modulation

Nonlinear Vibro-Acoustic Modulation (VAM) techniques rely upon the simultaneous application of two excitation signals of distinct frequency and amplitude: a low-frequency, high-amplitude wave (*pump signal*) that perturbs the material at the damaged region and a low-amplitude, high-frequency wave (*probe signal*) that captures the nonlinearities in the material behavior generated by the pump excitation. Under the hypothesis of perfect linearity, the response of the material varies proportionally to the applied excitation and the superposition principle holds. Therefore, since the input provided in a VAM test consists of two harmonic excitations, the response of the inspected sample $u\,(t)$ can be written as a linear combination of the two responses $u_{LF}\,(t)$ and $u_{HF}\,(t)$:

$$u\,(t) = u_{LF}\,(t) + u_{HF}\,(t) = U_{LF}\sin(\omega_{LF}\,t) + U_{HF}\sin(\omega_{HF}\,t) \tag{1}$$

Here U_{LF} and U_{HF} respectively indicate the amplitude of the system response at the pump (ω_{LF}) and the probe (ω_{HF}) frequency. Therefore, the spectrum of the response of a linear undamaged material exhibits only the two frequency components corresponding to the pump and the probe excitation signals (figure 1).

Conversely, the presence of internal damage (e.g., cracks, delamination, cracks, and kissing bonds) perturbates the propagation of elastic waves and triggers the emergence of modulation effects. Although the mechanisms underlying the modulation of an acoustic wave are still not completely understood, it is widely accepted that they rely upon the nonlinear contact dynamic. Phenomena occurring within the damaged area, such as contact, decohesion, clapping, and friction, cause the material to behave nonlinearly against the applied inputs, which lead to the generation of additional spectral components. Thus, the time-domain response of a nonlinear system can be expressed as [31]:

$$u\,(t) = U_{LF}\sin(\omega_{LF}\ t) + U_{HF}\sin(\omega_{HF}\ t) + \sum_i H_{LF,i}\sin(i\,\omega_{LF}\ t) + \sum_j H_{HF,j}\sin(j\,\omega_{HF}\ t) + \sum_i\sum_j U_{SB,i\pm j}\ \sin(i\,\omega_{LF}\ t\ \pm\ j\,\omega_{HF}\ t) \tag{2}$$

where $H_{LF,i}$ and $H_{HF,i}$ respectively represent the amplitude of the higher harmonics of the pump and probe signal, while $U_{SB,i\pm j}$ represent the amplitudes of multiple mixed frequency components, which are generally referred to as modulation sidebands (figure 1). The number of sidebands and their amplitude depends on the modulation intensity, which is related to damage severity. It is, however, worth remarking that non-damage-related nonlinear effects, such as those arising from boundaries, inherent material nonlinearities, or the adopted instrumentation, might trigger the emergence of similar nonlinear effects. Thus, a certain number of modulation sidebands will be generally present even in the spectrum of an intact structure, regardless of the employed equipment or the applied approach [21].

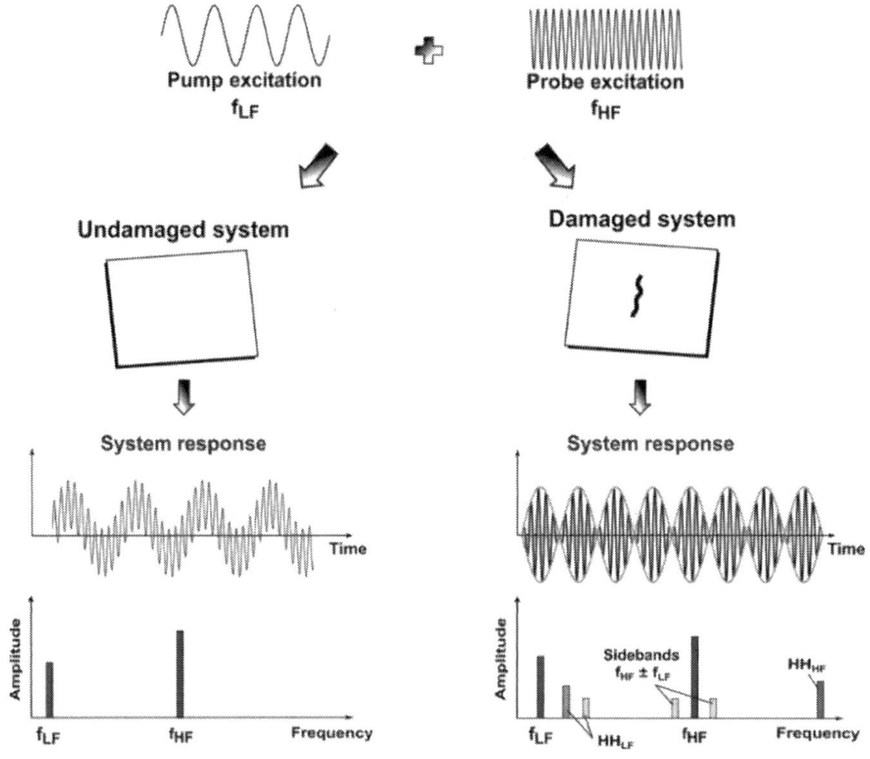

Figure 1. Schematic representation of the Nonlinear Vibro-Acoustic Modulation (VAM).

3. Experimental setup and preliminary analysis

Experimental tests were carried out on a laminated composite beam manufactured from *Seal Texipreg ® HS160/REM* carbon/epoxy unidirectional prepreg plies (0.17 mm nominal thickness). The sample was cut in the shape of a 530 mm × 59 mm x 2 mm beam (figure 2) from a panel laminated with a $[0/90]_{3s}$ stacking sequence and cured in autoclave at a maximum temperature of 160 °C and a pressure of 6 bar. Before testing, the specimen was ultrasonically C-scanned to assess the quality of the laminate and to exclude the presence of possible manufacturing defects.

The sample was mounted on a *Bruel and Kjaer 4809* electrodynamic shaker using a threaded aluminum stud glued at its bottom surface. Three *PI Ceramic PIC 151* low-profile piezoceramic transducers (10 mm diameter, 1 mm thickness), positioned as shown in figure 2, were surface bonded using a two-component epoxy adhesive and wired using additionally welded connectors. In particular, the piezoceramic transducer at location A1 was used as high-frequency actuator, while those at locations S1 and S2 were employed to sense the system response. During the tests, the shaker was driven by a signal provided by a *TTI-TGA1241* function generator and amplified through a *Bruel & Kjaer 2706* power amplifier, while a *TTI TG5011* function generator in association with a *FLC F20A* amplifier was used to control the PZT transducer bonded at location A1. The response of the sample was acquired at the two sensors (S1 and S2 in figure 2) using a 14-bit, 100 MSa/s, PC-controlled oscilloscope (*Cleverscope CS328A*). The instrumentation used to carry out the experimental investigations is shown in figure 3.

An instrumented drop-weight testing machine, whose 2.34 kg impactor was equipped with a hemispherical indenter of 12.5 mm diameter, was used to impact the sample with a 1.9 J energy. Non-destructive penetrant-enhanced X-ray images were taken to assess the nature and extent of internal damage. As illustrated in figure 4, the induced damage is mainly a combination of matrix cracks and delaminations, with a projected damage area of about 40 mm^2. Short fiber fracture paths can be also observed at the boundary of the indentation area.

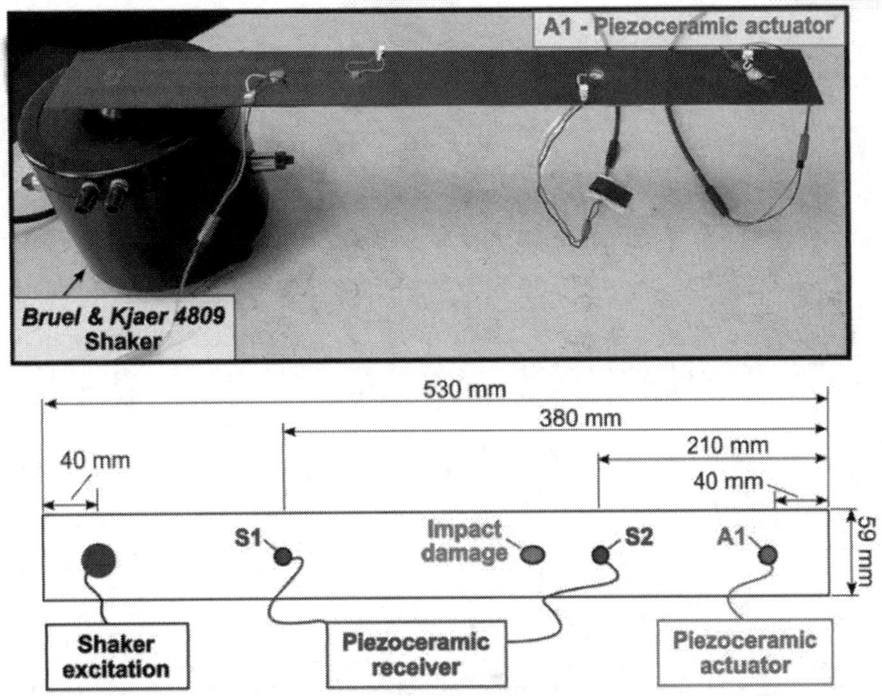

Figure 2. Schematic representation of the adopted experimental setup.

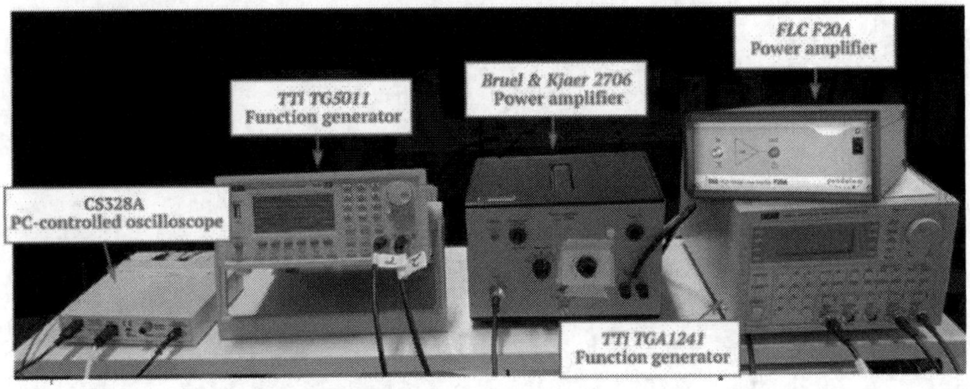

Figure 3. Instrumentation used for the experiments.

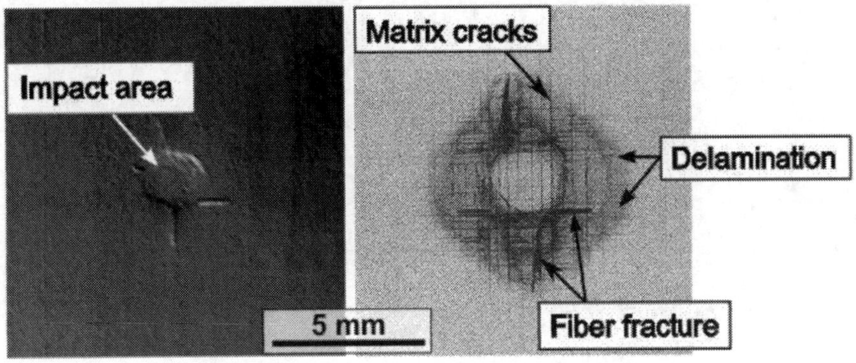

Figure 4. Impacted region (left) and X-ray image (right) of the internal damage induced by a low-velocity impact of 1.9 J energy.

To preliminarily characterize the modal response of the system, the composite sample was excited through a sweep sine signal ranging from 1 Hz to 1200 Hz in 4 s. The response of the inspected sample was acquired at locations S1 and S2. In order to increase the signal-to-noise ratio, forty data sets were recorded and post-processed in *Matlab* to calculate the average power spectrum. The averaged FFT amplitude, presented in figures 5 and 6, shows the first eigenfrequencies of the plate. Two frequency ranges around the resonance frequencies of 316 Hz and 510 Hz were chosen to explore the effectiveness of a broadband pump excitation in the VAM techique. A numerical modal analysis performed in *Abaqus* revealed that the mode shapes associated with these two selected frequencies are characterized by out-of-plane bending (figure 7).

A similar procedure was applied to select the probe frequency value. In this case, a linear sweep sine signal ranging from 5 to 40 kHz in 10 s was fed into the low-profile piezoceramic transducer bonded at location A1. The vibration patterns were again recorded at sensor locations S1 and S2 and then processed to obtain the corresponding spectra (figures 8 and 9). The 21070 Hz resonance frequency, characterized by a high relative amplitude at both S1 and S2 locations, was chosen for the subsequent VAM experiments.

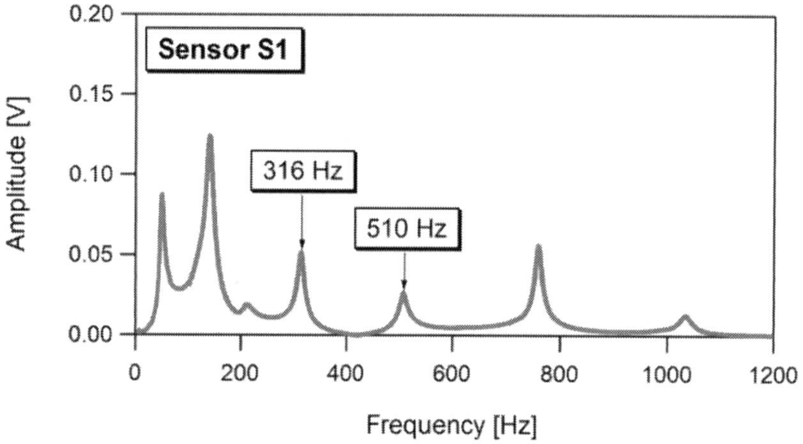

Figure 5. Power spectrum (low-frequency range) taken at sensor location S1.

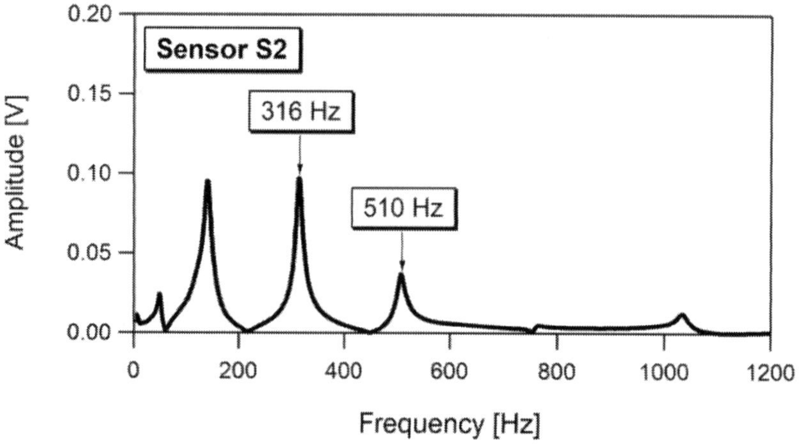

Figure 6. Power spectrum (low-frequency range) taken at sensor location S2.

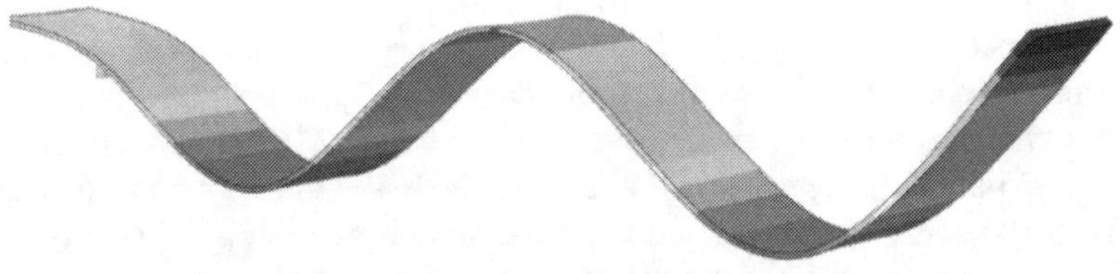

4th flexural mode (316 Hz)

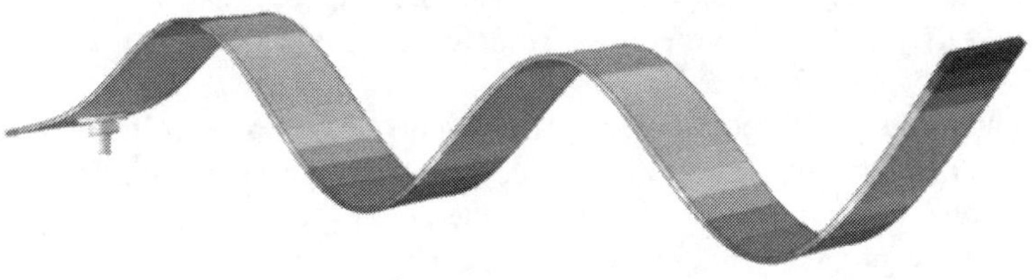

5th flexural mode (510 Hz)

Figure 7. Numerically predicted mode shapes corresponding to the selected resonance frequencies.

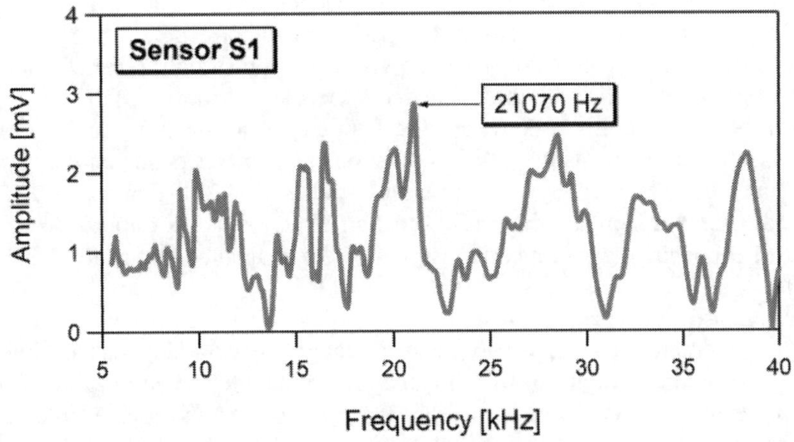

Figure 8. Power spectrum (high-frequency range) taken at sensor location S1.

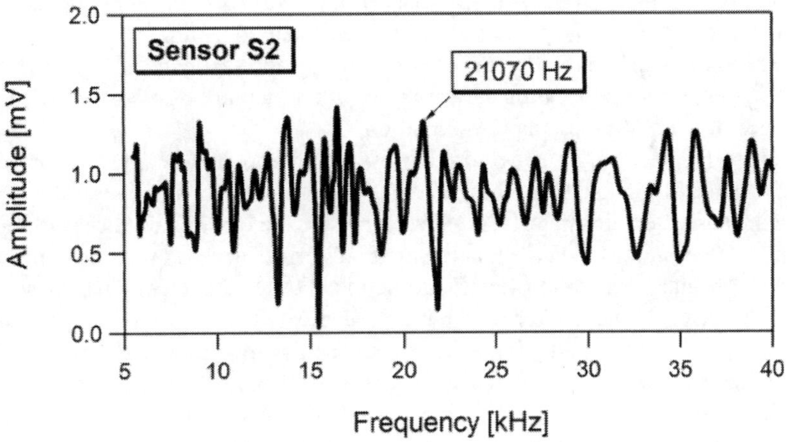

Figure 9. Power spectrum (high-frequency range) taken at sensor location S2.

IOP Conf. Series: Materials Science and Engineering 1275 (2023) 012008 doi:10.1088/1757-899X/1275/1/012008

4. Nonlinear Vibro-Acoustic Modulation tests and results

In the vibro-acoustic tests, the laminated composite beam was vibrated through a broadband pump excitation and a simultaneous mono-harmonic probe wave. The pump signals consisted of linear sine sweeps that included the two selected resonance frequencies (i.e., 316 Hz and 510 Hz). The frequency of the pump wave was initially swept across the range 240÷360 Hz in 10 s and, subsequently, over the range 440÷560 Hz in 10 s. Three pump excitation levels, namely 50 mVpp, 250 mVpp, and 1 Vpp, were considered for testing. Concurrently, a 21070 Hz pure-tone harmonic signal (probe excitation) was fed into the piezoceramic transducer at location A1. The system response was sensed at locations S1 and S2, and recorded at a sampling rate of 400 kSamples/s. The acquired data were averaged over ten data sets and post-processed by a Short-Time Fourier Transform (STFT) algorithm. To this purpose, 69 sub-windows with 30% overlapping were used.

Figures 10-13 show the STFT spectrograms of the acoustic response of the inspected sample over the two considered frequency ranges and for the three considered pump excitation levels (i.e., 50 mVpp, 250 mVpp, and 1 Vpp). Figures 10 and 11 refer to the signals acquired at sensor location S1, while figures 12 and 13 correspond to the response sensed at location S2. The spectrograms provide information on the amplitude (represented by the grey intensity) of the frequency content of the system response (horizontal axis) as it varies against the instantaneous pump frequency (vertical axis). In order to assess the emergence of modulation sidebands, the spectrograms were zoomed around the spectral line corresponding to the probe frequency (i.e., 21070 Hz). Examples of the averaged power spectra corresponding to a specific pump frequency value are shown below the associated spectrograms.

As expected, the higher amplitude contained in the spectrograms corresponds to the frequency of the probe excitation, while the modulation sidebands are visible as thin straight lines with a distance from the probe frequency varying linearly with the instantaneous frequency content of the swept pump signal. The comparison of the spectrograms obtained for the different sensor positions and the three levels of pump excitation shows that the intensity of the modulation sidebands depends on the pump excitation energy and frequency, as evidenced by other previous studies [21, 28]. For instance, for the pump excitation amplitude of 250 mVpp and the sensor location S2, no sidebands appear for the range 240-360 Hz (see figure 12b), while they are clearly visible in the damaged condition for the range 440-560 Hz (figure 13b).

The spectrograms also show that the modulation intensity is affected by the sensor position. For example, for the pump amplitude of 250 mVpp and the range 240-360 Hz, sidebands do not emerge when the sensor is in position S2 (figure 12b), while they are evident for sensor location S1 (figure 10b). The effect of the position of the sensor on the modulation intensity was investigated in other studies and was seen to depend on the distance of the sensor location from the nodal points of the deformed beam at the excitation frequency [39].

Examining, in particular, the response of the beam acquired at the position S1 we may see that when the amplitude of the provided pump input is increased from 50 mVpp to 250 mVpp, a clear pattern of modulation sidebands emerges in the spectrogram of the damaged sample (figure 10 and figure 11). These sidebands exhibit higher amplitudes in the proximity of the resonance frequency, thus showing the influence of the interrogating pumping frequency on the modulation intensity and, in turn, on the detection capabilities of the Vibro-Acoustic Modulation.

The comparison of the spectrograms acquired at 50 mVpp and 250 mVpp in damaged conditions highlights the importance of the excitation level on the VAM sensitivity and shows that enough energy is required to activate and detect the material nonlinearities. On the other hand, the spectrogram acquired at 1 Vpp (see figures 10c and 11c) indicate that the method's effectiveness might be compromised when comparatively too high pump excitation levels are adopted. As a matter of fact, these spectrograms reveal a significant increase in the background noise, as indicated by the change in the grey intensity, which makes it almost impossible to distinguish the presence of modulation sidebands. This evidence might be linked to the intrinsic characteristics of the adopted equipment. For example, when powered, the electrodynamic shaker generates an electromagnetic field that varies its intensity proportionally to the amplitude of the applied input. The induced electromagnetic interference reduces the signal-to-noise ratio of the signal received from the PZT sensors, compromising the VAM performance.

IOP Conf. Series: Materials Science and Engineering 1275 (2023) 012008 doi:10.1088/1757-899X/1275/1/012008

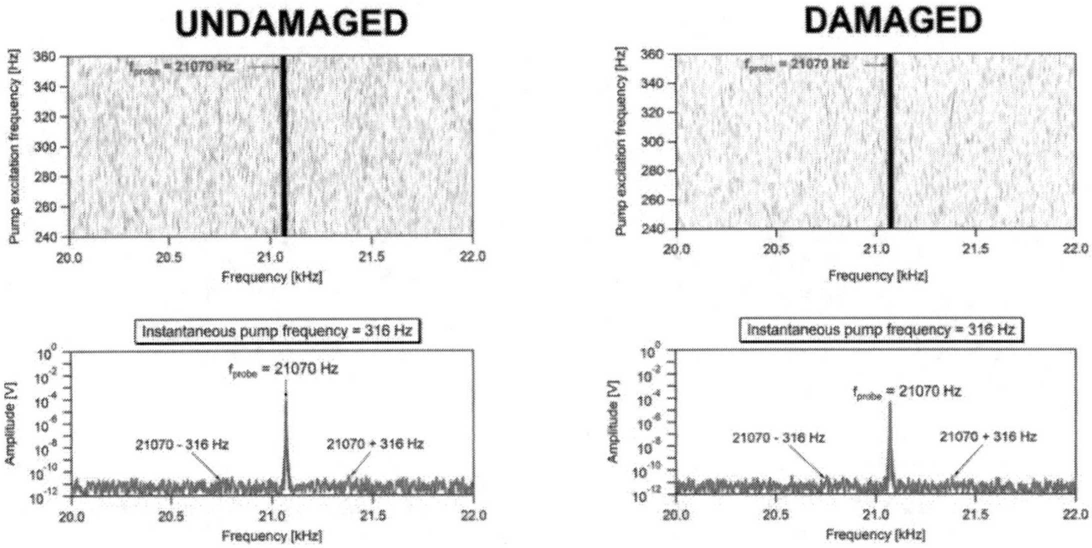

(a) Excitation amplitude 50 mVpp

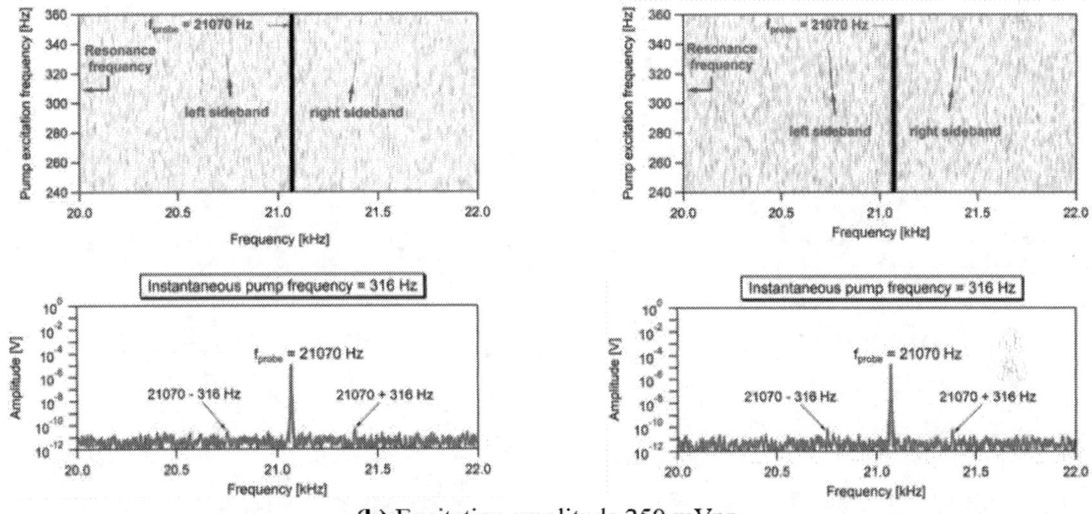

(b) Excitation amplitude 250 mVpp

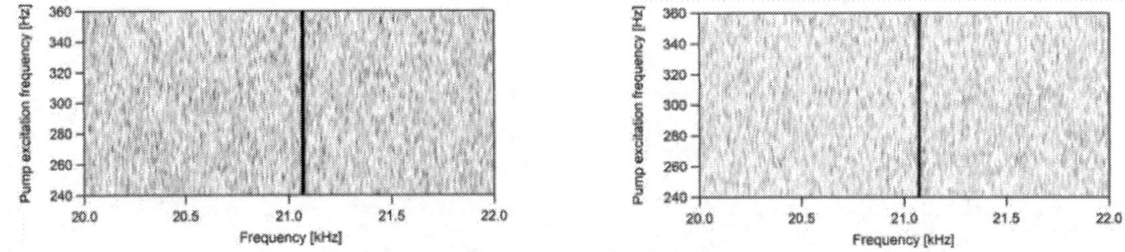

(c) Excitation amplitude 1Vpp

Figure 10. Spectrograms of the beam response to a swept pump excitation ranging from 240 Hz to 360 Hz at sensor location S1 in pristine (left) and damaged (right) conditions. The data were acquired at a pump excitation amplitude of 50 mVpp (a), 250 mVpp (b), 1Vpp (c).

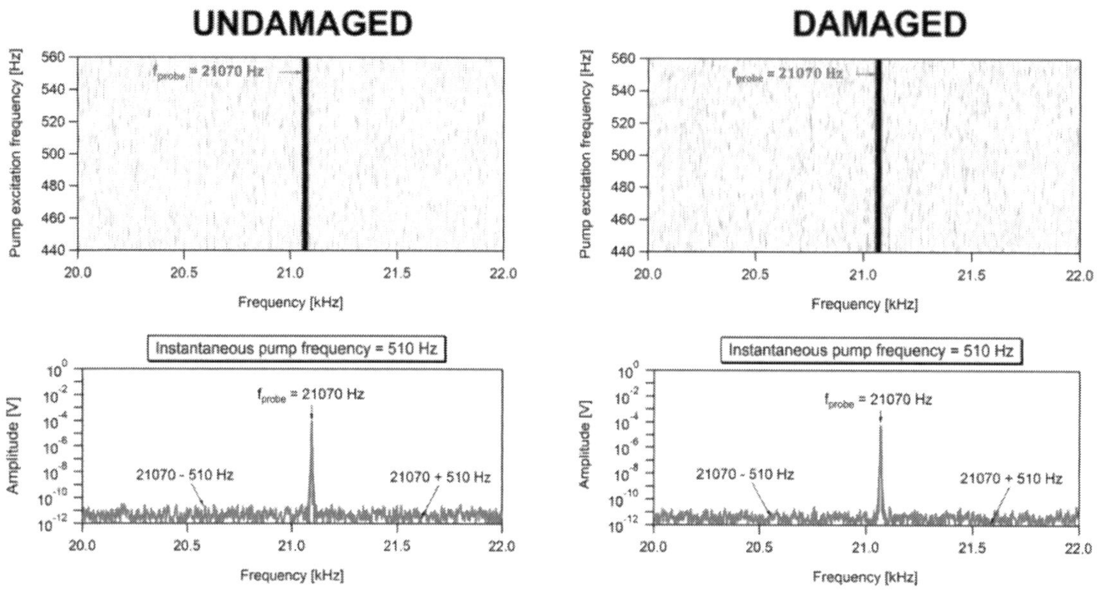

(a) Excitation amplitude 50 mVpp

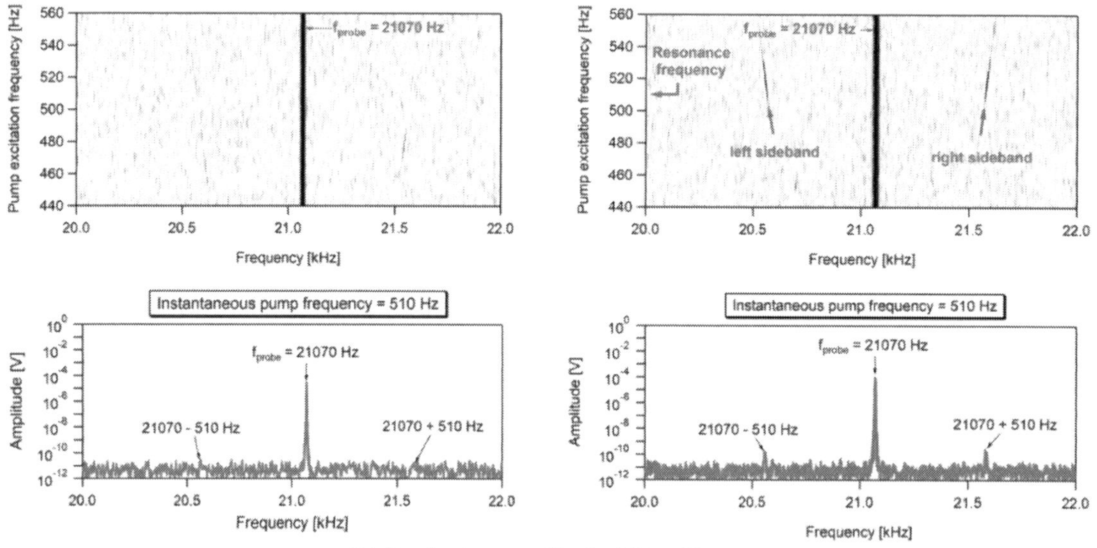

(b) Excitation amplitude 250 mVpp

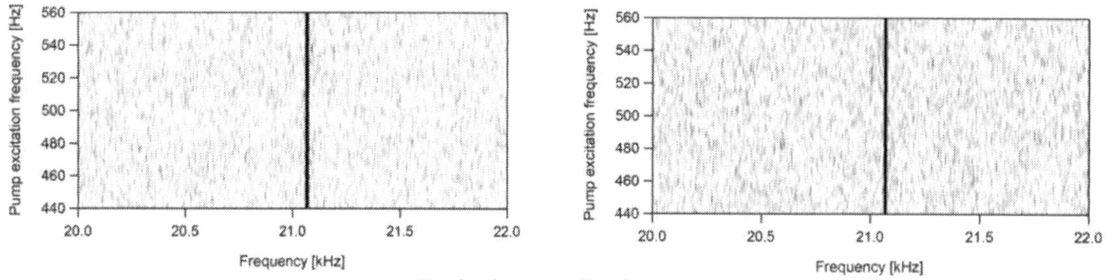

(c) Excitation amplitude 1 Vpp

Figure 11. Spectrograms of the beam response to a swept pump excitation ranging from 440 Hz to 560 Hz at sensor location S1 in pristine (left) and damaged (right) conditions. The data were acquired at a pump excitation amplitude of 50 mVpp (a), 250 mVpp (b), 1Vpp (c).

IOP Conf. Series: Materials Science and Engineering 1275 (2023) 012008 doi:10.1088/1757-899X/1275/1/012008

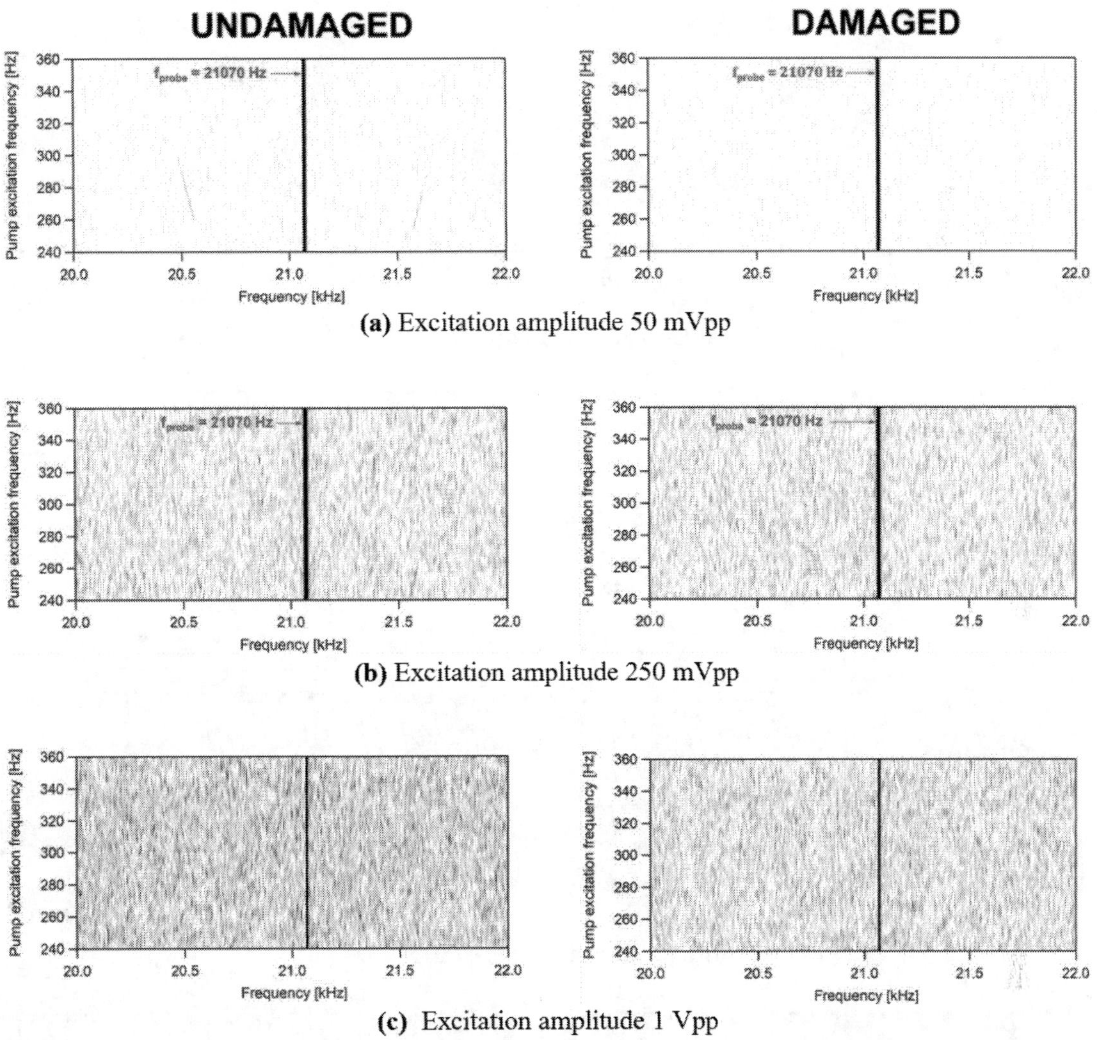

Figure 12. Spectrograms of the beam response to a swept pump excitation ranging from 240 Hz to 360 Hz at sensor location S2 in pristine (left) and damaged (right) conditions. The data were acquired at a pump excitation amplitude of 50 mVpp (a), 250 mVpp (b), 1Vpp (c).

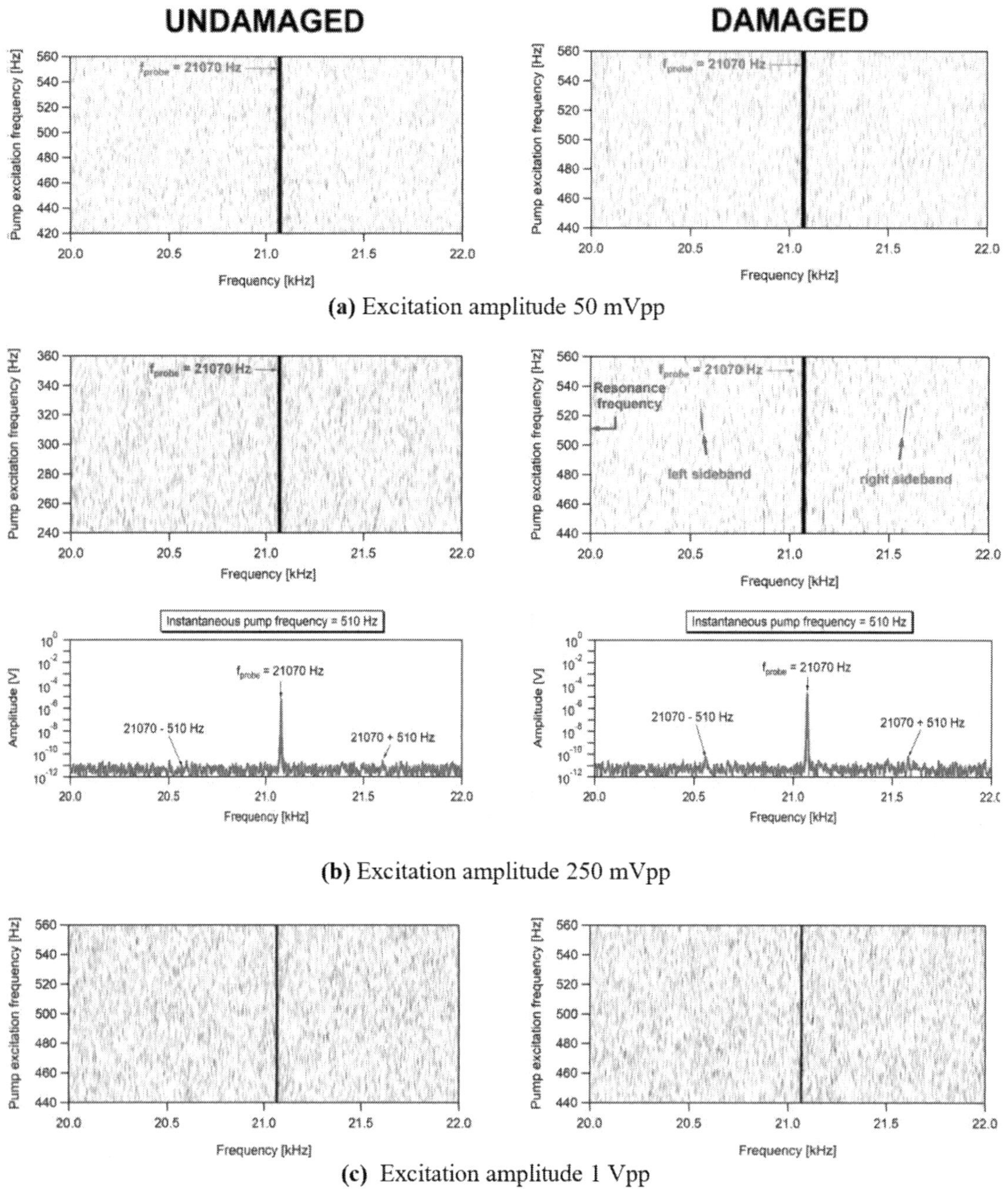

Figure 13. Spectrograms of the beam response to a swept pump excitation ranging from 440 Hz to 560 Hz at sensor location S2 in pristine (left) and damaged (right) conditions. The data were acquired at a pump excitation amplitude of 50 mVpp (a), 250 mVpp (b), 1Vpp (c).

In order to provide a quantitative indication of the effectiveness of the sweep-based VAM approach to the presence of Barely Visible Impact Damage (BVID) in composite samples, the values of the magnitude of the modulation sidebands were calculated from the STFT spectrograms for different levels of the pump signal. For this purpose, the two pump frequency ranges were divided into twelve sub-ranges with a frequency width of about 12 Hz. The average magnitudes of the sidebands within each sub-range were extracted, summed, and then plotted as a function of the pump excitation amplitude.

The graphs in figure 14 show, for each sensor location, the cumulative amplitude of the modulation sidebands over the two considered sweep frequency ranges (240÷360 Hz and 440÷460 Hz), which provides a global indication of the method's ability to detect changes in the nonlinear content of the system response. Because of the significant increase of the background noise retrieved at 1 Vpp, the graphs show only the results associated with 50 mVpp and 250 mVpp amplitude levels.

The results obtained for data acquired at position S1 (figure 14a) show modulation intensities always higher for the damaged beam than for the pristine beam, indicating the general ability of the Vibro-Acoustic Modulation to detect the presence of impact-induced damage. It is also seen that the increase in the sideband amplitude generated by the damage is more marked for the higher pump excitation level, confirming the importance of perturbing with a sufficient pump energy the linearities to be sensed by the probing signal. On the other hand, the data acquired by the sensor in S2 (figure 14b) show quite similar values for the sideband amplitudes recorded for the undamaged and damaged conditions, thus revealing the strong influence of sensor position on the performance of the method.

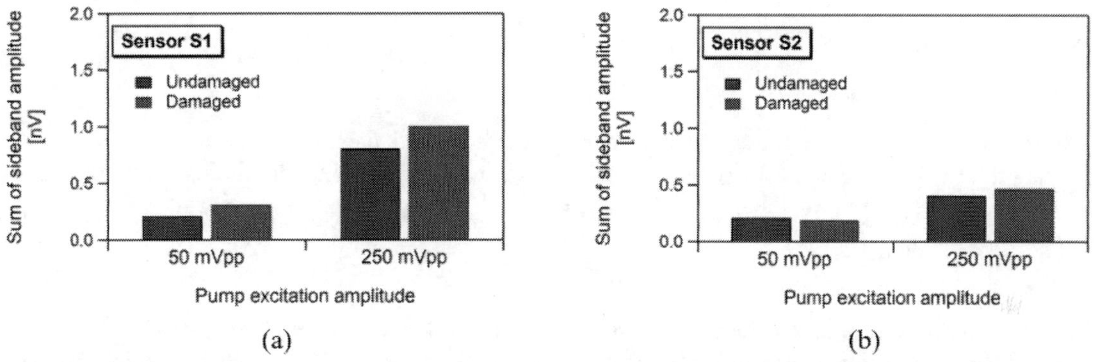

Figure 14. Cumulative sum of the average sideband amplitudes recorded in both the undamaged and damaged conditions at sensor S1 (a) and sensor S2 (b), for two excitation levels. Data are obtained by vibrating the inspected sample through a combination of a swept pump excitation and a pure-tone probe signal.

As a final attempt to explore the quality of the indications provided by the sweep-based VAM approach, a series of conventional vibro-acoustic tests was carried out exciting the inspected sample with a combination of two pure-tone driving signals. The resonance frequencies of 316 Hz and 510 Hz were chosen for the pump excitation, while the 21070 Hz frequency was selected for the probe excitation. In correspondence to each of the considered testing scenarios, twenty data sets were recorded and post-processed to determine their averaged Fast Fourier Transform. The amplitude of the first pair of modulation sideband was extracted and, subsequently, plotted against the pump excitation level (figure 15).

The graphs shown in figure 15 reveal general trends and provide indications like those obtained for the sweep pump excitation. In particular, the data of figure 15 show not only that the performance of the technique strongly depends on the position of the sensor and on the amplitude of the pump excitation, but also that the choice of a suitable pump frequency is required for a reliable detection of the introduced damage. These results clearly suggest the adoption of a sweep multi-mode pump excitation to increase the effectiveness and robustness of VAM-based approaches to detect impact damage at its early stage in composite laminates.

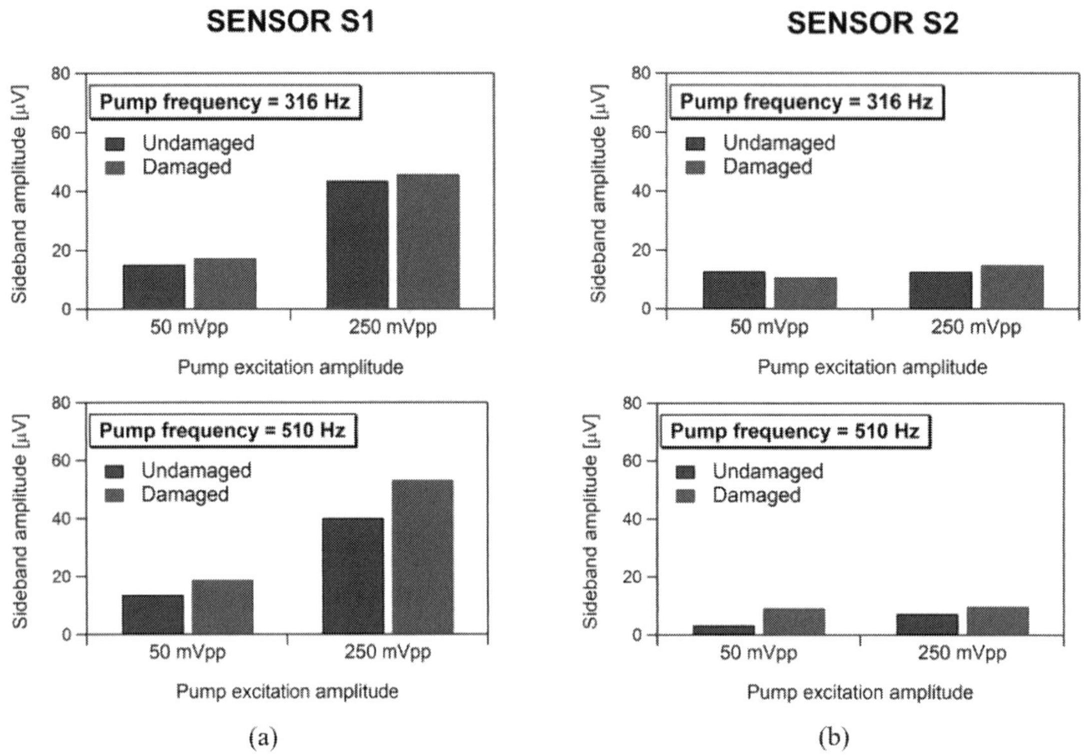

Figure 15. Sideband amplitudes recorded in undamaged and damaged conditions at the sensor location S1 (a) and sensor S2 (b) for two pump excitation levels. The data plotted in these graphs were obtained by vibrating the inspected sample through a combination of two pure-tone signals.

Conclusions

The present study provides a further analysis of the potential of a sweep-based VAM approach to identify the occurrence of barely visible impact damage in composite materials. For this purpose, a laminated composite beam, tested in pristine and damaged conditions, was vibrated through a combination of a frequency-swept pump signal and a mono-harmonic probe wave. The response of the system to the excitations were acquired by low-profile PZT sensors. The obtained results allow us to draw the following insights:

- The application of a broadband swept signal is a promising option to detect the presence of barely visible impact damage in composite laminated beam.
- The adopted approach has strong potential to improve the reliability and robustness of the VAM technique. In particular, the use of a broadband pump excitation: (1) avoids the need of a preliminary modal analysis to identify the system resonances; (2) reduces the risk of selecting pump-probe frequency combinations with low sensitivity to damage; (3) allows to excite effective resonance frequencies even when they shift because of damage or of possible changes of environmental or boundary conditions.
- The selection of the excitation level has a critical influence on both the effectiveness and the sensitivity to damage of the VAM technique. The experimental evidence shows that small internal damage may remain undetected when the energy provided to the sample is not sufficient to activate the damage induced nonlinearities.

Nonetheless, further investigations are required to broaden the scope of application and verify its performance when different types and severity of internal damage, as well as boundary and environmental conditions, are considered.

References

[1] Krautkramer J and Krautkramer H 1990 *Ultrasonic Testing of Materials* (New York: Springer-Verlag) p 677

[2] Masters J 1992 *Damage Detection in CompositeMaterials* ASTM STP 1128 (Philadelphia: ASTM) p 290

[3] Hellier C 2012 *Handbook of Nondestructive Testing* (New York: McGraw Hill Professional) p 720

[4] Wronkowicz A, Dragan K and Lis K 2018 Assessment of uncertainty in damage evaluation by ultrasonic testing of composite structures *Comp. Struct.* **203** 71-84.

[5] Gupta R, Mitchell D, Blanche J, Harper S, Tang W, Pancholi K, Baines L, Bucknall D G, Flynn D 2021 A Review of sensing technologies for non-destructive evaluation of structural composite materials *J. Comp. Sci.* **5**(12) 319

[6] Boller C, Chang F and Fujino Y 2009 *Encyclopedia of Structural Health Monitoring* (Chichester: Wiley) p 2960

[7] Stepinski T, Uhl T and Staszewski W J 2013 *Advanced Structural Damage Detection: from Theory to Engineering Applications* (Chichester: Wiley) p 352

[8] Giurgiutiu V 2008 *Structural Health Monitoring with Piezoelectric Wafer Active Sensors* (New York: Elsevier Accademic Press) p 1024

[9] Porcu M C, Patteri D M, Melis S and Aymerich F 2019 Effectiveness of the FRF curvature technique for structural health monitoring *Constr. Build. Mater.* **226** 173-87

[10] Su Z, Ye L and Lu Y 2006 Guided Lamb waves for identification of damage in composite structures: A review *J. Sound Vib.* **295**(3) 753-80

[11] Chona R, Suh C S, and Rabroker G A 2003 Characterizing defects in multilayer materials using guided ultrasonic waves *Opt. Lasers Eng.* **40**(4) 371-78

[12] Kundu T 2019 *Nonlinear Ultrasonic and Vibro-Acoustical Techniques for Nondestructive Evaluation* (Cham: Springer Nature Switzerland) p 759

[13] Morris W L, Buck O and Inman R V 1979 Acoustic harmonic generation due to fatigue damage in high strength aluminium *J. Appl. Phys.* **59**(11) 6737-41

[14] Richardson M 1979 Harmonic generation at an unbound interface. Part I: Planar interface between semi-infinite media *Int. J. Eng. Sci.* **13** 73-75

[15] Loi G, Porcu M C, Pieczonka L, Staszewski W J and Aymerich F 2019 Scaling Subtraction Method for damage detection in composite beams *Procedia Struct. Integr.* **24** (118-26)

[16] Loi G, Porcu M C and Aymerich F 2021 Impact damage detection in composite beams by analysis of non-linearity under pulse excitation *J. Comp. Sci.* **5**(2) 39

[17] Rauter N and Lammering R 2015 Impact damage detection in composite strucutres considering nonlinear Lamb wave propagation *Mech. Adv. Mater. Struct.* **22**(1-2) 1647-54

[18] Li W, Cho Y and Achenbach J D 2012 Detection of thermal fatigue in composites by second harmonic Lamb waves *Smart Mater. Struct.* **21** 085019

[19] Van Den Abeele K E A, Carmeliet J, Ten Cate J A and Johnson P A 2000 Nonlinear elastic wave spectroscopy (NEWS) techniques to discern material damage, Part II: Single mode nonlinear resonance acoustic spectroscopy *Res. Nondestr. Eval.* **12**(1) 31-42

[20] Van Den Abeele K E A, Johnson P A and Sutin A M 2000 Nonlinear elastic wave spectroscopy (NEWS) techniques to discern material damage, Part I: Nonlinear wave modulation spectroscopy (NWMS) *Res. Nondestr. Eval.* **12**(1) 17-30

[21] Duffour M, Morbidini M and Cawley P 2006 A study of the vibro-acoustic modulation technique for the detection od cracks in metals *J. Acoust. Soc. Am.* **119**(3) 1463-75

[22] Parsons Z and Staszewski W J 2006 Nonlinear acoustic with low-profile piezoceramic excitation for crack detection in metallic structures *Smart Mater. Struct.* **15** 1110 - 1118

[23] Klepka A, Pieczonka L, Staszewski W J and Aymerich F 2014 Impact damage detection in laminated composites by non-linear vibro-acosutic wave modulations *Compos. B Eng.* **65** 99-108

[24] Pieczonka L, Klepka A, Martowicz A and Staszewski W J 2016 Nonlinear vibroacoustic wave modualtion for structural damage detection: an overview *Opt. Eng.* **55**(1) 011005

[25] Loi G, Uras N, Porcu M C and Aymerich F 2022 Damage detection in composite materials by flexural dynamic excitation and accelerometer-based acquisition *IOP Conf. Series: Materials Science and Engineering* **1214** 012007

[26] Zaitsev V Y, Sutin A M, Belyaeva I Y and Nazarov V E 1995 Nonlinear interaction of acoustical waves due to cracks and its possible usage for cracks detection *J. Vib. Control* **1**(3), 335–44

[27] Donskoy D M and Sutin A M 1999 Vibro-Acoustic modulation nondestructive evaluation technique *J. Intell. Mater. Syst. Struct.* **9**(9) 765-71

[28] Sohn H, Lim H J, DeSimio M P, Brown K and Derriso M 2014 Nonlinear ultrasonic wave modulation for online fatigue crack detection, *J. Sound Vib.* **333**(5) 1473–84.

[29] Dziedziech K, Pieczonka L, Adamczyk M, Klepka A and Staszewski W J 2018 Efficient swept sine chirp excitation in the non-linear vibro-acoustic wave modulation technique used for damage detection *Struct. Health Monit.* **7**(3) 565–76.

[30] Zhang Z, Liu M, Su Z and Xiao Y 2016 Quantitative evaluation of residual torque of a loose bolt based on wave energy dissipation and vibro-acoustic modulation: A comparative study *J. Sound Vib.* **383** 156 –70

[31] Chen B Y, Soh S K, Lee H P, Tay T E and Tan V B C 2016 A vibro-acoustic modulation method for the detection of delamination and kissing bond in composites *J. Comp. Mater.* **50** 3089-104

[32] Oppermann P, Dorendorf L, Rutner M and Renner C 2021 Nonlinear modulation with low-power sensor network using undersampling *Struct. Health Monit.* **20**(6) 3252-64

[33] Meo Z and Zumpano G 2005 Nonlinear elastic wave spectroscopy identification of impact damage on a sandwich plate *Comp. Struct.* **71**(3) 469-74

[34] Aymerich F and Staszewski W J 2010 Impact damage detection in composite laminates using nonlinear acoustic *Comp. A Appl. Sci. Manuf.* **41**(9) 1084-92

[35] Liu B, Yang J and Gang T 2020 Analysis of sound and vibration interaction on a crack and its use in high-frequency parameter selection for vibro-acoustic modulation testing *Mech. Syst. Signal Process* **143**(106835)

[36] Liu B, Yang J, Zhang G and Gang T 2021 The choice of the sweep-frequency rate for the probing wave in nonlinear-modulation method of crack detection in resonant samples *NDT Eval. Int.* **123** (102524)

[37] Yoder N C and Adams D E Vibro-acoustic modulation utilizing a swept probing sgnal for robust crack detection *Struct. Health Monit.* **9**(3) 257-67

[38] Donskoy D M, Sutin A M and Ekimov A 2001 Nonlinear acoustic interaction on contact interfaces and its use for nondestructive testing *NDT Eval. Int.* 34(4) 231–38

[39] Loi G, Aymerich F and Porcu M C 2022 Influence of sensor position and low-frequency modal shape on the sensitivity of vibro-acoustic modulation for impact damage detection in composite materials *J. Comp. Sci.* 6(190)

IOP Conf. Series: Materials Science and Engineering 1275 (2023) 012009 doi:10.1088/1757-899X/1275/1/012009

Influence of Core Cellular Structures on Collapse Mechanisms Maps of Sandwich Beams

M Utzeri*, L Morichelli and A Lattanzi

Polytechnic University of Marche, Department of Industrial Engineering and Mathematical Sciences, 12 Breccie Banche, Ancona 60131, IT

E-mail: m.utzeri@pm.univpm.it

Abstract. The main advantage of 3D printing is manufacturing complex and innovative shapes which guarantee high mechanical properties. Therefore, it is necessary easily figure out the most suitable structure for the required design requirements. The well-known strategy to design sandwich panels is evaluating collapse maps as they determine the panel performances based on their geometrical features. The aim of this study is to update the traditional collapse maps by showing how the core shape can improve the sandwich beam performance. The collapse maps proposed are based on advanced analytical models than the traditional Gibson theories. The analytical modelling of the indentation phenomenon is based on Vlasov's model. The analytical modelling of the bending phenomenon is based on the First Shear Order Theory. The overall panel stress and strain maps are computed superposing both effects. A composite sandwich panels with Gyroid core based are analyzed to verify the proposed model consistency. A core failure criterion is chosen by experimental testing evidence on the representative core structure. Once the computed stress state overtakes failure criterion ones, the critical load is defined. In the end, the model is exploited to compare the performances of four sandwich panels with cores based on different lattice structures.

1. Introduction

Nowadays, Additive Manufacturing (AM) enables the manufacturing of intriguing material architectures inspired by the nature. These cellular materials achieve excellent mechanical properties and functional performances[1, 2]. However, it is a challenge to define the best cellular structure inside a complex mechanical system. The need to find out the most suitable structure for a broad range of engineering applications is getting more and more relevant. In aerospace and defence sectors, 3D printed cellular structures are deployed as a core inside a sandwich panel[3, 4].

The well-known strategy to design sandwich panels is evaluating collapse maps as they determine the panel performances based on their geometrical features. This study aims to update the traditional collapse maps by showing how the core shape can improve the sandwich beam performance. The collapse maps proposed are based on advanced analytical models than the traditional Gibson theories[5, 6]. The analytical modelling of the indentation phenomenon is based on Vlasov's model[7, 8]. Its main advantage is the great accuracy of the stress state inside the sandwich due to taking into account the transverse shear. Abrate et al. highlighted its applicability by comparing the Finite Element Method (FEM) results with the well-known Wrinkler's model[9]. Vlasov's model was modified by taking into account the axial

Content from this work may be used under the terms of the Creative Commons Attribution 3.0 licence. Any further distribution of this work must maintain attribution to the author(s) and the title of the work, journal citation and DOI.
Published under licence by IOP Publishing Ltd

core deformation. Furthermore, the core is considered an orthotropic material to evaluate any core lattice structure by its own homogenized stiffness matrix. The bending phenomenon is based on the First Shear Order Theory coupled with the shear correction parameter[10, 11].

The core failure is supposed to occur once the core stress state overtakes critical stress. The critical stress was determined by carrying out tensile, compression, shear, and combined shear-compression tests on the core lattice structure. Failure criteria for cellular materials should include the effect of hydrostatic pressure, anisotropic behaviour and strength dependent on the loading direction. Several criteria have been proposed in the literature and these criteria usually fall into four categories. Each category corresponds to a different shape of the failure surface in terms of stress and deviatoric stress invariants [12, 13]. So, the failure surface was defined by fitting a failure criterion, such as Von Mises, Desphande-Fleck, Gibson, Raghava and GATZT. Then, a composite sandwich beam with a 3D printed core was manufactured to verify the proposed model consistency. The model consistency was verified also through a comparison in terms of strain maps through a DIC analysis carried out on the core lateral side[14].

In the end, the analytical model is exploited to compare the performances of four panels with cores based on different structures, each one has been chosen for its specific feature. The conventional structure is the hexagonal structure as it is the most used cell inside the sandwich beam. The auxetic structure is the Re-Entrant structure as it is widely studied in literature and it is getting more and more used for energy absorption applications[1, 3]. The Gyroid structure is an extremely stiff structure with a shear modulus slightly higher than an isotropic material with the same Young's modulus and Poisson's ratio. In the end, a complex structure composed of two different structures is analyzed to have a macro-isotropic structure.

2. Theorical Analysis

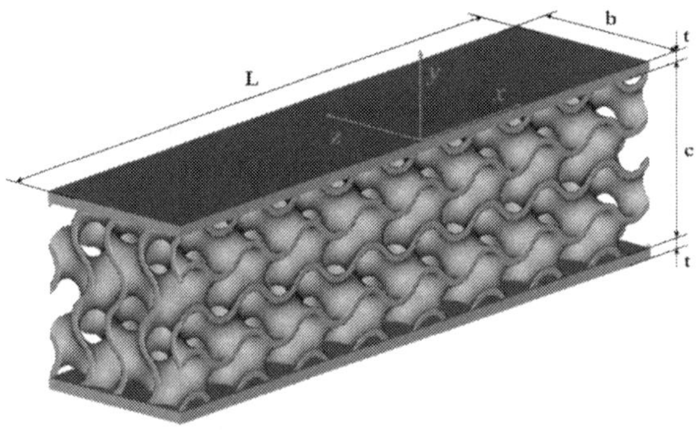

Figure 1: Geometry of sandwich beams.

A symmetric sandwich beam loaded under three-point bending test with the span length L and constant width b is sketched in figure 1. The external skins are assumed to be made of balanced composite lamintes with thickness t. This assumption ensures applicability of the isotropic material theory on the skins. Therefore, the skins material properties are defined by Young's modulus E_f and shear modulus G_f. The skins are perfectly bonded to a core with thickness c. The core mechanical properties are strictly related to the cellular structure properties that makes it up. As the core can have all kind of shape it is assumed to be an orthotropic material.

The core stiffness matrix is defined as follows

$$\begin{bmatrix}\sigma_{xx}\\ \sigma_{yy}\\ \sigma_{zz}\\ \sigma_{yz}\\ \sigma_{yx}\\ \sigma_{xy}\end{bmatrix} = \begin{bmatrix}C_{11} & C_{12} & C_{13} & 0 & 0 & 0\\ C_{21} & C_{22} & C_{23} & 0 & 0 & 0\\ C_{31} & C_{32} & C_{33} & 0 & 0 & 0\\ 0 & 0 & 0 & C_{44} & 0 & 0\\ 0 & 0 & 0 & 0 & C_{55} & 0\\ 0 & 0 & 0 & 0 & 0 & C_{66}\end{bmatrix} \begin{bmatrix}\varepsilon_{xx}\\ \varepsilon_{yy}\\ \varepsilon_{zz}\\ 2\varepsilon_{yz}\\ 2\varepsilon_{yx}\\ 2\varepsilon_{xy}\end{bmatrix} \tag{1}$$

The core stiffness matrix can be obtained by Representive Volume Element (RVE) theory. It can be computed trough a homogenization analysis of the representive unit cell on its elastic properties[15, 16]. Therefore, the cellular structure stiffness matrix will be the core one. In this study, the core elastic properties are defined following the absolute reference sistem showed on figure 1. The $x-axis$ dirtection is aligned to the sandwich beam length and $y-axis$ direction is aligned to the load direction. In order to have a quick comparison with the well-known analytical bending theory the E_c is the core Young's modulus along to $x-axis$ direction and it is computed by the core compliance matrix as $E_c = \frac{1}{S_{11}}$.

2.1. *First-Order Shear Deformation Theory*

Several papers deal with the sandwich beam loaded in three-point bending. It is a common knowledge the sandwich bending stiffness is coupled with the skins bending and the core shear. The equivalent flexural rigidity of sandwcih beam section is

$$(EI)_{eq} = E_f\frac{bt^3}{6} + \frac{E_f bt}{2}(c+t)^2 + E_c\frac{bc^3}{12} \tag{2}$$

The equivalent shear rigidity of sandwich beam section is

$$(AG)_{eq} = bdG_{xy} = bd^2\frac{G_f G_c}{2tG_c + cG_f} \tag{3}$$

where $d = 2t+c$. Thus, it can easily defined the maximum deflection δ reached by the three-point loaded beam on its half:

$$\delta = \frac{PL^3}{48(EI)_{eq}} + \frac{PL}{4(AG)_{eq}} \tag{4}$$

The Reissner-Mindlin theory, also called First-order Shear Deformation Theory (FSDT), is applied to determine the stress and strain state inside the loaded beam. The FSDT theory assumes the core plane section remains plane under bending, following the inextensibility of transverse normal hypotesis. However, the FSDT enables the cross-section to not be necessarly perpendicular to the midsurface after deformation. That means the slope of the middle surface doesn't coincide with the shear deformation as the Eulero-Bernuolly theory. The FSDT takes into account this effects with the rotation term ψ. Thus, the differential equation governing bending is

$$\frac{dw}{dx} = -\psi_x + \frac{\beta}{dG_{xy}}\frac{dM_x}{dx} \tag{5}$$

where $w(x)$ is the core mid-plane displacements and G_{xy} the equivalent shear modulus. Due to the FSDT assumes the shear stress contant along the thickness, the shear correction factor β must be considered. β can be determined by several assumptions, the most common is based on parabolic shear able to satisfy the stress free boundary conditions[17]. In this study, β is computed following the Huang and Kardomateas formulas[10]. The M_x is the moment variation along the beam and it well-known from static equilibrium considerations, namely

$$M_x = -\frac{Px}{2b} \qquad 0 \leq x \leq \frac{L}{2} \tag{6}$$

Actually, the FSDT neglects the core in-plane stresses, i.e. σ_x. In this study, the core stiffness is considered inside the sandwich flexural rigidy so it is taken into account. The shear strain, i.e. τ_B, and normal stress, i.e. $\sigma_{Bx}(x, y)$, are obtained by the compatibility equations. The core stress state tensor, i.e. σ_B, inside a three-pointed bended sandwich beam is described as follows

$$[\sigma_B] = \begin{bmatrix} \sigma_{Bx}(x,y) & \tau_B & 0 \\ \tau_B & 0 & 0 \\ 0 & 0 & 0 \end{bmatrix} \tag{7}$$

where

$$\begin{aligned} \sigma_{Bx}(x,y) &= E_c y \frac{d\psi_x}{dx} = -yx\frac{E_c P}{2(EI)_{eq}} \\ \tau_B &= G_c\left(\psi + \frac{dw}{dx}\right) = \beta\frac{G_c P}{2(AG)_{eq}} \end{aligned} \tag{8}$$

2.2. Indentation Phenomenom Theory

The analitical theories able to predict the indentation phenomenom are usually based on the governing differential equation for beams on two-parameters elastic foundation. Among them, the Vlasov's theory adopts a simplified continuum approach based on the varitional principle[9]. Vlasov assumes the foundation is elastic, isotropic and homogeneous, and he defines the potentianl energy of the skin/core system as follows

$$\Pi = \frac{1}{2}\int_{-\frac{L}{2}}^{\frac{L}{2}} E_f I_f \left(\frac{d^2w}{dx^2}\right)^2 dx + \frac{1}{2}\int_{-\infty}^{+\infty}\int_{-c}^{0} b(\sigma_{xx}\varepsilon_{xx} + \sigma_{yy}\varepsilon_{yy} + \tau_{xy}\gamma_{xy})dydx \tag{9}$$

where the σ_{xx}, σ_{yy} and τ_{xy} are the stresses inside the core instead the ε_{xx}, ε_{yy} and γ_{xy} are the core strains. In this study, the foundation has an orthotropic behaviour so the foundation elastic properties are defined following the core stiffness matrix, equation (1) . The total strain energy definition is unchanged. The Vlasov's model assumes the vertical displacement in the core, i.e. $v(x, y)$, are function of an appropriate mode shape, $\phi(z)$. Vlasov neglected the axial displacement whereas, in this study, they are taken into account:

$$v(x,y) = w(x)\phi(y) \qquad u(x,y,z) \neq 0 \tag{10}$$

The mode shape $\phi(z)$ imposes a restriction on the deflection within the foundation to guarantee the boundary contitions. Indeed, imposing the boundary conditions $\phi(0) = 1$ and $\phi(-c) = 0$ the vertical displacement are compatible with the depth of influence zone where the indentation occurs[18]. The axial displacements are computed imposing the plain strain condition along the sandwich beam $x - axis$. This assumption can be considered a acceptable strain state condition due to the higher core thickness and length compare to the local indentated zone[7]. The axial displacements are computed imposing the plain strain state inside the equilibrium equations and neglecting the volume forces, i.e. $b_y = b_x = 0$. Imposing the boundary conditions $u(0, y, z) = 0$ and $u(x, -c, z) = 0$. The axial displacements can be expressed as

$$u(x,y,z) = B\frac{\partial\phi}{\partial y}\int_{-\frac{L}{2}}^{\frac{L}{2}} w(x)\,dx + K_1(z,y) - K_1(z,-c) \qquad B = -\frac{C_{22}}{C_{66}+C_{12}} \tag{11}$$

That means the axial, transverse and shear strains are defined as follows:

$$\varepsilon_{xx} = \frac{\partial u}{\partial x} = Bw(x)\frac{\partial \phi(y)}{\partial y} \qquad \varepsilon_{yy} = \frac{\partial v}{\partial y} = w(x)\frac{\partial \phi(y)}{\partial y} \qquad \gamma_{xy} = \frac{\partial u}{\partial y} + \frac{\partial v}{\partial x} \approx \frac{\partial w}{\partial x}\phi y \tag{12}$$

The potential energy of the skin/core system is defined taking into account the updated strains and using the core constitutive relations of a plane elasticity.

$$\Pi = \frac{1}{2}\int_{-\frac{L}{2}}^{\frac{L}{2}} E_f I_f \left(\frac{d^2w}{dx^2}\right)^2 dx + \frac{1}{2}\int_{-\infty}^{+\infty}\int_{-c}^{0} b\left[C_{22}^* w(x)^2 \left(\frac{d\phi}{dy}\right)^2 + C_{66}\left(\frac{dw}{dx}\right)^2 \phi(y)^2\right] dydx \tag{13}$$

where $C_{22}^* = C_{22} + 2C_{12}B + C_{11}B^2$ and $C_{66} = G_{xy}$. If the axial dispalcement was neglected and the material was considered isotropic, the constants wuold have coincide with the Vlosv's model ones, namely

$$C_{22}^* = C_{22} = \frac{E_C(1-\nu_c)}{(1+\nu_c)(1-2\nu_c)} \qquad C_{66} = \frac{E_C}{2(1+\nu_c)} \tag{14}$$

The governing differential equation of the system was obtained minimizing the total energy function Π respect to w

$$E_f I_f \frac{d^4w}{dx^4} - k_1\frac{d^2w}{dx^2} + k_2 w = 0 \tag{15}$$

where

$$k_1 = C_{66} b\int_{-c}^{0}\phi^2 dy \qquad k_2 = C_{22}^* b\int_{-c}^{0}\left(\frac{d\phi}{dy}\right)^2 dy \tag{16}$$

The k_1 parameter is the shear foundation coefficient. The k_2 parameter represents the compression foundation behaviour and it is equivalent to all other core subgrade reaction coefficient as the Wrinkler spring constant. The mode shape ϕ was defined by Vallaban minimizing the potential energy respect to ϕ[7]. It is expressed as

$$\phi(y) = \frac{\sinh(\gamma(1+\frac{y}{c}))}{\sinh(\gamma)} \qquad \gamma^2 = c^2\frac{C_{66}\int_{-\infty}^{+\infty}\left(\frac{dw}{dx}\right)^2 dx}{C_{22}^*\int_{-\infty}^{+\infty} w^2 dx} \tag{17}$$

where γ is the core deformation parameter. γ depends to the sandwich structure features and has to be obtained using an iterative procedure since it is unknown a priori[7].

Solving the governing differential equation and imposing the boundary condition the skin deflection is determined. The skin deflection is dependent to the k_1 and k_2 so the general solution of the equation (15) can have 3 solutions. The complete equations are discussed by Abrate et all. as well as the boundary conditions imposed in the differential equations[9]. Once the skin deflection is determined the strains inside the elastic foundation can be determined using the equation (12). Then, the sigma and shear stresses can be computed imposing the constitutive relations of plane strain and elasticity. The core stress state tensor, i.e. σ_I, inside the indented sandwich beam is described as follows

$$[\sigma_I] = \begin{bmatrix} \sigma_{Ix}(x,y) & \tau_I(x,y) & 0 \\ \tau_I(x,y) & \sigma_{Iy}(x,y) & 0 \\ 0 & 0 & \sigma_{Iz}(x,y) \end{bmatrix} \tag{18}$$

where

$$\begin{aligned}
\sigma_{Ix}(x,y) &= (BC_{11}+C_{12})\varepsilon_{yy} = (BC_{11}+C_{12})\frac{\gamma\cosh\left(\gamma^2\left(1+\frac{y}{c}\right)\right)}{c\sinh(\gamma)}w(x)\\
\sigma_{Iy}(x,y) &= (BC_{21}+C_{22})\varepsilon_{yy} = (BC_{21}+C_{22})\frac{\gamma\cosh\left(\gamma^2\left(1+\frac{y}{c}\right)\right)}{c\sinh(\gamma)}w(x)\\
\sigma_{Iz}(x,y) &= (BC_{31}+C_{32})\varepsilon_{yy} = (BC_{31}+C_{32})\frac{\gamma\cosh\left(\gamma^2\left(1+\frac{y}{c}\right)\right)}{c\sinh(\gamma)}w(x)\\
\tau_I(x,y) &= C_{66}\gamma_{xy} = C_{66}\frac{\sinh(\gamma(1+\frac{y}{c}))}{\sinh(\gamma)}\frac{dw}{dx}
\end{aligned} \tag{19}$$

2.3. Core failure under coupled indentation-bending condition

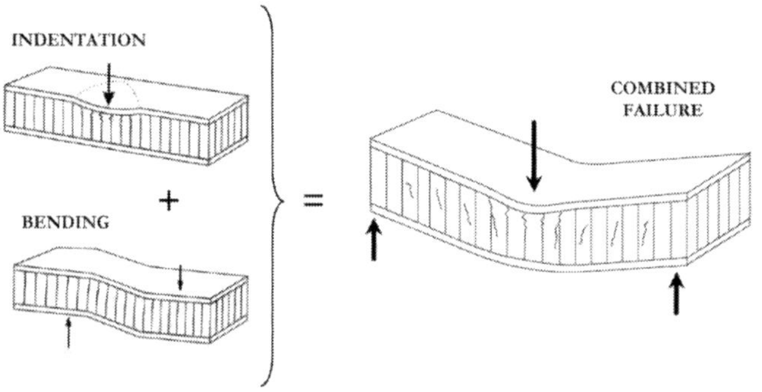

Figure 2: Combined failure mechanism on TPB sandwich beam.

As aforementioned on the previous sections, the three point bending can generate two main failure mechanisms: indentation and shear. A straightforward procedure to define a unique core stress state is combining both effects. The stress state generated by the indentation is added to the stress state comes from the bending, as depicted in figure 2. This assumptions is based on the statement that the applied load causes both effects separatly as two in-series springs system[19]. Consequently, the overall sandwich beam deflection is the sum of deflections generated by indentation and bending. The combined core stress state tensor, i.e. σ_C, inside the three point bended sandwich beam is described as follows

$$[\sigma_C] = \begin{bmatrix} \sigma_{Bx}(x,y)+\sigma_{Ix}(x,y) & \tau_I(x,y)+\tau_B & 0\\ \tau_I(x,y)+\tau_B & \sigma_{Iy}(x,y) & 0\\ 0 & 0 & \sigma_{Iz}(x,y)\end{bmatrix} \tag{20}$$

The core failure occurs when the stress state inside it overtake a certain condition. It is well-known that the failure criteria for cellular materials should include the effect of hydrostatic pressure, anisotropic behavior and strength dependent to the loading direction. Several criteria have been proposed in licterature and these criteria usually fall into four categories. Each category corresponds to a different shape of the failure surface in terms of stress and deviatoric stress invariants, i.e. I_1 and J_2. Since the failure criteria are defined in terms of stress invariants, the J_2 and I_1 are also defined for coupled indentation-bending case as follows

$$\begin{aligned}
J_2(x,y) &= \frac{1}{6}\left(\sigma_{Bx}(x,y)+\sigma_{Ix}(x,y)-\sigma_{Iy}(x,y)\right)^2+\frac{1}{6}\left(\sigma_{Iy}(x,y)-\sigma_{Iz}(x,y)\right)^2\\
&+\frac{1}{6}\left(\sigma_{Iz}(x,y)-\sigma_{Bx}(x,y)-\sigma_{Ix}(x,y)\right)^2+\left(\tau_I(x,y)+\tau_B\right)^2,\\
I_1(x,y) &= \sigma_I(x,y)+\sigma_B(x,y)
\end{aligned} \tag{21}$$

3. Material and Methods

Three point bending test is carreid out to control the accuracy of the developed model. Hybrid composite sandwich beams was chosen because of the core shape complexity. The sandwich core is made of 3D printed cellular structure and the skins are made of Carbon Fiber Reinforced Polymer (CFRP). The sandwich beam is 100 mm long and 16mm width. The core thickness is 16mm in order to be filled by 2x2 Gyroid unit cell along the thickness and width.

3D printed core was manufactured through a commercial Digital Light Printing (DLP) 3D printer, Anycubic Photon S. The raw material is a commercial high-quality resin named Value, Prima Creator product. The resin has a composition of 40-50 epoxy thermoset, 20-40 monomer, and 3-5 photoinitiators in wt.% (Prima creator datasheet). The resin has a crystal-clear transparency. Also, it has already charaterized through compression and tensile test along all 3D printing direction. The bulk material is isotropic with a Young's modulus $E_s = 2250$ MPa. The Gyroid structure with relative density $\overline{\rho} = 40\%$ was repeated filling the entire sandwich core by means of nTopology© software. The generated core was exported as STL format files and then imported to the slicing software Chitubox©. The process parameters were defined inside the slicing software. Based on our extensively testing in the laboratory, they ensure the best choice in strength, printability, and size accurancy[20, 16]. Once the printing is complete, the parts undergone two post-treatement to obtain ultimate and stable mechanical properties.

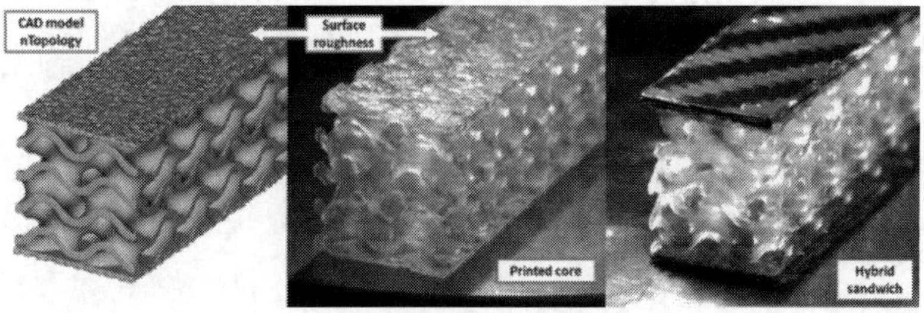

Figure 3: Hybrid sandwich beam made of CFRP skins and 3D printed core.

The sandwich skins are made of pre-preg T700 (Twill 4x4 fabric). The pre-preg sheets were laminated on the external surface of the printed Gyroid structure. The hybrid sandwich was put in oven at 40°C in order to ultimate the curing process of the CFRP skins. In addition, external pressure was applied on the skins through vacuum bag moulding technique. Since the Gyroid structure is a open-cell structure, the core surface close to the skins had to be close in order to avoid the epoxy resin flowing from the pre-preg laminates inside the core during the curing process. Therefore, the sandwich core was printed with two surface 0.2 mm thick. Those thin surfaces were also printed with a specific surface roughness as shown in figure 3. The roughness were textured on the the surfaces by means of nTopology© software. In this way, an high interface strength is guaranteed and the delamination phenomena avoided. Therefore, the core failure will occur under only indentation or shear mechanism.

In order to choice a core failure criterion, the tensile, compression, shear, and combined shear-compression experiments were carried on the core. Representative core portions were 3D printed in addition to the sandwich core. The representive core cells were a Gyroid structure with dimension 16x16x16 mm and 2x2x2 unit cell configuration. The compressive and tensile test were carried out on the Gyroid structures following the ASTM standards for sandwich contruction test methods, ASTM C365 and ASTM C297 respectively. The combined shear-compression tests were achieved rotating the structures of loading angle θ and then compressing it, as shown in figure 4 [21].

The Gyroid structures for combined shear-compression test were printed along with two parallel bevels, as shown in figure 4c. The bevels were 3D printed full density to guarantee the

(a) Tensile test. (b) Compression test. (c) Combined shear-compression test.

Figure 4: Representive core specimens based on the test type.

highest stiffness of the load device. That kind of shape allows to load the specimens vertically and then achieve a combined shear-compressive stress. This 3D printing strategy allows any load angle to be achieved and avoids problems caused by slippage. Six different loading angles were considered, namely 10°,20°,30°,40°,50°, and 60°. Following the standards C273 about the test method for shear properties of sandwich core materials, the specimen should hade been too bigger than the printing plate so the single lap shear test was carried out on the representive core cube, as shown in figure 5. The shear plates were printed full density along with the representive core to guarantee the highest stiffness in this case too.

All mechanical tests were performed on Zwick/Roell Z50 at a loading rate of 0.5 $\frac{mm}{min}$ and the experimental test set-up is showed in figure 5. Focusing on the coupled shear-compression tests, the normal and shear stress can be obtained supposing that the compressive load can be divided into two components, namely

$$F_s = F \sin\theta \qquad F_n = F \cos\theta \tag{22}$$

where F is the compressive load. F_s is the tangential compressive load component that generates the shear effects in the representative core. Instead, the F_n is the normal compressive load component that generates the normal stress effects. As the loading angle θ increases the shear effects became more and more relevants meanwhile the normal effects decrease.

(a) Single lap shear test. (b) Three point bending test.

Figure 5: Experimental tests set-up.

The results of the experimental tests show a brittle beahviour of the 3D printed Gyroid structure. Once the structures overtake the collapse or yielding stress, they explode without buckling, plasticity or crushing phenomena. The failure stress can be assumed to be similar to the collapse stress as the sandwich beam core will break without reaching plastic phase. That means, the aforementioned analytical model can predict the sandwich failure even if it is not take into account plastic deformation. The collapse stress were determined as the intersection between the elastic and hardening straight lines. The collapse stresses are plotted on the princpal stresses plane, as shown in figure 6. The first and fourth quadrant of $\sigma_1 - \sigma_2$ plane were completely filled. Those quadrants are crucial because the pure compressive, tensile, shear, and coupled shear-compressive stresses are the typical stress states inside the sandwich beam core. Subsequently, the best failure criterion was determined. Several failure criteria were evaluated,

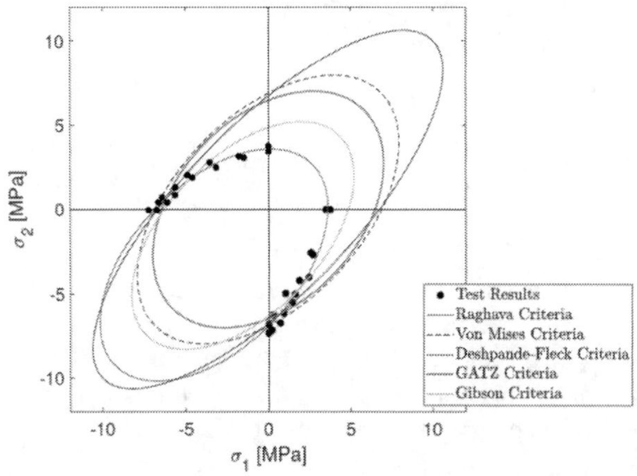

Figure 6: $\sigma_1 - \sigma_2$ plane.

such as Von Mises, Desphande-Fleck, Gibson, Raghava and GATZT[12, 22]. Among them, Raghava criterion revealed to be the best one because it takes into account the different failure stresses in compression and tensile condition, i.e. $3J_2 + I_1(C - T) - CT \leq 0$. The absolute compression stress C and tensile stress T were obtained by means a fitting algorithm and they are 7.0793 MPa and 3.6264 MPa, respectively.

The CFRP skins properties are evaluated under tensile test. The CFRP has a Young's modulus $E_f = 28$ GPa and a strength about 380 MPa. The skins were laminated separately with the same lamination parameters od the sandwich beam, i.e. temperature, pressure and time, due to the manifacturing process affects the CFRP mechanical properties.

4. Results and Discussion

Three-pointe bending test was carried out on the hybrid sandwich beam. The beam deformations and displacement were measured through DIC technique. The analitycal model was applied to obtain the critical loading and core strain maps. The core stiffness matrix was computed by Homogenization technique on the Gyroid structure. Then, the coupled indentation-bending equations are solved imposing a step-by-step load. Once the load imposed overtaked the core failure stress determined by Raghava criterion, the load was considered the critical loading.

Only half-length beam was recorded to take advantage of camera resolution, as shown in figure 7a. The strains were calculated only on the planar face since the Gyroid structure is a 3D structure with large open cells. The DIC analysis based on the global method was applied[23, 24]. The grid nodes were exactly located on the Gyroid "waves"on the lateral face and the speckled baseplates, as shown in figure 7a. Thus, the mesh size was dependent to the core unit cell. The quadrangulate mesh dimension was 4x4 mm as the Gyroid unit cell was 8x8x8 mm. The node highlighted by the red circle was located in the middle of the beam. It was used as a reference node for the beam vertical displacement. The force on the Three-Point bending test is

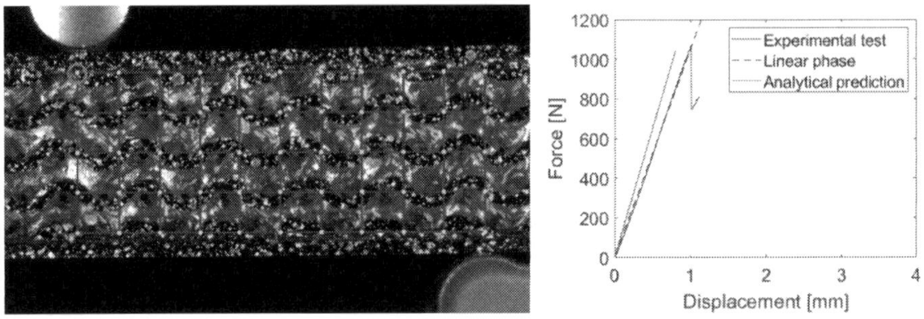

(a) Half-length sandwich beam with grid mesh superposed.

(b) Force-displacement curves.

Figure 7: Three point bending test results.

showed in figure 7b by the blue line and the analytical prediction by the red line. The sandwich beam unveiled an extremely brittle behaviour. It suddently broke without the plastic collapse phase and the cells underneath the indenter didn't crush, as predicted by core charaterization. The skins and core keep being bonded after the load dropping. The figure 7b also shows the analytical predicition shows a consistent result in terms of failure prediction. However, the panel stiffness is quite stiffer than the experimental result. The failure load reached the 1082 MPa and it was 4.6 % higher than the predicted one.

The DIC results are analyzed and compared with the analytical ones. The comparison of ε_{xx} maps is showed in figure 8. DIC result shows the core is stretched underneath the indenter as well as the analytical predictions. That means, the modified Vlasov' model is able to descibe coerently the axial displacement. The bending effect is irrelevant compared to indentation effect due to the large streched zone close to indented zone.

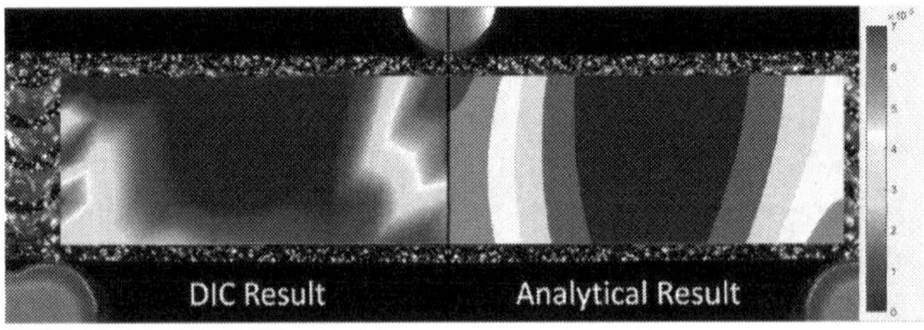

Figure 8: ε_{xx} distribution.

The comparison of ε_{yy} maps is showed in figure 9. Surely, the compression stress state prevailed below both the indenter and supports. In this case, the strain fields are slightly different on the bottom skins due to the boundary conditions on the Vlasov's indentation model. This mismatch doesn't affect the failure region due to the compressive strain under the indenter are quite similar.

The comparison of γ maps is showed in figure 10. The DIC results shows a peak of shear strain close to the indendeters as well as the analitycal one. As predicted by Vlasov's model the shear peak does not appear on the beam midlle but it is shifted. The Analitycal solution predicted the maximum shear stress is 3.2 MPa at 10.2 mm to the sandwich beam center, as depicted by the red X in figure 10. The experimental result confirms the failure occurs close

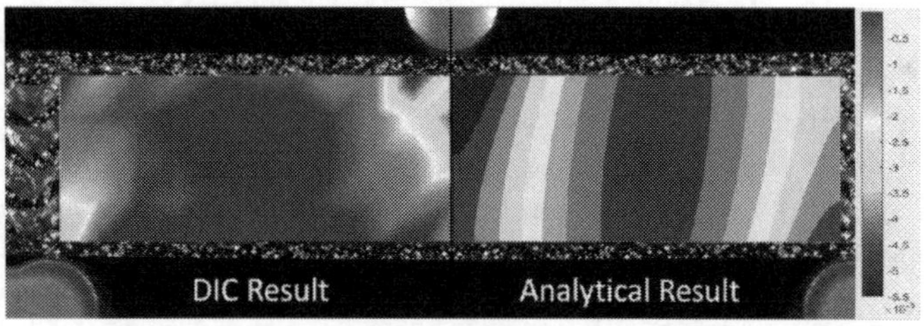

Figure 9: ε_{yy} distribution.

the analitycal prediction, about 11.4 mm. The overall shear strain trend is consistent to the analitycal one even in the sandwich midsurface.

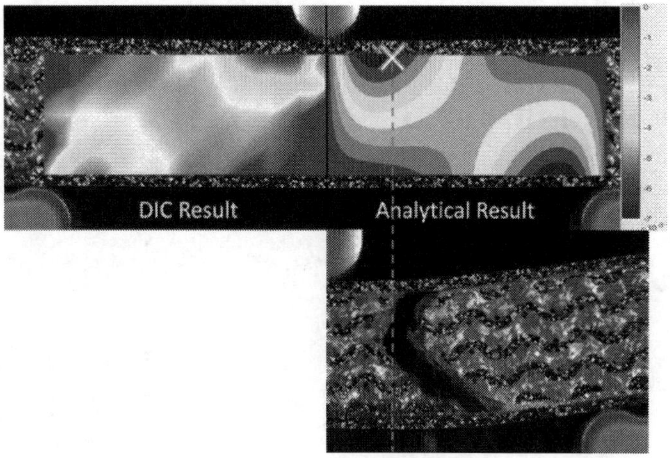

Figure 10: γ distribution.

All three DIC results show a sharp strain transition compared to the analitycal ones. They might be smothed decreasing the mesh size. However, it's not possible because of costrained mesh nodes position.

5. Advanced Collapse Mechanism Maps

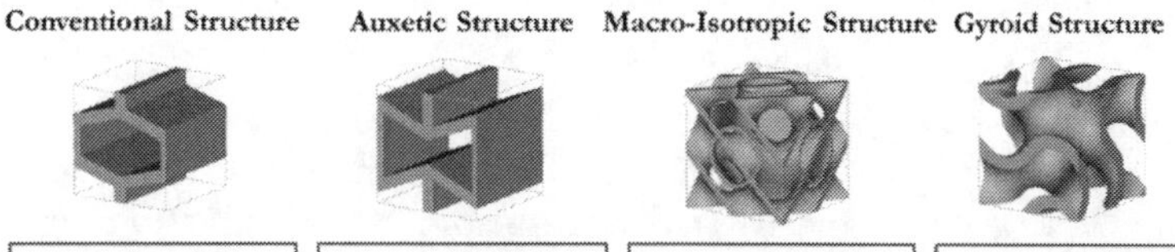

Conventional Structure	Auxetic Structure	Macro-Isotropic Structure	Gyroid Structure
$E_{xx} = 216\ MPa$	$E_{xx} = 100\ MPa$	$E_{xx} = 94\ MPa$	$E_{xx} = 239\ MPa$
$E_{yy} = 30\ MPa$	$E_{yy} = 17\ MPa$	$E_{yy} = 94\ MPa$	$E_{yy} = 239\ MPa$
$E_{zz} = 602\ MPa$	$E_{zz} = 598\ MPa$	$E_{zz} = 94\ MPa$	$E_{zz} = 239\ MPa$
$\nu_{xy} = 0.25$	$\nu_{xy} = -0.32$	$\nu_{xy} = 0.3$	$\nu_{xy} = 0.32$
$\nu_{xz} = 0.35$	$\nu_{xz} = 0.35$	$\nu_{xz} = 0.3$	$\nu_{xz} = 0.32$
$\nu_{yz} = 0.35$	$\nu_{yz} = 0.35$	$\nu_{yz} = 0.3$	$\nu_{yz} = 0.32$
$G_{xy} = 28\ MPa$	$G_{xy} = 5.8\ MPa$	$G_{xy} = 35\ MPa$	$G_{xy} = 104\ MPa$

Figure 11: Lattice structures with associated orthotropic elastic constants.

The analytical model discussed hitherto is able to take into account the core shape by means the general core stiffness matrix introduction. Therefore, four different core shape with the same relative denisities are studied. The unit cell analyzed are showed in figure 11 with the respective stiffness matrix computed by homogenization technique. Each lattice structure has been chosen for its specific feature. The conventional structure is the hexagonal structure as it is the most used cell inside sandwich beam. The auxetic structure is the Re-Entrant structure as it is widely studied in licterature and it is getting more and more used for energy absotption applications. The Gyroid structure is an extremely stiff structure with a shear modulus slightly higher than an isotropic material with the same Young's modulus and Poisson's ratio. At the end, a complex structure composed by two different structure is analyzed in order to have a macro-isotropic structure.

It is supposed a hybrid sandwich beam with the same skins made of CFRP and 3D printed core made of epoxy resin. So, it is supposed the same skins strength and core failure criterion for the sake of semplicity. The analytical model was applied to determine the critical loading and the collapse mechanism[25, 6]. The collapse mechanism maps were obtained investigating a wide range of possible sandwich dimensions[5]. The core thickess runs between 0 to 40% of sandwich beam length and the skin thickness runs between 0 to 20% of core height. Thus, the collapse maps are generated. The critical loading (CL) and the related absorbed energy (EA) up to this load are plotted as isocurves. The isocurves were computed taking into account the highest CL and EA values for each structures. Therefore, the highest performed sandwich beam are highlighted on the same collapse map, as shown in figure 12.

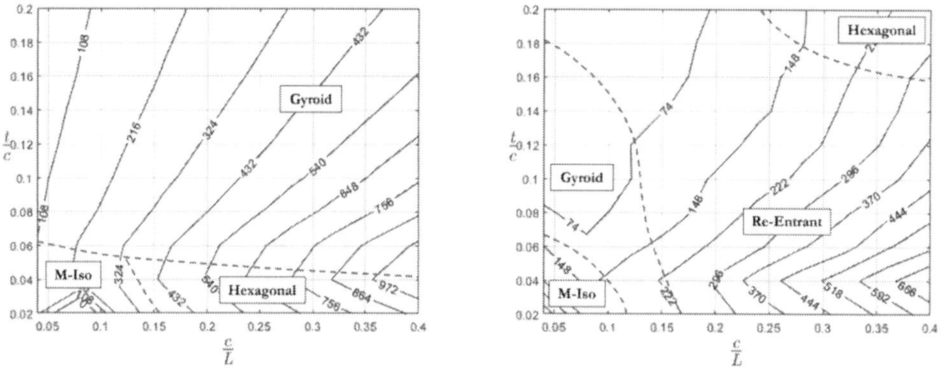

(a) Critical loading map. Isocurves show the load magnitude in N.

(b) Energy absorbtion map. Isocurves show the EA magnitude in mJ.

Figure 12: Collapse mechanism maps for hybrid sandwich with different core lattice structure.

The collapse maps do not show difference between the shear or indentation collapse mechanism even on the left side of the maps. That means the critical collapse mechanism for all hybrid sandwich beam is due to the indentation. However, the core lattice structure plays an important rule on the sandwich performance. The core filled with Gyroid structure guarantees the highest load for almost the entire sandwich sizes. On the contrary, the Re-entrant structure ensures the highest sandwich energy absorption. This results confirm the peculiarity of auxetic structures to absorb energy when they are compressed.

6. Conclusion

An advanced analytical theory on the three-point bending behaviour of sandwich beams has been proposed. The analytical model couples two important phenomena: sandwich indentation and bending. The indentation phenomenon is based on the Vlasov theory as it exploits a simplified approach to continuum mechanics. Vlasov's model was modified by taking into account the axial core deformation. Furthermore, the core is considered an orthotropic material to evaluate any core lattice structure by its own homogenized stiffness matrix. The bending phenomenon is based on the First Shear Order Theory coupled with the shear correction parameter.

The core failure occurs once the core stress state overtakes critical stress. The critical stress was determined by carrying out tensile, compression, shear, and combined shear-compression tests. So the failure surface was defined and fitted through Raghava failure criterion. A composite sandwich beam with a 3D printed core was manufactured to verify the proposed model consistency. The results showed a slight underestimation between the critical load predicted and the real one. The model consistency was verified also through a comparison in terms of strain maps.

In the end, the model is exploited to compare the performances of four panels with cores based on different structures. The Gyroid structure turns out to be the best one in terms of panel strength. On the contrary, the auxetic one ensures the highest sandwich energy absorption confirming the peculiarity of auxetic structures to absorb energy when they are compressed.

References

[1] Ubaid J, Schneider J, Deshpande V S, Wardle B L and Kumar S 2022 *Adv. Eng. Mater.* **24** 2200194
[2] Utzeri M, Farotti E, Coccia M, Mancini E and Sasso M 2021 *J. Mater. Res.* **36** 2083–2093
[3] Kumar S, Ubaid J, Abishera R, Schiffer A and Deshpande V S 2019 *ACS Appl. Mater. Interfaces* **11** 42549–42560
[4] Tancogne-Dejean T, Diamantopoulou M, Gorji M B, Bonatti C and Mohr D 2018 *Advanced Materials* **30** 1803334
[5] Ashby M F, Evans T, Fleck N A, Hutchinson J W, Wadley H N G and Gibson L J 2000 *Metal Foams: A Design Guide* (Oxford, England: Butterworth-Heinemann)
[6] Banghai J, Zhibin L and Fangyun L 2015 *Composite Structures* **133** 739–745 ISSN 0263-8223
[7] Vallabhan C and Das Y 1991 *International Journal of Solids and Structures* **27** 629–637 ISSN 0020-7683
[8] Hetényi M 1950 *Journal of Applied Physics* **21** 55–58
[9] Navarro P, Abrate S, Aubry J, Marguet S and Ferrero J F 2013 *Composite Structures* **100** 79–88 ISSN 0263-8223
[10] Carlsson L A and Kardomateas G A 2006 *Structural and failure mechanics of sandwich composites* Solid mechanics and its applications (New York, NY: Springer)
[11] Birman V and Bert C W 2002 *Journal of Sandwich Structures & Materials* **4** 83–95
[12] Abrate S 2008 *J. sandw. struct. mater.* **10** 5–51
[13] Deshpande V and Fleck N 2001 *Acta Materialia* **49** 1859–1866 ISSN 1359-6454
[14] Sasso M, Antonelli M G, Mancini E, Radoni M and Amodio D 2017 *International Journal of Impact Engineering* **103** 50–63 ISSN 0734-743X
[15] Mancini E, Utzeri M, Farotti E and Sasso M 2021 *ESAFORM 2021*
[16] Mancini E, Utzeri M and Sasso M 2022 *Key Eng. Mater.* **926** 2119–2126
[17] Hernández-Pérez A, Avilés F and Carlsson L A 2012 *J. sandw. struct. mater.* **14** 229–245
[18] Teodoru I B 2009 *Bulletin of the Polytechnic Institute of Jassy, CONSTRUCTIONS. ARCHITECTURE Section, Tomme LV (LIX), Fascicle 4, pages 37-46 (2009)* **LV (LIX)**
[19] Petras A and Sutcliffe M 1999 *Composite Structures* **44** 237–252 ISSN 0263-8223
[20] Utzeri M, Scapin M, Sasso M and Peroni L 2022 *Dynamic behaviour of additively manufactured structures & materials : DYMAT 2022, 26th Technical Meeting Conference Proceedings* ed Ganzenmüller G (Albert-Ludwigs-Universität Freiburg) pp 247–251
[21] Li Z B, Li X Y and Zheng Y X 2020 *Trans. Nonferrous Met. Soc. China* **30** 41–50
[22] Utzeri M, Bhagavatam A, Mancini E, Dinda G, Sasso M and Newaz G 2021 *Journal of Engineering Materials and Technology* **143**
[23] Sasso M, Sarasini F, Chiappini G, Mancini E and Tirillò J 2018 *Dynamic Behavior of Materials, Volume 1* ed Kimberley J, Lamberson L and Mates S (Cham: Springer International Publishing) pp 235–241
[24] Mancini E, Campana F, Pilone D, Amodio D and Sasso M 2022 *Materials Science and Engineering: A* **833** 142346 ISSN 0921-5093
[25] Steeves C A and Fleck N A 2004 *International Journal of Mechanical Sciences* **46** 561–583 ISSN 0020-7403

A contribute to the development of a design procedure for space-born sandwich panels with CFRP skins and an aluminum core

E Mancini[1,2], L Mussolin[3], M Palmieri[1], G Ambrosi[2] and F Cianetti[1]

[1] University of Perugia, Department of Engineering, Via G. Duranti 93, 06124 Perugia, Italy
[2] National Institute of Nuclear Physics (INFN), Section of Perugia, Via Alessandro Pascoli, 06123 Perugia, Italy
[3] University of Perugia, Department of Physics, Via Alessandro Pascoli, 06123 Perugia, Italy

E-mail: edoardo.mancini2@studenti.unipg.it

Abstract. Composite sandwich panels have a vast heritage and are commonly employed in various fields. The space industry is not an exception and often space structures and experiments use this type of panels. This paper represents an effort toward the development of a design procedure for space-born sandwich panels. The here-presented work has been performed on a real test case: the structural sandwich panel of a scientific space mission. In this preliminary phase, classical methodologies were used to ensure the satisfaction of the mechanical requirements and for the minimization of the panel mass and dimensions. Tools such as Finite Element (FE) were employed to help with the design. Increasingly detailed models were developed and used to design the space panels. In addition, this work describes the future roadmap of the project.

1 Introduction

1.1 HERD

The present work details the design status of the supporting sandwich panels of the Silicon Charge Detector (SCD), a particle physics detector part of the High Energy cosmic Radiation Detection facility (HERD).

HERD ([1], [2]) is a space-born particle physics experiment intended to be launched and installed on the Chinese space station in 2027. Being the experiment large and complex, a wide collaboration, with teams from different continents, is working on it.

As other similar projects currently operating in space (e.g. DArk Matter Particle Explorer - DAMPE[3]), HERD should characterize the space particle flux. To be more precise, the main scopes are the dark matter indirect search, the measurement of the cosmic rays spectrum and composition, the monitoring of gamma rays, and the full sky survey.

The actual design foresees a scientific spacecraft made of 4 different layered detectors each of which with a different function. Namely, the SCD, the Plastic Scintillator Detector (PSD), the FIbre Tracker (FIT), and the Calorimeter. By the actual design, the whole experiment should be $3 \times 2.3 \times 1.5$ m^3 large and have a mass of about 3200 kg. As shown in figure 1, the experiment is shaped as a rectangular prism with 5 detective surfaces while the sixth surface is used for the connection to the space station. The prism's lateral faces are smaller than the top face and are

Content from this work may be used under the terms of the Creative Commons Attribution 3.0 licence. Any further distribution of this work must maintain attribution to the author(s) and the title of the work, journal citation and DOI.
Published under licence by IOP Publishing Ltd

identical. By looking at figure 2, it should be clear that each of the five sides is a stack-up of an SCD, a PSD, and a FIT detector while the calorimeter is an octagonal prism placed in the center of the experiment.

The experiment collects the signals generated by the interaction between the incoming particles and the detectors and determines the particles' properties through post-processing and data crossing.

As already said, the HERD collaboration is large and scattered either in Italy, Europe, and the rest of the World. In particular, the INFN office in Perugia is responsible for the development of the outermost detector, the SCD, and for what concerns the detector's structural aspects is collaborating with the machine design group of the University of Perugia.

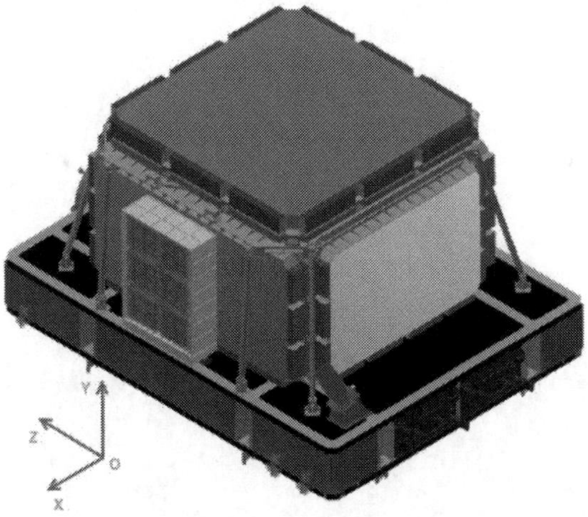

Figure 1: HERD experiment at the current design status.

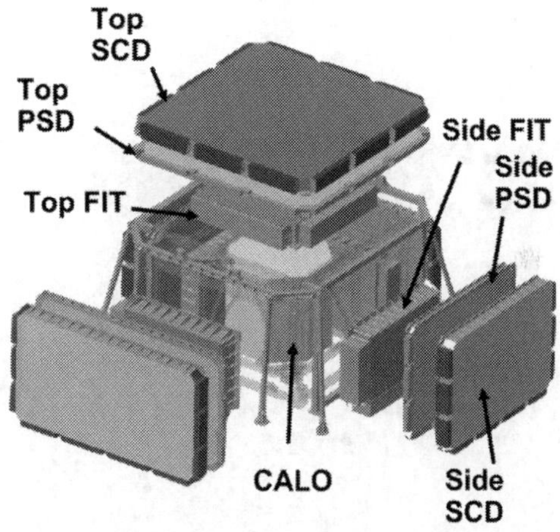

Figure 2: HERD experiment layout and detectors' arrangement inside the structure.

1.2 SCD

The SCD is a physics instrument made of 300 μm thick doped silicon layers with a superficial metallization. The superficial metalization provides the silicon layers with a strip-like pattern, this and the internal structure of the tile allow the particle detection mechanism. In seek of clarity, the silicon surfaces show this strip configuration with linear sensitive areas spaced-out by non-sensitive areas, while the internal structure of the silicon allows the motion of charges. In this way, the strips can collect information when crossed by a particle.

Such detective surfaces (figure 3), which dimensions can range from 50 x 50 mm^2 to 100 x 100 mm^2, are glued on a Kapton or glass fibre composite substrate and then wire-bonded one another to create what is called a "ladder" (figure 4). Then, the ladders are bonded to the read-out electronics (figure 5). Finally, the ladders are placed one next to the other to create a detective plane.

As a whole, the detector will be a stuck-up of planes made of ladders and oriented in different directions. It is understood that such fragile components can not undergo a space flight without experiencing mechanical failure thus, they must be installed on a stiff supporting structure.

As previously stated, the SCD detector is at the utmost part of the experiment, its allocated envelope is 1700 x 910 x 140 mm^3, for the side detectors, and 1700 x 1700 x 140 mm^3 for the top detector. Moreover, the SCD should have at least two mechanical planes capable of supporting 2 to 4 detective planes (with a total of 4 or 8 detective surfaces which is still under discussion). Being the silicon strips unidirectional, to measure the particle position two planes tilted by 90°

IOP Conf. Series: Materials Science and Engineering 1275 (2023) 012010 doi:10.1088/1757-899X/1275/1/012010

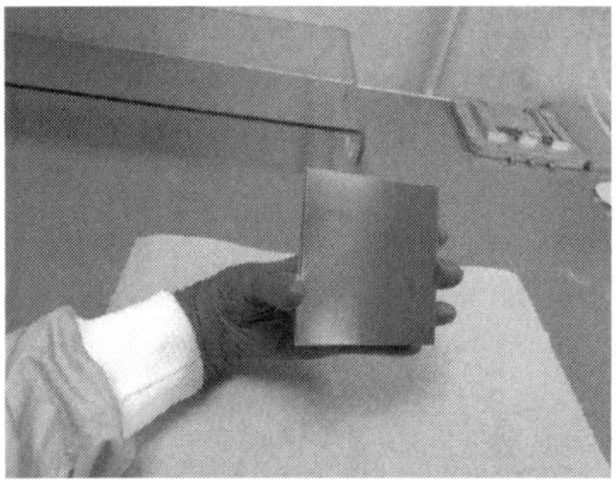

Figure 3: Single-sided Silicon Strip Detector (SSD).

Figure 4: A ladder composed of 10 SSD and the front-end electronic in the foreground.

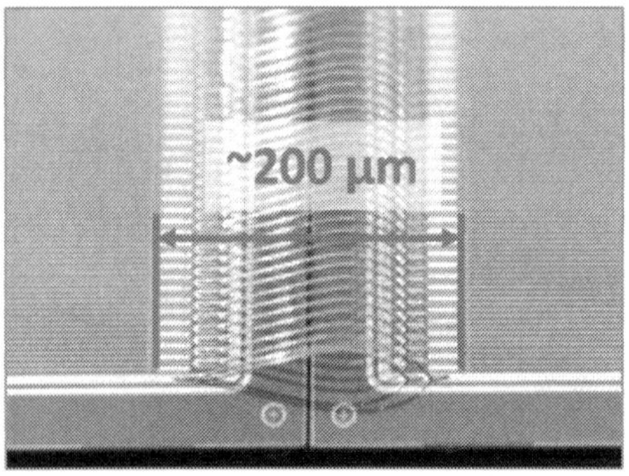

Figure 5: Detail showing the micro soldering (wire bonding) of two silicon strip detectors placed side by side. Moreover, alignment marks are shown in the foreground.

are necessary. Indeed, each plane provides information on which band has been crossed by a particle thus, to determine the exact position of a particle two parallel and tilted planes are necessary.

2 Loads and requirements

Before moving on with the description of the sandwich panel design, let us spend some words on the load profiles and the requirements imposed by the physics. The loads can be related to two different phases: the launch phase, characterized mainly by mechanical loads, especially of dynamic nature, and the space phase, much longer and during which the object undergoes thermal gradients. In addition, also the physics requirements have to be taken into account. Those are related to the achievement of the best performance of the experiment and are mainly associated with the position accuracy and the detector's geometrical acceptance.

2.1 Launch loads

As of today, satellites and spacecraft are placed in orbit with launchers. During the flight (typically lasting a few hours), the rocket transfers relevant loads to the transported payloads[4]. As a matter of fact, from a mechanical point of view, this is usually the most intense phase of the payload life. The mechanical loads experienced during this phase are mainly of four kinds: the quasi-static acceleration, the harmonic load, the random load, and the shocks (figure 6 show the load profiles provided in [4] as an example).

The quasi-static acceleration can be threaded as a static force since it is related to the instantaneous acceleration of the space vector. The harmonic loads are associated with the launcher combustion and internal processes, and they consist of an input of several g's in a frequency range that goes from a few Hz to about 100 Hz. Similarly, the random accelerations are due to the rocket-atmosphere fluid-dynamic interaction and are usually described with a PSD and its RMS. The magnitude of the RMS is usually comparable with the magnitude of the harmonic excitation while the frequency span is wider and could reach 2000 Hz. Finally, the impulsive shock loads come from the stages separations and concern a much higher frequency range (typically spanning up to 10 kHz).

It is relevant to state that the dynamic loads are much more severe than the quasi-static ones. Indeed, the first goal of the mechanical designer is to avoid low-frequency structural resonances, especially in the harmonic range (below 100 Hz). As a rule of thumb, the main requirement is to achieve a first resonance frequency value above $70 - 100$ Hz depending on the S/C (SpaceCraft) size.

2.2 Space Thermal loads

As previously stated, severe thermal gradients characterize the on-orbit phase. Indeed, the external surface temperature of a S/C such as the International Space Station (used here as an example due to its similarity with the foreseen on-orbit conditions of HERD) can range between about -160 °C and $+120$ °C [5]. Those cycles are quite fast since one orbit period lasts about 90 minutes.

Surely, thermal cycling is dependent on the time of the year and the motion of the Earth therefore, the thermal design of these objects which can be either hit by direct sunlight, radiate to deep space (about 3° K) or both is not an easy matter. Figure 7 is included to clarify the former discussion and to show the spacecraft's sun/umbra periods. Nevertheless, the thermally induced stresses are not critical as the thermal distortions since the latter affect the experiment's accuracy, as will be explained in subsection 2.3.

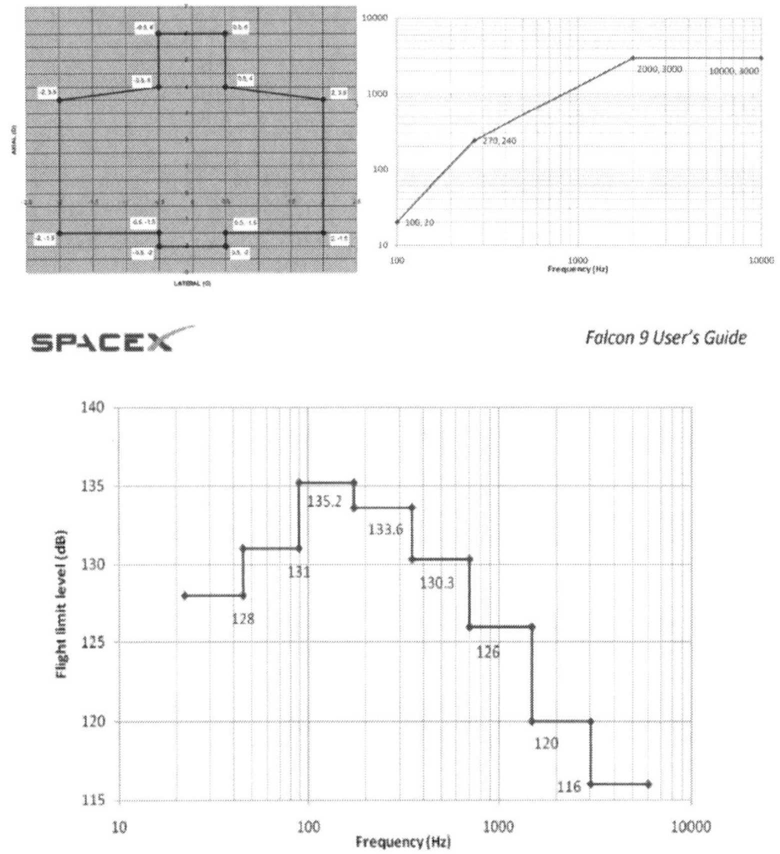

Figure 6: Falcon 9 declared profiles (from the User Manual).

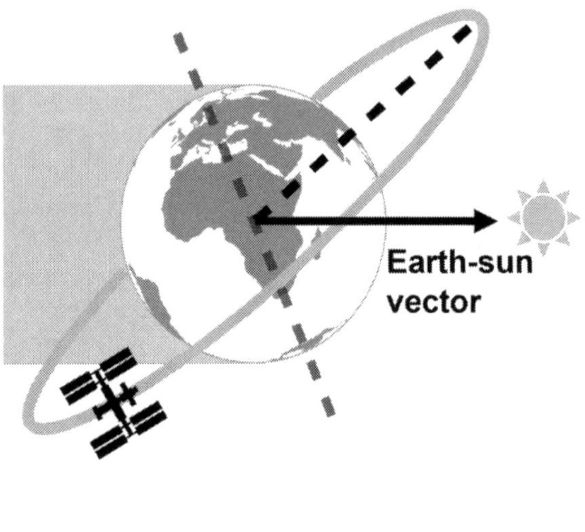

Figure 7: Picture showing the shadow cast by the Earth and the orbit of the ISS. The angle between the orbital plane and the Earth-Sun vector is called the beta angle.

2.3 Physics requirements
In the introduction, the experiment's working principle was briefly described. Namely, particles arrive, hit the various detectors, and produce an electric signal then, the information on the particles is retrieved through data post-processing.

It is understood that, for this matter, it is fundamental to know the exact position of the experiment surfaces to accurately measure the particle properties, especially the trajectory.

Hence from a manufacturing and operative point of view, it is important to have a precise assessment of the position of the detective surfaces. To this end, the on-orbit phase is much more critical than the on-ground phase where it is possible to soften the impact of silicon misplacement through metrology and calibration. Indeed, deformations and distortions can not be measured once in orbit, and therefore it is important to take smart design choices to minimize them.

Another physics requirement is the so-called "material budget". Indeed, not only the mass should be minimized (like in any space project) but also the quantity of material encountered by the incoming particles. To be more specific, the interaction of the particles with other materials perturbs the former thus, the density of the supporting structure affects the accuracy of the measurement. Moreover, if the object on the particle trajectory is too dense the perturbation is so high that the measure must be discarded. Therefore, the presence of thick objects, such as beams, on the particle's trajectory is not desirable and results in blind spots which reduce the detector's coverage.

3 SCD panels layout and approach to the design
3.1 Heritage and sandwich choice
As a whole, this activity aims to design a stiff (relatively high first resonance frequency) and thermally stable (minimized detector movement) sandwich structure with a minimal material budget.

To this end, the first step was the study of other similar projects and the heritage of the research group. During the years, the INFN section of Perugia participated in the design of various space-borne particle detectors such as DAMPE.

DAMPE is a Chinese project which has been operating in space since the end of 2015. As per HERD, in DAMPE there is a PSD, a stack-up of planes equipped with silicon detectors, and a calorimeter. The main difference between the two experiments is that DAMPE is mono-directional, meaning that it can only characterize the particle flux coming from a single direction, while HERD performs tri-dimensional mapping. In terms of dimensions and mass, DAMPE is smaller than HERD but it is still quite large (silicon panels 970 x 970 mm^2 large and an overall experiment mass of 1900 kg). Part of the design of DAMPE was carried out in Perugia and it is detailed in [6]. The former Ph.D. thesis describes the choice of the sandwich as the optimal solution for this kind of application, the design in terms of FEA models of different complexity, and the experimental activities performed on a scaled prototype to validate the models.

Concerning the sandwich choice, this solution has good performance when loaded in bending moreover, it is made of composite material which has good thermal stability (low CTE) and a low material budget. In addition, the sandwich has a large heritage in the space industry as proven by the fact that the European Space Agency (ESA) related standard is more than 10 years old [7]. This last fact is quite relevant because often space agencies and founding institutions are skeptical about using innovative solutions with low TRL, especially in large experiments with other goals.

3.2 The SCD supporting structure
Following the DAMPE approach, the planes are made of a central sandwich area (Carbon Fibre Reinforced Polymer - CFRP - skins and honeycomb core) with a perimetral stiffening CFRP frame. The silicon detectors are placed only in the central part (the material budget is too high

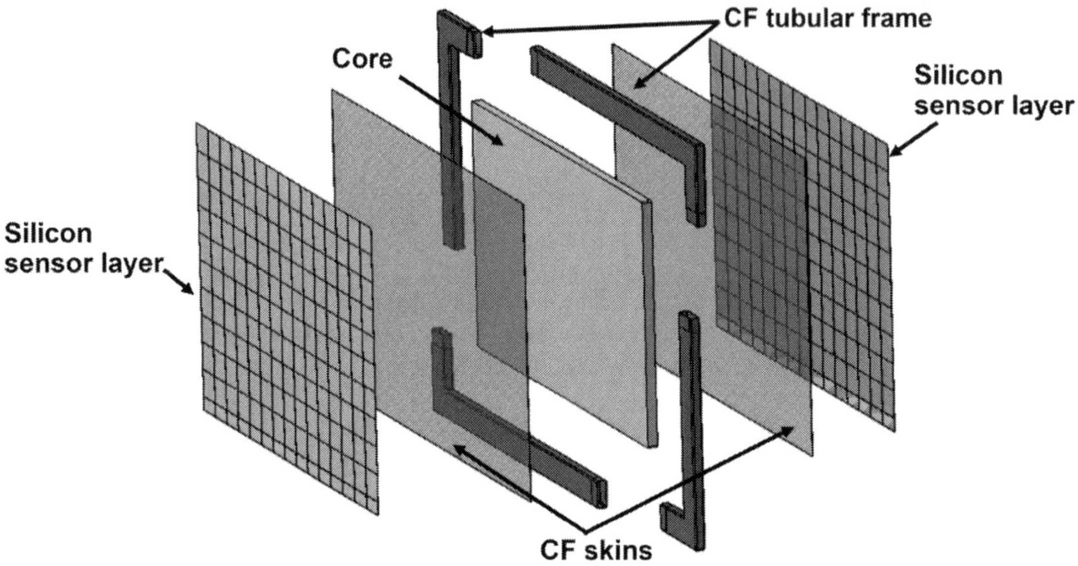

Figure 8: Layout of the SCD mechanical planes.

on the sides where the frame is) therefore, the particles passing through the plane encounter only the two skins and a low-density aluminum region. The sandwich plane is fixed only on its corners to avoid over-constraint and undesired mechanical distortions, the purposes of the CFRP frame are the stiffening of the lateral sides and the achievement of boundary conditions more similar to that of a plane fixed on its sides instead of a plane with free boundaries. Indeed, the first resonance frequency and the perpendicular deformation do not only depend on the plane bending stiffness but also on the boundary conditions. Hence, the presence of the frame results in elastic reactions on the planes' sides which are both torsional (bending about the frame axis) and flexural (vertical displacement of the panel side), reactions with amplitudes proportional to the frame stiffness. As a whole, the layout chosen for the SCD structural planes is shown in figure 8.

As for the material, the present design foresees sandwich skins made of high-modulus UD CFRP, an aluminum honeycomb core, and a woven CFRP frame (better for manufacturing). To this day, alternatives for the core are under evaluation as detailed in section 4.

The SCD mechanical design was divided into three phases. The first one, described in this paper and currently ongoing, consists of the development of preliminary models of the panels (section 5, section 6) and an experimental campaign on a set of prototypes (section 4) which aims to provide the correct inputs to the models. The second phase will consist of the design and realization of a prototype of the full object, the experimental activities to characterize it, and the numerical-experimental validation. Finally, the third phase will concern the development of the flight object and its qualification.

4 Prototypes for the model validation and materials characterization

As already stated in the previous section, to have precise predictive capabilities it is important to validate the models and input them with correct material properties.

Therefore, the preliminary design will be followed by a prototyping phase which has two goals: to determine the mechanical properties of the materials and the whole object and to verify the accuracy of the models.

IOP Conf. Series: Materials Science and Engineering 1275 (2023) 012010 doi:10.1088/1757-899X/1275/1/012010

To be more specific, the commissioned objects were: a CFRP sample of both the skin and of the frame material to be tested following the ASTM regulation, a complete frame not embedded in any sandwich, a miniaturized (500 x 500 mm^2) version of the SCD plane and some sample of a possible alternative for the sandwich core.

For what concerns the alternative core, a CFRP orthogrid core is currently under investigation and it could be a valid alternative to honeycomb for physics-related reasons. Indeed, the mismatch between the core and the skin CTE would be much smaller if the core was to be made of CFRP (even if it is not the skins CFRP) if opposed to the case in which the CFRP skins are coupled with an aluminum core; as a matter of facts, the thermal deformations will be less significant. Moreover, the shape of the orthogrid (rectangular cells of about 50 x 50 mm^2) would result in a simplification in the physics post-processing software which has to take into account the presence of the core and of the skin to correct the estimated particle's properties. All these benefits have to be further investigated and then balanced with the drawbacks of this solution such as the reduced gluing area and other problems which may arise during the experimental activities.

5 SCD preliminary model

The preliminary design of the plane was conducted through a parametric analysis performed with FEA software.

To be more specific, the study started with the development of a model of the top detector assembly (most critical due to its dimensions), where assembly means the whole top detector made of two mechanical planes connected to "ground" on their corners and one to each other on their sides (to increase the side stiffness without linking them to ground). The two planes were modeled in the same way: the CFRP components (skins and frame) were modeled as surface elements, the core was modeled as a solid object with equivalent properties, and the lateral fixation between the planes was assumed to be an aluminum-made plate 3 mm thick. A specific modeling tool was used to define the stack-up on CFRP parts. This model consisted of about 250000 nodes and 121000 elements.

Before starting with the parametric analysis, the authors considered if it was possible to develop a simpler model which would lead to similar results with a reduced computational cost. To attain this result, the sandwich was collapsed in a single surface with an imposed materials stack-up which included both the two skins and the honeycomb core; the lateral frame was modeled with beam elements and the side aluminum plates remained the same. This effort led to a much simpler model (about 7300 nodes and 6500 elements) which errors are quite negligible, especially in terms of the first resonance frequency. The accordance between the two models is shown in table 1.

Table 1: Comparison between the full and simplified models, static and dynamic results.

Model	1st natural freq. [Hz]	Error%	Max out-of-plane def. [mm]	Error%
`Full`	94.49	-	0.846	-
`Simplified`	94.37	0.12	0.866	2.36

6 Results

The parametric analysis was conducted on four geometrical parameters: the overall sandwich panel thickness, the frame in-plane width, the frame skin thickness, and the sandwich skin thickness. At the same time, the panels' in-plane dimensions were fixed and set equal to the envelope allocated to the SCD from the HERD experiment engineering team, the inter-plane distance was adjusted to equally space the planes in the allocated SCD thickness, and the boundary conditions were never changed since the design of the fixation is a concern of the next phase.

The choice of the design point was led by the requirements discussed in the former sections and by the allocated envelope dimensions. In seek of clarity, the skin thickness should be as low as possible to minimize the material budget; additionally, a large frame in-plane width negatively affects the detector's performance by reducing the exploitable area (detectors can not be placed over the frame); the sandwich thickness does not affect significantly the physics performances (at fixed skin thickness) nor the overall mass since the honeycomb density is low but it is anyway constrained by the defined envelope thickness which includes not only the two sandwich planes but also the inter-plane spacing (to be designed to avoid clashing) and the ladders thicknesses; finally, the frame skin thickness is constrained only by the state of the art of the manufacturing. As a whole, table 2 shows the chosen design point.

Table 2: Chosen design point.

Parameter	Value
`Skin thickness`	0.72 mm
`Sandwich panel thickness`	50 mm
`Frame in-plane width`	50 mm
`Frame skin thickness`	2 mm

7 Conclusions and prospects

In the present work, the authors described the status of the design of the SCD detector, part of the HERD experiment. As of today, the preliminary design of the object is concluded, and the group is moving toward a more refined design through an experimental phase that has already begun.

The design work was led by the physics requirements and constraints (especially in terms of accuracy and position stability), by the mechanical loads experienced during the space flight, and by the effects of the space environment.

The detector is made of five parts, four of which are identical (4 sides + 1 top). Apart from the dimensions, the layout is the same for all the parts and consists of two sandwich panels with aluminum core and high modulus CFRP skins which are fixed on their corners to the experiment supporting structure and stiffened by a CFRP frame that goes along the panels' perimeter. Finally, the two panels are laterally connected through aluminum plates.

As for the prospects, the main goal is to validate and assess the accuracy of the model in predicting stresses and deformations either when the object is subjected to dynamic, quasi-static, or thermal-loads loads. In this perspective, the closure of the experimental activities is a mandatory requirement.

The experimental activities will not only help to tune the model but also provide reliable data on the properties of the materials involved. In addition, a study of different core possibilities is foreseen.

In a later stage, all the outputs from the prototyping will allow the finalization of the detector's design and components and will lead to the production of the flight version of the SCD detector.

IOP Conf. Series: Materials Science and Engineering 1275 (2023) 012010 doi:10.1088/1757-899X/1275/1/012010

References

[1] S N Zhang et al. 2014 The high energy cosmic-radiation detection (HERD) facility onboard China's Space Station *Proc. of SPIE - The International Society for Optical Engineering Monreal*

[2] S N Zhang et al. 2017 Introduction to the high energy cosmic-radiation detection (HERD) facility onboard China's future space station *Proceedings of Science 35th International Cosmic Ray Conference, ICRC Busan July 2017*

[3] P Azzarello et al. 2016 The DAMPE silicon–tungsten tracker *Nuclear Instruments and Methods in Physics Research, Section A: Accelerators, Spectrometers, Detectors and Associated Equipment* vol. **831** pages 378-384

[4] SpaceX 2011 *Falcon 9 Launch Vehicle Payload User's Guide https://www.spacex.com/media/falcon-users-guide-2021-09.pdf*

[5] NASA website 2001 *https://science.nasa.gov/science-news/science-at-nasa/2001/ast21mar_1/*

[6] A Bolognini 2015 *Development and testing of the DAMPE Silicon Tracker mechanical structure*, Ph.D. program in physics and physical technologies, University of Perugia

[7] European Cooperation for Space Standardization (ECSS) 2011 *E-HB-32-20_Part3A Structural materials handbook*

A numerical study on the cold sprayability of carbon fibre reinforced composites

A Heydari Astaraee[1], S Bagherifard[1] and C Colombo[1]*

[1] Politecnico di Milano, Department of Mechanical Engineering, Via La Masa 1, 20156 Milano, Italy

* Corresponding author: chiara.colombo@polimi.it

Abstract. One of the open questions in cold spraying on fibre reinforced composites is the optimal thickness of the top layer to provide a suitable base for successful deposition of the metallic particles and at the same time to hinder the probable damage of the fibres. In this study, a detailed finite element model is developed to study the deformation of a single Cu particle deposition on to polyether ether ketone (PEEK) substrate reinforced with carbon fibres. A PEEK layer with 30, 40 or 60 μm thickness was considered on the top surface of the composite. The particle impact velocity was varied in the range of 300-600 m/s to analyse its effects on the induced deformations as well as the structural integrity of the critical carbon fibres. It is believed that the proposed model can provide a helpful tool for predicting the optimal conditions in the metallization of polymers using the cold spray technique.

1. Introduction

Cold gas dynamic spray or "cold spray" is a solid-state material deposition technique, where micron-sized particles of a powder bond to a substrate, as a result of the high-velocity impact and the associated severe plastic deformation. Acceleration of particles to high velocities is obtained via the expansion of a pressurized and "hot" gas through a de-Laval nozzle. Despite this, the feedstock remains in a solid state throughout the entire process; hence the name "cold" spraying.

Cold spray (CS) is primarily used to produce coatings on the surface of materials. However, it also can be adapted for repair applications and even additive manufacturing of parts [1]. The main applications deal with Metal-on-Metal coating [2], but recently attention is moving also toward Metal-on-Composites [3]. The application of CS to polymers and polymer matrix composites to produce metallic coatings is known as polymer metallization. Metallization of polymers improves their thermal and electrical properties, wear and erosion resistance, and can act as electromagnetic shielding, and as lightning strike protection for instance in the aerospace industry. Literature reports different polymer metallization methods such as metal sheet bonding, electroless plating, physical and chemical vapour deposition, and thermal spray techniques [4]. However, the metallization using CS has several advantages compared to some counterparts including: 1) reduction of processing costs, 2) possibility to get higher coating thicknesses, 3) reduction of metal oxidation, which occurs with any thermal process at high temperatures, 4) reduction of substrate degradation due to the temperature, and 4) avoid size limitations of the component, because the gun is mounted on a robot allowing for a full 3D deposition [5].

The main parameters affecting the process of coating deposition are the gas pressure, temperature and type, the powder feed rate and nozzle transverse speed, the spray distance or stand-off distance, the

Content from this work may be used under the terms of the Creative Commons Attribution 3.0 licence. Any further distribution of this work must maintain attribution to the author(s) and the title of the work, journal citation and DOI.
Published under licence by IOP Publishing Ltd

spray angle and the scanning step. These parameters need to be tuned properly to ensure on the one hand the cohesion between the first deposited layer and the substrate, and on the other hand the growth of the coating. In other words, there exists a valid deposition window whose selection is a challenge both from the experimental and the numerical viewpoints. Regarding the application to composites with polymeric matrix, the deposition on a thermoset matrix is extremely difficult due to erosion [6]; for this reason, most of the recent works are focusing on composites with thermoplastic matrices or including a purely thermoplastic layer at the top surface [7,8], also to prevent direct spraying over the brittle fibres. The optimal thickness of the thermoplastic top layer to provide a suitable base for a successful deposition of the metallic particles and at the same time to hinder the probable damage of the fibres is one of the open issues in cold spraying fibre reinforced composites. Indeed, limiting the mass of the overall coated structure is often a target to the designer, especially if this performing composite is applied in the aerospace field.

This strategy affects the manufacturing of the composite substrate, but definitively improves the applicability of the CS coating, especially at the first layer, where the adhesion between the polymeric substrate and the metallic coating is the weakest structural point [9]. Indeed, when dealing with the first layer (Metal-to-Composite), the particle velocity must be higher than the interlocking velocity of particles with the polymeric substrate, but lower than the erosion velocity of the polymer itself. On the other hand, dealing with the rest of the coating (Metal-to-Metal), the velocity must be higher than the critical velocity but lower than the one inducing coating erosion. For a successful deposition, the starting and key point of the coating is the first layer which is dictated by the interlocking velocity. Our previous research [10] focused on the deformation of a single Copper particle, deposited onto polyether ether ketone (PEEK). It was shown numerically that for the mechanical interlocking, a velocity threshold exists below which the particle detaches from the substrate. This numerical model predicted that the minimum particle velocity for the successful embedment and interlocking of Copper particles with 20 μm diameter in a PEEK substrate should be around 550-600 m/s, in agreement with experimental findings from the literature [11,12].

Starting from that work, we investigate here the feasibility of cold spraying Cu particles on Carbon Fibre Reinforced (CFR) PEEK with a numerical approach. A detailed finite element model is developed to study the interlocking particle velocity and to analyse the structural integrity of the fibres and the matrix. Two main manufacturing parameters are analysed with this numerical approach: the particle velocity and the thickness of the pure polymer at the impact surface. The aim is to investigate the optimal (or minimum) thickness of the PEEK layer in order to keep safe the underlying composite.

2. Numerical modelling

The three-dimensional numerical model was introduced in our previous work [10], considering the only PEEK substrate. Here, only the main points are recalled, introducing the different implementations due to the composite.

Within the FE framework of the software Abaqus/Explicit 2019, we selected a 3D Coupled Eulerian-Lagrangian formulation, widely used when dealing with high deformations as in impact problems. Here, the particle was modelled as a deformable Lagrangian section while the substrate was an Eulerian section, because of the high deformations induced during the impact. It is worth noting that generally, for Metal-to-Metal coating, the particle is simulated as Eulerian and the substrate as Lagrangian; hence, here we show a different application of these models, to the case of composites.

Figure 1.a shows the geometry as a function of the particle's radius equal to 10 μm and the boundary conditions. A friction coefficient value of 0.35 was assigned to the whole model, acting at the interface between the copper particle and the PEEK top layer. The carbon fibres were modelled explicitly as cylinders of 8 μm diameter, with a volume fraction equal to 50%. An ordered hexagonal unidirectional arrangement was considered for simplicity. The most critical fibres were those immediately below the particle. Thus, mesh size was refined only here to $1/25^{th}$ of the particle's radius, to decrease the computational cost and minimize the dependency of the results on the mesh size, see Figure 1.b,c. The particle and substrate were meshed using 8-node C3D8R and EC3D8R elements with reduced integration and hourglass control, respectively. Exploiting the symmetry of the problem, only a quarter of the model was built up. Thermal effects were considered using adiabatic heating conditions. A non-reflecting outflow boundary condition was applied to the bottom and side faces of the substrate, acting very similarly to infinite elements. Initial thermal boundary conditions were applied to the particles (T_p=150°C) and to the substrate (T_s=25°C).

Two input parameters were varied during the simulations, to understand their effect on the model's outputs: 1) the particles' velocity V_p, in the range 300-600 m/s, as from [10], and 2) the thickness of the PEEK layer at the top surface of the substrate, with the values 30, 40, and 60 μm taken in accordance to [7].

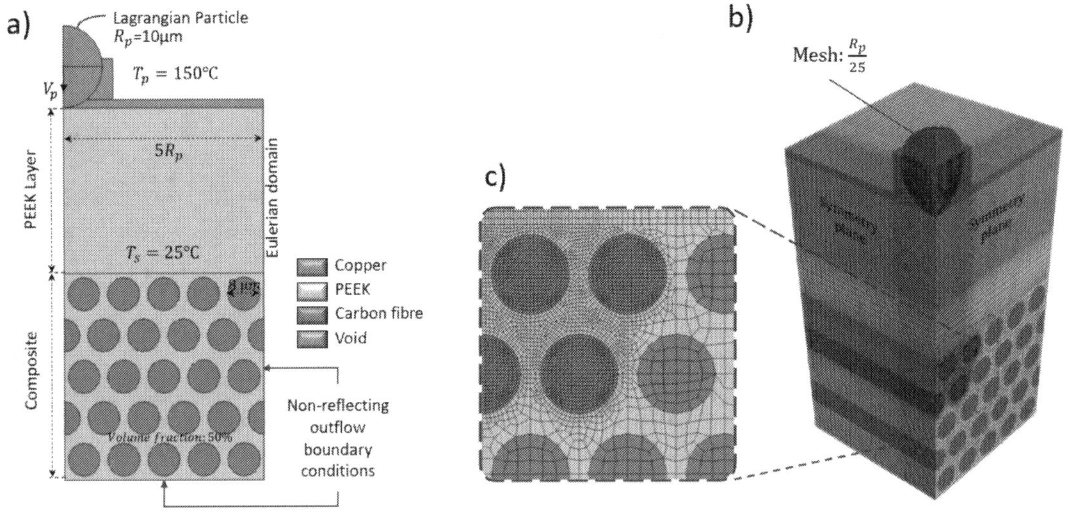

Figure 1. a) Geometry of the problem with boundary conditions; b) 3D view of the mesh and c) detail at the most refined region.

Due to the large strains, ultra-high strain rates, and thermal softening occurring during the CS, the particle and substrate materials were defined with the Johnson-Cook plasticity model [13]:

$$\sigma=\left[A+B\cdot\varepsilon_p^n\right]\left[1+C\cdot\ln\left(\frac{\dot{\varepsilon}}{\dot{\varepsilon}_0}\right)\right]\left[1-\left(\frac{T-T_0}{T_m-T_0}\right)^m\right] \quad (1)$$

where σ and ε_P are the true stress and true plastic strain, $\dot{\varepsilon}$ is the plastic strain rate, and T is the temperature variable, while A, B, n, C, m, $\dot{\varepsilon}_0$, T_0 and T_m are material constants listed in Table 1 and taken from literature works [14,15]. The constants C and m for PEEK were implemented into the software with a bilinear trend as a function of the strain rates, using the available data in the literature [16,17]. A VUHARD subroutine was used to implement this bilinear Johnson-Cook plasticity model in Abaqus. On the other hand, carbon fibres were supposed to have a linear elastic behaviour up to failure.

The use of an Eulerian domain limits high deformation of the finite elements for these cases of impact simulations. However, it is worth to point out that in this formulation, necessary for impact problems, it is not possible to include material anisotropy with the present software version.

Table 1. Material properties used in the numerical model.

Property	Copper	PEEK	Carbon Fibre
Density (kg/m^3)	8960	1300	1760
Young's modulus (GPa)	124	3.5	231
Poisson's ratio	0.34	0.40	0.28
Thermal conductivity (W/m°C)	386	0.25	8.0
Specific heat (J/kg°C)	383	Variable, see [18]	-
Elastic limit, A (MPa)	90	132	-
Hardening constant, B (MPa)	292	10	-
Hardening exponent, n	0.31	1.2	-
Strain rate constant, C	0.025	0.029 ($\dot{\varepsilon}<100\ s^{-1}$) 0.0834 ($\dot{\varepsilon}\geq100\ s^{-1}$)	-
Reference strain rate, $\dot{\varepsilon}_0$ (s^{-1})	1.0	0.001 ($\dot{\varepsilon}<100\ s^{-1}$) 1.0 ($\dot{\varepsilon}\geq100\ s^{-1}$)	-
Thermal exponent, m	1.09	0.634	-
Melting temperature, T_m (°C)	1083	341	-
Reference temperature, T_0 (°C)	25	23	-
Inelastic heat fraction	0.9	0.9	-

3. Results

This section presents the numerical results in terms of displacements and crater depth at the impact region, stress field in the fibres and plastic strain field in the PEEK matrix.

Figure 2.a identifies the steps during the impact of the particle over the substrate; after reaching the maximum penetration depth, when the maximum plastic strain is generated, the particle is pulled out due to the elastic reaction of the PEEK.

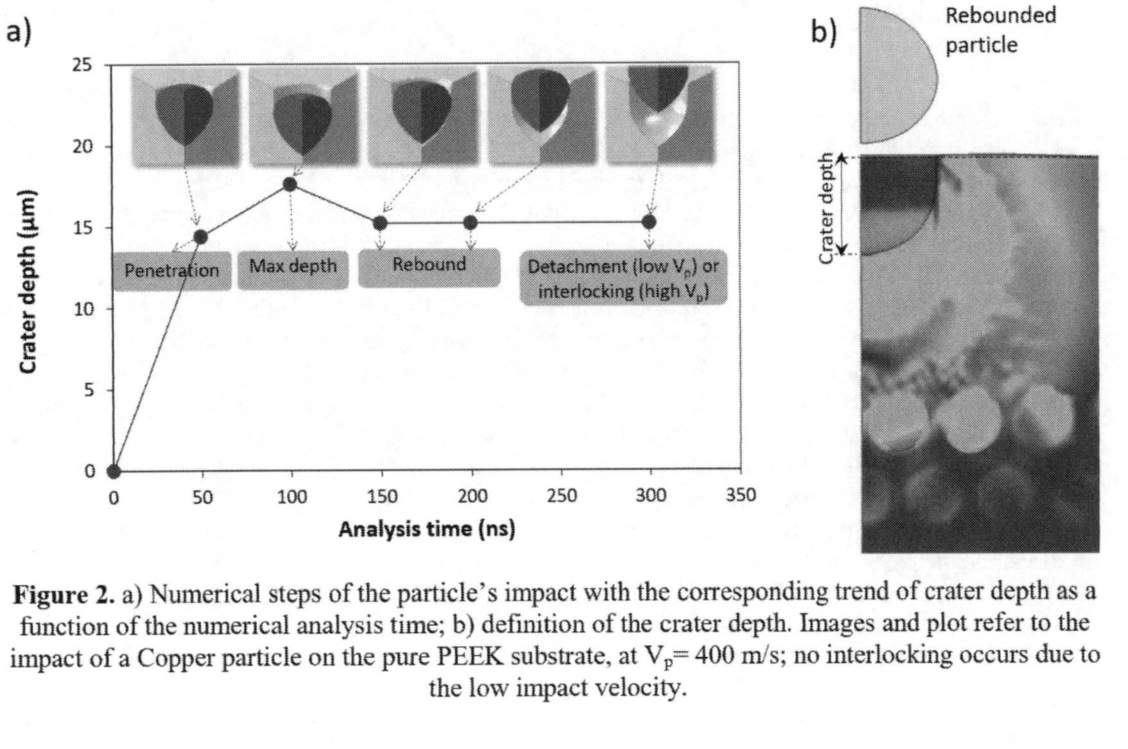

Figure 2. a) Numerical steps of the particle's impact with the corresponding trend of crater depth as a function of the numerical analysis time; b) definition of the crater depth. Images and plot refer to the impact of a Copper particle on the pure PEEK substrate, at V_p= 400 m/s; no interlocking occurs due to the low impact velocity.

Depending on the particle velocity imposed at the beginning of the simulation, the result can be the rebound with full detachment from the substrate, at $V_p < 550$ m/s, or the mechanical interlocking, at $V_p > 550$ m/s. The first result leaves an empty crater in the PEEK, resembling a hole; the second result shows a small empty region below the particle. Figure 2.b defines the crater depth as the distance between the undeformed shape and the maximum depth of the deformed shape after the elastic recovery of the PEEK, e.g., the central bottom point of the crater. The final crater depth is intended when either the particle is fully rebounded from the substrate (if the velocity is not sufficient for the embedment) or stays embedded (if the particle velocity is sufficiently high to reach mechanical interlocking). All the simulations clearly show the indentation, and the penetration depth is always not null.

Figure 3 summarizes the numerical results of the crater depth as a function of V_p and PEEK thickness at the top of the CFR polymer. The figure also shows the interlocking limit identified from the analyses.

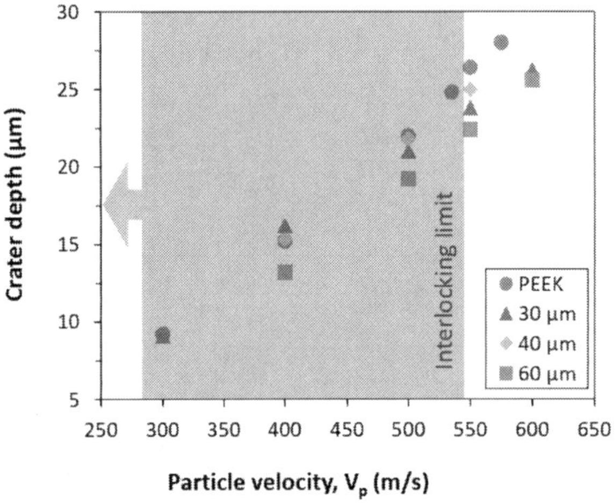

Figure 3. Trend of the crater depth as a function of the particle impact velocity; comparison between the case of pure PEEK substrate and the cases of a PEEK layer with different thicknesses at the top of the CFR polymer.

Before analysing the output of the model in terms of stress and strains, Figure 4 compares the peak stress in the fibres with two formulations, i.e., the coupled Eulerian-Lagrangian (CEL) model, described previously in Section 2, and the purely Lagrangian model with infinite boundary conditions. Indeed, literature reports that the nodal stress/strains of the Eulerian domain are not realistic, because the material flows through the mesh, fixed in the space [19]. Instead, it is suggested to use the volume-averaged variables. This direct comparison between the two models ensures that the averaged stresses are providing realistic values. The comparison is performed at the low velocity of 400 m/s, since the Lagrangian analysis does not abort, even if extremely distorted elements with a non-favourable aspect ratio in the crater can be observed in Figure 4.c. Figure 4.a shows that the averaged von Mises stress in the Eulerian model is very similar to the nodal von Mises stress in the Lagrangian model, suggesting the validity of the volume-averaged stress. The difference after the peak point in Figure 4.a between the two trends (CEL vs. L models) as a function of time could be attributed to the extremely distorted mesh of the Lagrangian model which deteriorates the results.

Once the validity of the outputs is confirmed, Figure 5 and Figure 6 show the stress and the equivalent plastic strain fields, respectively. Both these quantities are averaged volume over the finite element volume (SVAVG is the volume-averaged von Mises stress and PEEQAVG is the equivalent volume-averaged plastic strain). The most stressed region is the fibre placed below the impact (Figure 5.a), while the most strained one is the matrix at the crater, together with some intensification occurring in the region of the most stressed fibre (Figure 6.a,b).

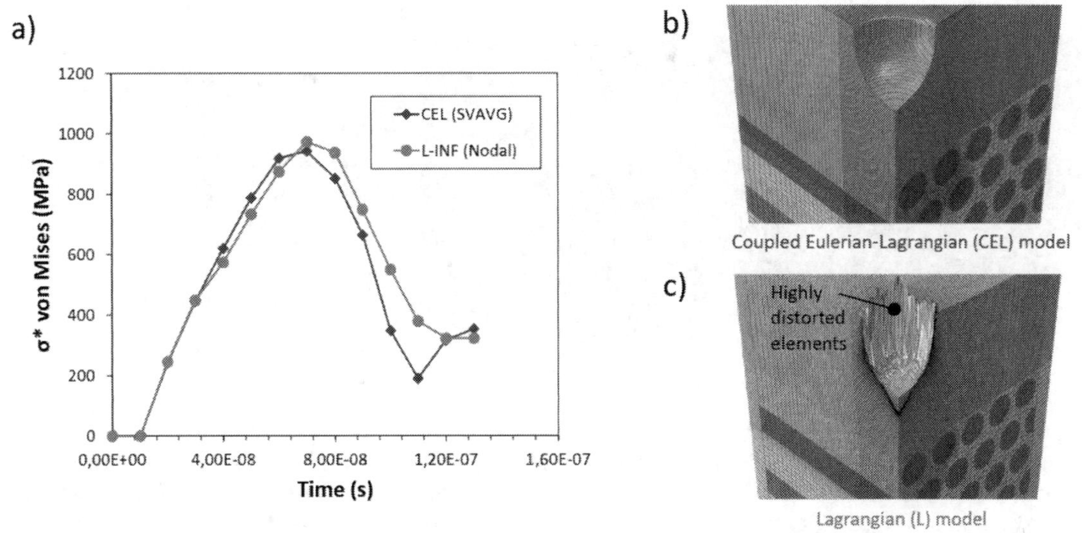

Figure 4. a) Trend of the equivalent von Mises stress at the most stressed carbon fibre as a function of the analysis time: comparison between a Coupled Eulerian-Lagrangian (CEL) model and a Lagrangian (L) model; b) CEL deformed shape and c) L deformed shape. Impact particle velocity: $V_p = 400$ m/s.

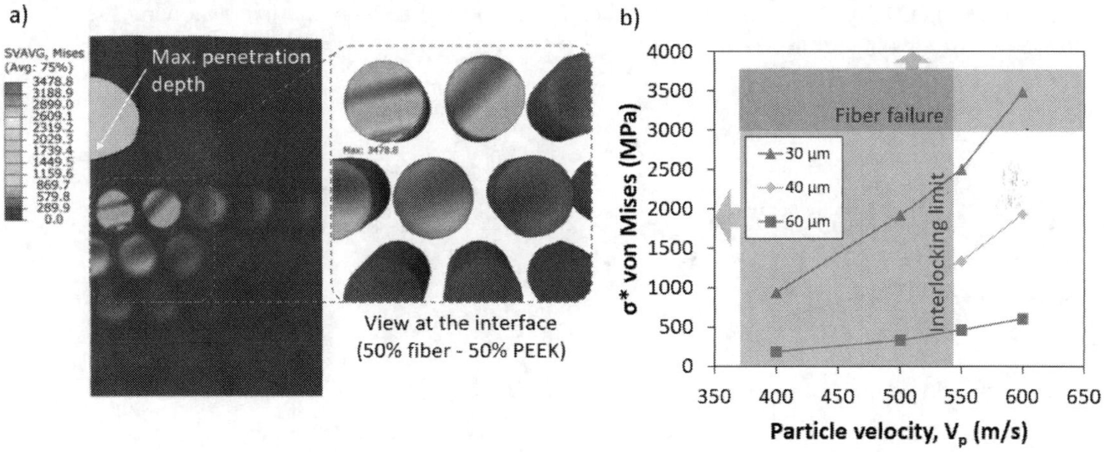

Figure 5. Numerical results into the fibres: a) von Mises equivalent stress field averaged over the element's volume (SVAVG), with a detail at the fibre region, for the case of a surface PEEK layer with 30 µm thickness and $V_p = 600$ m/s; b) trend of the maximum von Mises stress as a function of the particle velocity.

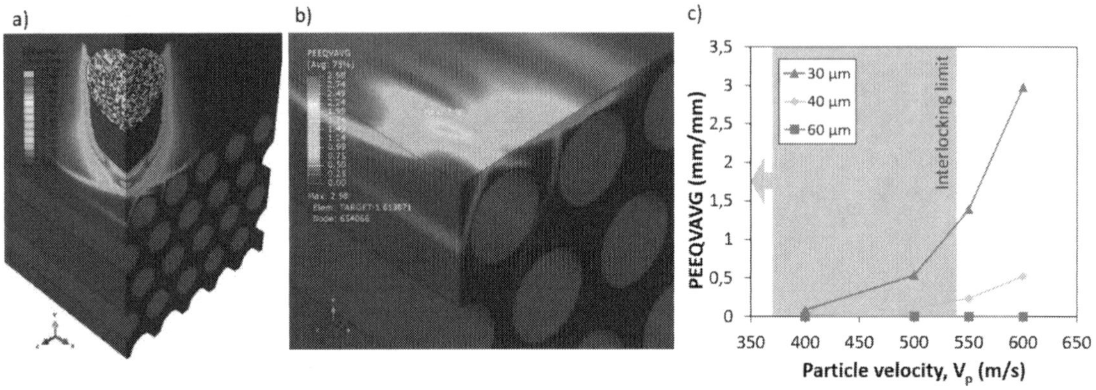

Figure 6. Numerical results in the PEEK: a) equivalent plastic strain field averaged over the element's volume (PEEQAVG), for the case of a surface PEEK layer with 30 µm thickness and V_p = 600 m/s; b) detail at the interface between the pure PEEK and the composite; c) trend of the maximum value of the PEEQAVG in the PEEK as a function of the particle velocity.

After showing these outputs of the simulation, the model was modified to study the dependence of the SVAVG in the fibres and of the PEEQAVG in the matrix as a function of the distance between the fibres, e.g. of the local fibre volume, that can locally increase also due to the randomness of the fibre positioning. Figure 7 shows the plots for the case of a pure polymeric layer with 30 µm thickness. Here, the results of inter-distance equal to 3 µm (solid line) correspond to the previous simulations, while a second case with 0.5 µm inter-distance (dashed line) is added as a comparison, showing the trend of stresses and strains in the substrate.

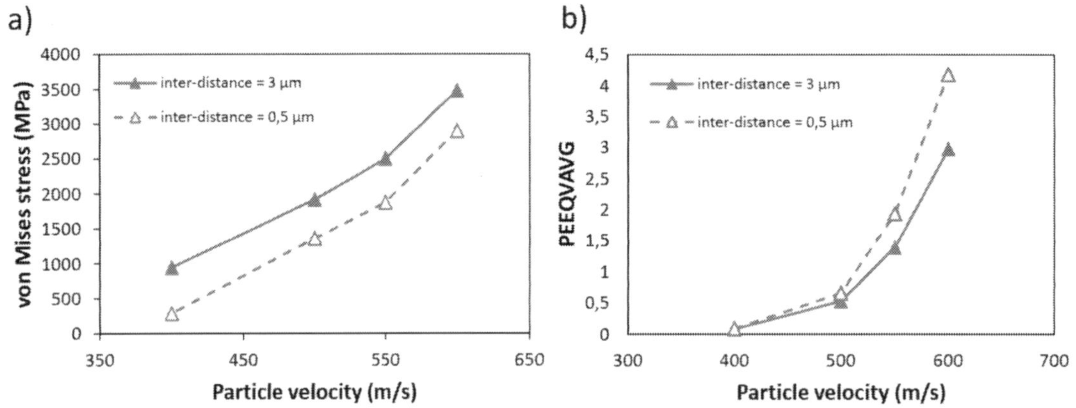

Figure 7. Effect of the distance between the fibres on: a) SVAVG in the fibres; b) PEEQAVG in the matrix. Case of 30 µm thickness, with variable particle velocity.

4. Discussion

Different interesting outputs can be obtained from the implemented numerical models, which are worth discussing with some focus on the impact region as well as at the composite level.

Focusing at first on the crater depth, Figure 3 shows linear trends in the investigated range of vertical particle velocities, both for pure PEEK as well as for PEEK with a composite substrate. The crater depths are lower for the cases with a composite substrate compared to the pure PEEK substrate, meaning that the underlying composite increases the stiffness of the structure and offers resistance to deformation. However, the model is not very sensitive to the underlying composite. Indeed, the difference can be appreciated only at sufficiently high particle velocities. In these cases, the difference is less than 20% with respect to the case of pure PEEK substrate. Figure 3 also reports the interlocking limit of 550 m/s, which is insensitive to the presence of the CFR region below the PEEK. Providing that this layer is placed at the surface, the models always identified the interlocking phenomenon occurring at the same velocity.

It is worth noting that the crater depth, which is an indicator of the coating penetration into the substrate, suggests the required minimum thickness of the PEEK bond layer to serve as protection for the fibres. In other words, the minimum PEEK thickness could not be lower than the final crater depth. This structural condition is necessary but not sufficient because also the failure of the fibres or the PEEK matrix can occur. For this reason, it is important to focus also on stresses and strains below the impact region.

Focusing on the fibres, the maximum equivalent stress is always in the fibre immediately below the impact region, with a gradient due to the induced bending (Figure 5.a), e.g., the most stressed area of the fibre is in its south pole region. Besides, the time interval when maximum stress occurs is close to the maximum penetration of the particle in the PEEK matrix. Although the other fibres are also stressed, the stress level decreases with increasing distance from the point of impact. The plot of Figure 5.b summarizes the results from the simulations of the single particle impact with different PEEK thicknesses. As expected, the maximum stress increases with a decrease in the thickness. However, the stress growth is more considerable at higher particle velocities, experiencing a nonlinear trend. Here, a successful window of deposition is identified, dictated by two conditions: 1) the upper limit of carbon fibre strength that is around 3000 MPa [20], and 2) the effective interlocking of the particle, generating the first layer, as seen in Figure 3. Figure 5.b suggests that a PEEK layer of 60 μm is a good choice because the developed stresses remain well below the fracture strength of the fibres even upon high velocities, while 30 μm thickness is too small to ensure a good safety factor in the fibres, especially at the velocities allowing for the interlocking.

On the other hand, it is necessary to examine also the PEEK matrix in terms of volume-averaged equivalent plastic strain (PEEQAVG). This field underlines some cruciality at the most critical fibres, acting as a strain raiser (Figure 6.a,b). It was found that the only region experiencing plasticity is above the most critical fibre. Besides, it should be noted that the highest PEEQAVG value (3 mm/mm) is due to the severe plasticity around the crater, which is not structurally interesting for the composite. Figure 6.c summarizes the peak PEEQAVG in the matrix at the fibre region. It is difficult to add a limitation in this plot because the strain limit depends on the triaxiality, which is not an available output of the simulation. However, it is clear that with 60 μm thickness no failure will occur.

Since the PEEQAVG are null for such a layer thickness, the case with 30 μm of pure PEEK as the top layer was selected to understand the effect of the different inter-fibre distance on the stresses in the fibres and strains in the matrix. Figure 7 gives a comparative trend and underlines that there is an almost constant difference in the peak stress values of about 600 MPa (Figure 7.a), while strains in the matrix progressively increase with the particle velocity reaching +40% at 600 m/s (Figure 7.b). Even if these strain values could be beyond the limit of the PEEK, the model can catch the effect of this parameter, underlying the weakest regions more prone to the failure.

All these numerical results agree with the experimental work by Gillet et al. [7], suggesting that in the case of cold sprayed Copper coating over a carbon reinforced composite, it is necessary to place a pure PEEK top layer with a thickness in the range of 50-100 μm. This direct comparison underlines the importance of numerical models in the selection of the optimal manufacturing strategies for such new coatings. Indeed, cold spray is a recent technology recently spreading; the implementation and validation of a numerical model can be extremely useful to save time and costs for experimental campaigns and allow for a quick selection of targeted experiments.

5. Conclusion

The implemented numerical model of a single Copper particle impact on a substrate made of a top layer of pure PEEK and an underlying carbon fibre reinforced composite was in good agreement with experimental literature data. For all the cold spray parameters set in this study, a 60 µm PEEK bond layer was found to be safe while a 30 µm is not recommended.

From a more general viewpoint, the model can:

- estimate the interlocking velocity of the particle in the PEEK layer, suggesting that the minimum PEEK thickness could not be lower than the final crater depth;
- evaluate the maximum stress in the most critical fibre;
- evaluate the maximum equivalent plastic strain in the matrix.

These three conditions identify structural limits to the mechanical resistance of the coated composite. As an application of this model, the effect of a change in the inter-fibre distance was analysed, showing the trends of stresses in the fibres and strains in the matrix. Future steps of the work will consider different distributions and arrangements of the fibres as well as study the effect of a multiple impact.

From a wider perspective, this numerical approach could be a valid tool to select the best deposition window during cold spray manufacturing, performing targeted experiments.

Acknowledgements

This research is part of the project COSMEC (Cold Spray of Metal to Composite), funded by the Italian Ministry of University and Research (MUR) under the call PRIN2017 (Grant 2017N4422T).

References

[1] Yin S, Cavaliere P, Aldwell B, Jenkins R, Liao H, Li W and Lupoi R 2018 Cold spray additive manufacturing and repair: Fundamentals and applications *Addit. Manuf.* **21** 628–50

[2] Prashar G and Vasudev H 2021 A comprehensive review on sustainable cold spray additive manufacturing: State of the art, challenges and future challenges *J. Clean. Prod.* **310**

[3] Sandhu H S, Phull G S, Anand U, Preet Singh J and Gulati P 2021 Deposition of Metallic Coating on Plastic Substrate Material Using Cold Spray *Lect. Notes Mech. Eng.* 61–70

[4] Melentiev R, Yudhanto A, Tao R, Vuchkov T and Lubineau G 2022 Metallization of polymers and composites: State-of-the-art approaches *Mater. Des.* **221**

[5] Parmar H, Tucci F, Carlone P and Sudarshan T S 2022 Metallisation of polymers and polymer matrix composites by cold spray: state of the art and research perspectives *Int. Mater. Rev.* **67** 385–409

[6] Che H, Chu X, Vo P and Yue S 2017 Cold spray of mixed metal powders on carbon fibre reinforced polymers *Surf. Coatings Technol.* **329** 232–43

[7] Gillet V, Aubignat E, Costil S, Courant B, Langlade C, Casari P, Knapp W and Planche M P 2019 Development of low pressure cold sprayed copper coatings on carbon fiber reinforced polymer (CFRP) *Surf. Coatings Technol.* **364** 306–16

[8] Astaraee A H, Salerno A, Bagherifard S, Carlone P, Parmar H, Astarita A, Viscusi A and Colombo C 2021 Thermographic analysis of composite metallization through cold spray *Metals (Basel).* **11**

[9] Raoelison R N, Lalu Koithara L and Costil S 2021 Cold spray coating of PEEK surface by copper deposition: Interfacial adhesion at high deposition efficiency and bonding strength *CIRP J. Manuf. Sci. Technol.* **35** 63–8

[10] Heydari Astaraee A, Colombo C and Bagherifard S 2021 Numerical Modeling of Bond Formation in Polymer Surface Metallization Using Cold Spray *J. Therm. Spray Technol.*

[11] Bortolussi V 2016 *Experimental and Numerical Study of the Electrical Conductivity of Cold Spray Metal-Polymer Composite Coatings on Carbon Fiber-Reinforced Polymer* (Université Paris sciences et lettres)

[12] Che H, Vo P and Yue S 2019 Investigation of Cold Spray on Polymers by Single Particle Impact Experiments *J. Therm. Spray Technol.* **28** 135–43

[13] Johnson, R. G 1983 A constitutive model and data for materials subjected to large strains, high strain rates, and high temperatures *Proc. 7th Inf. Sympo. Ballist.* 541–7
[14] Wang Q, Luo X, Tsutsumi S, Sasaki T, Li C and Ma N 2020 Measurement and analysis of cold spray residual stress using arbitrary Lagrangian–Eulerian method *Addit. Manuf.* **35**
[15] Garcia-Gonzalez D, Rusinek A, Jankowiak T and Arias A 2015 Mechanical impact behavior of polyether-ether-ketone (PEEK) *Compos. Struct.* **124** 88–99
[16] Rae P J, Brown E N and Orler E B 2007 The mechanical properties of poly(ether-ether-ketone) (PEEK) with emphasis on the large compressive strain response *Polymer (Guildf).* **48** 598–615
[17] El-Qoubaa Z and Othman R 2017 Temperature, Strain Rate and Pressure Sensitivity of the Polyetheretherketone's Yield Stress *Int. J. Appl. Mech.* **9**
[18] Koerdt M, Koerdt M, Grobrüg T, Skowronek M and Herrmann A S 2022 Modelling and analysis of the thermal characteristic of thermoplastic composites from hybrid textiles during compression moulding *J. Thermoplast. Compos. Mater.* **35** 127–46
[19] Oyinbo S T and Jen T C 2020 Feasibility of numerical simulation methods on the Cold Gas Dynamic Spray (CGDS) Deposition process for ductile materials *Manuf. Rev.* **7**
[20] Minus M L and Kumar S 2005 The processing, properties, and structure of carbon fibers *Jom* **57** 52–8

IOP Conf. Series: Materials Science and Engineering 1275 (2023) 012012 doi:10.1088/1757-899X/1275/1/012012

A test rig for the experimental characterization of the mechanical energy dissipation due to the propagation of elastic waves during the impact of cylinders

M Abruzzo

DICI, Department of Civil and Industrial Engineering, University of Pisa, Italy

E-mail: michele.abruzzo@phd.unipi.it

Abstract. The proposed work describes the design and construction of an experimental test rig used to quantify the mechanical energy dissipation due to the propagation of elastic waves during the impact of cylinders and validate the background theoretical models. Firstly, the impact of perfectly elastic cylinders is studied using finite element and lumped parameter models. The main result obtained is the coefficient of dissipation, which allows quantifying the total mechanical energy dissipation of the system. Then, the final layout of the test rig designed to reproduce the test conditions of the theoretical models is shown. Two kinds of tests are performed to obtain the energy dissipation of the system with and without impacts. The position of one cylinder during its motion is acquired with an industrial camera. Finally,the experimental results are compared with an analytical model of a nonlinear pendulum and used to quantify the coefficient of dissipation.

1. Introduction

In geared mechanical transmissions, the quantification of the total energy dissipation can be fundamental for the characterization of the dynamic behavior of the system. In particular, in high-performance applications, it is not possible to avoid the intersection between the working regimes and the natural frequencies of the transmission, and the total damping controls the shape and the intensity of the resonance peaks experienced. Moreover, the dissipative phenomena are kept to a minimum to maximize the transmission's efficiency, and, in near-resonant regimes, highly nonlinear phenomena (i.e., contact loss and backside contact) are experienced. Those two factors lead to the necessity to obtain a reasonable estimation of the damping to characterize the dynamic response of the system.

Usually, unless proper dissipative devices are implemented, the damping quantification is challenging and often uncertain, and full-scale tests have to be performed to characterize the transmission. As a consequence, several models have been proposed to study the energy dissipation due to not-conservative forces, such as friction on the tooth flanks, friction on the supports of the shafts, windage, oil spreading, dissipative properties of the gear materials [1], and the presence of small and limited plastic or viscous strains in the contact region [2]. All these models assess the mechanical energy dissipated and, even if several processes can be involved, the global effect is usually collected into a unique damping parameter [3, 4].

The main issue with those classic approaches is that the dissipated energy strongly depends on the operating conditions and all those dissipative phenomena are difficult to model (or absent) if

Content from this work may be used under the terms of the Creative Commons Attribution 3.0 licence. Any further distribution of this work must maintain attribution to the author(s) and the title of the work, journal citation and DOI.
Published under licence by IOP Publishing Ltd

contact loss is experienced, as it happens at the resonant regimes of the transmission. To obtain a damping estimation independent of the working conditions of the gears and that can be applied even when contact loss is present, a simplified approach based on the energy dissipation only due to the propagation of elastic waves inside the gear material during the impact was previously developed [5]. The dissipation due to the propagation of elastic waves is usually a phenomenon of little entity, but it may become relevant when the other dissipative sources are kept to a minimum, as it is always present. The main result of the study was the quantification of a coefficient of dissipation, which was used to obtain the total energy dissipation during the impact.

In the present work, the design of a test rig to quantify the energy dissipation due to the propagation of elastic waves during the impact of cylinders is proposed. In the first part, the background theoretical models and the main results obtained are briefly recalled. Then, the design specification and the final layout of the rig are presented. The preliminary experimental results are shown and compared with a nonlinear model of pendulum used to quantify the mechanical energy dissipation of the system using an inverse algorithm.

2. Theoretical models

2.1. *Propagation of elastic waves*

In previous work, the mechanical energy dissipation due to the propagation of elastic waves was quantified using two different models [5]. A dynamic finite element model was used to reproduce the perfectly symmetric impact of steel cylinders in plain strain conditions. The analyses were conducted with cylinders of different diameters ($D = 20, 40, 80$ mm) and at different initial impact speeds v_{imp}, ranging from 0 to 5 m/s. This computationally expensive model was used as a benchmark for the development of a more efficient dynamic lumped parameter model, used for parametric analyses.

It was demonstrated that the energy associated with the propagation of elastic waves after the impact E_{dis}, which will be inevitably dissipated due to the second principle of thermodynamics, can be obtained using only the center of mass speed calculated after the separation of the two bodies v_{fin}. In particular, v_{fin} was used to calculate the coefficient of dissipation d, which was defined as the fraction of the center of mass speed lost in the collision:

$$d = 1 - \frac{v_{\text{fin}}}{v_{\text{imp}}} \tag{1}$$

The energy dissipated E_{dis} by each cylinder can be calculated as:

$$E_{\text{dis}} = \frac{1}{2}m(v_{\text{imp}}^2 - v_{\text{fin}}^2) = \frac{1}{2}mv_{\text{imp}}^2(2d - d^2) \tag{2}$$

in which m is the mass of each cylinder.

The coefficient of dissipation obtained by the finite element model is represented in Fig. 1 as a function of the initial impact speed (v_{imp}) normalized with the speed of the longitudinal waves in the material (v_{w}). It can be noticed that for $v_{\text{imp}} > 0.1 \cdot 10^{-3} v_{\text{w}}$, the same curve was obtained for all the couples of impacting identical cylinders with a different mass and diameter. Moreover, d increases with a decreasing slope. For $v_{\text{imp}} < 0.1 \cdot 10^{-3} v_{\text{w}}$, d increases and seems to approach a finite value for a vanishing impact speed. The presence of a minimum at $v_{\text{imp}} \simeq 0.1 \cdot 10^{-3} v_{\text{w}}$ was supposedly due to the particular features of the Hertzian contact. At low impact speeds, the size of the contact region decreases more rapidly than the decrease in the impact energy and produces a more localized contact region which is a more efficient source of elastic waves. No solutions were obtained for $v_{\text{imp}} < 0.02 \cdot 10^{-3} v_{\text{w}}$ because numerical instabilities were observed as a consequence of the unilateral contact conditions.

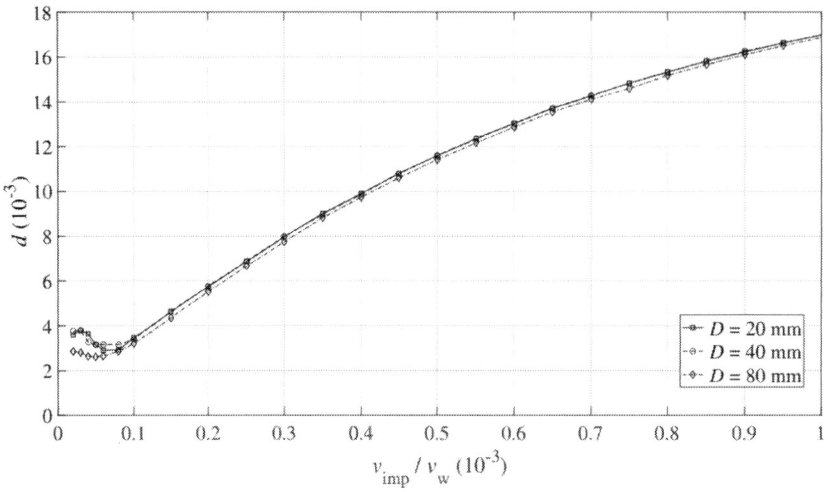

Figure 1: Finite element model coefficient of dissipation.

2.2. Nonlinear pendulum

The lumped parameter model used for the interpretation of the experimental results is shown in Fig. 2.

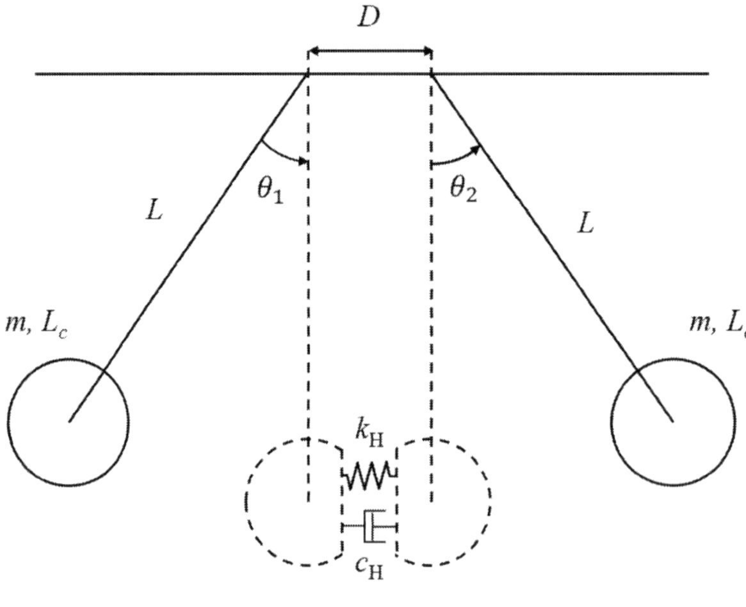

Figure 2: Nonlinear pendulum lumped parameter model.

The model was made of two identical ideally rigid cylinders connected to a fixed frame using two massless and inextensible cables of length L, so that each body has one degree of freedom, θ_1 and θ_2. The contact was modeled using a unilateral spring reproducing the Hertzian contact [6], whose elastic constant is equal to:

$$k_{\mathrm{H}} = \frac{\pi E L_{\mathrm{c}}}{8(1-\nu^2)} \tag{3}$$

in which L_c is the axial extension of the cylinders, E is the elastic modulus and ν is the Poisson's ratio.

The damping was included in the model by means of a quadratic function of the tangential speed of the cylinders, which represent the minimum order formulation able to include the air drag, which is the main source of dissipation. The damping force parameters c_1 and c_2 are to be determined based on the free pendulum tests. With these hypotheses, and considering the equivalent damping of the impact c_H, the equations of motion can be written as:

$$\begin{cases} mL^2\ddot{\theta}_1 + c_2L^3\dot{\theta}_1^{\,2}\mathrm{sgn}(\dot{\theta}_1) + c_1L^2\dot{\theta}_1 + mgL\sin(\theta_1) = -LF_H \\ mL^2\ddot{\theta}_2 + c_2L^3\dot{\theta}_2^{\,2}\mathrm{sgn}(\dot{\theta}_2) + c_1L^2\dot{\theta}_2 + mgL\sin(\theta_2) = LF_H \end{cases} \tag{4}$$

In equation 4, F_H represents the unilateral contact force:

$$F_H = \begin{cases} k_H(L\sin(\theta_1) - L\sin(\theta_2)) + c_H(L\dot{\theta}_1\cos(\theta_1) - L\dot{\theta}_2\cos(\theta_2)) & \text{if } \theta_1 - \theta_2 > 0 \\ 0 & \text{otherwise} \end{cases} \tag{5}$$

For the dynamic simulation of the pendulum, the length of the cables and the initial position of each cylinder were measured directly on the test rig and used as parameters and initial conditions of the nonlinear model. The simulation was used to simulate the first impact of the two cylinders.

3. Design of the test rig

The test rig design was based on Newton's pendulum principle, in which the spheres are substituted by cylinders. This layout allows a simple adjustment of the test parameters and reproduces a symmetric impact, as in the theoretical models.

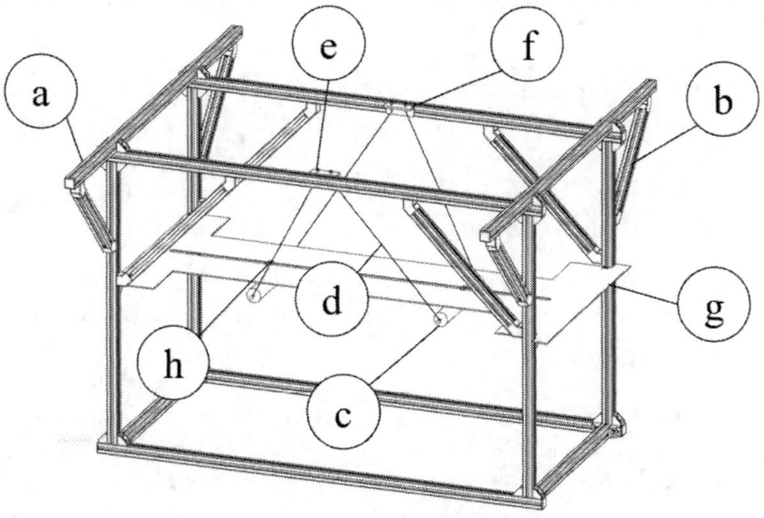

Figure 3: Test rig layout. (a) Main portal profiles. (b) Stiffening elements. (c) Cylinders. (d) Nylon cables. (e) Hollow screws. (f) Regulation plates. (g) Support plate. (h) Grasping system.

The test rig was built using a portal architecture made using lightweight (i.e., aluminum) Bosch profiles, and its final layout is shown in Fig. 3, in which all the elements of the rig are listed from (a) to (h). Besides the main portal profiles (a), stiffening elements (b) were used to avoid instabilities during the test and to increase the natural frequencies of the rig (which must

IOP Conf. Series: Materials Science and Engineering 1275 (2023) 012012 doi:10.1088/1757-899X/1275/1/012012

not intersect the frequencies of the tested cylinders). The cylinders (c) were connected to the upper part of the test rig using a system of four hollow screws and four nylon cables. The nylon cables (d) were used for their small inertial properties. The hollow screws (e) were used for the connection with the portal structure through a pair of regulation plates (f) and allowed to set the height of the cylinder faces independently so that their alignment can be guaranteed. A support plate (g) was used for choosing the initial impact speed and to obtain a symmetric impact. In particular, by moving the plate along the vertical profiles and the grasping systems (h) along the central guide, the initial height of each cylinder is modified to achieve the requested initial impact speed. The plate was machined to allow the passage of cables not exceeding the weight of 15 kg and the axial reference to the portal was guaranteed by a common contact surface at one of its ends. The grasping systems (h) were made using steel supports (which reproduce the external surface of the cylinders) connected to a pair of electromagnets used to hold the cylinders in position. These allow the simultaneous release of the cylinders and to reproduce a symmetrical impact.

The tested cylinders were made of steel with a yield strength of 500 MPa. Considering that the energy dissipation is diameter independent, cylinders with $D = 40$ mm were chosen for their greater simplicity of installation. In addition, the length of the cylinders must be high enough to reproduce the plain strain test condition. In the proposed design, the length of the cylinder was equal to the width of the plate (300 mm) to simplify the verification of the cylinder alignment.

The test rig dimensions (overall dimensions 1300x1300x900 mm) were designed to achieve an impact speed continuously varying from 0 to 2 m/s. This speed range allows the reproduction of the coefficient of dissipation curve without producing any plastic strain in the contact region, which would significantly influence the total energy dissipation during the impact.

4. Experimental results

Two different kind of tests were executed to separate the several sources of dissipation that are active during the motion of the cylinders (i.e., test rig damping, air damping, elastic waves damping):

- Free pendulum tests. Those tests were used to calculate the overall damping of the system during the free motion of a single cylinder, excluding the dissipation due to the propagation of elastic waves.
- Impact tests. Those tests were used to quantify the dissipation due to the propagation of elastic waves for a certain impact speed.

In each case, the experimental response of the system was compared to the results of the theoretical model (Section 2.2), and at least three repetitions of the test were performed.

4.1. Acquisition procedure

The position of one of the cylinders during its motion was acquired by using the "DMK 23UP031" produced by "The Imaging Source", an industrial camera capable to reach a maximum resolution of 2592x1944 pixels (at 15 frames per second) and a maximum frame rate of 120 frames per second at the minimum resolution. For the position tracking, the frontal faces of the cylinder were sprayed in white and two markers were applied to the cylinder face. Considering that the cylinder face must remain in focus from the release until the impact, the maximum number of frames per second was limited to 60 (corresponding to a resolution of 1280x720 pixels). After the acquisition, the frames were processed using a corner finder algorithm (based on Harris corner finder method), which was used to obtain the position of the center of each marker as a function of time.

4.2. Free pendulum test

The free pendulum test was conducted by removing one of the two cylinders from the test rig and acquiring the velocity of the two markers from the release of the cylinder until the motion is exhausted (Fig. 4).

Figure 4: Free pendulum test example.

The period of the system is equal to 1.4 s, so a sampling frequency of 60 frames per second was high enough to derive the velocity of the cylinder using the incremental ratio. The coefficients c_1 and c_2 characterizing the damping function (Equation 4) were obtained by minimizing the difference between the theoretical and the experimental dynamic response of the system (Fig. 5). It can be noticed that the quadratic term is prevalent at high tangential speeds and almost constant, whilst the linear damping term is prevalent at low tangential speeds. An example of comparison between the nonlinear pendulum and the experimental results is shown in Fig. 6, for a maximum theoretical speed of 1 m/s (at the minimum height).

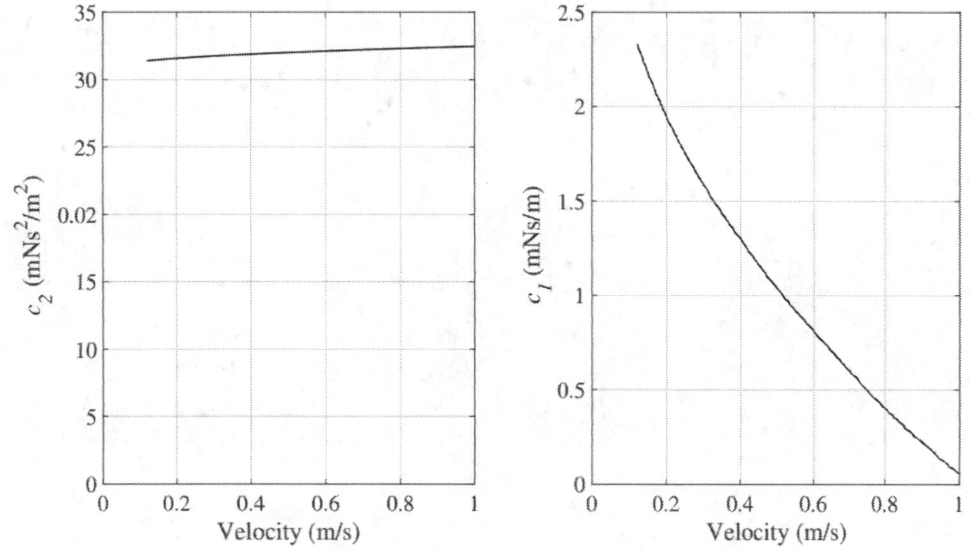

Figure 5: Damping function coefficients obtained basing on the free pendulum test.

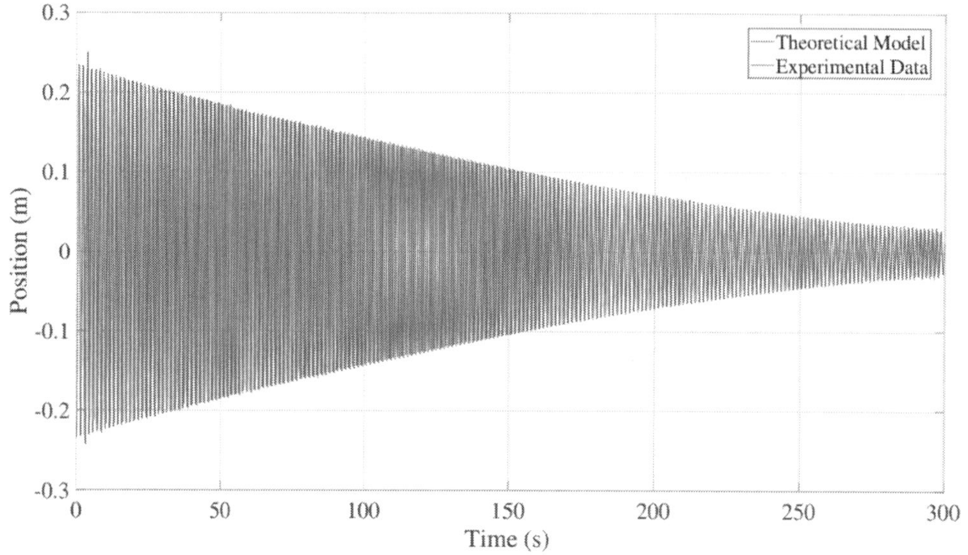

Figure 6: Free pendulum dynamic response comparison.

4.3. Impact test

The impact tests were conducted for a theoretical impact speed $v_{\text{imp}} = 1.5$ m/s and the motion of one of the cylinders from its release until the first impact is shown in Fig. 7. It can be observed that the first impact between the cylinders complies with the specification because it is symmetric and localized on a cylinder generatrix. As for the subsequent impacts, the unavoidable asymmetries present in the system cause the presence of a bending motion of the cylinders at a higher frequency that can be observed in the frames, in the distance between the consecutive response peaks, and the estimated energy dissipation. For this reason, only the first response peak of the system was used to calculate the coefficient of dissipation.

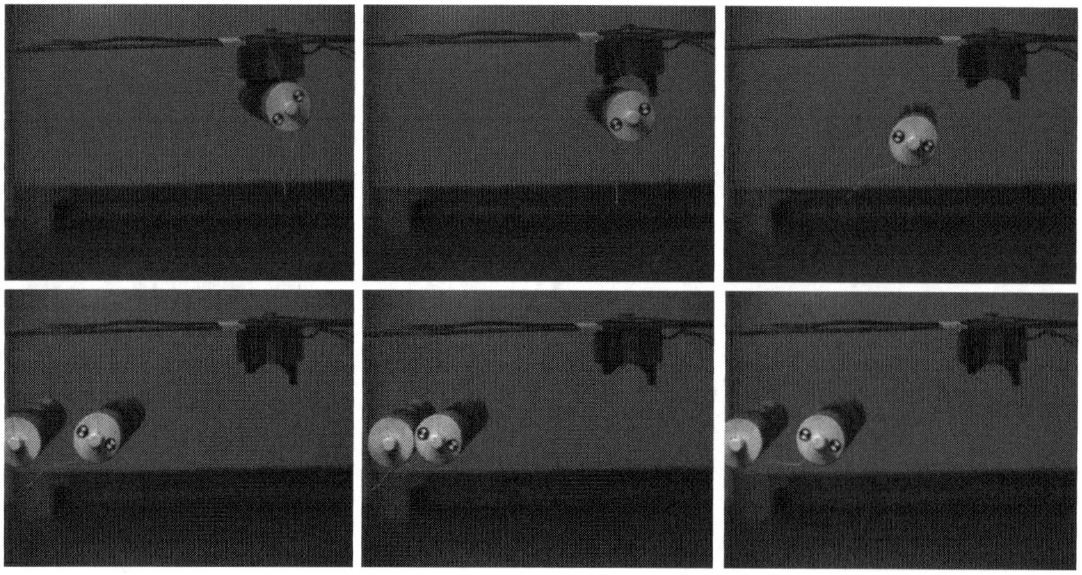

Figure 7: First impact at $v_{\text{imp}} = 1.5$ m/s.

In this case, the period of the system is 0.7 s, so the sampling frequency adopted was sufficient to characterize the global motion of the cylinder. Despite this, the impact time is in the order of tens of microseconds, so a sampling frequency of 60 frames per second is not high enough to calculate directly the coefficient of dissipation d.

4.4. Coefficient of dissipation calculation

An inverse algorithm was used to calculate the coefficient of dissipation without changing the acquisition devices. In particular:

- The free pendulum tests were used to calculate the overall damping of the system and this value was used as an input for the nonlinear pendulum model.
- The nonlinear pendulum model was used to simulate the global motion of the cylinders from the release until the impact, and the actual impact speed was derived.
- The nonlinear pendulum model was initialized again, assuming the speed of the cylinder after the impact as an unknown variable.
- The speed of the cylinder after the impact was determined to match the first response peak of the system.
- The coefficient of dissipation was calculated.

An example of application of the proposed algorithm is shown in Fig. 8. For $v_{imp} = 1.5$ m/s, the coefficient of dissipation was $d = 0.118$, which is twenty times higher than its theoretical value ($d = 0.006$). Even if the first impact occurs along a generatrix of the cylinders, it is plausible that other motions are active due to the asymmetries of the system, thus explaining the considerable increase obtained in the coefficient of dissipation compared to the theoretical results.

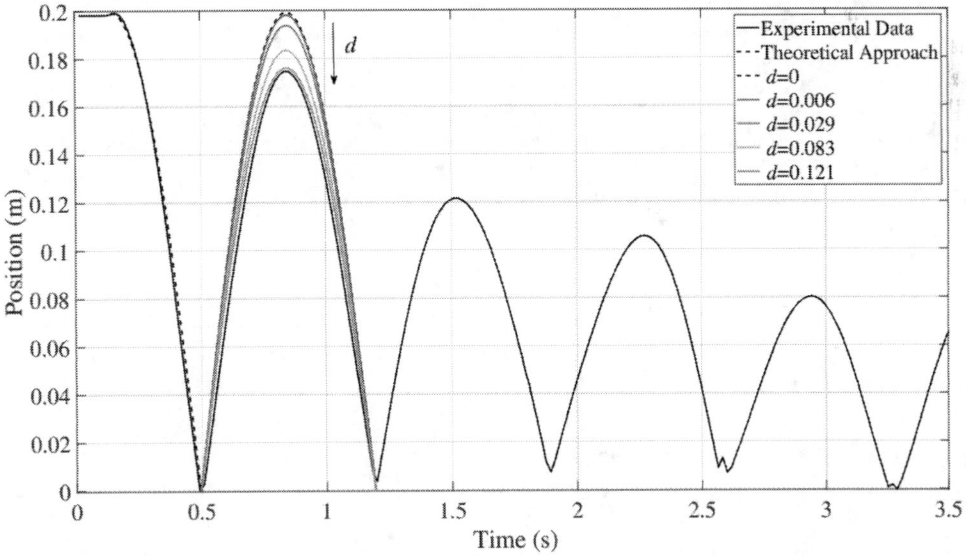

Figure 8: Impact test comparison and determination of d for $v_{imp} = 1.5$ m/s.

5. Conclusions

A test rig for the experimental quantification of the energy dissipation produced by the propagation of elastic waves during the impact of cylinders is proposed.

The final design of the test rig was based on Newton's pendulum principle, obtaining a simple and versatile layout. The results obtained show that the proposed test rig satisfied all the technical specifications. Different cylinders can be tested, and continuous regulation of the initial impact speed is guaranteed. Two different kinds of tests were proposed to separate the several sources of energy dissipation present in the system. The overall damping of the system was calculated by executing free pendulum tests, and an inverse algorithm was used to derive the coefficient of dissipation from the impact tests. The preliminary results show that the overall damping of the system can be calculated by using the free oscillations of one cylinder. In addition, for the tested impact speed used for the tuning of the test rig, the experimental coefficient of dissipation was found one order of magnitude higher than its theoretical value, meaning there may be other forms of dissipation to be investigated. These include the non-synchronous impact on the cylinder's generatrix, which causes the activation of other motions and a significant increase in dissipation. For this reason, we are preparing several modifications to the test.

In the near future, a series of experimental tests are planned to reproduce the theoretical coefficient of dissipation curve for different impact speeds by limiting the presence of out-of-plane motions and using laser interferometry to get a redundant measure and validate the inverse algorithm used. Then, cylinders of several dimensions or materials can be tested.

References

[1] A Laderou, M Mohammadpour, S Theodossiades, R Daubney and G Meeks 2020 On the Effect of DLC and WCC Coatings on the Efficiency of Manual Transmission Gear Pairs *Applied Sciences* **10** 3102
[2] C T Lim and W J Stronge 1999 Oblique elastic–plastic impact between rough cylinders in plane strain *International Journal of Engineering Science* **37** 97-122
[3] K H Hunt and F R E Crossley 1975 Coefficient of Restitution Interpreted as Damping in Vibroimpact *Journal of Applied Mechanics* **42** 440-445
[4] Hany A Sherif and Fahad A Almufad 2019 Models for Materials Damping, Loss Factor, and Coefficient of Restitution *Journal of Engineering Materials and Technology* **142** 011006
[5] M Abruzzo, M Beghini and C Santus 2022 Mechanical energy dissipation due to the propagation of elastic waves during the lateral impact of elastic cylinders *Journal of Sound and Vibration* **535** 117075
[6] V L Popov 2010 *Contact mechanics and Friction* Springer Berlin Heidelberg

IOP Conf. Series: Materials Science and Engineering 1275 (2023) 012013 doi:10.1088/1757-899X/1275/1/012013

Performance of hybrid supercapacitors in the presence of mechanical vibrations for applications of micro-mobility

M Laschi[1], M S Gulino[2] and D Vangi[3]

[1] PhD Student, University of Florence, Department of Industrial Engineering, via di Santa Marta 3, 50139, Florence, Italy
[2] Adjunct Professor, University of Florence, Department of Industrial Engineering, via di Santa Marta 3, 50139, Florence, Italy
[3] Associate Professor, University of Florence, Department of Industrial Engineering, via di Santa Marta 3, 50139, Florence, Italy

E-mail: maurizio.laschi@unifi.it

Abstract. Supercapacitors (SCs) have always been regarded as a source of electrical energy storage confined to equipment requiring high power, and particularly high currents, for a few moments. Hybrid SCs characterized by increasingly higher energy densities have been gaining ground. The main advantages of such accumulators are long service life, a high number of cycles which drastically reduces the problem of recycling lithium cells and good operation even at low temperatures. One of the most underestimated aspects, but with an impact on the service life of accumulators, is the mechanical vibration behavior. In micro-mobility vehicles, the battery pack is often attached to the charter, which in turn is not sufficiently dampened for normal road surface conditions. The literature mainly analyses the variation of ESR (internal resistance) as a function of the received vibration dose. In the following work, the impact of mechanical vibrations on the various parameters of an equivalent electrical circuit is assessed using impedance spectroscopy, and how the variation of individual components can influence the performance of a micro-mobility vehicle battery. The reference vibration cycle was realized by sampling a real driving cycle on an electric scooter on various types of urban terrain and was replicated on various types of cells using a shaker.

Content from this work may be used under the terms of the Creative Commons Attribution 3.0 licence. Any further distribution of this work must maintain attribution to the author(s) and the title of the work, journal citation and DOI.
Published under licence by IOP Publishing Ltd

1. Introduction

EDLC super-capacitors (SC) have always been a source of energy storage for all those applications where large powers are required for small time intervals, such as the movement of rudders of large vessels or actuators for industrial use [1]. Two of the most stringent limitations of these SCs are their low energy density and high daily self-discharge. This is a major limitation if one wants to extend their use as an energy storage source for mobility, especially for light electric vehicles. Various SC Li-ion hybrid systems can be found in the literature to exploit the advantages of both technologies [2]. Recently, however, new SC technologies are being commercialized: hybrid super-capacitors. There are various subcategories of them; the ones examined in the article have these two main characteristics:

- The cathode is like that of a battery, thus providing high capacitance
- The anode is an evolved version of EDLC SCs.

The most important positive aspect certainly is the reduction of the detrimental effect of current fluctuations, and a higher lifespan [3, 4].

Of course, this technology still has limitations, such as the low specific power density slightly higher than lithium-ion cells.

The application of such SCs in an urban micro-mobility context would bring undoubted advantages:

- The increased useful life of one's vehicle
- Lower environmental impact due to less battery recycling
- Less use of metals such as cobalt
- Increased vehicle safety since both the electrolyte and the internal components are less flammable than a traditional Li-ion cell

A very important aspect, but one that is not yet in the literature, is the study of the shock and/or vibration resistance of these SCs; in fact [5] it is shown that vibrations in the medium-term cause irreversible damage to the internal jelly roll structure of Li-ion cells. Accumulators in light electric vehicles for micro-mobility are often poorly fixed and almost without dampening, especially along the vertical axis. This causes important stresses in the form of mechanical vibrations to affect the individual cells. The purpose of this work will be to investigate how and how much damage is done to these SCs and to make a comparison with conventional lithium cells. An impedance spectroscopy technique will be used since this technique allows valuable information to be extracted regarding the complex impedance and then health status as a function of the frequency of discharge of use. [6]

2. Experimental setup

To best simulate the mechanical stresses to which a light electric vehicle battery pack is subjected, an electric scooter battery pack was first instrumented using a single-axial accelerometer as in Figure 1 below.

IOP Conf. Series: Materials Science and Engineering 1275 (2023) 012013 doi:10.1088/1757-899X/1275/1/012013

Figure 1. Instrumented battery pack on e-kick scooter used for measuring vibration along the vertical axis

The battery was laterally locked to minimize the effects of transverse vibrations and left free to move along the vertical axis. The vehicle also has solid tires and no damping system. Several minutes of driving was done on various types of terrain that a user may commonly find. From the data collected, three sections where the greatest accelerations are recorded were isolated:

- A section of road with a speed bump
- A section of bumpy road
- A stretch of road on cobblestones

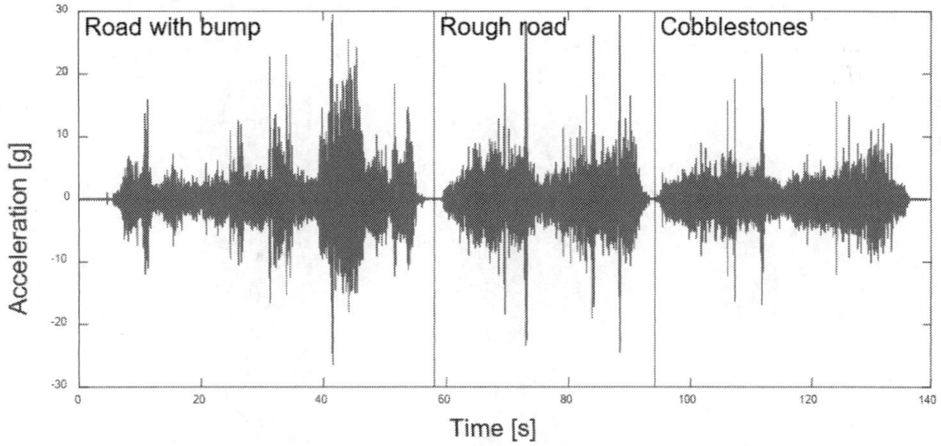

Figure 2. Profile of chosen mechanical vibration to use to stress SC and Li-Ion cells

Before proceeding to stress the SCs and lithium cells, the selected vibration profile was tested on a 4000F hybrid SC using a shaker. The response of the SC was recorded using a linear accelerometer, mounted on the SC jacket, all to verify that the vibrations were congruent with what was acquired.

Figure 3. Experimental setup used to validate mechanical vibration profile. The same setup is then used to apply the profile in figure 2 to all the cells

Once the mechanical stress cycle was validated, the basic measurement of the complex impedance of three types of cells was carried out:

- A 4000F SC
- An 18650 lithium-ion cell
- A 21700 lithium-ion cell

The cells used are in figure 4 below.

Figure 4. From left to right: the 18650 Li-Ion cells, the 4000F SC, the 21700 cell

2.1. The complex impedance measurement

The following tools were used for the complex impedance acquisition and calculation procedure:

- An electrical load of DC EA 9080 45T PS
- A PicoScope 5444D oscilloscope.
- A 0.001875mOhm current deviator, accuracy class 1.0%

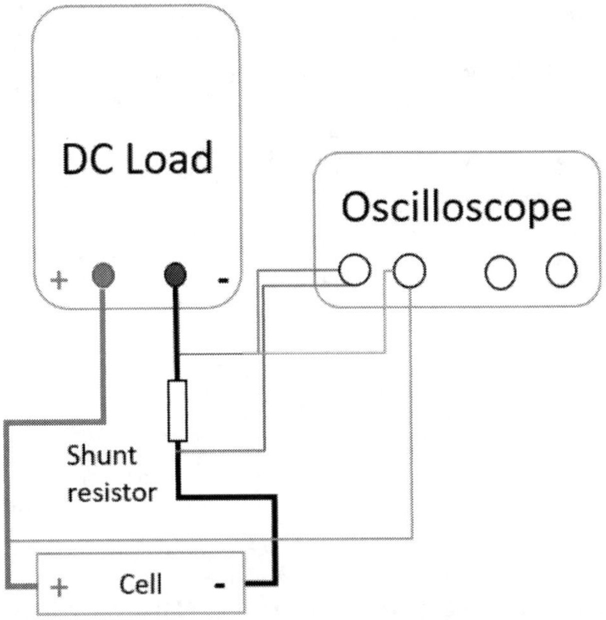

Figure 5. Experimental setup used to measure the complex impedance

The load was set to draw a sinusoidal current at 2, 4, 6, 8, 20, 40, 60, 80, 200, 400, 600, 800, 1000, 2000, 4000, 6000, and 8000 Hz and a peak intensity of 2A.

Following the diagram in Figure 5, two channels of the oscilloscope were used:

- One probe is located with its ends attached to the ends of the shunt.
- The other, however, has the positive end attached to the anode of the battery and the ground in common with the other probe.

For the lithium cells, two pure nickel foils were spot-welded to the poles, onto which the electrical connection cables were then soldered. For the SC, on the other hand, cables were soldered directly to the two ends

2.2. *Signal processing*

MATLAB 2022a software was used to process the acquired signals. The signals thus acquired are referred to as V_{TOT} between shunt and SC and V_{SH} that of the current shunt. The V_{TOT} signal is composed of the sum of the voltage signal at the ends of the shunt and that at the ends of the SC (1)

$$V_{TOT}=V_{SH}+V_{SC} \quad (1)$$

This signal must then be subtracted from the shunt component to obtain the voltage at the ends of the SC (V_{SC}).

Next, the current flowing through the shunt was calculated by applying Ohm's first law (2).

$$I=\frac{V_{SH}}{0.001875\Omega} \quad (2)$$

The signals then were then filtered from external and network noise by an IIR passband zero-phase digital filtering that preserves phase information and then smoothed by a Savitzky-Golay filter with a smoothing factor of 0.35 to obtain nearly noise-free waves. To eliminate de-polarization problems due to overpotential effects [6] of the cells, the signal was cut from 30 per cent onward.

2.3. *Calculation of the phase*

The V_{SC} peaks for each discharge frequency were identified and the time step difference with the corresponding I. The phase was obtained through this formula (3):

$$\vartheta=2*\pi*fr*\Delta p*t_p \quad (3)$$

Where Δp is the time step difference, t_p the sampling period and fr the frequency of the current of discharge

The phase shifts of each peak were calculated and averaged

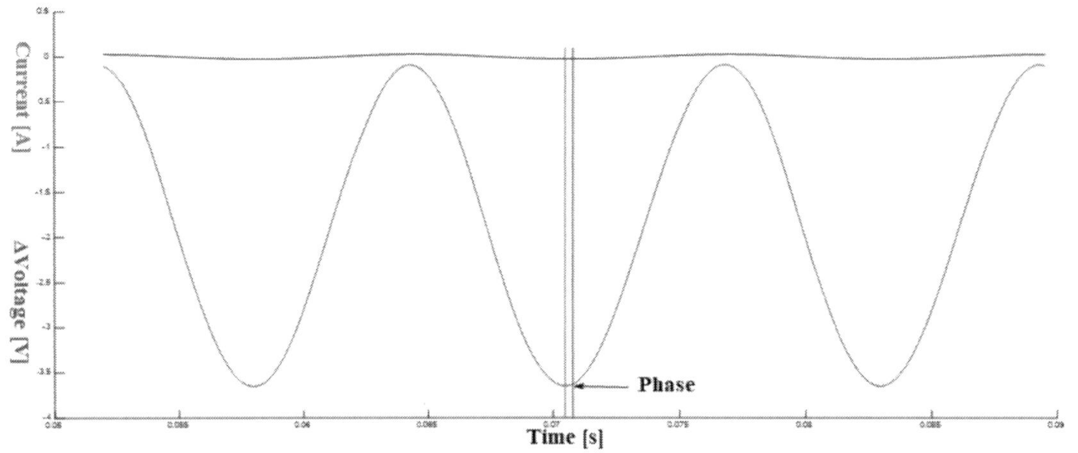

Figure 6. Filtered sinewaves obtained from measurement of current, red, and voltage, blue, of a cell

2.4. Calculation of complex impedance

The average component was removed from each signal. Then peak-to-peak was calculated for the V_{SC} signal, thus obtaining the total voltage change. The calculation of the complex impedance was obtained through the following formula (4):

$$Z=\frac{\Delta V}{I_{mean}*(1+\cos(\vartheta)+i*\sin(\vartheta))} \quad (4)$$

ΔV's phase is set to 0 because a cell could be primarily treated as a constant voltage generator if discharged with a short pulse of current

These steps are then repeated for the remaining acquired waveforms. The following Figure 7 shows the impedances of the lithium-ion cells and the SC before they were subjected to vibration cycling

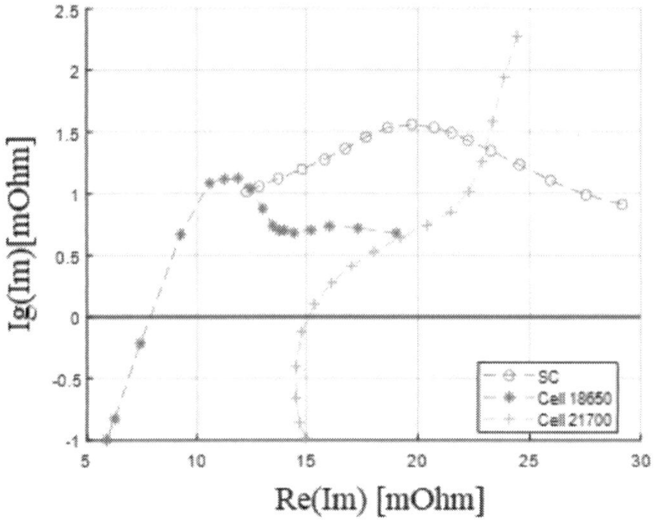

Figure 7. Complex impedance of the three cells tested before any vibration cycle

IOP Conf. Series: Materials Science and Engineering 1275 (2023) 012013 doi:10.1088/1757-899X/1275/1/012013

2.5. Application of vibration

At this point, you move on to apply vibrations to each cell. The same shaker as in Figure 3 was used. The vibration profile in Figure 2 was cyclically applied for twenty hours, and then the complex impedance was measured as described above. This procedure was repeated nine more times, for a total of 180 hours of vibrations. This series of vibrations corresponds to about 400 hours of actual use since, through the applied vibration profile, limited situations were reproduced, against which the driver goes against about 10% of the total driving time.

3. Results

The purpose of the work is to see how the complex impedance changes. Attention will be placed on the real part because it gives us more information about the health and power status of a single cell, as well as is the quantity to vary most significantly [8].

We begin by analysing the Argand diagrams, also called Nyquist diagrams. They are shown in Figure 8 below:

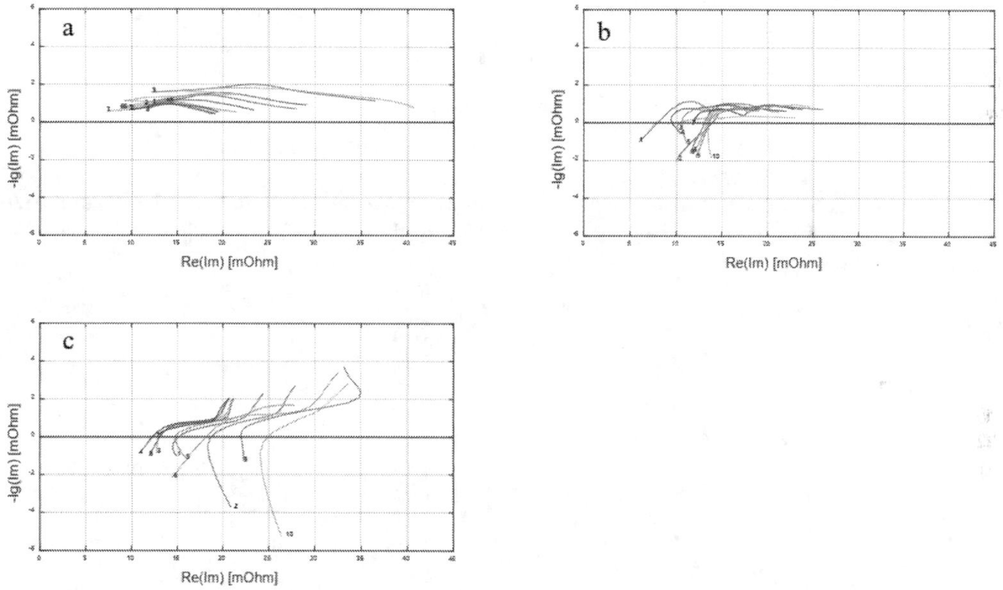

Figure 8. Complex impedance of (a) SC, (b) 18650 cell and (c) 21700 cell. The numbers at the right of each curve are the number of the vibration cycle

The first thing that stands out is the different behaviour of SCs compared with Li-Ion cells. This is mainly due to the porous electrodes of the elements constituting the anode, which promote a capacitive effect for a wide range of discharge frequencies [8]. Discontinuities of the first kind or concavity changes are noted; these correspond with the different charge transport modes in the cells and/or from the change in the graphite lattice due to intercalation phenomena. Each mechanism can be associated with an equivalent electrical circuit net [9].

The chart in Figure 9 shows the impedances at the ninth discharge cycle of the cells under consideration. The SC is the one whose real part varies the most due mainly to a de-intercalation process that is not particularly favourable at low frequencies. [10]

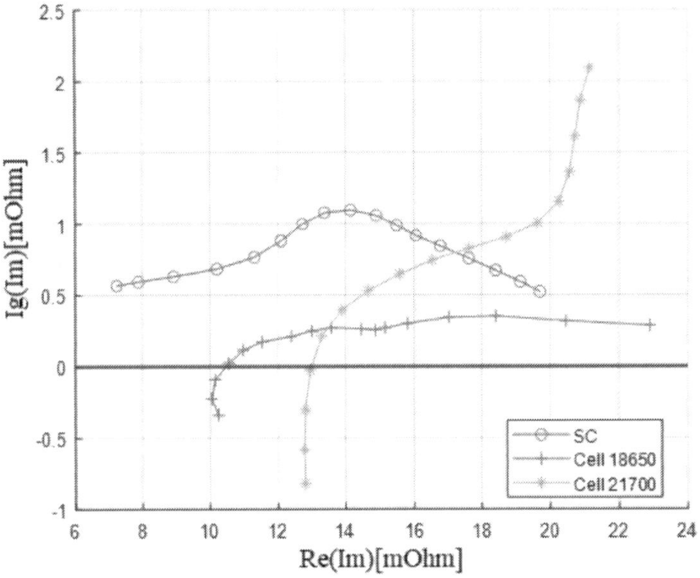

Figure 9. Complex impedance after 9 cycles of vibration

Analysing the real part of the impedance as a function of the vibration cycle (Figure 10 a and b), they always tend to first compress and shift to the left and then dilate and shift to the right. SCs show much more stable behaviour.

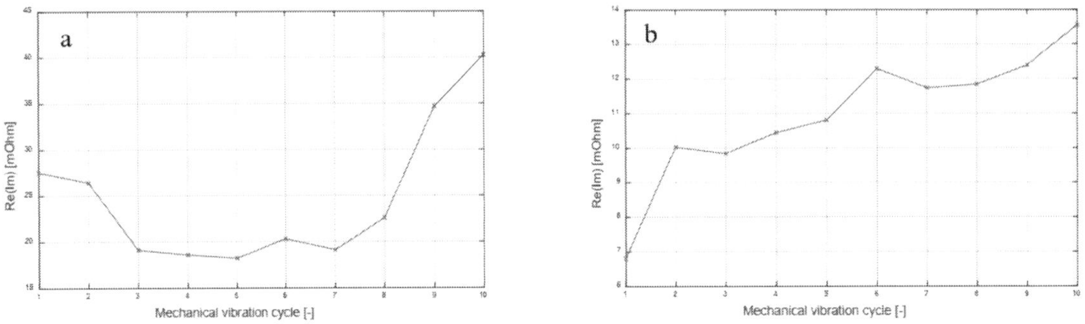

Figure 10. Example of the evolution of complex impedance of SC and the 18650 cell in the time. The (a) is the SC and (b) the 18650 cell both at 8000Hz

This is best seen by studying the change over time of the real part of the impedance. In fact, in the graphs of Figure 10, one can see that at high discharge frequencies SCs tend to damage much less than lithium ion cells; on the other hand, at low discharge frequencies, a greater effect of mechanical vibrations is appreciated. The central plateau phase, always present about halfway through cycles, is due to a stabilization phase of the internal structure

The variations between beginning and end are summarized in this table and the following graph in Figure 11

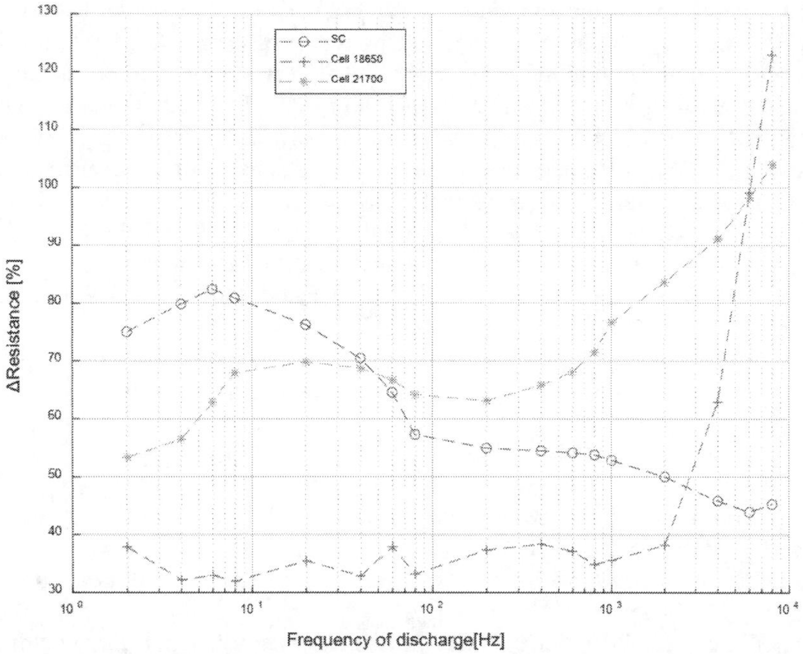

Figure 11. Variation of internal resistance of each cell plotted versus the frequency of discharge

We now turn to the Bode plot. In the graph, in Figure 12 the SCs show a monotonous and strong decrease in impedance as a function of frequency, something that for lithium-ion cells stops above 6000Hz. Almost certainly the SCs also have an inversion point, but given the limited DC load it was not possible to investigate further

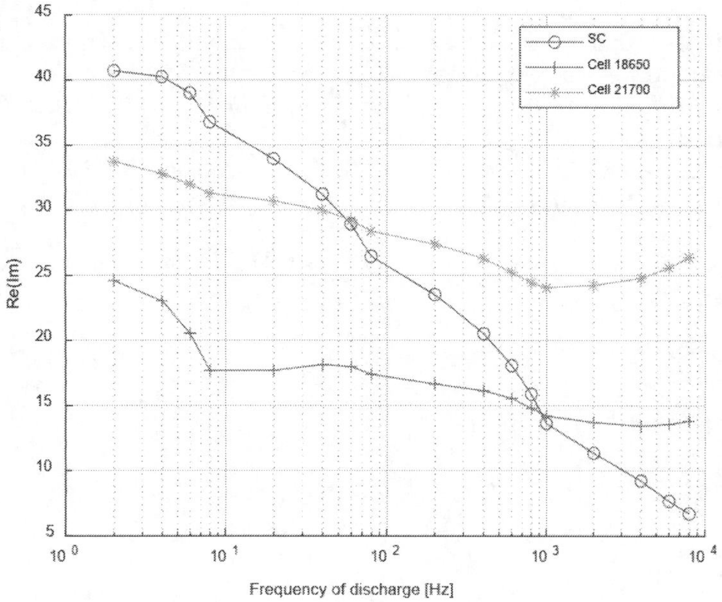

Figure 12. Real part of impedance in function of discharge frequency

4. Conclusion

Tests have shown that the mechanical vibrations of an urban driving cycle over various types of rough terrain damage SCs, however, because the internal structure is different, they have a different response to damage than conventional lithium cells. Both electrotechnically and mechanically, SCs lend themselves to being discharged at frequencies higher than 1000 Hz. In addition, the relative damage is less than lithium batteries, thus ensuring a longer life. This lays the foundation for the development of ESC boards for micro-mobility vehicles that operate in a switching mode rather than continuously as is traditionally the case. Moreover, the measurement done are very useful to create a base for an algorithm for estimating both the State of Power and State of Health of the battery pack, both useful to grant easier maintenance. In the future, vibration cycles will be increased to see how the resistance curve evolves, and these results will be implemented for onboard evaluation systems to diagnose damaged batteries for micro-mobility vehicles for sharing platforms in time

In the future, testing will be improved by moving to a potentiostatic measurement that will allow fewer data to be processed and ensure faster and more accurate measurements

References

[1] Miller JM, 2011 Jan 1, *Ultracapacitor Applications,* (IET Power and Energy series 59, London, United Kingdom) pp 189 -211

[2] Desphande RP, *Ultracapacitors future of energy storage,* (McGraw Hill Education (India) Private Limited) pp 267 - 285

[3] Kouchachvili L, Yaïci W, Entchev E. Hybrid battery/supercapacitor energy storage system for the electric vehicles. Journal of Power Sources [Internet]. 2018 Jan;374:237–48

[4] Zuo W, Li R, Zhou C, Li Y, Xia J, Liu J. Battery-Supercapacitor Hybrid Devices: Recent Progress and Future Prospects. Advanced Science [Internet]. 2017 Feb 21;4(7):1600539

[5] Siddique ARM, Tasnim S, Mahmud S, Van Heyst B. Thermal field investigation of lithium-ion battery with porous medium under vibration. Journal of Energy Storage [Internet]. 2022 Mar;47:103632

[6] Barsoukov E, Macdonald JR, editors. Impedance Spectroscopy. 2005 Jan 27

[7] Chen Z, Danilov DL, Raijmakers LHJ, Chayambuka K, Jiang M, Zhou L, et al. Overpotential analysis of graphite-based Li-ion batteries seen from a porous electrode modelling perspective. Journal of Power Sources [Internet]. 2021 Oct;509:230345

[8] Andrea D, 2010 Jan 1, *Battery Management Systems for Large Lithium-Ion Battery Packs*, Artec House, London, 14 – 28

[9] Barsoukov E, Macdonald JR, editors. Impedance Spectroscopy. 2005 Jan 27, pp 444 - 446

[10] Barsoukov E, Macdonald JR, editors. Impedance Spectroscopy. 2005 Jan 27 pp 469 - 496

[11] Sun J, Li G, Wang Z, Guo H, Li X, Yan G, et al. Balancing the anions adsorption and intercalation in carbon cathode enables high energy density dual-carbon lithium-ion capacitors. Carbon [Internet]. 2022 Nov;200:28–37

IOP Conf. Series: Materials Science and Engineering 1275 (2023) 012014 doi:10.1088/1757-899X/1275/1/012014

Active thermography characterization of aerogel materials for vehicle electrification

F Curà[1], R Sesana[1], M Dugand[2] and L Corsaro[1]

[1] Dimeas Politecnico di Torino, corso Duca degli Abruzzi 24, 10129 Turin – Italy
[2] CRF - Materials Engineering Methods & Tools, E/E & Multifunctional Materials - Opto-Electronic & Glazing, C.so Settembrini 40, 10235 Turin – Italy

E-mail: francesca.cura@polito.it

Abstract. Aerogel materials are very important in the actual scenario of both vehicle electrification and green mobility. Temperature control and cell physical state of battery packs are relevant for the correct operating conditions of electric systems. As a matter of fact, the achieved temperature inside the battery is a critical parameter to increase the battery performance. Thanks to its physical properties, due to the low thermal conductivity, aerogel materials can store a portion of the thermal energy released during discharging operation and then use it during the runaway one. This way, a passive thermal control increases the whole system performance without any further device. The evaluation of both thermal diffusivity and conductivity of aerogel materials becomes significant from an industrial point of view in order to test different aerogel typologies. Aim of this work was to develop an Active Thermography approach for aerogel testing to determine thermal parameters as diffusivity and conductivity. Experimental data were processed basing on dedicated ISO Standards. These results were also compared to those available in literature.

1. Introduction

Carbon dioxide emissions play an important role in global climate change. The roadmap outlined to reduce these effects operates on an economical and sustainable energy supply [1]. Among challenges to reduce carbon dioxide emissions, vehicle electrification is one of the most significant. To satisfy this goal, the use of internal combustion engines in vehicles has to be progressively replaced with electric motors or hybrid systems [2]. Advantages in adopting these solutions may produce environmental benefits and, on the other hand, may guarantee high performance and comfort of vehicles [3]. As a matter of fact, the actual scenario of green mobility requires continuous improvements in the sustainable energy production field and in the capability in both storing and supplying the energy.

In vehicle electrification contest, storing energy has a great impact on the vehicle performance. The battery design is focusing the attention on three aspects to store energy in an efficient way. These aspects consist in storing sustainable energy with long stability, in carrying on high life cycle products and in keeping environmental constraints [4]. Nowadays, an optimal solution may be represented by lithium-ion batteries, due to their high power/energy density, low energy consumption and low weight. However, these advantages are highly influenced by the operating temperature, as energy and the power densities decrease for low and high temperature conditions. A study from the American Automobile Association stated a reduction of 57% in the electric vehicle autonomy because of external temperature [5].

Content from this work may be used under the terms of the Creative Commons Attribution 3.0 licence. Any further distribution of this work must maintain attribution to the author(s) and the title of the work, journal citation and DOI.
Published under licence by IOP Publishing Ltd

Different industrial solutions may be adopted to reduce this problem, firstly by using additives. As described in [6] [7] [8], additives improve the effective properties of the electrolyte, but this advantage deteriorates at high temperature with a reduction in terms of cell performance.

The second solution may be a pre-heating of the battery up to a certain optimal temperature before use. This technique allows to increase the battery power with an increasing in vehicle autonomy. The disadvantage of this solution consists in the required time of the heating phase, which is approximately 40 minutes. Another possibility to apply this technique is described in [9], where the heating phase is realized by using regenerative braking.

A further adopted solution may be the passive thermal control, as thermal insulating materials are able to keep the battery internal temperature to a certain value. One of the most common material is aerogel, that is characterized by low values of thermal diffusivity and conductivity [10]. Thanks to these properties, the thermal dispersion of the batteries through the environment is reduced. In [11], an experimental study was conducted to evaluate the advantages of aerogel materials in the passive thermal control technique. This way, aerogel materials may keep the heat generated during the discharging process and then use it during the runaway operation.

For these reasons, a complete knowledge of physical and thermal properties of aerogel materials is necessary to improve their performance. To determine both thermal diffusivity and conductivity of plastic insulating materials some Standards have to be taken into account, as ISO 8301 (Heat Flow Meter method, HFM) [12], ISO 8302 (Guarded Hot Plate method, GHP) [13] and ISO 22007-4 (Laser Flash method, LF) [14]. For what concerns physical and thermal characterization of aerogel materials, specific Standards have to be followed. In particular, ISO 8301 [12] and ISO 8302 [13] Standards provide formula to obtain thermal diffusivity and conductivity values, while ISO/DIS 22482.2 Standard [15] equations aim to compute physical properties.

An example of thermal characterization of aerogel materials may be found in [16] [17] [18], basing on the Laser Flash technique that is useful to provide thermal diffusivity and conductivity values in a very fast way.

Also the classical Active Thermography approach may be used to characterize materials from the thermal point of view, instead of the more common Laser Flash. Adapting classical Standard procedures ([19], [20]), Active Thermography technique in transmission configuration was used to analyze Thermal Barrier Coatings peculiarities [21].

Aim of this paper is to develop an Active Thermography approach for aerogel materials testing in order to determine thermal parameters as diffusivity and conductivity. Adequate aerogel specimens were prepared for thermal characterization.

Experimental data were processed basing on dedicated ISO Standards. More in detail, half rise and areal heat diffusion time methods were adopted to obtain apparent thermal diffusivity and thermal conductivity. These results were also compared to those available in literature.

2. Materials and Methods

The complete sample analyzed in this work (see Figure 1) is composed of two different parts, a black plastic coating (external part) and an aerogel material (internal part). Physical (density) and geometrical parameters of both external (black plastic coating) and internal (aerogel material) part were determined. Three specimens for the black plastic coating and three specimens for the aerogel material were adopted (surface dimensions of each specimen were about 15x15mm).

Firstly, an optical microscope analysis was done to get the thickness of the two parts (black plastic coating and aerogel material) and to analyze more in detail the aerogel material structure. Figure 2 shows the optical microscope images (black plastic coating on the left, aerogel material on the right); silica fibers in the aerogel material are clearly distinguishable. Then, weight measurements of black plastic coating and aerogel material were performed by a precision balance.

IOP Conf. Series: Materials Science and Engineering 1275 (2023) 012014 doi:10.1088/1757-899X/1275/1/012014

Figure 1. Complete aerogel sample.

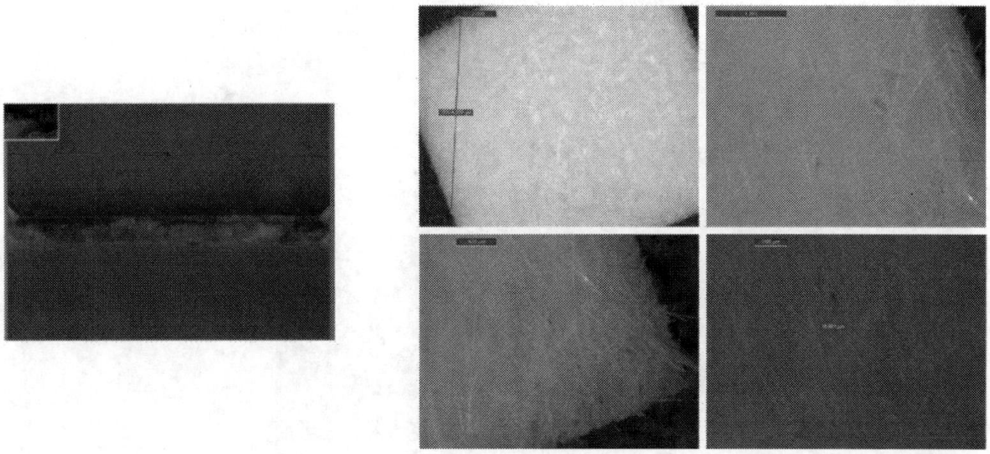

Figure 2. Optical microscope images; left side: black plastic coating - right side: aerogel material.

Once thicknesses of black plastic coating and aerogel material were known, both volume and mean density of each sample were calculated. The black plastic coating density value is in good agreement with that of a PTFE plastic material (2140 - 2180 kg/m^3), as reported in [22]. Specific heat value of both samples was found in literature ([22], black plastic coating and [23], fibrous aerogel material).

Table 1 summarizes all estimated parameters (thickness d, mean density ρ and specific heat c).

Table 1. Material characterization.

Material	d [mm]	ρ [kg/m^3]	c [J/kg °C]
Black plastic coating	0.05	2079.15	1000
Aerogel material	1.38	204.95	1000

Active Thermography (AT) equipment is composed of a IR thermo camera, a laser excitation source and a PC control unit. The IR thermo camera is a FLIR A6751sc with sensitivity lower than 20 mK and 3-5 μm spectral range, while the laser source can generate a maximum power of 50 W concentrated in a small surface. Thermal excitation is defined by power entity (% of the maximum power, %Pmax) and duration time (time period). The "*transmission mode*" configuration was adopted for the thermal characterization. A dedicated sample holder was designed to reduce the heat dissipations during the test. Each specimen was clamped in the sample holder with plastic wires, at the level of the laser source excitation, at 160 mm distance from the laser source and 940 mm from the thermal camera. The frame rate acquisition was 785.67 Hz. Figure 3 shows the experimental setup and all devices. Thermal characterisation is similar to that of Laser Flash approach, but here a classical AT was used.

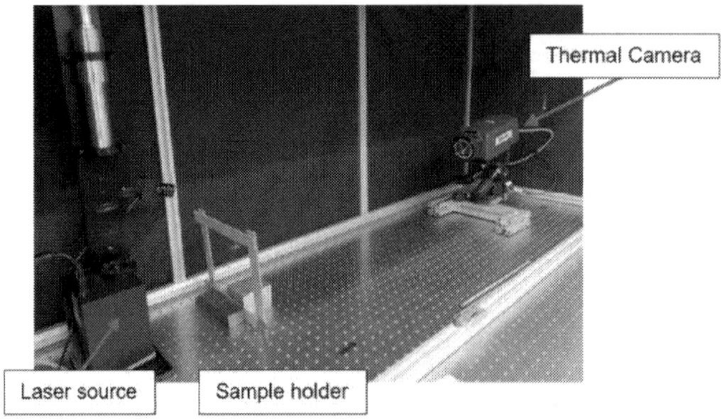

Figure 3. Active Thermography setup in transmission mode.

Figure 4. ROI on the specimen.

Heating profiles were acquired during Active Thermography tests. More in detail, absolute temperature over time was extracted referring to a Region Of Interest (ROI) located in the centre of the specimen (see Figure 4). The ROI size was about 900 pixel and its mean temperature was considered. In all cases, the emissivity value was set to 0.85 based on the specimen surface aspect. Experimental signals were denoised by using a Gaussian filter with 50 points to reduce the noise due to the high frame rate acquisition (785.67 Hz). This procedure allowed us to better identify the starting point of each heating profile. At the end, the relative temperature ΔT (absolute temperature - environmental temperature) was considered for the thermal characterization. The maximum relative temperature ΔT_{max} was also monitored during each test and then the material thermal response in terms of maximum temperature increment was analyzed with respect to the energy excitation.

Thermal parameters (thermal diffusivity and apparent thermal diffusivity) were obtained by the methods proposed in Standards [19] [20] starting from relative temperature ΔT. In particular, ISO 18755 Standard [19] allows to determine the thermal parameters (thermal diffusivity and thermal conductivity) by using the *half rise time* method and ISO 18555 Standard [20] may be used for multilayer materials (as metallic and inorganic coatings), allowing to determine the thermal parameters (thermal diffusivity and thermal conductivity) of each layer by using the *areal heat diffusion time* method. In this case, it was used to estimate the thermal parameters of only one layer with the heat diffusion time τ of the tested sample.

A preliminary testing campaign was carried on to verify Standards requirements on thermal profiles. In particular, preliminary measurements emphasised that thermal profiles generated by both materials (black plastic coating and aerogel) didn't respect ISO 18755 [19] and ISO 18555 Standards [20] requirements. Requested conditions about half rise time and thermal stabilization of thermal profiles were not verified due to the small thickness of the black plastic coating and both low density and surface

IOP Conf. Series: Materials Science and Engineering 1275 (2023) 012014 doi:10.1088/1757-899X/1275/1/012014

aspect of the aerogel material (see Table 1). Standards [19][20] propose a general solution to increase the thermal energy absorption by simply applying a coating on samples. In that case, the condition to be verified is that the half rise time of the coating has to be lower than the 2% of the half rise time of the whole specimen, covered by the coating.

In this work, a small layer of graphitized grease (about 0.05 mm) was chosen, following two different architectures.

In the first case (Specimen 1, see Figure 5), the layer of graphitized grease was interposed between plastic and aerogel to guarantee their adhesion. This solution provided an apparent thermal diffusivity for the aerogel due to the presence of the black plastic coating.

In the second case (Specimen 2, see Figure 5), the aerogel material was simply covered by the graphitized grease layer providing a classic aerogel thermal characterization with thermal diffusivity and thermal conductivity parameters.

So, a theoretical thermal characterization of each coating was carried on. As a matter of fact, the small thickness of the black plastic coating (Specimen 1) and the semi-solid state of the graphitized grease coating (Specimen 2) make impossible the application of the Laser Flash Method separately to estimate their half rise time. A calculation of the "theoretical" half rise time $t_{0.5}^{th}$ [19] was then performed for black plastic coating, graphitized grease, aerogel material and corresponding values were reported in Table 2, that illustrates all parameters useful to compute the thermal diffusivity α. To do that, aerogel thermal conductivity k value was found in [1], while some information about physical properties of graphitized grease was found in [24] [25]. The graphitized grease thickness was considered equal to the black plastic coating one (0.05 mm).

As it can be observed in Table 2, obtained "theoretical" half rise time $t_{0.5}^{th}$ values for black plastic coating and graphitized grease are quite negligible respect to the theoretical value computed for a generic aerogel material ($t_{0.5}^{th}$ = 1.23 s).

This point, the actual experimental campaign was carried on, following some testing parameters resumed in Table 3. In particular, for both specimens (Specimen 1 and Specimen 2), nine testing conditions were chosen (Test 1, 2, ..., 9). Laser excitations involved different energy conditions (time period and power) to perform a good tuning of thermal profiles (10 - 90 °C acquisition window of the thermal camera). Three replications were considered for each thermal excitation.

Figure 5. Specimen 1 and Specimen 2.

Table 2. Theoretical half rise time $t_{0.5}^{th}$.

Material	k [W/m°C]	ρ [kg/m³]	c [J/kg °C]	α [m²/s]	d [mm]	$t_{0.5}^{th}$ [s]
Black plastic coating	0.025	2079.15	1000	1.20E-08	0.05	0.029
Graphitized grease	0.23	907.27	2.15	1.18E-04	0.05	2.94E-06
Aerogel material	0.044	204.95	1000	2.14E-07	1.38	1.23

Table 3. Thermal excitations for Specimen 1 and Specimen 2.

Test	Specimen 1			Specimen 2		
	Time period [ms]	Power [%Pmax]	Energy [J]	Time period [ms]	Power [%Pmax]	Energy [J]
Test 1	20	45	0.45	5	100	0.25
Test 2	20	30	0.30	5	80	0.20
Test 3	20	15	0.15	10	100	0.50
Test 4	5	15	0.04	10	90	0.45
Test 5	5	30	0.08	10	70	0.35
Test 6	5	45	0.11	10	60	0.30
Test 7	10	45	0.23	20	60	0.60
Test 8	10	40	0.20	20	55	0.55
Test 9	10	35	0.18	20	40	0.40

3. Results

Figure 6 (left side) shows the relative temperature profiles ΔT of Specimen 1. One replication of tests due to its repeatability (Test 1, 2, ….,9), described in Table 3, was considered for that analysis. The relative temperature profiles ΔT are plotted with different color to emphasize the energy corresponding to the laser excitation. As an example, the green relative temperature ΔT (Test4) corresponds to the lower energy excitation (0.04 J) and the red relative temperature ΔT (Test1) corresponds to the higher one (0.45 J). Figure 6 (right side) may relates the energy excitation of each test with the corresponding relative maximum temperature increment ΔT_{max}. The material thermal response increases linearly with the energy excitation.

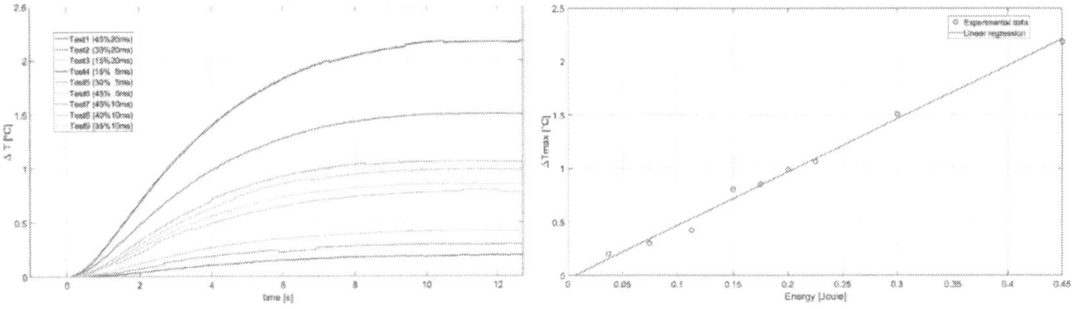

Figure 6. Relative temperature ΔT (left) and material thermal response (right) for Specimen 1.

Figure 7 shows the normalized relative temperature profiles with respect to their maximum value. This elaboration is the starting point to characterize the material by applying ISO 18755 and ISO 18555 Standards [19] [20]. Half rise time $t_{0.5}$ and heat diffusion time τ parameters are computed to evaluate the apparent thermal diffusivity.

Table 4 resumes for all tests and for Specimen 1 the following data: half rise time $t_{0.5}$, heat diffusion time τ, apparent thermal diffusivity $\alpha_{t0.5}$ (see ISO 18755 Standard [19], half rise time method) apparent thermal diffusivity α_{area} (see ISO 18555 Standard [20], areal heat diffusion time method). The low energy excitation test (Test 4) is not aligned with the other ones, due to its low thermal increment ΔT, so it wasn't considered in processing these mean values.

Table 4. Thermal characterization for Specimen 1.

	$t_{0.5}$ [s]	τ [s]	$\alpha_{t0.5}$ [m²/s]	α_{area} [m²/s]
Specimen 1	3.12	22.11	9.75E-08	9.93E-08

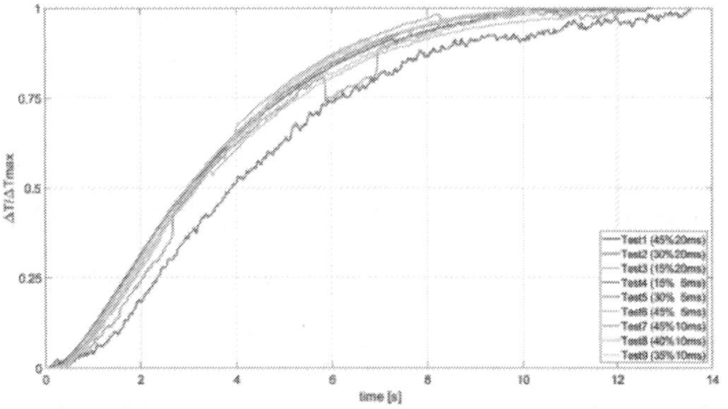

Figure 7. Normalized relative temperature of each profile for Specimen 1.

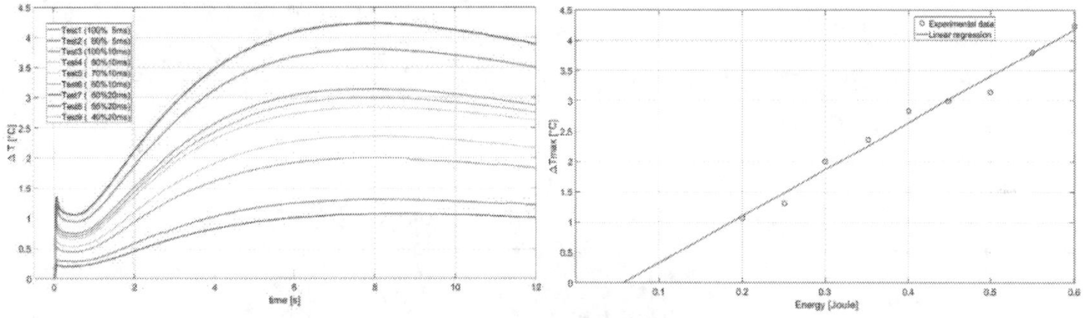

Figure 8. Relative temperature ΔT (left) and material thermal response (right) for Specimen 2

Similar results are shown in Figures 8 and 9 and in Table 5 for Specimen 2.

In particular, Figure 8 (left) shows the relative temperature ΔT, related to all tests and referred to one replication. The relative temperature ΔT was plotted, as for Figure 6, with different colours to appreciate the energy variation corresponding to each thermal excitation.

In this case, relative temperature profiles are different to those generated by Specimen 1. As a matter of fact, the initial heating phase of all the tests presents a spike, as already pointed out in ISO 18755 [19] Standard. Thermal response of Specimen 2 (see Figure 8, right) shows also in this case a linear behaviour with the supplied energy.

Figure 9 shows the normalized relative temperature profiles with respect to its maximum value. Thermal characterization for Specimen 2 was done by applying ISO 18755 and ISO 18555 Standards [19] [20]. For Specimen 2, Test 2 wasn't considered in data processing due to its low thermal increment ΔT. Half rise time $t_{0.5}$ and areal diffusion time τ parameters were computed to evaluate both thermal diffusivity and thermal conductivity.

Similarly to Table 4, Table 5 resumes the mean value of all tests for: half rise time $t_{0.5}$, heat diffusion time τ, thermal diffusivity $\alpha_{t0.5}$ (ISO 18755 Standard [19], half rise time method), thermal diffusivity α_{area} (ISO 18555 Standard [20], areal heat diffusion time method) and the corresponding two thermal conductivities $k_{t0.5}$ and k_{area} values obtained from the two thermal diffusivities ($\alpha_{t0.5}$ and α_{area}). Test 2 wasn't considered in the mean values computation.

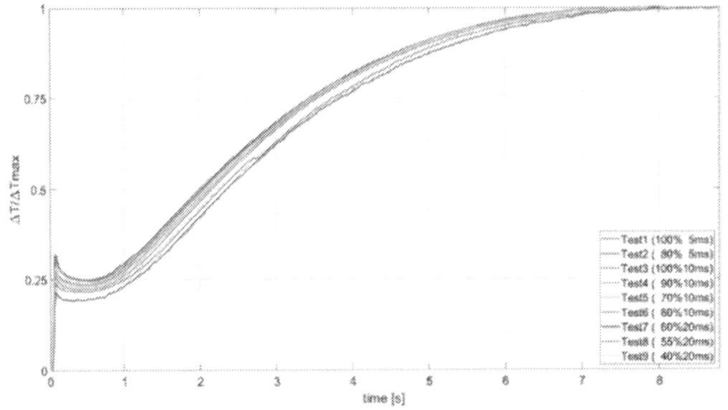

Figure 9. Normalized relative temperature of each profile for Specimen 2.

Table 5. Thermal characterization for Specimen 2.

	$t_{0.5}$ [s]	τ [s]	$\alpha_{t0.5}$ [m²/s]	α_{area} [m²/s]	$k_{t0.5}$ [W/m°C]	k_{area} [W/m°C]
Specimen 2	2.12	13.78	1.25E-07	1.38E-07	0.026	0.028

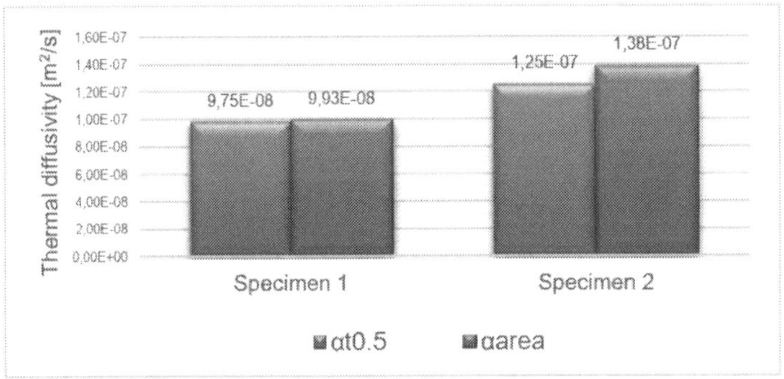

Figure 10. Thermal diffusivity for Specimen 1 and Specimen 2.

As already pointed out, from the analysis of Figures 6 (right side) and Figure 8 (right side), it can be observed that adopted solutions to test Aerogel materials generate a linear thermal behaviour in the response. This way, a possible forecast of the maximum thermal response ΔT_{max} can be estimated to improve the setup of the laser energy excitations.

Figure 10 compares the results (apparent thermal diffusivity and thermal diffusivity values) obtained for Specimens 1 and 2 (ISO 18755 ($\alpha_{t0.5}$) and ISO 18555 (α_{area}) Standards [19] [20]).

It can be noted that the results generated for both thermal characterizations (ISO 18755 and ISO 18555 Standards [19] [20]) show the same value. Thermal conductivity values (see Table 5) are in a good agreement with those provided by literature for nanocellulose aerogels [10] (0.018 - 0.044 W/m°C).

4. Conclusions

The results obtained in the present work allow to draw the following conclusions.

Active Thermography technique may be satisfactory utilised for thermal characterisation of aerogel materials provided that testing set up was chosen to guarantee the applicability of both procedures and formula involved in classical Standards.

From the experimental point of view, the focus is to realise a sample able to achieve a good warm-up and to stabilise the thermal response to the given excitation.

During the preliminary testing campaign, some measurements were performed to investigate separately plastic and aerogel materials, but their thermal characteristics make impossible to absorb the provided energy and then to produce appreciable thermal profiles. As a matter of fact, laser power and corresponding time period always depends on the thermal response of a system. In other words, suitable laser excitations may be chosen only after the evaluation of the response in terms of system thermal profile.

In order to overcome this impasse and to increase the thermal energy absorption required by Standards, two different architectures for specimens were defined, keeping the same plastic and aerogel materials, but introducing in a different way a conductive coating. Specimen 1, involving the use of a graphitized grease layer interposed between plastic and aerogel, provided apparent thermal diffusivity values. In Specimen 2 the aerogel material was simply covered by the graphitized grease layer and its characterisation provided us a response in terms of thermal conductivity values.

Anyhow, thermal diffusivity and conductivity values obtained from the two different specimens and by two different approaches are comparable and in a good agreement to those available in literature.

Finally, it may be concluded that, in this research activity, design and specimen preparation become a crucial phase to characterize aerogel materials from the thermal point of view with an Active Thermography technique.

References

[1] Armand M and Tarascon J-M 2008 Building better batteries *Nature* **451** n 7179 pp 652-657

[2] Etacheri V, Marom R, Elazari R, Salitra G and Aurbach D 2011 Challenges in the development of advanced Li-ion batteries: a review *Energy &Environmental Science* **4** n 9 pp 3243-3262

[3] Manzetti S and Mariasiu F 2015 Electric vehicle battery technologies: from present state to future systems *Renewable and Sustainable Energy Reviews* **51** pp 1004-1012

[4] Liu H, Wei Z, He W and Zhao J 2017 Thermal issues about Li-ion batteries and recent progress in battery thermal management systems: a review *Energy Conversion and Management* **150** pp 304-330

[5] https://newsroom.aaa.com/2014/03/extreme-temperatures-affect-electric-vehicle-driving-range-aaa-says/

[6] Xu G et al 2019 Functional additives assisted ester-carbonate electrolyte enables wide temperature operation of a high-voltage (5 V-Class) Li-ion battery *Journal of Power Sources* **416** pp 29-36

[7] Tornheim A, Peebles C , Gilbert J A, Sahore R, Garcia J C, Bareño J, Iddir H, Abraham D P and Liao C 2017 Evaluating electrolyte additives for lithiumion cells: a new figure of merit approach *Journal of Power Sources* **365** n C pp 201-209

[8] Senyshyn A, Mühlbauer M, Dolotko O and Ehrenberg H 2015 Low-temperature performance of Li-ion batteries: the behavior of lithiated graphite *Journal of Power Sources* **282** pp 235-240

[9] Zhang G, Ge S, Yang X G, Leng Y, Marple D and Wanet C Y 2017 Rapid restoration of electric vehicle battery performance while driving at cold temperatures *Journal of Power Sources* **371** pp 35-40

[10] Chen Y, Zhang L, Yang Y, Pang B, Xu W, Duan G, Jiang S and Zhang K 2021 Recent progress on Nanocellulose naerogels: preparation, modification, composite fabrication, applications *Advanced Materials* **33** n 11

[11] Wu H, Zhang X, Wang C, Cao R and Yang C 2020 Experimental study on aerogel passive thermal control method for cylindrical lithium-ion batteries at low temperature *Applied Thermal Engineering* **169** 114946
[12] ISO 8301 Thermal insulation - Determination of steady-state thermal resistance and related properties - Heat flow meter apparatus
[13] ISO 8302 Thermal insulation - Determination of steady-state thermal resistance and related properties-Guarded hot plate apparatus
[14] ISO 22007-4 Plastics - Determination of thermal conductivity and thermal diffusivity
[15] ISO/DIS 22482.2 Thermal insulation products-Aerogel blanket for buildings - Determination of phisical properties
[16] Liu G, Li J, Li X, Pan X and Qian C 2021 Preparation and properties of novel superhydrophobic cellulose nanofiber aerogels *Journal of Nanomaterials* **2021** pp 1-8
[17] Zhong Y, Zhou M, Huang F, Lin T and Wan D 2013 Effect of graphene aerogel on thermal behavior of phase change materials for thermal management *Solar Energy Materials and Solar Cells* **113** pp 195-200
[18] Yoo J K, Wagle R, Lee C W and Lee E Y 2019 Synthesis of silica aerogel thin sheets and evaluation of its thermal, electrical, and mechanical properties *International Journal of Applied Ceramic Technology* **16** n 2 pp 832-842
[19] ISO 18755 Fine ceramics (advanced ceramics, advanced technical ceramics) - Determination of thermal diffusivity of monolithic ceramics by laser flash method
[20] ISO 18555 Metallic and other inorganic coatings - Determination of thermal conductivity of thermal barrier coatings
[21] Curà F, Mura A and Corsaro L 2022 Active thermography technique for barrier coatings characterization *IOP Conf. Ser.: Mater. Sci*. Eng. **1214** 012034
[22] http://www.indat.it/plastics/wpcontent/uploads/2016/12/TekPTFE.pdf. (2022/03/20)
[23] Lakatos Á, Csík A, Trník A and Budai I 2019 Effects of the heat treatment in the properties of fibrous aerogel thermal insulation *Energies* **12** n 10
[24] https://files.canon-europe.com/files/webcontent/safety-data-sheets/EU_Service-aterials/gy2-0741_multemp_et-r_pdf.pdf
[25] https://pubs.rsc.org/en/content/articlepdf/2019/ra/c9ra09201c

The use of induction heating for thermographic inspection of composite laminates

R Nobile[1], A Saponaro[1*] and A Pirinu[1]

[1]University of Salento, Department of Engineering for Innovation, via per Monteroni, Lecce 73100, Italy

[*]Corresponding author's e-mail: andrea.saponaro@unisalento.it

Abstract In the present paper, an active thermographic technique is proposed for the inspection of a unidirectional GFRP panel with integrated artificial internal defects and an aeronautical CFRP panel with real small production defects.
Different inductor geometries were considered by varying the length and the number of spires, finally allowing to identify the optimal solution. Several tests allowed to identify the optimal experimental setup for defect detection as a function of the location of the defect.
The acquired thermal maps were processed in Matlab with a suitable analysis algorithm for the determination of the absolute and normalized thermal contrasts. Another important consideration concerns the possibility of obtaining a qualitative indication on the defect depth. Although the thermal contrast depends on the choice of the intact and defective area, the maximum value of the absolute contrast was higher for the more superficial defects than for the deeper ones. Furthermore, the absolute contrast was correlated with the size of the defects. In the study of CFRP panel, the defects characterization was conducted in a comparable way to evaluate the contrasts with a suitable choice of undamaged reference area.
The comparison of the proposed technique with the traditional pulsed technique has demonstrated the real potential of application on SMArt GFRP and aeronautical CFRP elements with the same level of reliability as traditional thermographic techniques. However, the induction technique has shown better ability to detect small real defects more accurately in CFRP panels. However, the limitation of the power generator and the size of the inductor could limit the detectability of defects.

Keywords: Composite material, SMArt thermography, Pulsed thermography, Defect detection, Electromagnetic Induction

1. Introduction

Defects can be present in composite material deriving from the production process or they can occur during the normal service life due to the continuous stresses or other factors that determine their degradation over time. Therefore, it becomes essential to develop new techniques for controlling the structural health state. In this regard, the authors have proposed a valid in-situ inspection technique based on the integration of a grid of electrically conductive Shape Memory Alloy (SMA) wires into the GFRP glass fiber composite to create a multifunctional SMArt material [1]. The integrated wire grid allows a lightning protection function, which is fundamental for the application of this material for the realization of wind blade. Moreover, using the integrated wire grid as an internal heat source constitutes

Content from this work may be used under the terms of the Creative Commons Attribution 3.0 licence. Any further distribution of this work must maintain attribution to the author(s) and the title of the work, journal citation and DOI.
Published under licence by IOP Publishing Ltd

an alternative for the application of the NDT control using an active thermographic technique, which usually requires halogen or flash lamps as external heat sources.

The induction heating process is a no-contact process to heat an electrically conductive metal or CFRP composite by generating eddy currents through the physical principle of electromagnetic induction. Since the generated eddy current flows against the resistivity of the metal, due to the Joule heating principle, heat is generated in the material [2].

The first industrial application of induction heating as heat source for thermographic control was the detection of surface cracks on steel bars and billets during the movement of the latter through an induction coil and evaluating the signal by means of four infrared scanners placed around the bars [3].

In the literature, inductive thermography is one of the non-destructive methods of active thermography studied and developed over the years by various research groups [4-10]. For example, Busses' research group studied induction thermography to detect defects in steel and CFRP composite specimens with both the Lock-in and Burst techniques, based on pulse heating [11]. Induction heating is a very efficient technique as it is fast and accurately controlled [12, 13]. The limitation of the technique is that the material must be electrically conductive, excluding for example the examination of GFRP panels.

In this work, the application of the induction heating within an active thermographic control has been considered for two case study that are particularly critical: the first one is a unidirectional GFRP panel with integrated artificial defects and a grid of electrically conductive Shape Memory Alloy wires, while the second one is an aeronautical CFRP panel with real production defects with a particularly limited characteristic dimension.

Different inductors geometries were considered varying the shape, size, and number of turns until an optimal solution was found for the correct functioning of the induction heating. The thermal maps were acquired with an experimental setup obtained after several tests carried out by varying the heating time, the acquisition time and the power of the induction generator.

The acquired thermal maps were processed in Matlab environment by means of an appropriate custom analysis algorithm for the computed of the classical thermal parameters of absolute and normalized contrast.

In the case of the CFRP panel, the characterization of the defects was studied by evaluating the thermal contrasts with an appropriate choice of the reference area not damaged during the transient cooling phase, evaluating the influence of the heating time on the results of the thermal contrast.

2. Materials and experimental methods

The inductor used for the experimental tests consists of a 3 mm diameter copper tube that has been designed specifically for each field of application. The geometry, length and number of turns are the main parameters that affect the efficiency of the technique. Initially, given the extension of the panel, it was decided to model the inductor with a "coil" type geometry that was able to heat a larger area of the panel, but this last geometry was not found to be functional because the control system detected a series of errors. Subsequently, the optimal solution was found by making a spiral inductor consisting of only 3 turns, with a width of 25 mm comparable with the greater size of the defects incorporated within the GFRP panel. The latter was subsequently welded to the heating head as shown in Figure 1.

Since the limited dimensions of the inductor allow only localized heating, a horizontal and vertical movement system for the panel was provided in the experimental setups to ensure complete inspection of the same. By moving the panel horizontally and then vertically, several tests have been performed; the first ones are aimed at defining the optimal setup for the detection of defects, improving the setup in terms of power generator, the distance between coil and panel and between IR camera and panel.

Full-field thermographic inspections of the SMArt GFRP and CFRP panels were carried out using a FLIR A655sc microbolometer thermal camera with a highly reliable uncooled sensor and a resolution of 640 × 480 pixels of high quality and stability of measurement, having a thermal resolution of 30 mK.

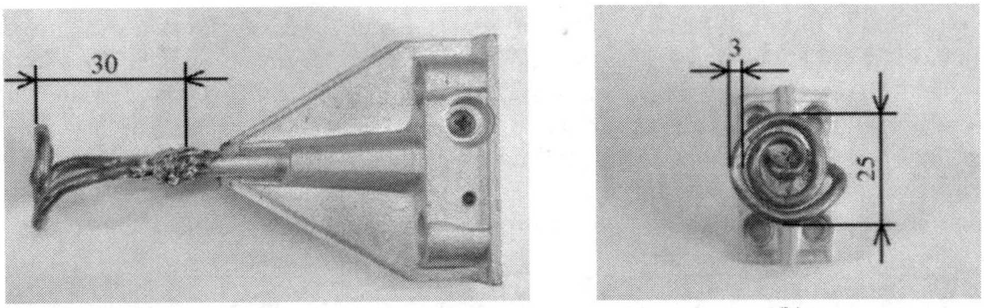

(a) (b)

Figure 1. Heating head and inductor geometry (Dimensions in mm): Front view (a) and lateral view (b).

Each acquisition was subsequently exported in .mat format to make it manipulable in the Matlab environment, through which the post-processing of the thermal maps recorded in each test was carried out. Absolute C_a and normalized C_n thermal contrasts were used to study the detectability of defects by means of induction thermography.

A customized post-processing algorithm was developed in Matlab environment for the characterization of defects by loading the three-dimensional matrix of thermal data, returning the selected thermal maps of the sample under investigation for various tests, highlighting various sites of heat accumulation over time. The algorithm provides the temperature variations referred to the defective areas and to the chosen intact area. Subsequently, the routine provides the plots obtained during the cooling phase for the calculation of the normalized contrast C_n and the absolute contrast C_a.

Since it is difficult to obtain uniform heating between the defective zone and the intact one, a single known intact zone has been chosen for the calculation of the absolute and normalized contrast, which is heated for induction under the same conditions as the defective one.

The classical absolute and normalized contrasts [14] are defined by the following equations (1) e (2):

$$C_a(t) = T_{DZ}(t) - T_{IZ}(t) \tag{1}$$

$$C_n(t) = \left(\frac{T_{DZ}(t)}{T_{DZ}(t_0+1)}\right) - \left(\frac{T_{IZ}(t)}{T_{IZ}(t_0+1)}\right) \tag{2}$$

The average temperatures used in these formulas are the temperature in the reference area associated with the defective (DZ) and intact (IZ) zones at a given instant (t) during the cooling phase and (t_0+1) represents the temperature at start of cooling phase.

2.1. First case study: artificial defect detection on SMArt GFRP panel

The first case study concerned a GFRP glass fiber panel representative of the material commonly used for the construction of wind blade. It consists of 8 unidirectional layers with a common orientation of 0°, laminated using a hand lay-up process [15, 16]. The panel has a dimension of 210 x 300 mm^2 and a total thickness of 8 mm. The unidirectional glass fiber available in coils had a grammage of 600 kg / m^2 while the matrix was made of epoxy resin.

Table 1 shows the mechanical and physical properties of the fiber and matrix.

Table 1. Mechanical and physical properties of glass fibre and epoxy resin.

	E-glass fibre	**Epoxy resin EC 130 LV + hardener W340**
Mean diameter [µm]	14	1.14÷1.16
Density [kg/m^2]	600	2900÷3100
Young modulus [MPa]	72500	75÷80
Ultimate stress [MPa]	2150	8.5÷9
Ultimate strain [%]	3.75	1.14÷1.16

The panel has a grid of SMA Shape Memory Alloy) wires made of Flexinol having a diameter of 0.25 mm, positioned between the first and second layer incorporated in the laminate. A series of 8 defects was inserted between the plies. Defects d1, d3, d5 and d7 are located between the 6th and 7th layer at a depth of 4.5 mm from the inspection surface, while the defects d2, d4, d6 and d8 have been inserted between the 4th and 5th layer to a depth of 3 mm from the inspection side (Figure 2b). The artificial defects have been realized in polystyrene and placed between the layers during the lamination process. The dimensions of the defects placed at different depths are indicated in Table 1.

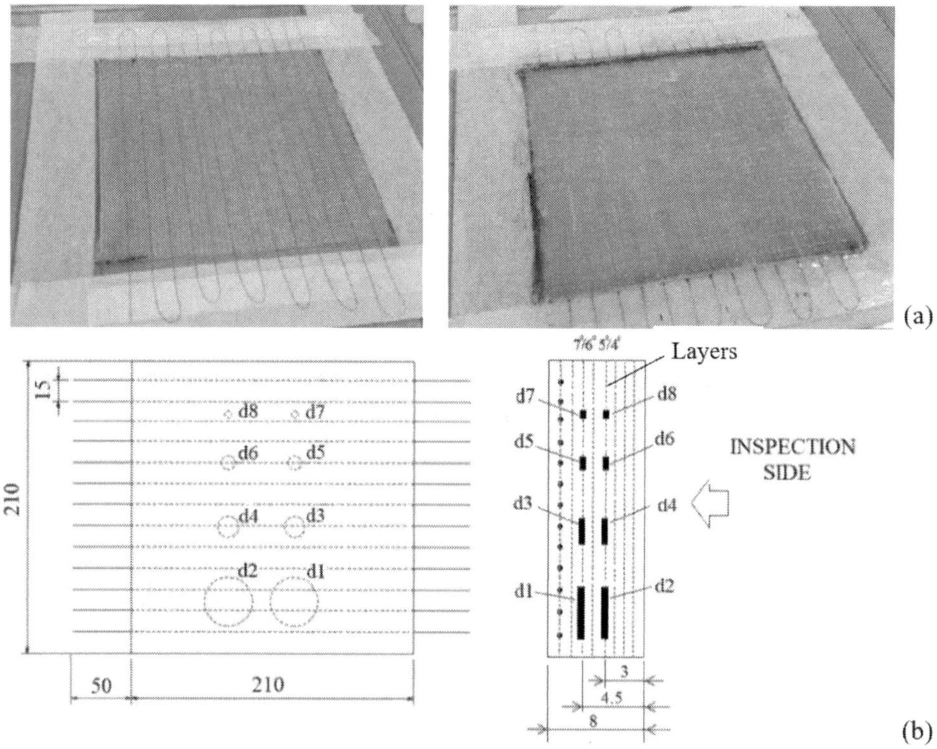

Figure 2. Manufactured GFRP panel with embedded SMA wires and artificial defects (a); geometry and location of defects (dimensions in mm) (b).

Table 2. Geometric data of the defects introduced in the panel.

Defect ID	d1	d2	d3	d4	d5	d6	d7	d8
Diameter d [mm]	25	25	15	15	10	10	5	5
Depth h [mm]	4.5	3	4.5	3	4.5	3	4.5	3

The thermal stimulation system used in the thermographic inspection of SMArt GFRP panel is based on the electromagnetic induction principle: the goal is to generate eddy currents within the SMA wire grid and thus obtain heating by the Joule effect.

The proposed setup for induction heating includes an induction generator (CEIA Power cube 45-A model), a control unit for the programming and automatic management of the heating (Power controller v3 model), with the possibility of setting the power supplied by the generator, the frequency, the voltage and the current on the inductor, a cooling system to prevent the inductor from overheating due to the passage of electric current and a heating head that allows the connection with the inductor.

Adopting the experimental setup show in Figure 3, several tests were carried out with the induction method, identifying the optimal setup for inspections. The best configuration was characterized by a heating time t_h of 60 s, keeping the acquisition time t_{acq} constant at 120 s. Considering a frame rate of 6.25 Hz, 750 total frames were recorded at the end of each test.

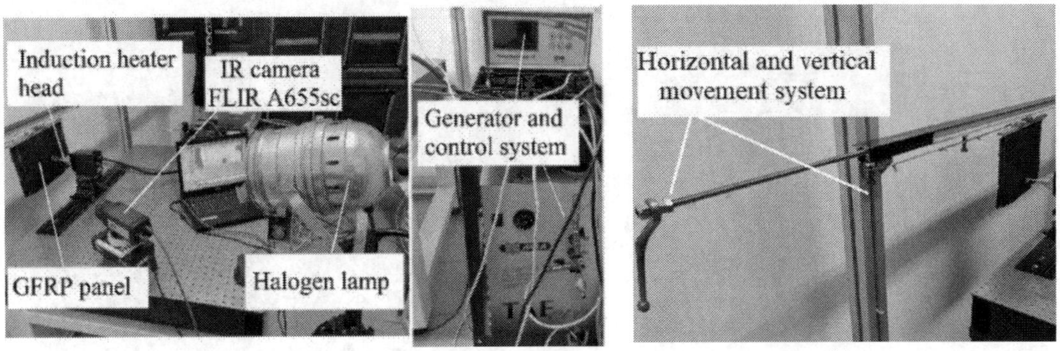

(a) (b)

Figure 3. (a) Experimental setup for induction and pulsed thermography to inspect the SMArt GFRP panel; (b) horizontal and vertical movement system.

The infrared camera and the induction heat source were placed on the same side of the panel to be inspected in reflection mode. In addition, an optimized inductor-panel distance of 2-3 mm and a generator power set at 82% ensure a significant heating of the panel. The IR camera, on the other hand, was placed at 76 cm from the panel to be inspected. The acquisition of thermal images was carried out only during the cooling phase, and the side of the panel to be inspected was painted with a matte black paint to avoid annoying reflections.

In order to compare the results obtained with the induction method, in terms of detecting defects, the traditional pulsed technique was applied using a 1kW halogen lamp located at a distance of 1m from the panel as an external heat source, orienting it in order to have a uniform heating of the panel, maintaining the same experimental setup used in the induction method (Figure 3a). With this last experimental setup, a test with pulsed thermography was carried out using a heating time of 10 s and an acquisition time of 180 s, acquiring 1125 total frames.

The heat input of both techniques is different. In the induction method, the heating takes place from the inside through the heat generated by the SMA wires or by the carbon fiber in the case of the investigation of CFRP aeronautical panels. In the pulsed technique, on the other hand, the heating takes place by adding heat from the external source constituted by the halogen lamp.

2.2. Second case study: real defect detection on aeronautical CFRP panel

A panel of aeronautical derivation laminated in CFRP with high porosity values were investigated with the induction method. The application of the technique, in this case, is possible thanks to the high electrical conductivity of the carbon fibers. The panel with dimension of 400 x 400 mm^2 have been manufactured in a vacuum bag autoclave, inducing porosity through pressure variation during curing cycles (Figure 4a). The cure parameters, stacking sequence and the thickness of the laminated panel are shown in Table 3.

Table 3. Lay-up, number of plies and cure pressure of laminated panel.

Specimen	Thickness [mm]	No. of plies	Pressure [bar]	Lay-up
CFRP-panel	3.023	16	0.4	$[0, +45, -45, 90, 0, +45, -45, 90]_s$

Figure 4b shows the experimental setup used for thermographic tests both with the induction and traditional pulsed thermography method. Also, in this second case of study, the experimental equipment used to conduct thermographic controls consist of halogen lamp, an induction generator, a control unit for managing the heating and a heating head with spiral geometry as illustrated in Figure 1.

(a) (b)

Figure 4. Laminate plate specimen 16 plies (a); experimental setup used in the tests with induction and pulsed thermography (b).

The different tests were carried out with the induction method using two different heating times t_h (3 s and 5 s) and acquisition time of 120 s in different central areas of the panel. The tests were performed using a frame rate of 6.25 Hz. Table 4 summarizes the test parameters used for both tests on 16 plies panel with the induction method.

Table 4. Scheduled of the tests carried out with the induction thermographic method.

Specimen	t_h [s]	t_{acq} [s]	Frame rate [Hz]	Total frames
Panel – 16 plies	3	120	6.25	750
	5	120	6.25	750

The results obtained with the induction method were qualitatively compared with those obtained by applying the pulsed technique, in terms of detectability of the real defects present in the panel. For the tests, a heating time (t_h) of 3s on the 16 plies panel was performed.

The experimental equipment used to conduct pulsed thermographic inspection is the same of the thermographic induction method and consist of halogen lamp 1 kW as an external heat source to obtain a uniform heating of the surface. In thermal tests, the CFRP panel surface was positioned at 1m from the lamp as shown in the setup of Figure 4b. Table 5 summarizes the tests performed on the panel with the pulsed technique.

Table 5. Scheduled of the tests carried out with the traditional thermographic method.

Specimen	t_h [s]	t_{acq} [s]	Frame rate [Hz]	Total frames
Panel – 16 plies	3	180	6.25	1125

3. Results and discussion

3.1. *First case study: SMArt GFRP panel*

Using the experimental setup shown in Figure 3a, a heating time of 60 s was sufficient to detect all defects present in the panel. Figures 5 and 6 show as an example the processing of thermal maps after 8s of cooling from the start of acquisition for defects d2 and d3 with respect to the selected intact area (I_1). From the graphs of Figures 5b and 6b it can be seen, during the cooling phase, that the defect releases heat more slowly than the chosen intact area which instead cools faster even if the latter was subjected to the same heating conditions.

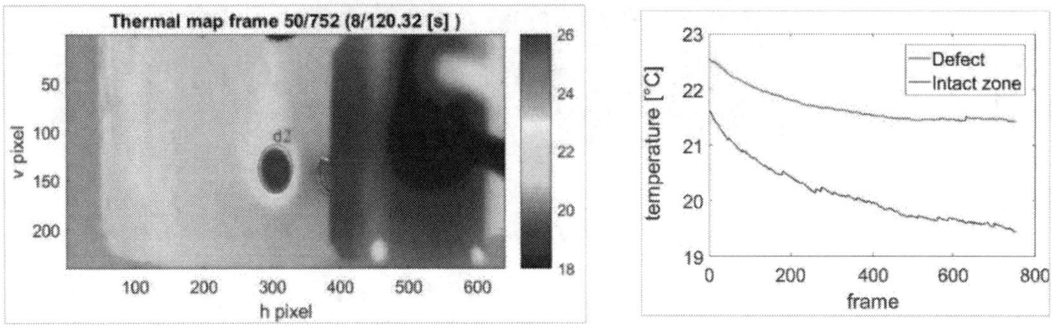

(a) (b)

Figure 5. Example of thermal map processing for defect d2 with induction technique and heating time of 60 s (a); Temperature vs. frames for defect d2 and intact zone (b).

By moving the panel vertically, with the movable crosspiece, it was possible from time to time to inspect the panel horizontally from bottom to top. The defect d3 of 15 mm diameter also showed a lower cooling rate compared to the intact area, heated for the same time of 60 s as can be seen from the thermograms of Figures 6 and 7.

From Figures 5a and 6a it is possible to observe that at the same cooling time of 8 s there was a greater accumulation of heat for the defect d2 compared to the defect d3. This result is justified by the fact that the latter has a smaller size than the d2 defect and is located deeper.

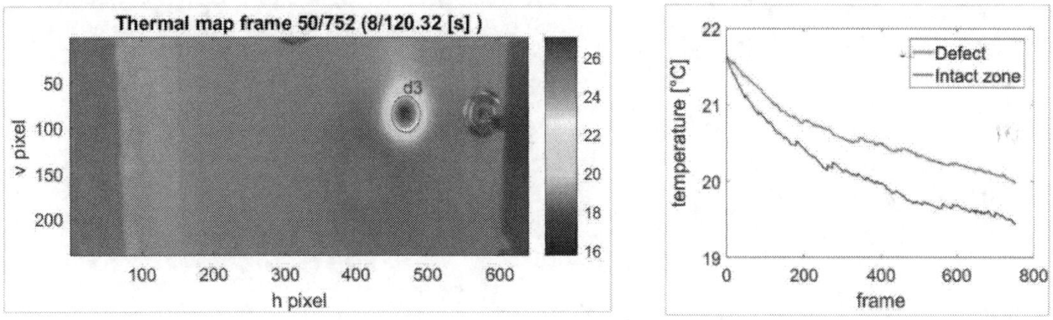

(a) (b)

Figure 6. Example of thermal processing defect d3 with induction technique and heating time of 60 s (a); Temperature vs. frame for defect d3 and intact zone.

By observing the temperature trend of the defective and intact area for defect d3 for the entire acquisition time (Figure 6b), the intact area cools down faster than the defective area. Defect d2 instead shows a significant accumulation of heat in the defect area compared to the undamaged zone already from the beginning of the acquisition, in the cooling phase, as can be seen from the graph in Figure 5b.

Figure 7. Thermal map of the intact area used for processing all defects.

The results obtained by processing the thermal data of all the defects show that the maximum values of the absolute contrast are reached just before the end of the observation time (see Figure 8a) and for this reason it was preferred to avoid longer acquisition times. Furthermore, due to the limited heating of the panel with integrated SMA wires, there is a slow heating and cooling phase. The normalized contrast curves instead show a trend similar to that of absolute contrasts as shown in Figure 8b. The values of the maximum absolute and normalized thermal contrasts C_{a_MAX} e C_{n_MAX} and the respective observation times t_{a_MAX} e t_{n_MAX} have been reported in Table 6.

Table 6. Absolute and normalized contrast for the induction thermography.

			Heating time 60 s			
Defect ID	**Diameter d [mm]**	**Depth h [mm]**	**Absolute contrast C_{a_MAX} [°C]**	**t_{a_MAX} [s]**	**Normalized contrast C_{n_MAX}**	**t_{n_MAX} [s]**
d1	25	3	1.1978	119.36	0.0275	119.36
d2	25	4.5	1.9852	119.04	0.0447	119.04
d3	15	3	0.5112	74.44	0.0112	74.44
d4	15	4.5	0.6676	106.82	0.0001367	106.82
d5	10	3	0.9725	80.95	0.0010	80.95
d6	10	4.5	1.3695	119.04	0.0037	119.04
d7	5	3	1.833	119.04	0.0016	119.04
d8	5	4.5	1.9075	118.88	0.0119	118.88

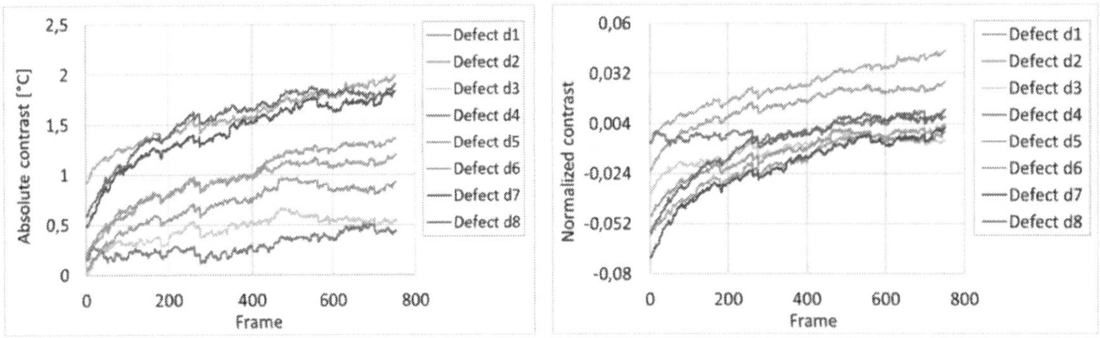

(a) (b)

Figure 8. Absolute contrasts (a) and normalized contrasts (b) (heating time 60s).

Figure 9a shows the thermal map used for thermal processing the defects present in the panel by means of pulsed technique with a heating time of 10 s as indicated above. The choice of the intact zone was made based on the Source Distribution Image (SDI) method generally used in this case to select a defect-free area of the panel that receives the same heat as the corresponding damaged area minimizing errors in the determination of the thermal contrast [17].

The thermal map for the processing of defects was selected based on the best visualization of the defects detected.

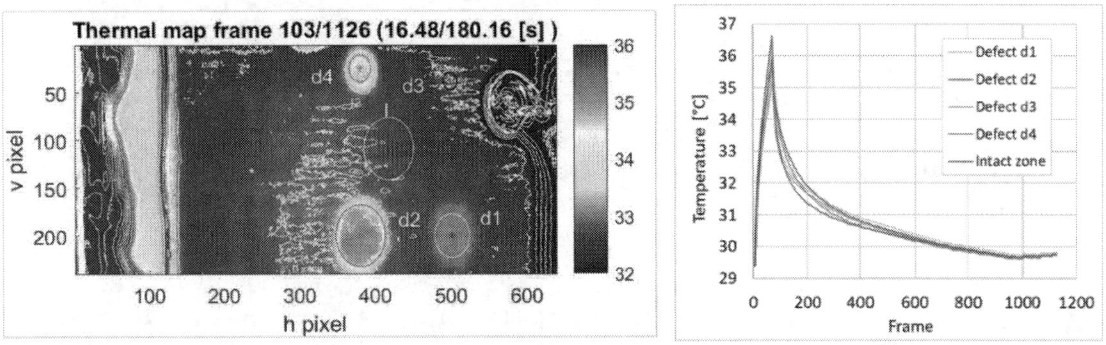

(a) (b)

Figure 9. (a) Thermal map processing of the defects d1, d2, d3 and d4 with pulsed thermography (heating time of 10 s); (b) Trend of the temperature vs. frame for processed defects and intact zone.

By analysing the thermogram of Figure 9a, it is possible to observe that the pulsed technique is able to easily detect artificial defects with a single test because it offers a larger control area of the panel than the induction method for which, given the inductor dimension comparable with the largest defect size of the largest size (25 mm in diameter), has a field of observation reduced to a smaller area. However, it has been observed that despite this limitation, the proposed induction technique allows all defects present in the panel to be detected with the same accuracy and precision. Furthermore, from the tests it was observed that the pulsed technique allows a greater and faster heating of the panel with higher temperatures than the induction technique and this also due to the small size of the SMA wires and the grid pitch. The temperature trend as the number of frames varies is shown in Figure 9b. The graph shows that in the cooling phase, the defects cool more slowly than the only intact area chosen.

The graphs of the absolute and normalized contrasts shown in Figure 10 a-b, show similar trends and are characterized by a low thermal noise compared to that detected with the induction technique. Table 7 reports the absolute and normalized contrast values determined with the traditional technique.

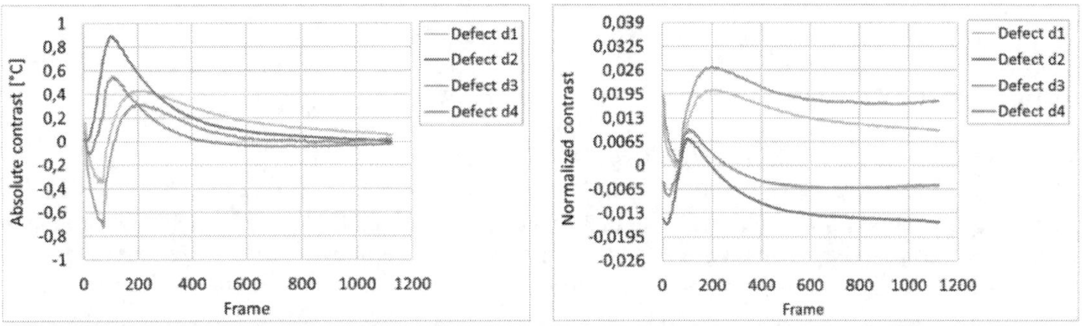

(a) (b)

Figure 10. Trends of absolute (a) and normalized (b) contrasts vs. frames for the artificial defects d1, d2, d3 and d4 determined by processing the thermal map with the pulsed technique.

An interesting observation concerns the possibility of obtaining a qualitative assessment of the depth of the defects. Although the values of the thermal contrast strongly depend on the choice of the defective and intact area, the maximum value of the absolute contrast was higher for the defects placed more on the surface than for those positioned deeper as expected. Furthermore, from the graph of Figure 11a it is observed that for defects of diameter 5, 10 and 15 mm, the absolute contrast seems to decrease with the size of the defects, while for defects with a larger diameter (25 mm) an increase was observed of the absolute contrast values. The general trend of the experimental data suggests that there is no real dependence of the absolute contrast with the size of the defect.

Table 7. Absolute and normalized contrast for the traditional pulsed thermography.

			Heating time 10 s			
Defect ID	**Diameter d [mm]**	**Depth h [mm]**	**Absolute contrast C_{a_MAX} [°C]**	**t_{a_MAX} [s]**	**Normalized contrast C_{n_MAX}**	**t_{n_MAX} [s]**
d1	25	4.5	0.4333	32	0.0206	32
d2	25	3	0.8959	15.84	0.0074	16.16
d3	15	4.5	0.3248	31.68	0.0272	31.68
d4	15	3	0.5588	17.12	0.0101	17.12

By comparing the absolute contrast values obtained with the induction and pulsed method, it is noted that for both techniques the contrast increases as the size of the defects increases for defects of 15 and 25 mm (Figure 11b). Furthermore, for these latter defects, higher absolute contrast values were observed with the induction technique compared to the traditional technique and this could be due to fact that for the same heating time, the induction technique provides a greater heat localized input to the material.

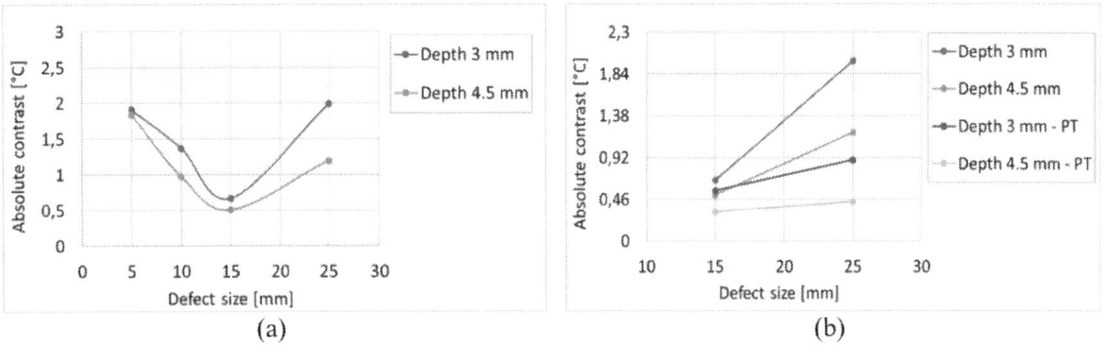

Figure 11. (a) Absolute contrasts vs. defect size for all defects (induction thermography) and (b) comparison absolute contrast vs. defect size for defects d1, d2, d3 and d4 computed with the induction and traditional thermography method (PT: Pulsed Thermography).

3.2. Second case study: CFRP panel with real small defects

In this second study, the induction method was also applied for the analysis of the CFRP aeronautical panel with real defects to evaluate the potential of the proposed method to detect small real production defects. For this purpose, the experimental setup shown in Figure 4b was used. The experimental data were processed using the numerical procedure described above, which provides for the identification of defective areas and those intact appropriately chosen based on the observation of the thermogram at the beginning of the acquisition and based on the SDA method. The processing of the thermal data of the defects detected with the induction method for the 16 plies panel and a heating time of 3s is shown in Figure 12a. The thermal map highlights small real defects concentrated in a limited central area of the panel as shown in Figure 12b.

The thermographic survey with the proposed method was subsequently extended to another area of the 16 plies panel (see Figure 12d). From the thermal map shown in Figure 12c, a small porosity is observed after a cooling time of 5.6 s from the start of the acquisition.

IOP Conf. Series: Materials Science and Engineering 1275 (2023) 012015 doi:10.1088/1757-899X/1275/1/012015

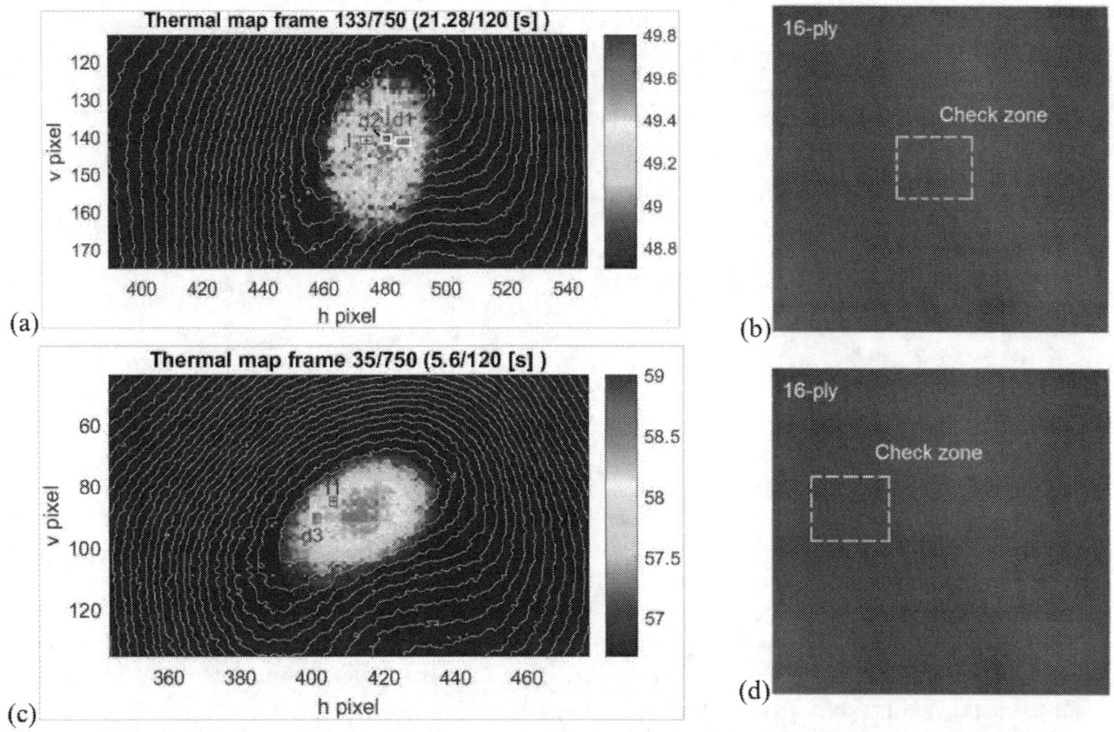

Figure 12. Defects processing at 21.28 s (a) and 5.6 s (c) of cooling with the induction method with a heating time of 3 s and 5 s respectively on the 16 plies panel; (b), (c) Location of the control area on the 16 plies panel.

Observing the graphs showing the temperature trends as the number of frames varies (Figure 13a-b) for defects d1, d2 and d3, it is possible to observe the slight differences in the temperature profile during the cooling phase between the defective areas and those intact choices. From the graphs it is also observed that the defects tend to cool down more slowly than the respective intact areas.

The trends of the absolute and normalized contrasts over the entire acquisition time for the processed defects are shown in Figure 14a-b. From the trend of the absolute contrast graph, it was observed that the defects d2 and d3 have the maximum values of the absolute contrast compared to that detected for the defect d1. For the defect d3, the maximum absolute thermal contrast, equal to 0.48 °C is obtained at 14.24 s from the start of the acquisition, in the cooling phase.

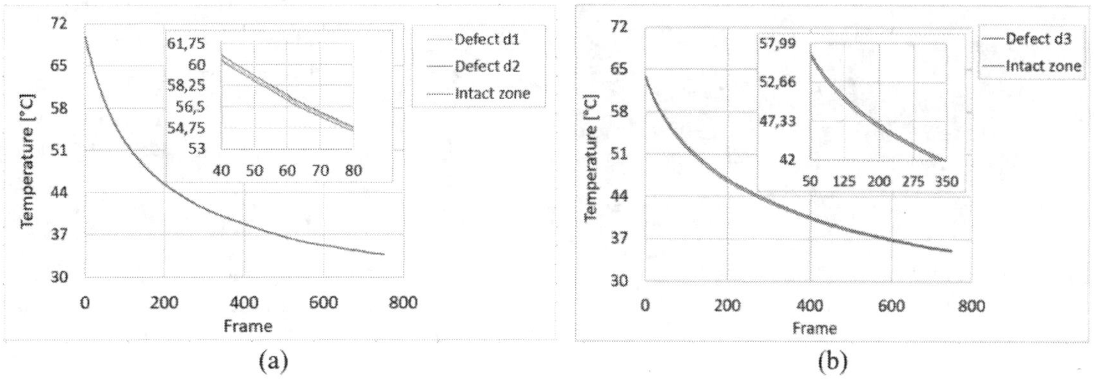

Figure 13. Temperature trends vs. frames in different zones of the 16 plies panel for defects d1 and d2 (a) and for defect d3 (b) (induction method).

The thermal contrasts showed a smooth behavior for all small real defects processed. The trend of the absolute and normalized thermal contrast curves is also affected also in this case by a thermal noise due to small contrast values generated by the defects compared to the intact area which are close to the sensitivity of the infrared camera sensor (Figura 14a-b).

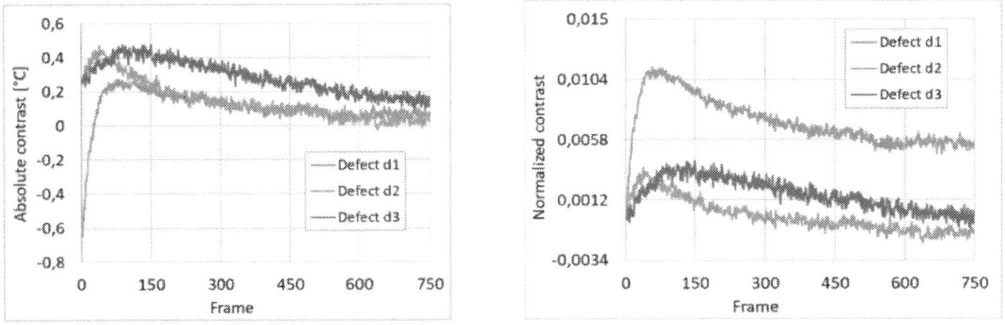

(a) (b)

Figure 14. Trends of the absolute (a) and normalized (b) contrasts for the real defects d1, d2 and d3 with the induction thermographic method for the 16 plies panel.

Finally, a qualitative comparison was made with the pulsed thermographic technique for a heating time of 3s to highlight the potential of both techniques in terms of detecting defects, always using the same thermographic setup, as shown in Figure 4b.

From the experimental results obtained by processing the defects detected on the basis of the SDI method, it was observed that unlike the GFRP panel with integrated wires and artificial defects, the pulsed technique (see Figure 15) highlights areas of heat concentration in patches by placing in evidence macro-areas of diffuse defects distributed on the surface of the panel in a large area, while the proposed induction technique seems to be more effective by locating smaller defect areas more accurately and therefore concentrating the control on small areas of interest, obtaining in the inspected areas a greater resolution of smaller defects such as porosities, as evidenced by the thermal maps shown in Figures 12a-b and 15 relating to the processing of real defects with both techniques.

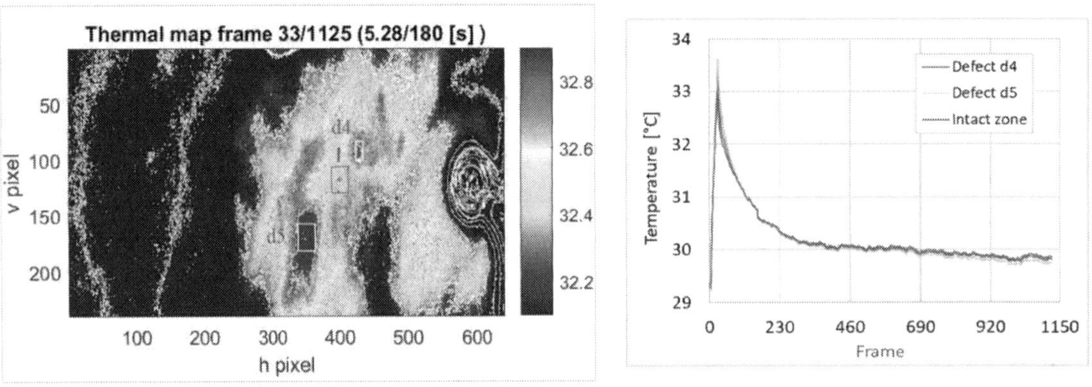

(a) (b)

Figure 15. Defect processing on the thermal map at 5.28 s from the beginning of cooling phase with the traditional method (heating for 3s) on the 16 plies panel; (b) Trend of the temperature vs. frame for processed defects and intact zone.

IOP Conf. Series: Materials Science and Engineering 1275 (2023) 012015 doi:10.1088/1757-899X/1275/1/012015

The trends of the absolute thermal contrasts detected with the pulsed technique are shown in Figure 16a-b. From the graph of Figure 16a, it is observed that the defect d5 has a higher absolute contrast than the maximum contrast detected for the defect d4.

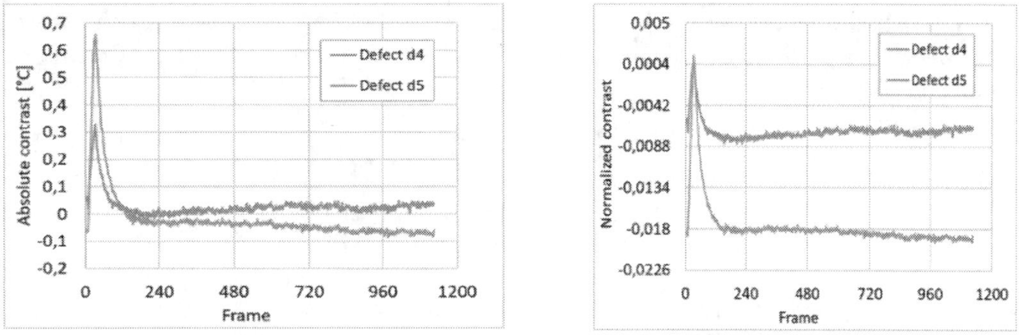

(a) (b)

Figure 16. Absolute (a) and normalized (b) contrasts as the number of frames varies for defects d4 and d5 determined with the pulsed thermography method for the 16 plies panel (heating for 3s).

The values of the absolute and normalized contrasts, experimentally determined for both the induction and pulsed techniques are summarized in Table 8 for the case of the 16 plies panel.

Table 8. Absolute and normalized contrast for the induction and pulsed thermography (16 plies panel).

INDUCTION THERMOGRAPHIC TECHNIQUE (CFRP - 16 plies – t_h = 3 s)				
Defect ID	**Absolute contrast C_{a_MAX} [°C]**	**t_{a_MAX} [s]**	**Normalized contrast C_{n_MAX}**	**t_{n_MAX} [s]**
d1	0.2724	12.96	0.0114	9.28
d2	0.4730	6.24	0.0037	6.24
d3	0.4779	14.24	0.0042	24
PULSED THERMOGRAPHIC TECHNIQUE (CFRP – 16 plies – t_h = 3 s)				
Defect ID	**Absolute contrast C_{a_MAX} [°C]**	**t_{a_MAX} [s]**	**Normalized contrast C_{n_MAX}**	**t_{n_MAX} [s]**
d4	0.3290	4.96	0.014	4.96
d5	0.6572	5.12	0.0011	5.12

4. Conclusions

In this work, the induction thermographic technique was proposed for the non-destructive investigations of a glass fiber panel, with an integrated grid of shape memory alloy wires and artificial defects at different depths. Moreover, an aeronautical CFRP carbon fiber panel containing small real manufacturing defects were also analyzed.

The experimental results showed that the induction thermographic technique allows to detect the artificial defects on the SMArt panel with the same level of reliability as the pulsed technique, even if the latter in this case had the advantage of being able to inspect a larger area of the panel with a single test in relation to the size of the defects. In the case of the investigation of CFRP aeronautical panel with small real production defects, induction thermography has presented a considerable advantage by obtaining on the one hand a higher heating of the panel with the same heating time compared to the pulsed one thanks also to the high electrical conductivity of the carbon fiber and on the other hand demonstrated a greater sensitivity of defect detection, with a higher resolution than the pulsed technique, by localizing the heating in a smaller area of the panel. In conclusion, induction thermography has shown good application potential on multifunctional fiberglass composites and CFRP aeronautical elements with the same reliability as traditional thermographic techniques, although the limitation of the generator and the size of the inductor could limit the detectability of the defects. Furthermore, the proposed technique showed improved sensitivity of accurately detecting real small defects in CFRP panels.

Acknowledgments
The authors thank the student Maria Rosaria Petracca for her contribution to the research carried out during the experimental thesis in machine construction at the University of Salento - Department of Engineering for Innovation.

References

[1] Gibson R F 2010 A review of recent research on mechanics of multifunctional composite materials and structures *Compos. Struct.* **92 (12)** 2793–2810
[2] Bayerl T, Duhovic M, Mitschang P, Bhattacharyya D 2014 The heating of polymer composites by electromagnetic induction *A review Composites Part A: Applied Science and Manufacturing* **57** 27-40
[3] Wally G, Oswald-Tranta B 2007 The influence of crack shapes and geometries on the results of thermos inductive crack detection *In Proceedings of the SPIE Thermosense XXIX* **6541** 9-12
[4] Shiratori M, Miyoshi T, Nakanishi T, Noda T, Harada M 1990 Detection of Cracks and Measurement of Stress Intensity Factors by Infrared Video System *JSME Int. J.* **33** 400-408
[5] Riegert G, Zweschper Th, Busse G 2004 Lockin Thermography with Eddy Current Excitation *QIRT Journal* **1** 21-32
[6] Zenzinger G, Bamberg J, Dumm M, Nutz P 2005 Crack Detection using EddyTherm AIP *Conference Proceedings* **760** 1646-1653
[7] Oswald-Tranta B, Wally G 2005 Thermo-Inductive Investigations of Steel Wires for Surface Cracks SPIE **5782** 245-254
[8] Benzerrouk S, Ludwig R 2007 Infrared Detection of Defects in Powder-Metallic Compacts Journal of NDE **26**, 1-9
[9] Vrana J, Goldammer M, Baumann J, Rothenfusser M, Arnold W 2008 Mechanisms and Models for CrackDetection with Induction Thermography *AIP Conference Proceedings* **975** 475-482
[10] Balaji L, Balasubramanian K, Krishnamurty C 2013 Induction thermography for non-destructive evaluation of adhesive bonds *Proc. 39th RQNDE, AIP conf. proc.* **1511** 579-586
[11] Srajbr C 2016 Induction Excited Thermography in Industrial Application *in Proceedings of the 19th Word Conference on Non-Destructive Testing*, Munich, Germany
[12] Lucía O, Maussion P, Dede E, Burdío J M 2014 Induction Heating Technology and its Applications: Past Developments, Current Technology, and Future Challenges *IEEE Trans. Ind. Electron.* **61** 2509-2520
[13] Oswald-Tranta B 2018 Induction Thermography for Surface Crack Detection and Depth Determination *Applied sciences* **8(2)** 257
[14] Dattoma V, Nobile R, Panella F W, Saponaro A 2018 NDT thermographic techniques on CFRP structuralcomponents for aeronautical application *Procedia Structural Integrity* **8**, 452–461
[15] De Giorgi M, Nobile R, Saponaro A 2020 Numerical and experimental validation of SMArt thermography for the inspection of wind blade composite laminate *SN Applied Science* **2** 1662
[16] De Giorgi M, Nobile R, Saponaro A 2019 Experimental evaluation of artificial defects using SMArt thermography *Procedia Structural Integrity* **24** 866-874
[17] Susa M, Maldague X, Boras I 2010 Improved Method for Absolute Thermal Contrast Evaluation Using Source Distribution Image (SDI) *Infrared Physics & Technology* **53 (3)** 197–203

Thermal analysis under static and fatigue conditions on HDPE

D Santonocito*, F Cucinotta, E Guglielmino and G Risitano

University of Messina, Department of Engineering, Contrada di Dio, 98166 Messina Italy

*email: dsantonocito@unime.it

Abstract. High Density Polyethylene (HDPE) represents one of the best choices for water and gas distribution pipeline thanks to its light weight and good mechanical performance. One of the strengths of this plastic material is the possibility to join it via welding adopting a procedure according to the standard. However, the welding procedure can severally alter the mechanical performance of the material under service loads. In the present work comparison between virgin and welded HDPE specimens is carried out adopting rapid approaches, such as the Risitano's Thermographic Method and the Static Thermographic Method, monitoring the evolution of the specimen's superficial temperature under static tensile and fatigue tests. Mechanical performance has been compared, as well as the fatigue limit obtained with traditional fatigue tests, showing good agreement. Reliable fatigue data have been obtained in a very short amount of time and with few specimens by adopting Energy Methods.

1. Introduction

High Density Polyethylene (HDPE) is one of the most adopted materials for the realization of pipeline for gas and water distribution. Its strength points are the low cost and light weight if compared to steel pipes. The joining of two different sections of pipe, thanks to their viscoplastic nature, is very easy. Indeed, they can be welded by heating the pipe extremities and applying a pressure for a prescribed time.

In literature, several authors have investigated the mechanical properties of HDPE. In particular, for this kind of material the correlation between the slow crack growth and the chemical composition has been studied [1]. Some authors studied the fatigue properties of HDPE developing damage models [2,3]. To shorten the required test time, Risitano and Santonocito applied infrared thermography to investigate the fatigue life of PE100.

Few studies regarding the mechanical properties of welded HDPE exist in literature. These studies demonstrated how the welding process lead to a restructuring of the crystalline phases, with an increment of the mechanical properties [4–7].

To assess the fatigue properties of welded details compared to the base material, in this work infrared thermography has been applied to specimens of HPDE tested under static tensile and fatigue loading conditions. The Static Thermographic Method (STM) [8] has been applied to identify the limit stress of the material during static tensile tests. It is the macroscopic stress level that causes irreversible damage within the material. The Risitano's Thermographic Method (RTM) [9,10] has been applied to derive the fatigue limit of the HDPE rapidly and with few specimens. Moreover, to verify the findings of the

Content from this work may be used under the terms of the Creative Commons Attribution 3.0 licence. Any further distribution of this work must maintain attribution to the author(s) and the title of the work, journal citation and DOI.
Published under licence by IOP Publishing Ltd

Thermographic Methods, constant amplitude fatigue tests have been performed on the same specimen's geometry in order to obtain the SN curve and the fatigue limit of the plain and welded HPDE.

2. Theoretical background

In this section a short overview of the main Thermographic Methods is provided. For more information we remand to specific paper.

2.1. Risitano's Thermographic Method

By observing the temperature evolution of a specimen subjected to cyclic loading condition above its fatigue limit, it is possible to observe three phases [9]. The first phase is characterized by the increment of the temperature. The second phase is characterized by a constant value of the temperature signal in the time domain (Figure 1a). Risitano correlated this temperature value to the onset of fatigue damage within the material. As the applied stress level increase, the stabilization temperature, ΔT_{st}, increase, but the integral of the temperature vs. number of cycles curve remain almost constant. This value, the Energy Parameter Φ, can be related to the dissipated energy of the material.

It is possible to apply on a single specimen, in a stepwise way, a series of stress levels and record the relative stabilization temperatures (Figure 1b). By plotting the stabilization temperatures vs. the applied stress levels, it is possible to recognize the fatigue limit of the material as the knee region of that curve. Exploiting the constancy of the Energy Parameter, it is also possible to rapidly evaluate the SN curve of the material adopting a single specimen. For more information about the Risitano's Thermographic Method we remand to [9,10].

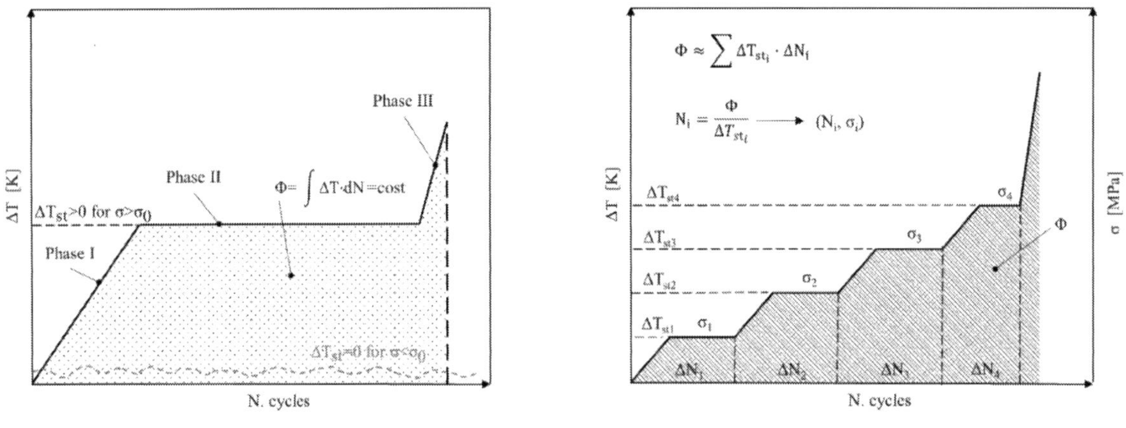

(a) (b)

Figure 1. a) Temperature trend during a constant amplitude fatigue test; b) Temperature trend during a stepwise fatigue test.

2.2. Static Thermographic Method

The main idea of the Static Thermographic Method is to assess the end of the thermoelastic behaviour of a material during a static tensile test.

Risitano and Risitano [8], in 2013, observed how the temperature signal during a static tensile test shows three different phases. Phase I is characterized by a linear decrement of the temperature, due to the thermoelastic law of Lord Kelvin. Phase II is characterized by a temperature decrement but with a different trend compared to the Phase I. At the end of Phase II, the temperature reaches a plateau region, which correspond to the yielding stress of the material. In Phase III, the plastic damage is more predominant compared to the elastic behavior, hence the temperature begins to rise up to the specimen failure.

The transition point between Phase I and Phase II correspond to the onset of irreversible damage within the material. If we are able to assess this point in the temperature signal, we can correlate it to a macroscopic stress, defined as the "limit stress", σ_{lim}, which produces an irreversible energy release due to the increment of plastic damage. For more information about the STM, we remand to [8,11–13].

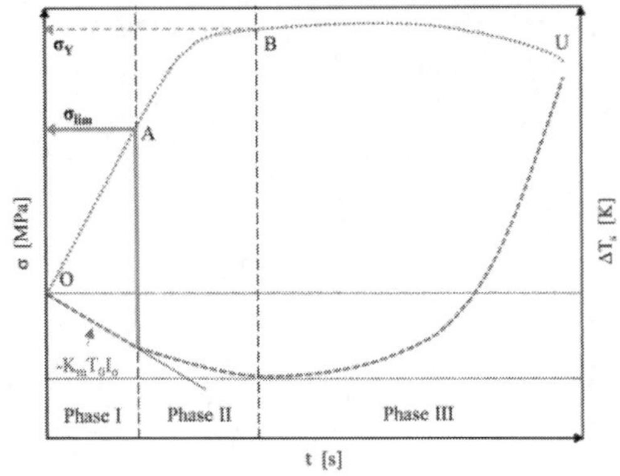

Figure 2. Temperature trend during a static tensile test. Phase I is characterized by a linear decrement due to the thermoelastic constant K_m and the first stress invariant I_σ.

3. Materials and Method

Specimens of HDPE (class PE100, Hostalen CRP 100 Black) were obtained from a pipe section by cutting them according to the ISO 527-2 standard. Two kinds of specimens were tested: the "Plain" one and the "Welded" one. Welded specimens were retrieved from pipe section were the welding procedure, according to the Italian standard UNI 10520, was performed. The welding bead was removed by milling. The nominal cross section of the specimens is equal to 10x4 mm^2.

The specimens were tested adopting a servo hydraulic machine ITALSIGMA 25 kN. Static tensile tests were performed on three specimens per type, adopting a crosshead speed of v= 5 mm/min. Engineering strains were measured adopting an extensometer with an initial gauge length of L_0= 50 mm.

Fatigue tests, both Constant Amplitude (CA) and Stepwise, were performed adopting a stress ratio of R= 0.1 and testing frequency of f= 1Hz, to prevent an excessive self-heating of the specimens. The tests were performed under stress control, starting from 10 MPa up to 17 MPa, with a number of cycles per block equal to 3000. The infrared camera FLIR A40 was adopted to monitor the temperature evolution during the static tensile tests and the fatigue tests. A sampling frequency of 5 Hz was adopted.

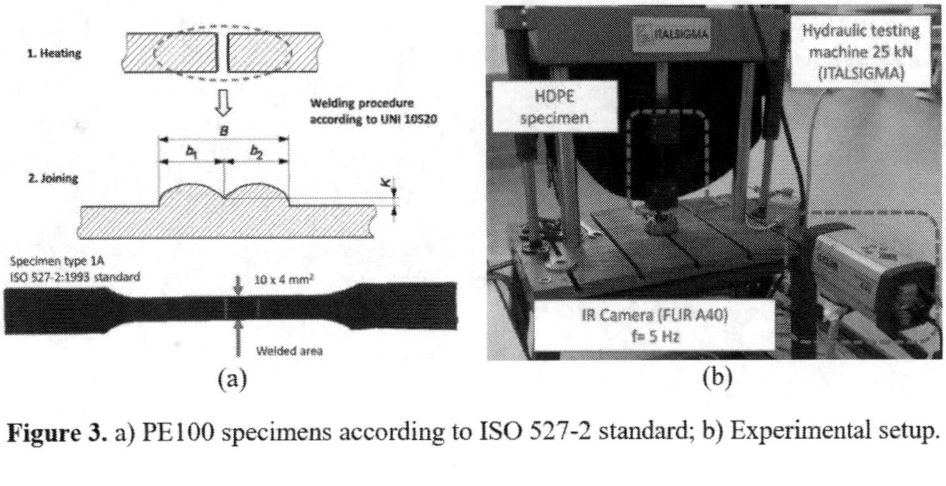

Figure 3. a) PE100 specimens according to ISO 527-2 standard; b) Experimental setup.

4. Results and discussion

4.1. Static tensile test – Mechanical properties

Static tensile tests were performed under displacement control, adopting a speed of 5 mm/min, as prescribed from ISO527-1 standard.

The stress has been evaluated as the ratio between the force and the specimen's cross section. Figure 4 reports the engineering curves of the tested specimens, both plain and welded. The plain PE100 (Figure 4a) exhibits a typical behaviour of plastic materials with a yielding point (i.e. the maximum stress value). After that point, necking begins in the specimen and the stress reduces as the strain increases. For a strain value of $\varepsilon \approx 0.34$, the stress level became almost constant. This stress level is called plastic flow stress.

Welded PE100 (Figure 4b) shows the same behaviour of the plain PE100. The yielding stress is reached for the same stress and strain levels; however, the plastic flow stress has been reached for lower strain values compared to the plain PE100 ($\varepsilon \approx 0.26$).

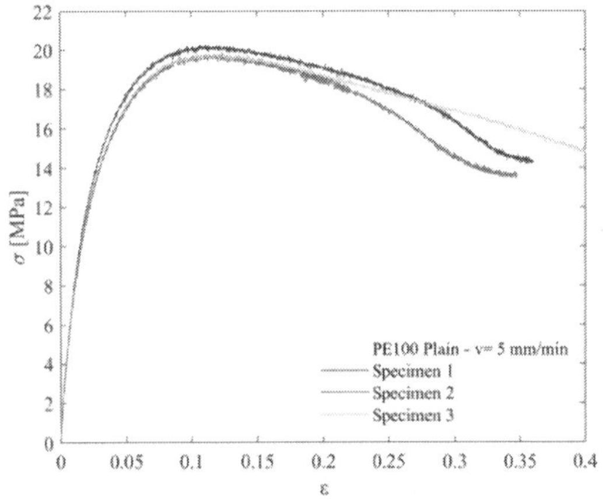

(a)

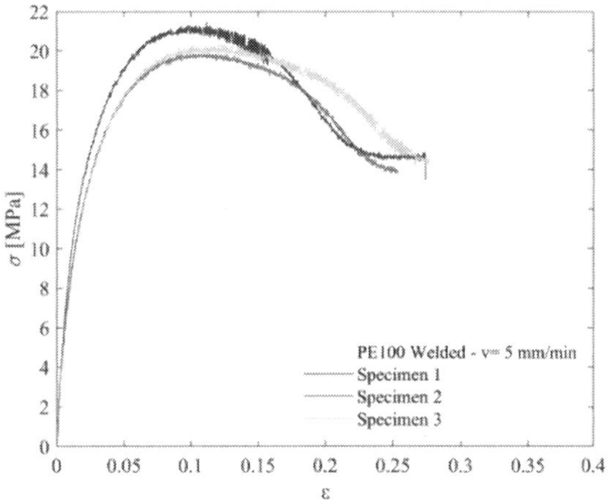

(b)

Figure 4. Engineering curve of: a) Plain PE100; b) Welded PE100.

Young's Modulus has been evaluated, according to the ISO527 standard, as the linear regression of the stress vs. strain within the strain levels of $\varepsilon_1 = 0.0005$ e $\varepsilon_2 = 0.0025$. Table 1 reports the value of the Young's Modulus and the yielding stress of PE100. For the plain PE100, average values of $E = 1018.9 \pm 33.9$ MPa and $\sigma_Y = 20.0 \pm 0.3$MPa have been calculated. These values are in good agreement with the manufacturer datasheet ($E = 850$ MPa and $\sigma_Y = 22$ MPa), even if Young's Modulus is higher. The same considerations apply for the welded PE100 ($E = 1155 \pm 124$ MPa and $\sigma_Y = 20.6 \pm 0.8$ MPa).

There are no excessive deviations of the mechanical properties of welded PE100 compared to the plain PE100, although a greater standard deviation is noticed with respect to it. The values obtained are consistent with the datasheet values of the material as regards the Young's Modulus, while they are slightly lower as regards the yield stress.

The welded PE100 specimens showed the formation of the necking always in areas of the gauge section not included in the zone involved in the welding process, to indicate that the welding zone does not represent a possible site trigger of failure.

IOP Conf. Series: Materials Science and Engineering 1275 (2023) 012016 doi:10.1088/1757-899X/1275/1/012016

Table 1. Comparison of the mechanical properties of PE100 specimens.

Specimen type	No. Specimen	E [MPa]	σ_y [MPa]	E [MPa]	σ_y[MPa]
	1	1055	20.3		
PE100 Plain	2	1014	19.9	1019 ± 34	20.0 ± 0.3
	3	988	19.8		
	1	1288	21.5		
PE100 Welded	2	1133	19.9	1155 ± 124	20.6 ± 0.8
	3	1044	20.4		

4.2. Static tensile test – Temperature trend

During the static tensile tests, the evolution of the superficial temperature of the specimens has been performed adopting an infrared camera. The temperature of a rectangular spot, placed on the specimen's gauge section, has been monitored and the maximum value has been retrieved and filtered with a *rlowess* filter (data span of 5%). The filter has been adopted to highlight the linear trend of the temperature signal and to remove possible outlier.

The temperature signal has been plotted vs. the applied stress level respect the test time. Figure 5 reports the results for a test performed on plain and welded PE100 specimens. It is possible to distinguish the first linear elastic phase (phase I), followed by a different phase (phase II), where the temperature signal reaches a minimum value and then it begins to increase. By performing the linear regression of the temperature points of phase I (ΔT_1 points) and phase II (ΔT_2 points), and making their intersection, it is possible to assess the limit stress of the material.

For the plain PE100, an average value of σ_{lim}=15.2 MPa has been estimated, while σ_{lim}= 14.2 MPa has been estimated for the welded PE100. Such macroscopic stress level activates in the material some process that lead to irreversible energy release compared to the thermoelastic behaviour, where all the energy is recovered.

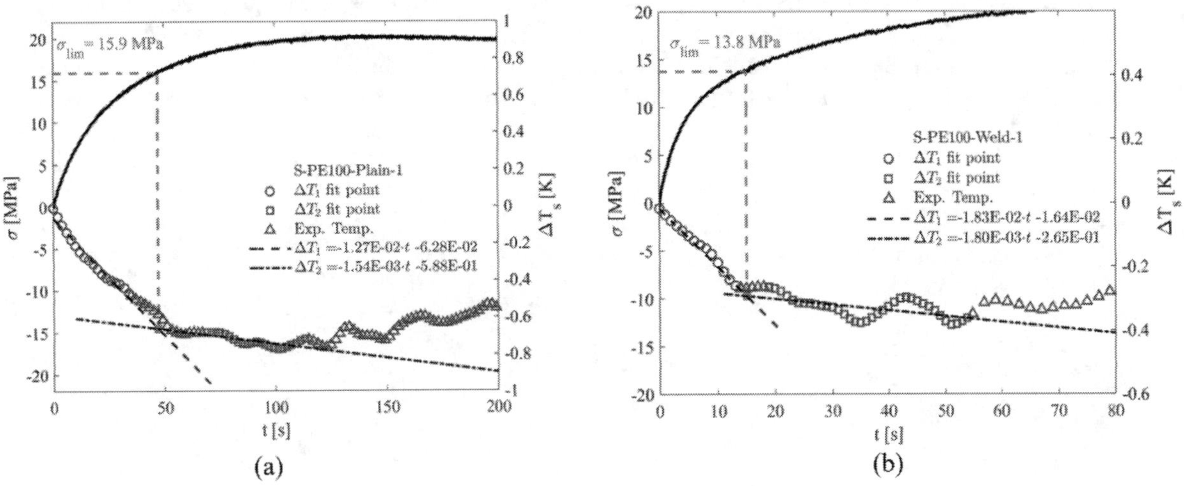

Figure 5. Temperature trend vs. applied stress level during a static tensile test of: a) Plain PE100; b) Welded PE100.

4.3. Stepwise fatigue test

Stepwise fatigue tests have been performed on a specimens per type to assess the fatigue limit with the RTM. The stress ratio has been R= 0.1, while the testing frequency has been set to f= 1 Hz to prevent excessive self-heating of the polyethylene. The superficial temperature signal has been recorded respect the number of cycles and the applied stress level as already done for static tensile tests.

Figure 6 shows the temperature evolution of a stepwise fatigue test, both for plain and welded PE100. At the beginning of the test, the temperature rapidly increases and then stabilizes. It is possible to assess an average value of the temperature signal for the superior applied stress level (σ_{sup}). For the plain PE100 specimen, a further increase of the temperature can be noticed after that 15 MPa stress level has been passed, while, for the welded specimen, it can be noticed after that 14.3 MPa stress level has been passed.

After that these stress levels have been passed, fatigue damage is produced within the material. For lower stress levels, the temperature increments are strictly related to the intrinsic self-heating of the material due to its viscoplastic behaviour.

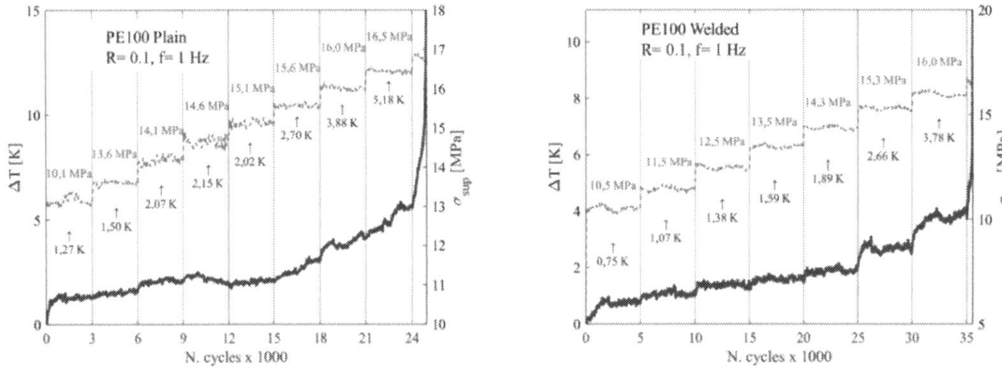

(a) (b)

Figure 6. Stepwise fatigue test of: a) Plain PE100; b) Welded PE100.

By reporting the stabilization temperature vs. the applied stress levels, it is possible to observe a bilinear curve (Figure 7). The first region is characterized by stress levels below the fatigue limit of the material; on the other hand, the second region is characterized by stress levels above the fatigue limit. The knee region is where fatigue damage begins. By performing two linear regressions, it is possible to assess the fatigue limit of the PE100 by RTM. For the plain PE100, a value of 15.5 MPa has been estimated, while 14.4 MPa has been estimated for the welded specimen.

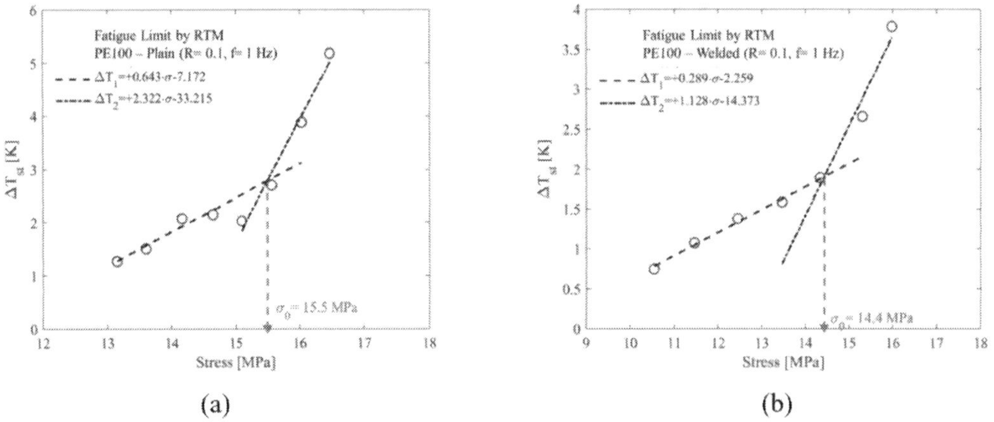

(a) (b)

Figure 7. Fatigue limit by RTM of: a) Plain PE100; b) Welded PE100.

IOP Conf. Series: Materials Science and Engineering 1275 (2023) 012016 doi:10.1088/1757-899X/1275/1/012016

Figure 8 shows the evolution of the damage in a welded PE100 specimen during the stepwise fatigue test. The extensometer has been placed within the welding zone. In the first frame (Figure 8a) no damage is highlighted by the IR camera, while in the second frame (Figure 8b) a slight temperature increment has been noticed which evolutes up to the specimen's failure (Figure 8c).

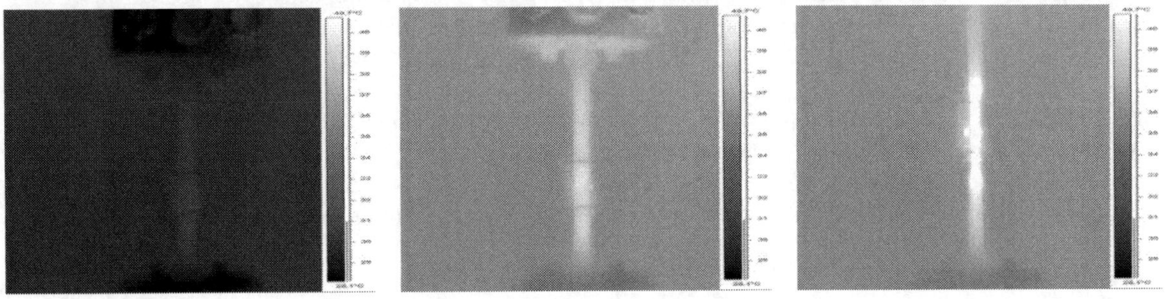

(a) (b) (c)

Figure 8. Damage evolution during a stepwise fatigue test on a welded PE100 specimens. The IR scale is the same for all the frames.

4.4. Constant Amplitude fatigue test

To verify the findings of the Thermographic Methods, a CA fatigue test campaign has been performed on plain and welded PE100 specimens adopting the same stress ratio and test frequency of the stepwise tests. The applied stress levels have been reported vs. the number of cycles to failure in a bi-logarithmic diagram. The scatter band has been estimated with probability of survival (PS) of 10 and 90%, as well as the inverse slope k and the scatter index T_σ. Run-out tests have been considered after that the specimens reaches $1x10^5$ cycles without any kind of failure.

The S-N curves of the plain PE100 are shown in Figure 9. It is possible to estimate the inverse slope of this curve by performing a linear regression of the log of the stresses against the log of the number of cycles. For this material type, a value of k= 10.72 has been obtained, while k= 13.78 for the welded PE100. The fatigue limit estimated at $4x10^4$ cycles for a probability of survival equal to 50%, is $\sigma_{0,\,50\%}$= 15 MPa for the plain PE100, while it is $\sigma_{0,\,50\%}$= 14.7 MPa for the welded PE100. The other two fatigue limit values, at 10 and 90% of PS, are adopted to estimate the band scatter index ($T_\sigma = \sigma_{0,\,10\%}/\sigma_{0,\,90\%}$).

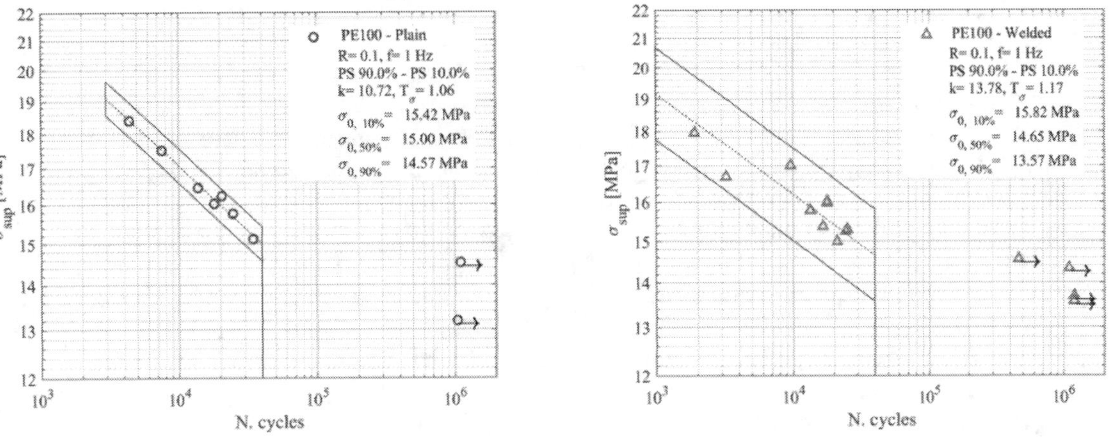

(a) (b)

Figure 9. S-N curve for: a) Plain PE100; b) Welded PE100.

By observing the graph with the same level of applied stress, welded PE100 has a shorter fatigue life than plain PE100. This phenomenon is particularly evident for higher stress levels. As regards the values of the scatter index T_σ (T_σ= 1.06 for plain and 1.17 for welded specimens), they are consistent with the values typically found in fatigue tests for engineering materials (T_σ = 1 ÷ 1.2).

However, it is therefore possible to observe how the presence of welding has not a detrimental effect on the fatigue limit of the material, but introduces an higher scatter compared to the plain material.

5. Conclusion

In this work, mechanical properties of high density polyethylene have been assessed adopting static tensile and fatigue test. Comparison between plain and welded PE100 has been performed showing how the welding process does not affect the mechanical performances. On the other hand, fatigue damage has shown how the welded PE100 has shorten fatigue life compared to the plain PE100.

To shorten the testing time, rapid energy methods, such as the Static Thermographic Method and the Risitano's Thermographic Method, have been adopted. By monitoring the energy release during a static tensile test or during a stepwise fatigue test, it is possible to severally shorten the testing time, from month to hours, obtaining reliable fatigue information, comparable to time expensive constant amplitude fatigue tests.

References

[1] Deveci S and Fang D 2017 Correlation of molecular parameters, strain hardening modulus and cyclic fatigue test performances of polyethylene materials for pressure pipe applications *Polym. Test.* **62** 246–53

[2] Bourchak M and Aid A 2017 PE-HD fatigue damage accumulation under variable loading based on various damage models *Express Polym. Lett.* **11** 117–26

[3] Djebli A, Bendouba M, Aid A, Talha A, Benseddiq N and Benguediab M 2016 Experimental Analysis and Damage Modeling of High-Density Polyethylene under Fatigue Loading *Acta Mech. Solida Sin.* **29** 133–44

[4] Galchun A, Korab N, Kondratenko V, Demchenko V, Shadrin A, Anistratenko V and Iurzhenko M 2015 Nanostructurization and thermal properties of polyethylenes' welds *Nanoscale Res. Lett.* **10** 1–6

[5] Kovalchuk M, Iurzhenko M, Demchenko V and Senchenkov I 2019 *The investigation of the welding process of different-type polyethylenes* (Springer Singapore)

[6] Demchenko V, Iurzhenko M, Shadrin A and Galchun A 2017 Relaxation behavior of polyethylene welded joints *Nanoscale Res. Lett.* **12** 1–6

[7] Dai H and Peng J 2017 The effects of welded joint characteristics on its properties in HDPE thermal fusion welding *Mod. Phys. Lett. B* **31** 1–11

[8] Risitano A and Risitano G 2013 Determining fatigue limits with thermal analysis of static traction tests *Fatigue Fract. Eng. Mater. Struct.* **36** 631–9

[9] La Rosa G and Risitano A 2000 Thermographic methodology for rapid determination of the fatigue limit of materials and mechanical components *Int. J. Fatigue* **22** 65–73

[10] Fargione G, Geraci A, La Rosa G and Risitano A 2002 Rapid determination of the fatigue curve by the thermographic method *Int. J. Fatigue* **24** 11–9

[11] Clienti C, Fargione G, La Rosa G, Risitano A and Risitano G 2010 A first approach to the analysis of fatigue parameters by thermal variations in static tests on plastics *Eng. Fract. Mech.* **77** 2158–67

[12] Corigliano P, Cucinotta F, Guglielmino E, Risitano G and Santonocito D 2020 Fatigue assessment of a marine structural steel and comparison with Thermographic Method and Static Thermographic Method *Fatigue Fract. Eng. Mater. Struct.* **43** 734–43

[13] D'Andrea D, Guglielmino E, Risitano G and Santonocito D 2022 A comparison on the Energy Release between traditional and Additive Manufactured AISI 316L steel during static tensile test *IOP Conf. Ser. Mater. Sci. Eng.* **1214** 012013

IOP Conf. Series: Materials Science and Engineering 1275 (2023) 012017 doi:10.1088/1757-899X/1275/1/012017

Fatigue health monitoring of AISI 304 notched specimens by means of thermographic analysis and UT-based measurements

F W Panella[1], A Pirinu[1*], A Saponaro[1], R Nobile[1], R Rucco[1] and V Renna[1]

[1]University of Salento, Department of Engineering for Innovation, Via per Monteroni, Lecce (LE), 73100, Italy

*alessandra.pirinu@unisalento.it

Abstract. The damage accumulation and the cracks development due to fatigue loads are the main failure mechanisms that occur in metal structures. Therefore, the prediction of structural damage is critically important for the safe and reliable operation of complex engineered systems. In the present work, the evolution of fatigue damage was monitored by infrared thermographic and ultrasonic measurements by means of piezoelectric sensors (PZTs) applied to a batch of notched specimens in AISI 304 stainless steel. Concerning the ultrasonic data, a linear and non-linear frequency study was carried out on the data acquired at various steps during the fatigue life through an appropriate MATLAB analysis algorithm based on the Fast Fourier Transform (FFT). In details, the proposed damage parameters related to thermal variations and the peak-to-peak signal voltage (Vpp) were considered to evaluate the fatigue damage process. Regarding the IRT technique, the thermographic monitoring inspection is implemented with measurements processed in MATLAB to estimate damage indices related to the fatigue damage of the same specimens. The results of the thermal analyses were finally compared with the data of the ultrasound measurements obtained to provide complementary information on the prediction of fatigue damage. The experimental results indicate an interesting increase in the proposed damage parameters that could be associated with an irreversible change due to fatigue damage at the notch tip. The final propagation phase of the crack, instead, is characterized by a quick data variation in agreement with the stiffness reduction. The applied Monitoring strategies has proved efficient for detecting damage induced by fatigue in metal materials.

Keywords: Damage Prediction, Fatigue, Stainless Steel, Nonlinear Ultrasonic, Structural Health Monitoring, Thermographic Monitoring.

1. Introduction

In the metallic structures, the fatigue failure represents a common damage mode that could occur under cyclic stresses during service life, generally lower than the nominal yield strength of the material [1].

Specifically, as reported in Table 1, the fatigue failure seems to reveal the greater percentage of failures both in engineering and in aircraft component, highlighting that the incidence of fatigue failure clearly dominates the distribution in aircraft [1]. However, higher frequency of failure mechanism is reached by corrosion damage in the industrial components, even if this failure mode substantially shows a comparable percentage frequency to fatigue damage [1]. Moreover, three separate stages of crack growth are observed on metallic components that fail under fatigue cycles [1, 2], where the first phase consists of the initiation of a fatigue crack that could be influenced by stress concentrations due to

Content from this work may be used under the terms of the Creative Commons Attribution 3.0 licence. Any further distribution of this work must maintain attribution to the author(s) and the title of the work, journal citation and DOI.
Published under licence by IOP Publishing Ltd

material defects or design features [3 - 5]. The second and third stages are characterized by the propagation of the fatigue crack and final sudden failure, respectively, where a progressive cyclic of crack growth is observed until the propagating crack reaches a critical size at which the damaged component could not support the service loads, thus a sudden rupture occurs [1, 3].

However, the main research efforts are focused on the prediction of the first stage that clearly represents the most critical stage of fatigue crack grown [6, 7]. In fact, the design state and mechanical behaviour of material component are substantially altered in the last two phases due to the damage accumulation and the rapid propagation of fatigue cracks, where a progressive degradation of mechanical properties are observed before the final structural failure [7]. In particular, the Structural Health Monitoring (SHM) aims to provide a specific analysis of the integrity state of the constituent materials, of the different parts, and of the full assembly of these parts constituting the whole structure during the service life of a component [7, 8].

Therefore, a proper strategy through non-destructive and low expensive monitoring techniques represents a continuous challenge of Structural control and Health monitoring for the prevention and detection of crack initiation on industrial metallic components under fatigue loading to ensure stable and reliable operating condition during service life [9 - 11]. In literature, different and alternative methods are proposed as advanced possibilities of several SHM indicators, such as smart materials [15], wireless sensors [14], infrared thermography [12, 20], piezo-electric (PZT) sensors [13], radiographic techniques [16], 3D digital image correlation [17], and so on. These solutions and clearly proper combination of different approaches should enhance the intervention time or maintenance cost, which would prevent major collapses of monitored structure under fatigue loads [6].

Table 1. Frequency of failure mechanisms [1, 2].

Failure modes		Corrosion	Fatigue	Brittle fracture	Overload	High temperature corrosion	SCC/ Corrosion fatigue/ HE	Creep	Wear/ abrasion/ erosion
Percentage of Failures [%]	Engineering Components	29	25	16	11	7	6	3	3
	Aircraft Components	16	55	–	14	2	7	–	6

In this work, an experimental campaign is performed on metallic specimens under sinusoidal cyclic loading to evaluate the structural integrity of tested elements through two non-destructive monitoring techniques. In particular, the crack initiation and propagation were studied on five notched AISI 304 samples using simultaneously ultrasonic PZT sensors and Infrared Thermography (IRT).

Specifically, this research work highlights the main advantages of these promising and low-cost approaches to assess and analyse the fatigue damage initiation and monitoring of inspected specimens [18, 19]. In fact, for UT technique, only one side of the surface specimen is required for the placement of two PZT sensors that detect the UT signals emitted by the material under cyclic stress that could be correlated a part directly to the crack propagation phenomenon. At the same time, the tested sample is monitored by a bolometric thermal camera placed in front of the specimen to acquire thermal sequences.

Clearly, the thermographic monitoring provides a full-field and real time information that highlight the temperature increase around the crack tip for a rapid evaluation of the crack initiation and progression during fatigue test [6]. Therefore, continuous damage accumulations are monitored both through infrared thermal camera and UT measurements at specified intervals during fatigue life, coupled with mechanical stiffness degradation.

Finally, a comparative analysis was performed between the two monitoring methods to assess the relative capability of each technique. Specifically, different damage parameters are proposed to evaluate damage initiation and crack progression on each notched element based on the continuous degradation of UT signals, dissipative and reversible data and stiffness behaviour, respectively.

2. Materials and Methods

2.1. AISI samples and experimental procedures

The fatigue tests are conducted on flat specimens AISI 304 austenitic stainless steel with nominal thickness equal to 2.5 mm, whose geometry and dimensions are reported in Figure 1 (a).

Generally, the AISI 304 is a widely employed austenitic chromium-nickel stainless steel characterized by good formability and also highly corrosion-resistance for different industrial applications such as food handling, processing equipment, screws, machinery parts, utensils, tubing or exhaust manifolds [6, 7].

The physical and mechanical properties of material samples are reported in Table 2.

Experimental cyclic tests are carried out on a servo - hydraulic axial-torsion test machine INSTRON model 8850, equipped with a load cell of ± 250 kN and governed by a computer equipped with INSTRON WAVE MATRIX software for the remote control, as shown in Figure 2.

In Table 3, test parameters and resulting data in terms of fatigue endurance are reported, where the fatigue parameter denoted as σ_{amp_norm} represents the percentage stress amplitude σ_{amp} of each fatigue test normalized to reference yield stress σ_{yield} of AISI 304.

As indicated in Table 3, the AISI specimens were subjected to the fatigue tests under the load-control mode with sine wave shape employing a constant frequency of 9 Hz and a stress ratio of 0.1.

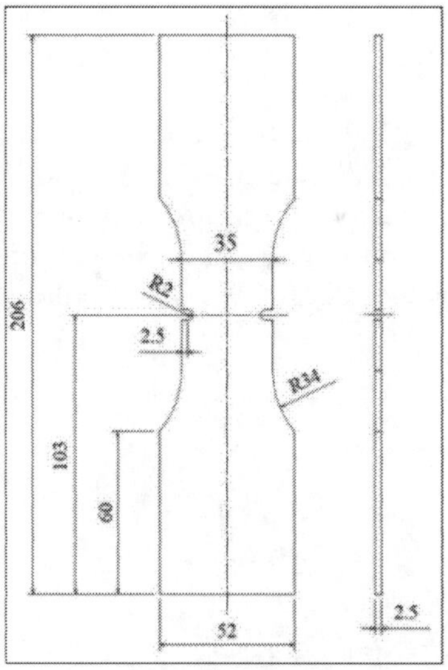

(a)

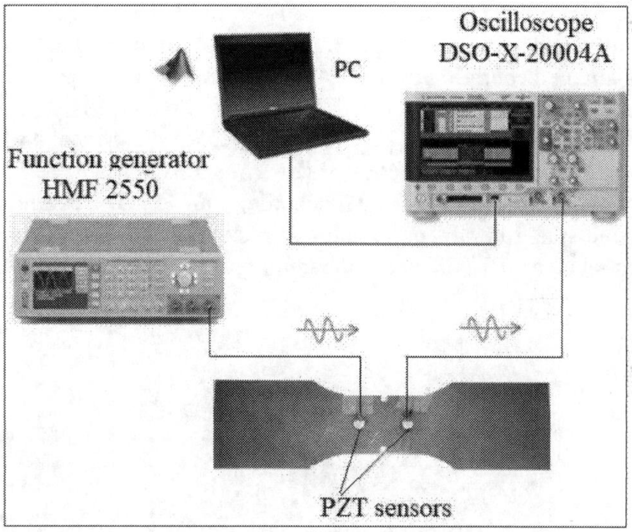

(b)

Figure 1. (a) Geometry and dimensions [mm] and (b) schematic view of connection mode and location of PZT on the AISI 304 specimens.

Table 2. Physical and mechanical properties of AISI 304 alloy.

Density (ρ) [Kg/m^3]	**Young Modulus (*E*)** [GPa]	**Poisson Modulus (υ)**	**Ultimate tensile strength (σ_R)** [MPa]	**Yield stress (σ_{yield})** [MPa]	**Ultimate elongation (*A*)** [%]
8000	193 ÷ 200	0.29	505	215	70

As proposed in a previous work [12], a cumulative damage $D_K(N)$ parameter related to stiffness variation of specific specimen is evaluated through experimental data recorded during fatigue test as defined by the following equation:

$$D_K(N) = \left[1 - \frac{K(N)}{K_0(N_0)}\right] * 100 \quad (1)$$

where the terms $K(N)$ and $K_0(N_0)$ represent the stiffness data related to a generic instant of fatigue life (N) and to the reference initial state (N_0), respectively.

Additionally, the experimental results include the compliance analysis where the stiffness behaviour is evaluated during all fatigue life until final failure [12]. In the next paragraph, the monitoring strategy consists of PZT sensors and thermographic setup and relative processing approaches are presented.

Table 3. Experimental parameters of fatigue tests for AISI 304 specimens.

ID sample	Stress ratio (R)	Load frequency [Hz]	Normalized Stress amplitude ($\sigma_{amp\_norm} = \sigma_{amp}/\sigma_{yield}$) [%]	Number of cycles (N_f)
P1			50.23	97,534
P2			55.81	53,000
P3	0.1	9	47.44	80,850
P4			58.60	30,905
P5			54.42	73,680

2.2. PZT monitoring procedure

A promising SHM technique is related to the use of piezoelectric sensors (PZTs) that are based on the phenomenon of guided Lamb waves, which propagate on the surface of the material under examination and generated by particular excitation types [13, 19]. Basically, the crack initiation and propagation of generally determines a voltage reduction of UT signal in the time domain respect to the reference signal at undamaged condition of the tested element [10, 11, 19].

In the present work, the PZT sensors with a frequency of 300 kHz have been bonded on sample surface and symmetrically positioned respect to the critical section at a distance of 17.5 mm along the loading axis, as shown in Figure 1 (b). Thus, this experimental arrangement of sensors seems to enable a suitable monitoring of whole notched region [13]. The continuous sinusoidal Lamb wave was applied on tested steel sample employing two piezoelectric sensors, one as transducer (PZT T1) and the other one as receiver (PZT R1), as shown in Figure 2.

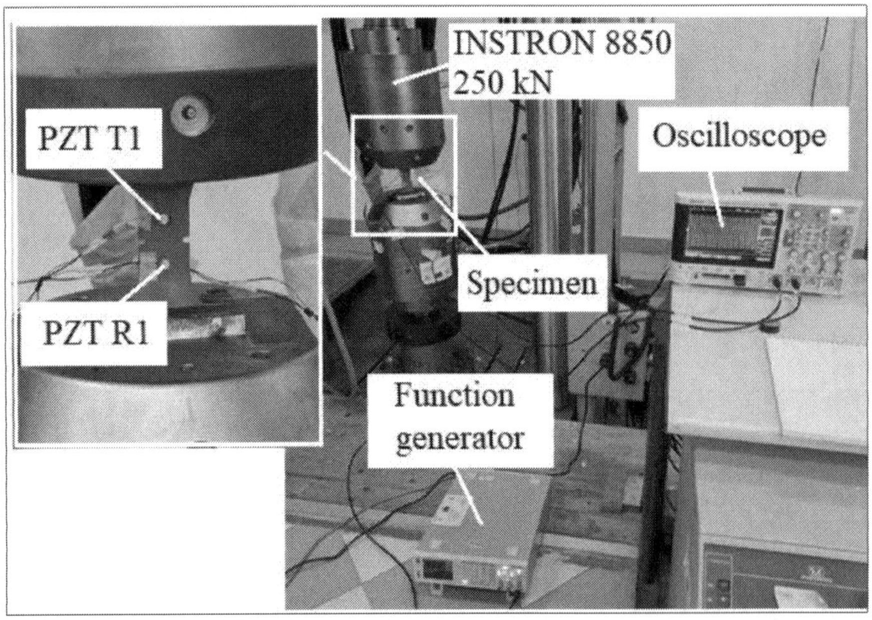

Figure 2. PZT experimental setup for fatigue tests on the AISI 304 specimen.

Thus, the voltage signal data are acquired in the time domain by the oscilloscope in a '.txt' format and represent the sinusoid UT signal at specific instant captured by the receiving sensor with a frequency equal to the fundamental one of the sensors, as shown in Figure 3 (a). Specifically, both linear and nonlinear ultrasonic analyses are performed processing the UT signals acquired during fatigue tests in the time domain or in the frequency domain employing the Fast Fourier Transform (FFT) algorithm, respectively [13]. Specifically, for the linear ultrasonic analysis, a MATLAB code is performed to assess the average value of signal amplitude V_{pp} at a specific fatigue instant t, calculated as following

$$V_{pp}(t) = V_{max}(t) - V_{min}(t) \qquad (2)$$

Where the terms $V_{max}(t)$ and $V_{min}(t)$ represent the maximum and the minimum voltage values of UT signal related to the peak and valley of the waveform at the specific instant t^{th} of fatigue life, respectively, as shown in Figure 3 (a).

Consecutively, the UT sequences are processed in MATLAB environment to evaluate the data results of average signal amplitude V_{pp} at function of fatigue life for each tested specimen. In Figure 3 (b), the diagram shows an example of comparative analysis between three different UT signal curves in the time domain recorded at the 0 % (0 cycle), 40 % (26000 cycles) and the 80% (61000 cycles) of fatigue life for the AISI 304 P3 sample. In addition, based on previous D_K parameter, an Ultrasonic Damage Index D_{UT} is proposed and calculated during whole fatigue life as follows

$$D_{UT}(t) = \left| \frac{V_{pp}(t) - V_{pp0}(t_0)}{V_{pp0}(t_0)} \right| \times 100 \qquad (3)$$

Where the term $V_{pp0}(t_0)$ represents the reference average value of signal amplitude at the initial undamaged instant t_0 (0 cycle).

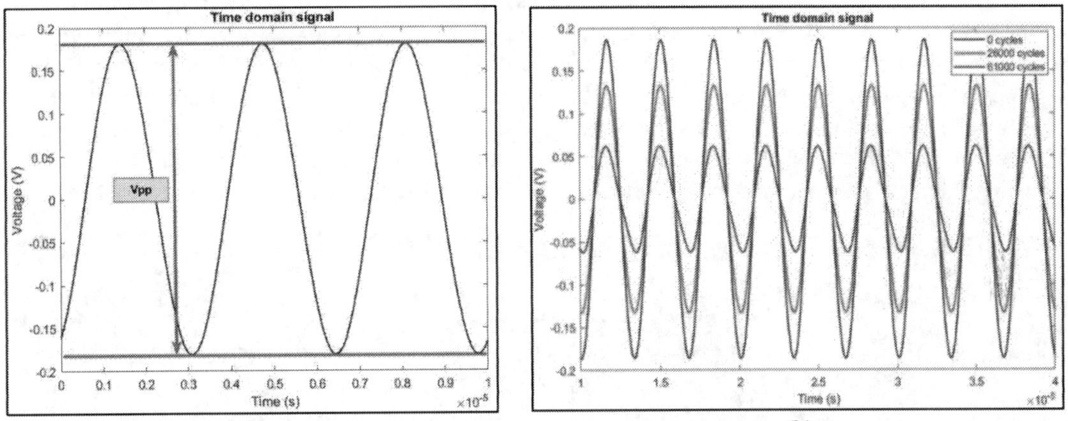

(a) (b)

Figure 3. (a) Example of V_{pp} measurement on a sine UT wave and (b) UT signal curves in the time domain recorded at the 0 %, 40 % and the 80 % of fatigue life related to the AISI P3 specimen.

2.3. Thermographic monitoring procedure

During fatigue tests, thermal acquisitions are recorded to evaluate crack initiation and propagation. Metallic samples are monitored at regular intervals with a micro-bolometric infrared thermal camera, as shown in the thermographic setup of Figure 4 (a) [12, 18].

Monitored surfaces are preventively prepared with the same high-temperature black paint proper for thermal analysis. The FLIR A655sc thermal camera is employed, equipped with an uncooled vanadium oxide micro-bolometer detector that provides 14-bit data up to 50 frames per second. Thermal sequences are acquired at regular intervals of 5000 cycles for all specimens as a function of the expected fatigue life with a frame of 640 × 240 pixels, using a frame rate of 100 Hz and observation time of 10s, in controlled environment temperature between 23÷26 °C. The Research-IR is the software employed for remote control of thermal camera, that could convert the thermal sequences from '.seq' in '.mat' or '.csv' file for the post processing analysis.

In previous work [12], the authors proposed two thermal contrast parameters based on reversible and irreversible thermoelastic sources for the cumulative fatigue damage identification and quantification in selected zones. Assuming initially undamaged conditions, the first damage index is denoted as Thermo-elastic Contrast parameter D_{rev} related to reversible heating sources that occur during the fatigue life (N_{tot}) and calculated as:

$$D_{rev}(N) = \left[\frac{\Delta T}{T} / \left(\frac{\Delta T}{T}\right)_0\right]_d - \left[\frac{\Delta T}{T} / \left(\frac{\Delta T}{T}\right)_0\right]_i \tag{4}$$

Where ΔT [K] represents the temperature oscillation, T [K] is the absolute temperature at the mean load level on inspected specimen under fatigue loading, the terms $(\Delta T/T)_d$ and $(\Delta T/T)_i$ represent the TSA ratio for damaged and undamaged regions (ROIs) at a given fatigue instant N, normalized respect to the same relative TSA ratio at initial fatigue life, respectively.

The second damage factor is the Dissipative Contrast parameter D_{diss} related to dissipative heat release that provide interesting information on material degradation as function of fatigue life (N / N_{tot}), calculated as:

$$D_{diss}(N) = \left[\left.\frac{T - T_0}{T_0}\right|_d - \left.\frac{T - T_0}{T_0}\right|_i\right] = \left[\left.\frac{T}{T_0}\right|_d - \left.\frac{T}{T_0}\right|_i\right] * 100 \tag{5}$$

Where the T_0 [K] represents the absolute temperature at the mean load level at initial undamaged conditions. The first parameter is used to examine of the damage progression based on a contrast measurement related to thermal oscillation and normalized to actual temperature, whilst the second D_{diss} parameter represents the cumulative damage evolution associated to a normalized contrast evaluated at the mean load level in selected damaged and undamaged zones, respectively. In this work, only the second damage thermal parameter is employed for data processing.

Specifically, several Regions of Interest (or ROIs) should be selected in different undamaged and damaged zones in the raw temperature sequences [12]. Then, the second processing step consists of post-processed phase where the 3D thermal sequences of each ROI are elaborated in MATLAB environment following Equations (4) and/ or (5) as function of fatigue life and compared with relative stiffness behaviours. Since the crack initiation could occur in one or both notched zones during fatigue life, multiple defective spots are considered only the defective ROIs close to the critical zones where interesting thermoelastic and dissipative effects could be observed, as illustrated in Figures 4 (b).

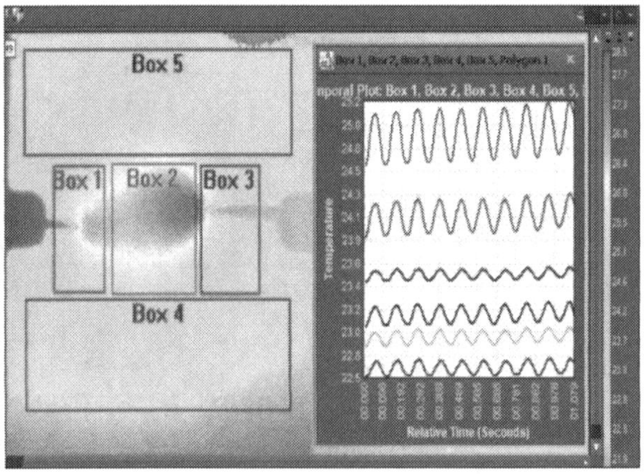

(a) (b)

Figure 4. (a) Thermographic monitoring setup and (b) thermal map with thermal profiles referred to selected ROIs at 99 % of fatigue life.

3. Results and discussion

3.1. Experimental fatigue results

In Figure 5 (a), stiffness data K normalized to specific maximum initial value (K_0) are reported as function of fatigue life for tested AISI samples. The compliance analysis highlights similar stiffness behaviours in each fatigue test, where an almost flat stabilized trend with small K deviations is observed in the initial and middle phases of fatigue life. In Figure 5 (a), a slightly different behaviour is observed in the experimental data of P3 and P4 specimens, where a slow decreasing trend is noted between the 70 ÷ 80 % of fatigue life, reaching a stiffness degradation around 4 ÷ 8 %.

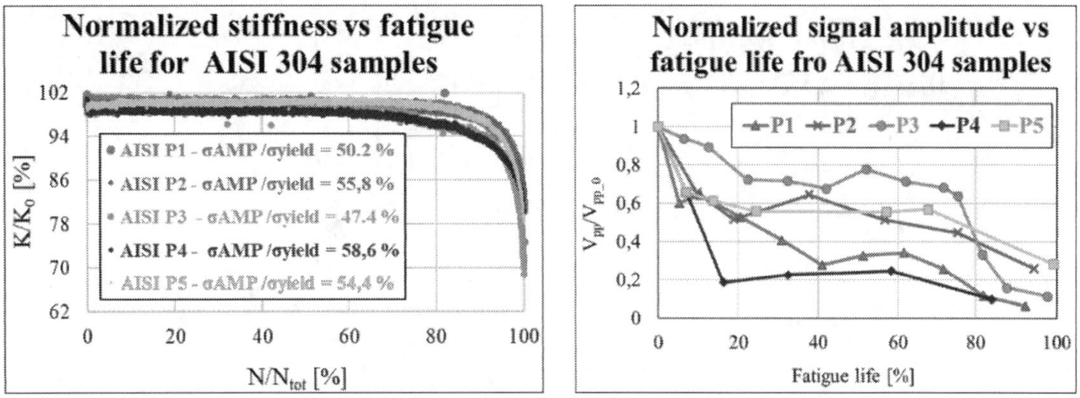

(a) (b)

Figure 5. (a) Normalized stiffness vs normalized fatigue life and (b) normalized signal amplitude vs normalized fatigue life for AISI 304 samples.

However, in this series, a slight stiffness reduction of 5 ÷ 7% is observed at the 85 – 95 % of total fatigue life when evident rapid crack progression occurs around the two notched zones. After 90 % of fatigue life, crack growing seems to accelerate the propagation until final failure when a sudden stiffness degradation is observed with a final decrease of 20 ÷ 30 at 99% of fatigue life.

In addition, visual inspections of damaged sections are performed by means of a stereo - microscope Nikon SMZ 1000 to assess the damage zones due to the applied fatigue loads. Examples of fracture surface analysis are reported in Figures 6, where similar damage zones are observed for AISI 304 P2 and P5 specimens respectively after relative fatigue failure. Specifically, the optical microscopic images of fracture surfaces indicate the first crack initiation and relative propagation sites on notched zone.

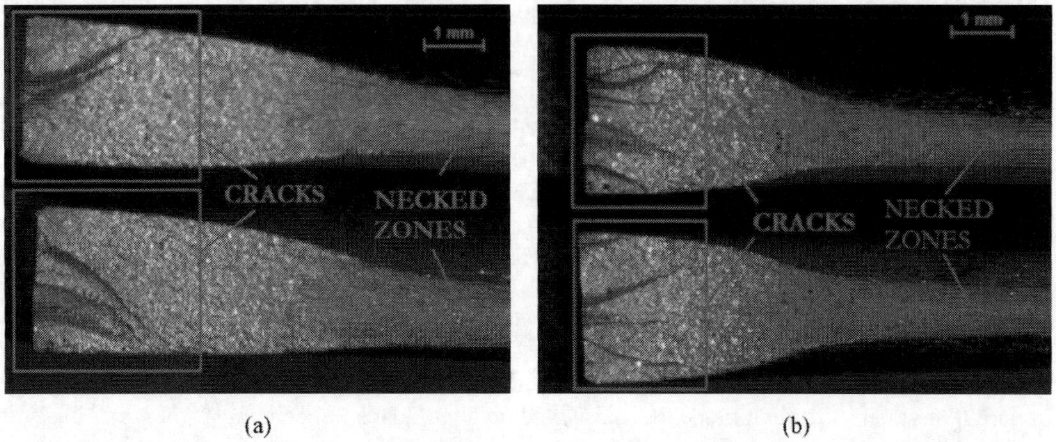

(a) (b)

Figure 6. (a) Fracture surfaces at notched zone on AISI - P2 and (b) on AISI - P5 samples after relative fatigue failure.

3.2. PZT monitoring results

For the linear ultrasonic analysis, in the Figure 7 (a), the relative curves of ultrasonic damage index described in previous Equation (3) are reported as function of fatigue life for different AISI samples.

The D_{UT} data of P3 and P4 samples seem to increase up to 20% of fatigue life, when an almost constant flat trend is observed between 20 ÷ 60% of fatigue life and followed by a sudden increase up to final failure. The D_{UT} data of AISI P5 specimen show a wide steady state behaviour between the 15 ÷ 70% of fatigue life, when a rapid increase is observed until fatigue failure.

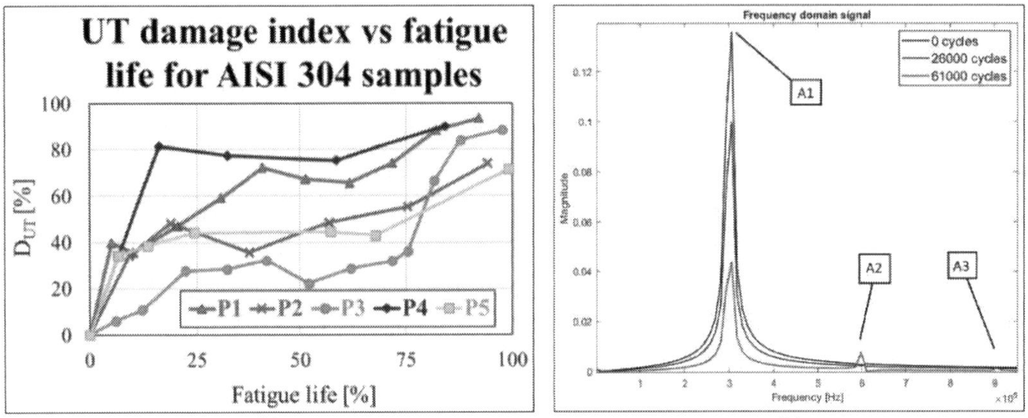

(a) (b)

Figure 7. (a) Ultrasonic damage index data vs fatigue life for AISI 304 samples and (b) FFT profiles at the 0%, 40 % and the 80 % of fatigue life for AISI P3 specimen.

The AISI P1 element seems to provide a growing trend up to 40% of fatigue life, followed by a plateau behaviour between the 40 - 60 % and by a final sudden increase. However, the almost total of increasing behaviour of P4 sample is observed until the 20 % of fatigue life, after that a constant trend occurs up to 60 % of fatigue life, when a slight increase is noted.

Therefore, a similar behaviour of damage factor D_{UT} [%] seems to be observed for each specimen, with a constant trend between the 20% and the 60% of relative fatigue life, when a sudden increase arises at 60% - 70% of fatigue life.

Concerning the nonlinear ultrasonic analysis, the Figure 7 (b) illustrates a representative example of frequency spectrum with different FFT profiles obtained at the 0%, 40 % and the 80 % of fatigue life for AISI P3 specimen, where the terms A_1, A_2 and A_3 represent the first, the second and the third harmonics, respectively.

The spectrum analysis could be studied to evaluate the non-linear phenomena that occur during the damage progression that interacts with the nonlinear behaviour of material. As noted in Figure 7 (a), the fundamental frequency related to the first higher peak arises at a frequency equal to 300 kHz, that corresponds to excitation frequency of the PZT sensors.

In addition, a continuous decrease of the data values is observed in the first and third frequencies, whilst a different behaviour seem to occur for the second harmonic data, where the last selected curve clearly shoes a higher value respect to the two previous ones.

For a nonlinear comparative analysis, the first A_1 and the third A_3 harmonics normalized respect to the relative initial data for different AISI samples as function of fatigue life are reported in the Figures 8 (a) and 8 (b), respectively.

In the Figure 8 (a), the diagram of normalized 1st harmonics seems to highlight an interesting correlation between first amplitude decrease and applied loads. Specifically, the first harmonics decreases are observed at lower fatigue life stages for higher loading conditions.

Finally, the Figure 8 (b) shows the normalized data of third harmonics as function of percentage fatigue life for each AISI specimens. Concerning the third harmonics, a similar behaviour is distinguished in the diagram of Figure 8 (b), where a slight relationship with applied loading is noted.

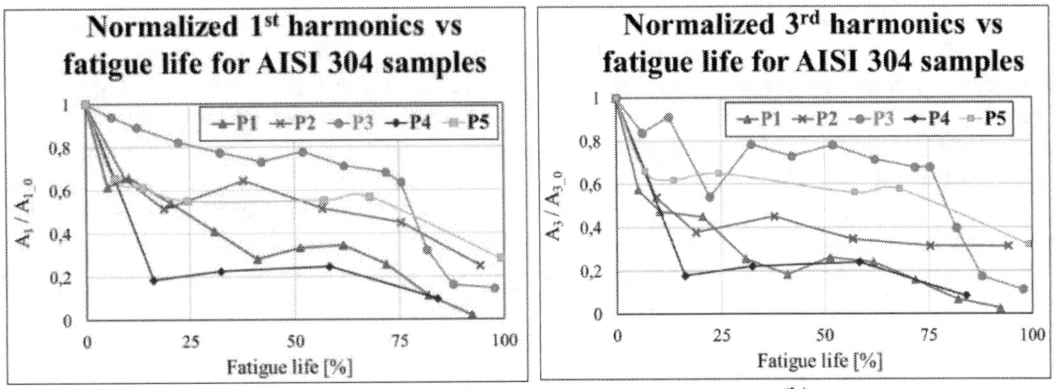

(a) (b)

Figure 8. (a) Normalized behaviour of first and (b) third FFT harmonics vs normalized fatigue life for AISI 304 samples.

3.3. Thermographic monitoring results

The thermal damage parameters defined as in Equation (4) and (5) are analysed for crack initiation and evolution, selecting specific ROIs for processing analysis for each specimen by maximizing the damage parameter [12, 20]. Specifically, the intact area was calculated in several zones to evaluate the influence of ROI selection on damage factors values.

In Figures 9 and 10, the processed thermal profiles of selected ROIs related to thermo-elastic (D_{rev}), or dissipative heating sources (D_{diss}) are reported as function of normalized fatigue life for each AISI sample and compared with stiffness degradation factor (D_K).

Specifically, both thermal damage parameters are compared in two selected ROIs, where one ROI is located neighbouring the notched zones (B1 or B3) where the first crack initiation is observed in the relative samples and the other one represents the central ROI (B2) where the final crack progression is observed.

In Figure 9 (a) and 10 (b), a high level of noise (scattering) and a certain trend variability is observed in the D_{rev} curves between 0 ÷ 50 % of fatigue life. In addition, the D_{rev} thermal data seem to provide greater values in the ROIs located neighbouring the notched zones (B1 or B3) respect to the central ROI (B2). A similar behaviour is observed in the D_{diss} curves, but lower percentage values respect to D_{rev} and D_K factors are noted for each specimen.

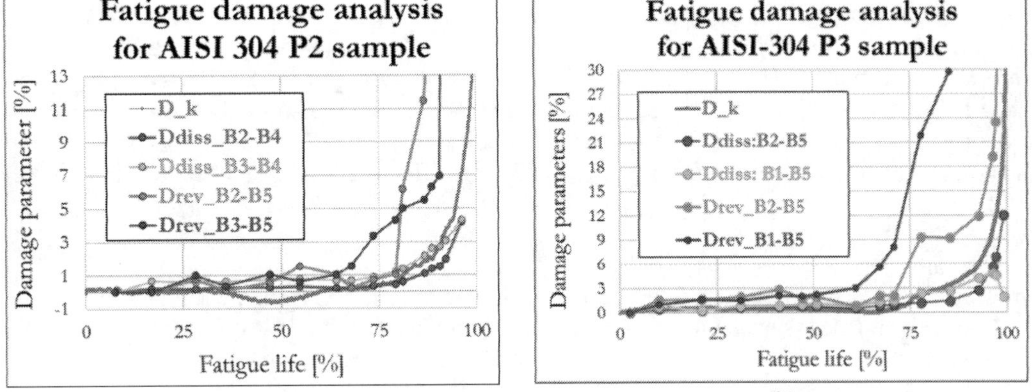

(a) (b)

Figure 9. (a) Comparative analysis between D_K and dissipative damage D_{diss} parameter vs fatigue life for AISI P2 and (b) P3 samples.

The general behaviour of all fatigue damage parameters appears with an approximately exponential trend characterized by an almost steady state at the initial phase without an index trend increase of followed by a second phase, where an evident curve growth is observed in all the AISI specimens.

Finally, a rapid parameter rise seems to follow this increasing trend, leading to the final rupture, except for the D_{diss} parameters near the notched zone of AISI P3 and P5, as shown in Figures 9 (b) and 10 (b). Specifically, both D_K and D_{diss} profiles show a similar damage evolution and an interesting comparable behaviour with foible data variability until the 70 ÷ 75 % of fatigue life, appearing clearly interlocking and intersecting each other.

Then, both D_K and D_{diss} data exhibit a gradually increasing growth between 70 ÷ 75 % of fatigue life where critical damage propagation starts to occur up to failure.

However, in the final stage after 75 ÷ 85 % of fatigue life, the stiffness variation seems to slight differ much from D_{diss} curves, which look very similar in case of AISI P2 and P3 specimens. In general, a very similar correspondence and a proper agreement are observed between the proposed damage parameter evaluated by D_{diss} data and the reference cumulative damage D_K profiles related to stiffness degradation.

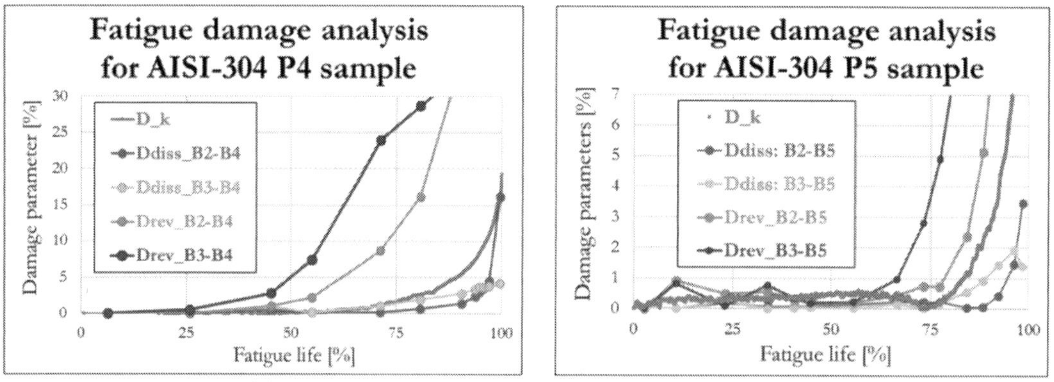

(a) (b)

Figure 10. (a) Comparative analysis between D_K and dissipative damage D_{diss} parameter vs fatigue life for AISI P4 and (b) P5 samples.

Concerning the third damage factor, even if the D_{rev} data seem to suffer of a higher level of noise, this thermal factor results clearly more sensitive to crack damage initiation and progression, providing an evident data variation between 50 ÷ 60 % of fatigue life and highlighting similar behaviour between AISI P2, P3 and P5 specimens.

In the Figure 10 (a), a sudden growth of D_{rev} data is noted after 25% of fatigue life for the AISI P4 sample, probably due to the higher loads applied on this sample, as indicated in Table 3. Therefore, the more critical loading conditions is clearly distinguished with the D_{rev} parameter.

In addition, a more rapid variation trend of D_{rev} data is observed for the ROIs neighbouring the notched zones (B1 or B3) respect to the central ROI (B2). However, also the D_{rev} variation of ROI B2 clearly provides a more rapid prediction of crack initiation respect to the D_K and D_{diss} parameters in all AISI samples. Therefore, the D_{rev} factor seems to be more effective for the notched metallic specimens.

4. Conclusions

In the present work, a suitable combined approach of ultrasonic PZT and thermographic monitoring procedures is performed to evaluate the crack initiation and propagation of AISI 304 notched specimens under fatigue loads. Specifically, different fatigue damage parameters based on processed ultrasonic and thermal data are proposed and compared to the stiffness degradation.

Concerning the UT procedure, the proper data results are examined in the curves of peak-to-peak signal voltage that seems to result more strictly correlated to fatigue damage initiation and propagation of metallic samples. Comparing to relative stiffness data, a general decreasing trend is observed in all tested samples starting from approximately 70% of fatigue life, previously anticipated by a slight increase of signal. In particular, the processed data highlighted an inverse behaviour of the signal increase respect to the applied load. Lastly, the non-linear analysis provided interesting results employing damage factors based on the first, the second and the third order harmonics, where the parameter related to the third order seem to be more sensitive to damage effects.

Finally, concerning the thermographic procedure, two thermal damage parameters are evaluated in specific processing ROIs for crack initiation and damage progression of each AISI sample. Specifically, a comparative analysis is performed between three proposed damage factors consisting of two thermal contrast-based approaches consisting of the D_{rev} parameter based on the thermoelastic effects and the alternative thermal dissipation D_{diss} index, and the reference damage factor D_K based on stiffness degradation.

In general, a very similar behaviour is observed between the proposed damage D_{diss} parameter and the reference D_K, whilst the D_{rev} data results clearly more sensitive to crack damage initiation and progression, also highlighting a correlation to critical loading conditions.

References

[1] Findlay S J and Harrison N D, 2002. Why aircraft fail. Materials Today 5 (11), Pages 18-25.

[2] Bernabei M, Allegrucci L, Amura M, 2016. Chapter 5 - Fatigue failures of aeronautical items: Trainer aircraft canopy lever reverse, rescue helicopter main rotor blade and fighter-bomber aircraft ground-attack main wheel. Handbook of Materials Failure Analysis with Case Studies from the Aerospace and Automotive Industries, Pages 87-116.

[3] Dowling N E, 2012.Mechanical Behavior of Materials, Engineering Methods for Deformation, Fracture, and Fatigue. Pearson.

[4] Correia J A F O, Blasón S, De Jesus A M P, Canteli A F, Moreira P M G P, Tavares P J, 2016. Fatigue life prediction based on an equivalent initial flaw size approach and a new normalized fatigue crack growth model. Engineering Failure Analysis 69, Pages 15 - 28. https://doi.org/10.1016/j.engfailanal.2016.04.003

[5] Milella P P, 2013. Fatigue and Corrosion in Metal. Springer-Verlag, Italia.

[6] Barile C, Casavola C, Pappalettera G, Pappalettere C, 2016. Analysis of crack propagation in stainless steel by comparing acoustic emissions and infrared thermography data. Engineering Failure Analysis 69, Pages 35 - 42. https://doi.org/10.1016/j.engfailanal.2016.02.022.

[7] Deng G J, Tu S T, Wang Q Q, Zhang X C, Xuan F Z, 2014. Small fatigue crack growth mechanisms of 304 stainless steel under different stress levels. International Journal of Fatigue 64, Pages 14 - 21. https://doi.org/10.1016/j.ijfatigue.2014.01.027.

[8] Daniel Balageas D, 2006. Chapter 1: Introduction to Structural Health Monitoring. In: Structural Health Monitoring. Editors Daniel Balageas and Claus-Peter Fritzen, Pages 1 - 21.

[9] Yang W, Gao Pi, 2019. Lamb Wave-Minimum Sampling Variance Particle Filter-Based Fatigue Crack Prognosis. Sensors 19, 1070, pages 1 – 20.

[10] Chen H, Zhang G, Fan D, Fang L, Huang L, 2020. Nonlinear Lamb wave analysis for microdefect identification in mechanical structural health assessment. Measurement 164 (2), pages 1 - 11.

[11] Pozo F, Tibaduiza D A, Vidal Y, 2021. Sensors for Structural Health Monitoring and Condition Monitoring. Sensors, 21 (5), 1558; https://doi.org/10.3390/s21051558.

[12] Nobile R, Panella F W, Pirinu A, Saponaro A, 2022. Full-field monitoring methods for damage analysis on aeronautical CFRP specimens under fatigue loads, IOP Conference Series Materials Science and Engineering 1214 (1). https://doi.org/10.1088/1757-899X/1214/1/012011.

[13] Nobile R, Panella F W, Antonazzo R, Pirinu A, Saponaro A, 2022. Fatigue damage evolution using NDT techniques for Health Monitoring of CFRP laminates. IOP Conference Series Materials Science and Engineering 1214 (1): 012008. https://doi.org/10.1088/1757-899X/1214/1/012008.

[14] Noel B A, Abderrazak A, Tarek E, Mohamed Hossam A, Badawy A, Mohamed S S, 2017. Structural health monitoring using wireless sensor networks: a comprehensive survey. IEEE Commun. Surv. Tutorials, 19 (3), Pages 1403-1423.

[15] Chee-Kiong S, Yaowen Y, Suresh B, 2012. Smart Materials in Structural Health Monitoring, Control and Biomechanics. Springer Science & Business Media.

[16] Wenhui Hou, Dashan Zhang, Wei Ye, Jie Guo, Xiaolong Zhang, 2020. Review on computer aided weld defect detection from radiography images. Appl. Sci., 10 (5), p. 1878.

[17] Reagan D, Sabato A, Niezrecki C, 2018. Feasibility of using digital image correlation for unmanned aerial vehicle structural health monitoring of bridges. Structural Health Monitoring 17 (5), Pages 1056-1072.

[18] Thatcher J E, Crump D A, Devivier C, Bailey P B S, Dulieu-Barton J M, 2020. Low cost infrared thermography for automated crack monitoring in fatigue testing. Optics and Lasers in Engineering 126, 105914. https://doi.org/10.1016/j.optlaseng.2019.105914.

[19] Rong Wang, Qi Wu, Fengming Yu, Yoji Okabe, Ke Xiong, 2018. Nonlinear ultrasonic detection for evaluating fatigue crack in metal plate. Structural Health Monitoring 18 (1).

[20] Pirinu A, Panella F W, 2022. Full-Field Thermographic Analysis for Fatigue Damage Detection of Composite Specimens. Book: European Workshop on Structural Health Monitoring, Chapter No: 59. https://doi.org/10.1007/978-3-031-07258-1_59.

An innovative experimental energy-based approach for the life prediction of notched details

D Crisafulli[a]*, P Foti[b] and D Santonocito[a]

[a]University of Messina, Department of Engineering, Contrada di Dio, 98166 Messina, Italy

[b]Norwegian University of Science and Technology, MTP Gloshaugen, Richard Birkelands vei 2B, Trondheim 7491, Norway

*Corresponding author: davide.crisafulli@studenti.unime.it

Abstract. Fatigue assessment of mechanical components can be very challenging due to the presence of macroscopic defects and notches. However, local approaches as the Stain Energy Density, have shown how the fatigue resistance is strictly related to the local stress field near the notch tip and not on the overall dimensions of the component. It is necessary to know some peculiar material's parameters to apply those local approaches. A short way to obtain such parameters can be to apply Thermographic Methods, such as the Risitano's Thermographic Method and the Static Thermographic Method, monitoring the energy release during stepwise fatigue and static tensile tests. In this work, experimental tests were performed on plain and double edge steel notched specimens. The SED method has been applied through finite element models, with the easy application of the volume-free procedure, to evaluate its trend at several stress levels. The analysis of the energy release and of the local averaged strain energy density during a static tensile test may be adopted as a novel approach to relate the local stress state with a macroscopic stress value that introduce in the material the first micro plasticization leading to the failure of the component.

1. Introduction

Estimating the fatigue properties of materials is of fundamental importance, however the process is extremely time consuming. During fatigue test, the material is involved into internal transformations that cause a heat dissipation into the surrounding environment.

By monitoring the energetic release of a stressed material adopting an infrared camera, it is possible to correlate the temperature trend with its fatigue life. Thermographic Methods, such as the Risitano's Thermographic Method (RTM) and the Static Thermographic Method (STM), are used as rapid testing protocol to correlate the temperature trend with the fatigue properties of the component during, respectively, a stepwise fatigue test or static tensile test, using only few specimens [1–5].

To assess the fatigue life of notched components, local approaches, as the Strain Energy Density (SED) method, evaluate the strain energy in a control volume (CV) centered at the notch tip. Fatigue resistance depends only on local energy conditions, not on the nominal stress applied to the component

Content from this work may be used under the terms of the Creative Commons Attribution 3.0 licence. Any further distribution of this work must maintain attribution to the author(s) and the title of the work, journal citation and DOI.
Published under licence by IOP Publishing Ltd

[6–11]. To be practically applied, this local approach requires an important parameter: the radius of the CV R_0, which is an intrinsic property for each material.

Generally, the CV radius is assessed performing several numbers of tests to estimate the fatigue limit of the material. Then, by knowing some fracture mechanic parameters and the notch conditions, it is possible to estimate it applying formulations from fracture mechanic.

In the present paper, for the first time, a rapid way to assess the CV radius is proposed. It is based on the adoption of the RTM and STM to assess the fatigue limit of the material. With a numerical procedure, it is possible to compare the critical SED value, estimated with the fatigue limit of the material, and to obtain the CV radius.

2. Theoretical background

2.1. Risitano's Thermographic Method

By observing the temperature evolution of a specimen subjected to cyclic loading condition above its fatigue limit, it is possible to observe three phases [1]. The second phase is characterized by a constant value of the temperature signal in the time domain (Figure 1a). Risitano correlated this temperature value to the onset of fatigue damage within the material. As the applied stress level increase, the stabilization temperature increase, but the integral of the temperature vs. number of cycles curve remain almost constant. This value, the Energy Parameter Φ, can be related to the dissipated energy of the material.

It is possible to apply on a single specimen, in a stepwise way, the applied stress level and record the stabilization temperature. By plotting the stabilization temperature vs. the applied stress level, it is possible to recognize the fatigue limit of the material as the knee region of that curve. Exploiting the constancy of the Energy Parameter, it is also possible to rapidly evaluate the SN curve of the material adopting a single specimen. For more information about the Risitano's Thermographic Method we remand to [1,2].

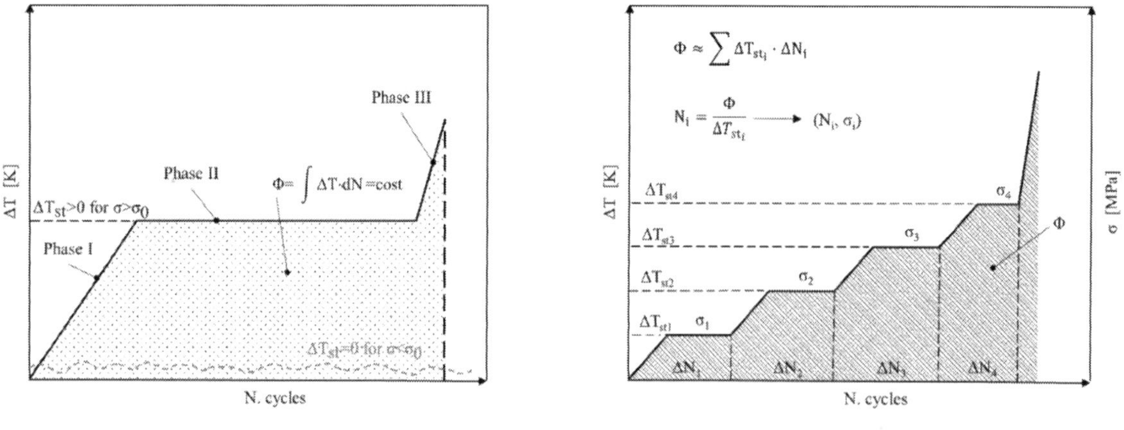

(a) (b)

Figure 1. a) Temperature trend during a constant amplitude fatigue test; b) Temperature trend during a stepwise fatigue test.

2.2. Static Thermographic Method

The main idea of the Static Thermographic Method is to assess the end of the thermoelastic behavior of a material during a static tensile test.

Risitano and Risitano [12], in 2013, observed how the temperature signal during a static tensile test, show three different phases. Phase I is characterized by a linear decrement of the temperature, due to the thermoelastic law of Lord Kelvin. Phase II is characterized by a temperature decrement but with a

IOP Conf. Series: Materials Science and Engineering 1275 (2023) 012018 doi:10.1088/1757-899X/1275/1/012018

different trend compared to the Phase I. At the end of Phase II, the temperature reaches a plateau region, which correspond to the yielding stress of the material. In Phase III, the plastic damage is more predominant compared to the elastic behavior, hence the temperature begins to rise up to the specimen failure.

The transition point between Phase I and Phase II correspond to the onset of irreversible damage within the material. If we are able to assess this point in the temperature signal, we can correlate it to a macroscopic stress, defined as the "limit stress", σ_{lim}, which produce an irreversible energy release due to the increment of plastic damage. For more information about the STM, we remand to [4,12,13].

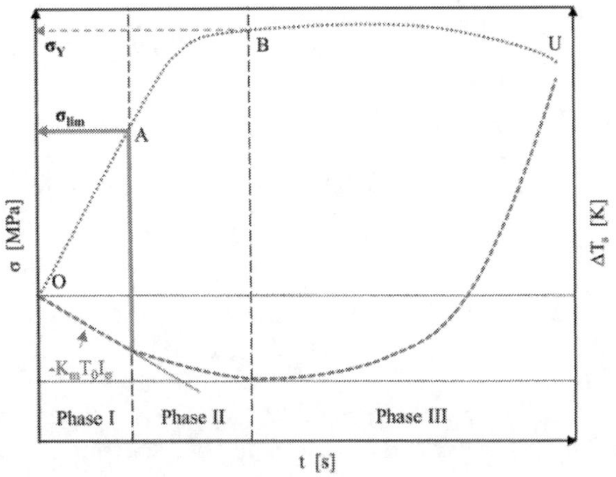

Figure 2. Temperature trend during a static tensile test. Phase I is characterized by a linear decrement due to the thermoelastic constant K_m and the first stress invariant I_σ.

2.3. *Strain Energy Density method*

The SED method is a local approach where the strain energy density near the notch tip is evaluated and compared to a critical value. If the material, stressed under fatigue load, locally reaches the critical condition, it will fail. This method is independent from the size of the analysed component. Different large or small components can be in the same local energetic critical conditions if the geometry of the notch is the same. In mechanical components, failure happen when the local SED reaches a critical value[8,9]. For an isotropic elastic material, the critical value of SED is defined as follow [6–8,14,15]:

$$W_c = \frac{\sigma_t^2}{2E} \tag{1}$$

In this equation W_c is the critical value of strain energy density and σ_t is the value of the ultimate stress.

This limit value is evaluated on the CV that is centered on the notch tip, according to the theory proposed by Lazzarin [8,9]. The fundamental parameter of this method is the value of the CV radius which, according to Lazzarin's theory, is an intrinsic property of the material, recalling the Neuber's elementary material volume theory. It is therefore possible, starting from the energy evaluated on the entire CV, to make the ratio between calculated energy and volume to express an average strain energy density.

According to the theory of fracture mechanics (Figure 3) [6,7,9,11] the value of the CV radius has been estimated.

IOP Conf. Series: Materials Science and Engineering 1275 (2023) 012018 doi:10.1088/1757-899X/1275/1/012018

Recently, the application of the SED method has been proposed adopting the free mesh procedure [10,16] to reduce the high time simulations effort.

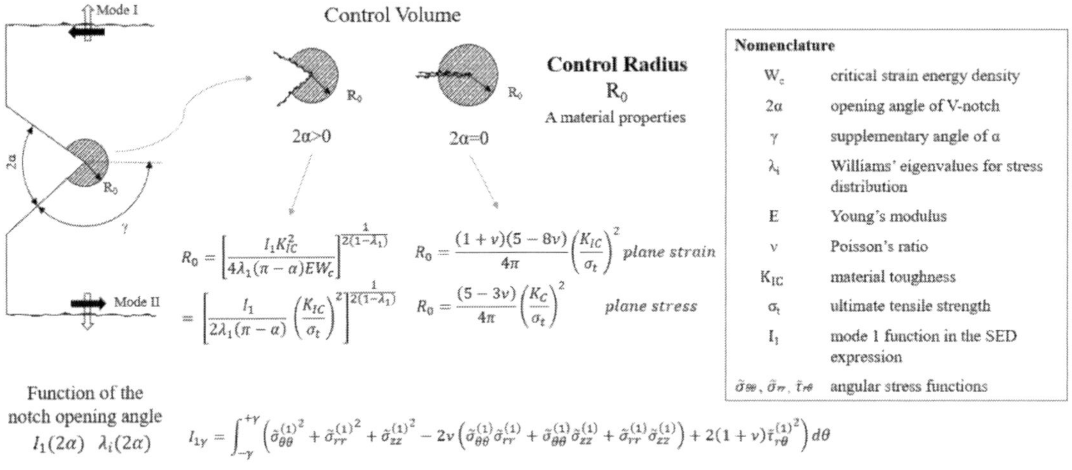

Figure 3. CV radius estimation by fracture mechanics theory.

3. Materials and methods

The experimental campaign has been performed on two kind of specimen geometries of stainless steel AISI 304L (Figure 4). A plain geometry and a blunt V-notched specimens with opening angle 2α= 135° and fillet radius ρ= 2 mm.

Static and fatigue tests were performed using a servo-hydraulic load machine MTS 810 with maximum load capacity of 250 kN. To evaluate the energy release of the material and in order to apply the thermographic methods, the infrared camera FLIR A40 was used during all tests. In particular for V-notched specimen's tests, an extra optical zoom was adopted to analyse the energy release near the notch area in a more accurate way. Before carrying out the tests, all the specimens were coated with a black paint to increase the thermal emissivity of the material up to 0.98.

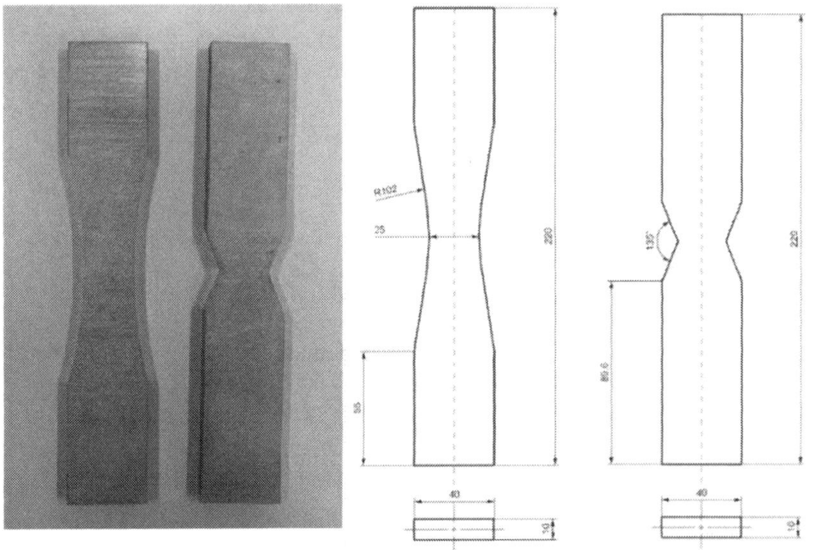

Figure 4. Plain and V-notch specimen: real and 2D CAD representation.

The acquisition of the temperature evolution of the specimen surface temperature was made through a rectangular measurement spot placed near the restricted resistant section of the specimen with a temperature measurement range between -40°C and +120°C.

Three static tensile tests were performed under stress control with a stress rate of 240 MPa/min and 360 MPa/min. Some preliminary tests have been carried out to choose the stress rate, to highlight the thermal phenomenon. Indeed, it is very important that the specimen must not have time to exchange heat with the external environment.

For the stepwise fatigue tests, the chosen load range was from 100 up to 300 MPa, with a stress step of $\Delta\sigma$=10 MPa and a number of cycles per block of ΔN= 10000. Four fatigue tests were performed, two of each type of geometry, adopting a stress ratio R= -1 and a frequency f= 5 Hz.

During the static and fatigue tests the evolution of the specimen surface temperature was evaluated with an infrared camera FLIR A40, with a sample rate of 5 Hz.

To apply the SED method, a set of numerical simulations have been performed in Ansys® APDL. The parametric model of the V-notched specimen was realized but only 1/4th of the geometry was modelled, taking advantage of the specimen symmetries (Figure 5). A free-mesh procedure for finite element analysis was adopted, considering the 8-nodes element PLANE183 with a global mesh size of 1 mm and three level of refinement around the notch tip.

The simulations were carried out using the plain strain conditions and the elastic material behavior was considered (E = 205 GPa, ν = 0.3). Symmetric boundary conditions were applied to the geometry, while the tensile force was equally distributed on the quarter grip section nodes.

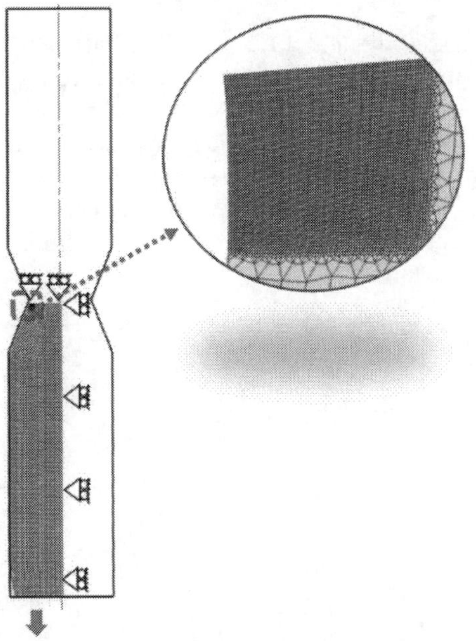

Figure 5. Quarter V-notch specimen structural FE 2D model.

4. Results and discussion

4.1. Static tensile tests

During static tensile test, the evolution of the specimen's surface temperature has been analysed to apply the STM. An optical zoom has been applied to the IR camera to better visualize the energy release near the notch tip. Two rectangular measurement areas, placed on the specimen's reduced section near the notch tip, have been adopted to measure the maximum value of temperature (Figure 6). The temperature signal of the measurement area with the highest temperature increment has been considered in the following elaborations.

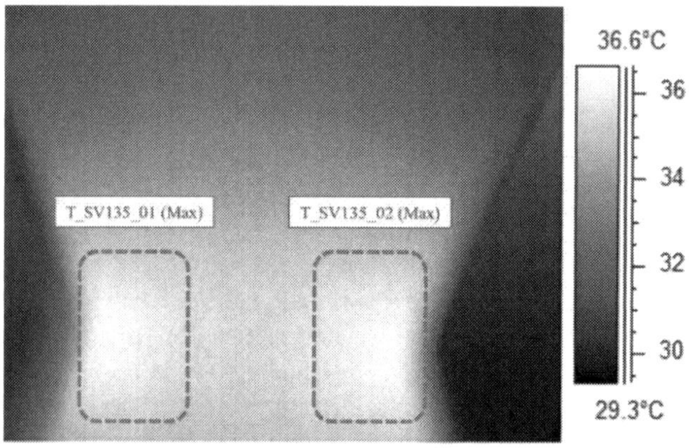

Figure 6. Measurement areas to monitor the evolution of the maximum surface temperature during the static test.

The temperature data was processed thought a *rlowess* filter with a 10% range to eliminate off-scale peaks and highlight the temperature trend. The applied stress level has been reported vs. the temperature signal.

In the initial part, the linear trend is distinguished and, when it deviates from linearity, it is possible to observe a plateau region (Figure 7). It is possible to perform linear regression of the temperature point of the first and second phase and identify the coordinates of the intersection point between the lines. The corresponding stress is equivalent to the macroscopic stress, the limit stress σ_{lim} that induces micro plasticization phenomena in the material.

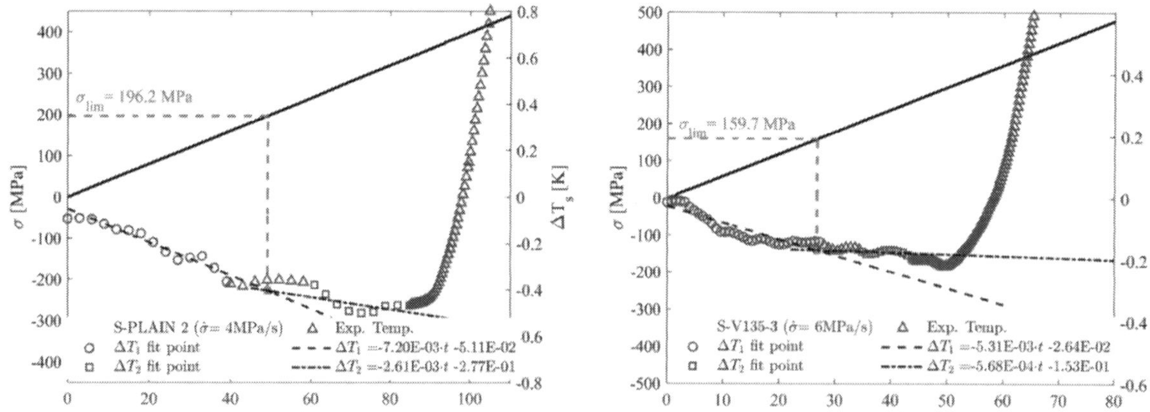

(a) (b)

Figure 7. Temperature trend vs. applied stress of (a) Plain and (b) V-notch specimen.

In Table 1 are reported, for each tested specimen, the value of the ultimate strength σ_U and the limit stress σ_{lim} as assessed by STM, with their average estimation. The average ultimate strength stress is higher for the V-notch specimens. This result is probably related to the hardening of the material due to the notch presence which did not bring the specimen to break during the test.

While the average limit stress is higher in the plain specimen, so the presence of notch causes a worsening of the fatigue properties.

Table 1. Mechanical properties of AISI304L specimens.

Specimen type	No. Specimen	σ_U [MPa]	σ_{lim} [MPa]	$\sigma_{U\ ave}$[MPa]	$\sigma_{lim\ ave}$[MPa]
Plain	1	603	156	612 ± 10	179 ± 21
	2	622	196		
	3	612	186		
V-notch 135°	1	647	121	646 ± 1	152 ± 27
	2	646	174		
	3	645	160		

4.2. Fatigue tests

During the fatigue tests, a stepwise stress increment has been applied. The increment of temperature has been acquired in the same way as done for the static tensile tests (Figure 6). The different stabilization temperature phases have been detected during each load step. In Figure 8, it is possible to observe the three different phases of thermographic method for the V-notch specimen recorded during the fatigue test.

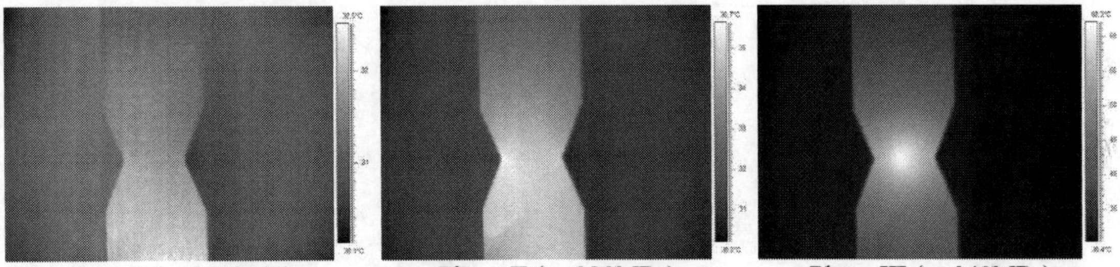
Phase I (σ=140MPa) Phase II (σ=200MPa) Phase III (σ=240MPa)

Figure 8. V-notch specimen's surface temperature during fatigue test.

For stress levels below 110 MPa, no significative increase in the temperature trend can be appreciated and the temperature signal can be assumed noisy. On the other hand, when the stress level reaches 150 MPa, a rapid temperature increment can be noticed.

For the stress level higher than 220 MPa, a very high further temperature increment is exhibited until the specimen failure.

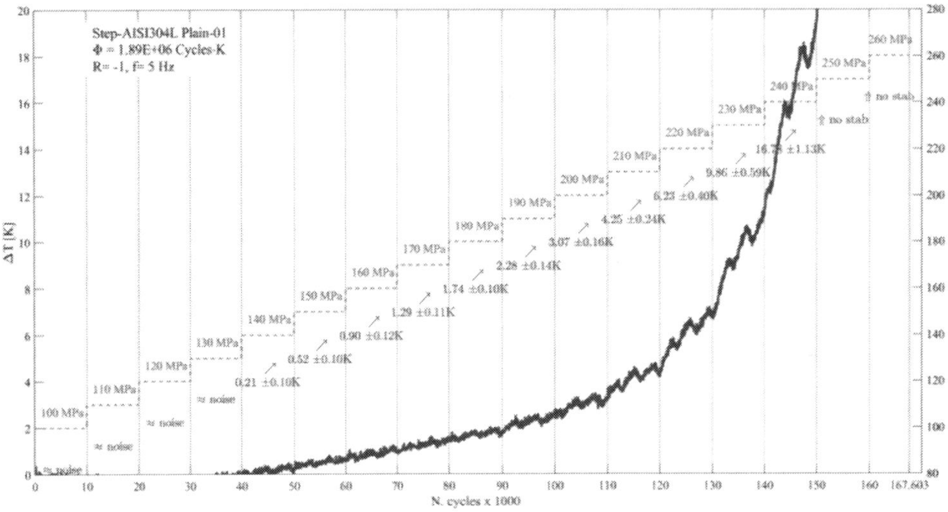

(a)

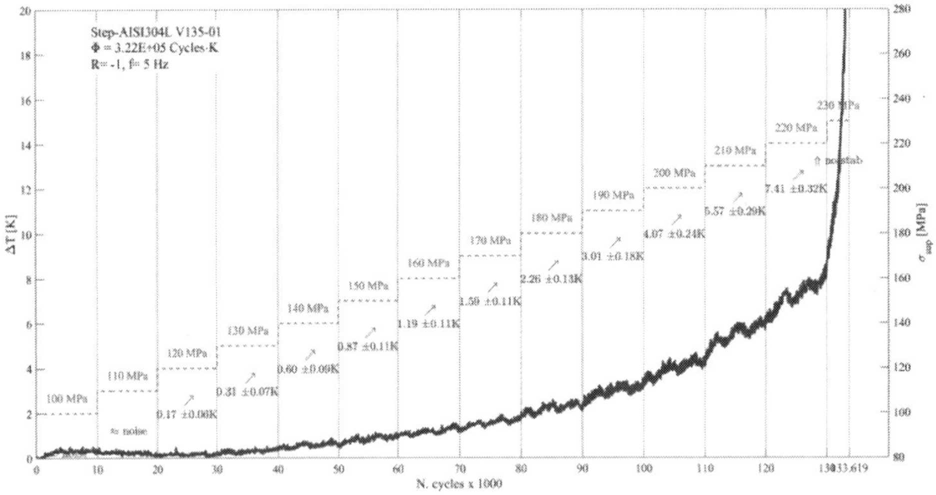

(b)

Figure 9. Temperature trend related to stepwise stress level for (a) Plain and (b) V-notch specimen.

The variation of temperature trend of both specimens is shown in Figure 9. The significant temperature increase is noticed moving from 110-120 MPa to 220-240 MPa for the two types of specimens. It is reasonable to think that the fatigue limit of the material under study falls within these stress levels.

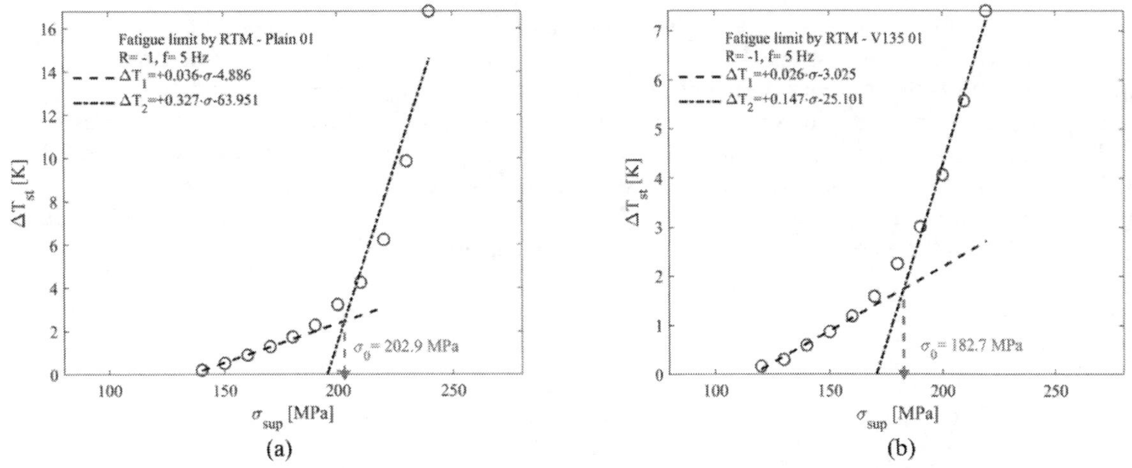

Figure 10. Fatigue limit of (a) Plain and (b) V-notch specimen by RTM.

For the estimation of the fatigue limit adopting the RTM, by representing the different stabilization temperatures and the applied stress levels (Figure 10), it is possible to observe two linear trends with different slopes, the first trend is typical of the stress level below the fatigue limit of the material and may be due to noise in the temperature signal or to the damping of the material. The second trend is characteristic of the energy release caused by the fatigue damage. From the intersection of the linear trends, it is possible to quickly define the fatigue limit of the material.

The results of stepwise fatigue tests performed to obtain the fatigue limits by RTM were of $\sigma_{0\ R\text{-}1}$ = 199 ± 4 MPa for Plain and $\sigma_{0\ R\text{-}1}$= 186 ± 3 MPa for V-Notch.

4.3. Rapid estimation of critical CV radius for SED method application

A simulation for the calculation of the SED values around the notch tip has been performed with a 2D elastic finite element model. The average limit stress detected by STM for V-notch specimens (152 MPa) has been applied as loading condition in the FE model.

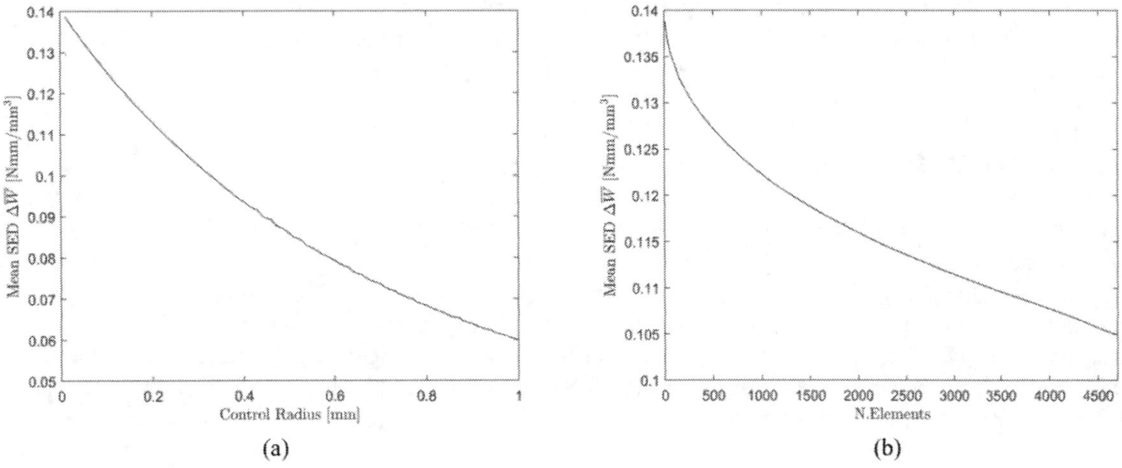

Figure 11. a) Average SED trend vs. the CV radius; b) Average SED trend vs. the number of elements.

In the post processing phase, the SED values were obtained with a polar selection from the notch tip by the variation of the dimension of the CV radius from 0.01 mm to 1 mm with steps of 0.01 mm.

The average SED trend vs. the CV radius has been reported in Figure 11. As the control radius increases, as well as the number of elements, the SED values follow a decreasing trend.

In Figure 12, the constant line shows the critical value of SED obtained analytically by Eq. (1) using the limit stress resulted from the STM for Plain specimen (179 MPa). The intersection between the SED critical value for Plain specimen and the average SED trend obtained from the FE model of V-notch specimen over the CV radius identify a critical CV radius characteristic of the material. A value of 0.36 mm has been found as a critical CV radius for the SED evaluation of AISI 304L material.

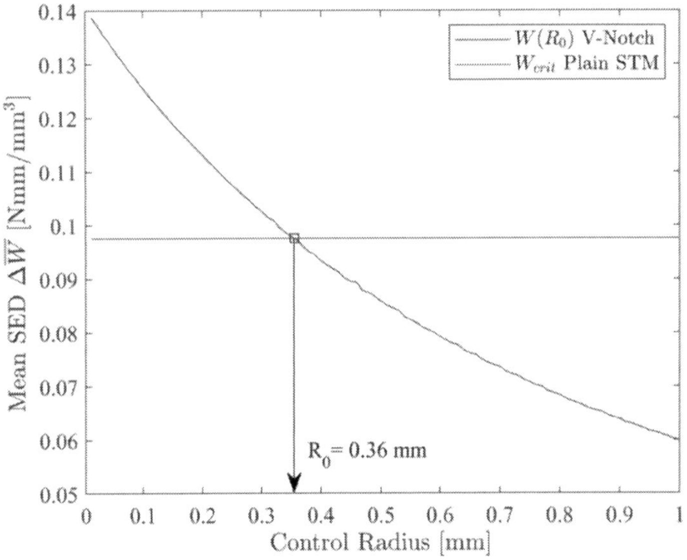

Figure 12. Identification of the critical CV radius using SED trend.

5. Conclusion

In this paper, for the first time, the energetic release during tensile tests of AISI 304L steel has been evaluated to rapidly estimate the CV radius of the SED method. For the STM the average value of the limit stress has been evaluated as the stress level at which the temperature deviates from its linear trend, obtaining a value of 179 ± 21 MPa for Plain specimens and of 152 ± 27 MPa for V-Notch specimens.

Stepwise fatigue tests have been performed to obtain the fatigue limits by RTM of $\sigma_{0\,R\text{-}1}$= 199 ± 4 MPa for Plain and $\sigma_{0\,R\text{-}1}$= 186 ± 3 MPa for V-Notch specimen.

Experimental campaigns have shown that the presence of the notch cause an increase of the mechanical properties of the AISI 304L on static field, while the fatigue performances of V-Notch specimen are lower compared to the Plain one.

An innovative method for the rapid estimation of the CV radius for SED through thermographic methods it was presented. The trend of SED value versus CV radius for V-notch specimen of AISI 304L was evaluated with the easy way of the volume-free procedure. The experimental critical SED value for AISI 304L Plain specimen was estimated through STM. Comparing these two different approach the critical CV radius for AISI 304L material was assessed.

References

[1] Rosa, G. La and Risitano, A. (2000) Thermographic methodology for rapid determination of the fatigue limit of materials and mechanical components. 22, 65–73.

[2] Fargione, G., Geraci, A., La Rosa, G. and Risitano, A. (2002) Rapid determination of the fatigue curve by the thermographic method. *International Journal of Fatigue*, 24, 11–9. https://doi.org/10.1016/S0142-1123(01)00107-4

[3] Risitano, A., Risitano, G. and Clienti, C. (2010) Fatigue limit by thermal analysis of specimen surface in mono axial traction test. *EPJ Web of Conferences*, 6. https://doi.org/10.1051/epjconf/20100638010

[4] Corigliano, P., Cucinotta, F., Guglielmino, E., Risitano, G. and Santonocito, D. (2020) Fatigue assessment of a marine structural steel and comparison with Thermographic Method and Static Thermographic Method. *Fatigue and Fracture of Engineering Materials and Structures*, 43, 734–43. https://doi.org/10.1111/ffe.13158

[5] Foti, P., Risitano, G., Berto, F. and Santonocito, D. (2021) Evaluation of the Energetic Release During Tensile tests in Notched Specimens by means of Experimental and Numerical Techniques. *IOP Conference Series Materials Science and Engineering*,. https://doi.org/10.1088/1757-899X/1038/1/012038

[6] Lazzarin, P., Berto, F. and Ayatollahi, M.R. (2013) Brittle failure of inclined key-hole notches in isostatic graphite under in-plane mixed mode loading. 942–55. https://doi.org/10.1111/ffe.12057

[7] Foti, P., Santonocito, D., Risitano, G. and Berto, F. (2021) Fatigue assessment of cruciform joints : Comparison between Strain Energy Density predictions and current standards and recommendations. *Engineering Structures*, Elsevier Ltd. 230, 111708. https://doi.org/10.1016/j.engstruct.2020.111708

[8] Lazzarin, P. and Zambardi, R. (2002) The equivalent strain energy density approach re-formulated and applied to sharp V-shaped notches under localized and generalized plasticity. *Fatigue and Fracture of Engineering Materials and Structures*, 25, 917–28. https://doi.org/10.1046/j.1460-2695.2002.00543.x

[9] Lazzarin, P. and Zambardi, R. (2001) A finite-volume-energy based approach to predict the static and fatigue behavior of components with sharp V-shaped notches. 275–98.

[10] Foti, P., Crisafulli, D., Santonocito, D., Risitano, G. and Berto, F. (2022) Effect of misalignments and welding penetration on the fatigue strength of a common welded detail : SED method predictions and comparisons with codes. *International Journal of Fatigue*, Elsevier Ltd. 164, 107135. https://doi.org/10.1016/j.ijfatigue.2022.107135

[11] Fischer, C., Fricke, W. and Rizzo, C.M. (2016) Experiences and recommendations for numerical analyses of notch stress intensity factor and averaged strain energy density. *Engineering Fracture Mechanics*, Elsevier Ltd. https://doi.org/10.1016/j.engfracmech.2016.08.012

[12] Risitano, A. and Risitano, G. (2013) Determining fatigue limits with thermal analysis of static traction tests. 631–9. https://doi.org/10.1111/ffe.12030

[13] Clienti, C., Fargione, G., La Rosa, G., Risitano, A. and Risitano, G. (2010) A first approach to the analysis of fatigue parameters by thermal variations in static tests on plastics. *Engineering Fracture Mechanics*, 77, 2158–67. https://doi.org/10.1016/j.engfracmech.2010.04.028

[14] Beltrami, E. (1885) Sulle condizioni di resistenza dei corpi elastici. *Il Nuovo Cimento*, 18, 145–55. https://doi.org/10.1007/BF02824697

[15] Atzori, B., Berto, F., Lazzarin, P. and Quaresimin, M. (2006) Multi-axial fatigue behaviour of a severely notched carbon steel. 28, 485–93. https://doi.org/10.1016/j.ijfatigue.2005.05.010

[16] Foti, P. and Berto, F. (2022) Some useful expressions and a proof of the validity of the volume free procedure for the SED method application. *Engineering Fracture Mechanics*, Elsevier Ltd. 108818. https://doi.org/10.1016/j.engfracmech.2022.108818

An analysis of the dependence of the material length parameter a_0 on the stress ratio

B Atzori, D Rigon and G Meneghetti

University of Padova – Department of Industrial Engineering
Via Venezia, 1, 35131 Padova, Italy

bruno.atzori@unipd.it

Abstract. Safe life approaches in notch fatigue of metallic components rely on the knowledge of the material response to fatigue stresses at notches, which is typically quantified by a material-dependent length parameter. As a general material behaviour, such material parameter depends on several factors and one of the most important of them is the mean stress. The present study is focused on the El Haddad-Smith-Topper material length parameter a_0 for analysing under which circumstances and to which extent it may vary with the applied mean stress or stress ratio. The analysis starts by considering the broader problem of the mean stress influence on the fatigue limit of plain material, blunt and sharp notches and cracks. Eventually, an analysis of previous sets of experimental data on wrought steels led to simple relations for estimating the a_0 parameter at different stress ratios on the basis of the sole Vickers Hardness

1. Introduction

Geometrical stress concentration effects in mechanical components and structures started to be quantified analytically between the end of XIX and the beginning of the XX century, leading to the definition of the stress concentration factor K_t. On the other hand, the effects of stress raisers on the fatigue behaviors, and particularly on the fatigue limit, has been analysed systematically only starting from approximately the mid of the previous century thanks to the investigations due to Neuber [1] and Peterson [2]. They led to the definition of the fatigue strength reduction factor K_f, defined as the ratio between the plain and notched fatigue limits for zero mean stress. When K_f and K_t factors are compared, then notch sensitivity arises, which is made to depend on material fatigue properties and, to a large extent, on the stress gradient at notches, which in turn strongly depends on the notch tip radius. The material-dependent constant governing its notch sensitivity is usually made to depend on the tensile strength of the material [3].

Besides the aforementioned classical approach, Fracture Mechanics discipline brought a significant advancement of fatigue design methods, in that enabled to estimate the fatigue strength in presence of nucleated cracks, which cannot be treated with the existing approaches since both the notch acuity (defined as the ratio between the notch depth and the notch tip radius $\zeta=a/\rho$) and the theoretical stress concentration factors growth to infinite. Kitagawa e Takahashi [4] highlighted the dependence of the fatigue limit on the crack size for short cracks and in their original diagram showed that for very short crack the fatigue limit is given by a constant stress condition (i.e. the plain material fatigue limit) while for sufficiently long cracks the fatigue limit is given by a constant stress intensity factor condition (namely the threshold value of the stress intensity factor range). Schematically reasoning, the critical

Content from this work may be used under the terms of the Creative Commons Attribution 3.0 licence. Any further distribution of this work must maintain attribution to the author(s) and the title of the work, journal citation and DOI.
Published under licence by IOP Publishing Ltd

crack size separating the two criteria is a material parameter, dependent on the mean stress, which can be expressed on the basis of the threshold range of the stress intensity factor ΔK_{th} and the plain material fatigue limit $\Delta\sigma_{Mth}$:

$$a_0 = \frac{1}{\pi}\left(\frac{\Delta K_{th}}{\Delta\sigma_{M,th}}\right)^2 \tag{1}$$

Crack sizes around a_0 proved to have a fatigue limit lower than predicted by both criteria, namely $\Delta\sigma = \Delta\sigma_{M,th}$ and $\Delta K = \Delta K_{th}$.

The use of Fracture Mechanics-based concept in notch fatigue was ever increasing in subsequent years (e.g. Smith e Miller [5]) and Atzori e Lazzarin formalised the use of LEFM for fatigue analysis of notches [6] by analysing the variation of the fatigue limit of notches at constant K_t-values (figure 1). Isibasi [7] introduced the concept of "branch point" as the characteristic K_t value above which the critical stress for crack propagation is different from and greater than the critical stress for crack initiation. This concept was clarified by Atzori and Lazzarin, which proposed a 3D diagram for assessing the fatigue strength of defects/cracks/notches characterised by a "branch line" (rather than a point) separating the fatigue behaviour of sharp and blunt notches [8]. The characteristic size separating blunt and sharp notch fatigue behaviour resulted $a^* = a_0\,K_t^2$, and notch size around a* proved to exhibit a higher fatigue strength than both the sharp and blunt notch behaviour. Subsequently [9,10], Atzori, Lazzarin and Meneghetti systematically validated the diagram and also extended the diagram to open V-notches using non-conventional advances of the LEFM. Figure 1 shows that the sharp notch fatigue behaviour is governed by ΔK_{th}, while K_t does not play any role. On the other hand, K_t defines the range of sizes where the notch exhibits a sharp behaviour, i.e., the same fatigue behaviour of a crack having the same size.

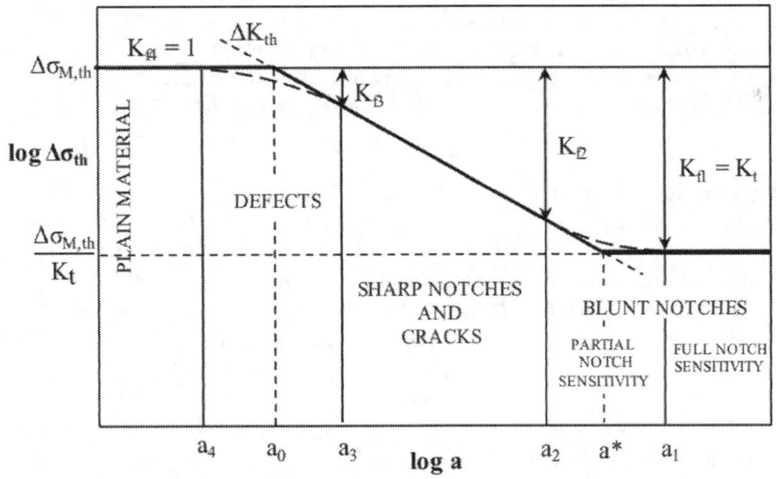

Figure 1. Scale effect in notch fatigue at constant K_t-values.

Regarding the mean stress effect on the fatigue limit, it is convenient to recall separately the criteria for classic notch fatigue and for the LEFM domain. In notch fatigue, mean stress sensitivity is analysed by using the Haigh diagram ($\Delta\sigma_{th}/2=f(\sigma_m)$) [11] and the most simple rule considers a linear dependence of the fatigue limit on the mean stress, with an upper bound of the mean stress equal to the material strength (σ_{uts} according to Goodman [12], σ_f o σ_f' according to Morrow [13]) which is coupled with a null fatigue strength:

$$\frac{\Delta\sigma_{th}}{\Delta\sigma_{th,-1}}+\frac{\sigma_m}{\sigma_{UTS}}=1 \tag{2}$$

In LEFM studies it is common practise to study the dependence $\Delta K_{th}=f(R)$ and synthesize the crack growth data by combining range and maximum values according to Walker [14]:

$$\Delta K^m K_{max}{}^{1-m}=\Delta K_{R=0} \tag{3}$$

or Klesnil and Lukas [15]:

$$K_a{}^{1-\gamma}K_{max}{}^{\gamma}=K_{a,R=-1} \tag{4}$$

with $\gamma=1$-m.

Equations 3 and 4 can both be expressed as a function of R, leading to the one usual in Fracture Mechanics:

$$\Delta K_{th}=\Delta K_{th,0}(1-R)^{\gamma} \tag{5}$$

referred to R=0 (as equation 3), where $\gamma = 1$-m. To allow and easy comparison of the fatigue limit of plain materials and cracks, besides equation (5), in what follows reference will be made to another formulation of ΔK_{th} dependence on the K_m value as proposed by Radhakrishnan [16].

Two remarks are worth mentioning. Firstly, the following analysis considers both R-ratio and mean stress dependence, as in practical application those situations are commonly encountered. Secondly, it is convenient in what follows to disregard the material strength in Goodman Eq (1) (or material toughness in Radhakrishnan equation) for the sake of better fitting experimental results. However, the compatibility of the newly proposed equations with well-established mean stress or R-ratio equations (Masounave [17], McEvily [18] and Davenport [19]) will be analysed.

The analysis of R-ratio dependence of the plain material fatigue limit and of the threshold range of the stress intensity factor will lead to analyse the R-ratio dependence of a_0 by virtue of Eq. (1). As a first application, simple, though approximate, estimations of a_0 for wrought steels, subjected to pure tensile mode I fatigue loading, will be proposed to be used in a preliminary design phase particularly when recent advanced local approaches to fatigue lifetime assessment are adopted [20,21].

2. Analysis of blunt and sharp notches

Although the terms 'blunt' and 'sharp' are widely used in notch fatigue, it may be useful to clarify their definition in the present context and reference will be made to the diagram proposed by Atzori and Lazzarin [8] previously introduced in figure 1. The diagram can be referred to the R-ratio -1 and illustrates the variation of the fatigue limit of a notch having an increasing size and constant K_t value; alternatively interpreted, the diagram shows the fatigue limit reduction factor K_f ($K_f = \Delta\sigma_{Mth}/\Delta\sigma_{Nth}$) for a variation to scale of a notch having constant acuity ζ (and hence K_t). Three well-defined fatigue behaviours are highlighted in the diagram: for vanishingly small and for sufficiently large notch sizes the fatigue behaviour is governed by the plain material fatigue limit $\Delta\sigma_{Mth}$, while for intermediate notch sizes it is governed by the threshold range of the SIF for long crack. For large notch sizes, the fatigue behaviour corresponds to the full notch sensitivity, the fatigue limit can be estimated by accounting for the K_t value, $\Delta\sigma_{Nth} = \Delta\sigma_{Mth}/K_t$ (with $K_f=K_t$) and it is referred to here as blunt notch behaviour; in the central zone of the diagram the fatigue limit can be estimated using the LEFM, $\Delta\sigma_{SNth} = \Delta K_{th}/\sqrt{\pi a}$ and the fatigue behaviour is referred to here as sharp notch behaviour. The range of notch sizes involving the sharp notch behaviour is wider the greater the notch acuity ζ and then the stress concentration factor K_t. The central zone of the sharp notch behaviour smoothly connects to the other two zones through transition behaviours called 'defect sensitivity' and 'notch sensitivity' [8,9].

3. Influence of stress cycle asymmetry on ΔK_{th}

Although the stress cycle asymmetry is generally expressed in terms of R-ratio, Radhakrishnan [16] proposed an expression similar to the Goodman line:

$$\frac{\Delta K_{th}}{\Delta K_{th,-1}}+\frac{K_m}{K_C}=1 \tag{6}$$

However, linking the sensitivity to stress cycle asymmetry of materials and cracks to strength (σ_{UTS}, equation (2)) and toughness (K_c, equation (6)) is questionable; an appropriate example is the comparison between the intercept σ_f'/E of the Manson-Coffin strain-life curve and the tensile strength of the stabilised stress-strain curve (see. for example table 1 of ref. [22]). Therefore, both equations (2) and (6) are modified by substituting the strict dependence on strength and toughness, respectively, with an angular coefficient α, to be calibrated on experimental results. More in detail, the angular coefficient appropriate for equation (6) is $\alpha_K = (\Delta K_{th,-1}/2)/K_0$, where K_C is thought of a free parameter capable of fitting the experimental data, in light of the previous argument. Therefore, equation (6) can be re-written either for constant K_m or constant R-ratio problems:

$$\frac{\Delta K_{th}}{\Delta K_{th,-1}}=1-\alpha_K\frac{2K_m}{\Delta K_{th,-1}} \quad \text{or} \quad =\frac{1}{1+\alpha_K\frac{(1+R)}{(1-R)}} \tag{7}$$

By comparing equations (5) and (7), both indicate that greater values of parameters γ ed α, respectively, translate into greater sensitivity to stress cycle asymmetry; moreover, the same fatigue behaviour is predicted in the limit cases $\gamma = \alpha = 0$ and $\gamma = \alpha = 1$.

Dealing with equation (2), the angular coefficient can be referred to the plain material (α_M), blunt notches (α_{BN}) and sharp notches (α_{SN}) with constant mean stress σ_m or R-ratio:

$$\frac{\Delta\sigma_{th}}{\Delta\sigma_{th,-1}}=1-\alpha\frac{2\sigma_m}{\Delta\sigma_{th,-1}} \quad \text{or} \quad =\frac{1}{1+\alpha\frac{(1+R)}{(1-R)}} \tag{8}$$

where α means α_M, α_{BN} or α_{SN}.

Figure 2 illustrates equation (8). The reduction in the push-pull fatigue limit ($R = -1$, $\sigma_m = 0$) for blunt notches starting from the plain material, the static strength being unchanged, implies lower angular coefficients α_{BN} than α_M: the greater K_f, the lower α_{BN} as compared to α_M. On the other hand, a different argument must be put forward for sharp notches. Since their fatigue limit is predicted by the LEFM equation (7), the angular coefficient α_{SN} must equal α_K. A similar situation should also occur for the inverse slope k of the Woehler curve: an increasing K_f implies a reduction of k for blunt notches (lower k values meaning steeper Woehler curves), for sharp notches, higher K_f values translate into lower Woehler curves with inverse slope unchanged.

Figures 3 and 4 compare equation (7) with well-established formulae of the literature (Masounave [17], equation (9), McEvily [18], equation (10), e Davenport [19], equation (11)), for constant R-ratio and constant K_m, respectively. Since in fracture mechanics studies the reference R-ratio is typically R = 0 (see for example [20]), all equations refer to the pulsating tension case, where equation (12) updates the previous equation (7):

$$\frac{\Delta K_{th}}{\Delta K_{th,0}}=(1-R) \quad \text{or} \quad =2-\frac{2K_m}{\Delta K_{th,0}} \tag{9}$$

$$\frac{\Delta K_{th}}{\Delta K_{th,0}}=\left(\frac{1-R}{1+R}\right)^{0.5} \quad \text{or} \quad =\left(\frac{2K_m}{\Delta K_{th,0}}\right)^{-1} \tag{10}$$

$$\frac{\Delta K_{th}}{\Delta K_{th,0}}=(1-R)^{0.5} \quad \text{or} \quad =\frac{\sqrt{\left(\frac{2K_m}{\Delta K_{th,0}}\right)^2+8}-\frac{2K_m}{\Delta K_{th,0}}}{2} \tag{11}$$

IOP Conf. Series: Materials Science and Engineering 1275 (2023) 012019 doi:10.1088/1757-899X/1275/1/012019

$$\frac{\Delta K_{th}}{\Delta K_{th,0}} = \frac{1+\alpha_k}{1+\alpha_k\left(\frac{1+R}{1-R}\right)} \quad \text{or} \quad =(1+\alpha_k) - \alpha_K \frac{2K_m}{\Delta K_{th,0}} \tag{12}$$

Figures (3) and (4) demonstrate the compatibility of the proposed equation (12) with existing formulations and also highlight its flexibility to fit experimental data thanks to the calibration parameter α.

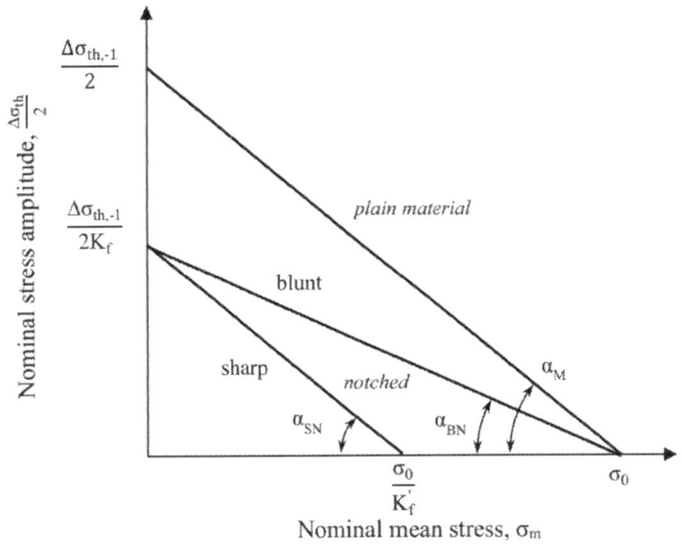

Figure 2. Different mean stress sensitivities of blunt (α_{BN}=f(K_f)) and sharp (α_{SN}=constant) notches for ductile metals.

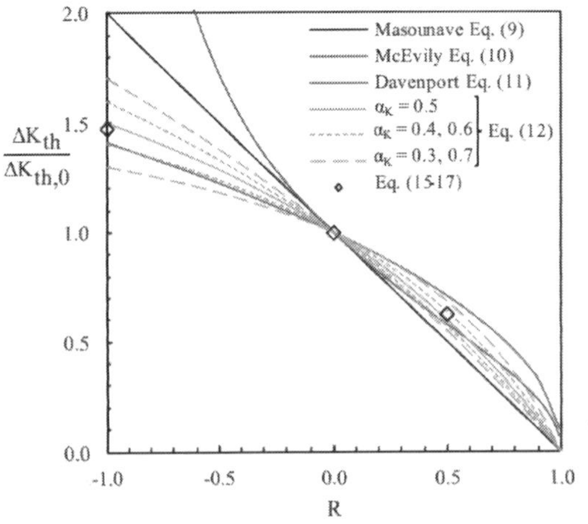

Figure 3. Comparison between the proposed equation (12) and existing formulae for describing the dependence ΔK_{th}=f(R).

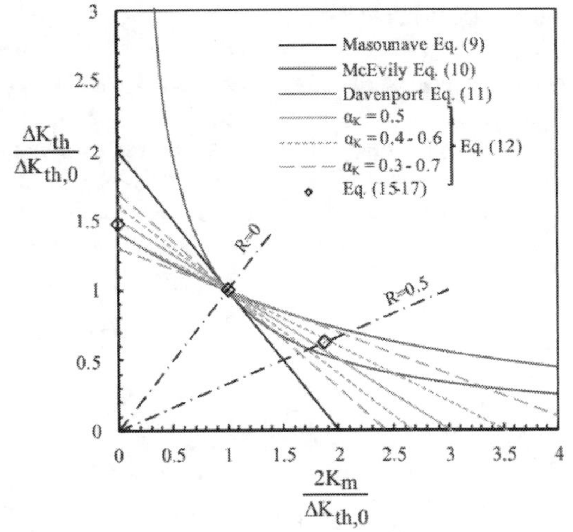

Figure 4. Comparison between the proposed equation (12) and existing formulae for describing the dependence $\Delta K_{th}=f(K_m)$.

4. Influence of stress cycle asymmetry on a_0

Using equations (7) and (8), it is straightforward to evaluate the influence of R on the El Haddad-Smith-Topper length parameter, $a_{0(R)}$, compared to $a_{0(R=-1)}$:

$$\frac{a_{0(R)}}{a_{0(R=-1)}}=\left(\frac{\Delta K_{th(R)}}{\Delta\sigma_{M,th(R)}}\cdot\frac{\Delta\sigma_{M,th(R=-1)}}{\Delta K_{th(R=-1)}}\right)^2=\left(\frac{1+\alpha_M\left(\frac{1+R}{1-R}\right)}{1+\alpha_K\left(\frac{1+R}{1-R}\right)}\right)^2 \tag{13}$$

or, by using equation (5) applied to both ΔK_{th} and $\Delta\sigma_{M,th}$, the same dependence referred to $a_{0(R=0)}$:

$$\frac{a_{0(R)}}{a_{0(R=0)}}=\left(\frac{(1-R)^{\gamma_K}}{(1-R)^{\gamma_M}}\right)=(1-R)^{\gamma_K-\gamma_M} \tag{14}$$

According to equations (13) and (14) a_0 is insensitive to the R-ratio if $\alpha_K=\alpha_M$ or $\gamma_K=\gamma_M$; in all other cases there is a R-sensitivity illustrated in figure 5 for different exemplary values of (γ_K-γ_M).

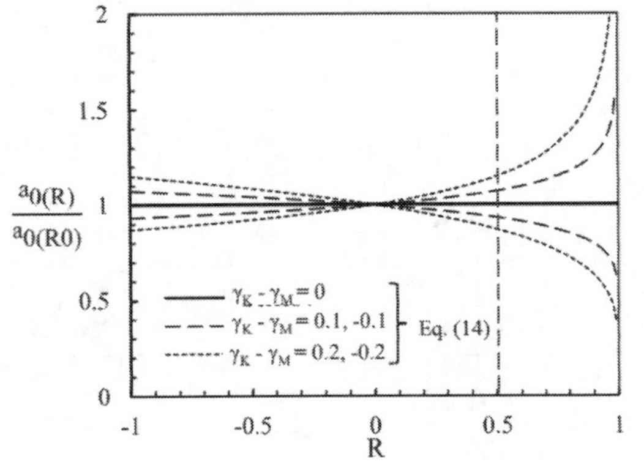

Figure 5. R-ratio dependence of the material length parameter a_0 (equation (1)) for different (γ_K- γ_M) values according to equation (14).

5. R dependence of ΔK_{th} and a_0 for wrought steels

Recently, Meneghetti and Rigon [23,24] have proposed empirical equations for estimating the R-dependence of ΔK_{th} for steels, which are based on the Vickers hardness HV and a microstructural length parameter *l*. In the present study, a simplification is attempted to make the ΔK_{th} of steels dependent solely on HV, similarly to the equation proposed by Murakami for estimating the threshold stress of defects [25]. By considering the experimental data published previously [23,24], to which the reader is referred, the reanalysis has been conducted in the present study by simply eliminating the dependence on L and by analysing the sole influence of HV on ΔK_{th} for long cracks for wrought steels and for R-ratios equal to -1, 0 and 0.5. The data were interpolated by a power function with fixed exponent equal to 0.5 for all R ratios only accounting HV values between 100 and 400.

The resulting expressions for different R values and valid for 100<HV<400, which best fit the available experimental results are as follows:

$$R=-1:\quad \Delta K_{th}=135HV^{-0.5}\quad [MPa\sqrt{m}] \tag{15}$$

$$R=0:\quad \Delta K_{th}=93HV^{-0.5}\quad [MPa\sqrt{m}] \tag{16}$$

$$R=0.5:\quad \Delta K_{th}=58HV^{-0.5}\quad [MPa\sqrt{m}] \tag{17}$$

It is worth noting that such equations lead to estimating errors higher than ±20% (obtained by using equations proposed in [23,24]). However, equations 15-17 provide a rough estimation of ΔK_{th} necessary to analyse the α_K value for wrought steels.

ΔK_{th} values obtained from equations (15-17) are reported in previous figures (3) and (4). By using ΔK_{th} evaluated at R = 0 (equation (16)) into equation (7), α_K results 0.45, while using ΔK_{th} evaluated at R=0.5 (equation (17)) into equation (7), α_K results 0.44, while for the fatigue limit $\Delta\sigma_{Mth}$ the value of α_M resulted to be 0.5. Alternatively, by using equation (5), the corresponding values are $\gamma = 0.53$ for R=-1 and $\gamma = 0.68$ for R = 0.5. It is interesting to note that in case of Al alloys Walker [14] found $\gamma = 0.5$ for 2024-T3 and 0.575 for 7075-T6, such exponents being valid for both $\Delta\sigma_{Mth}$ and ΔK_{th}. Regarding the SWT parameter [26] γ is invariably taken as 0.5, independently of the steel material considered.

Due to the similar values of α_K and α_M for the analysed test results, by plotting the available experimental results on a chart reporting a_0 versus HV for different R values (figure 6), R-independence can be assumed as a first approximation. Consequently, since the data are linearly distributed in a double log scale (figure 6), by considering all data belonging to a single statistical population, a power function has been fitted by considering HV values ranging from 100 and 400, and calculating the scatter band relevant to 10-90% of estimating probabilities. Finally, the resulting equation is the following:

$$a_0=\left(\frac{893}{HV}\right)^{2.93}\quad [\mu m] \tag{18}$$

which will be approximated as in next equation (19).

By considering the analytical relations between Vickers hardness and fatigue limit of steels [25] and between fatigue limit and static strength [27,28] ($\Delta\sigma_{th,-1}=\sigma_{UTS}=3.2HV$), a_0 can be estimated, starting from equation (19), also with equations (20) and (21):

$$a_{0,10\%\div90\%}=\left(\frac{500\div1600}{HV}\right)^{3}\ [\mu m]\ \rightarrow\ a_{0,50\%}=\left(\frac{900}{HV}\right)^{3}\ [\mu m] \tag{19}$$

$$a_{0,10\%\div90\%}=\left(\frac{1600\div5000}{\Delta\sigma_{th,R-1}}\right)^{3}\ [\mu m]\ \rightarrow\ a_{0,50\%}=\left(\frac{3000}{\Delta\sigma_{th,R-1}}\right)^{3}\ [\mu m] \tag{20}$$

$$a_{0,10\%\div90\%}=\left(\frac{1600\div5000}{\sigma_{UTS}}\right)^{3}\ [\mu m]\ \rightarrow\ a_{0,50\%}=\left(\frac{3000}{\sigma_{UTS}}\right)^{3}\ [\mu m] \tag{21}$$

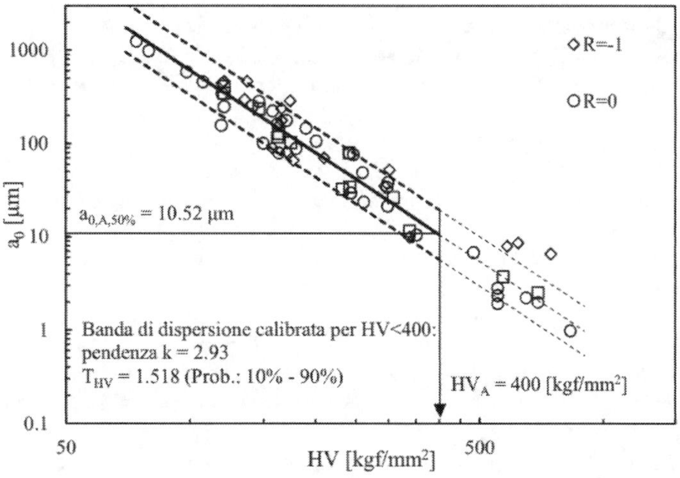

Figure 6. El Haddad-Smith-Topper length parameter a_0 versus Vickers hardness for R-ratios between -1 and 0.5.

6. Conclusions

The El Haddad-Smith-Topper length parameter a_0 is known to be widely used in advanced local approaches for fatigue design. The present study has focused on the R-ratio dependence of a_0 and, on the definition of simple, though approximate, analytical expressions to estimate a_0 in a preliminary design phase.

Since a_0 depends on two material properties, namely the threshold range of the stress intensity factor for long cracks and the fatigue limit of plain material (equation (1)), new equations have been proposed to express their stress cycle asymmetry dependence. In such a way, the mean stress sensitivity is taken up by a parameter which best fits the experimental results, while the explicit dependence on material static properties like strength and toughness is disregarded. The newly proposed equations have been checked for consistency with similar equations existing in the literature. Finally, this study reports the simple empirical equations to estimate a_0 for wrought steels at different stress ratios, which have been determined by analysing previously published experimental data taken from the literature.

The following conclusions can be drawn:

- The mean stress sensitivity of blunt notches is different from sharp notches, in that the latter behave like brittle materials in the Haigh diagram, even though the material behaviour is ductile;
- The influence of the R ratio on the a_0 can be captured by the equation proposed in this study (equation (13)), which consist in a modified Goodman-type equation, and by the traditional equation used in fracture mechanics (equation (14));
- The degree of R-ratio sensitivity of a_0 depends on the difference between the parameters that describe the mean stress sensitivity of the plain and of the cracked material (α_M and α_K in equation (13), γ_M and γ_K in equation (14));
- The empirical simple equations for estimating a_0 have been expressed as a function of (i) Vickers hardness HV or (ii) plain material push-pull fatigue limit or (iii) material static strength and may be useful in a preliminary design phase

References

[1] Neuber H 1937 *Kerbspannungslehre* (Berlin: Springer-Verlag)
[2] Peterson R E 1959 Notch Sensitivity *Metal Fatigue*, (New York: McGraw Hill)
[3] Pilkey W D and Pilkey D F 2008 *Peterson's stress concentration factors* (John Wiley & Sons, Ltd)
[4] Kitagawa H and Takahashi S 1976 Applicability of fracture mechanics to very small cracks or the cracks in the early stage *Proc. 2nd Int. Conf. Mech. Behav. Mater. - ICM2* pp 627–31.
[5] Smith R A and Miller K J 1978 Prediction of fatigue regimes in notched components *Int. J. Mech. Sci.* **20** 201–206
[6] Atzori B and Lazzarin P 2000 Notch sensitivity and defect sensitivity under fatigue loading: two sides of the same medal *Int. J. Fract.* **107** L3–L8
[7] Isibasi T 1954 *Prevention of fatigue and fracture of metals*, (Tokyo)
[8] Atzori B and Lazzarin P 2002 A three-dimensional graphical aid to analyze fatigue crack nucleation and propagation phases under fatigue limit conditions *Int. J. Fract.* **118** 271–84
[9] Atzori B, Lazzarin P and Meneghetti G 2003 Fracture mechanics and notch sensitivity *Fatigue Fract. Eng. Mater. Struct.* **26** 257–67
[10] Atzori B, Lazzarin P and Meneghetti G 2005 A unified treatment of the mode I fatigue limit of components containing notches or defects, *Int. J. Fract.* **133** 61–87
[11] Haigh B P 1917 Experiments on the Fatigue of Brasses *J. Inst. Metals* **18** 55-86
[12] Goodman J 1899 *Mechanics applied to Engineering* (London: Longman, Green and Company).
[13] Morrow J 1968 Fatigue Design Handbook (Pasadena: Pub. N. AE-4, SAE) Section 3.2 Fatigue Properties of Metals
[14] Walker K 1970 The Effect of Stress Ratio During Crack Propagation and Fatigue for 2024-T3 and 7075-T6 Aluminum *Effects of Environment and Complex Load History on Fatigue Life*, ASTM STP **462** 1-14.
[15] Klesnil M and Lukáš P 1972 Effect of stress cycle asymmetry on fatigue crack growth *Mater. Sci. Eng.* **9** 231–240
[16] Radhakrishnan V M 1990 Endurance diagram *Int. J. Fatigue* **12** 513–17
[17] Masounave J and Bailon J P 1975 The dependence of the threshold stress intensity factor on the cyclic stress ratio in fatigue of ferritic-pearlitic steels *Scr. Metall.* **9** 723–30
[18] McEvily A J 1977 Current aspects of fatigue *Met. Sci.* **11** 274–284
[19] Davenport R T and Brook R 1979 The threshold stress intensity range in fatigue *Fatigue Fract. Eng. Mater. Struct.* **1** 151–58
[20] Taylor D 2007 *The theory of critical distances a new perspective in fracture mechanics (Elsevier)*
[21] Radaj D and Vormwald M 2013 *Advanced Methods of Fatigue Assessment*,(Berlin: Springer, Berlin, Heidelberg)
[22] Dowling N E, Calhoun C A and Arcari A 2009 Mean Stress Effects in Stress-Life Fatigue and the Walker epquation *Fatigue Fract. Engng. Mater. Struct.* **32** 163-179
[23] Rigon D and Meneghetti G 2020 An engineering estimation of fatigue thresholds from a microstructural size and Vickers hardness: application to wrought and additively manufactured metals *Int. J. Fatigue.* **139** 105796.
[24] Rigon D and Meneghetti G 2022 Engineering estimation of the fatigue limit of wrought and defective additively manufactured metals for different load ratios *Int. J. Fatigue* **154** 106530
[25] Murakami Y 2002 *Metal Fatigue: Effects of Small Defects and Nonmetallic Inclusions* (Elsevier)
[26] Smith K N, Watson P and Topper T H 1970 A Stress-Strain Function for the fatigue of metals, *J. mater.* **5** 767-778
[27] Forrest P G 1962 *Fatigue of Metals* (London: Pergamon Press)
[28] Atzori B, Meneghetti G and Susmel L 2005 Material fatigue properties for assessing mechanical components weakened by notches and defects *Fatigue Fract. Eng. Mater. Struct.* **28** 83–97

IOP Conf. Series: Materials Science and Engineering 1275 (2023) 012020 doi:10.1088/1757-899X/1275/1/012020

The use of fractional order statistics for estimating nonparametric confidence intervals for quantiles of the fatigue damage computed in service random loadings

D Benasciutti[1] and J M E Marques[2]

[1] University of Ferrara, Department of Engineering, via Saragat 1, 44122 Ferrara, Italy
[2] Czech Technical University in Prague, Department of Mechanics, Biomechanics and Mechatronics, Faculty of Mechanical Engineering, Technická 4, 166 36 Prague, Czech Republic.

Corresponding author's email address: denis.benasciutti@unife.it

Abstract. This paper deals with the statistical variability of fatigue damage in random loadings as caused by the inherent randomness of the loading. It examines the use of a nonparametric method based on fractional order statistics as a tool for constructing confidence intervals for the quantile of fatigue damage, regardless of the knowledge of the damage probability distribution. The approach here discussed relies on a statistical method existing in the literature. After a short review of this method, with also an emphasis on possible advantages and limitations, the paper presents a numerical benchmark, followed by a practical application to time-histories measured on a mountain-bike in off-road tracks. The paper concludes with some critical remarks and an outlook.

1. Introduction

The statistical variability of the fatigue damage $D(T)$ of a random loading of length T can be characterized by the variance of damage $\mathrm{Var}[D(T)]$, which – along with the expected damage $E[D(T)]$ – is used to define the coefficient of variation (CoV) $\sqrt{\mathrm{Var}[D(T)]}/E[D(T)]$ as a dimensionless measure of uncertainty.

The interest on the study of the variance (or CoV) of fatigue damage in random loadings dates to the sixties, with the pioneering works of Bendat [1] and Crandall et al. [2,3]. They devised analytical formulas for calculating the variance $\mathrm{Var}[D(T)]$ in the case where the random time-history $z(t)$ is the output of a stationary Gaussian linear oscillator. The obtained variance expressions depend upon a few statistical properties of the random loading (e.g. variance, mean up-crossing frequency) and the oscillator (e.g. damping ratio), other than upon SN parameters.

After Bendat's and Crandall et al.'s works, the interest on the damage variance continued in subsequent decades, for example in Madsen et al. [4] and, more recently, in the studies [5] and [6,7], which extended Bendat's and Crandall et al.'s solutions to any type of narrowband random loading, either Gaussian [4-6] or non-Gaussian [7].

All the above variance solutions rely on the knowledge of the statistical properties characterizing the random loading, e.g. autocorrelation function or power spectral density. Often, these quantities are not known exactly by mathematical expressions and they must be estimated from measured data. This

Content from this work may be used under the terms of the Creative Commons Attribution 3.0 licence. Any further distribution of this work must maintain attribution to the author(s) and the title of the work, journal citation and DOI.
Published under licence by IOP Publishing Ltd

introduces a sampling variability into the computational model of the damage variance, the effect of which is not considered by the aforementioned methods.

This aspect motivated our attempt to develop a different approach in which the statistical variability of fatigue damage is estimated by means of confidence intervals, constructed after a direct analysis of one or more time-histories [6]. First developed for the case of stationary random loadings, the approach has been extended to non-stationary loadings of "switching type" (i.e., formed by a sequence of stationary states) [8,9]. The method has been benchmarked, first against numerically simulated time-histories, and then against more realistic time-histories measured in a mountain-bike in off-road tracks; in either case, a quite good accuracy was observed [8,9].

It is the purpose of the present article to improve further this approach by abandoning two of its hypotheses, in which the confidence interval is: i) constructed for the expected damage $E[D(T)]$ and ii) under the assumption of $D(T)$ following a normal distribution, a hypothesis introduced in early studies [3, 5,10] but recently called into question [11].

With the aim to overcome the two previous hypotheses, the method proposed herein makes use of nonparametric fractional order statistics to construct confidence intervals for any quantile of damage $D(T)$, regardless of the specific form of its probability distribution. The method thus allows for the construction of confidence intervals other than the median or the expected value. In this article, numerical and experimental examples are used to emphasize advantages and limitations of the proposed method.

It has finally to be noted that, in the following, only the statistical variability coming from the randomness of the loading will be considered. It is assumed that the scatter of constant amplitude fatigue data is taken into account e.g. by the definition of a characteristic SN line referred to prescribed levels of failure probability and confidence (e.g. P_f=2.3% with 95% confidence).

2. The uncertainty in fatigue damage

2.1. Fatigue damage and its expected value

Let $z(t)$, $0<t<T$, be a random time-history of finite length T. The fatigue damage under Palmgren-Miner (PM) rule is:

$$D(T)=\sum_{i=1}^{l}\frac{n_i}{N(s_i)} \tag{1}$$

where $N(s_i)$ is the number of cycles to failure at constant amplitude s_i, n_i is the number of cycles within the i-th class with amplitude s_i, and l is the number of amplitude classes.

Equation (1) considers a discrete set of amplitudes s_i referred to cycles counted in $z(t)$; another set is obtained if another time-history $z(t)$ is considered. In the limit of continuously distributed amplitudes, classes s_i become infinitely narrow and the summation approaches an integral [6]:

$$E[D(T)]=E[n]\int_0^{\infty}\frac{1}{N(s)}p(s)\mathrm{d}s \tag{2}$$

where $E[D(T)]$=expected damage, $E[n]$=expected number of cycles in $z(t)$, $p(s)$=probability distribution of amplitudes. In Eq. (2), the probability distribution $p(s)$ describes the randomness of the amplitudes of the cycles counted in $z(t)$.

The previous formulas are very general as they do not imply any mathematical expression to relate s and $N(s)$, nor do they require $z(t)$ being stationary – $z(t)$ can also be nonstationary. Often, it is assumed that amplitudes and cycles to failure are related as $N(s)=As^{-k}$ (elementary PM rule), which in a log-log diagram is a straight line with slope k and intercept $A=s_A^kN_A$, where the reference amplitude s_A is computed at N_A cycles (for example, N_A=2×10^6 cycles), for a predetermined failure probability. For this single-slope SN curve, the fatigue damage in Eq. (1) and (2) writes:

$$D(T)=\frac{1}{A}\left(\sum_{i=1}^{l} n_i\, s_i^k\right) \quad , \quad E[D(T)]=E[n]\frac{E[s^k]}{A} \tag{3}$$

Where $E[s^k]$ is the k-th moment of the amplitude distribution $p(s)$. Closed-form expressions can be obtained if the expression of $p(s)$ is known; for example, if $z(t)$ is a so-called narrowband loading, $p(s)$ is Rayleigh and Eq. (2) yields $E[D(T)]=(\nu_0^+ T/A)\left(\sqrt{2\mathrm{Var}[z(t)]}\right)^k \Gamma(1+k/2)$, where ν_0^+ is the number of mean value upcrossings of $z(t)$ [6].

When a single-slope SN equation is believed to be too conservative, two-slope SN models are used in which a knee point is followed by a less steep line with slope k'=2k–1 or 2k–2 (Haibach model [12]) or even by a horizontal line that represents a "endurance (fatigue) limit". In this case, it is straightforward to modify the previous damage formulas accordingly.

3. Confidence intervals

Let us consider a random variable X with cumulative distribution function $F(x)$. Further, assume that θ is a parameter of $F(x)$, e.g. it may be the expected value of a normal distribution. A two-sided confidence interval (CI) for an unknown parameter θ is an interval, defined by two end points θ_l and θ_u, that is expected to enclose the unknown parameter θ a specified percentage of cases after that an experiment is repeated under identical conditions. If the specified percentage is 100(1 − α)%, a CI is defined as [13]:

$$\Pr[\theta_l<\theta<\theta_u]=(1-\alpha) \quad \text{for } 0<\alpha<1 \tag{4}$$

(subscripts l and u mean "lower" and "upper"). The values θ_l and θ_u, which are constructed around the estimator $\hat{\theta}$ of θ, are called the lower and upper confidence limits, respectively, and the probability (1 − α) is named coverage probability or confidence level.

Often, the parameter θ of interest is the expected value of a Gaussian probability distribution. In this case, the two-sided 100(1 − α)% CI for μ is [13]:

$$\bar{x} - t_{\alpha/2;n-1}\frac{\hat{\sigma}}{\sqrt{n}} < \mu < \bar{x} + t_{\alpha/2;n-1}\frac{\hat{\sigma}}{\sqrt{n}} \tag{5}$$

where $t_{\alpha/2;n-1}$ is the upper $100\,\alpha/2$ percentage point of the Student t-distribution with n–1 degrees of freedom, while $\bar{x}$ and $\hat{\sigma}^2$ are the sample mean and sample variance, respectively. In other engineering applications, the parameter θ of interest is a quantile of the probability distribution. In the following, we provide an overview of some confidence interval expressions.

3.1. Confidence intervals for the expected damage E[D(T)]

The last result of previous section forms the basis of the confidence interval for the expected damage. Let $z(t)$ be a non-stationary "switching" random loading, formed by a finite number of stationary states N_S. Each stationary state has a total time length T_j, so that $T=\sum_{j=1}^{N_S} T_j$ is the total duration of $z(t)$. Note that each state needs not to appear in $z(t)$ only once in its full length T_j, but it can appear a multitude of times, each with duration shorter than T_j, provided that its total time length is T_j. It is irrelevant in which time order and how many times a state appear in $z(t)$, so at the beginning of the analysis, the states are reordered so that they appear only once in their full length T_j [8,9].

Methods for identifying different states in $z(t)$ (fast Fourier transform, run test) are described in [14]. After state identification, each state is further subdivided into N_B blocks of equal length T_B. For each block, the fatigue damage is finally computed. In summary, state identification and block subdivision yield the set of damage values $D_{B,ij}(T_B)$, with i=1,2,…,N_B and j=1,2,…,N_S. Damage $D_{B,ij}(T_B)$ refers to block i in state j. Under the assumption of $D(T)$ normally distributed, the confidence the confidence interval for the expected damage $E[D(T)]$ then is [8,9]:

$$D(T) - t_{\eta,\beta/2}\sum_{j=1}^{N_S}\sqrt{N_B \cdot \hat{\sigma}^2_{D_B j}} \leq E[D(T)] \leq D(T) + t_{\eta,\beta/2}\sum_{j=1}^{N_S}\sqrt{N_B \cdot \hat{\sigma}^2_{D_B j}} \quad (6)$$

where $\hat{\sigma}^2_{D_B j}$ is the sample variance of the damage of blocks for the j-th stationary state, and $t_{\eta,\beta/2}$ is the quantile of the Student t-distribution with an equivalent number of degrees of freedom $\eta \cong (N_B - 1)\left(\sum_{j=1}^{N_S}\hat{\sigma}^2_{D_B j}\right)^2 / \sum_{j=1}^{N_S}\left(\hat{\sigma}^2_{D_B j}\right)^2$. When the switching time-history $z(t)$ has only one stationary state (N_S=1), the previous confidence interval expression converges to the solution given in [6] for the stationary case. Numerical and experimental results have confirmed the validity of the above confidence interval for both stationary and non-stationary random loadings [6,8,9].

It has to be emphasized that Eq. (6) relies on the hypothesis of normally distributed damage $D(T)$, which can be assumed under the validity of the central limit theorem in the limit $T\to\infty$. This hypothesis, first introduced in [3], is also mentioned in other studies [5,10], though it was not until the works [5,11] that its validity has been scrutinized by numerical simulations. In [5], we read: "*the departure from the Gaussian distribution is only marginal. These results suggest that the Gaussian hypothesis made in prior studies is valid when the CoV is small, which usually indicates that T is long enough for the central limit theorem to apply. Fortunately, this is true for the majority of practical situations.*" Further investigations [11] showed that larger values of Cov (e.g. large k, very narrow spectral bandwidth of random loading) lead to larger deviations from gaussianity, with the damage distribution being skewed. In this case, the assumption of normally distributed damage is under conservative [11].

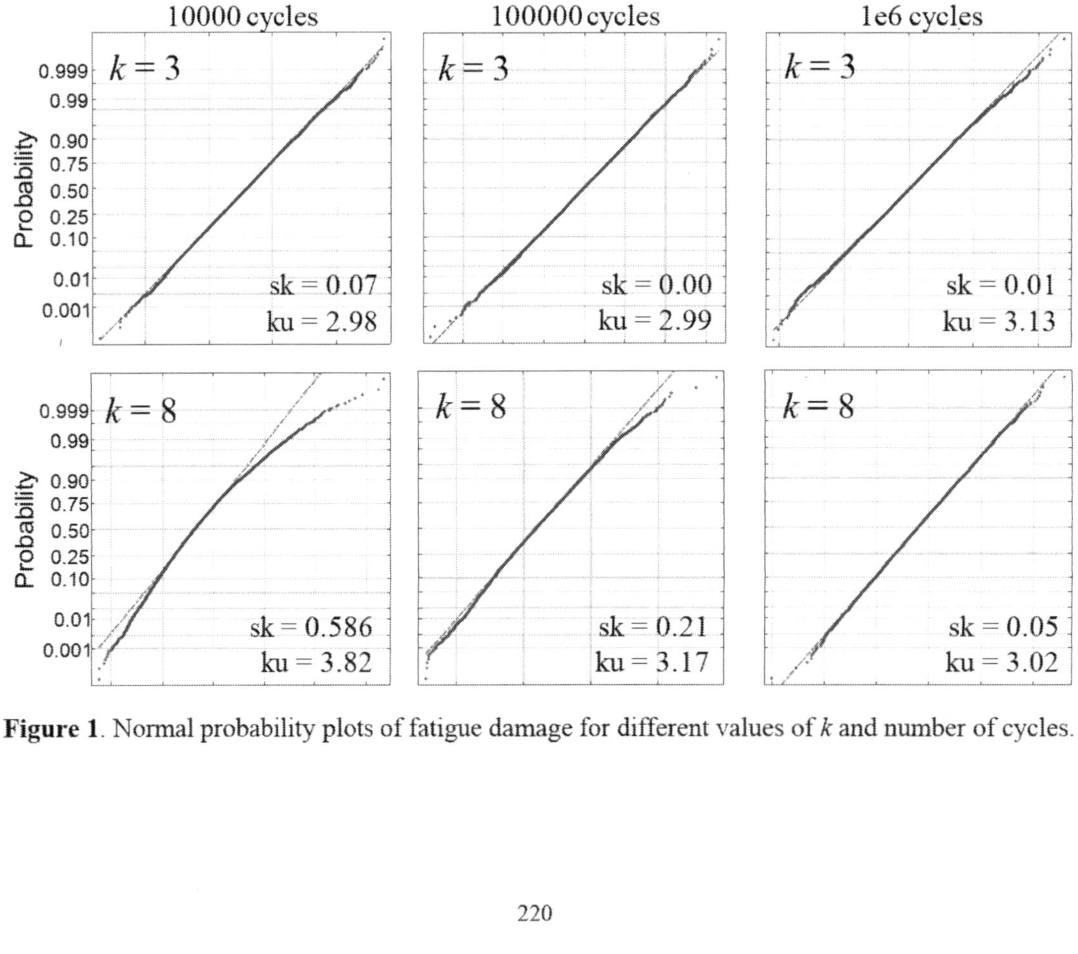

Figure 1. Normal probability plots of fatigue damage for different values of k and number of cycles.

It has to be noticed, though, that the numerical example in [11] considered a small number of cycles (3200 cycles) that, while accentuating the skewness of damage distribution, appears to be too small for any engineering application. On the other hand, the rate of convergence to the normal distribution for the damage depends on the number of random variables being summed (i.e. number of counted cycles) and to which extent each random variable departs from the normal distribution (i.e. effect of inverse slope k). This trend is exemplified in Figure 1, which compares the normal probability papers for different combinations of k and number of counted cycles. The graphs refer to 10000 damage values computed from samples of independent Rayleigh distributed stress amplitudes (corresponding to narrowband loadings); independence among amplitudes neglects the time correlation between adjacent cycles, which in fact corresponds to the output of any counting method in which the relative time position of cycles is lost. From Figure 1, it is apparent that, for high k, far more cycles are required for the damage to become normally distributed.

3.2. *Confidence intervals for quantile estimates based on normal approximation*

For an assigned probability P, let $\hat{x}_P=\widehat{F}^{-1}(P)$ be the estimator of the Pth quantile x_P, where $\widehat{F}^{-1}$ denotes the inverse of the probability distribution $\widehat{F}(x)$ that approximates $F(x)$. A method for constructing CIs for quantiles is to assume that $\hat{x}_P$ is normally distributed, with mean x_P and variance $\mathrm{Var}[\hat{x}_P]=\frac{1}{n}\frac{P(1-P)}{\hat{f}^2(\hat{x}_P)}$ [15,16], where $\hat{f}(x)$ is the probability density corresponding to $\widehat{F}(x)$, and n is the sample size. For large sample size (validity of central limit theorem), an approximate CI is:

$$\hat{x}_P - z_{1-\alpha/2}\sqrt{\mathrm{Var}[\hat{x}_P]} \le x_P \le \hat{x}_P + z_{1-\alpha/2}\sqrt{\mathrm{Var}[\hat{x}_P]} \tag{7}$$

where $z_{1-\alpha/2}$ is the upper $1-\alpha/2$ percentage point of the standard normal random variable. This approach is simple as it only requires the knowledge of the estimator variance $\mathrm{Var}[\hat{x}_P]$, which in turn depends on the probability density $\hat{f}(x)$ – in literature, analytical expressions are available for the most common probability distributions [15,16].

3.3. *Confidence intervals for quantiles based on order statistics*

An alternative approach to construct CIs for quantiles, which is especially useful when $F(x)$ or its estimation $\widehat{F}(x)$ are not known in advance, is by means of order statistics. Let $X_{1:n}\le X_{2:n}\le\ldots\le X_{n:n}$ be the order statistics from a sample of size n of a random variable X with probability distribution $F(x)$. It is well known that [13]:

$$\pi(q,r,n,P)=\Pr\left[X_{q:n}\le x_P\le X_{r:n}\right]=\sum_{i=q}^{r-1}\binom{n}{i}P^i(1-P)^{n-i} \quad , \quad q<r \tag{8}$$

where $P=F(x_P)$ is the probability associated to the quantile. Thus, any choice of q and r such that $\pi(q,r,n,P)=1-\alpha$ provides a $100(1-\alpha)\%$ nonparametric CI for the Pth quantile x_P. The approach followed in practice is to try different combinations of q and r that make π as much close to $(1-\alpha)$ as possible. However, if the sample size n is small – as usually happens in practice – the probability π may substantially deviate from the prescribed coverage probability $(1-\alpha)$.

The use of fraction order statistics [17] allows the definition of an *exact* confidence interval for x_P for any value of confidence. The first step is to rewrite Eq. (8) as [18]:

$$\Pr[x_l\le x_P\le x_u]=I_P\big(n'P_l,n'(1-P_l)\big)-I_P\big(n'P_u,n'(1-P_u)\big) \tag{9}$$

where $0<P_l<P_u<1$, $n'=n+1$ and $I_P(a,b)=\int_0^P t^{a-1}(1-t)^{b-1}\mathrm{d}t/I(a,b)$ denotes the incomplete beta function ratio (it is the beta cumulative distribution function). The *exact* $100(1-\alpha)\%$ CI for the Pth quantile x_P

is defined by $(x_l, x_u)=(X_{n'P_l:n}, X_{n'P_u:n})$ by using fractional order statistics, where the two probabilities P_l and P_u are determined numerically by solving:

$$I_P(n'P_l, n'(1-P_l))=1-\alpha/2 \quad , \quad I_P(n'P_u, n'(1-P_u)) = \alpha/2 \tag{10}$$

Fractional order statistics is also used to define the estimator of the Pth quantile as $\hat{x}_P=X_{n'P:n}$.

As pointed out in [18], the fractional quantile $X_{n'P:n}$ represents only a technical device for defining a continuum of order statistics and it cannot be calculated from the sample, since $n'P$ may not be integer. The same drawback also applies to $X_{n'P_l:n}$ and $X_{n'P_u:n}$. They are then approximated by a linear interpolation, as proposed in [18]. Indeed, it has been shown that the distribution of $X_{n'P:n}$ is well approximated by the distribution of the following linear combination of order statistics:

$$\hat{x}_P^{int}=(1-\varepsilon)X_{\lfloor n'P\rfloor:n}+\varepsilon X_{\lfloor n'P\rfloor+1:n} \quad , \quad \frac{1}{n+1}<P<\frac{n}{n+1} \tag{11}$$

where $\lfloor\cdot\rfloor$ denotes the floor function, and $\varepsilon=n'P-\lfloor n'P\rfloor$, with $0<\varepsilon<1$. The same linear interpolation in Eq. (11) is also used to compute the values x_l^{int} and x_u^{int}, provided that P is replaced by P_l and P_u, respectively; this introduces two new interpolation weights $\varepsilon_l=n'P_l-\lfloor n'P_l\rfloor$ and $\varepsilon_u=n'P_u-\lfloor n'P_u\rfloor$. The two latter values define the lower and upper limits of the interval (x_l^{int}, x_u^{int}), which represents the *approximate* $100(1-\alpha)\%$ nonparametric CI for the Pth quantile x_P.

An illustrative example is now discussed. Suppose to have an ordered sample $X_{i:n}$ of size $n=10$ and consider the quantile $P=0.75$. Since $Pn'=0.75\times11=8.25$, the fractional order statistics gives the quantile estimator $\hat{x}_{0.75}=X_{8.25:10}$, which clearly does not exist in the ordered sample. The nearest sample values, $X_{8:10}$ and $X_{9:10}$, are then used to compute the linear interpolation estimator as $\hat{x}_P^{int}=0.75\cdot X_{8:10}+0.25\cdot X_{9:10}$, weight parameter being $\varepsilon=n'P-\lfloor n'P\rfloor=8.25-8=0.25$.

A point worth to mention is that the linear interpolation in Eq. (11) is only valid for quantiles P in the region between $1/(n+1)$ and $n/(n+1)$, which are the lowest and highest quantiles corresponding to a sample size n. Owing to this condition, the lower and upper confidence limits cannot be computed if $\lfloor n'P_l\rfloor=0$ and $\lfloor n'P_u\rfloor=n$ (see [19]), which is equivalent to the conditions $P_l<1/(n+1)$ and $P_u>n/(n+1)$. For example, assume we want to construct a 99% CI for a sample of size $n=50$, and for two extreme quantiles $P=0.1$ and $P=0.9$. According to Eq. (10), for $P=0.1$ it is $P_l=0.019371$ and so $\lfloor n'P_l\rfloor=0$ (note that $P_l=0.019371<1/51=0.0196$). Similarly, for $P=0.9$, it is $P_u=0.980629$ and $\lfloor n'P_u\rfloor=50$, so the required value $\lfloor n'P_u\rfloor+1=51$ exceeds the last available sample value (again, note that $P_u=0.980629>50/51=0.9804$). In both cases, the CI limits do not exist.

These conditions restrict the admissible combinations of sample size, confidence and quantile levels. As an example, Table 1 lists the minimum sample size n required for some combinations of quantile and confidence levels. For $P=0.75$ and $\alpha=0.05$ (confidence 95%), a minimum sample size of 13 is needed, but this threshold almost triples if the quantile level becomes 0.90, the confidence being kept unchanged. The table therefore emphasizes how a simultaneous large quantile level and confidence require an equally large sample size.

Table 1. Minimum sample size n for several combinations of quantile and confidence

	Quantile level, P		
$(1-\alpha)$	P=0.5	P=0.75	P=0.90
0.90	5	11	29
0.95	6	13	36
0.99	8	19	50

IOP Conf. Series: Materials Science and Engineering 1275 (2023) 012020 doi:10.1088/1757-899X/1275/1/012020

A possibility to partly overcome the above limitations is to use the extrapolation formula proposed in [20], which extends the definition in Eq. (10) to extreme quantiles:

$$\hat{x}_P(P)=\begin{cases} X_{1:n}+(X_{2:n}-X_{1:n})\ln(n'P) & , \quad 0<P\leq\frac{1}{n+1} \\ (1-\varepsilon)X_{\lfloor n'P\rfloor:n}+\varepsilon X_{\lfloor n'P\rfloor+1:n} & , \quad \frac{1}{n+1}<P<\frac{n}{n+1} \\ X_{n:n}+(X_{n:n}-X_{n-1:n})\ln\big(n'(1-P)\big) & , \quad \frac{n}{n+1}\leq P<1 \end{cases} \tag{12}$$

where the second expression in Eq. (12) is the linear interpolation introduced previously. For a random variable with values $X>0$, and sufficiently near zero, the first expression modifies as $\varepsilon X_{1:n}$ [20].

4. Numerical examples

The accuracy of the nonparametric confidence interval described in Sec. 3.3 has already been checked in [18] for the median and third quartile, by considering five different standardized probability distributions (uniform, normal, Cauchy, exponential, Gumbel). The method was proved to yield coverage probabilities very close to the prescribed nominal value of 95%.

Here, the aim is to perform another simulation study that considers a higher quantile (P=0.90) and also relatively "small" sample sizes for which $P_u>n/(n+1)$, a condition that requires the use of the extrapolation formula in Eq. (12). In the present numerical study, 100000 pseudo-random samples are generated from four standardized probability distributions (uniform, normal, exponential, Gumbel); for each sample, the CI for the quantile x_P of the distribution is constructed by the method in Sec. 3.3. The coverage probability is estimated by counting the percentage fraction of confidence intervals that enclose the true quantile x_P, and it is finally compared with the required nominal value 95%. Note that for the considered combination of quantile level (P=0.90) and confidence (95%), the limit sample size is n=36; for smaller values, the extrapolation in Eq. (12) must be invoked. The comparison of results is summarized shown in Figure 2.

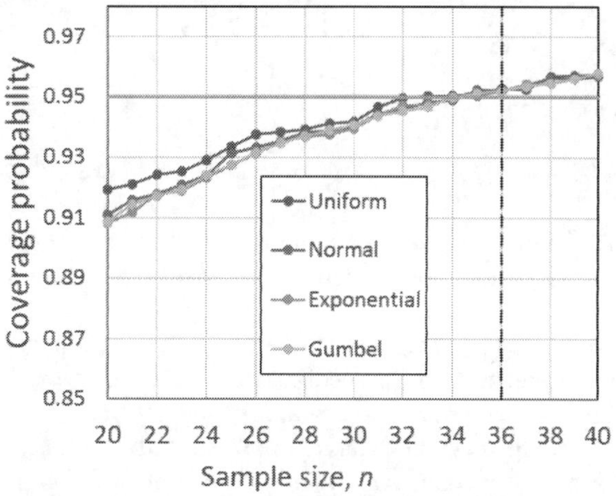

Figure 2. Coverage probability as a function of sample size for different probability distributions. Quantile P=0.90, nominal confidence 95%.

The figure emphasizes that, for a sample size n larger than the limit value 36 (i.e. extrapolation formula not used), the coverage probability is close to the nominal value 0.95, with a slightly increasing trend against n (a tendency already observed in [15]), towards a horizontal asymptote for very large n (result not shown in Figure 2). By contrast, the coverage probability diminishes as n becomes lower than 36, a condition in which $P_u \gg n/(n+1)$ and the upper confidence limit is approximated by the extrapolation formula Eq. (12). The figure nevertheless shows that it is possible to use a sample size smaller than that dictated by the condition $P_u<n/(n+1)$, if a small loss of confidence is accepted.

5. Case study: Mountain-bike data

This section considers a set of time-histories recorded in a mountain-bike on off-road tracks. The aim is not that of evaluating the safety of the bicycle, but to show an engineering application of the statistical methods described in the previous Sections.

The mountain-bike is a 2010 Scott Sportster P6, with a frame in 6061 aluminum alloy and a Unicrown front fork in carbon steel. Two Rigida Cyber 10 size 700C wheels are coupled with 700×37c S207 semi-slick tires. Handlebar and saddle are from Scott Sports; transmission, chain, and crankset are from Shimano, Inc.

The loadings acting on the bicycle front fork were measured by two strain gages, placed symmetrically on the left tube and connected in half-bridge configuration; the system was calibrated in laboratory static loading tests. Strain gage are from HBM, model LY Linear strain gauge with one measuring grid (only one direction). Time-histories were recorded by a Dewesoft data acquisition system (model Minitaurs Dewe-101 with 8 channels). This model has an industrial power computer built directly into the unit. Sampling frequency was 1000 Hz. A filter was set up with cut-off frequency of 300 Hz, which was above the maximum frequency of interest. The data acquisition system was fixed at the inclined tube of the main "triangle" frame of the bicycle: the rechargeable supply battery was allocated behind the seat, on a welded support [21].

The bicycle speed was monitored by a speedometer (Marwi Group, model Union 8 Cycling Computer), formed by a sensor and magnet fixed on a wheel rod. When fully equipped, the mountain-bike weighed about 12.2 kg.

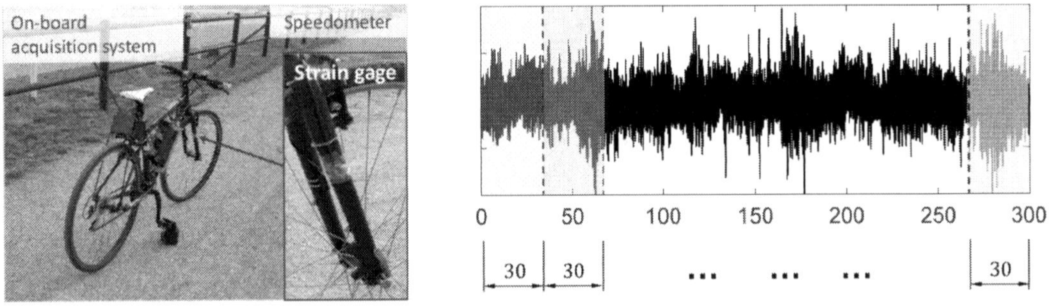

Figure 3. Mountain-bike with strain gages and onboard acquisition system, and example of measured time-history subdivided into segments of 30 seconds.

A total of n=40 time-histories was measured and used in the subsequent analysis. Each time-history is first subdivided into 10 segments of 30 seconds each. The fatigue damage is computed for each segment, by assuming an SN curve $N(s)=As^{-k}$, with A=1 and k=3 and 5. Therefore, 10 damage values were obtained from each time-history; their maximum was finally computed. In summary, the initial set of measured time-histories returned a sample of 40 maxima of damage $D_{max,i}$, i=1,...,40, which represents the random sample X_i used to construct the nonparametric confidence interval of quantile described in Sec. 3.3. Subsets with n=10, 20, 30 damage values were also considered to study the behavior of the confidence interval computed from smaller samples.

Figure 4(a) compares the 95% confidence intervals for 0.90 quantile, obtained for different sample sizes. Damage values are normalized to the sample mean. Black circles (•) represent the sample damage values, the star (✱) is the quantile estimator, while the vertical lines are the quantile confidence intervals. For a sample size n<36 (limit value), the extrapolation formula Eq. (12) is used (confidence intervals are red colored). Note that the nonparametric confidence interval needs not to be symmetric around the quantile estimator. For n=40, it is $P_u<n/(n+1)$ and therefore the upper limit of CI falls below the last

sample value in the ordered statistics. Instead, for sample size <36, the confidence interval has an increasing width, while its upper limit exceeds the last samples value (also by a large amount for small n) and falls in the region of validity of the extrapolation formula.

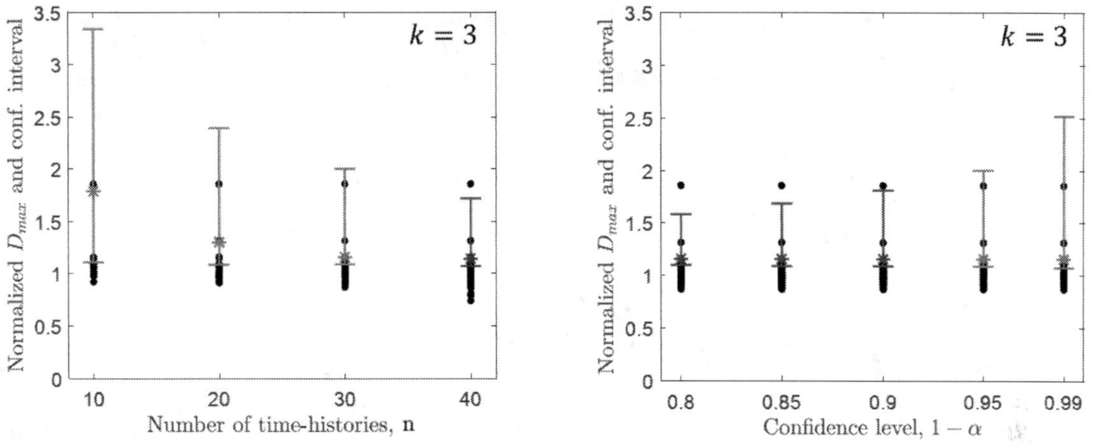

Figure 4. Confidence intervals of the 0.90 quantile, in the case of (a) fixed confidence 95% and different sample size, or (b) fixed sample size n=40 and different confidence.

Figure 4(b) replots the same results of Figure 4(a), where CIs are now shown as a function of the confidence level. It can be seen that the extrapolation formula is used only for confidence levels higher than 0.90; in this case, the upper confidence limit exceeds the highest value in the ordered statistics.

6. Conclusions

The paper studied the applicability of a nonparametric method based on fractional order statistics – and developed in two literature papers [18,20] – as a tool for estimating the confidence interval of quantiles of fatigue damage computed in random loadings. After a brief theoretical summary of the method, a numerical study and an experimental case study have been examined, the results of which allow for the following considerations:

- the nonparametric method has the undoubtful advantage of making no prior assumption on the probability distribution of fatigue damage, contrary to our previous method that hypothesized a normally distributed damage to construct the confidence interval;
- in the original formulation of the nonparametric method [18], there is a constraint in the admissible combinations of sample size n, quantile level P and confidence $(1-\alpha)$, which is dictated by the limit quantile levels, $1/(n+1)$ and $n/(n+1)$, established by sample size n;
- in a subsequent formulation [20], an approximated extrapolation formula allows the previous constraint to be overcome, which however occurs – as shown by our simulation results – at the expense of a loss in confidence, which becomes more pronounced if the sample size decreases far below the "limit" value;
- in order for the extrapolation formula not to be applied, a large sample size n must be used. However, for very high quantiles (e.g. >0.99) and confidence levels (e.g. 95%), the required n may be so large that the method could become impractical (indeed, n represents the number of time-histories from which the sample of damage values are computed). For very high quantile levels (e.g. >0.99), the use of the extreme value theory may be more appropriate [22].

Acknowledgments

Research has partially been funded by CTU Global Postdoc Fellowship Program (research topic #2-11).

References

[1] Bendat J S 1964. *Probability functions for random responses: Prediction of peaks, fatigue damage, and catastrophic failures. National Aeronautics and Space Administration contract report 33* (Washington, D.C., USA) p 57

[2] Mark W D 1961 *The inherent variation in fatigue damage resulting from random vibration. Ph.D. Thesis Department of Mechanical Engineering MIT* (Cambridge, Massachusetts, USA) p 93

[3] Crandall S H, Mark W D and Khabbaz G R 1962 *The variance in Palmgren-Miner damage due to random vibration. Proc. of 4th US National Congress of Applied Mechanics* vol 1 (California, Berkeley, USA) pp 119–26

[4] Madsen H O, Krenk S and Lind N C 1986 *Methods of Structural Safety* (Englewood Cliffs, NJ, USA: Prentice-Hall) p 416

[5] Low Y M 2012 Variance of the fatigue damage due to a Gaussian narrowband process. *Struct. Saf.* **34**(1) 381–9

[6] Marques J M E, Benasciutti D and Tovo R 2020 Variability of the fatigue damage due to the randomness of a stationary vibration loading. *Int. J. Fatigue* **141** 105891

[7] Marques J M E, Benasciutti D and Tovo R 2021 Variance of the fatigue damage in non-Gaussian stochastic processes with narrow-band power spectrum. *Struct. Saf.* **93** 102131

[8] Benasciutti D and Marques J M E 2021 Methods for estimating the statistical variability of the fatigue damage computed in one single stationary or non-stationary random time-history record. *Proc. of 3rd Int. Conf. on Structural Integrity for Offshore Energy Industry STRUCTURAL INTEGRITY 2021* (15–16 November 2021, Aberdeen, UK)

[9] Marques J M E and Benasciutti D 2021 Evaluating confidence interval of fatigue damage from one single measured non-stationary time-history. *Proc. of VIRTUAL 20th Int. Colloq. on Mechanical Fatigue of Metals* (15–17 September 2021, Wrocław, Poland)

[10] Desmond A F 1987 On the distribution of the time to fatigue failure for the simple linear oscillator. *Probab. Eng. Mech.* **2**(4) 214–8

[11] Low Y M 2012 An analytical formulation for the fatigue damage skewness relating to a narrowband process. *Struct. Saf.* **35**(1) 18–28

[12] Haibach E 2002 *Betriebsfestigkeit* 2006 (Heidelberg, Germany: Springer-Verlag Berlin)

[13] Montgomery D C and Runger G C 2014. *Applied Statistics and Probability for Engineers* 6th ed. (Hoboken, NJ, USA: John Wiley & Sons) p 284

[14] Marques J M E, Solazzi L and Benasciutti D 2022 Fatigue analysis of nonstationary random loadings measured in an industrial vehicle wheel: Uncertainty of fatigue damage. *Metals* **12**(4) 616

[15] Serinaldi F 2009 Assessing the applicability of fractional order statistics for computing confidence intervals for extreme quantiles. *J. Hydrol.* **376**(3–4) 528–41

[16] Su S 2009 Confidence intervals for quantiles using generalized lambda distributions. *Comput. Stat. Data Anal.* **53**(9) 3324–33

[17] Stigler S M 1977 Fractional order statistics, with applications. *J. Am. Stat. Assoc.* **72**(359) 544–50

[18] Hutson A D 1999 Calculating nonparametric confdence intervals for quantiles using fractional order statistics. *J. Appl. Stat.* **26**(3) 343–53

[19] Nagaraja C H and Nagaraja H N 2020 Distribution-free approximate methods for constructing confidence intervals for quantiles. *Int. Stat. Rev.* **88**(1) 75–100

[20] Hutson A D 2002 A semi-parametric quantile function estimator for use in bootstrap estimation procedures. *Stat. Comput.* **12** 331–8

[21] Marques J M E 2021 Confidence intervals for the expected damage in random loadings: Application to measured time-history records from a mountain-bike. *IOP Conf. Ser. Mater. Sci. Eng.* **1038**, 012025

[22] Castillo E *Extreme value theory in engineering* 1988 (San Diego, USA: Academic Press)

IOP Conf. Series: Materials Science and Engineering 1275 (2023) 012021 doi:10.1088/1757-899X/1275/1/012021

Development, realization and validation of a device for real-time monitoring of fatigue behavior

M Palmieri, F Morganti and F Cianetti

University of Perugia, Department of Engineering, Via G. Duranti 93, Perugia 06125, Italy

E-mail: massimiliano.palmieri@unipg.it

Abstract. The health monitoring of mechanical components and systems is a technique known and applied for decades in almost all fields of engineering since it allows a drastic reduction of the risks associated with any unexpected failure of components and at the same time allows reducing the plant maintenance and management costs. Several monitoring techniques exists and all of them are based on the acquisition and processing on-field measurements aimed to evaluate possible changes in the behavior of the component. Among these, in recent years a method able to monitor in real-time the fatigue damage or the potential damage of a structure with few information has been developed. In this paper, this method, implemented and tested only numerically, has been implemented in an acquisition board, expressly developed, and its ability to operate in real-time by monitoring the failure of a real component subjected to random vibrations has been tested.

1. Introduction

The use of techniques aimed to monitor the structural behavior of mechanical components and systems has been widespread for decades in any field of engineering. The fundamental advantage deriving from the application of health monitoring techniques is mainly linked to the reduction of the risk of unexpected failures. This aspect brings additional benefits such as safety and the reduction of maintenance and repairing costs. Due to the increasing use of mechanical systems in particularly aggressive environmental conditions, active control techniques have recently been developed to minimize fatigue damage and for alleviating the loads acting on a structure [1]. Several techniques for health monitoring exist in literature but the most common involve analyzing the loads acting on the structure, or exploiting acoustic emissions or analyzing vibrations [2,3].

In this context, in a recent work Cianetti et al. [4,5] proposed a damage estimation method, mainly developed for wind turbines but extendable to any mechanical system, able to monitor in real time the accumulated fatigue damage in a structure or the potential damage (if the technique is applied on any quantities and not on the stress states). The proposed technique involves applying a cycle counting method (i.e. rainflow counting method) and the Palmgreen-Miner damage accumulation rule [6,7] not to the entire time history (which would make it inapplicable in real-time) but to a moving windows of the signal, keeping in memory only the computed damage and rejecting the rests of data.

The main advantage of using this monitoring technique is linked to the fact that a single measurement is enough to determine the cumulative damage on the entire structure, both real and potential, through a relationship numerically determined.

Content from this work may be used under the terms of the Creative Commons Attribution 3.0 licence. Any further distribution of this work must maintain attribution to the author(s) and the title of the work, journal citation and DOI.
Published under licence by IOP Publishing Ltd

This technique, which has only been tested numerically, in this activity has been implemented within an electronic device created ad hoc. The developed device and the algorithm implemented are based on few low-costs components. The device was then experimentally tested trying to predict the failure of a simple Y-shaped component [8], subject to random loads. The obtained results showed that the developed device is perfectly able both to operate in real time but also to provide reliable results in terms of predicted fatigue life with an error always lower than 20% compared to a classical fatigue-failure control technique.

2. Evaluation of fatigue damage in real time

The evaluation of the fatigue damage of structures subjected to random vibrations in real time can be carried out by exploiting the classical methodologies in the time domain. That is, given the measurement of any signal (such as accelerations, forces, stress), this can be processed with a cycle counting method, typically the rainflow counting method, and once the load spectrum and the S-N curve expressed in terms of the measured quantity are known, a damage accumulation law can be applied, i.e. Palmgreen-Miner [6,7], through which determining the fatigue life. However, using this approach means applying the cycle counting method over the entire time history, from the moment the component was installed to the moment the calculation is made. However, this approach does not go well with a real-time analysis as it would involve managing an enormous amount of data (especially for high-frequency signals), and therefore is de-facto inapplicable in real-time.

In this context, Cianetti et al. [4,5] proposed a method that exploits what stated in previous line, thus rainflow counting algorithm and the Palmgreen-Miner rule, but instead of calculating the damage on the entire acquired time history, the calculation is carried out over a window, with appropriate characteristics (time length, sampling frequency) and therefore the computational effort is substantially reduced. Figure 1 shows the proposed procedure.

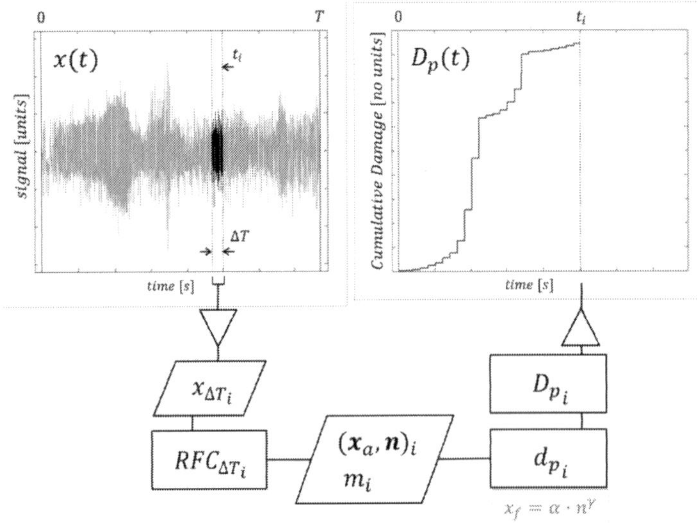

Figure 1. Flowchart of proposed damage evaluation technique

As visible from figure 1, only a small window of length ΔT is processed by the rainflow counting algorithm and what previously happened is discarded. On the available time window, it is possible to apply the rainflow counting and the Palmgren-Miner rule determining the potential damage associate with the ith window, namely instantaneous damage in this activity, as shown in equation 1.

$$d_{p_i} = \sum_{k=1}^{m_i} \frac{(n_k)_i}{\left(\frac{(x_{a_k})_i}{a}\right)^{1/b}} \tag{1}$$

In equation 1 a and b are the Wholer curve parameters, n_k is the extracted number of cycles in the ith window where k is the generic spectrum cycle counted, x_{a_k} represents the alternate value of the acquired signal while m_i is the total number of cycles counted in the window.
Once the potential fatigue damage associated to the ith window is known, the total accumulate damage at a predefined time instant D_p can be easily computed by summing the instantaneous damages d_{p_i} as follow:

$$D_p = \sum_{r=1}^{i} d_{p_i} \tag{2}$$

In this paper the authors define the damage as "potential" since it can be computed not only on stress process but on generic signals and in such case, the computed fatigue damage is potential and useful more for comparison than to estimate the fatigue life.

3. The device

In order to have a device able to evaluate the real or potential cumulative damage within a component, it was necessary both to create an ad-hoc device capable of performing all the necessary operations and to implement an algorithm aimed to data acquisition and processing operation.

3.1. The hardware

The realized device is based on 5 elements, detailed in the following:

- Sensor = Needed to acquire whichever signals
- Signal conditioner = Needed to amplify the acquired signal
- AD Converter = necessary to convert the analog signal into a digital one
- Buffer board = Electronic board aimed to memorize the acquire data
- Processor = Electronic board aimed to get data and compute them simultaneously

An accelerometer produced by PBC is used as sensor in this activity while the signal conditioner model 480E09 produced by PBC is adopted. The Raspberry Pi AD/DA Expansion Board ADS1256 with 8 channels at 24bit and sampling rate equal to 30ksps is used as ADC. For the buffer board, necessary to memorize the acquired data, the WeAct Black Pill V2.0 board has been used while the Raspberry PI4 is used as processor. The choice of these components has been made in order to minimize the total cost, that excluding the sensor and signal conditioner, is lower than 150 Euro. The final device is shown in figure 2.

As visible from figure 2, an additional resistor was needed since the ADC converter can read only positive values in a range 0 - 3.3 Volts. This means for examples that for a zero mean signal, the negative values would be lost. By introducing an additional resistance, it was possible to insert a bias, thus a known mean value of voltage equal to 1.75V, to be removed after the acquisition.

3.2. The software

The realized device should be able to acquire a signal and processing it simultaneously. To this aim, a Python algorithm has been implemented within the Raspberry PI4. This algorithm exploits the multi-threads logic. In such a way two threads operate at the same time. The

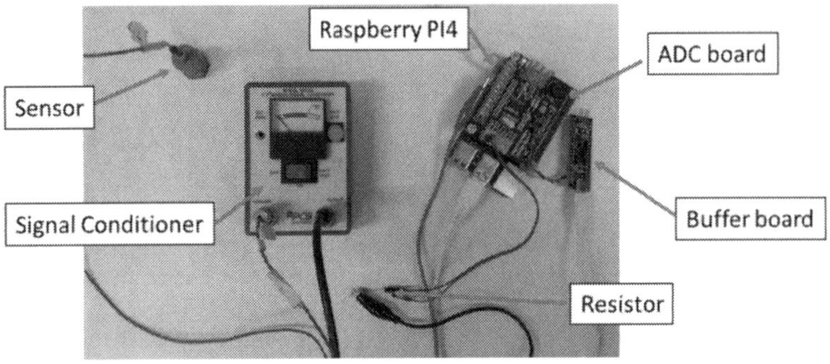

Figure 2. The realized device with all necessary components

first one is used to acquire portion (window) of the signal in real time while the second thread computes the fatigue damage, according to what stated in Sec. 2, of the previously acquired window. Figure 3 shows the multi-threads logic used to compute the fatigue damage in real time.

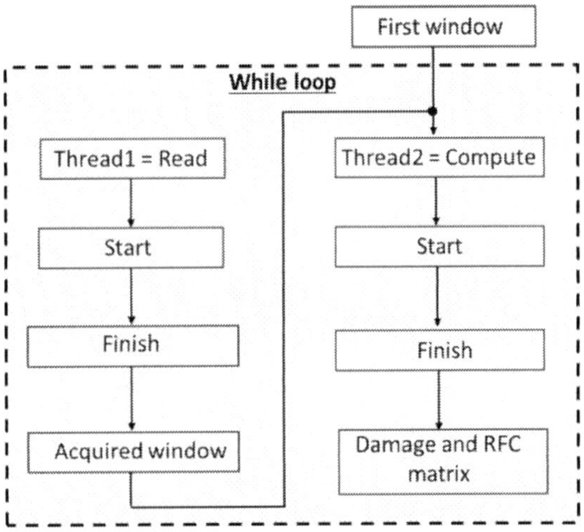

Figure 3. Multi-threads logic used to compute the fatigue damage in real time

As visible from figure 3, the algorithm foresees to acquire a first window, secondly the two threads operate simultaneously. Both threads are inserted in a while loop that ends when the accumulated damage is equal or close to a predefined threshold defined critical for the analyzed component.

4. Validation of the device through vibration tests

In order to evaluate the ability of the device to operate in real time to monitor a fatigue failure induced by vibrations, vibration tests were performed on a simple laboratory component. The specimens were realized in additive manufacturing in Polylactic acid (White-Pearl PLA produced by ultimaker) for which the slope and the intercept of the S-N curve, defined as $S = aN^b$ had been determined in a previous activity [8]. These turn out to be: $a = 204.7$ MPa and $b = -0.1874$. Figure 4 shows the used experimental setup.

As shown in Figure 4, the chosen parameter to measure and then related to the fatigue damage was the acceleration measured on one arm of the specimen. However, in order to use this type of signal and at the same time use the known parameters of the fatigue curve, it was necessary to determine a scale factor able to transform the acceleration measured on the arm of the specimen to the stress component at the failure point. The latter, given the simple geometry of the specimen, is in fact known and is located near the hole (Figure 4). To determine

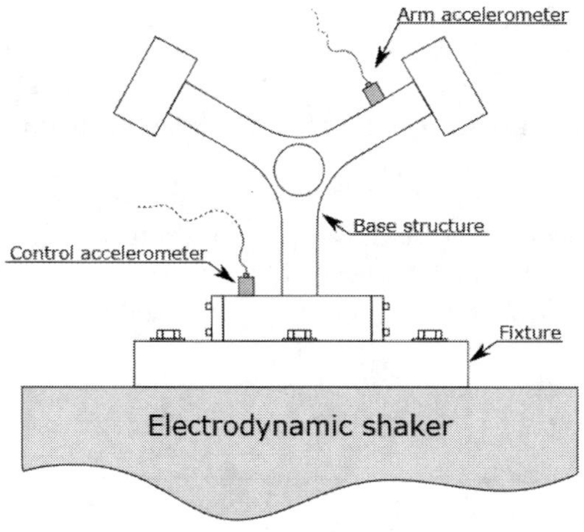

Figure 4. Experimental setup used to validate the accuracy of the proposed device through vibration tests.

the scale factor between acceleration measured on the arm and the stress at the failure point, the hypothesis that there is a linear correlation between the two physical quantities was made. This can be acceptable in this case since the component behaves like a single degree of freedom system and therefore there are no phase issues and there is only one stress component that causes the failure. The scale factor between acceleration and stress was determined by taking the maximum value of the frequency response function in terms of stress/acceleration, in correspondence of the resonance frequency and it results to be equal to 0.146 MPa/g. The frequency response function was determined by an experimentally calibrated finite element model. To obtain an experimental validated model, a set of vibration tests were performed and the results, in terms of frequency response function between arm and base acceleration, were compared to those numerically obtained. In such a way it was possible to adjust the material parameters such as Young modulus, density e and Poisson ratio, aimed to obtain a good fit between experimental and numerical frequency response functions [8].

Before proceeding with experimental tests, an intense numerical analysis has been performed in order to define the best setting parameters for the analysis. The parameters to be determine are the length of the window ΔT, the frequency sampling and the residue processing methodology [9]. The latter is indeed much important in this case since, applying the rainflow counting method to short time window there are many residues to be managed in order to determine the correct load spectrum. In particular four different residue processing methodologies have been considered:

- Counting the residues as half-cycles
- Considering all the residues as closed cycles
- All the residues are discarded
- Carrying all the residues to the next window

The analysis has been conducted numerically using several random signals such as non stationary signals, wide band stationary signals, variable mean value signals and bimodal processes. The results of the analysis led to define that the window time length should be not lower than 30 times the lowest frequency of the process, while (as expected) the sampling frequency should be at least ten times the maximum frequency of the signal. Moreover it comes out the best residues processing methodologies to obtain a fatigue damage comparable to the actual one, are to consider the residues as half cycle or to carry them to the next window over which the rainflow counting is further re-applied. The actual fatigue damage, used as benchmark for the fatigue damage obtained with the proposed algorithm, was computed applying the rainflow counting algorithm to the entire process, for each considered random signal, adopting the Cloorman-Seeger hypothesis [10-11].

Once the analysis parameters were determined, it was possible to perform the experimental test campaign.

The specimen was tested with three different random loads, with a different frequency content and a different RMS. The properties of the signals used are shown in table 1. According to these three different inputs, the failure of the components were monitored with the device proposed in this activity. The results obtained are shown for each input in table 1.

Table 1. Comparison between fatigue life obtained with the proposed device and standard methods for three different loading conditions

Test ID	Frequency range [Hz]	Input RMS [g]	Estimated life [s]	Real life [s]	Percentage error [%]
1	90-170	1	2020	2350	16.3
2	80-250	3	620	770	19.4
3	100-400	2	980	1180	17.1

To validate the accuracy of the results obtained with the proposed device, the acceleration signals were also acquired with a commercial instrument (Siemens Testlab) and were post-processed with a well-known technique in vibration fatigue to determine the failure of a component. The latter, known as frequency drop, foresees to monitor the resonance frequency of the specimen declaring a failure when the resonance frequency drop down below 5 % compared to its initial value [12]. The results obtained with the proposed device and by monitoring the position of the natural frequency are compared in table 1. As visible, there is a good agreement between the obtained fatigue life, indicating a good accuracy of the realized device. The committed error is in fact always lower than 20 %. However, it should be noted that the device always anticipates the prediction of the component failure but this can be attributed to the stress/acceleration ratio numerically obtained.

5. Conclusions

In this activity a device has been developed, both from a hardware and software point of view, within which a procedure aimed at calculating the fatigue damage in real time of structures subjected to random loads was implemented. The forecasting method used involves calculating the fatigue damage by applying the rainflow counting method for the extraction of fatigue cycles and the Miner's rule not on the entire time history but on short-term windows, thus making it applicable in real time.

The device consists of few low-cost elements available on the market while the implemented algorithm exploits the multi-threads logic so as to be able to simultaneously acquire and compute the signal. The effectiveness of the device to operate in real time and to provide a correct prediction of the fatigue life of a mechanical component was verified by means of vibration tests carried out through an electrodynamic shaker, on a Y-shaped specimen until failure occurred. The obtained results show that the developed device is viable since the obtained results, in terms of fatigue life, are in line with those obtained by monitoring the frequency drop.

References

[1] Zieglera L, Gonzalez E, Rubert T, Smolka U and Melero J 2018 Lifetime extension of onshore wind turbines: A review covering Germany,Spain, Denmark, and the UK. *Renewable and Sustainable Energy Reviews* **82** 1261-1271

[2] Yang Y, Zhang Y and Tan X 2021 Review on Vibration-Based Structural Health Monitoring Techniques and Technical Codes. *Symmetry* **13**

[3] Yuequan B, Zhicheng C, Shiyin W, Yang X, Zhiyi T and Hui L. 2019 The State of the Art of Data Science and Engineering in Structural Health Monitoring. *Engineering* **5** 234-242

[4] Cianetti F, Cetrini A, Corradini ML, Ippoliti G and Orlando G 2018 Dynamic behavior of wind turbines. An on-board evaluation technique to monitor fatigue. *Procedia Structural Integrity* **12** 102-112

[5] Cetrini A, Cianetti F, Corradini ML, Ippoliti G and Orlando G 2019 On-line fatigue alleviation for wind turbines by a robust control approach. *International Journal of Electrical Power & Energy Systems* **109** 384-394

[6] Collins JA 1981 Failure of Materials in Mechanical Design *Wiley - New York*

[7] ASTM Standard 2011 Standard Practices for Cycle Counting in Fatigue Analysis. *West Conshohocken, PA: ASTM International* **E 1049**

[8] Palmieri M, Zucca G, Morettini G, Landi L and Cianetti F. 2022 Vibration Fatigue of FDM 3D Printed Structures: The Use of Frequency Domain Approach *Materials* **15**

[9] Marsh G, Colin W, Philipp R, Nigel B, Atilla I, Vengatesan V and Lars J 2016 Review and application of Rainflow residue processing techniques for accurate fatigue damage estimation. *International Journal of Fatigue* **82** 757-765

[10] Clormann U and Seeger T 1986 Rainflow – HCM – Ein Zählverfahren für Betriebsfestigkeitsnachweise auf werkstoffmechanischer Grundlage. *IStahlbau* **55** 65–117

[11] ASTM Standard E 1049 1985 Standard Practices for Cycle Counting in Fatigue Analysis *ASTM International, West Conshohocken*

[12] Palmieri M, Slavič J and Cianetti F 2021 Single-process 3D-printed structures with vibration durability self-awareness. *Additive manufacturing* **47** 102303

IOP Conf. Series: Materials Science and Engineering 1275 (2023) 012022 doi:10.1088/1757-899X/1275/1/012022

Strength of notched Ti-6Al-4V specimens not subjected to solution treatment and over-aging under cyclic loading

E V Arcieri[1] and S Baragetti[1]

[1]Department of Management, Information and Production Engineering, University of Bergamo, Viale Marconi 5, Dalmine 24044, Italy

sergio.baragetti@unibg.it

Abstract. Ti-6Al-4V alloy is a popular titanium alloy due to its high strength / density ratio, high toughness, superior corrosion resistance and good biocompatibility. After forming operations, Ti-6Al-4V is typically subjected to Solution Treatment and Over-Aging (STOA). Fatigue is one of the most frequent causes of failure in mechanical components. The prediction and optimization of the strength of structural components under cyclic loads therefore plays a strategic role and must take into consideration the presence of defects. The aim of the work is to investigate the axial fatigue behaviour of Ti-6Al-4V alloy not subjected to STOA in inert environment in the presence of notches. The experimental results agree well with the results obtained using the Haigh diagram.

1. Introduction

Ti-6Al-4V titanium alloy is commonly adopted in the aeronautical, maritime and biomedical industries owing to its high strength / mass ratio, high toughness, high resistance to corrosive environments, and good biocompatibility [1-4].

Cyclic loading is a common cause of failure in the above-mentioned fields [5]. The structural integrity of the components must be guaranteed during their service under different loading and environmental conditions [6, 7], even in the presence of defects, which are caused by the production process or by the impact of objects. Some components are today produced by additive manufacturing techniques [8] which are feasible to produce complex geometries, even in titanium alloys. The poor surface quality obtained with these techniques contributes to the reduction of fatigue strength [9]. Even the conventional processes can induce defects that need to be monitored. Foreign objects can collide with the aircraft engine parts, frequently made of titanium alloys. The occurrence of high tensile stresses associated with the impact-damage may reduce service life [10, 11].

After forming operations, Ti-6Al-4V is commonly Solution Treated and Over-Aged (STOA), i.e. subjected to solution treatment and vacuum annealing [12]. The behavior of STOA-treated Ti-6Al-4V in inert and aggressive environments and under quasi-static and fatigue loads is presented in [13-17]. In [18, 19] the behavior of the Ti-6Al-4V alloy not subjected to the STOA treatment under quasi-static loading is described for different notches and different environments.

The aim of this work is to study the fatigue behavior of Ti-6Al-4V without STOA treatment in inert environment [20]. The lack of the treatment could reduce the manufacturing cost. Notched Ti-6Al-4V specimens not subjected to STOA were tested under cyclic axial loading, with a load ratio $R = 0$. The experimental results are in good agreement with those provided by the analysis using the Haigh diagram.

Content from this work may be used under the terms of the Creative Commons Attribution 3.0 licence. Any further distribution of this work must maintain attribution to the author(s) and the title of the work, journal citation and DOI.
Published under licence by IOP Publishing Ltd

2. Materials and methods

Authors applied in this work the same research method adopted in past works on applications in different mechanical engineering sectors [21-24].

The geometry of the tested specimens is shown in Figure 1, where *d* is the depth of the notch. The following values of *d* were analyzed: 0.5, 1.0 and 2.0 mm. The chemical composition of the specimens is reported in Table 1 [14]. The Ti-6Al-4V alloy was not subjected to STOA, with a consequent small improvement in mechanical properties compared to the alloy subjected to STOA as shown in Table 2 [12, 13, 25, 26], where *UTS* indicates the ultimate tensile strength, *YS* the yield stress and *E* the Young's modulus. The STOA treatment applied to obtain the mechanical properties of Table 2 is solution treatment at 925 °C for 1 h and vacuum annealing at 700°C for 2 h. The notches of the specimens were created by milling at low cutting speed in order to induce low residual stresses. No stress relief was conducted after the production of the notches.

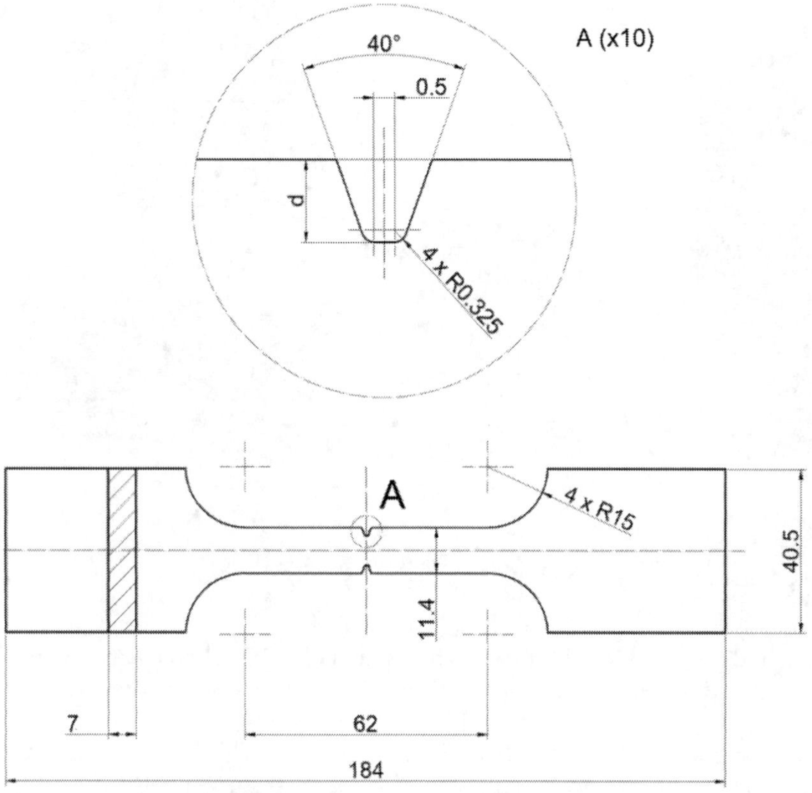

Figure 1. Geometry of the tested specimens.

Table 1. Chemical composition of Ti-6Al-4V (%wt.) [14].

Ti	Al	V	Fe	O	N	H	C
Bal.	5.97	4.07	0.20	0.19	0.05	0.015	0.003

Table 2. Mechanical properties of Ti-6Al-4V with and without STOA [12, 13, 25, 26].

	UTS (MPa)	*YS* (MPa)	*E* (MPa)	ν
With STOA	947-990	900-945	110000	0.342
Without STOA	1000-1100	958-1050	110000	0.342

To determine the stress concentration factor *SCF* at the notch a plane stress linear elastic finite element analysis was performed on the model of a quarter of the specimens using the Abaqus Standard code [27], Figure 2 [28]. A homogeneous isotropic elastic material model was used for the analyses, with the properties of Table 2. Figure 2a shows the applied boundary and load conditions. Symmetry boundary conditions were implemented. A uniform load of 1 MPa was applied to the upper edge of the specimens. Figure 2b shows a detail of the mesh created at the notch. For each notched specimen, finite element model convergence was achieved in order to accurately capture the stress gradient at the notch. Figure 2c shows the resulting stress distribution, i.e. the stresses in the y direction according to the coordinate system in Figure 2a, for one of the investigated geometries, $d = 2.0$ mm.

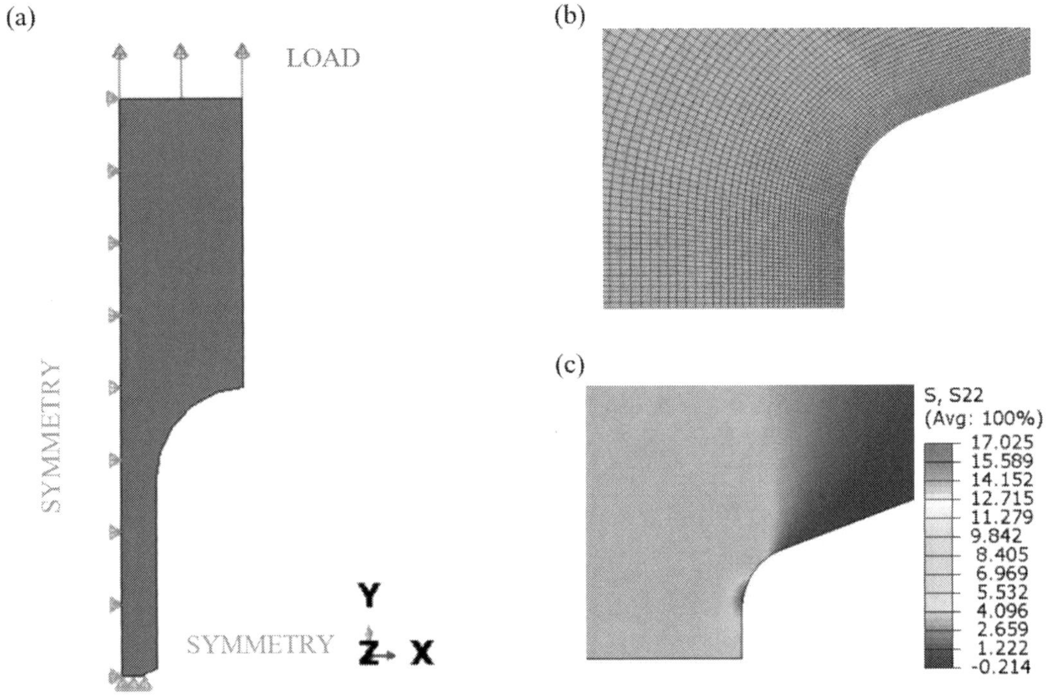

Figure 2. Finite element model for the determination of *SCF*: (a) boundary and load conditions, (b) mesh at the notch, (c) axial stresses at the notch (MPa) [28].

Before the experimental tests, the surface of the specimens, including the notch surface, was polished with grit paper and diamond paste. Fatigue axial tests were conducted with a frequency $f = 5$ Hz and a load ratio $R = 0$. The three specimens were tested in air environment, but a layer of insulating tape was placed adherent to the notches of the two specimens with $d = 0.5$ and 1.0 mm to simulate vacuum conditions. Another technique for simulating vacuum conditions is the application of beeswax or paraffin oil [15]. The fatigue tests were performed on the notched specimens according to the step loading procedure described by Nicholas in [29], which enables obtaining preliminary results in a fast way, similarly to the Locati method [30, 31]. The step loading procedure consists of the sequential implementation of load blocks on the specimen. In each load block, the specimen is subjected to N_{life} loading cycles, where N_{life} is the number of cycles at which the stress to failure must be determined. The load is incrementally increased in the following load blocks until the specimen fails, at a number of cycles N_{fail} less than or equal to N_{life}. According to Equation 1, the stress range σ_{SL} that causes the failure of the specimen after N_{life} loading cycles is determined as the linear interpolation between the stress range applied in the last unfailed load block, σ_{prev}, and the stress range applied in the load block in which the failure occurs, σ_{fail}:

$$\sigma_{SL}=\sigma_{prev}+\frac{N_{fail}}{N_{life}}\left(\sigma_{fail}-\sigma_{prev}\right) \quad (1)$$

The step loading procedure provides reliable results if the crack propagation is very fast in the experiments. Any crack propagation in previous load blocks could influence the results, as the specimen strength would be changed. The ideal would be to have the entire crack propagation phase in the load block in which the specimen fails [29].

3. Results and discussion

Table 3 shows the results of the analysis conducted in this work. F_{fail} is the axial load applied to the specimen in the load block in which the failure occurred in the experimental test, σ_H is the stress to failure (stress range) calculated using the Haigh (modified-Goodman) diagram and Δ is the discrepancy between the two admissible stress ranges σ_{SL} and σ_H. The reported stresses are nominal stresses referred to the net area of the specimens. The first specimen, with a notch depth of 0.5 mm, failed at a maximum load in the fatigue cycle of 17.0 kN after 135544 cycles applied in the last load block. The second specimen, with a 1.0 mm notch depth, failed at a maximum load in the fatigue cycle of 16.8 kN after 160650 cycles in the block. The third specimen, with a notch depth of 2.0 mm, failed in the first load block, when a maximum load in the fatigue cycle of 13.0 kN was applied; the failure occurred after 139458 cycles. The step loading methodology provides a stress range to failure of 220 and 248 MPa for the specimens with a notch depth of 0.5 and 1.0 mm respectively at a fatigue life of 200000 cycle. For the third specimen Equation 1 was not used because the specimen failed in the first applied load block. The stress range to failure for the specimen with the notch depth 2.0 mm is 251 MPa at a fatigue life of 139458 cycles [28]. The σ_H value was calculated according to the procedure illustrated in Figure 3. The figure is an example and refers to the case of the specimen with d = 0.5. The first step was the determination of the alternating stress to failure σ_a (N_{life}, R = -1) at the fatigue life N_{life} starting from the S-N (σ_a-N_{cycles}) curves for each notched specimen not subjected to STOA (Figure 3a). The S-N curves for the three notched specimens with the three corresponding values of stress concentration factor *SCF* were built taking into consideration the brittle behavior of the alloy observed in [19]. In the analysis it was assumed that the alternating stress to failure at a fatigue life of 1000 loading cycles corresponds to the stress which induces static failure, i.e. *UTS* / *SCF*. The alternating stress to failure at a fatigue life of 2 million cycles for the notched specimens was assumed to be 0.7 x 0.33 x *UTS* / *SCF* [28]. The average of the two *UTS* values reported in Table 2 for Ti-6Al-4V not subjected to STOA treatment was used to create the S-N curves. Given the σ_a (N_{life}, R = -1) value calculated in the first step, a Haigh diagram was built for each notched specimen. Even in this diagram the brittle behavior of the alloy observed in [19] was considered, dividing *YS* by *SCF* to define the yield line and *UTS* by *SCF* to define the fatigue limit curve (Goodman line). The yield domain is defined by the area between the vertical axis of the diagram, the horizontal axis, and the yield line. The failure domain is defined by the area between the vertical axis, the horizontal axis and the Goodman line. The intersection of these two domains provides the strength domain. By intersecting the curve corresponding to R = 0 with the limit curve of the strength domain, it is possible to identify the mean and alternating stresses to failure, $\sigma_{m,lim}$ (N_{life}, R = 0) and $\sigma_{a,lim}$ (N_{life}, R = 0) at the fatigue life N_{life} when the load ratio R is equal to 0. In all the three analysed cases, the failure domain is stricter than the yield domain. The sum of $\sigma_{m,lim}$ (N_{life}, R = 0) and $\sigma_{a,lim}$ (N_{life}, R = 0) is σ_H. The analysis with the Haigh diagram provides a stress range to failure of 268, 229 and 225 MPa for the tested specimens. The corresponding discrepancies Δ are 18%, 8% and 12%. Scatter is an intrinsic characteristic of fatigue properties of materials and components. The fatigue strengths of similar specimens or components subjected to the same fatigue load may be very different [32]. For this reason, it is possible that the repetition of the experimental tests provides slightly different results, with slightly different discrepancies compared with those reported in Table 3. Since all the values of discrepancies obtained in this work are less than 20%, they can be considered acceptable. The discrepancies may also be due to the assumptions made in the calculation including the absence of residual stresses after the creation of the notches. As known, residual stresses affect the fatigue life of components [33]. Even if the results of the step loading procedure and the estimation of the stress to failure with the Haigh diagram are quite similar, further investigations are needed.

Table 3. Results of the experimental tests (data of the third specimen from [28]).

d (mm)	SCF	F_{fail} (kN)	N_{fail}	σ_{SL} (MPa)	σ_H (MPa)	Δ (%)
0.5	2.48	17.0	135544	220	268	18
1.0	2.91	16.8	160650	248	229	8
2.0	3.11	13.0	139458	251	225	12

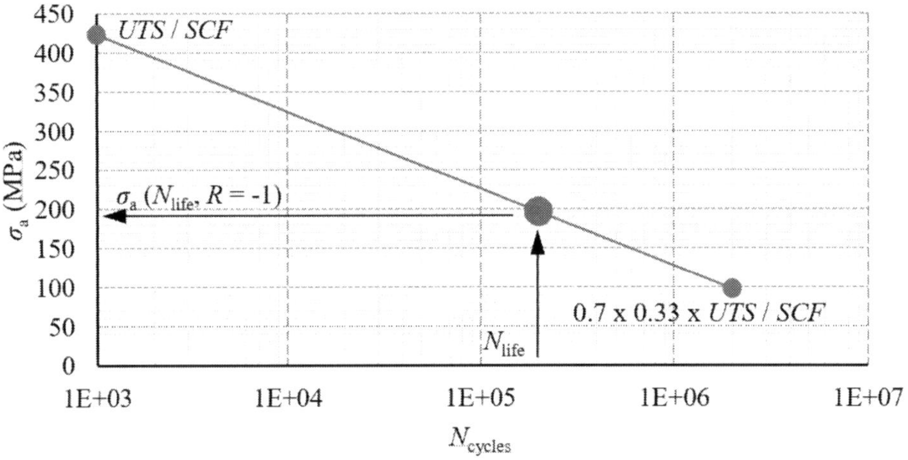

(a)

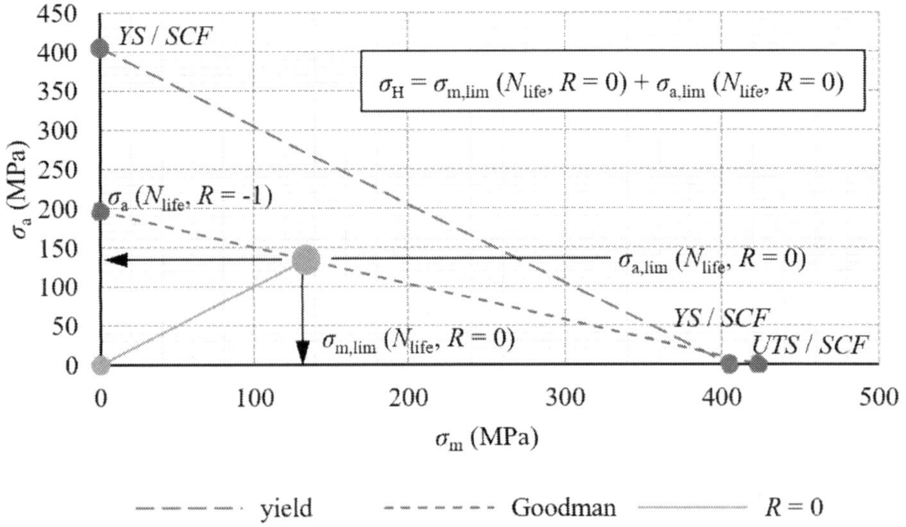

(b)

Figure 3. Calculation of σ_H: (a) S-N curve, (b) Haigh diagram.

Figure 4 shows the fracture surfaces of the tested specimens, observed with a stereo microscope. The nucleation areas are marked. In all the cases the fatigue cracks propagated from the notches due to the stress concentration induced by them. In the specimen with notch depth d = 0.5 mm, the crack nucleation area lies between a vertex of the fracture surface and the midplane of the specimen normal to the notch edge (Figure 4a). In the specimen with d = 1.0 mm, the crack propagated from a point between the two vertices of the fracture surface (Figure 4b). In the specimen with d = 2.0 mm, the crack propagated from the area close to one vertex of the fracture surface (Figure 4c, [28]). There are no symmetry planes in the three fracture surfaces. Only the fracture surface of Figure 4b tends to be symmetric, with the crack nucleation point quite equidistant from the vertices lying on one notch. In this area the theoretical stress state is plane stress. The general lack of symmetry in the fracture surface can be justified by possible mounting misalignments of the specimen with respect to the axis of the testing machine, presence of defects in the specimen, and lack of symmetry in the geometry of the specimens due to the machining of the notches by milling, which is relatively inaccurate on small scales. These aspects could be responsible for a partial overload of some areas of the specimens, where premature cracking occurred.

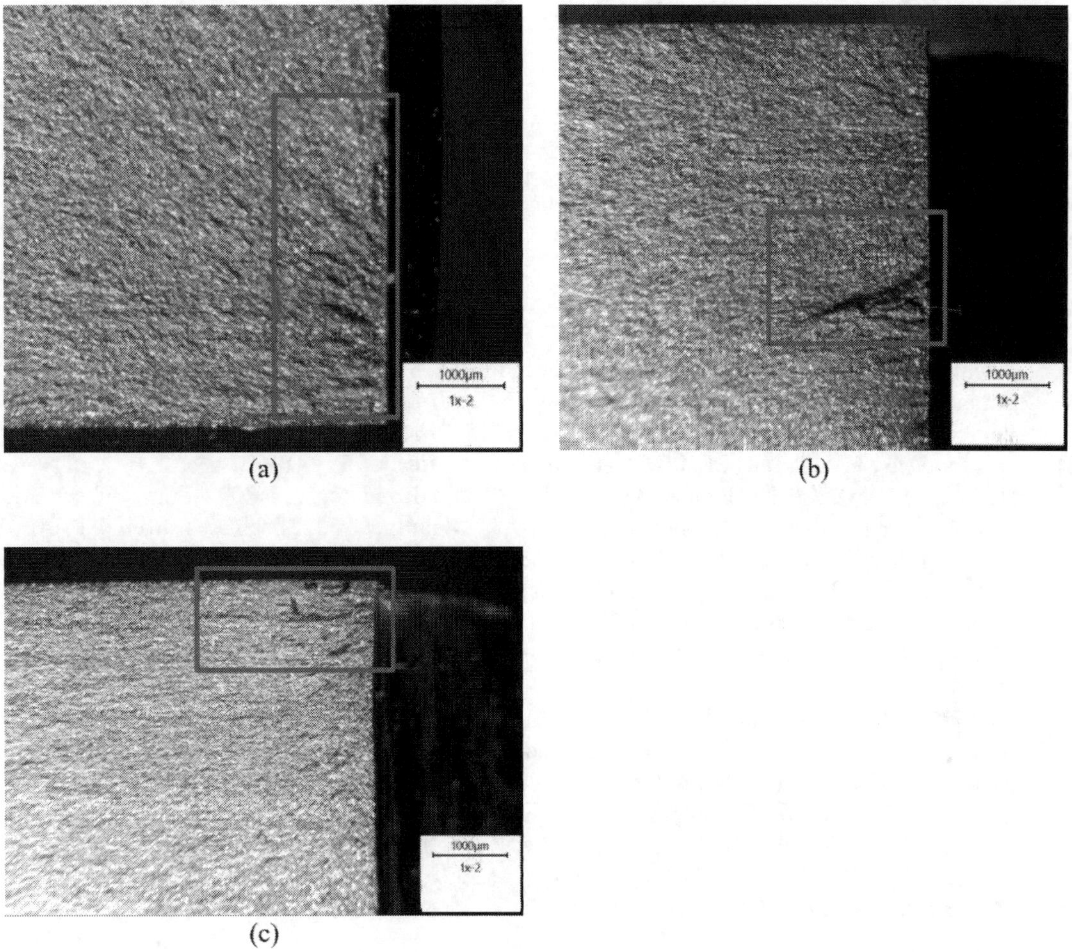

Figure 4. Fracture surfaces of the tested specimens: (a) d = 0.5 mm, (b) d = 1.0 mm, (c) d = 2.0 mm (third fracture surface adapted from [28]).

4. Conclusions and future developments

This work investigated the axial fatigue behavior of some notched specimens made of Ti-6Al-4V not subjected to STOA in inert environment. The values of *SCF* of the tested specimens were calculated using the finite element method. The specimens were tested according to the step loading procedure. The experimental results are quite in agreement with the analysis conducted with the Haigh (modified-Goodman) diagram assuming a brittle behavior of the material and an alternating stress to failure equal to 0.7 x 0.33 x *UTS* at a fatigue life of 2 million loading cycles. The analysis of the fracture surfaces with a stereo microscope provided some preliminary information on the mechanism that led to failure the tested specimens. Further experimental tests would be useful to fully understand the fatigue behavior of Ti-6Al-4V without the STOA treatment and to compare the fatigue strength of the untreated alloy with the treated one. The analysis of the fracture surfaces of the failed specimens at high magnification could help to clarify the failure mechanism. The application of strain gauges on the specimens to monitor the stress state in the specimen during the experimental tests could enable the identification of eventual parasitic stresses.

Acknowledgments

The authors wish to acknowledge NEVI S.r.l. for the machining of the specimens.

References

[1] Zheng C, Wang F, Cheng X, Liu J, Fu K, Liu T, Zhu Z, Yang K, Peng M and Jin D 2015 *Int. J. Impact Eng.* **85** 161-169
[2] Lütjering G and Williams J C 2007 Alpha + Beta Alloys *Titanium Engineering Materials* Springer (Berlin) pp 203–258
[3] Peirs J, Tirry W, Amin-Ahmadi B, Coghe F, Verleysen P, Rabet L, Schryvers D and Degrieck J 2013 *Mater. Charact.* **75** 79-92
[4] Kakogiannis D, Verleysen P, Belkassem B, Coghe F and Rabet T 2018 *Int. J. Impact Eng.* **119** 1-13
[5] Altenberger I, Nalla R K, Sano Y, Wagner L and Ritchie R O 2012 *Int. J. Fatigue* **44** 292–302
[6] Baragetti S, Guagliano M and Vergani L 2000 *Int. J. Mater. Prod. Technol* **15** 91–103
[7] Baragetti S and Tordini F 2007 *Int. J. Fatigue* **29** 1832–1838
[8] Lewandowski J J and Seifi M 2016 *Annu. Rev. Mater. Res.* **46** 151–186
[9] Nesládek M, Matušů M, Papuga J, Mžourek M and Roudnická M 2022 Some Observation Concerning Fatigue Response of Additively Manufactured Specimens from Ti-6Al-4V *Fatigue and Fracture of Mateials and Structures ed* (vol 24) eds G Lesiuk, S Duda, J.A.F.O. Correia, A.M.P. De Jesus (Cham: Springer) pp 251-358
[10] Arcieri E V, Baragetti S and Božić Ž 2021 *Procedia Struct. Integr.* **31** 22-27
[11] Nicholas T 2006 Foreign Object Damage *High Cycle Fatigue* Elsevier Science Ltd (Oxford) pp 322–376
[12] Baragetti S, Cavalleri S and Tordini F 2012 *Key Eng. Mater.* **488-489** 502-505
[13] Baragetti S 2013 *Surf. Interface Anal.* **45** 1654-1658
[14] Baragetti S, Foglia C and Gerosa R 2012 *Key Eng. Mater.* **525–526** 505–508
[15] Baragetti S 2014 *Mater.* **7** 4349-4366
[16] Baragetti S and Villa F 2014 *Frat. ed Integrita Strutt.* **30** 84–94
[17] Baragetti S and Arcieri E V 2018 *Int. J. Fatigue* **112** 301-307
[18] Baragetti S, Borzini E and Arcieri E V 2018 *Procedia Struct. Integr.* **12** 173-182
[19] Baragetti S, Borzini E and Arcieri E V 2019 *Corr. Rev.*
[20] Muttoni S and Legrenzi N 2022 Progettazione di telai per veicoli ibridi e caratterizzazione di materiali ad alto rapporto resistenza / massa (leghe di titanio e alluminio) (bachelor thesis)
[21] Baragetti S and Baryshnikov A 2001 *J. Mech. Design* **123** 456-463
[22] Srinivasan N, Bhaskar L K, Kumar R and Baragetti S 2018 *Mater. Des.* **160** 303-312
[23] Baragetti S 2006 *Meccanica* **41** 443-458

[24] Baragetti S and Arcieri E V 2019 *Procedia Struct. Integr.* **24** 91-100
[25] Yan G, Crivoi A, Sun Y, Maharjan N, Song X, Li F and Tan M J 2018 *Manuf. Process.* **32** 763-772
[26] Baragetti S and Medolago A 2013 *Key Eng Mater.* **525-526** 501-504
[27] ABAQUS, ABAQUS Documentation, Dassault Systèmes, Providence, RI, USA, 2017
[28] Arcieri E V and Baragetti S 2022 *AIP Conf. Proc.* (submitted)
[29] Nicholas T 2002 *Fatigue Fract. Eng. Mater. Struct.* **25** 861-869
[30] Locati L 1955 *Metall. Ital.* **47** 301–308
[31] Braut S, Tevčić M, Butković M, Božić Ž and Žigulić R 2021 *Procedia Struct. Integr.* **31** 33-37
[32] Schjve J 2009 Fatigue and Scatter *Fatigue of Structures and Materials* Springer (Dodrecht) pp 373-394
[33] Baragetti S and Arcieri E V 2022 *Proc. Inst. Mech. Eng, Part C* **236** 10713-10722

DIC analysis of mechanical response of tooth aligners under simulated swallowing acts

C Casavola[1], G Pappalettera[1], C Pappalettere[1], M Patronelli[1], G Renna[1], D Ciavarella[2] and M Laurenziello[2]

[1]Dipartimento di Meccanica, Matematica e Management, Politecnico di Bari, Via Orabona 4, 70126 – Bari, Italy

[2]Dipartimento di Medicina Sperimentale e Clinica, Università di Foggia, Foggia, Italy

Corresponding author: Gilda Renna email: gilda.renna@poliba.it

Abstract. In this work, the mechanical and deformation behavior of clear Polyethylene Terephthalate-glycol (PET-g) aligners, under cyclic loading was investigated using a full-field optical technique: the Digital Image Correlation. In particular, the PET-g aligners thermoformed from 0.88 mm thick discs, were subjected to cyclic compression tests for 13000 load cycles from 0 to 50 N in the atmospheric environment (~25°C). This number of cycles was chosen because it simulates, on average, the intraoral load associated with the swallowing acts that an aligner is subjected to during the time of use of 1 week. At the same time, the results from the analysis of hysteresis loops obtained by the DIC technique were compared with those obtained by the testing machine. The mechanical response of clear aligners was evaluated in terms of maximum displacement, energy loss and relative stiffness along the load direction to seven different stages of the 13000 load cycles. A comparable trend was found between the measurements obtained by Digital Image Correlation analysis and the analysis of the hysteresis loops obtained from the cyclic compression tests. Furthermore, the morphological features of the PET-g aligner at the end of the tests were analyzed by optical microscopy (OM). The OM analyses showed that the surface of PET-g aligner was affected by morphological variations such as high depressions and cracks.

Keywords: PET-g aligners, mechanical behaviour, digital image correlation technique, DIC technique, compression tests, deformation behavior

1. Introduction

In the last decades, the rising demand among adult patients, for dental misalignment problems or the treatment of malocclusion, of aesthetic devices has led to an exponential growth in the market of clear aligners [1-5]. The orthodontic treatment with these devices consists in the sequential use of clear aligners characterized by appropriate geometry and made with a thermoplastic polymeric material. Based on the specific orthodontic treatment plan, these aligners, through a controlled distribution of forces, allow gradual repositioning and reorientation of the teeth [6].

Since, the forces exerted by clear aligners on the dental arch, depend substantially on their geometry, it is necessary to make orthodontic devices that do not undergo important deformation throughout the treatment period. In addition, among the parameters that influence the quality of the force applied to the teeth, and thus the final outcome of orthodontic treatment, the mechanical properties of the manufacturing material are of fundamental importance. Generally, the desired properties for an aligner

Content from this work may be used under the terms of the Creative Commons Attribution 3.0 licence. Any further distribution of this work must maintain attribution to the author(s) and the title of the work, journal citation and DOI.
Published under licence by IOP Publishing Ltd

material are: biocompatibility, transparency, low rigidity, good formability, environmental stability and resistance to aging [7].
To date, PolyPropylene, PolyCarbonate, Polyethylene Terephthalate glycol, Ethylene Vinyl Acetate and Thermoplastic PolyUrethanes are just some of the thermoplastic materials available on the market for this type of application [1,8-10]. Nonetheless, only some materials among those available exhibits both before and after use of the same chemical characteristics. As a result, the mechanical and especially the optical properties of many aligner materials will be affected differently by factors, such as food colourings, masticatory stress and salivary enzymes, to which they are normally exposed in the oral cavity during the period of use [10]. Several scholars have shown that factors such as temperature, humidity, thickness, characteristics of the forming process, and time elapsed after elastic deformation can influence the deformation of aligner material under load [11,12]. Therefore, knowledge of deformation behavior is of fundamental importance not only for the proper design of aligners made from these materials but also to guarantee correct functionality during their use. Besides, the knowledge of the mechanical and deformation behaviour of the aligner can help the orthodontist to accurately define the most suitable strategy to use, simultaneously improving the patient's comfort and treatment time.
At the present time, thanks to its exceptional formability and its excellent aesthetic characteristics, the polyethylene terephthalate-glycol (PET-g) is forenumber used plastic material for the production of such devices [11,13]. Nowadays, the majority of researchers that have focused on aligners' mechanical behaviour and the effect of the intraoral environment perform tests on rectangular strips of material cut from the discs used to thermoform these devices. However, this way of evaluating the performance of aligners is entirely questionable because it does not capture the complexity of the behavior of a geometrically complex device such as a dental aligner. Therefore, unlike the works in the literature, in this research work, the effect of intraoral loading on PET-g aligners was evaluated by performing tests directly on the thermoformed aligners from 0.88 mm thick discs. Particularly, PET-g aligners were subject to cyclic compression tests for 13000 load cycles in the atmospheric environment (~25°C). Each load cycle was increased from 0 to 50 N. 13000 cycles were chosen because it simulates, on average, the intraoral load associated with the swallowing acts that an aligner is subjected to during the time of use of 1 week. Digital image correlation (DIC) [14,15] was used to investigate the cyclic compression behavior of clear aligners. This is an innovative full-field measurement technique for this type of application. This, being a contactless optical technique, allows to follow the evolution of the aligner behavior, instant by instant, taking into consideration the entire arch-aligner set of teeth.
In this research, the mechanical response of PET-g aligners was evaluated in terms of maximum displacement, energy loss (irreversible work) and relative stiffness along the load direction (y axis). These terms were derived from the analysis of the load-displacement curves obtained by the DIC technique. At the same time, a comparison of these quantities with those obtained from the analysis of the hysteresis loop of the test machine was performed to assess whether this technique is suitable to describe the aligners' behavior. Finally, the morphological features of the aligners surfaces at the end of the loading test were investigated under an optical microscope (OM).

2. Experimental Procedure

2.1. Materials
The clear aligners samples were manufactured from PET-g discs of 0.88 mm nominal thicknesses using an Erkodent® 3D vacuum machine (Figure 1b). Especially, Polyethylene Terephthalate-glycol was an Ace® Plastic produced by GAC International, made of 95% copolyester and 5% trade secret. Before thermoforming, it was necessary to make a digital dental record and then print a resin dental cast of a patient's mouth (Figure 1c). The clear thermoformed aligner is shown in Figure 1a and the overview of the occlusal plane of the 1st molar before performing the cyclic tests is reported in Figure 1d.

IOP Conf. Series: Materials Science and Engineering 1275 (2023) 012023 doi:10.1088/1757-899X/1275/1/012023

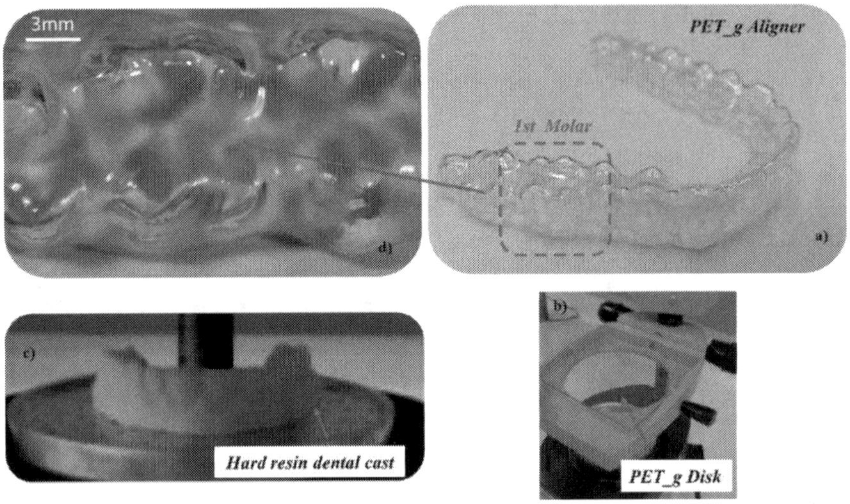

Figure 1. a) PET-g aligner sample after thermoforming, b) PET-g disc of 0,88 mm thick inserted in the support plate of the Erkodent® 3D vacuum machine used for termoformed the clear aligners c) resin dental cast of a patient's mouth and d) view macrographic of 1st molar

2.2. *Setup of Cyclic Compression Test and Full-field displacement measurement*

Cyclic compression test was performed at room temperature (~25°C) using an Instron 3344 (Norwood, MA, United States) electro-mechanic machine with a single column, shown in Figure 2a, equipped with the software Bluehill 2.15 and a 1 kN loading cell. The aligner was fitted on a proper hard resin dental cast which reproduces the upper dental arc and then the system was put in occlusion with the corresponding lower dental cast, as shown in Figure 2b. The cyclic load, under load control conditions, includes four stages each one lasting 1 s, visible in Figure 2c. The total number of cyclic load was calculated considering the average number of daily dental contacts during the swallowing act (1500 swallowing acts) and the time of swallowing act (1 s). So, one load cycle comprises four stages: the first in which the load increases from 0 to 50 N; the second in which the maximum load is held for 1 s; the third in which a descending ramp brings the load to 0 N; the fourth in which the sample is kept at 0 N for 1 s. The clear aligner was subjected to 14250 load cycles consecutively for 16 hours. This load history corresponds to a one-week usage of a clear aligner, considering that at certain times of the day is not worn for eating or hygiene operations. Measures were taken using a contactless full-field optical technique: Digital Image Correlation (DIC). The measurement setup, comprised the DIC system Dantec Dynamics Q400 (Dantec Dynamics A/S, Skovlunde, Denmark), the INSTRA 4D software (Dantec Dynamics A/S, Skovlunde, Denmark), the Timebox, one GigE CCD Manta cameras of 1628 x 1436 pixels (Allied Vision Technologies GmbH, Stadtroda, Germany) equipped with a high-resolution lens Xenoplan 2.8/50-0902 (Schneider) and the LED light, is represented in Figure 2a. The speckle pattern of black random dots on a white background was realised on the half of dental arc of the clear aligner as is shown in Figure 2d. Before the test the quality of the pattern was evaluated through a Matlab code, so a black/white ratio of 31% was obtained. During the test four acquisitions were taken: the first at the beginning of the test and the others respectively after 1000, 5000 and 13000 load cycles. Thanks to the full-field measurements, local displacements values were extrapolated by placing a virtual strain gauge (small area A) on the 1st molar as shown in Figure 2e. The 1st molar was chosen as the most stressed tooth because of the specific clinical conditions of the patient who has a regular dentition. DIC measurements are based on the differences between the area selected on the reference image, taken at zero load, and the subsequently images captured during the test. The INSTRA software provides the average of the displacements, $\overline{D_F}$, given by the displacement $D_{P_{i,F}}$ of each single pixel, i, contained in area A, for each F image recorded during the test:

$$\overline{D_F} = \frac{\sum_{i=0}^{n} D_{P_{i,F}}}{n} \qquad 0 \leq F \leq N . \tag{1}$$

Where $P_{i,F}$ is referred to the point P associated with the pixel i of the image F, n indicates the number of pixels contained in the area A and N is the number of frames recorded by the camera. Before the test, the DIC system was calibrated to reduce measurement errors as much as possible. Displacement noise-floor of 0.785 ± 0.734 μm and 0.193 ± 1.518 μm were recorded respectively along the y and the x direction using a subset size of 17 px, a step size of 13 px and ad SSD matching criterion. All the DIC hardware parameters and the DIC analysis parameters used for this experiment are well schematized in [16].

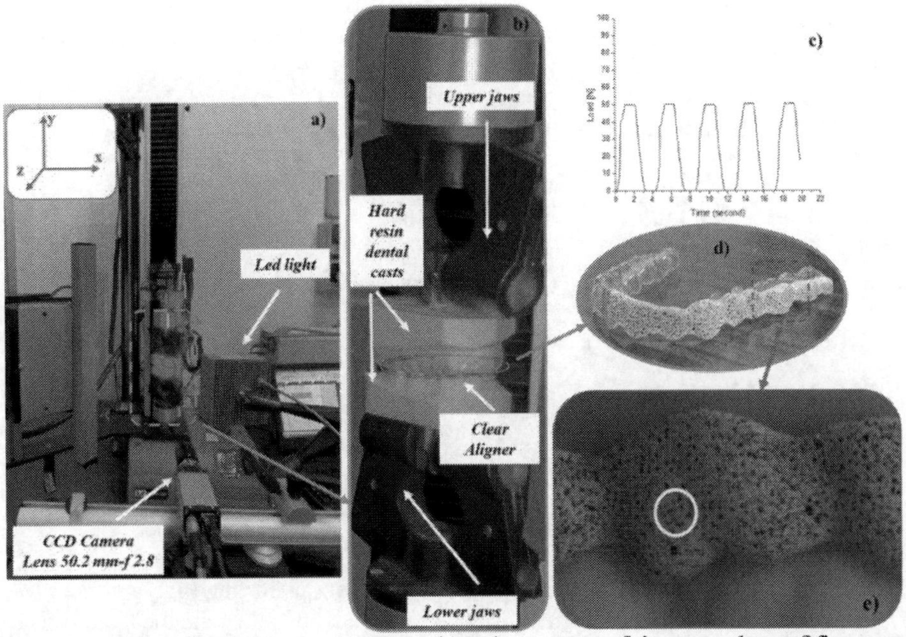

Figure 2. a) Setup of Cyclic Compression Test, b) enlargement of the green box of figure a showing the entire resin dental cast-aligner assembly, c) representation of load as a function of time during cyclic loading, d) image of aligner with speckle pattern of black random dots on a white background and e) enlargement of 1st molar showing the data analysis circular area.

2.3 Optical Microscopy (OM) Analysis
Surface morphology analysis was performed on the occlusal plane of the 1st molar by ZEISS - Axio Vert A1 optical microscopy (OM) under transmitted light. The surface' morphology of the aligners was imaged at the end of the cyclic compression test.

3. Results and Discussions

3.1. Comparison of the load–displacement hysteresis loops
Figure 3a and b show the evolution of the load–displacement hysteresis loops in the y direction (along load axis) obtained from the crosshead displacement of the testing machine and from the full-field (DIC) analysis, respectively. Specifically, seven curves describing load-displacement behaviour, are shown for each analysis technique: 2, 6, 10, 20, 1000, 5000 and 13000 loading cycles. For clarity data reported for DIC analysis, cycles from 2 to 20 were acquired in the 1st acquisition, the 1000th cycle was acquired during the 2nd acquisition, the 5000th cycle was acquired during the 3rd acquisition and the 13000th cycle

was acquired during the 4th and least acquisition. Observing Figure 3a, it can be seen that as the number of cycles n increases, the hysteresis loops tend to shift to the left toward lower displacements. Especially is interesting to note that the amount of displacement reduction is very small during the very first few cycles. After of 20th cycle, the amount of displacement reduction tends to grow progressively up to 1000 cycles and then decrease significantly again until it stabilizes toward the last cycles. Additionally, the curves remain highly nonlinear until the end of the test and become more narrow (reduction of area of the hysteresis loops) at high cycle numbers [16-18]. With reference to the load–displacement hysteresis loops obtained from DIC technique (Figure 3b), trends comparable to those obtained from the test machine are found up to 1000 cycles. After which the curves tend to return back towards greater displacements (right shift), indicating that after the 1000th cycle the contact point of the dental crown gradually moves upwards. This is due to the fact that as cycles progress, the areas of the dental crown do not always touch in the same way but vary continuously. Therefore, the aligner undergoes the same rigid movement of the dental arches [19]. Also, unlike the loops obtained by the test machine, those obtained by the DIC technique are more linear throughout the test and tend to become narrower as the test progresses.
Therefore, it is clear that the local analysis performed with the DIC allows us to highlight phenomena that cannot be observed from the analysis of the data provided by the test machine.

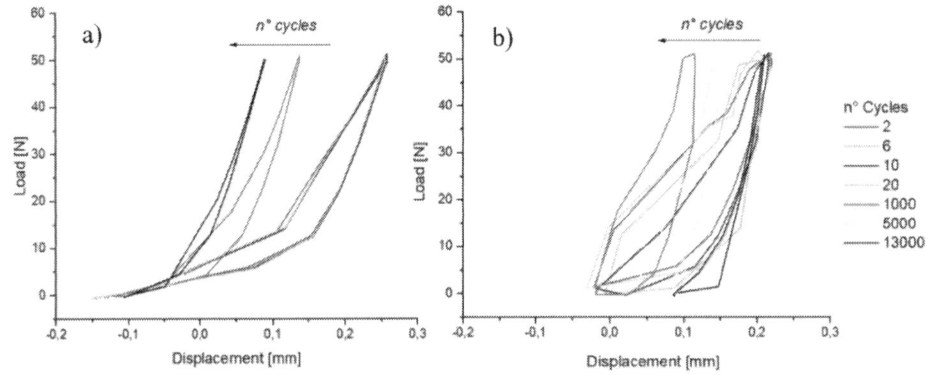

Figure 3. Comparison of the load-displacement hysteresis loops for the PET-g aligner (y direction) obtained from: a) analysis of the displacement of the crossheadand and b) the local analysis carried out with the DIC technique

3.2. PET_g mechanical behavior

The mechanical behavior of the PET_g aligner can be detailed better by determining the maximum displacement reached, the relative stiffness and the energy loss per cycle (Figure 4a-c). Furthermore, this last parameter also allows evaluating the viscoelastic nature of the thermoplastic material. In particular, the energy (N*mm) was determined by integrating the load-displacement hysteresis loops while the stiffness (N/mm) was determined as the slope of a linear regression fitting of the hysteresis loops. It can be observed from Figure 4a-c that the measurements derived from the analysis of the load-displacement loops (Figure 3) show similar trends. In particular, both curves show substantial changes from the 20th cycle up to the achievement of the 5000th cycle, after which these values tend to remain almost constant. More specifically, it can be seen that the maximum displacement decreases as $n°$ increases (Figure 4a); while the rigidity of the aligners tends to increase significantly with increasing cycles (Figure 4b) and the energy loss (irreversible work) decreases (Figure 4c). Therefore, the reduction in max. displacement is associated with a significant increase in aligner stiffness as it is fatigued. In other words, the orthodontic thermoplastic material becomes stiffer and less viscoelastic as cycles increase. The observed stiffening effects may be due to several factors. First of all, the stiffening can be attributed to the entanglement of the polymer chain that occurs as an effect of the application of the load, and secondly to the stress generated during the mechanical compression cycle. It is known that the stresses are generated in orthodontic device by the initial deformation and they tend to cumulate with increasing the loading cycles [19]. Other hypotheses that could explain the increase in stiffness are related to the absorption of water molecules from the air (environmental humidity) during the cyclic compression test [16,20,21]. In addition, comparing the stiffness curves shown in Figure 4b, it can be seen that the stiffening values obtained by the DIC technique are greater than those provided by the testing machine. Particularly, with reference to the DIC technique, a maximum stiffness of 451 N/mm was found, while the maximum stiffness value obtained from the cyclic tests is 256 N/mm. However, the increase in stiffness measured with the DIC is 131.4% compared to that measured by the test machine which is 107.4%. On the other hand, this is completely normal as the DIC technique, based on local analysis, provides us with information regarding the actual behavior of the aligner. While the results obtained from the crosshead displacement analysis of the testing machine represent only an average behavior of the aligner, thus not grasping in detail, the behavior of the aligner that is characterized by a rather complex geometry.

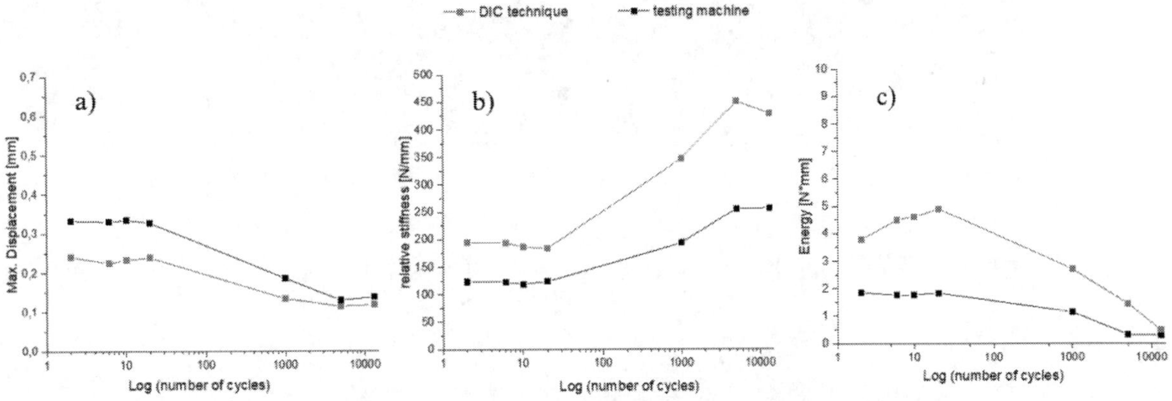

Figure 4. Comparison of the maximum displacement (a), energy loss (b) and relative stiffness (c) as a function of the log of number of cycles obtained by the two different techniques

If the displacement data recorded by the testing machine give information only along the y direction, the DIC technique allows us to obtain full-field measurements, and therefore also to know the total displacement component (which is directly related to the direction of the load distribution) and the principal compression strains. Figures 5 and 6 show the results deriving from the analysis of the total displacement component and the deformation chromatic maps (i.e. compressive strain), respectively. By analyzing Figure 5, it can be observed that the first 20 cycles are characterized by a constant total displacement component developing in the same verse and direction. In accordance with what was seen above (Figure 3), during the very first cycles, the aligner has a displacement component of about 300 μm. This value tends to decrease to about 100 μm at 1000 cycles. Interestingly, at 1000 load cycles the total displacement component is mirrored along the vertical axis. After this number of cycles, the displacement components rise again up to 240 μm always maintaining the same verse and direction. In other words, during these first cycles, the aligner undergoes a displacement relative to the cast fitting due to initial settlement. As the cycles progress, this displacement decreases, indicating that the aligner adheres to the cast. After that, for high numbers of cycles, the aligner again loses adherence with the resin dental cast walls exactly as it occurs during the stages. This loss of adherence could be associated with the aligner wear which can occur with cracks initiation and propagation.

The experimental results reported in Figure 6 show instead, that the most deformed portion of the material was the one close to the contact area with the opposite tooth (note that the most deformed bands are the purple and blue ones); whereas, on the lateral part of the tooth the entity of strain is quite slight.

It is also interesting to note that the thickness of the observed strain band corresponds perfectly to the thickness of the aligner (0.88 mm). This behavior is clearly highlighted by the graph in Figure 6e

IOP Conf. Series: Materials Science and Engineering 1275 (2023) 012023 doi:10.1088/1757-899X/1275/1/012023

which shows the trend of the principal compressive strain along the height of the 1st molar. The curves show an almost constant value along the entire height of the tooth, from 0, close to the gum, up to about 4.5 mm, after which the curves become steep in correspondence with the contact area with the opposite tooth. In addition, it is possible to observe that as the number of cycles increases *n*, the strain band moves along the profile of the tooth, as a function of the area of contact that occurs between the crowns of the upper tooth with the lower one as the number of cycles increase.

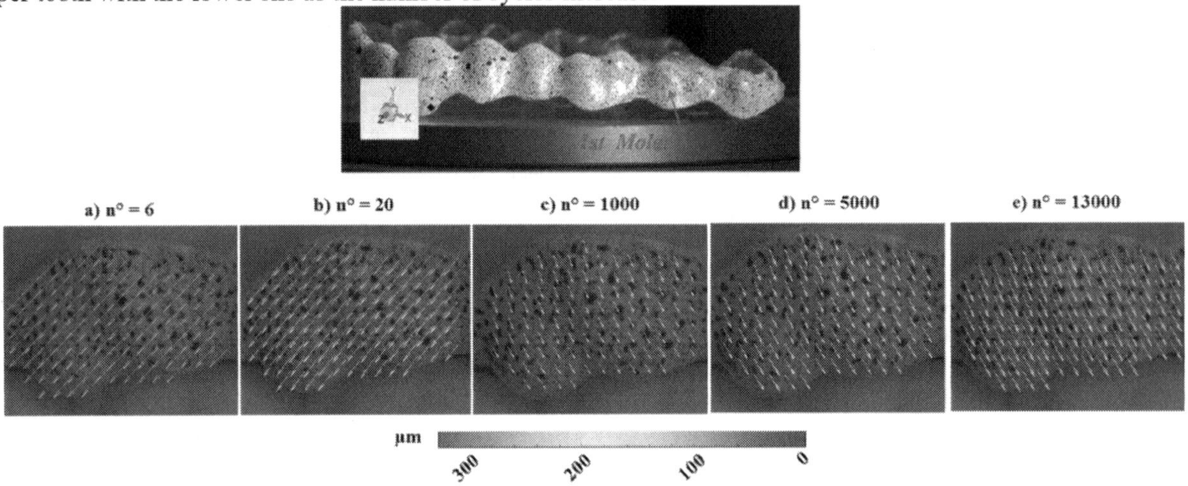

Figure 5. Total displacement direction analysis during the; a) 6nd, b) 20th, c) 1000th, d) 5000th and f) 13000th cycle.

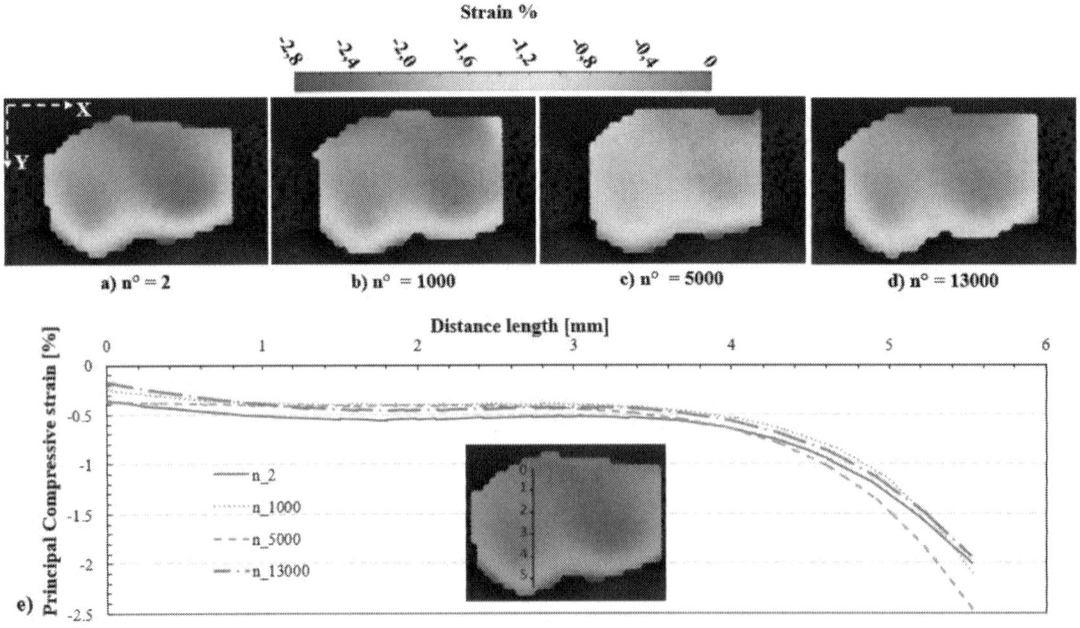

Figure 6. Principal compressive strain performed by DIC technique for the PET-g aligner during the a) 2nd, b) 1000th, c) 5000th. d) 13000th load cycle and e) trend of the principal compressive strain along the length of the tooth in the Y-direction.

Finally, for completeness, Figure 7 shows the results of the displacement analysis along the y and x-axis for the entire aligner. In order to immediately highlight the differences, a monochrome scale was used.

Note that for the chromatic scale along x, the positive direction increases to the left, while, for the one along y, the positive direction increases to the right. By observing Figure 7a-d, it is possible to infer that the largest displacements occur in the anterior part of the teeth: in the incisors. In fact, this is what really happens when the two dental arches, upper and lower, move apart and reconnect; this again confirms that the experiment is faithfully reproduced from start to finish. Furthermore, it can be observed that the displacement ranges affected by the aligner change with the cycles confirming that the crowns of the teeth touch differently during the test. Especially, in the 2nd cycle a range of 168÷468 μm can be observed, at the 1000th cycle a range of 90-327 μm (Figure 7a), at the 5000th cycle a range of 100 ÷ 366 μm (Figure 7b) and finally at the 13000th cycle a range of 185 ÷ 440 μm (Figure 7c). While with regard to the x axis (Figure 7e-h), it can be observed that the displacement is fairly constant for the entire aligner. Particularly, in the contact area of the aligner there is an average displacement of -235 μm for the 2nd cycle (Figure 7e), 86 μm for the 1000th cycle (Figure 7f), 114 μm for the 5000th (Figure 7g) and 144 μm for the 13000th cycle (Figure 7h). It should be specified that these data also include the rigid displacement component, which is important precisely in correspondence with the incisors. If the rigid displacement component is excluded, along X there is a displacement range from +40 μm to -35 μm, while along Y there is a range from 110 μm to -90 μm, with the greatest displacements reached in correspondence of the molars. As a result of the analyses seen above, it is therefore clear how the analysis performed with the DIC is able to detect and describe the real behavior of the aligner allowing at the same time to verify that the test continues according to the imposed operating conditions

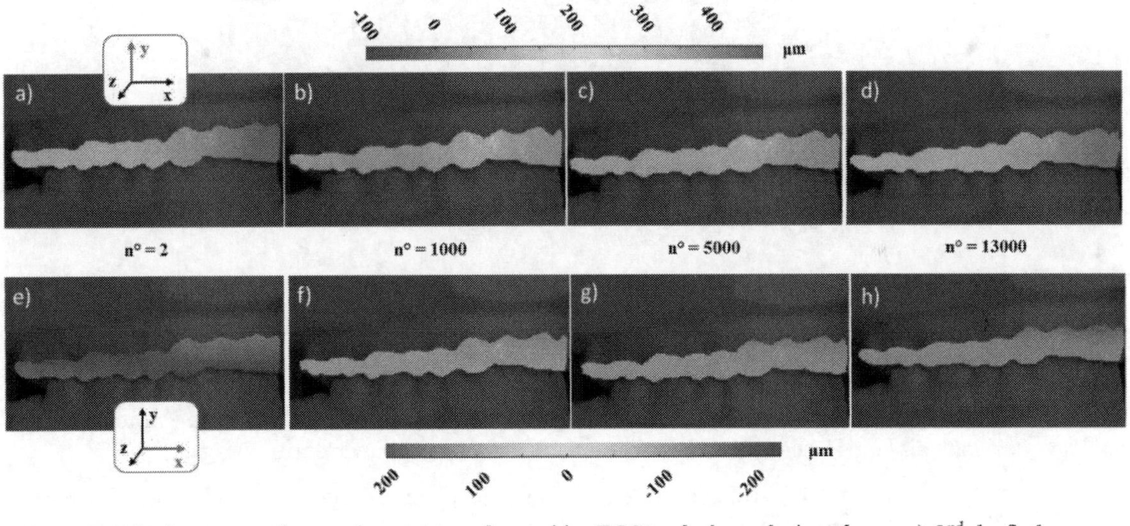

Figure 7. Displacement chromatic maps performed by DIC technique during the: a, e) 2nd, b, f) the 1000th, c, g) the 5000th and d, h) the 13000th load cycle along the load axis y and x, respectively.

3.3. Optical Microscopy characterization

The OM analysis allowed to corroborate the above results. Figure 8 shows the micrographs of the occlusal plane of the 1st molar. It can be observed that the surface of the aligner is strongly degraded following the 13000 load cycles. It is interesting to note that the most worn area is the central one of the tooth. In particular, in this area, the surface is characterized by the formation of an elevated depression (fracture) and cracks that branch out towards the dental crown starting from the fractured area. The presence of these cracks indicates high degradation kinetics in the central part of the tooth. In addition, it is worth noting that all around the crack surface there is the presence of numerous crazes. It was also observed that the density and size of these crazes decrease as one moves away from the crack surface. While along the crown of the tooth small micro-voids of different morphology and with sizes of 20-50 μm were observed. The micro-voids with acicular morphology observed as well as large voids are generally preferential sites for the nucleation of the fatigue crack, due to the local intensification of the stress dependent on the size and morphology of the voids. The origin of this damage lies in the fact that the elasticity degree of the PET-g thermoplastic material is not enough to counteract the stresses induced by the occlusal load for an exercise period of 13 days. The presence of this degree of wear could compromise the correct transmission of orthodontic forces. Furthermore, the observations under the light microscope show important chromatic variations all around the fractured area. This phenomenon is known as birefringence and is due to the cyclic compressive load which induces internal stresses and important deformations within the aligner.

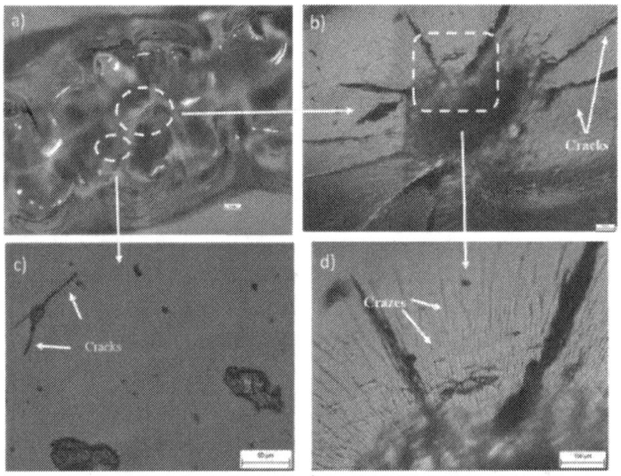

Figure 8. a) overview of the occlusal plane of 1st molar of PET-g aligner at the end of the 13000 load cycles, b) enlargement of the central area of the tooth showing high depressions and cracks, c) enlarged image of the spot fringe of Figure a and d) enlarged image of the spot fringe of Figure b showing the crazes around the crack surface

4. Conclusion

The study of mechanical response of 0.88 mm thick PET-g thermoformed aligners using the DIC technique has brought the following evidences:

- A good correlation was found between the measurements obtained by full-field (DIC) analysis and the analysis of the hysteresis loops obtained from the cyclical tests. Moreover, the DIC analysis helps to understand phenomena that occur in aligner during usage.
- The hysteresis loops reflect a ratcheting displacement reduction in the aligners during loading cycles. At the same time, the orthodontic thermoplastic material become stiffer and less viscoelastic as the aligner was fatigued. The loss of the viscoelastic behavior of thermoplastic material is evidenced by the decrease in viscoelastic energy loss per cycle.
- Local displacement maps show that the quantities derived from the load-displacement cyclic curves, such as energy and stiffness, are influenced not only by the mechanical behavior of the material itself, but also by the complexity of the loading distribution.
- The DIC technique has been found to be an excellent optical method for determining local displacements of an aligner under load, which is fairly new in this field
- The OM analyses showed that the surface of PET-G aligner was affected by significant morphological variations such as high depressions and cracks indicating a significant reduction of functionality after 7 consecutive days of wear.
- From a medical point of view, it is necessary to carefully evaluate possible use of the aligners for a further week due to considerable degradation of the surface

IOP Conf. Series: Materials Science and Engineering 1275 (2023) 012023 doi:10.1088/1757-899X/1275/1/012023

References

[1] Lombardo L, Arreghini A, Maccarrone R, Bianchi A, Scalia S and Siciliani G 2015 Progress in Orthodontics 16:41
[2] Boyd R, Miller RJ and Vlaskalic V 2000 J Clin Orthod. 34:13–5.
[3] Kohda N, Iijima M, Muguruma T, Brantley WA, Ahluwalia KS and Mizoguchi I 2013 Angle Orthod. 83(3):476–83.
[4] Rosvall MD, Field HW, Ziuchkovski J, Rosenstiel SF and Johnston WM 2009 Am J Orthod Dentofacial Orthop 135:276–88.
[5] Kim T W and Echarri P 2007 World J Orthod. 8:13–8
[6] Elkholy F, Schmidt F, Jäger R, and Lapatki BG 2016 Angle Orthodontist, vol. 86, no. 6, pp. 883–890
[7] Zhang N, Bai Y, Ding X and Zhang Y 2011 Dental Materials Journal, vol. 30, no. 6, pp. 954–959
[8] Krzypow DJ and Rimnac CM 2000 Biomaterials, vol. 21, no. 20, pp. 2081–2087
[9] Kwon J-S, Lee Y-K, Lim B-S and Lim Y-K 2008 American journal of orthodontics and dentofacial orthopedics : official publication of the American Association of Orthodontists, its constituent societies, and the American Board of Orthodontics, vol. 133, pp. 228–34
[10] Gracco A, Mazzoli A, Favoni O, Conti C, Ferraris P and Tosi G 2009 Aust Orthod J. 25(1):34–40.
[11] Lombardo L, Martines E, Mazzanti V, Arreghini A, Mollica F and Siciliani G 2017 Angle Orthodontist, vol. 87, no. 1, pp. 11–18
[12] Eliades T, Eliades G and Watts DC 1999 EurJ Orthod. 21:649–58.
[13] Hassan MH Omar AM, Daskalakis E, Hou Y, Huang B, Strashnov I, Grieve B D and Bártolo P 2020 Polymers, vol. 12, no. 12, pp. 1–12
[14] Chu TC, Ranson W F, Sutton M A and Peters W H 1985 Exp. Mech. 25, 232–244
[15] Schreier H, Orteu J -J and Sutton M A 2009 Image Correlation for Shape, Motion and Deformation Measurements. Boston, MA: Springer US
[16] Casavola C, Pappalettera G, Pappalettere C, Patronelli M, Renna G, Laurenziello M and Ciavarella D 2022 Journal of the Mechanical Behavior of Biomedical Materials Volume 134, October, 105391
[17] Shen X, Xia Z and Ellyin F 2004 Polym. Eng. Sci., vol 44 (12), pp. 2240-2246
[18] Cianci C, Pappalettera G, Renna G, Casavola C, Laurenziello M, Pappalettere C and Ciavarella D 2020 Frontiers in Materials, vol. 7
[19] Cianci C, Pappalettera G, Renna G, Casavola C, Laurenziello M, Pappalettere C and Ciavarella D 2020 IEEE International Symposium on Medical Measurements and Applications (MeMeA)
[20] Schrauwen B A G 2003 Deformation and failure of semicrystalline polymer systems: influence of micro and molecular structure, Eindhoven, NL, Technische Universiteit Eindhoven
[21] Dupaixa R B and Boyce M C 2005 Polymer vol. 46, pp. 4827–4838

IOP Conf. Series: Materials Science and Engineering 1275 (2023) 012024 doi:10.1088/1757-899X/1275/1/012024

Effect of roughness small scales on the adhesion of randomly rough surfaces with high fractal dimension

G Violano [1], D Dini [2], A Di Bari [3] and L Afferrante [1]

[1]Department of Mechanics, Mathematics and Management, Polytechnic University of Bari, Via E. Orabona, 4, 70125, Bari, Italy
[2]Department of Mechanical Engineering, Imperial College London, London SW7 2AZ, United Kingdom
[3]Bosch Italia, Via delle Ortensie, 15, 70026, Modugno (BA), Italy

E-mail: luciano.afferrante@poliba.it

Abstract. Surface roughness affects several tribological phenomena like adhesion. In this work, JKR-like experimental tests were performed between Polydimethylsiloxane (PDMS) spherical samples and flat glass substrates with different surface roughness properties. Data were interpreted in the light of classical JKR theory. We find that adhesion is destroyed by roughness in agreement with the most common and well-known results in the literature. However, from the analysis of the Power Spectral Density (PSD) of the surfaces, it emerges that the smaller scales of roughness have a greater influence on the surface adhesion energy in the range of the high fractal dimensions ($D > 2.5$). In this range, for surfaces with the same root mean square (RMS) roughness amplitude h_{rms}, which is mainly affected by large scale roughness, the pull-off force (i.e., the maximum tensile force reached during the retracting phase) is observed reducing by increasing the RMS slope of the surface roughness h'_{rms}, that is by changing the roughness content at the small scales. This result agrees with recent theoretical findings and has never been shown experimentally in the literature to the best of our knowledge.

1. Introduction

Adhesion plays a crucial role in many important biological and engineering applications. For example, adhesion governs the mechanism of locomotion of several insects, lizards and beetles [1, 2, 3, 4], and is of extreme importance in the cell adhesion [5]. Moreover, there are many other applications exploiting adhesion including pressure sensitive adhesives ([6]), adhesive gloves and suits ([7]), thin coatings in orthopedic implants ([8, 9]), medical adhesive bands ([10]), just to mention someone. Adhesive interactions affect also the tribological performance of mechanical devices like gears, bearings, seals [11, 12, 13]. However, adhesion may be deleterious in micro- and nano-devices, e.g. in electromechanical switchers [14]. In other cases, adhesion can regularize the "*ill posed nature of contact even for perfectly fractal surfaces*", as argued by Gao & Bower [15] on the basis of results given in Ref. [13]. Such result finds corroboration in Refs. [16], which showed that adhesion has a great effect on fractally rough surfaces. In particular, in absence of adhesion, no load is large enough to ensure complete contact, and there are not even any contact areas of finite dimension. In presence of adhesion and for low fractal dimension, regularization of the contact area is observed at a large magnification where the contact area consists of regions of finite size.

Content from this work may be used under the terms of the Creative Commons Attribution 3.0 licence. Any further distribution of this work must maintain attribution to the author(s) and the title of the work, journal citation and DOI.
Published under licence by IOP Publishing Ltd

IOP Conf. Series: Materials Science and Engineering 1275 (2023) 012024 doi:10.1088/1757-899X/1275/1/012024

It is well known that the strength of adhesion depends on the surface topography. In particular, as explained in Ref. [13], the effective interfacial adhesive mechanism is a result of two competitive mechanisms: (i) the short-range inter-molecular force responsible for adhesion energy and (ii) the repulsive force due to the elastic energy stored in the material.

The first scientifically organized work aimed at studying the effect of surface roughness on adhesion is due to Fuller & Tabor [17], which made evident that roughness can produce a large reduction in adhesion. In particular, even a relatively small surface roughness can completely remove the adhesion. A different result was obtained by Briggs & Briscoe [18], which found that making a rigid smooth surface slightly rough may enhance adhesion when a rubber is brought into contact. In fact, they found that adhesion can be maximized at an optimum value of roughness. However, in a successive work Fuller & Roberts [19] found that this effect was associated with the use of silicone rubber, which shows a clear viscoelastic behaviour. In particular, the presence of a viscous film would result in low adhesion against a smooth substrate, but "*against a slightly rough substrate asperity puncture of the film would increase the solid-solid contact*", thus leading to higher adhesion (see also Ref. [20]). Moreover, one should consider that a soft rubber is capable to flow into the valleys of the rough substrate increasing the interfacial area of contact, without high stresses form around asperities.

Fuller & Tabor explained results of their experiments with a very simple model based on the assumption of real contact area much smaller than the nominal one, and contact regions independent of each other. Starting from Greenwood & Williamson (GW) model [21], they applied JKR theory [22] to the surface asperities which were approximated by spherical cups.

Peressadko et al. [23] proposed a different approach to explain the effect of roughness on adhesion on the basis of the Persson's theory [24, 25]. They considered the limit of nearly complete contact and similarly found roughness destroys adhesion. However, no relation between adhesion properties and surface RMS roughness amplitude h_{rms} was identified. They found the main roughness quantity affecting adhesion is the second moment of the surface power spectral density (PSD), which is related to the RMS of the surface slopes h'_{rms}. However, the effect of h_{rms} on adhesion is still an open issue as discussed in Ref. [26], where it is shown the effective energy of adhesion can increase or decrease with the roughness amplitude depending on the applied boundary conditions. Recently, Jacobs et al. [27] showed roughness reduces the portion of atoms that are close enough to adhere. Moreover, the presence of a layer of contamination molecules, like water and hydrocarbons, on the surface of hard solids should be taken into account when studying the problem at the nanoscale [28].

There are in the literature many theoretical models aimed at modeling adhesion between rough surfaces [25, 29, 30, 31, 32, 33, 34]. However, the debate on the effect of roughness on adhesion is still open. For example, Fuller & Tabor [17] in their pioneristic work, found adhesion is mainly destroyed by long wavelengths of roughness. This result has been theoretically confirmed by Violano et al. [35] and experimentally by Tiwari et al. [36], which showed that "*adhesion in most cases is killed by the longest-wavelength roughness*". On the contrary, Pastewka & Robbins [37] and Muser [38] defined local stickiness criteria in which adhesion is also affected by the small scale roughness through h'_{rms}. In a recent work, Wang and Muser [39] emphasized the importance of a global energy analysis for the assessment of stickiness over purely local stickiness criteria.

The recent literature seems to agree that stickiness does not depend on the upper truncation frequency of roughness spectrum for fractal surfaces with low fractal dimension D; however, the question is open for the case at high D. In this respect, Violano et al. [40], showed theoretically that for high D, "*stickiness is more influenced by short wavelength roughness with respect to the low D case*".

For this reason, in this paper, we performed experimental JKR-like tests on glass substrates with different surface roughness properties focusing on the case at high fractal dimension, which has not been experimentally investigated in the literature. We found the surface roughness Power Spectral Density PSD is strongly correlated to the adhesion properties of the surface, and results suggest the smallest scales mainly affect adhesion.

2. Materials and methods

2.1. Materials

2.1.1. PDMS samples. Polidimetilsiloxane (*PDMS, Sylgard 184, Dow Corning*) samples (Figure 1) with standard mixing ratio (10 : 1) of base and cross-linking agent were used in all the experiments. Polymerization of the samples was performed in a vacuum dryer to remove air bubbles formed during the phases of components mixing and casting in the mold.

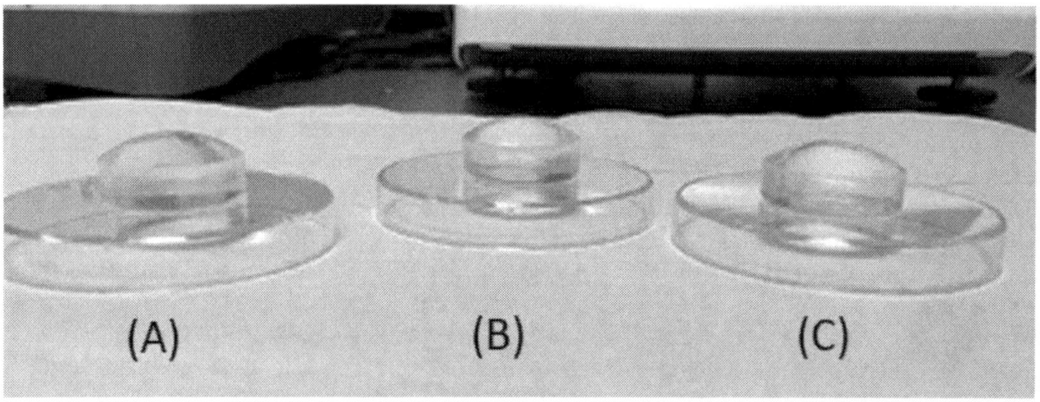

Figure 1. The Polidimetilsiloxane (PDMS) samples used in the experiments.

Plano-concave lens were used as mold to obtain specimens with spherical shape and radius $R = 19.62$ mm. Notice samples have a cylindrical base of thickness 8 mm to avoid the material stiffness is influenced by the steel plate on which samples are attached during the tests (Ref. [41]).

2.1.2. Glass substrates. Glass plates (*TECHSPEC-4 N-BK7 Precision Windows*) with thickness $t = 4$ mm, Young's modulus $E = 82$ GPa and Poisson's ratio $\nu = 0.206$ were used as rigid substrate (Figure 2).

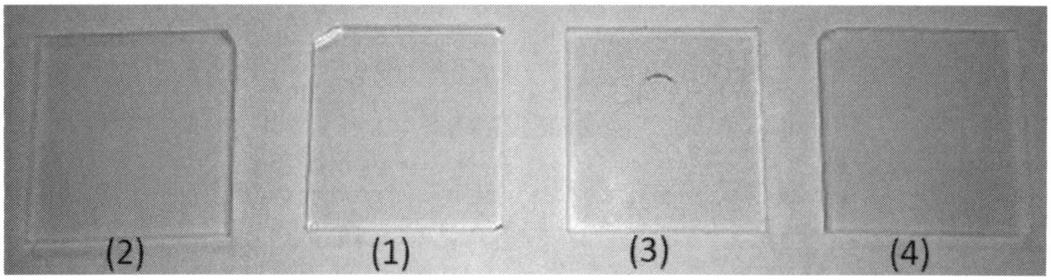

Figure 2. The glass plates used as substrate in the experiments. Each plate is characterized by a different surface roughness to allow to study the effect of roughness on adhesion.

To investigate the effect of roughness on adhesion, tests were performed on substrates with different surface properties. Specifically, the surface of specimen 1 in Figure 2 can be considered smooth being the root mean square (RMS) roughness amplitude very small (h_{rms} = 19.57 nm). For the other specimens the surface roughness properties are modified by using different processes: (i) manual sanding with sandpaper P400, for specimen 2, (ii) fluid polishing, for specimen 3, performed with diamond particles of size 9 μm, and (iii) sandblasting, for specimen 4, performed with a jet of sand and air at pressure of about 5 bar and angle of 45° with respect to the horizontal plane.

2.2. Surface analysis

The surface characteristics of the PDMS samples were measured indirectly by performing roughness measurements on the plano-concave lens with the aid of the profilometer *Talysurf Series 2 – Taylor Hobson*. Four measurements were carried out obtaining an averaged value of the radius $R = 19.62$ mm, and a RMS roughness amplitude $h_{rms} = 9.9$ nm (a value sufficiently low to consider smooth the surface).

About the glass plates, surface roughness was characterized by using the white light interferometer *Wyko NT9100 - Veeco.*

Table 1. Average roughness R_a, root mean square roughness amplitude h_{rms} and root mean square roughness slope h'_{rms} of the glass substrates surface used in the experiments.

Substrate	R_a [μm]	h_{rms} [μm]	$h\prime_{rms}$
1	0.0089	0.0112	0.02
2	0.2946	0.5456	0.36
3	0.4588	0.5972	0.90
4	4.3	5.24	2.29

Figure 3 shows contour plots of the glass specimens surfaces heights. The roughest substrate is clearly the number 4, which is characterized by the highest values of R_a and h_{rms}, as shown in Table 1. The RMS values of the surface slopes $h\prime_{rms}$ are also given. Specimen 3 shows an almost uniform surface roughness. On the contrary, specimen 2 shows evident scratches produced by the manual sanding; however, roughness parameters are smaller than those of specimen 3.

2.3. Experimental set-up

Figure 4 shows the experimental equipment used in the tests. The PDMS sample is fixed to a steel circular plate tightened to an uniaxial load cell (*Honeywell Model 34*) with load range of 150 g. The latter is in turn screwed to the crosspiece of the mechanical tester (*Biomomentum Mach-1 V500CST*). Under the glass substrate is placed a mirror forming an angle of 45° on the horizontal plane. In this way, the video camera (a *Phantom MIRO with lens Mitutoyo M PLAN APO 2X*) located in front of the mirror may capture the reflected image of the contact area. An halogen light source (*Schott KL1500 Electronic*) has been also used to obtain high quality images of the contact spot.

Before each test, the load cell was calibrated using a standard weight of 50 g (*MA3261210A-7*) and an ultrasonic cleaning process was carried out on samples and substrates. Notice the

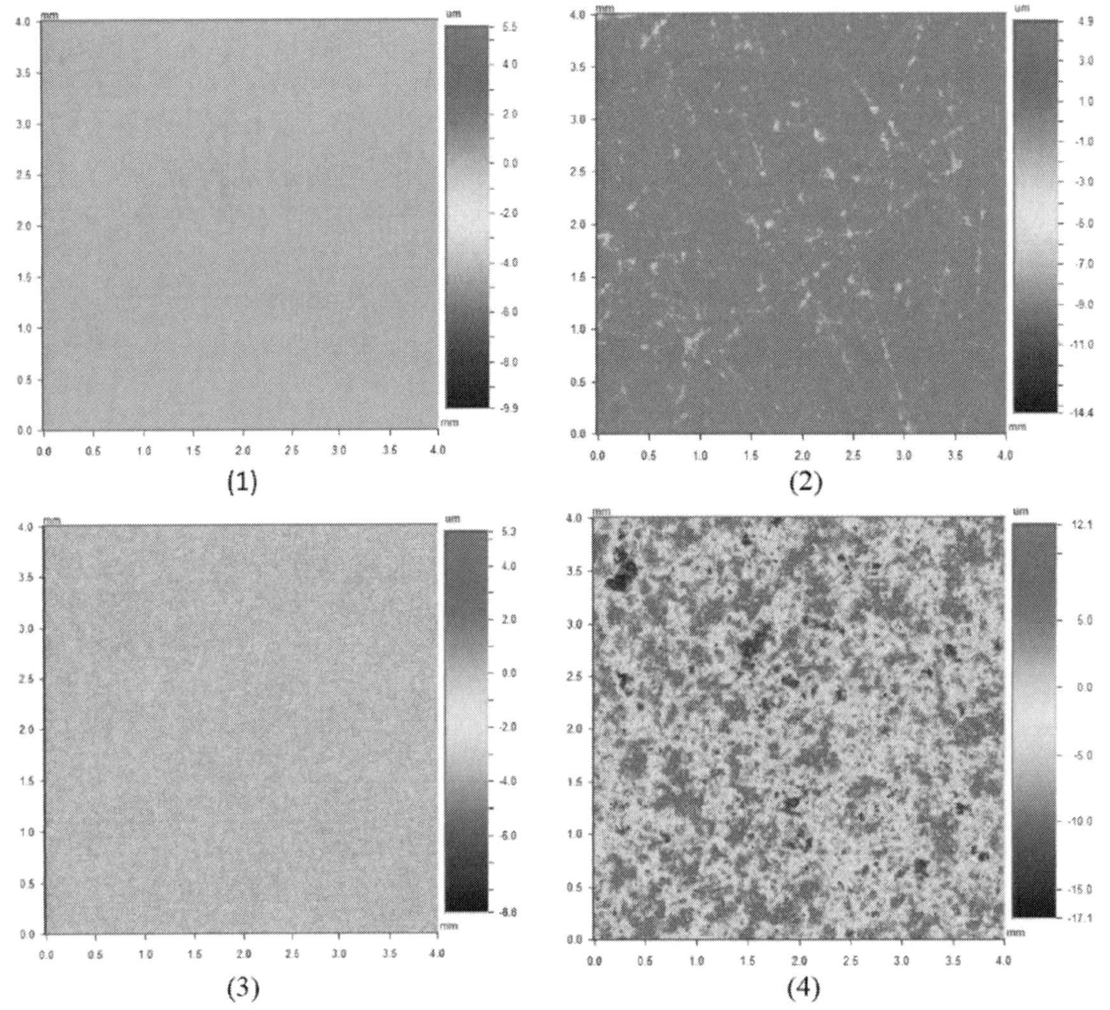

Figure 3. Images of the surface roughness of the glass substrates: (1) smooth substrate; (2) substrate treated with manual sanding; (3) substrate treated with fluid polishing; (4) substrate treated with sandblasting.

frame structure, which supports the glass substrate, was covered with matt black ribbon to avoid light reflections which can induce problems in the visualization of the contact spots.

3. Results and discussion

Adhesion tests were performed on the four glass substrates with each of the three PDMS samples (A, B, C). Tests on the smooth glass specimen 1 were aimed at determining the Young's modulus of the PDMS spherical cups and the work of adhesion $\Delta\gamma$, also referred as the Duprè energy of adhesion, by fitting experimental data with classical JKR theory [22]. Once the PDMS elastic modulus is known, the effect of roughness on adhesion was quantified by calculating the pull-off force on rough substrates, i.e., the maximum tensile load reached during the unloading phase.

Figure 5 shows, for each PDMS sample, the experimental load-penetration curves obtained squeezing and then retracting the PDMS samples from the four glass substrates.

For all samples we observe the same dependence of data on roughness. However, samples

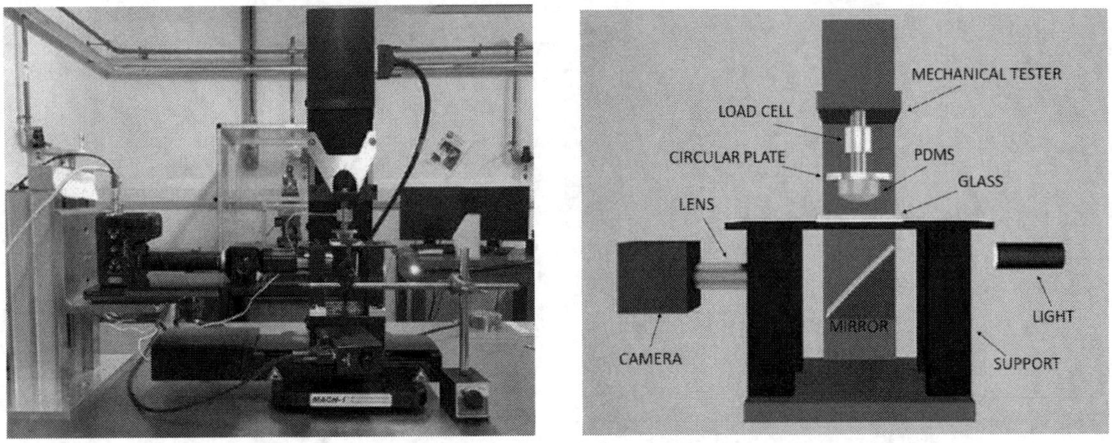

(a) (b)

Figure 4. The experimental set-up used in the tests.

A and C gave very similar results, while sample B showed significantly smaller values of the adhesion force, probably as a consequence of different elastic properties and adhesion surface energy. We indeed, for samples A and C, found very similar values of Young's moduli ($E = 1.916$ MPa and $E = 1.903$ MPa, respectively) and surface energies ($\Delta\gamma = 591$ mJ/m^2 and $\Delta\gamma = 608$ mJ/m^2, respectively) and lower values for sample B ($E = 1.746$ MPa and $\Delta\gamma = 382$ mJ/m^2). As a result, we may speculate that unexpected differences occurred during the components mixing and casting led to getting different properties.

In all cases, a clear difference is observed between loading and unloading phases, demonstrating the existence of a hysteretic behaviour related to the rate dependent nature of the adhesion which manifests during the unloading process. Specifically, the hysteresis can be an indication of interactions inside the contact area [41, 42]. In fact, after the loading process, the elastomer remained in contact with the substrate for a fixed period of time. During this time interval, the evolution of physical and chemical interactions may lead to an enhancement in the adhesion energy [43]. Moreover, it can also be a physical manifestation of the incomplete mechanical relaxation of the PDMS during the unloading if its rate is faster than the relaxation time [44].

The effect of roughness on adhesion can be clearly identified by observing that, at given penetration, smaller forces are required to detach the indenter as the roughness of the substrate increases. In particular, tests on glass specimen 4 (black triangles), whose surface was treated with sandblasting, show negligible hysteresis and very low adhesive interactions. In such case, roughness is high enough to destroy adhesion almost completely.

Table 2 shows the pull-off force calculated on each substrate and obtained as the mean of the values measured in the tests performed with the three PDMS samples. Its significant reduction with the roughness, which increases moving from substrate 1 to substrate 4, is explained by arguing that, adding roughness to the surface, the macro-contact regions are constituted by an increasing number of asperities which experience contact in different moments, thus poorly contributing to the increase in the total load.

From the Power Spectral Density (PSD) of the surface height distribution shown in Figure 6, we can observe that rough substrates 2, 3, and 4 show a trend with the modulus q of the wave vector that is typical of self-affine fractal surfaces. Specifically, such surfaces are characterized by a fractal dimension D of the order of about 2.6 (i.e., Hurst exponent $H \simeq 0.4$).

In this case, it is interesting to notice that the surface adhesion is mainly affected by the

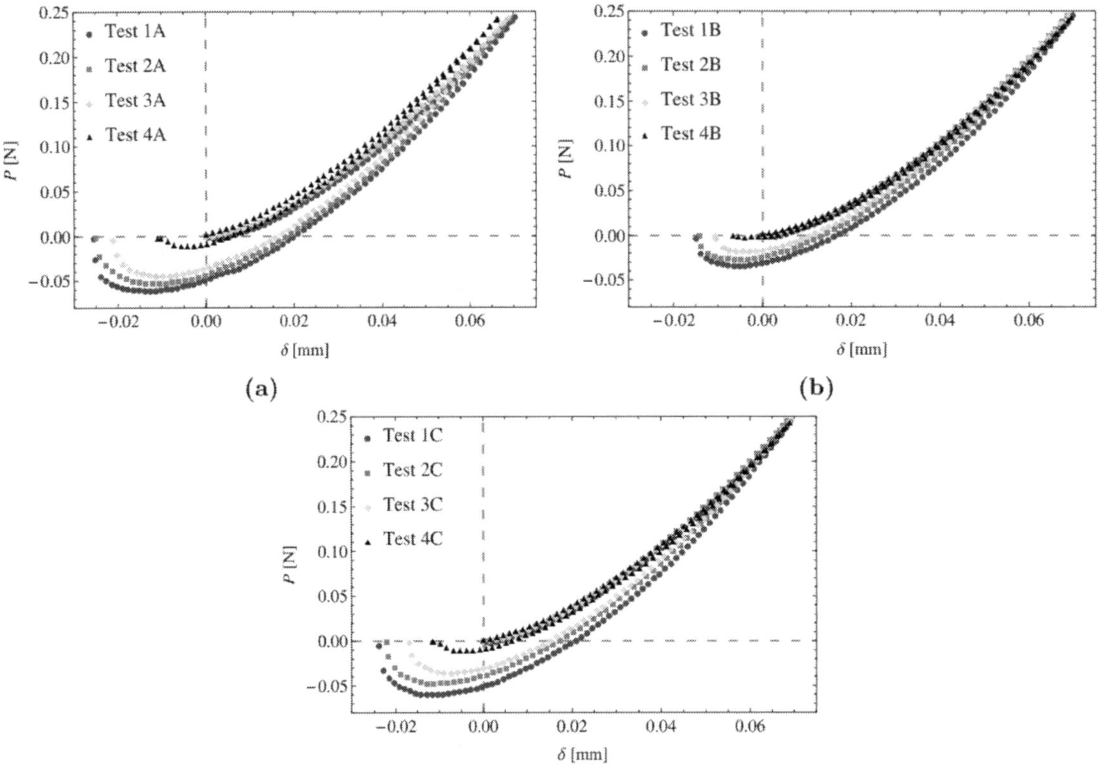

Figure 5. The experimental load vs. penetration curves obtained for each PDMS sample on the substrates with different roughness characteristics: smooth glass substrate 1 (blue circles), glass substrate 2 treated with manual sanding (red squares), glass substrate 3 treated with fluid polishing (green rhombus), and glass substrate 4 treated with sandblasting (black triangles). Figure 5a refers to results obtained with the PDMS sample A, Figure 5b with the sample B, and Figure 5c with the sample C.

Table 2. Average values of the pull-off force registered during the experiments on each substrate.

Substrate	$P_{\text{pull-off}}$ $[N]$
1	−0.0518
2	−0.0428
3	−0.0330
4	−0.0076

high spatial frequencies (finer scales). In fact, comparing for example the substrates 2 and 3, the latter shows lower surface adhesion as its PSD has larger values at the higher wave vectors despite the lower values detected at the smaller spatial frequencies (i.e., larger wavelengths).

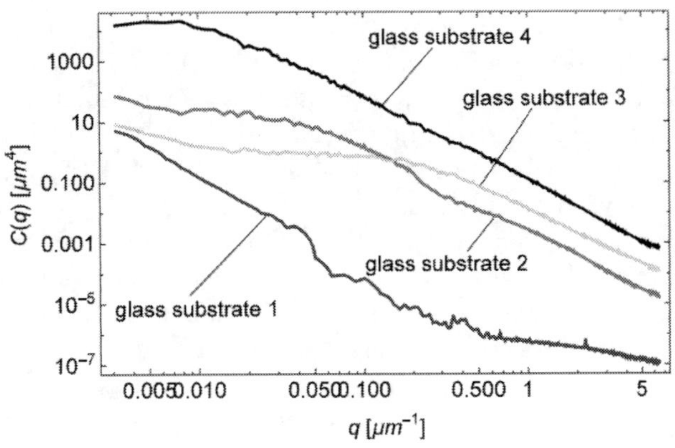

Figure 6. The power spectral density $C(q)$ of the surface height distribution for the glass substrates used in the experiments. Notice substrate 1 corresponds to the nominally smooth sample.

This effect appears clearer in Figure 7, where the pull-off force is plotted as a function of the RMS roughness amplitude h_{rms} (Figure 7a) and RMS surface slope h'_{rms} (Figure 7b). We recall that h_{rms} mainly depends on long wavelengths roughness, while h'_{rms} also depends on small wavelengths. As a result, the reduction of the pull-off force occurring in Figure 7a at about $h_{rms} \sim 0.55$ μm is unrelated to the long scale roughness. However, from Figure 7b, we observe that such reduction of the pull-off force is accompanied to an increase in the small wavelength component of the roughness as h'_{rms} is increasing. This suggests that, at high fractal dimensions $D > 2.5$ (low Hurst exponent $H < 0.5$), the effect of small wavelength components on stickiness is no longer negligible in agreement with the theoretical predictions given in Violano et al. [40].

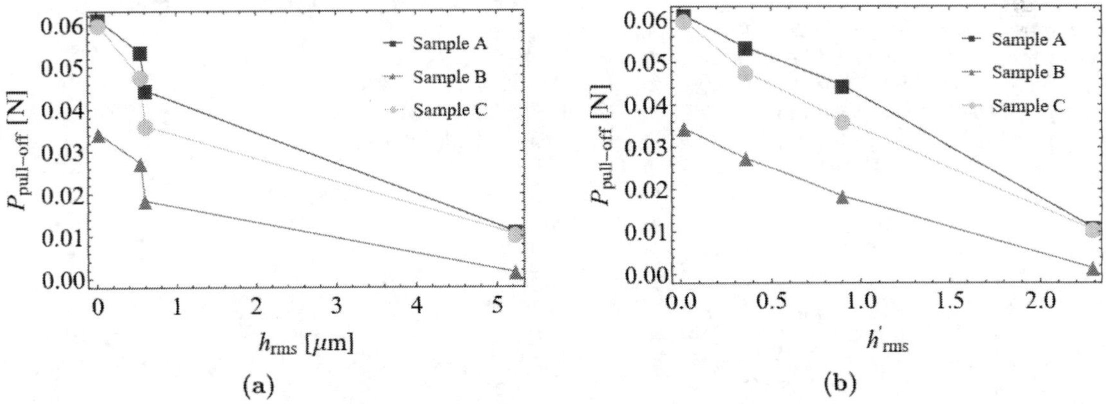

Figure 7. The pull-off force as a function of the (a) RMS roughness amplitude and (b) RMS surface slope, for the three PDMS samples considered in the tests.

4. Conclusions

In this work, we have carried out JKR-like tests on four glass substrates with three PDMS samples to investigate the effect of roughness on the surface adhesion properties. The experimental load-penetration curves were obtained squeezing and then retracting the PDMS samples from each of the four considered specimens with different roughness characteristics.

Specifically, tests were performed on a smooth specimen with the aim of determining the Young's modulus of the PDMS spherical cups and the work of adhesion. The influence of the roughness on the surface adhesive properties was estimated by calculating the pull-off force, i.e., the maximum tensile force registered during the unloading phase.

All rough specimens show a Power Spectral Density (PSD) with characteristics typical of self-affine fractal surfaces with fractal dimension $D > 2.5$ ($D \simeq 2.6$). We find that adhesion is destroyed by roughness in agreement with the well-known literature on the topic. However, experimental data show that, for these surfaces characterized by high fractal dimension, adhesive stickiness is mainly affected by the smaller scales of roughness. Indeed, samples with a very similar h_{rms}, which is mainly affected by long scale roughness, show a different pull-off force because of a difference in the small scale roughness, which affects the value of h'_{rms}. In the authors' knowledge, such a result, which confirms previous theoretical predictions, has been experimentally observed for the first time.

References

[1] Gorb S, Jiao Y and Scherge M 2000*J. Comp. Physiol.* A **186** 821-31
[2] Gorb S, Varenberg M, Peressadko A and Tuma J 2007 *J. R. Soc. Interface* **4** 271-75
[3] Afferrante L and Carbone G 2012 *J. R. Soc. Interface* **9** 3359-65
[4] Dening K, Heepe L, Afferrante L, Carbone G and Gorb S N 2014 *Appl. Phys.* A **116** 567-73
[5] Lackie J M 2013 *The Dictionary of Cell & Molecular Biology* (Academic Press)
[6] Krahn J, Liu Y, Sadeghi A and Menon C 2011 *Smart Mater. Struct.* **20** 115021
[7] Pugno N M 2008 *Nanotoday* **3** 35
[8] Allen M, Myer B and Rushton N 2001 *J. Biomed. Mater. Res.* **58(3)** 319-28
[9] Piveteau L D, Girona M I, Schlapbach L, Barboux P, Boilot J P and Gasser B 1999 *J. Mater. Sci. Mater. Med.* **10(3)** 161-7
[10] Kwak M K, Jeong H-E and Suh K Y 2011 *Adv. Mater.* **23** 3949-53
[11] Ping Y, Bo L N, Daoguo Y and Ernst L J 2006 *Microsystem Technologies* **12(12)** 1125-31
[12] Rymuza Z, Kusznierewicz Z, Solarski T, Kwacz M, Chizhik S A and Goldade A V 2000 *Wear* **238** 56-69
[13] Persson B N J, Albohr O, Tartaglino U, Volokitin A I and Tosatti E 2005 *J. Phys.: Condens. Matter* **17** R1-R62
[14] Katehi L P B, Harvey J F and Brown E. 2002 *IEEE Trans. Microwave Theory Tech.* **50** 858-66
[15] Gao Y F and Bower A F 2006 *Proc. R. Soc.* A **462**, 319-48
[16] Carbone G, Mangialardi L M and Persson B N J 2004 *Phys. Rev.* B **70** 125407
[17] Fuller K N G and Tabor D 1975 *Proc. Roy. Soc. London* A **345(1642)** 327-42
[18] Briggs G A D and Briscoe B J 1977 *J. Phys. D: Appl. Phys.* **10** 2453
[19] Fuller K N G and Roberts A D 1981 *J. Phys. D: Appl. Phys.* **14** 221
[20] Roberts A D and Othman A B 1977 *Wear* **42** 119-33
[21] Greenwood J A and Williamson J B P 1966 *Proc. Roy. Soc. London* A **295(1442)** 300-19
[22] Johnson K L, Kendall K and Roberts A D 1971 *Proc. Roy. Soc. London* A **324** 301-13
[23] Peressadko A G, Hosoda N and Persson B N J 2005 *PRL* **95** 124301
[24] Persson B N J 2001 *The Journal of Chemical Physics* **115(8)** 3840-61
[25] Persson B N J 2002 *Eur. Phys. J.* E **8** 385-401
[26] Menga N, Afferrante L and Carbone G 2016 *International Journal of Solids and Structures* **88-89** 101-09
[27] Jacobs T D B, Ryan K E, Keating P L, Grierson D S, Lefever J A, Turner K T and Carpick R W 2013 *Tribology Letters* **50(1)**, 81-93
[28] Samoilov V N, Sivebaek I M and Persson B N J 2004 *J. Chem. Phys.* **121** 9639
[29] Persson B N J and Scaraggi M 2014 *Journal of Chemical Physics* **141(12)** 124701
[30] Medina S and Dini D 2014 *International Journal of Solids and Structures* **51(14)** 2620-32
[31] Rey V, Anciaux G and Molinari J F 2017 *Computational Mechanics* **60(1)** 69-81
[32] Campana C and Muser M H 2006 *Phys. Rev.* B **74(7)** 075420
[33] Violano G and Afferrante L 2019 *Tribology International* **130** 36-42
[34] Violano G and Afferrante L 2019 *Tribology Letters* **67** 119
[35] Violano G, Afferrante L, Papangelo A and Ciavarella M 2019 *The Journal of Adhesion* **97(6)** 509-27
[36] Tiwari A, Wang J and Persson B N J 2020 *Phys.Rev.* E **102** 042803
[37] Pastewka L and Robbins M O 2014 *Proceedings of the National Academy of Sciences* **111(9)** 3298-303
[38] Müser M H 2016 *Tribology International* **100** 41-7

[39] Wang A and Müser M H 2022 *Friction* (doi:10.1007/s40544-022-0644-3)
[40] Violano G, Papangelo A and Ciavarella M 2021 *Tribology International* **159** 106971
[41] Deruelle M, Hervet H, Jandeau G and Léger L 1998 *Journal of Adhesion Science and Technology* **12** 225-47
[42] Ahn D and Shull K R 1996 *Macromolecules* **29** 43814390
[43] Woerdeman D L, Amouroux N, Ponsinet V, Jandeau G, Hervet H and Léger L 1999 *Composites: Part* A **30** 95109
[44] Chin P 1996 *PhD thesis* (University of Delaware)

Implementation of the Peak Stress Method for the automated FEA-assisted design of aluminium welded joints subjected to constant amplitude multiaxial fatigue loads

A Visentin[1], A Campagnolo[1], F Simonaggio[1] and G Meneghetti[1]

[1] University of Padova, Department of Industrial Engineering, 35131 Padova, Italy

giovanni.meneghetti@unipd.it

Abstract. The Peak Stress Method (PSM) is a FE-oriented local approach to the fatigue strength assessment of welded structures subjected to fatigue loading. Starting from the peak stresses calculated at the V-notch tip nodes defining weld toes or the weld roots, the PSM defines an equivalent peak stress which allows to estimate the fatigue failure location and fatigue lifetime of welded structures, in compliance with appropriate fatigue design curves. An Ansys® Mechanical extension has been developed to achieve full automated implementation of the tasks and calculations necessary to apply the PSM to welded structures. The tool allows to identify and analyse all the V-notch tip edges of the structure and perform fatigue life estimation on each analysed node. As an output, fatigue life results can be visualized through dedicated tabular data, graphs and contour results generated over the edges of the model. In this work, common-to-complex 3D geometries taken from the literature and related to aluminium alloys welded joints subjected to uniaxial as well as multiaxial fatigue loads have been analysed by comparing two design approaches: (i) manual application of the PSM, (ii) automated implementation of the PSM. The tool developed in Ansys® Mechanical allows to significantly contain the time and effort required to analyse welded structures according to the PSM.

1. Introduction

Dealing with the fatigue strength assessment of welded joints, Standards and Recommendations [1,2] suggest several methods: the nominal stress approach [3,4], the structural hot-spot stress approach [3–7], the notch stress approach [3,4,7–14] and the Linear Elastic Fracture Mechanics (LEFM) approach [3,12,13,15–17]. The nominal stress approach allows to employ nominal stresses, calculated according to the classic beam theory, for assessing the fatigue strength of welded components, in compliance with fatigue strength categories available in standards and depending on the welded details geometry and loading conditions. However, standards do not necessarily provide accurate reference categories when dealing with complex un-common welded joints, i.e. for instance welded components commonly installed on amusement park structures (see [18,19]). Adding to the complexity, the aforementioned joints are also generally subjected to real-life multiaxial constant amplitude as well as variable amplitude loadings. This requires analysts to find accurate methods to assess the fatigue strength of complex welded joints. In this perspective, local approaches, i.e. Notch Stress Intensity Factors (NSIFs) approach [11,20,21], the averaged Strain Energy Density (SED) criterion [12,22–26], the critical plane approach [27–29] or the Peak Stress Method (PSM) [30–37], provide design criteria based on local quantities, i.e. stresses, strains or strain energy parameters, intrinsically tied to the local geometry of the welded details.

Content from this work may be used under the terms of the Creative Commons Attribution 3.0 licence. Any further distribution of this work must maintain attribution to the author(s) and the title of the work, journal citation and DOI.
Published under licence by IOP Publishing Ltd

Modern local approaches, like the averaged SED approach or the PSM, take advantage of the capabilities offered by Finite Element Analysis (FEA) and provide computer-aided numerical procedures for evaluating the fatigue strength of welded joints subjected to both uniaxial and multiaxial constant amplitude [35,38,39] as well as variable amplitude [37] loadings. Methods like the PSM also allow to employ relatively coarse FE meshes if compared to those required by other local approaches, as for instance the NSIFs approach, in order to contain computational requirements for application. On the other hand, a significant effort is however generally requested to the analyst when manually applying local methods to complex joint geometries. To support the FE analyst in every-day fatigue design of welded structures, a new analysis tool has been developed in Ansys® Mechanical, which is able to automate the implementation tasks and calculations required to apply the PSM. Up to now, the tool has been employed to assess the fatigue strength of lab-scale steel joints [40] as well as more complex large-scale steel details [19]. In this work, seven geometries of common-to-complex welded joints made of different aluminium alloys and subjected to constant amplitude uniaxial as well as multiaxial loading have been re-analysed by comparing two design approaches: (i) manual application of the PSM and (ii) automated application of the PSM, taking advantage of the developed tool. Also, two case studies will be presented in which the PSM tool has been employed to assess the fatigue strength of simple and complex real-life aluminium welded details.

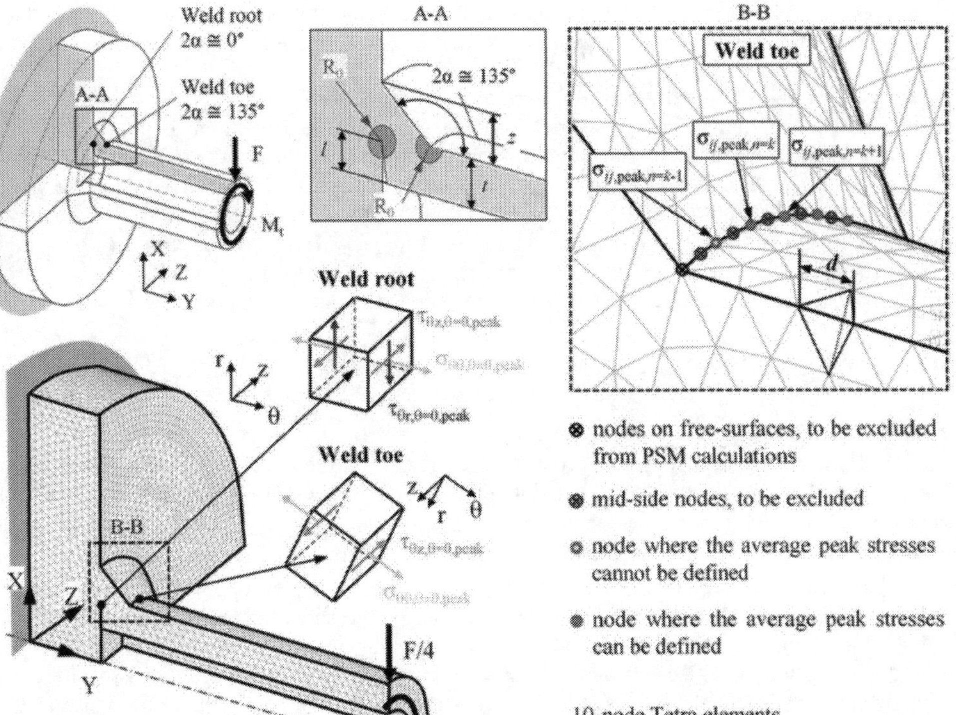

Figure 1. Definition of the characteristic size a from geometrical dimensions (l, z, t) in a partial-penetration tube-to-flange welded joint under combined bending and torsion loadings. Example of a FE model to apply the PSM using 10-node tetrahedral elements (SOLID 187 of Ansys® element library). Definition of cylindrical coordinate systems (r, θ, z) centered at the weld toe and at the weld root and relevant local elastic peak stresses [36].

2. The Peak Stress Method (PSM)

The Peak Stress Method is a FE-oriented local approach which allows to rapidly estimate the NSIF-terms K_1, K_2 and K_3 from the local opening, in-plane shear and out-of-plane shear peak stresses calculated at V-notch tip nodes from a linear elastic Finite Element Analysis [35]. Therefore, the weld toe and the weld root must be modeled as sharp V-notches having notch opening angle 2α and null tip radius (see figure 1). The estimated NSIF-terms can be evaluated from the following expressions [30–32]:

$$\text{(a) } K_1 \cong K^*_{FE} \cdot \sigma_{\theta\theta,\theta=0,peak} \cdot d^{1-\lambda_1}; \quad \text{(b) } K_2 \cong K^{**}_{FE} \cdot \tau_{r\theta,\theta=0,peak} \cdot d^{1-\lambda_2}; \quad \text{(c) } K_3 \cong K^{***}_{FE} \cdot \tau_{\theta z,\theta=0,peak} \cdot d^{1-\lambda_3} \qquad (1)$$

The peak stresses $\sigma_{\theta\theta,\theta=0,peak}$, $\tau_{r\theta,\theta=0,peak}$ and $\tau_{\theta z,\theta=0,peak}$ in equation (1) are defined with respect to a local cylindrical coordinate system centered at the V-notch tip node; z-direction is tangential to the notch tip edge, r-direction is defined by the notch bisector line and θ refers to the radial direction according to figure 1. Parameter d in equation (1) refers to the average element size used as input to generate a proper free finite element mesh according to the PSM [35]. Parameters λ_1, λ_2 and λ_3 are the stress singularity degrees depending on the opening angle 2α of the notch. Parameters K^*_{FE}, K^{**}_{FE} and K^{***}_{FE} in equation (1) depend on: (i) formulation of the element type; (ii) finite element mesh pattern; (iii) finite element commercial code employed to generate the mesh and perform the analysis. The reader is referred to the current state-of-the-art review of the PSM [35] for additional information.

2.1. Applying the Peak Stress Method with tetrahedral finite elements

Adopting tetrahedral finite elements allows to achieve a rapid discretization of both common and complex 3D geometries, thus avoiding sub-modeling techniques. However, the intrinsic irregularity of the free tetra mesh pattern produces variability of peak stress values along the notch tip profile. In order to compensate such variability and apply the PSM with tetrahedral elements, an average peak stress formulation [34,35] has been proposed according to equation (2). More in detail, the moving average of the peak stresses calculated on three adjacent vertex nodes, i.e. $n = k$-1, $n = k$ and $n = k$+1, must be performed and the calculated average stress must be assigned to the central node, that is $n = k$.

$$\bar{\sigma}_{ij,peak,n=k} = \left.\frac{\sigma_{ij,peak,n=k-1} + \sigma_{ij,peak,n=k} + \sigma_{ij,peak,n=k+1}}{3}\right|_{n=node} \quad \text{where } ij = \theta\theta, r\theta, \theta z \qquad (2)$$

The notch tip nodes belonging to the free surfaces of the analysed structure are usually affected by distorted mesh patterns and must be excluded. Eventually, only peak stresses calculated at vertex nodes located along V-notch tip edges must be accounted, while those belonging to mid-side nodes are to be discarded. PSM parameters K^*_{FE}, K^{**}_{FE} and K^{***}_{FE} have been calibrated using either 10-node [33] and 4-node [34] tetra elements (see table 1) in Ansys® and more FE commercial codes [36]. K^*_{FE}, K^{**}_{FE} and K^{***}_{FE} parameters calibration for tetra element types has been performed under the following assumptions [35]:

a. in the case of mode I and mode III loadings, notch opening angles 2α ranging from 0° to 135° have been investigated, being 2α = 0° and 2α = 135° the typical notch opening angles at weld root and weld toe respectively. Weld toe profiles having opening angles up to 150° have been also analysed adopting the same calibration constants [41]. In the case of mode II loading, only notch opening angles 2α = 0° [33,34] and 2α = 90° [42] have been investigated, being the typical cases of weld roots without or with a gap. A more in-depth investigation has been recently performed in order to extend the validity of the calibrated parameters K^*_{FE}, K^{**}_{FE} and K^{***}_{FE} for all the notch opening angle values between 2α = 0° and 2α = 180°, in the case of 4-node as well as 10-node tetra elements [43].
b. The free mesh pattern at the notch tip has been defined according to a minimum mesh density ratio $(a/d)_{min}$, a being the characteristic size of the considered notch [35]. As stated in table 1, the value

of the minimum mesh density ratio $(a/d)_{min}$ to fulfil depends on: (i) the adopted FE type, (ii) the loading mode (I, II or III) and (iii) the opening angle 2α of the analysed V-notch.

Table 1. Parameters K^{*}_{FE}, K^{**}_{FE} and K^{***}_{FE} and mesh density ratio *a/d* requirements to apply the PSM with 10-node tetra elements (SOLID 187) in Ansys® FE commercial code [35].

Loading	PSM parameters	2α = 0°	2α = 90°	2α = 120°	2α = 135°	*a* weld root	*a* weld toe
Mode I	K^{*}_{FE}	1.05±15%	1.05±15%	1.05±15%	1.21±10%	min{*l*, *z*}	*t*
Eq. (1a)	$(a/d)_{min}$	3	3	3	1		
Mode II	K^{**}_{FE}	1.63±20%	2.65±10%	n.a.	n.a.	min{*l*, *z*}	*t*
Eq. (1b)	$(a/d)_{min}$	1	1	n.a.	n.a.		
Mode III	K^{***}_{FE}	1.37±15%	1.37±15%	1.70±10%	1.70±10%	min{*l*, *z*}	*t*
Eq. (1c)	$(a/d)_{min}$	3	3	3	3		

[a] The 'Full Graphics' setting must be activated before extracting peak stresses according to the PSM

2.2. *Fatigue strength assessment according to the PSM*

The PSM adopts the Strain Energy Density (SED) averaged over a structural volume surrounding the V-notch tip as a fatigue strength criterion for assessing weld toe and weld root fatigue failures [24–26]. The averaged SED can be expressed as a function of the relevant NSIF-terms K_1, K_2 and K_3, according to equation (3).

$$\Delta\bar{W}=\left\{c_{w1}\frac{e_1}{E}\left[\frac{\Delta K_1}{R_0^{1-\lambda_1}}\right]^2+c_{w2}\frac{e_2}{E}\left[\frac{\Delta K_2}{R_0^{1-\lambda_2}}\right]^2+c_{w3}\frac{e_3}{E}\left[\frac{\Delta K_3}{R_0^{1-\lambda_3}}\right]^2\right\} \quad (3)$$

R_0 = 0.12 mm refers to the size of the structural volume in the case of welded joints made of aluminium alloys [35]. *E* refers to the material Young's modulus while coefficients e_i (© = 1, 2 and 3) depend on the notch opening angle 2α and on material Poisson's ratio *ν*. ΔK_1, ΔK_2 and ΔK_3 are the NSIFs range values tied to mode I, mode II and mode III, respectively. C_{wi} terms account for the mean stresses in the material and depend on the load ratio R_i of the *i*-th loading mode (© = 1, 2 or 3), according to equation (4). Differently from stress-relieved welded joints, welded joints tested in as-welded conditions are almost insensitive to mean stresses according to design standards [1]; therefore, c_{wi} = 1 should be adopted in this case.

$$C_{wi}(R_i) = \begin{cases} \dfrac{1+R_i^2}{(1-R_i)^2} & \text{if stress-relieved and } -1\le R_i\le 0 \\ \dfrac{1-R_i^2}{(1-R_i)^2} & \text{if stress-relieved and } 0\le R_i<1 \\ 1 & \text{if as-welded for any R value} \end{cases} \quad \text{where } i=1,2,3 \quad (4)$$

The PSM allows to estimate the NSIF-terms K_1, K_2 and K_3 taking advantage of the relevant peak stresses calculated using coarse FE meshes, according to equation (1). Therefore, the averaged SED can be re-formulated as a function of the relevant peak stresses too. Eventually, an equivalent peak stress quantity can be defined according to equation (5), which has been obtained by enforcing the SED expression for an equivalent uniaxial strain state $W=(1-\nu^2)\cdot\sigma^2_{eq,peak}/2E$ [35]. The equivalent peak stress represents a design stress quantity for the fatigue strength assessment of welded joints.

$$\Delta\sigma_{eq,peak} = \sqrt{c_{w1}\cdot f_{w1}^2\cdot\Delta\bar{\sigma}_{\theta\theta,\theta=0,peak}^2 + c_{w2}\cdot f_{w2}^2\cdot\Delta\bar{\tau}_{r\theta,\theta=0,peak}^2 + c_{w3}\cdot f_{w3}^2\cdot\Delta\bar{\tau}_{\theta z,\theta=0,peak}^2} \quad (5)$$

The average peak stress terms $\bar{\sigma}_{\theta\theta,\theta=0,peak}$, $\bar{\tau}_{r\theta,\theta=0,peak}$ and $\bar{\tau}_{\theta z,\theta=0,peak}$ are defined according to equation (2) in the case of 4-node as well as 10-node tetrahedral elements. Parameters f_{wi} (I = 1, 2 and 3, depending on the loading mode) account for peak stress averaging inside the material structural volume [35] and are defined in equation (6).

$$F_{wi} = K_{FE}\cdot\sqrt{\frac{2e_i}{1-\nu^2}}\cdot\left(\frac{d}{R_0}\right)^{1-\lambda_i} \qquad \text{where } i = 1, 2, 3 \quad (6)$$

The peak stresses involved in equation (5), as well as f_{wi} parameters in equation (6), depend on the average element size d adopted in the FE model (see figure 1).

Moreover, dealing with the fatigue strength assessment of welded joints subjected to multiaxial loadings, the following general approach has been adopted in this work. The average peak stresses $\bar{\sigma}_{ij,peak}$ ($ij = \theta\theta, r\theta, \theta z$) have been calculated according to equation (2) by applying each load case c separately. Afterwards, all average peak stress contributions relevant to the same loading mode (I, II and III) have been algebraically summed according to equation (7), in order to determine the average peak stresses accounting for the multiaxial combination of the different loadings.

$$\bar{\Sigma}_{ij,peak,M} = \sum_c \bar{\sigma}_{ij,peak_c} \qquad \text{where } ij = \theta\theta, r\theta, \theta z \quad (7)$$

Eventually, the peak stress ranges $\Delta\bar{\sigma}_{ij,peak,M}$ ($ij = \theta\theta, r\theta, \theta z$) have been evaluated and introduced into equation (5) in place of the $\Delta\bar{\sigma}_{ij,peak}$ ($ij = \theta\theta, r\theta, \theta z$) peak stress ranges, in order to calculate the equivalent peak stress range accounting for the multiaxial loading scenario on each analysed node.

The equivalent peak stress can finally be used to estimate the fatigue lifetime of welded joints, by comparing it with a proper PSM-based design curve. A local biaxiality ratio λ has been introduced as the ratio between the strain energy density contributions tied to mode II/III shear and mode I stresses, in order to correctly select which design curve to address [35]. Λ can be expressed as a function of the relevant averaged peak stresses according to equation (8), which is valid or 4-node as well as 10-node tetrahedral elements.

$$\lambda = \frac{c_{w2}\cdot f_{w2}^2\cdot\Delta\bar{\tau}_{r\theta,\theta=0,peak}^2 + c_{w3}\cdot f_{w3}^2\cdot\Delta\bar{\tau}_{\theta z,\theta=0,peak}^2}{c_{w1}\cdot f_{w1}^2\cdot\Delta\bar{\sigma}_{\theta\theta,\theta=0,peak}^2} \quad (8)$$

The proper PSM-based design curve must be selected according to the criteria reported in table 2 for the case of joints made of aluminium alloys [35]. $\Delta\sigma_{eq,peak,A,50\%}$ refers to the fatigue class at $2\cdot10^6$ cycles (for a survival probability of 50%) while k refers to the slope of the fatigue design curve.

Table 2. Criterion for selecting the PSM fatigue design curve for arc-welded aluminium joints [35].

Thickness, T [mm]	λ Eq. (10)	$\Delta\sigma_{eq,peak,A,50\%}$ [MPa]	k	T_σ
T ≥ 5 mm	λ = 0	123	3.8	1.80
	λ > 0	123	6.5	1.80
3 mm ≤ T < 5 mm	λ ≥ 0	123	6.5	1.80

The reader is also referred to [35] and [40] for more details on the manual procedure to apply the PSM to a generic welded structure.

3. Automated implementation of the PSM in Ansys® Mechanical

The automated application of the PSM to the welded details presented in this work has been carried out taking advantage of an interactive tool, originally developed in Ansys® Mechanical APDL [42] and eventually re-designed in Ansys® Mechanical. The PSM tool has been developed taking advantage of Ansys® Customization Toolkit (ACT), a development toolkit based on Application Programming Interfaces (APIs) which allow the developer to employ different modern programming languages (i.e. IronPython, Microsoft® C#, Microsoft® Jscript, XML) to create wizards and custom features and integrate them directly within Ansys® FE environment. Moreover, the APIs allow to easily automate different modeling, simulation and results post-processing tasks through IronPython scripting. Core calculation algorithms and graphical user interfaces (GUIs), e.g. the recently re-designed setup menu reported in figure 2, have been developed using Microsoft® C# and XML code, in order to achieve better code steadiness and integration with Microsoft® Windows environment. On the other hand, all custom results functions (see figure 2 and figure 6) and post-processing features have been developed taking advantage of IronPython scripts, in order to enhance code versatility and speed up debugging when required.

The PSM tool is designed to support the FE analyst in the fatigue strength assessment of real-life welded structures by automating the different post-processing tasks required to apply the PSM to a generic solved FE model and finally providing fatigue lifetime estimations according to paragraph 2. The developed tool features routines able to automate all the following PSM application tasks:

a. check FE mesh compatibility for PSM application by comparing the mesh density ratio a/d computed according to the user-inputs on a, d and T_{min} (i.e. the minimum welded thickness in the structure) with the minimum mesh density ratio $(a/d)_{min}$ prescribed for the weld toe and the weld root (see table 1). The *Model Characteristics* panel on the setup menu (see figure 2) allows the analyst to provide user-inputs and visualize mesh compatibility info, which provide the outcome of the automated mesh compatibility checks.
b. Identify weld toes and weld roots, by locally calculating the notch opening angle 2α on each node along the V-notch tip edges of the model. By default, the PSM tool evaluates all model edges in order to identify the V-notches to be analysed; however, the analyst can also decide to manually select a set of target edges the app will focus on during the analysis, in order to reduce computational requirements and analysis time. The panel *Edges Selection* on the setup menu (figure 2) is designed to assist the analyst in this selection.
c. Reference and compute the parameters required to apply the PSM. Stress singularity exponents λ_i and SED coefficients e_i are calculated taking advantage of recently developed polynomial expressions, which are function of the notch opening angle 2α and material Poisson's ratio ν [43]. The panel *Material Properties* provides the analyst information on the material properties of the model, which are automatically detected and filed by the PSM tool at launch. The option "STRESS-RELIEVED Material" allows the user to specify a stress-relieved material condition to be taken into account during the PSM analysis.
d. Define and orientate local coordinate systems on the analysed peak nodes defining weld toe and weld root, in order to retrieve the relevant mode I, mode II and mode III peak stresses, according to figure 1.

e. In the case of 4-node as well as 10-node tetrahedral elements, calculate average peak stresses by performing the moving average elaboration on adjacent vertex nodes as reported in figure 1 and discussed in paragraph 2.1.
f. Calculate c_{wi} and f_{wi} (i = 1, 2 and 3) parameters according to equations (4, 6).
g. In the case of multiaxial loading, algebraically sum the peak stresses retrieved from the separate analysed load cases into average peak stresses accounting for the multiaxial loads combination, according to equation (7). The panel *Analysis Settings* on the setup menu (see figure 2) allows the analyst to express preferences on the settings of the current PSM analysis. In the case of a structure subjected to multiaxial loading, the different loads must indeed be applied using separate load steps and the user can enable the option "Multiaxial PSM Analysis" in order for the app to automatically extract the relevant peak stresses due to each separate load step and combine them.
h. Calculate the equivalent peak stress on each analysed node, according to equation (5).
i. Calculate the local biaxiality ratio λ on each analysed node, according to equation (8).
j. On each analysed node, perform fatigue life estimation by automatically referencing the proper PSM-based design curve among those proposed in [35] according to the material type, local biaxiality ratio and survival probability. The panel *Fatigue Design Curve Settings* allows indeed the analyst to specify the set of PSM design curves (in terms of material type) and the survival probability to be adopted to estimate the fatigue lifetime of the structure.

As an output, the developed tool allows the analyst to access and visualize different PSM analysis results with the following levels of detail.

a. *Edge contour plots*. Equivalent peak stress, local biaxiality ratio and fatigue life results can be plotted directly over the analysed edges of the model, as shown in figure 2 and in figure 6. Edge contour plots allow the analyst to rapidly single out and compare the different competing crack initiation points in the analysed structure.
b. Results charts. Equivalent peak stress, local biaxiality ratio and fatigue life distributions along one or more analysed model edges can be visualized and compared taking advantage of dedicated charts, generated directly in Ansys® Mechanical. In conjunction with a relevant edge contour plot, custom charts provide a wider insight on the results set.
c. Numerical analysis reports. Comprehensive results reports are generated during the automated analysis. All numerical values relevant to the analysed edges and nodes, i.e. edge ids, nodal coordinates, local peak stresses, average peak stresses, local biaxiality ratios as well as equivalent peak stresses and fatigue life values are collected and made available in a formatted text file for the analyst to consult.

Up to now, the PSM tool has been employed to investigate several common lab-scale steel welded joint geometries subjected to constant amplitude as well as variable amplitude multiaxial loading [40,44] as well as complex amusement park steel welded details subjected to constant amplitude multiaxial loadings [19]. Moreover, the PSM tool has also been recently extended and improved in order to implement the new variable amplitude (VA) formulation of the PSM [37]. The outcome of this work has been presented in [44].

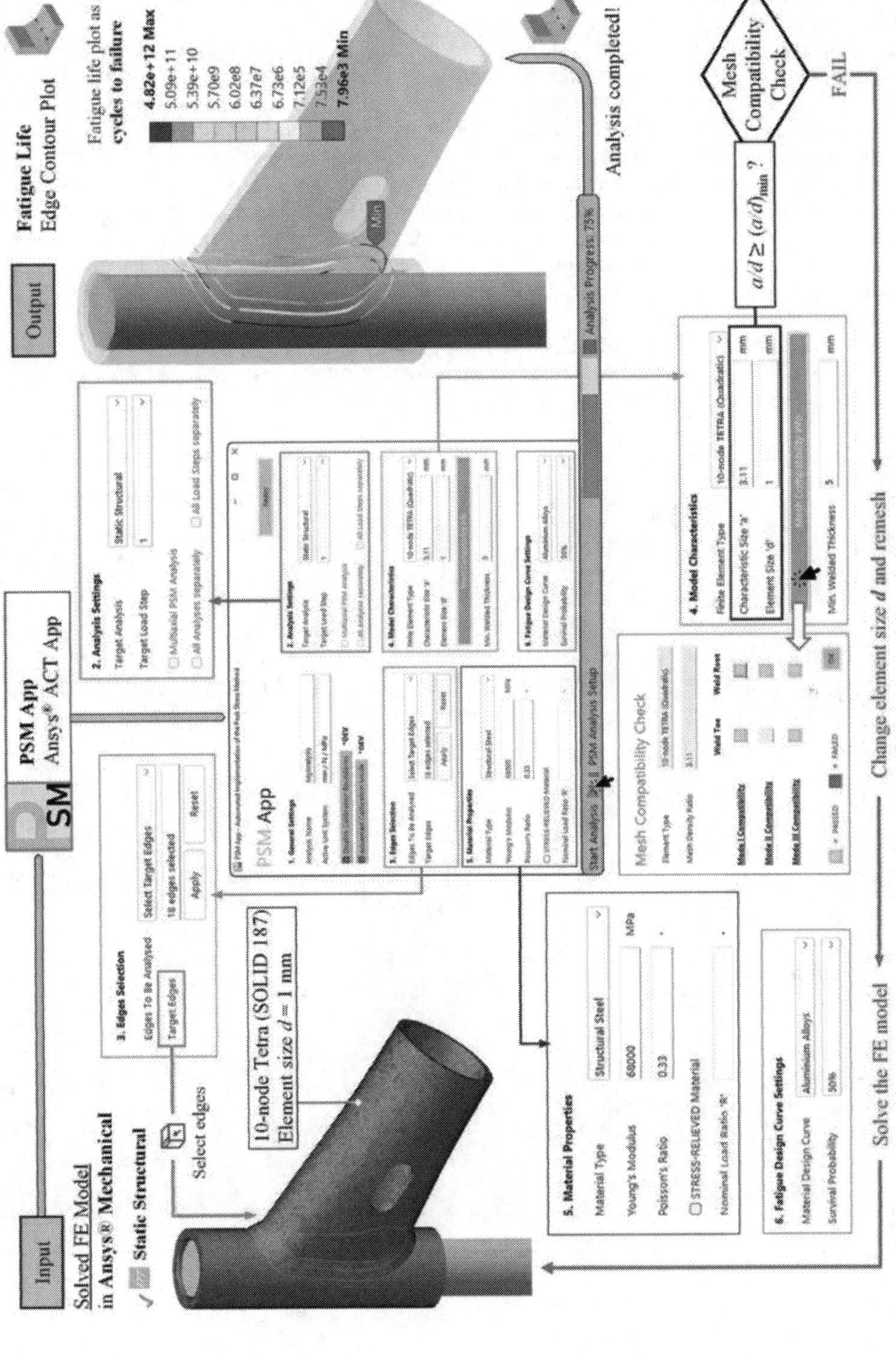

Figure 2. Automated application of the PSM using the tool developed in Ansys® Mechanical. Details of the analysis settings and custom results.

4. Analysed geometries of aluminium welded joints subjected to multiaxial loads and FE analyses according to the PSM

Various experimental data available in literature, dealing with weld toe and weld root fatigue failures in aluminium alloys welded joints subjected to both uniaxial and multiaxial fatigue loading [45–51] have been re-analysed according to the PSM. The considered joint geometries and relevant loading conditions are reported in table 3. It is worth noting that some of the considered geometries had already been analysed according to the PSM in previous works [38,52]. In these previous contributions [38,52], the joint geometries characterized by an axis-symmetric geometry were simulated using 2D 4-node plane harmonic elements (PLANE 25 of Ansys® element library), while the remaining geometries were simulated using 3D 8-node brick elements with K-option 2 set to 3 (SOLID 185 of Ansys® element library). In this latter case, the sub-modeling technique offered by Ansys® Mechanical APDL was employed to isolate and analyze only the most essential and critical regions of the joint dealing with the weld toes and the weld roots.

Material properties, as well as information on the welding process and on testing conditions for the considered geometries are reported in table 4. Details on the fatigue tests and analyses performed according to the PSM are available in table 5.

Table 3. Arc-welded aluminium joint geometries and load cases re-analysed according to the PSM.

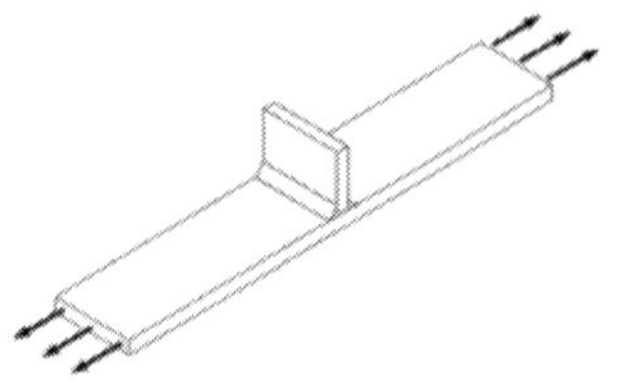
Meneghetti *et al.* [45]

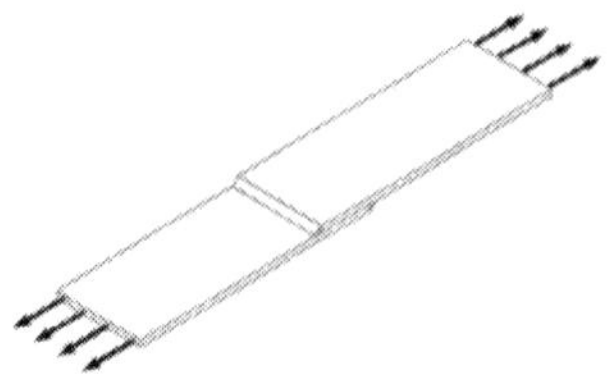
Da Cruz *et al.* [46]

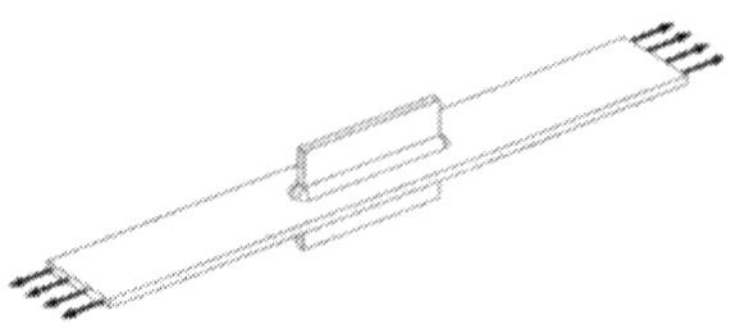
Haagesen *et al.* [47]

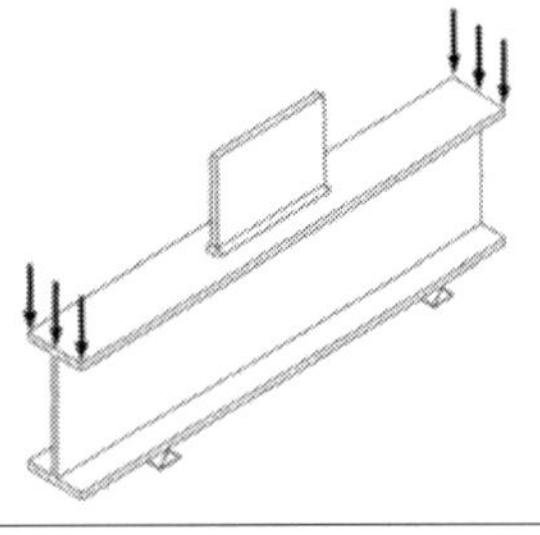
Voutaz *et al.* [48]

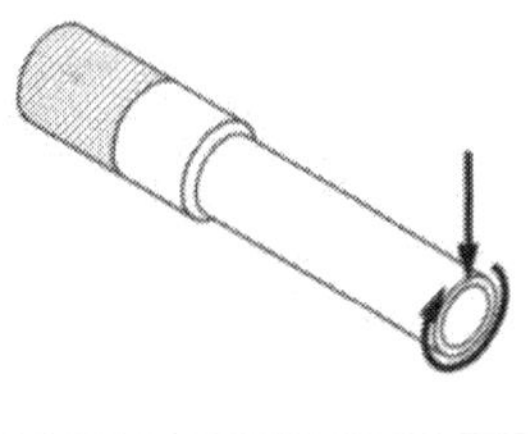
Costa *et al.* [49]

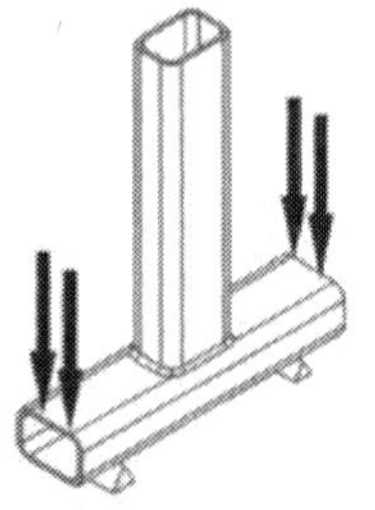
Macdonald and Haagensen [50]

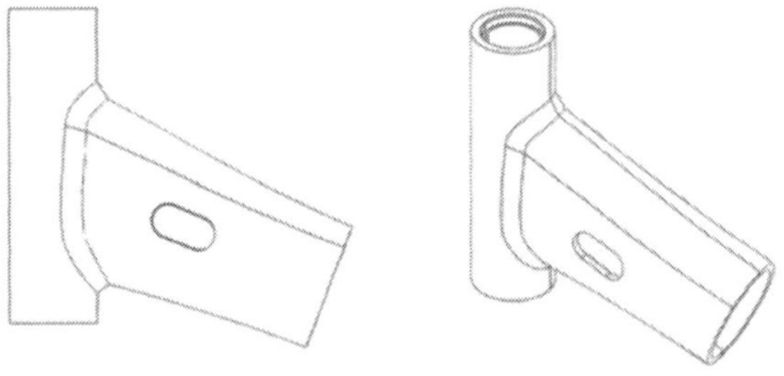
Petrone and Meneghetti [51]

In this work, the joint geometries reported in table 3 have been analysed by comparing the results coming from the following assessment approaches: (i) manual application of the PSM in Ansys® Mechanical APDL, as described in paragraph 2.2, and (ii) automated implementation of the PSM, taking advantage of the tool developed in Ansys® Mechanical and discussed in paragraph 3. More in detail:

a. a free 3D mesh has been generated for each considered geometry taking advantage of 10-node tetrahedral elements (SOLID187 of the Ansys® element library) both in Ansys® Mechanical APDL and Ansys® Mechanical. Given the joint geometry, the same global element size d has been adopted for both FE models.
b. Loads and constraints have been applied coherently for both FE models.
c. No sub-modeling techniques were employed, owing to the high discretization capability offered by 10-node tetrahedral elements with respect to plane and brick elements.

Table 4. Summary of material properties, type of welding process and testing conditions of the considered aluminium alloys welded joints in table 3 [38,52].

Reference	Material	Yield strength [Mpa]	Load type	Testing conditions
Meneghetti *et al.* [45]	5083 H321	250	Axial	As-welded
Da Cruz *et al.* [46]	AlMgSi1-T6	245	Axial	As-welded
Haagensen *et al.* [47]	5083-H22	250	Axial	As-welded
Voutaz et al. [48]	6082	260	Bending	As-welded
Costa *et al.* [49]	EN AW 6060-T6	215	Bending & Torsion	Stress-relieved
Macdonald and Haagensen [50]	6082-T5	290	Bending	As-welded
Petrone and Meneghetti [51]	6061-T6	-	Bending	As-welded

4.1. Case study: Costa et al. [49], *overlap-tubes joint*

In the original contribution by Costa *et al.* [49], the aluminium fillet-welded joints between tubes (see figure 3) were tested under rotating bending, reversed (R = -1) and pulsating (R = 0) in-phase bending (M_b) and torsion (M_t) fatigue loadings. The applied nominal stress ranges and the adopted biaxiality ratios Λ are reported in table 5. The number of cycles reported in the original contribution [49] is referred to a 10% stiffness loss of the joint under testing; the same number of cycles has been considered as the fatigue life in the current work.

The geometry has been modelled as a quarter of the full welded joint, taking advantage of the axis-symmetry. A 3D mesh of 10-node tetrahedral elements has been generated both in Ansys® Mechanical APDL and Ansys® Mechanical taking advantage of the free meshing algorithms provided by the FE programs (see figure 3). In order to grant PSM applicability at both the weld toe and the weld root accounting mode I, mode II (at the weld root only) and mode III loadings, the mesh density ratio requirement $a/d \geq 3$ needs to be fulfilled, according to table 1. The characteristic notch size a equals the tubes thickness at weld toe, namely $a = t = 3$ mm, which also corresponds to the weld leg length z (see figure 3a). Consequently, $d = 3/3 = 1$ mm must be assumed at least. Therefore, a 10-node tetrahedral mesh having global element size equal to 1 mm has been generated over the entire model (see figure 3c and figure 3d), achieving PSM applicability at both weld toe and weld root concurrently.

Moreover, all welded joints were originally tested in stress-relieved conditions. Therefore, the equivalent peak stress has been evaluated on each analysed node according to equation (5), where c_{wi} coefficients, depending on the load ratio R_i of the i-th loading mode (I = 1, 2 or 3), have been calculated according to equation (4).

According to Costa *et al.* [49], fatigue cracks always initiated at the weld toe side. As shown in figure 3 and reported in table 5, both the manual and automated PSM approaches allow to correctly highlight the weld toe as the estimated crack initiation location of the joint.

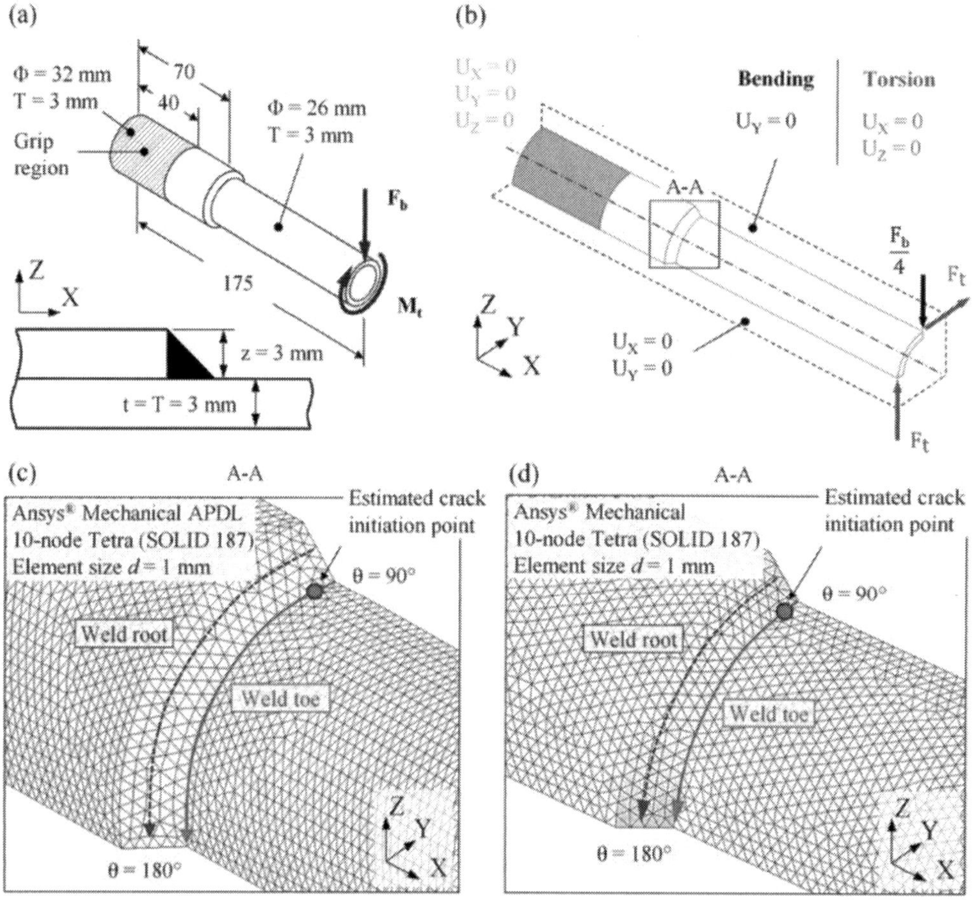

Figure 3. Overlap-tubes joint geometry by Costa *et al.* [49]. (a) detail of the joint geometry. (b) loads, constraints and FE analyses performed according to the PSM. (c) FE mesh generated in Ansys® Mechanical APDL for the manual application of the PSM; (d) FE mesh generated in Ansys® Mechanical for the automated application of the PSM.

4.2. Case study: Petrone and Meneghetti [51], *welded steering node of electric moped frame*

The aluminium fillet-welded steering node under exam is part of a lightweight two-wheeler electric moped frame made of 6061-T6 aluminium alloy (see figure 4). Constant amplitude fatigue tests were originally performed on several frames taking advantage of a servo-hydraulic test rig designed to replicate the most significant stress fields at the frame critical locations, which the vehicle is subjected to in exercise [51]. In this perspective, two main test configurations were employed: (i) a vertical load configuration and (ii) an horizontal load configuration. The latter test configuration only has been accounted in this work. The horizontal load configuration involves the support of a horizontal linear bearing slide to connect the front wheel axle of the frame to a 15 kN MTS 242 servo-hydraulic cylinder (controlled by a MTS Testar Iim). Moreover, the rear wheel has been clamped to the bench with respect to the wheel centre in order to fully constrain the frame (see test scheme in figure 4a). It is important to note that due to the misalignment between the front axle constraint and the rear support, a load reaction F_V is exerted on the frame by the linear slide, in addition to the horizontal load F_X applied by the cylinder.

Both loading components have been taken into account in developing the FE model in figure 4.

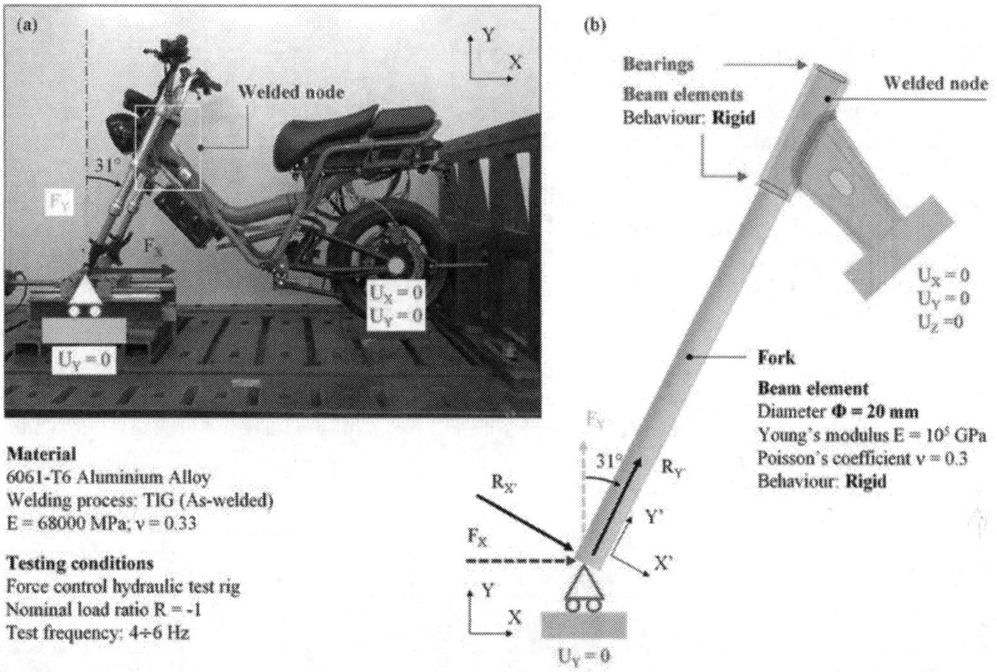

Figure 4. Welded steering node of electric moped frame by Petrone and Meneghetti [51]. (a) detail of the horizontal test configuration and loading scheme. (b) detail of the FE model and FE analyses performed according to the PSM.

The approximated FE model reported in figure 4 has been generated adopting the following assumptions:

(i) only the welded node between the steering tube (chord) and the main frame tube (brace) has been considered. The joint has been isolated from the frame structure, which in turn has been approximated by applying a fixed support to the free end of the welded component, as reported in figure 4b.

(ii) The front fork component has been simulated taking advantage of a single rigid beam element. A Young's modulus $E = 10^5$ GPa has been adopted in order to provide the beam fork a higher stiffness with respect to that of the welded node. The beam fork has been fastened on the steering tube by means of rigid beam elements set in a radial pattern in order to simulate the bearings seats.

(iii) The linear slide has been simulated applying a frictionless support constraint in Ansys® and the loading actions F_X and F_Y have been converted and applied as $R_{X'}$ and $R_{Y'}$ to the free end of the beam fork (see figure 4).

The joint model of the welded node proposed in figures 5a-5c has been generated with the aim of replicating the real geometry of the weld beads, starting from experimental measurements on real components (see some examples in figures 5d-5e). The weld roots have been modelled taking account of the gap between the steering tube (chord) and the main frame tube (brace), which has also been detected in the real components. Moreover, the so-modelled geometry allowed to exploit all the four critical locations of the welded joint under exam, i.e. the weld toe and the weld root on the brace-side; the weld toe and the weld root on the chord-side (see figure 5b and figure 5c).

A 3D mesh of 10-node tetrahedral elements has been generated both in Ansys® Mechanical APDL and Ansys® Mechanical taking advantage of the free meshing algorithms provided by the FE programs (see figure 2). In order to grant PSM applicability at both the weld toes and the weld roots accounting mode I, mode II and mode III loadings, the mesh density ratio requirement $a/d \geq 3$ must be fulfilled, according to table 1. The characteristic notch size a equals the minimum tube thickness at the weld toe, namely t_1 = 5 mm, while the minimum throat section width along the weld bead, namely $s \sim 3.11$ mm, has been adopted as characteristic reference size for the weld root (see figure 5), in order to obtain a more regular mesh pattern. Eventually, d = 3.11/3 = 1.04 mm must be adopted at least; therefore, a 10-node tetrahedral mesh having global element size equal to 1 mm has been generated on the entire model (see figure 2), achieving PSM applicability at both weld toe and weld root concurrently.

Horizontal reversed (R=-1) fatigue test were performed on eight available frames, employing test frequencies between 4 Hz and 6 Hz. Fatigue failure was defined as the number of cycles at which a 10% reduction of the initial frame stiffness (evaluated as the mean value of the first 5000 cycles) occurred [51]. According to the original contribution [51], at this stiffness threshold visible cracks were always detected at the joint between the steering tube (chord) and the main frame tube (brace). Six tested frames exhibited crack initiation at the lower-fibre weld toe (see figure 5c and figure 5e), while only 2 tested frames presented a final visible crack at the upper-fibre of the joint, initiated and propagated from the weld root (see figure 5b and figure 5d).

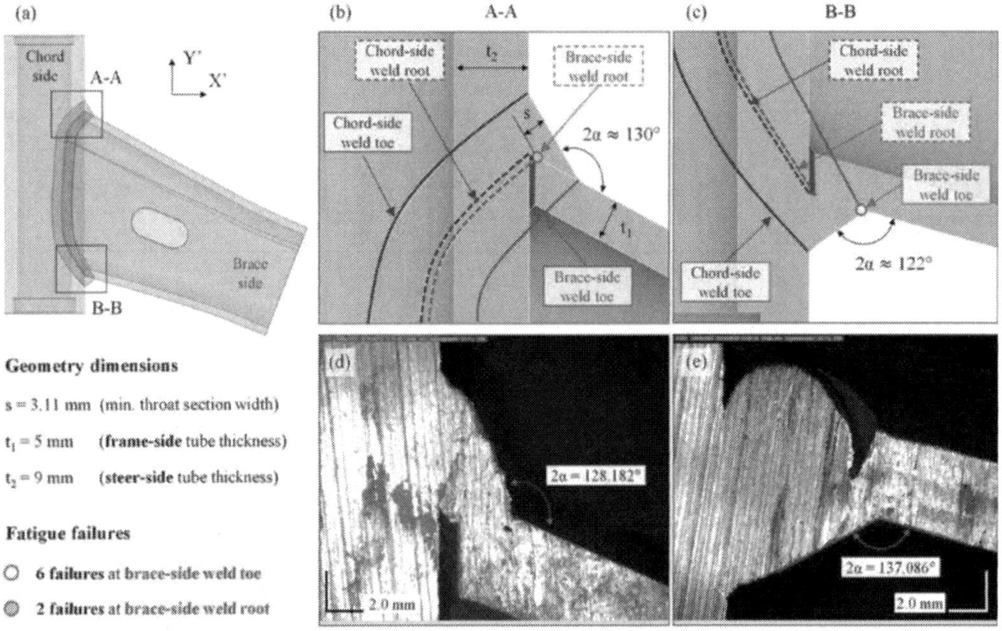

Figure 5. Steering welded node of electric moped frame by Petrone and Meneghetti [51]. (a) a wireframe view of the joint geometry. (b) local detail of the modelled weld bead at A-A. (c) local detail of the modelled weld bead at B-B. (d) micrograph of the actual weld bead geometry at A-A in a tested specimen. (e) micrograph of the actual weld bead geometry at B-B in a tested specimen.

The results coming from the manual and automated application of the PSM on the welded node are reported in figure 6. A good agreement can be observed between the equivalent peak stress distributions calculated by the PSM tool in Ansys® Mechanical and the equivalent peak stress distributions obtained by manually applying the PSM in Ansys® Mechanical APDL (see figure 6c and figure 6d). Moreover, the notch opening angle 2α has been locally evaluated along the brace-side weld toe and brace-side weld root and has been also reported in figure 6c and figure 6d. The brace-side weld root exhibits notch opening angles ranging between 70° and 150° (figure 6c) while the brace-side weld toe exhibits notch opening angles around 150° with a minimum opening angle $2\alpha \sim 122°$ at the estimated crack initiation point (node B in figure 6). As stated in paragraph 2.1, in a previous work [41], weld toe profiles having notch opening angles up to 150° have been analysed taking advantage of the same calibration constants valid for $2\alpha = 135°$. The same approach has been adopted in this work when manually applying the PSM to analyse both the weld toes and the weld roots of the model. On the other hand, the automated PSM tool implements polynomial expressions which allow to calculate accurate calibration constants depending on the specific value of the notch opening angle 2α [43]. The differences in terms of equivalent peak stress due to these two different approaches can also be appreciated in figure 6c (for the brace-side weld root) and in figure 6d (for the brace-side weld toe), especially when $2\alpha > 135°$.

Moreover, it is worth noting that both the manual and the automated PSM approaches allow to correctly estimate the most critical point at the brace-side weld toe (point B in figure 6), i.e. the location from which cracks initiated and propagated in six of the eight tested frames. On the other side, the fatigue results also highlight a secondary competing critical point at the brace-side weld root (point A in figure 6), i.e. the location from which cracks initiated and propagated in the remaining two tested frames. In this perspective, figures 6a-6b provide an example of fatigue life *edge contour plots* generated by the PSM tool, which allow to easily identify and compare the above mentioned competing critical points among all analysed weld toes and weld roots of the model. Indeed, it can be noted that point A and point B, relevant to the brace-side weld root and brace-side weld toe respectively, appear as competing critical points in terms of minimum estimated fatigue life (see figure 6b). In this perspective, experimental crack initiation in real tested components can be expected to occur alternatively either at the brace-side weld toe (point B in figure 6) or at the brace-side weld root (point A in figure 6), depending on the differences between the approximated modelled geometry (see figure 5a and figure 5b) and the actual geometry of the weld beads, which can be characterized by a remarkable geometric variability as proved by figure 5c and figure 5d.

5. Results of manual and automated assessment of fatigue failures according to the PSM

The original experimental results were provided in terms of nominal stress range and corresponding number of cycles to failure for each one of the considered welded joints (see tables 5-6). These data have been re-evaluated in terms of equivalent peak stress range and corresponding number of cycles to failure by means of both manual and automated application of the PSM, in order to be compared with the relevant PSM design scatter bands (see figures 7-8). The nominal stress ranges used in the experimental fatigue tests have been scaled by the equivalent peak stress calculated at the crack initiation point on the FE model applying unitary nominal stresses. The equivalent peak stress results have been reported in table 5 and table 6 for each one of the joint geometry in table 3.

When manually applying the PSM to welded joints subjected to multiaxial loadings, each load case has been individually simulated by means of a separate FE analysis in order to retrieve the relevant peak stresses on each analysed node. Afterwards, the peak stresses have been combined according to equation (7) and introduced in equation (5) in order to calculate the equivalent peak stress accounting for the multiaxial loading scenario (see paragraph 2.2). On the other hand, the PSM tool automatically performs the previous tasks when analyzing models subjected to multiaxial loads, as discussed in paragraph 3.

Table 5. Fatigue tests and PSM analyses results for the aluminium alloys joint geometries of table 3.

Reference	Load case[a]	R_σ	R_τ	Phase ϕ [°]	Λ^b	Failure location (N. failures)	$\Delta\sigma_{eq,peak}$ [MPa][c] Manual PSM	Automated PSM	Nominal stresses $\Delta\sigma_{nom}$ [MPa]	$\Delta\tau_{nom}$ [MPa]
Meneghetti *et al.* [45]	A	0.1	-	-	-	Weld toe (11)	2.157	2.152	80÷150	-
Da Cruz *et al.* [46]	A	0	-	-	-	Weld toe (14)	4.880	4.670	29÷98	-
Haagensen *et al.* [47]	A	0.1	-	-	-	Weld toe (6)	3.645	3.525	35÷100	-
Voutaz *et al.* [48]	B	0.1	-	-	-	Weld toe (18)	4.991	4.996	26÷50	-
Costa *et al.* [49]	Rot. B	-1	-1	-	∞	Weld toe (8)	2.330	2.410	65÷107	-
	B+T	-1	-1	0	2	Weld toe (5)	5.097	5.237	56÷91	28÷45
	B+T	-1	-1	0	0.167	Weld toe (3)	2.101	2.088	12÷15	72÷88
	B+T	0	0	0	1.33	Weld toe (5)	5.538	5.656	40÷64	30÷48
	B+T	0	0	0	2	Weld toe (8)	7.578	7.787	47÷73	24÷36
	B+T	0	0	0	3	Weld toe (4)	10.836	11.172	52÷70	17÷23
	Rot. B + const. T	-1	-1	-	1	Weld toe (8)	2.330	2.410	64÷86	64÷86
Macdonald and Haagensen [50]	B	0.1	-	-	-	Weld toe (20)	2.839	2.736	51÷107	-

[a] A = axial, B = bending, Rot. B = rotating bending, T = torsion, const. T = constant torsion
[b] Nominal biaxiality ratio is defined as $\Lambda = \sigma_{nom}/\tau_{nom}$
[c] Calculated at the crack initiation point with $\Delta\sigma_{nom}$ = 1 MPa and/or $\Delta\tau_{nom}$ = 1 MPa (Λ as indicated, when applicable).

Table 6. Fatigue test details and PSM analyses results for the welded node geometry by Petrone and Meneghetti [51].

Reference	Load case[a]	R_σ	Failure location (N° failures)	$\Delta\sigma_{eq,peak}$ [MPa][c] Manual PSM	$\Delta\sigma_{eq,peak}$ [MPa][c] Automated PSM	Loading action ΔF_X [N]
Petrone and Meneghetti [51]	B	-1	Weld toe (6)	183.1	184.2	2625÷3940
	B	-1	Weld root (2)	120.3	120.4	2625÷2801

[a] B = bending
[c] Calculated with ΔF_X = 2625 N (see figure 6)

Figures 7-8 show that a good agreement has been obtained between the experimental fatigue results, expressed in terms of equivalent peak stress, and the proposed PSM-based design scatter bands for aluminium alloys welded joints. Moreover, the analysed fatigue failures, occurring at the weld toe or at the weld root of the considered joints, have been correctly predicted by the PSM. Eventually, figures 6-8 also allow to compare the results obtained by means of the two adopted assessment approaches, i.e. manual and automated application of the PSM, the latter case involving the tool developed in Ansys® Mechanical and discussed in paragraph 3.

Percent variations below 5% were always detected between the equivalent peak stresses calculated manually and the equivalent peak stresses estimated by the automated tool.

Such differences account for:

(i) the different meshes generated by the free meshing algorithms available in Ansys® Mechanical APDL and Ansys® Mechanical, given the joint geometry.

(ii) The different procedure implemented by the automated PSM tool to define the local coordinate systems and retrieve the relevant peak stresses on the analysed nodes, with respect to the workflow adopted when manually applying the PSM.

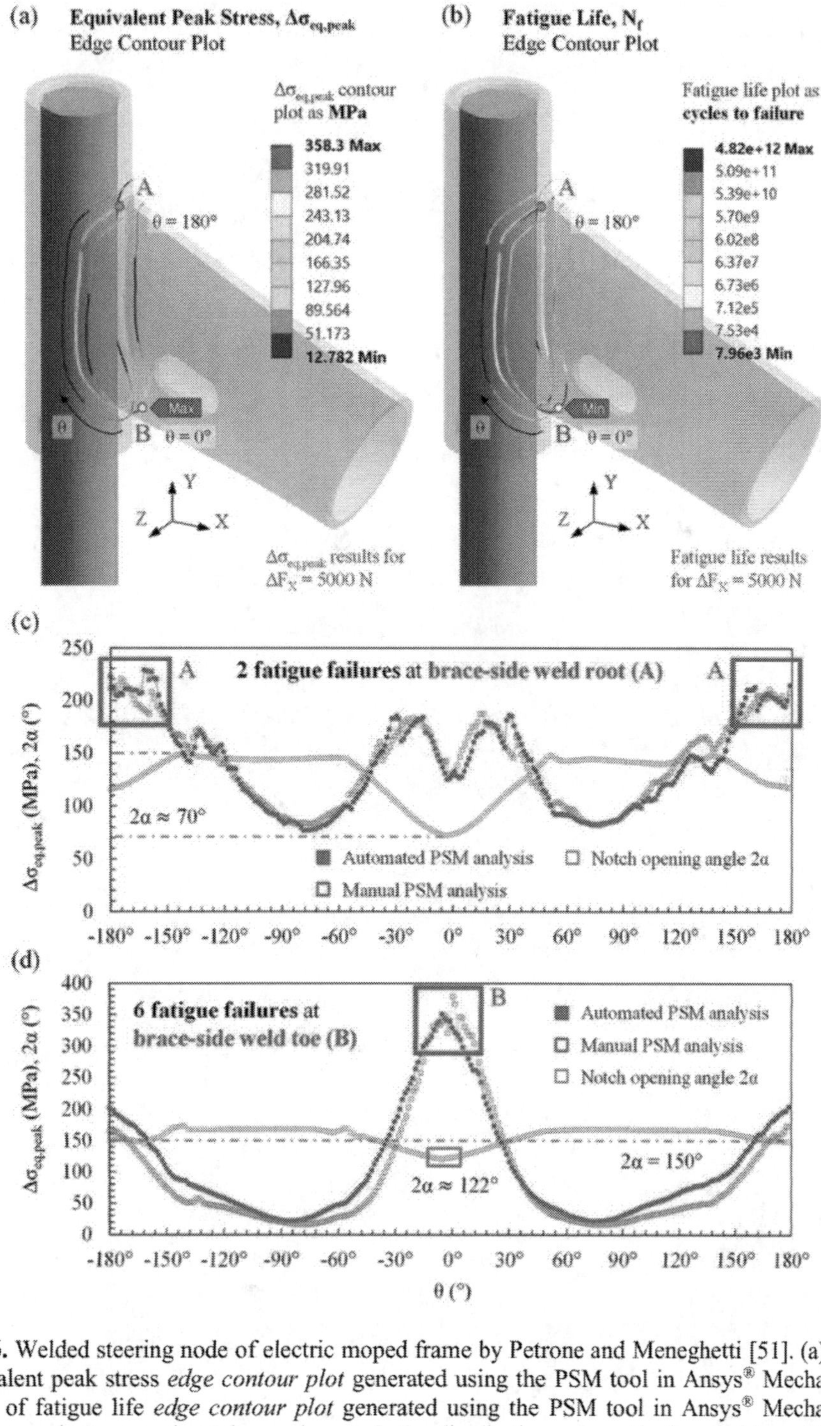

Figure 6. Welded steering node of electric moped frame by Petrone and Meneghetti [51]. (a) example of equivalent peak stress *edge contour plot* generated using the PSM tool in Ansys® Mechanical. (b) example of fatigue life *edge contour plot* generated using the PSM tool in Ansys® Mechanical. (c) equivalent peak stress and notch opening angle 2α distributions along the brace-side weld root. (d) equivalent peak stress and notch opening angle 2α distributions along the brace-side weld toe.

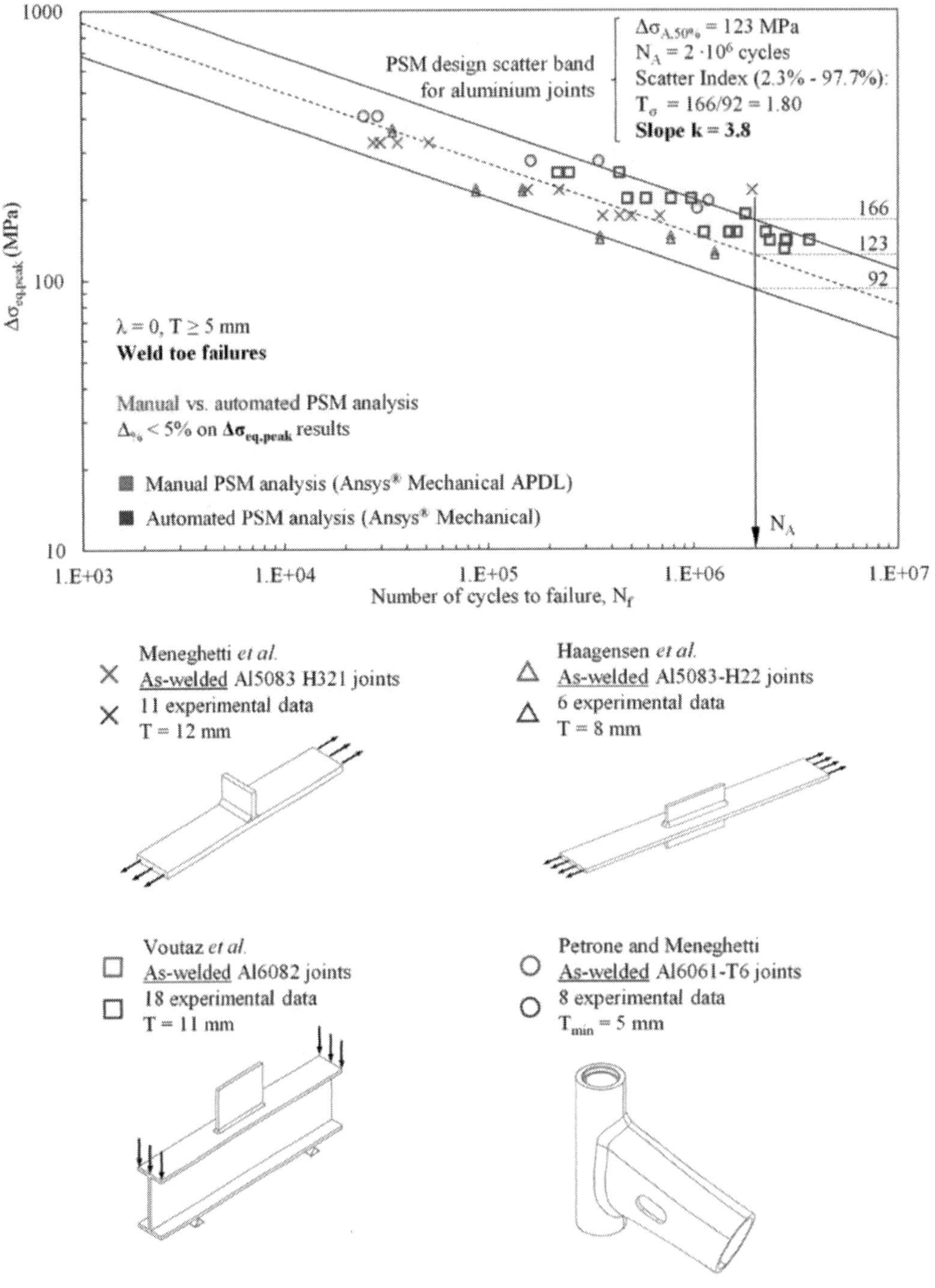

Figure 7. Assessment of weld toe fatigue failures according to the PSM. Comparison between the fatigue design scatter band for aluminium welded joints having $T \geq 5$ mm and $\lambda = 0$ at the crack initiation point [35] and fatigue data from literature [45,47,48,51], evaluated in terms of equivalent peak stress by means of manual and automated application of the PSM.

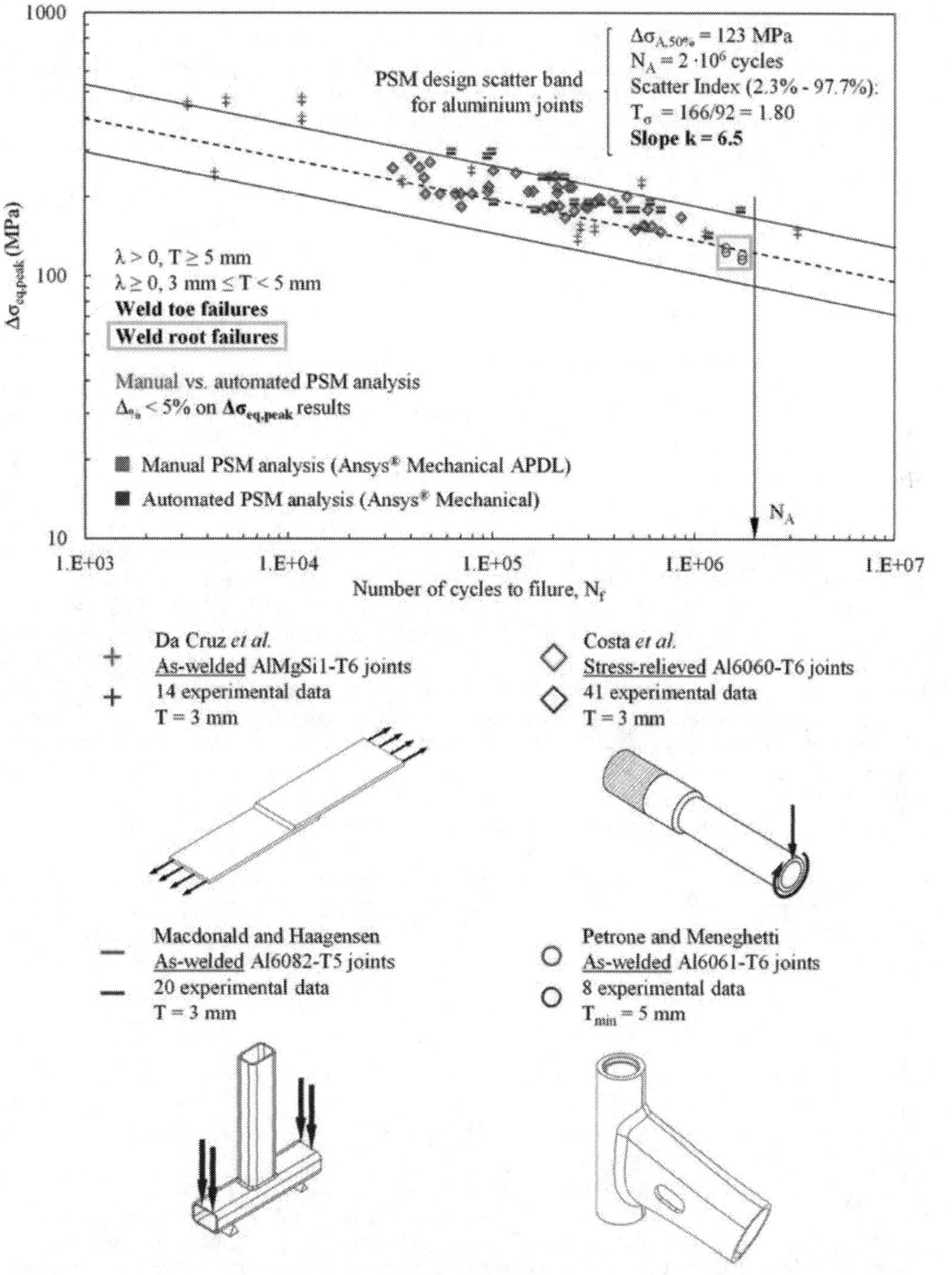

Figure 8. Assessment of weld toe and weld root fatigue failures according to the PSM. Comparison between the fatigue design scatter band for aluminium joints having $T \geq 5$ mm and $\lambda > 0$ or 3 mm $\leq T \leq 5$ mm and $\lambda = 0$ at the crack initiation point [35] and fatigue data from literature [46,49–51], evaluated in terms of equivalent peak stress by means of manual and automated application of the PSM.

6. Discussion and conclusions

In this paper, seven geometries of welded joints made of different aluminium alloys and subjected to constant amplitude uniaxial as well as multiaxial loadings have been re-analysed by comparing the results coming from manual and automated application of the PSM. As a result, 118 fatigue data of weld toe and weld root fatigue failures have been evaluated according to the PSM.

It has been proven that the Peak Stress Method (PSM) allows to assess fatigue failures in aluminium welded joints subjected to constant amplitude uniaxial as well as multiaxial loadings. In conjunction with a proper PSM-based design curve, the equivalent peak stress provides a reliable design quantity which allows the analyst to estimate the fatigue failure location when multiple competing crack initiation points are involved (see figure 3 and figure 6). The PSM allowed to successfully assess the crack initiation point in all the fatigue failures occurring at the weld toe, i.e. ~98% of the considered cases. Moreover, about 105 of the total 118 experimental data, i.e. ~89% of the considered data, fall inside the 2.3%-97.7% bands, proving a good agreement with the proposed PSM-based fatigue design curves. The remaining ~ 11% of the total considered data are on the safe side.

The new tool developed in Ansys® Mechanical allows to significantly contain the calculation effort and time required to apply the PSM on generic welded joints and structures. In detail, table 7 provides some estimations on manual versus automated analysis time, required to apply the PSM to analyze all the weld toes and weld roots of each joint geometry reported in table 3. It is worth noting that a remarkable result was obtained in the case of the welded steering node geometry proposed in [51] and discussed in paragraph 4.2. Dealing with this case study, the manual application of the PSM was indeed significantly hindered by the complexity of the welded detail, whose weld toes and weld roots are characterized by continuously variable notch opening angles. As a result, the analyst had to spend a long time in iterating the manual tasks required to manually estimate the notch opening angle 2α value at each node location along the analysed edges, properly define the local coordinate systems and extract the peak stresses. On the other hand, the PSM tool relies on algorithms designed to locally estimate the notch opening angles and orientate the local coordinate systems starting from node coordinates only; therefore, the automated routine is not dependent on the local complexity of the welded joint.

Eventually, owing to the possibility to adopt coarse FE meshes, which helps to reduce the computational resources required to perform compatible FE analyses according to the PSM, the tool developed in Ansys® Mechanical might provide a useful support to the FE analyst in every-day design practice of both simple and complex welded components and structures.

Table 7. Comparison of the time required to apply the PSM between manual and automated approach.

Reference	Analysed notches [a]	**Manual** PSM analysis [b] estimated time [s]	**Automated** PSM analysis [c] estimated time [s]
Meneghetti *et al.* [45]	2 WT; 1 WR	~ 600 s (~ 10 minutes)	~ 30 s
Da Cruz *et al.* [46]	2 WT; 2 WR	~ 600 s (~ 10 minutes)	~ 30 s
Haagensen *et al.* [47]	2 WT; 1 WR	~ 900 s (~ 15 minutes)	~ 90 s
Voutaz *et al.* [48]	2 WT; 1 WR	~ 900 s (~ 15 minutes)	~ 60 s
Costa *et al.* [49]	1 WT; 1 WR	~ 600 s (~ 10 minutes)	~ 60 s
Macdonald and Haagensen [50]	2 WT; 1 WR	~ 1800 s (~ 30 minutes)	~ 120 s
Petrone and Meneghetti [51]	2 WT; 2 WR	~ 7200 s (~ 2 hours)	~ 90 s

All FE models generated as ¼ of the entire joint, exploiting model symmetries
Time estimations refer to a one-time application of the PSM to analyse all weld toes and weld roots of the model
[a] Notch edges analysed in the FE model (taking account of symmetries). WT = weld toes; WR = weld roots
[b] Time required to: (i) calculate PSM-related parameters from literature, (ii) locally evaluate notch opening angles, (iii) define and orientate local coordinate systems,(iv) retrieve the relevant peak stresses and elaborate them, (v) calculate the equivalent peak stresses
[c] Time required to perform complete automated application of the PSM, including steps (i)-(v) and fatigue life estimation (assuming target edges have been pre-selected by the user, according to Paragraph 3)

References

[1] Hobbacher A F 2016 *Recommendations for Fatigue Design of Welded Joints and Components* (Cham: Springer International Publishing)
[2] Anon 2005 Eurocode 3. Design of steel structures – Part 1–9: Fatigue. CEN
[3] Fricke W 2003 Fatigue analysis of welded joints: state of development *Marine Structures* **16** 185–200
[4] Bruder T, Störzel K, Baumgartner J and Hanselka H 2012 Evaluation of nominal and local stress based approaches for the fatigue assessment of seam welds *Int J Fatigue* **34** 86–102
[5] Poutiainen I and Marquis G 2006 A fatigue assessment method based on weld stress *Int J Fatigue* **28** 1037–46
[6] Niemi E, Fricke W and Maddox S J 2018 *Structural Hot-Spot Stress Approach to Fatigue Analysis of Welded Components* (Singapore: Springer Singapore)
[7] Yıldırım H C 2014 Fatigue strength assessment of HFMI-treated butt welds by the effective notch stress method *Welding in the World* **58** 279–88
[8] Leitner M, Stoschka M and Ottersböck M 2017 Fatigue assessment of welded and high frequency mechanical impact (HFMI) treated joints by master notch stress approach *Int J Fatigue* **101** 232–43
[9] Baumgartner J 2017 Review and considerations on the fatigue assessment of welded joints using reference radii *Int J Fatigue* **101** 459–68
[10] Baumgartner J, Schmidt H, Ince E, Melz T and Dilger K 2015 Fatigue assessment of welded joints using stress averaging and critical distance approaches *Welding in the World* **59** 731–42
[11] Radaj D, Sonsino C M and Fricke W 2006 *Fatigue assessment of welded joints by local approaches* (Woodhead Publishing Limited)
[12] Remes H, Gallo P, Jelovica J, Romanoff J and Lehto P 2020 Fatigue strength modelling of high-performing welded joints *Int J Fatigue* **135** 105555
[13] Shams E, Malikoutsakis M, Savaidis G and Vormwald M 2014 Notch stress and fracture mechanics based assessment of fatigue of seam weld ends under shear loading *Fatigue Fract Eng Mater Struct* **37** 740–50
[14] Shams E and Vormwald M 2017 Fatigue of weld ends under combined loading *Int J Fatigue* **100** 627–38
[15] Zerbst U, Madia M and Vormwald M 2019 Applying fracture mechanics to fatigue strength determination – Some basic considerations *Int J Fatigue* **126** 188–201
[16] Nykanen T, Marquis G and Bjork T 2009 A simplified fatigue assessment method for high quality welded cruciform joints *Int J Fatigue* **31** 79–87
[17] Mikkola E, Murakami Y and Marquis G 2015 Equivalent crack approach for fatigue life assessment of welded joints *Eng Fract Mech* **149** 144–55
[18] Zanetti M, Babini V and Meneghetti G 2021 Fat classes of welded steel details derived from the master design curve of the peak stress method *Welding in the World* **65** 653–65
[19] Visentin A, Campagnolo A, Babini V and Meneghetti G 2022 Automated implementation of the Peak Stress Method for the fatigue assessment of complex welded structures *Forces in Mechanics* 100072
[20] Lazzarin P and Tovo R 1998 A notch intensity factor approach to the stress analysis of welds *Fatigue Fract Eng Mater Struct* **21** 1089–103
[21] Lazzarin P, Sonsino C M and Zambardi R 2004 A notch stress intensity approach to assess the multiaxial fatigue strength of welded tube-to-flange joints subjected to combined loadings *Fatigue Fract Eng Mater Struct* **27** 127–40
[22] Fischer C, Fricke W and Rizzo C M 2016 Experiences and recommendations for numerical analyses of notch stress intensity factor and averaged strain energy density *Eng Fract Mech* **165** 98–113
[23] Fischer C, Fricke W and Rizzo C M 2016 Review of the fatigue strength of welded joints based on the notch stress intensity factor and SED approaches *Int J Fatigue* **84** 59–66

[24] Lazzarin P and Zambardi R 2001 A finite-volume-energy based approach to predict the static and fatigue behavior of components with sharp V-shaped notches *Int J Fract* **112** 275–98

[25] Livieri P and Lazzarin P 2005 Fatigue strength of steel and aluminium welded joints based on generalised stress intensity factors and local strain energy values *Int J Fract* **133** 247–76

[26] Lazzarin P, Livieri P, Berto F and Zappalorto M 2008 Local strain energy density and fatigue strength of welded joints under uniaxial and multiaxial loading *Eng Fract Mech* **75** 1875–89

[27] Sonsino C 1995 Multiaxial fatigue of welded joints under in-phase and out-of-phase local strains and stresses *Int J Fatigue* **17** 55–70

[28] Susmel L 2009 *Multiaxial Notch Fatigue* (Cambridge, UK: Woodhead Publishing)

[29] Carpinteri A, Spagnoli A and Vantadori S 2009 Multiaxial fatigue life estimation in welded joints using the critical plane approach *Int J Fatigue* **31** 188–96

[30] Meneghetti G and Lazzarin P 2007 Significance of the elastic peak stress evaluated by FE analyses at the point of singularity of sharp V-notched components *Fatigue Fract Eng Mater Struct* **30** 95–106

[31] Meneghetti G 2012 The use of peak stresses for fatigue strength assessments of welded lap joints and cover plates with toe and root failures *Eng Fract Mech* **89** 40–51

[32] Meneghetti G 2013 The peak stress method for fatigue strength assessment of tube-to-flange welded joints under torsion loading *Welding in the World* **57** 265–75

[33] Campagnolo A and Meneghetti G 2018 Rapid estimation of notch stress intensity factors in 3D large-scale welded structures using the peak stress method ed G Hénaff *MATEC Web of Conferences* **165** 17004

[34] Campagnolo A, Roveda I and Meneghetti G 2019 The Peak Stress Method combined with 3D finite element models to assess the fatigue strength of complex welded structures *Procedia Structural Integrity* **19** 617–26

[35] Meneghetti G and Campagnolo A 2020 State-of-the-art review of peak stress method for fatigue strength assessment of welded joints *Int J Fatigue* **139** 105705

[36] Meneghetti G, Campagnolo A, Visentin A, Avalle M, Benedetti M, Bighelli A, Castagnetti D, Chiocca A, Collini L, de Agostinis M, de Luca A, Dragoni E, Fini S, Fontanari V, Frendo F, Greco A, Marannano G, Moroni F, Pantano A, Pirondi A, Rebora A, Scattina A, Sepe R, Spaggiari A and Zuccarello B 2022 Rapid evaluation of notch stress intensity factors using the peak stress method with 3D tetrahedral finite element models: Comparison of commercial codes *Fatigue Fract Eng Mater Struct* **45** 1005–34

[37] Campagnolo A, Vecchiato L and Meneghetti G 2022 Multiaxial variable amplitude fatigue strength assessment of steel welded joints using the peak stress method *Int J Fatigue* **163** 107089

[38] Meneghetti G, Campagnolo A and Rigon D 2017 Multiaxial fatigue strength assessment of welded joints using the Peak Stress Method – Part I: Approach and application to aluminium joints *Int J Fatigue* **101** 328–42

[39] Meneghetti G, Campagnolo A and Rigon D 2017 Multiaxial fatigue strength assessment of welded joints using the Peak Stress Method – Part II: Application to structural steel joints *Int J Fatigue* **101** 343–62

[40] Visentin A, Campagnolo A, Sacchet F and Meneghetti G 2022 Implementation of the Peak Stress Method for the automated FEM-assisted design of welded joints subjected to constant amplitude multiaxial fatigue loads *IOP Conf Ser Mater Sci Eng* **1214** 012022

[41] Meneghetti G, Campagnolo A and Berto F 2015 Fatigue strength assessment of partial and full-penetration steel and aluminium butt-welded joints according to the peak stress method *Fatigue Fract Eng Mater Struct* **38** 1419–31

[42] Meneghetti G, Campagnolo A and Visentin A 2020 Automated fatigue strength assessment of arc-welded structures according to the Peak Stress Method *Procedia Structural Integrity* **28** 1062–83

[43] Visentin A, Campagnolo A and Meneghetti G 2022 Analytical expressions for the rapid estimation of the NSIFs at V-notch tips using the Peak Stress Method *Fatigue Fract Eng Mater Struct* **Soon to be submitted**

[44] Visentin A, Vecchiato L, Campagnolo A and Meneghetti G 2022 Automated FEA-assisted fatigue design of welded structures subjected to variable amplitude multiaxial loads according to the Peak Stress Method *European Conference on Fracture ECF23, Funchal, Madeira (Portugal), 27 June - 1 July 2022*

[45] Meneghetti G, Tovo R and Volpone M Experimental analysis of the fatigue strength of welded structural details in aluminium alloy *Proceedings of the 26th AIAS National Conference, Catania (Italy)* (Catania, Italy) pp 393–400

[46] Pinho da Cruz J A M, Costa J D M, Borrego L F P and Ferreira J A M 2000 Fatigue life prediction in AlMgSi1 lap joint weldments *Int J Fatigue* **22** 601–10

[47] Haagensen P J and Oma S 2001 Fatigue life improvement of welded aluminium joints by peening techniques *Proceedings of 8th International Conference on Joints in Aluminium INALCO, 2001, Munich, Germany* (Munich, Germany) pp 1–14

[48] Voutaz B, Smith I F and Hirt M A 1995 Behaviour of Aluminium Beams withWelded Attachments *Proceedings of the 3rd International Conference on Steel and Aluminium Structures ICSAS* (Istanbul, Turkey) pp 213–20

[49] Costa J D M, Abreu L M P, Pinho A C M and M. F J A 2005 Fatigue behaviour of tubular AlMgSi welded specimens subjected to bending–torsion loading *Fatigue Fract Eng Mater Struct* **28** 399–407

[50] Macdonald K a. and Haagensen P J 1999 Fatigue design of welded aluminum rectangular hollow section joints *Eng Fail Anal* **6** 113–30

[51] Petrone N and Meneghetti G 2019 Fatigue life prediction of lightweight electric moped frames after field load spectra collection and constant amplitude fatigue bench tests *Int J Fatigue* **127** 564–75

[52] Meneghetti G 2008 The peak stress method applied to fatigue assessments of steel and aluminium fillet-welded joints subjected to mode I loading *Fatigue Fract Eng Mater Struct* **31** 346–69

Numerical modelling of the mechanical behaviour of Aluminosilicate Glass: A comparison between two simulation approaches

A Manes[1*], M N Tartaglia[1] and Z Wang[2]

[1] Politecnico di Milano, Department of Mechanical Engineering, Milan 20156, Italy
[2] Faculty of Materials and Manufacturing, Beijing University of Technology, Beijing 100124, China

Email: andrea.manes@polimi.it

Abstract. This paper presents a comparison between two numerical methods for modelling the mechanical and failure behaviour of aluminosilicate glass: Cohesive Elements Method (CEM) and Finite Element Method coupled to Smooth Particle Hydrodynamics (FEM-SPH). The failure behaviours, provided by these two approaches, are herein compared under i) quasi-static tests on material coupons, ii) dynamic tests (compressive and tensile tests) also on material coupons and iii) structural impact loading conditions in which the ballistic perforation of aluminosilicate glass tiles due to the impact of a flat-nosed steel projectile are considered. Both methods provide comparable results under quasi-static loading conditions, while more significant differences arise in the ballistic impact part, in which some limitations of CEM are shown.

1. Introduction

Aluminosilicate glass is a brittle material widely employed as windshields of vehicles, aircrafts, and also transparent touchscreens of electronic devices like smartphones and tablets. [1] Thanks to the presence of about 20% of Aluminum oxide in the composition of the material, Aluminosilicate glass possesses higher mechanical strength and chemical stability with respect to soda-lime glass. In general, the mechanical properties of Aluminosilicate glass differ quite a lot in compression and in tension loading conditions. As the compression strength and hardness are relatively high, the main weaknesses of the Aluminosilicate glass are its low tensile strength and its failure mechanisms characterized by large fragmentation. Accordingly, better understanding of the failure properties of the material under different loading conditions is necessary for accurate design of a component made of this material. Considering the efforts required to perform real experimental tests and the complex failure mechanism, the development of reliable numerical models able to predict the behaviour of glass structures under different loading conditions would simplify this task.

Studying the behaviour of a loaded glass structure is not simple and, especially, due to the complexity of crack patterns, the large number of factors the failure behaviour depends on (concentration of defects inside the material, stress state, loading speed, temperature, geometry, etc.) and the discontinuity of the failure process (crack initiation, propagation, branching and failure). All these factors make the prediction of the failure behaviour very challenging and complicated to be performed both in qualitative and quantitative terms.

Content from this work may be used under the terms of the Creative Commons Attribution 3.0 licence. Any further distribution of this work must maintain attribution to the author(s) and the title of the work, journal citation and DOI.
Published under licence by IOP Publishing Ltd

Finite Elements Method (FEM) remains one of the most used numerical methods for the analysis of stresses and strains due to its capability to mimic complex and specific physical behaviour and its calculation efficiency. However, in the analysis of non-linear problems such as failure simulation, some drawbacks arise [2]. In fact, FEM alone is quite unsuitable for the description of both cracking and fragmentation processes. Element erosion is a common choice to represent the formation of cracks in brittle solids inside the FEM framework. Element erosion implies that when an element of the model reaches the failure limit, according to a certain failure criterion, it is handled differently with respect to the remaining non-eroded elements. Several algorithms that handle failed elements exist, but usually eroded elements are either deleted and removed from the calculations, or, alternatively, their stresses are set to zero. However, such an operation is clearly unphysical since it involves the sudden disappearance of a part of the material and, thus, is associated with mass and energy loss [3]. Even if this approach may deliver reasonable results for a limited extent of failure, it becomes impractical for the reproduction of large fragmentations. Following these considerations, two approaches are herein considered by the authors to overcome this problem:

1. The insertion of Cohesive Elements in-between all the solid elements.
2. The transformation of the failed solid elements into Smooth Particle Hydrodynamics particles.

The Cohesive Elements Method (CEM) consists in the insertion of a zero-thickness solid element at the interface between every couple of elements (independent of their shape). Such an added element (which is properly a cohesive element) has different properties from the bulk material, but rather possesses the properties of the process zone ahead of the crack tip, following the Cohesive Zone Model theory by Dugdale and Barenblatt [4,5]. When a cohesive element reaches a certain loading condition (described by a cohesive law), the element is deleted, leaving the neighbouring solid elements as separated elements. Consequently, a realistic situation is reached, in which a disconnection between two neighboring solid elements, left by the deleted cohesive element, is formed. This disconnection represents the formation of a crack. Being zero-thickness, the deletion of cohesive elements does not cause any kind of mass or energy loss. A work by Pandolfi and Ortiz [6] used this technique with the aim to reproduce the tridimensional fragmentation of brittle materials in a three-point bending test. In other works, the fragmentation of rocks [5,7] and glass [3,8] was studied. However, very few studies on the ballistic impact of glass are present in the literature.

Pure Smooth Particle Hydrodynamics (SPH) is one of the earliest meshless method available and one of the most popular for a wide range of applications. SPH has proven to be effective in modelling the fracture of brittle materials [9–11]. In fact, lacking a predefined connection between nodes, this simulation technique predicts the behaviour of non-continuum materials well, like pieces of glass after failure. However, even though the SPH method is generally more suitable, efficient, and robust for problems involving severe distortions, its accuracy, stability, and efficiency is not as good as that of FEM for problems with mild distortions [12]. For this reason, the idea of investigating the feasibility of using traditional finite elements for the field of small distortion and eventually convert element subjected involved into large distortion into SPH particles. The newly generated SPH particles inherit all the mechanical properties of the eroded solid elements, including mass, energy, and constitutive properties. Thereby in a hybrid method the benefits of the classical FEM (easiness, fastness, simplicity) and the pure SPH method (conservation of mass and energy, good reproduction of pulverization of brittle materials) can be combined.

In this paper the Cohesive Elements Method (CEM) and the FEM-SPH (Finite Element coupled to Smooth Particle Hydrodynamics) simulation techniques were employed to run comparative analyses about compressive, tensile and impact behaviour of Aluminosilicate glass. Specifically, a Uniaxial Compression model and a Brazilian Disc Tensile model were used to calibrate and verify the two numerical techniques in different loading conditions, according to the experimental results. Finally, a numerical ballistic impact model was built to test the capability of the two methods to simulate the mechanical behaviour of Aluminosilicate glass. All calculations present in this work were developed on the commercial software LS-DYNA.

2. Materials and experimental results

2.1. Material description

Aluminosilicate glass is a kind of inorganic silicate glass that contains 15% to 25% aluminum oxide. This addition leads to an increase of the mechanical properties in terms of hardness and strength. In Table 1 the detailed chemical composition of Aluminosilicate glass is listed.

Table 1. Chemical composition of Aluminosilicate glass specimens considered in this work.

Oxides	SiO_2	Al_2O_3	Na_2O	MgO	K_2O	CaO	Others
Wt%	64,3	17,7	10,4	3,8	2,5	0,5	0,8

2.2. Specimens and experiments

All experimental data used herein were taken from previous works by Wang et al. [1] and no experimental activity was specifically conducted for the present work. The experimental data include the quantitative results for Uniaxial Compression, Brazilian Disc Tension and Flat-nosed projectile ballistic impact tests. These values are meaningful to provide a quantitative overview of the compressive and tensile properties as well as the impact response of the glass structure. The detailed geometry of the used specimens is shown in Figure 1. The quantitative results of the experimental tests are listed in Table 2.

Table 2. Material's mechanical properties under different loading conditions.

Loading condition	Loading Speed (mm/s)	Elastic Modulus (GPa)	Failure Strength (MPa)	Failure Strain
Uniaxial Compression	0,032	77,3 ± 10,9	579,22 ± 52,48	$8{,}31\cdot10^{-3} \pm 0{,}0009$
	1600	75,9 ± 27,5	897,03 ± 45,47	$1{,}32\cdot10^{-2} \pm 0{,}0020$
	3200	76,6 ± 27,2	1073,53 ± 66,66	$1{,}42\cdot10^{-2} \pm 0{,}0005$
Brazilian Disc Tension	0,002	59,2 ± 5,0	44,25 ± 8,15	$7{,}50\cdot10^{-4} \pm 0{,}0002$
	4000	61,3 ± 6,4	81 ± 11,92	$1{,}35\cdot10^{-3} \pm 0{,}0003$

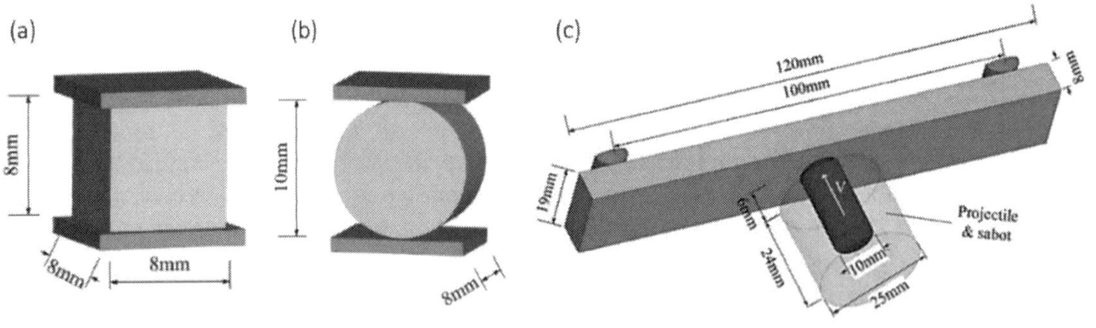

Figure 1. Sketch of the specimens' geometries: (a) Uniaxial Compression, (b) Brazilian Tension, (c) Ballistic Impact.

3. Numerical analyses

In order to develop a reliabl model for the prediction of the impact behaviour of Aluminosilicate glass, a Uniaxial Compression model and a Brazilian Disc Tensile model were built to calibrate and verify the two numerical techniques under different loading conditions (specifically, respectively under compressive and tensile loadings). The aim of this part was to trace the limits of validity of the considered numerical methods, in terms for instance of mesh size, mesh geometry and application of the loads. These are fundamental data for the correct development of a reliable model, able to provide realistic results in a wide range of situations.

3.1. Cohesive Element Method

3.1.1. Numerical model

For all the numerical models making use of CEM, only two types of material' models were employed: solid elastic elements for Aluminosilicate glass specimens and steel plates (whenever they were present) and cohesive elements in-between the solid ones. The default constant stress solid element defined in LS-Dyna by ELFORM=1 was considered for solid elements. Instead, ELFORM=19 elements were used as cohesive elements. They are flat zero-thickness 8-node solid elements with 4 integration points.

These elements were given the following materials' models in LS-Dyna: 001_Mat_Elastic for glass solid elements, 020_Mat_Rigid for the elements belonging to the steel parts and 138_Mat_Cohesive_Mixed_Mode or 184_Mat_Cohesive_Elastic for cohesive elements. The materials models' parameters are listed in Table 3. The properties of steel and Aluminosilicate glass were taken respectively from references [13] and [14].

Table 3. Materials parameters used in CEM analyses.

Steel 020_Mat_Rigid	Density	ρ	7,8	g/cm^3
	Elastic Modulus	E	210	GPa
	Poisson Ratio	ν	0,31	
Aluminosilicate Glass 001_Mat_Elastic	Density	ρ	2,546	g/cm^3
	Elastic Modulus	E	75	GPa
	Poisson Ratio	ν	0,22	
Cohesive Elements 138_Mat_Cohesive Mixed_Mode	Density	ρ	2,546	g/cm^3
	Stiffnesses in/normal to the plane of the cohesive element	E_T	1e+04	GPa/mm
		E_N	1e+04	GPa/mm
	Peak traction in normal/tangential direction	T	45	MPa
		S	225	MPa
	Energy release rate for mode I and mode II	G_{IC}	0,02	MPa/mm
		G_{IIC}	0,1	MPa/mm
	Ultimate displacements in normal/tangential direction	UND	0,002	mm
		UTD	0,002	mm
	Exponent of the model	XMU	1	
Cohesive Elements 184_Mat_Cohesive Elastic	Density	ρ	2,546	g/cm^3
	Stiffness in/normal to the plane of the cohesive element	E_T	1e+04	GPa/mm
		E_N	1e+04	GPa/mm
	Peak traction in normal/tangential direction	FN_FAIL	45	MPa
		FT_FAIL	225	MPa

For cohesive elements, a previous work by Wang et al. was considered [3], however the values of T and FN_FAIL were set equal to the maximum tensile strength of Aluminosilicate glass and the value of S and FT_FAIL were five times larger than T, as stated in the citated paper.

In the simulation of Uniaxial compression and Brazilian tension, two possible configurations of models were built with or without the compressing plates. In either case, the parts corresponding to the solid and cohesive elements of the specimens were always considered as a unique set of parts. For this reason, for the models in which the pressing steel plates were present, the AUTOMATIC_SURFACE_TO_SURFACE contact option was used, in which a contact between the plates and a set of parts (glass solid elements and cohesive elements) was imposed and the upper plate was subjected to a displacement-controlled loading scheme. Conversely, in the models without the presence of pressing plates, all the constraints of the boundary conditions were applied directly onto sets of nodes positioned at the top and bottom of the specimen.

When the LS-DYNA/Explicit solver is used for quasi-static loading conditions, the numerical loading time must be long enough to avoid dynamic effects, which can cause stress wave propagation in the specimen and dynamic oscillations in the loading curves. However, a longer loading time results in higher computational cost. Different loading times were, thus tried and the loading speed of 100 mm/s was used for the quasi-static simulations to obtain both good simulation results and acceptable computational efficiency.

3.1.2. Analyses and results

Several investigations were conducted using CEM both on the Uniaxial Compression model (UC) and Brazilian Disc Tensile model (BDT) to test the feasibility of this technique. Accordingly, the predefined displacement loading was applied through two steel compressive plates. To avoid non-negligible fragmentation of the elements situated in contact between the specimen and the plates due to stress concentration, two modifications were performed on the model:

1. Modification of the geometry of the specimen, in order to flatten and enlarge the contact surfaces
2. Introduction of some stronger cohesive elements near the contact zone, with the aim of reinforcing the contact area and reducing the effects of stress concentration

It was possible to employ stronger cohesive elements due to the fact that in Brazilian discs failure starts from the central part of the specimen and then propagates towards the contact areas. Hence, no direct contribution of the added stronger elements to the strength of the specimen can be expected. Specifically, the inserted values for T and S cards (peak traction values in normal and tangential direction) of the stronger cohesive elements were typically two or three times larger than those indicated in Table 3.

In Figure 2 a representation of a so-obtained Brazilian Disc Tensile model is shown. In particular, in the image on the right a representation of the model after having hidden the solid elements of the specimen is reported. As can be imagined, the remaining finite elements of the disc are only the cohesive elements. Due to their null thickness, they were not visible in the complete representation with all the elements of the model. The brown parts (near the contact zone) represent the stronger cohesive elements' part.

With this BDT model, the behaviour of different cohesive models was firstly explored. In fact, different cohesive models are present in LS-Dyna, such as Mat_138 (Cohesive Mixed Mode) and Mat_184 (Cohesive Elastic), differing in their cohesive law. The aim of this preliminary investigation was to explore the upper mesh size limit for the material's model below which a convergence of results could be obtained. The analysis was performed by building several models with different mesh sizes to identify at which point the desired output quantities (for instance strength and maximum strain) converge to the same value. Here three output quantities were considered for the comparison:

1. Maximum reaction force
2. Maximum normal stress of the firstly failed cohesive element

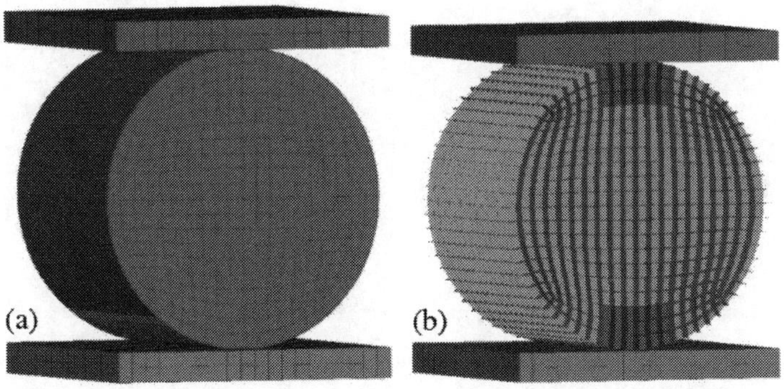

Figure 2. (a) Representation of the obtained Brazilian Disc Tensile model. (b) Representation of the model in which the cohesive elements (yellow and brown) are the only elements belonging to the specimen that are left visible.

The first analysis consists in the representation of the maximum reaction force exhibited by each models using different mesh sizes. In a similar way, the maximum normal stress of the first cohesive element reaching the failure condition was used as output magnitude to explore the mesh sensitivity of the models. In this analysis, for several models with different mesh sizes, the first cohesive element to reach the imposed deletion condition was retraced and its maximum stress directed orthogonally with respect to its plane (also referred to as "Z-Stress") was inserted in the diagram with the corresponding mesh size. These two analyses were useful to determine the dependence of each material's model on the mesh size and the plots obtained are reported in Figure 3. As visible the CEM models revealed not to be mesh size sensitive in tension (BDT).

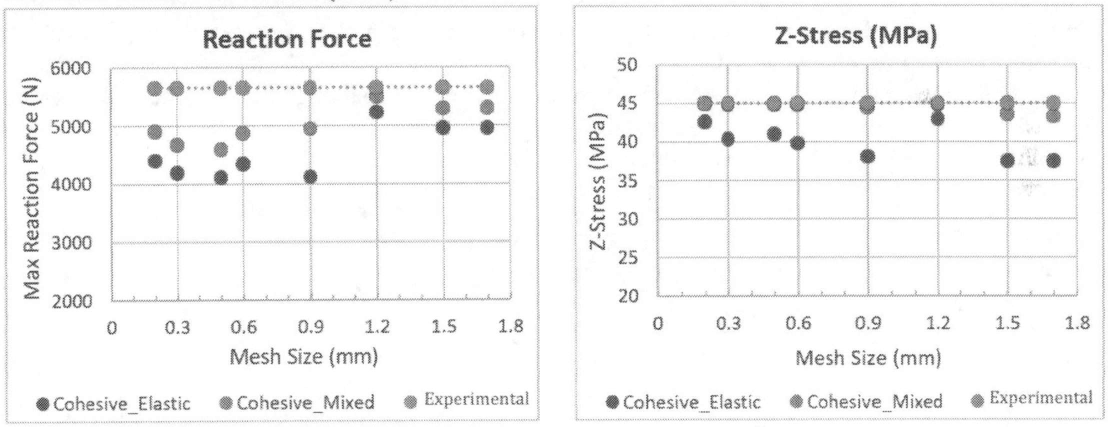

Figure 3. BDT simulations: comparison of the mesh size sensitivities of Cohesive_Elastic and Cohesive_Mixed_Mode cohesive models in terms of Maximum Reaction Force and Maximum Orthogonal Stress exiting the plane of cohesive elements ("Z-Stress").

According to the obtained results, Cohesive_Mixed_Mode demonstrated to be the cohesive model that provided the best results. Consequently, this model was employed for further analyses, including a further mesh size sensitivity, mesh geometry dependence and strain rate effect.

As highlighted before, no mesh size sensitivity could be outlined under tensile loading conditions for CEM. In an analogous way, the dependence from mesh size of the UC model was explored by tracing for different mesh sizes the maximum compressive stress determined from the values of maximum reaction force. The loading speed was chosen again equal to 100mm/s. As can be seen in Figure 4, no clear convergence was obtained but, despite some oscillations, a reasonable convergence of results can be noticed for mesh size finer than 0,6mm.

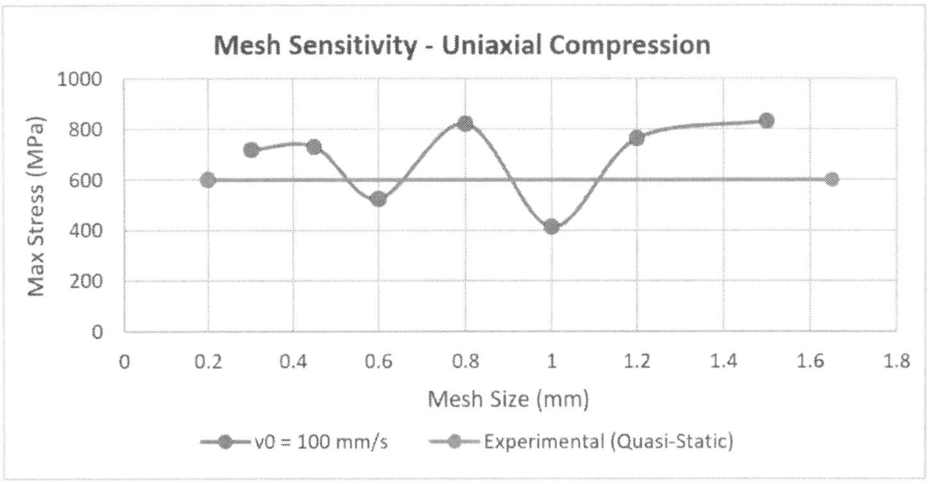

Figure 4. Mesh size sensitivity for the UC model using CEM.

Finally, the possibility to predict the strain rate dependence of Aluminosilicate glass using CEM was explored. In fact, as shown in Table 2, for increasingly higher strain rates, the specimens made with this material increase their strength and the maximum deformation that is reached before failure. The analysis was conducted again in parallel for BDT and UC models with mesh sizes respectively of 1,2 mm and 0,6 mm. It should be pointed out first that the exact quasi-static loading conditions (0,002 mm/s) could not really be reproduced as the corresponding loading speed would result in too long computational time. However, in contrast to the UC model, no strain rate dependence was outlined for the model in tensile loading conditions whereas with the UC model a strain rate effect were obtained for compression behaviour. The trends of the maximum stress and maximum strain are reported in Figure 5 and the numerical values were compared with experimental ones (already reported in Table 2). As visible, an increase in the maximum stress and strain was clearly detected.

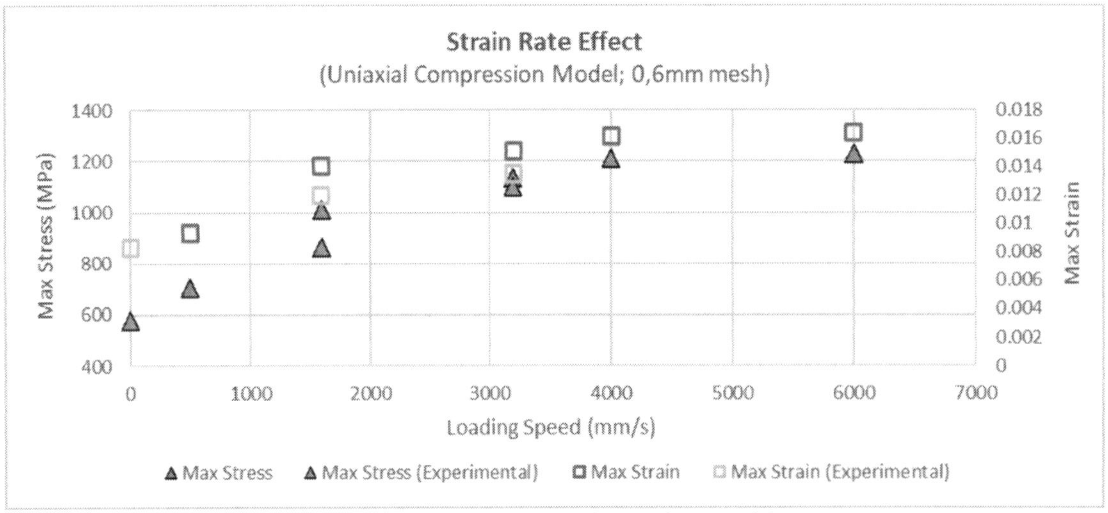

Figure 5. Strain rate dependence of UC model.

These results are not only good in qualitative terms, but also the quantitative results of the experimental tests were confirmed. The most interesting fact is that such a quantitative confirmation of the experimental results was obtained not only without parametrizing the material parameters with respect strain rate effect, i.e. without inserting any direct loading speed dependence in the constitutive parameters of the material. Such strain rate-dependent numerical strength could be explained by the fact that in the UC specimen multiple cracks initiate and propagate. After the deletion of cohesive elements, the interaction between the glass debris (corresponding in this simulation to the solid elements, remained free and unbounded after the failure of the cohesive elements, but still possessed a certain inertial contribution) provides some contribution to the final strength of the numerical specimen. This contribution was not present in tensile loading conditions, where following the elimination of cohesive elements, solid elements are driven far away from each other by the acting state of stress.

3.2. FEM-SPH

3.2.1. Numerical model

As mentioned above in the FEM-SPH approach the glass specimens were built at first as usual finite element parts and were converted into SPH particles after failure. Such a conversion process is regulated by the definition of the keyword Define_Adaptive_Solid_to_SPH already present in LS-Dyna. There are two user-input parameters ICPL and IOPT related to this keyword that were set equal to 1, meaning that coupling occurs when the solid element fails. Also, every finite element was adapted to one SPH element, by putting the NQ card equal to 1.

The SPH part is defined by the keyword card SECTION_SPH. In the selection of the parameters for the SPH part of the models, the default options were used: CSLH (a constant applied to the smoothing length of the particles) was set equal to 1,2, while the scale factors for the maximum and minimum smoothing length were respectively HMAX = 2 and HMIN = 0,2. Also, the default cubic spline kernel function was used (SPHKERN = 0).

Table 4. JH2 model parameters for Aluminosilicate glass.

Parameter	Notation	Value	Units
Density	ρ_0	2545,62	Kg/m^3
Poisson's Ratio	ν	0,22	
Elastic Modulus	E	75,13	GPa
Equivalent Stress at Hugoniot Elastic Limit	σ_{HEL}	5,95	GPa
Pressure at Hugoniot Elastic Limit	P_{HEL}	3,07	GPa
Maximum hydrostatic tensile stress	$\sigma_{t,max}$	45	MPa
Intact Strength coefficient	A	0,93	
Intact strength exponent	N	0,76	
Strain rate coefficient	C	0,036	
Fractured strength coefficient	B	0,2	
Fractured strength exponent	M	1	
Elastic bulk modulus	K_1	44,72	GPa
Coefficient for 2nd degree term in EOS	K_2	-67,45	GPa
Coefficient for 3rd degree term in EOS	K_3	141,6	GPa
Energy conversion coefficient	β	1	
Damage coefficient	D_1	0,043	
Damage exponent	D_2	0,85	

As far as the FEM part is concerned, the considered parameters are listed in Table 4. In this case, for the FEM part, 110_MAT_Johnson_Holmquist_Ceramics corresponding to Johnson-Holmquist 2 constitutive model [15] (often also abbreviated with JH2) was employed to describe the mechanical behaviour of Aluminosilicate glass before failure. The material parameters were taken from previous works of Wang et al. [1,14].

3.2.2. Analyses and results
The first conducted analysis was a mesh size sensitivity analysis, to find whether and in which conditions the models showed a convergence of results. Furthermore, the possibility to remove the compressing plates was explored for the UC model. The models and the results are shown in Figure 10. For the BDT model a proper mesh size can be outlined, while the UC model is mesh size insensitive. However, the presence of compressing plates decreases the reliability of the model. In the end, a 0,5 mm mesh was used for all remaining analyses, and no investigation on the behaviour of the model without the presence of plates was conducted, gigure 6.

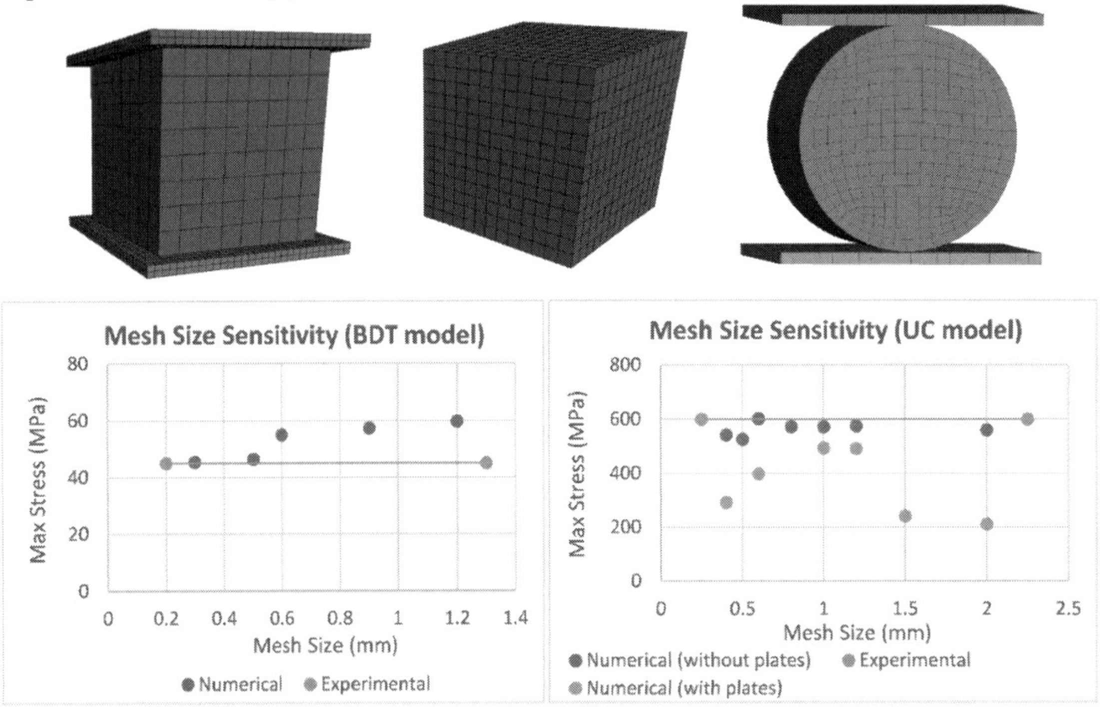

Figure 6. Models for mesh sensitivity analysis and results for FEM-SPH hexahedral mesh.

Different is the case of the strain rate effect. In fact, the strength included in JH-2 model parameter and indicated with the letter C in the constitutive equation in LS-Dyna, directly relates the compressive strength with the strain rate effect of the model. Coherently with this, only UC specimens showed a strain rate dependence in their strengths. As reported in Table 4, a value of 0,036 was calibrated for the parameter. In Figure 7 the strain rate effect of the UC model is depicted.

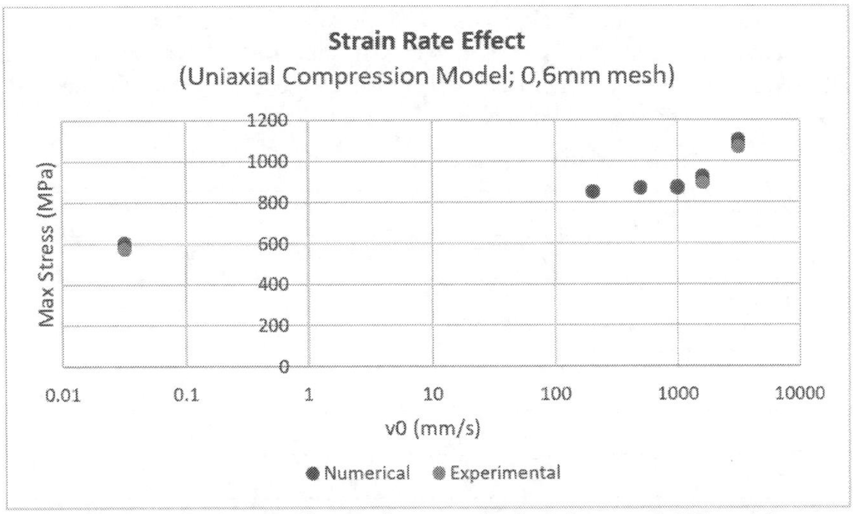

Figure 7. Strain rate effect on the strength of a 0,6mm mesh UC model.

The discussion of the last sections showed the capabilities of CEM and FEM-SPH models. CEM was studied by comparing 184_Cohesive_Elastic and 138_Cohesive_Mixed_Mode cohesive laws and following the demonstration of the Cohesive_Mixed_Mode model's reliability, this cohesive model was selected to investigate, the mesh size sensitivity of the CEM and its capability to predict the strain rate effect. Mesh size dependence above 0,6mm mesh and strain rate sensitivity only in compression were highlighted. Solid elements described by 110_MAT_Johnson_Holmquist_Ceramics constitutive model were used to build tensile and compressive specimens that turned into SPH particles whenever a failure condition was met. This model resulted to be mesh size sensitive in tension for meshes coarser than 0,5 mm and strain rate sensitive in compression. In this framework, for the following structural analysis of impact loadings, the two approaches were tested and compared.

4. Ballistic impact simulation

After having explored the potential of the two simulation techniques to reproduce the behaviour of the material in simple loading conditions, the attention was moved to the simulation of a ballistic impact onto an Aluminosilicate glass tile. The impact behaviour of Aluminosilicate glass was tested experimentally with a flat-nosed steel projectile at two impact velocities: 84 m/s and 139 m/s. A polymeric sabot covering the bullet was adopted to improve the trajectory during the test. The obtained experimental residual speeds of the bullet are 66 m/s and 127 m/s respectively, as shown by Wang et al. [16].

4.1. Numerical model

In this analysis, a glass tile modelled with the two numerical approaches considered so far was impacted by a flat-nosed projectile. In the numerical models, the only part subject to modifications was the glass tile, which was modelled with different mesh sizes (1mm for CEM, 0,5mm for FEM-SPH) and exploiting the two different approaches previously introduced: i) SPH conversion rule ii) cohesive elements embedded between the solid ones. All the remaining parts were kept unchanged and full size models were adopted with dimensions already shown in Figure 1.

The aluminium supporting bars were designed using the simple 001_Mat_Elastic material model with a density of 2,7 g/cm^3, an elastic modulus of 72 GPa and a Poisson ratio of 0,3 (with the consistent units of measure shown in section 4.1). The same material model was used for the polymeric sabot of the projectile (density = 1,13 g/cm^3, elastic modulus = 1,511 GPa, Poisson ratio = 0,3). The properties of the Cohesive elements and FEM-SPH parts were the same as already presented in the previous sections. The carbon steel projectile was modelled using the material model

098_Simplified_Johnson_Cook model (JC) with a mesh size of a 0,15mm. The projectile was put initially in contact with the glass tile and, together with the sabot, it was given an initial velocity according to the experimental tests using the Initial_Velocity_Node card. The parameters evaluated by Abouridouane et al. [13] were adopted here, as listed in Table 5.

Table 5. JC model parameters for carbon steel projectile

Parameter	Notation	Value	Units
Density	ρ_0	7850	Kg/m^3
Poisson's Ratio	v	0,31	
Elastic Modulus	E	186	GPa
-	A	546	MPa
-	B	487	MPa
-	n	0,25	
-	C	0,015	
Effective plastic strain rate	$EPS0$	1	

Also, Tied_Surface_to_Surface contact condition was used to attach the projectile to its polymeric sabot, while for the contact between the end of the supports and the glass tile as well as the contact between the tile and the projectile, Automatic_Surface_to_Surface option was selected. In the case of FEM-SPH, also other conditions were set. In fact, apart from the conversion of solid elements into SPH particles, the SPH part was put in contact using the Automatic_Nodes_to_Surface contact condition with all the parts initially in contact with the tile (projectile and ends of the supports).

4.2. *Results and discussion*

The effect of the impact velocity was explored as a consequence of the different modelling choices. The velocity history of the projectile for the two simulation techniques is shown in Figure 8. As can be seen, in general the FEM-SPH is able to more accurately approximate the experimental results. Specifically, a particularly higher residual speed is obtained for CEM with respect to the experimental values. The Aluminosilicate glass tile replicated by CEM seems to be unable to sufficiently slow down the projectile, absorbing its kinetic energy. Such discrepancy from the experimental results tends to decrease if the impact velocity increases. Also, for both techniques a wavy trend due to the formation of longitudinal mechanical waves on the bullet after the impact is exhibited in the curves. This behaviour is replicated in both simulation techniques with the same phase.

The two simulation techniques were compared in terms of the glass tile's failure mode as well. This analysis highlighted some important differences between the obtained results. First of all, the graphical representation of a crack using CEM immediately shows the difficulty in finding a suitable representation of the failure pattern, because with this technique the crack path is the result of the deletion of the cohesive elements that have a null thickness and, consequently, don't leave any significant and visible empty space in the model. In Figure 9 the failure process is shown both with half-shaded solid elements, in order to make the deletion of cohesive elements visible, and only from the cohesive elements point of view.

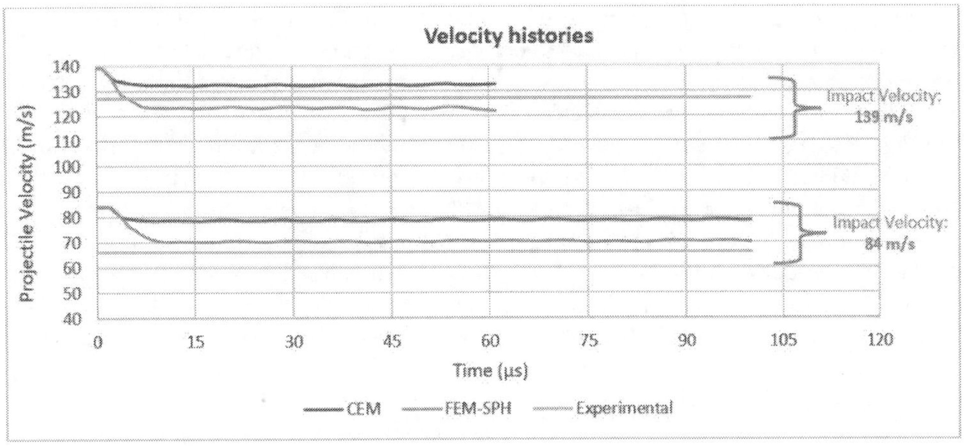

Impact Velocity (m/s)	Residual Velocity (m/s)		
	Experimental	CEM	FEM-SPH
84	66	79 (+19,6%)	69 (+4,5%)
139	127	132 (+3,9%)	123 (-3,1%)

Figure 8. Summary of the obtained velocity histories for CEM and FEM-SPH models at different impact velocities.

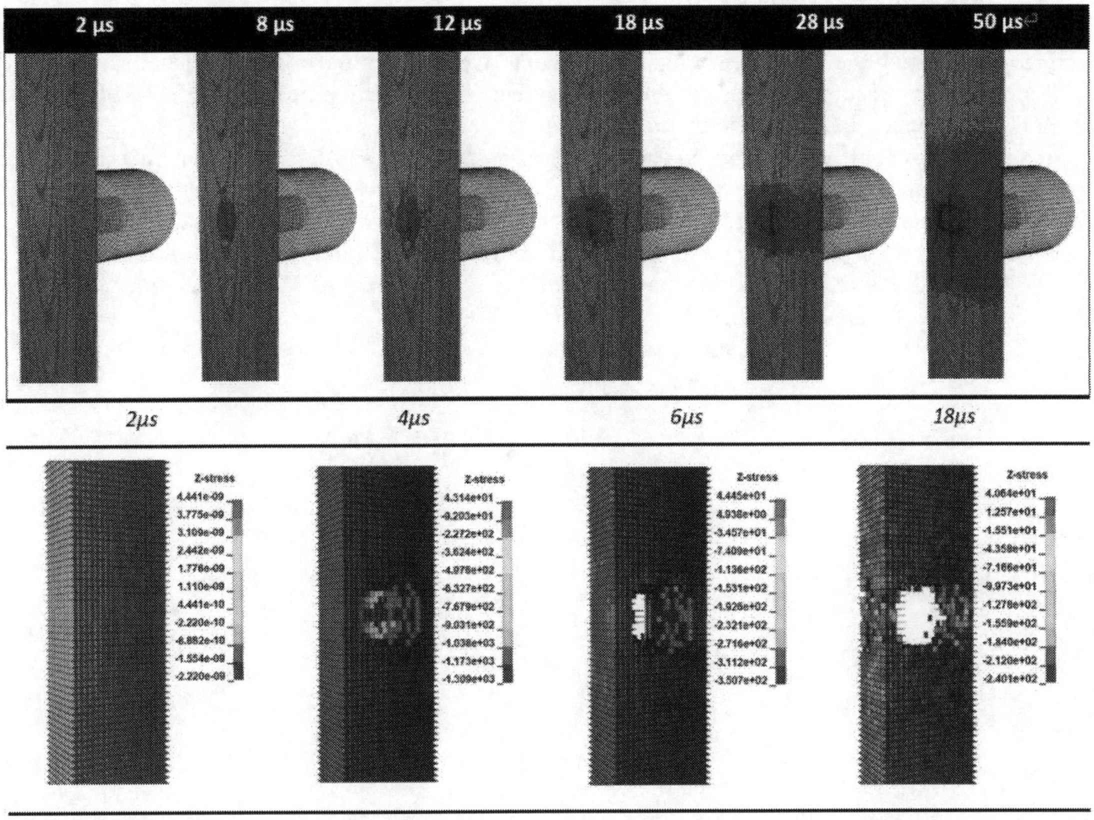

Figure 9. Above: Failure progressions for 1mm mesh impact CEM models. Cohesive elements are coloured in green. Below: Deletion progression of cohesive elements of the glass tile with an indication about the level of stress of cohesive elements exiting from the plane.

During the experiments, an ultra-high-speed camera with a frame rate of 500000 fps (time interval 2μs between neighbouring images) was used to capture the dynamic fragmentation process of the glass tiles. Due to the transparency property of glass, intact specimens were black in the image. When cracks initiated, the light from the flash was reflected by the newly formed crack surface and detected by the camera. In Figure 10 a comparison between the mid-section view of the failure is reported for the CEM and FEM-SPH models, while in Figure 11 an improved graphical representation shows the failure process using the Damage quantity defined for the JH2 model for different instants of time.

As can be seen from Figure 10, the FEM-SPH model provides a better failure mode, especially in terms of extent of the deformation of the broken glass tile. In fact, CEM is hardly able to transfer the loads for a long distance, and, rather, cohesive elements fail locally where the solicitation is applied. The consequence of this behaviour in the simulation of a ballistic impact is the formation of a circular hole in correspondence with the passage of the projectile, as visible clearly in Figure 11.

The incapability of CEM to transfer the loads far from the impacted area is reflected also by the fact that immediately after the impact, no visible crack or cohesive elements deletion is shown far from the projectile. Instead, the deletion of elements starts from the impacted area and then proceeds towards the extremes of the glass tile. Instead, in the case of FEM-SPH model, a larger fragmentation of the glass target follows the initial formation of a circular damage, and, as in the experimental images, some cracks are formed also far from the impacted area.

Accordingly, in CEM a smaller portion of the target material is solicited during the impact and, consequently, a smaller amount of kinetic energy is absorbed by the impacted material. This fact is consistent with the higher residual velocity that was obtained previously in the residual velocity plot for the CEM model with respect to FEM-SPH.

From a purely graphical point of view, also, though an improvement in the representation was obtained by adjusting the transparency of solid elements, CEM never allows a clear visualization of the crack propagation. Accordingly, also in the visual aspect of the failure process, the FEM-SPH method is more reliable both far and close to the impacted area.

Finally, one last comparison between the two techniques can be made in terms of computational time. With a 4 cores and 8 logical processors CPU @3.4GHz, the FEM-SPH model for ballistic impact simulation took about 1 hour and 12 minutes to process the data, while for CEM the same quantity took about 50 minutes if the mesh size was chosen equal to 0,5 mm (the same used also for the FEM-SPH model). Thus, it is possible to conclude that the reliability shown by the FEM-SPH technique requires a higher computational time than CEM.

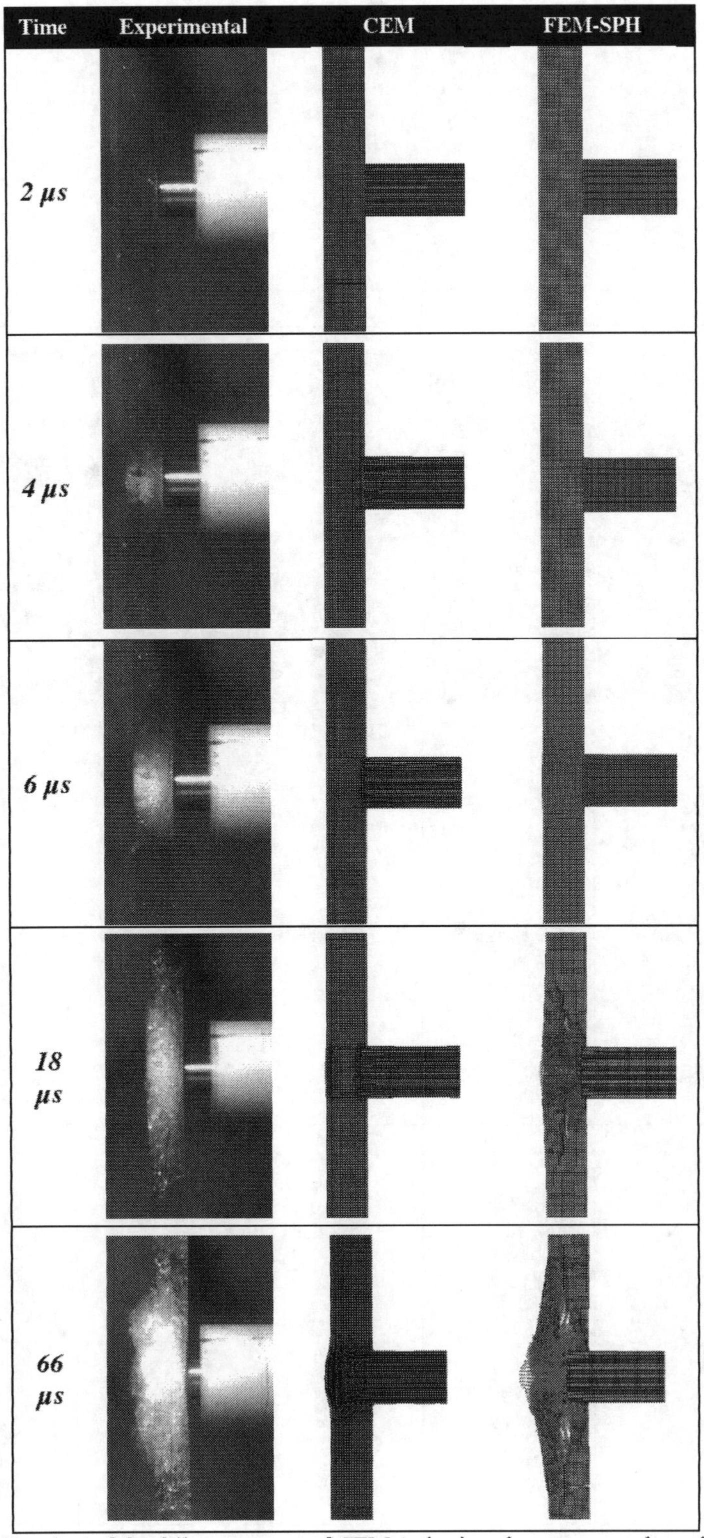

Figure 10. Mid-section view of the failure patterns of CEM (cohesive elements are coloured in green) and FEM-SPH models.

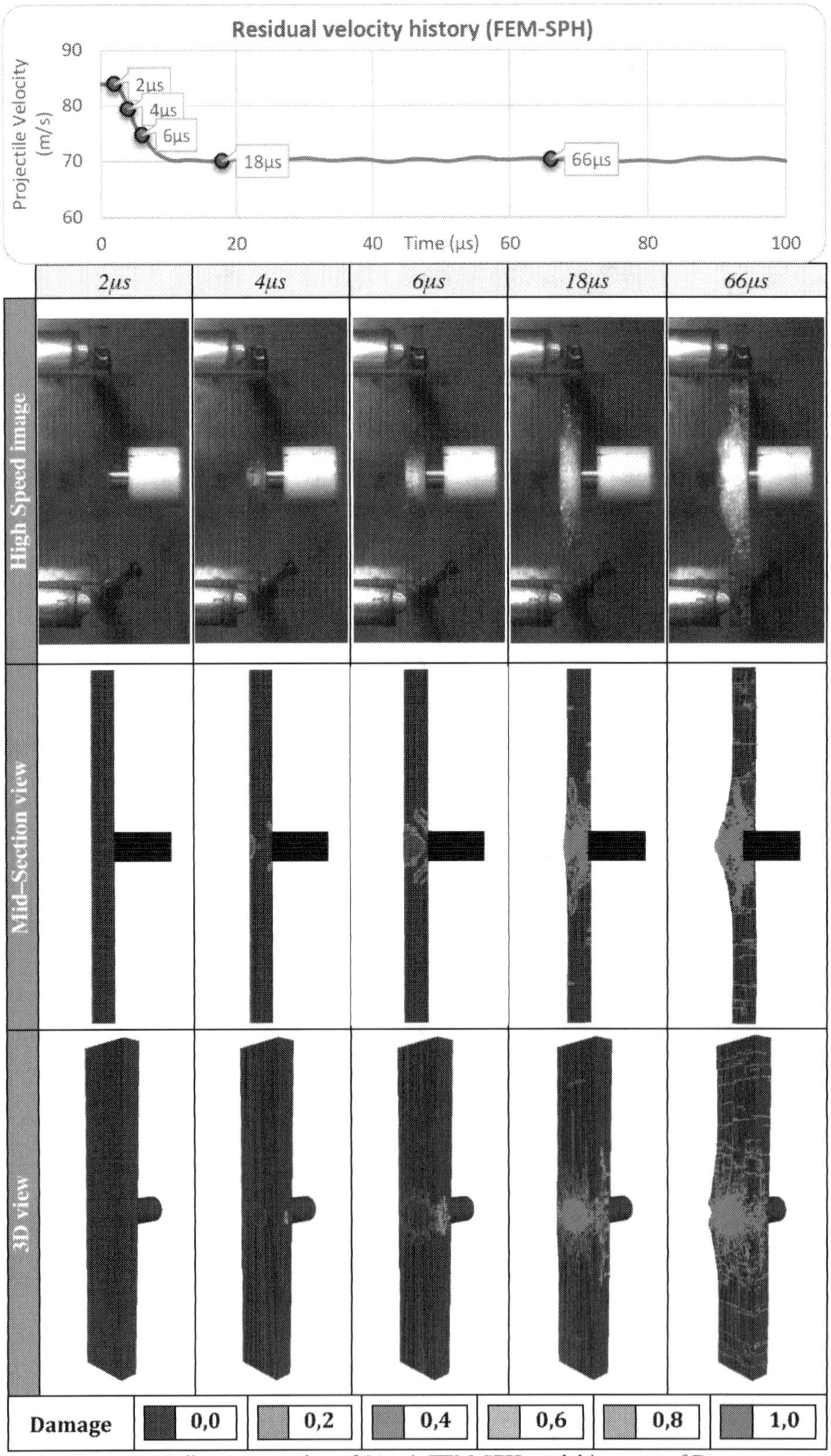

Figure 11. Failure progression of 84 m/s FEM-SPH model in terms of Damage.

5. Conclusions

In this paper two different simulation approaches (CEM and FEM-SPH) were employed in LS-Dyna and tested in the prediction of the mechanical and failure behaviour of Aluminosilicate glass. Two loading situations were considered: quasi-static and dynamic tests and impact loading conditions.

In the first case, in general both simulation techniques demonstrated to be valid and suitable to obtain reasonable results. In fact, under both tensile and compressive loading conditions, good results were obtained, and it was always possible to find solutions to all the characteristic problems that arose (like stress concentration). One interesting analogy between the two methods is that both techniques are able to predict the strain rate effect in compression, but not in tension. However, this fact is the result of two different and totally independent causes. In particular, CEM provided good strain rate sensitivity without inserting any direct correlation between the mechanical behaviour of the model and the loading rate. Conversely, in the other technique proper tuning of the strain rate sensitivity of the model needs to be performed. In fact, an advantage of CEM is in general the smaller number of parameters the model relies on, with respect to FEM-SPH method, that requires ad-hoc and time-consuming tuning processes.

As far as the ballistic impact simulations are concerned, in general terms the difference in the results between CEM and FEM-SPH increases. In fact, the reported results evidenced that FEM-SPH method behaves better than CEM practically always in terms of correctness of the results and capability to reproduce the crack aspect of the experimental specimen. However, this technique has a drawback which is the increased required computational time.

To conclude, none of the two simulation techniques demonstrated to be totally convenient with respect to the other, since both evidenced weaknesses and strengths. However, the simulation of impacts developed by the FEM-SPH approach and JH2 constitutive model provide better results, both in quantitative (residual speed) and qualitative terms (failure pattern).

References

[1] Wang Z, Ma D, Suo T, Li Y and Manes A 2021 *Int. J. Mech. Sci.* **196**
[2] Ma D, Esmaeili A, Manes A, Sbarufatti C, Jiménez-Suárez A, Giglio M and Hamouda A M 2020 *Mech. Mater.* **140** 103214
[3] Wang Z, Fu J and Manes A 2021 *Theor. Appl. Fract. Mech.* **111**
[4] Zhang W 2019 *Numerical Representation of Crack Propagation within the Framework of Finite Element Method Using Cohesive Zone Model* (University of Cincinnati)
[5] Wu X, Wang G, Li G, Han W, Sun S, Zhang S and Bi W 2020 *Symmetry (Basel).* **12** 1314
[6] Pandolfi A and Ortiz M 2002 *Eng. Comput.* **18** 148–59
[7] Jiang H and Meng D 2018 *Eng. Fract. Mech.* **199** 280–93
[8] Xu X, Chen S, Wang D and Zang M 2020 *Theor. Appl. Fract. Mech.* **108** 102660
[9] Liu M B and Liu G R 2010 *Arch. Comput. methods Eng.* **17** 25–76
[10] Liu G-R and Liu M B 2003 *Smoothed particle hydrodynamics: a meshfree particle method* (World scientific)
[11] Vignjevic R and Campbell J 2009 *Predictive modeling of dynamic processes* (Springer) pp 367–96
[12] Jankowiak T and Łodygowski T 2013 *Bull. Polish Acad. Sci. Tech. Sci.* 111–21
[13] Abouridouane M, Klocke F, Lung D and Adams O 2012 *CIRP Ann.* **61** 71–4
[14] Wang Z, Ren T, Suo T and Manes A 2021 *Thin-Walled Struct.* **165**
[15] Johnson G R and Holmquist T J 1994 *AIP conference proceedings* vol 309 (American Institute of Physics) pp 981–4
[16] Wang Z, Li Y, Ma D, Li Y, Suo T and Manes A 2022 *Submitt. to Mech. Mater.*

IOP Conf. Series: Materials Science and Engineering 1275 (2023) 012027 doi:10.1088/1757-899X/1275/1/012027

Experimental and postprocessing procedures for the response of sheet metals to high strain rate

G Mirone[1*], R Barbagallo[1], M M Tedesco[2], D De Caro[2] and M Ferrea[2]

[1] Department of Civil Engineering and Architecture, University of Catania, Via Santa Sofia 64, 95123, Catania, Italy
[2] C.R.F. S.C.p.A, Metals, Corso Settembrini 40, 10135, Turin, Italy

*Correspondence: gmirone@dii.uncit.it

Abstract. Metals subjected to high strain rates in Hopkinson bar testing remarkably increase their temperature, so that thermal and dynamic effects are always interleaved. Experimental procedures in such field are not clearly standardized and the most common methods for tensile testing and postprocessing cannot highlight some crucial aspects of the dynamic response of metals. When sheet metal specimens are tested instead of bulk specimens, the accurate derivation of the flow curves are further complicated due to intrinsic strain nonuniformities induced by the specimen geometry and to possible material anisotropy. The local strain peaks calculated by digital image correlation, together with the specimen elongation / shrinking evaluated on the deforming specimen by optical methods, deliver remarkably different estimates of the stress-strain curves and of the strain rate histories for the same given test. Such differences are discussed in this paper together with the assessment of the anisotropic response of the material at static, intermediate and high strain rates.

1. Introduction

Stress-strain characterization of sheet metals at high strains is still an open issue. The material characterization in the industrial sector typically relies on the classical dog bone specimens and on the technical specifications based on the engineering elongation-based approach. However, it is well known that this approach leads to great approximations after the necking onset, especially for ductile metals with early necking onset.

Many authors rely on FEM-based inverse approaches [1][2] and Virtual Fields Method [3][4]. This kind of approaches has many advantages, but the obtained flow curve can include some preliminary modeling choices. Direct approaches are logically simpler but have different difficulties to be taken into account. Other than the classical engineering approach, thanks to optical setups, it is possible to evaluate, with different approximations depending on the type of specimen and the specific used optical technique, the minimum cross-section of the specimen and, in turn, directly the true curve of the material. The latter is identical to the flow curve only until the necking inception. Then, many authors have worked on the extrapolation of the flow curve from the true section-based curve in the post-necking phase, starting from the well-known Bridgman method [5] for cylindrical specimens based on the optical evaluation of the curvature of the necked specimen. Mirone [6] proposed the MLR method, which can be used to transform true curves obtained with cylindrical or rectangular specimens into flow curves without the need for complex curvature evaluation. However, in the case of thick rectangular specimens, it is difficult to correctly evaluate the actual minimum cross-section, especially for anisotropic materials [7]. Indeed, if with isotropic materials it is possible to evaluate the thickness of the specimen assuming the constancy of the specimen's cross-section shape, with

Content from this work may be used under the terms of the Creative Commons Attribution 3.0 licence. Any further distribution of this work must maintain attribution to the author(s) and the title of the work, journal citation and DOI.
Published under licence by IOP Publishing Ltd

anisotropic materials it would be necessary a 3D DIC setup or trying to obtain useful information from post mortem specimen analysis [8]-[10]. Peirs et al. [11] used DIC to correct the experimentally obtained engineering stress–strain curves from high strain rate SHTB tensile tests on Ti6Al4V specimens. Abedini et al. [12] made a comparison between different constitutive characterization techniques for sheet metals based on virtual tensile experiments highlighting critical issues of common techniques. Mirone et al. [13] proposed a procedure able to transform the elongating gage length of flat specimens into their shrinking cross section, provided that the initial geometry of the specimen ensures a slenderness (length/width ratio) above 5 or 6 depending on the material.
The stress-strain characterization procedure gets more complicated if it is necessary to take into account also strain rate and temperature effects on the mechanical behavior of the material. Indeed, temperature and strain rate could have combined and correlated effects, which are difficult to correctly evaluate. In particular, in high strain rate tests, it is important to take into account a phenomenon that the authors and other researchers have studied in great detail, i.e., the spontaneous increase of the strain rate after the necking inception and its actual effect on the behavior of the material [14]-[19]. Indeed, the correlation between the actual strain rate undergoing in the material and the corresponding stress amplification is not straightforward, as also confirmed by Jacques and Rodríguez-Martínez [20].
In this work, an accurate analysis of the tensile behavior of the AL6061-T6 alloy at different strain rates has been made. Different experimental approaches have been considered, comparing the corresponding different results. Moreover, the material anisotropy has been analyzed and discussed.

2. Materials and methods

Thin sheet tensile specimens 2 mm thick were tested according to figure 1, with 4 mm width and 20 mm uniform section length between the fillets.

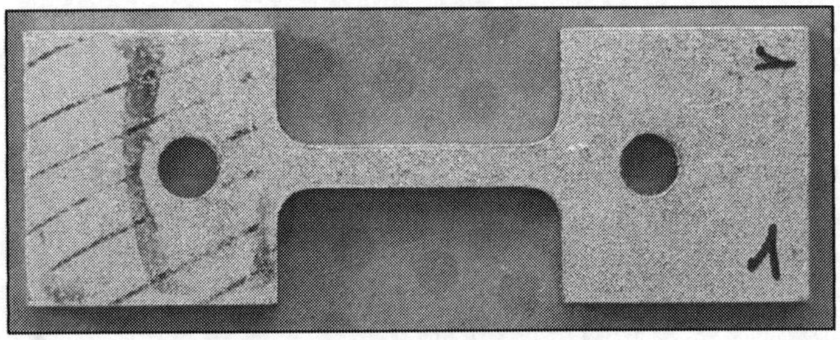

Figure 1. Undeformed specimen.

The experimental campaign includes static rates (S-series) and dynamic tests at intermediate rates of 1 and 10 s^{-1} (D01 and D10 series) and at high rates of 250 and 500 s^{-1} (D250 and D500 series).
Video acquisitions were taken for DIC measurements of evolving distances and strains over the specimen. 5000 fps at HD resolution were used for D01 and D10 tests, while 185000 fps at resolution of 256x104 were adopted for D250 and D500 tests.
The specimen width w of the neck section (initially equal to 5 mm) and the elongation of the gage length L (initially equal to 6 mm) are measured by DIC together with the axial strain ε_{xx} and transverse strain ε_{yy} in the midpoint of the neck section.
The stress-strain curves are calculated according to three procedures, respectively named as "elongation-based", "width-based" and "DIC peak-based".

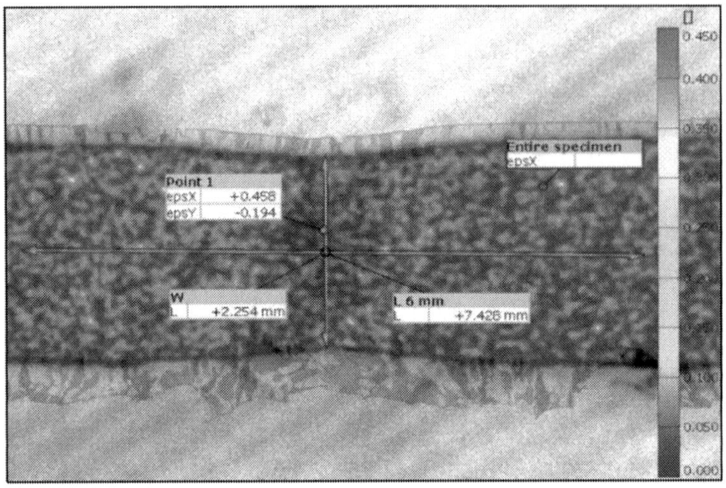

Figure 2. Example of DIC analysis.

Elongation-based true stress-strain curve

The true stress and true strain rate based on the elongation are defined as in equations (1), (2), and (3).

$$\varepsilon_{true_L} = ln\left(\frac{L}{L_0}\right) \tag{1}$$

$$\sigma_{true_L} = \frac{F}{A_0} e^{\varepsilon_{true_L}} \tag{2}$$

$$\dot{\varepsilon}_{true_L} = \frac{\partial \varepsilon_{true_L}}{\partial t} \tag{3}$$

After the onset of diffuse necking, equations (1) to (3) quit delivering the effective flow curve of the material.

Width-based true stress-strain curve

The classical derivation of true stress-strain data from engineering, elongation-based data is made according to equations (4), (5), and (6).

$$\varepsilon_{true_A} = ln\left(\frac{A_0}{A}\right) \tag{4}$$

$$\sigma_{true_A} = \frac{F}{A} \tag{5}$$

$$\dot{\varepsilon}_{true_A} = \frac{\partial \varepsilon_{true_A}}{\partial t} \tag{6}$$

Special care is taken for calculating the current resisting area A from the width w of the neck section, by making the assumption that the cross section shrinks homothetically, e.g., maintains the same shape and the same ratio between its edges all over each test.

So that the current area is expressed as in equation (7).

$$t_0/_{w_0} = t/_{w} \rightarrow A = w^2 \cdot t_0/_{w_0} \tag{7}$$

Local strain-based true stress-strain curve

True stress, strain and strain rate are also estimated by using the local DIC peak strain readings (at the midpoint of the neck section) for transforming engineering data into true data as in equations (8), (9), and (10), according to Peirs et al. [11] or Mirone et al. [10].

$$\varepsilon_{true_{local}} = \varepsilon_{xx_{DIC}} \tag{8}$$

$$\sigma_{true_{local}} = \frac{F}{A_0} e^{\varepsilon_{xx_{DIC}}} \tag{9}$$

$$\dot{\varepsilon}_{true_{local}} = \frac{\partial \varepsilon_{true_A}}{\partial t} \tag{10}$$

3. Results and discussion

Static tests

The true stress-strain curves from static tests according to the three different approaches are plotted in Figure 3.

The elongation-based curve, (nominal gage length 6 mm) cannot reflect the localizing deformation after the necking onset and quits delivering useful data at strains beyond 0.1.

The width-based true curve extends the range of the elongation-based curve is accurate up to plastic strains close to 0.3, then it decreases as ductile damage and/or localized necking decrease the load carrying attitude of the neck section, making the true curve not representative anymore of the hardening response.

The DIC peak-based true curve further extends the range of the area-based curve up to strains in excess of 0.4. In principle this curve reflects the strain nonuniformity within the neck section, although in an approximate way because, at late stages, it can vary depending on the choice of the material point where the local DIC measurements are taken.

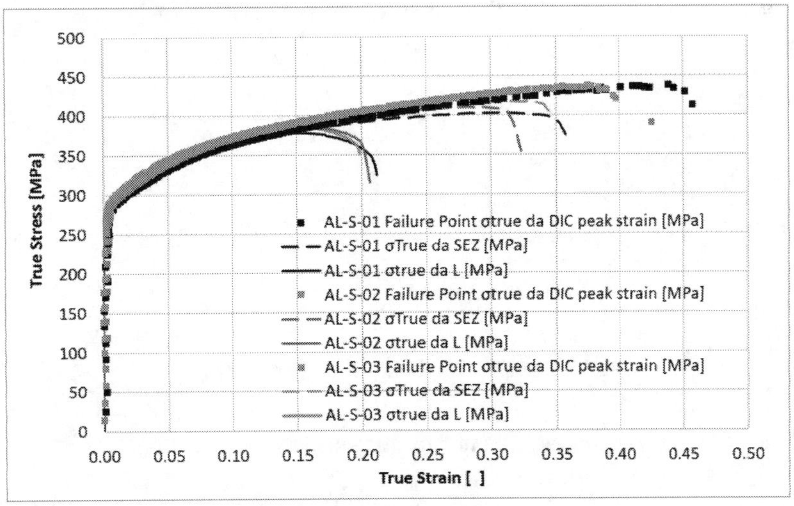

Figure 3. True stress-strain curves from the static tests.

Dynamic tests

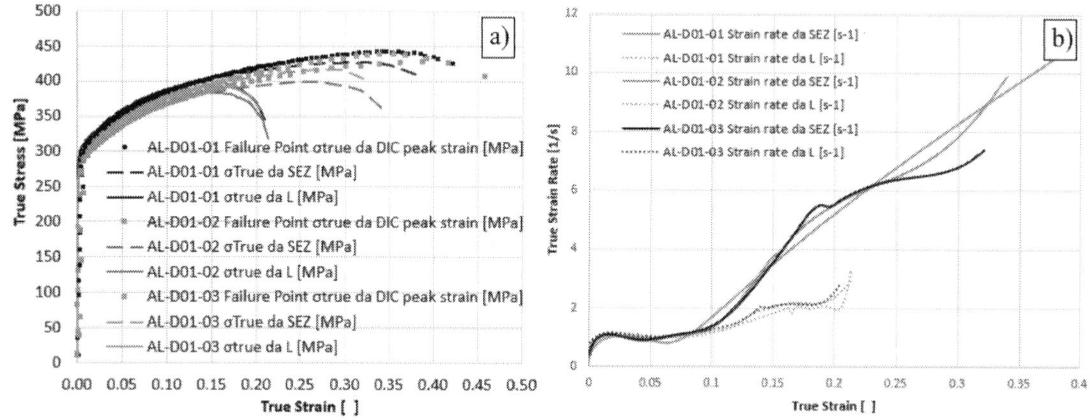

Figure 4. True stress-strain curves (a) and true strain rate vs strain curves (b) from the dynamic experiments at 1 s^{-1}.

The D01 tests in Figure 4 nearly reflect the same outcome of static tests, although a larger scattering occurs between the width-based curves.

Figure 5 shows the results of the D10 tests series; the D10-02 specimen exhibited more undulated edges parallel to the specimen axis (e.g., variability of w_{max} and w_{min} along the axis remarkably greater than other specimens), constituting micro-notches capable of slightly anticipating the onset of strain localization and failure.

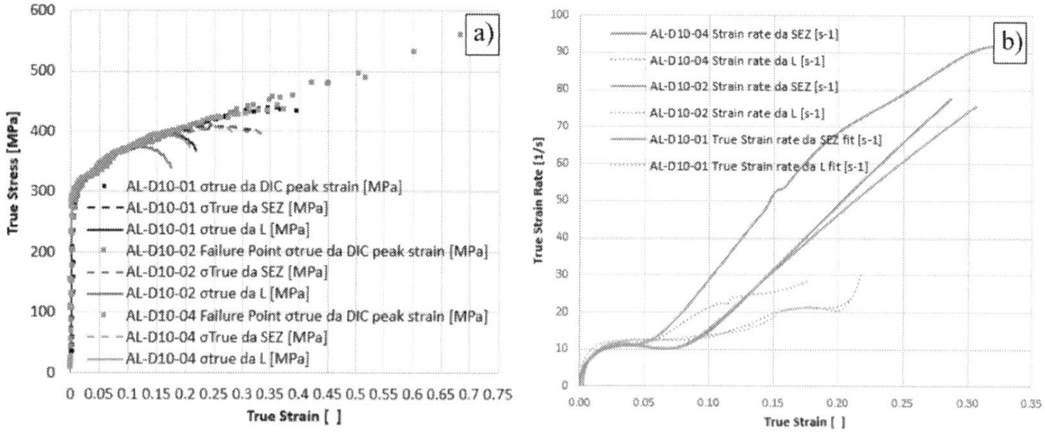

Figure 5. True stress-strain curves (a) and true strain rate vs strain curves (b) from the dynamic experiments at 10 s^{-1}.

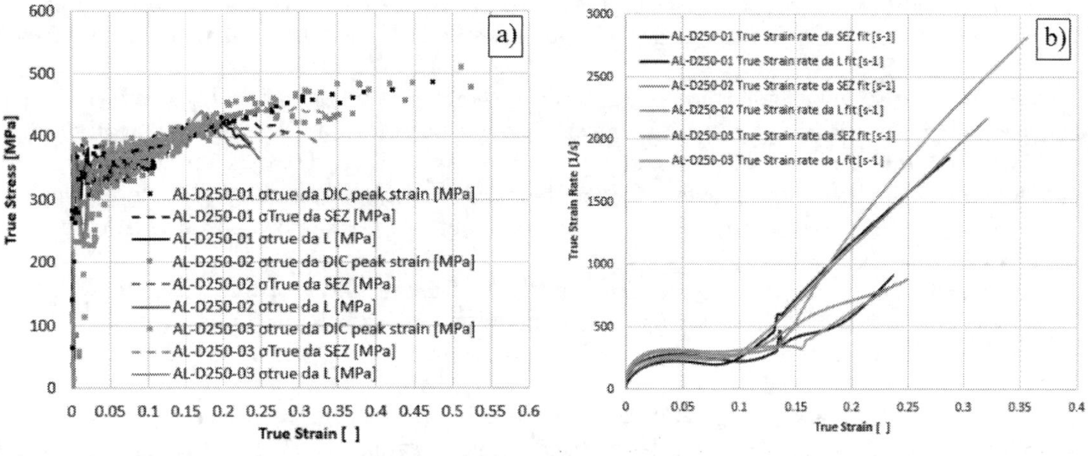

Figure 6. True stress-strain curves (a) and true strain rate vs strain curves (b) from the dynamic experiments at 250 s^{-1}.

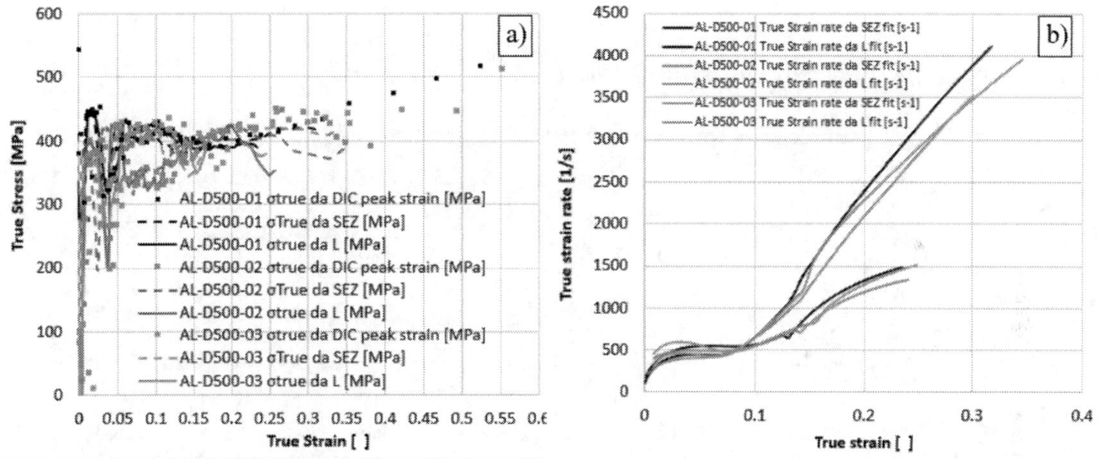

Figure 7. True stress-strain curves (a) and true strain rate vs strain curves (b) from the dynamic experiments at 500 s^{-1}.

Dynamic tests at the Hopkinson bar exhibited more pronounced oscillations than the slower dynamic tests, especially at early plastic strains within 0.15; however a reasonable overall repeatability is maintained.

Table 1 reports the relevant strains averaged for each test series.

Table 1. Necking and fracture strains from experiments.

	$\varepsilon_{true_{neck}}$	$\varepsilon_{true_{L}}{}_{fracture}$	$\varepsilon_{true_{A}}{}_{fracture}$	$\varepsilon_{true_{local}}{}_{fracture}$
S (avg. 3 tests)	0.10	0.21	0.29	0.43
D01 (avg. 3 tests)	0.10	0.21	0.35	0.42
D10 (avg. 3 tests)	0.10	0.20	0.31	0.49
D250 (avg. 3 tests)	0.12	0.24	0.32	0.50
D500 (avg. 3 tests)	0.11	0.25	0.33	0.52

The strain rate histories of dynamic tests exhibit the steep increase after the onset of the diffused necking, as already shown and discussed in [14]-[17]; the elongation-based approach almost misses such phenomena, especially when the gage length adopted is longer than 3-4 times the specimen width.

The true curves based on the elongation of a 6 mm gage length are only valid until the diffuse necking takes place, at plastic strains around 0.1; the specific elongation at fracture is close to 0.2 at all rates.

The width-based true curves steadily increase up to plastic strains around 0.3, then the ductile damage and localized necking rapidly take over affecting the macroscopic specimen response and making the last segments of such curves not representative anymore of the material hardening, up to failure strains around 0.35.

The DIC peak-based true curves fully confirm the trends of the width-based true curves up to strains of 0.3, then continue steadily increasing up to local failure strains in the range 0.4 - 0.5.

Recalling that the DIC peak true curves and the width-based true curves respectively tend to overestimate and to underestimate the effective true curve (that is load/resisting area vs. logarithm of initial/current resisting areas), the good agreement found between such series of curves demonstrates that both them are accurate (relative scattering band within a few percents) up to strain levels around 0.3.

Beyond plastic strains of 0.3 the DIC peak-based curves deliver an upper bound to the effective true curves, although no reference is available for comparison and only indirect validation (e.g., finite elements simulations) can be ran for assessing their accuracy in such strain range

Comparison of experimental results at all strain rates

The area-based stress-strain curves form various test series are then compared in Figure 8.

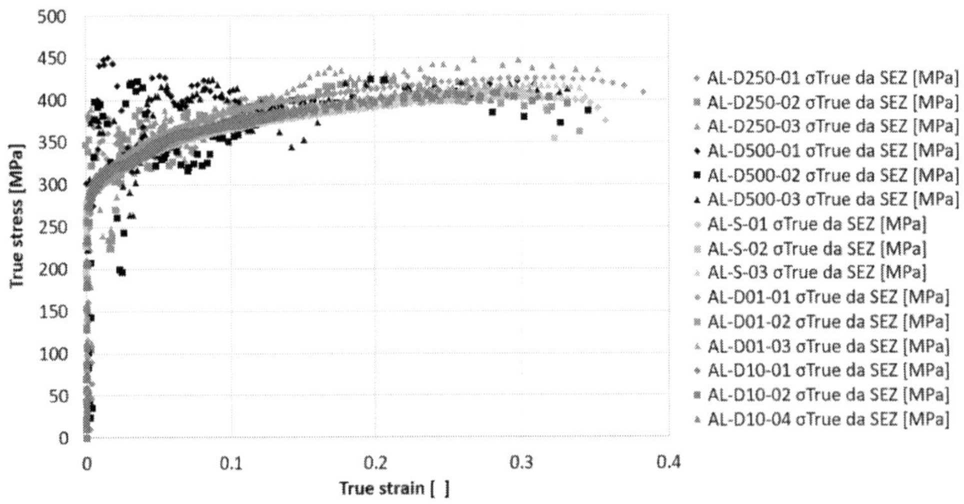

Figure 8. Comparison of the width-based True curves from all static and dynamic tests.

The true curves after the necking onset would require a postnecking correction for continuing to express the material flow curves, however their comparison at different rates is actually suitable for comparatively assessing the eventual presence of strain rate effects up to failure.

Figure 8 shows that the dynamic true curves at 1 s^{-1} and 10 s^{-1} are identical to the static ones, while those at 250 s^{-1} and 500 s^{-1} are quite overlapped to the same curves despite exhibit some oscillations at low plastic strains, due to late specimen equilibrium in Hopkinson bar testing.

Initially all specimens are at room temperature, then the fast dissipation of plastic work causes their progressive heating up to temperatures estimated around 100 C at incipient failure.

IOP Conf. Series: Materials Science and Engineering 1275 (2023) 012027 doi:10.1088/1757-899X/1275/1/012027

Therefore, the agreement of all curves at low-intermediate plastic strains, where eventual thermal effects are prevented by the room temperature, proves that the AL6061 at hand is insensitive to the strain rate at such strain levels.

The same true curves also agree to each other at large plastic strains, where the temperature increases up to around 100 C, this implies that either one of the two following possibilities must be true:

- the strain rate sensitivity of the material, initially null, progressively develops with the plastic strain in a fashion that perfectly compensates the thermal softening of the material in the temperature range 20 – 100 C°.
- the strain rate sensitivity, certainly null at low-intermediate strains, also remains null up to failure; then the agreement of static and dynamic true curves at late strains means that the AL6061 sheet is also insensitive to the temperature.

Although from a mathematical viewpoint both options are possible, the first one is rather unlikely because it implies that both the strain rate sensitivity and the thermal sensitivity of the AL6061 depend on the plastic strain in such way that they perfectly compensate to each other at all strains, from first yield to failure.

Therefore, the most reasonable outcome of such analysis is that the agreement of static and dynamic true stress-strain curves implies that the plastic response of the AL6061 is almost insensitive to both strain rate and temperature.

Static tests at high temperature were beyond the scopes of this work but are scheduled for future developments in order to assess which one of the above options really applies.

3.1. Anisotropy check

As already discussed, the width-based (or area-based) true curves are obtained assuming that the shape of the cross-section of the specimen deforms homothetically during the entire test.

In order to check this point, the width-based plastic strains are plotted against the elongation-based ones in Figure 9, for all static and dynamic tests.

Before the necking onset, the conservation of the prismatic volume of the specimen corresponding to the gage length ($L_0 \cdot w_0 \cdot t_0 = L \cdot w \cdot t$), coupled to the isotropy of the specimen under uniaxial tension ($\varepsilon_{yy} = \varepsilon_{zz} = -{}^{\varepsilon_{xx}}/_{2}$, homothetic shrinking of the cross section), would make the width-based and the elongation-based estimates of the equivalent plastic strains identical to each other ($2Ln\left(\frac{w_0}{w}\right) = Ln\left(\frac{L}{L_0}\right)$), implying that all the curves in Figure 9 should be overlapped to the 45° red dashed line.

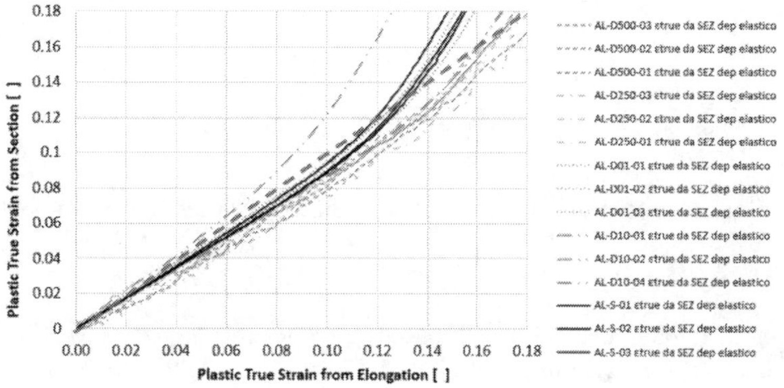

Figure 9. Plastic strain obtained from section vs plastic strain obtained from elongation for all tests.

Instead, Figure 9 shows that $2Ln(w_0 w) < Ln\left(\frac{L}{L_0}\right)$, evidencing a certain anisotropy of the AL6061 alloy at hand. Therefore, the conservation of the above specimen volume yields:

$$|\varepsilon_{zz}| = |\varepsilon_{xx}| - |\varepsilon_{yy}| > |\varepsilon_{yy}|; \\ Ln\left({}^{t_0}/_{t}\right) > Ln\left({}^{w_0}/_{w}\right); \tag{11}$$

Then the thickness, and in turn the effective resisting area, are both smaller than those estimated by equation (7), Then the effective true stress and true strain are slightly greater than those estimated by the specimen width according to equations (8) and (9).
However, the substantial identity discussed before between the width-based true curves and the DIC peak-based ones, ensures that neglecting the anisotropy introduced an error in the true curves which is below any detectable quantity, up to large postnecking plastic strains around 0.3.
The Lankford coefficient R according to equation (12) is then plotted in Figure 10 against the elongation-based plastic strain, where ε_{xx} and ε_{yy} are obtained by removing the elastic strains from the total strains respectively based on L and w,.

$$R_{xy} = \frac{\varepsilon_{yy}}{\varepsilon_{zz}} \tag{12}$$

The magnification in Figure 10b highlights a nearly constant R_{xy} after the elastic-plastic transition is completed and before the onset of diffused necking (strains between 0.05 and 0.10).
After the necking onset, the curves in Figure 10 lose their validity because the specimen volume within the gage length misses its prismatic shape and no conservation of any finite prismatic volume can be anymore enforced.

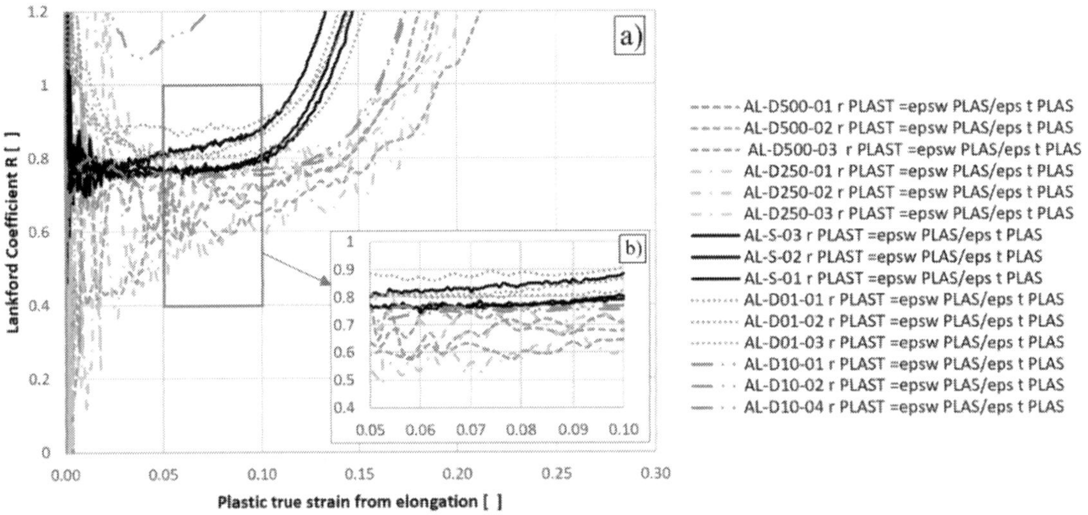

Figure 10. Lankford coefficient vs plastic true strain from elongation from all tests (a) and magnification (b)

The anisotropy found implies that the width-based cross section (equations (8) and (9)) is slightly overestimated, hence the corresponding width-based true curve is slightly underestimated.
Therefore if the area-based true curve would be calculated by taking into account for the above anisotropy (e.g. thickness derived from $Ln\left({}^{t_0}/_{t}\right) = Ln\left({}^{L}/_{L_0}\right) - Ln\left({}^{w_0}/_{w}\right)$ instead of equation (7)) , then the already good agreement with the DIC peak-based curves should further improve and also the length of the former curve should become closer to that of the latter one (e.g., the area-based failure strain should be closer to the DIC peak-based one).

4. Conclusions

An AL6061 sheet metal has been characterized by tensile tests at static, intermediate and high strain rates. Three different methods are adopted for deriving the true curves, demonstrating that:
- the length-based ones quit being useful at the onset of diffuse necking (plastic strain 0.1);
- both the width-based and the DIC peak-based curves deliver good estimates of the effective true stress up to plastic strain around 0.3 (however a post-necking correction would be necessary for converting such true stress into a flow stress over the strain range between 0.1 and 0.3);
- the DIC peak-based true curve delivers an upper bound of the effective true stress at strains higher than 0.3:

The true curves at all strain rates resulted to be almost identical to each other, at both low strains (where no thermal effects can exist because of room temperature) and high strains (where the temperature increases up to 100 C and thermal softening is eventually possible): this suggested that the AL6061 tested does not remarkably exhibit strain rate sensitivity nor thermal softening.
Length-based, width-based and DIC peak-based plastic strains at failure resulted to be almost independent of the strain rate and respectively close to 0.1, 0.3 and 0.45.
The comparison of width-based and length-based strains also evidenced the anisotropy of the alloy at hand, delivering a Lankford coefficient R_{xy} around 0.8 between first yield and diffused necking.

References

[1] Cortis G, Nalli F, Sasso M, Cortese L and Mancini E 2022 Effects of Temperature and Strain Rate on the Ductility of an API X65 Grade Steel. *Appl. Sci.*, **12**, 2444.

[2] Joun M, Eom J G and Lee M C 2008 A New Method for Acquiring True Stress–Strain Curves over a Large Range of Strains Using a Tensile Test and Finite Element Method. *Mech. Mater.* **40**, 586–593.

[3] Lattanzi A, Barlat F, Pierron F, Marek A and Rossi, M 2020 Inverse Identification Strategies for the Characterization of Transformation-Based Anisotropic Plasticity Models with the Non-Linear VFM. *Int. J. Mech. Sci.*, **173**, 105422.

[4] Rossi M, Lattanzi A and Barlat F 2018 A General Linear Method to Evaluate the Hardening Behaviour of Metals at Large Strain with Full-Field Measurements. *Strain*, **54**, e12265.

[5] Bridgman P W 1952 *Studies in Large Plastic Flow and Fracture: With Special Emphasis on the Effects of Hydrostatic Pressure* (Harvard University Press: Cambridge, MA, USA).

[6] Mirone G 2004 Approximate Model of the Necking Behaviour and Application to the Void Growth Prediction. *Int. J. Damage Mech.*, **13**, 241–261.

[7] Habraken A M, Aksen T A, Alves J L, Amaral R L, Betaieb E, Chandola N, ... and Vincze G 2022. Analysis of ESAFORM 2021 cup drawing benchmark of an Al alloy, critical factors for accuracy and efficiency of FE simulations. *Int. J. Mater. Form.*, **15**, 1-96.

[8] Larour P, Freudenthaler J and Weissböck T 2017 Reduction of cross section area at fracture in tensile test: measurement and applications for flat sheet steels. *J. Phys. Conf. Ser.* **896**, 1, 012073.

[9] Larour P, Wagner L, Felbinger A and Angeli J 2019 Local fracture strain measurement in AHSS uniaxial flat tensile tests considering specimen geometry and fracture morphology. *IOP Conf. Ser.: Mater. Sci. Eng.* **651**, 1, 012016.

[10] Mirone G, Barbagallo R, Tedesco M M, De Caro D and Ferrea M 2022 Extended Stress–Strain Characterization of Automotive Steels at Dynamic Rates. *Metals*, **12**, 960.

[11] Peirs J, Verleysen P, Van Paepegem W and Degrieck J 2011 Determining the stress–strain behaviour at large strains from high strain rate tensile and shear experiments. *Int. J. Impact Eng.*, **38**, 406-415.

[12] Abedini A, Narayanan A and Butcher C 2022 An Investigation into the Characterization of the Hardening Response of Sheet Metals Using Tensile and Shear Tests with Surface Strain Measurement. *Forces Mech.*, **7**, 100090.

[13] Mirone G, Verleysen P, and Barbagallo R 2019 Tensile Testing of Metals: Relationship between Macroscopic Engineering Data and Hardening Variables at the Semi-Local Scale. *Int. J. Mech. Sci.*, **150**, 154–167.

[14] Mirone G, Barbagallo R, Corallo D and Di Bella S 2016 Static and Dynamic Response of Titanium Alloy Produced by Electron Beam Melting. *Procedia Struct. Integr.*, **2**, 2355–2366.

[15] Mirone G, Barbagallo R, Giudice F and Di Bella S 2020 Analysis and Modelling of Tensile and Torsional Behaviour at Different Strain Rates of Ti6Al4V Alloy Additive Manufactured by Electron Beam Melting (EBM). *Mater. Sci. Eng.* A, **793**, 139916.

[16] Mirone G and Barbagallo R 2021 How Sensitivity of Metals to Strain, Strain Rate and Temperature Affects Necking Onset and Hardening in Dynamic Tests. *Int. J. Mech. Sci.*, **195**, 106249.

[17] Mirone G 2013 The Dynamic Effect of Necking in Hopkinson Bar Tension Tests. *Mech. Mater.*, **58**, 84-96.

[18] Zhang L, Pellegrino A and Petrinic N 2021 Dynamic Necking of a near α Titanium Alloy at High Strain Rates: Experiments and Modelling. *Def. Technol.*, **17**, 1126–1134.

[19] Zhang L, Pellegrino A, Townsend D and Petrinic N 2020 Strain Rate and Temperature Dependent Strain Localization of a near α Titanium Alloy. *Int. J. Impact Eng.*, **145**, 103676.

[20] Jacques N and Rodríguez-Martínez J A 2021 Influence on Strain-Rate History Effects on the Development of Necking Instabilities under Dynamic Loading Conditions. *Int. J. Solids Struct.*, **230–231**, 111152.

Computational modeling of confined blast waves with focus on interaction with structures

E Shehu, L Lomazzi, M Giglio and A Manes*

Politecnico di Milano, Department of Mechanical Engineering, Via La Masa n.1, Milan 20156, Italy

*Email: andrea.manes@polimi.it

Abstract. Blast loading is a critical extreme loading condition for most engineering structures. Modeling such scenarios is challenging due to the intrinsic non-linearities. Recent numerical methods can capture the physics governing blast waves and their interaction with structures in a more accurate way than established empirical methods. This article is intended as a proof of concept that state-of-the-art CFD and FE software can be combined to set up high-fidelity uncoupled simulations. Computational fluid dynamics is exploited in this work to map the pressure field developed during a detonation reaction in a Eulerian domain. Then, such pressure time history is applied in the finite element framework to perform Lagrangian simulations. The methodology is used to compare the structural response of blast-loaded plates in fully confined environments to that in free-field scenarios. It turns out that confined blast waves are more severe than blast waves in unconfined scenarios, mainly due to the multiple reflections and residual quasi-static pressure. Moreover, on the one hand, the quasi-static pressure appears to contribute to further increasing the plate out-of-plane deflection, while on the other hand, it prevents plates from undergoing reverse buckling or oscillations around the initial equilibrium configuration. Experimental tests are required to provide further evidence about the latter contribution.

1. Introduction

The failure of safety-critical components in oil & gas or chemical plant facilities, bad handling of dangerous materials and terrorist attacks can cause blast waves. Such high-pressure waves represent critical loading conditions to most engineering structures. Thus, several methods have been developed and proposed in the literature to evaluate the response of blast loaded structures, with the purpose of increasing their blast resistance capability. Historically, empirical methods were developed, based on data from experimental tests, both to characterize the unconfined propagation of blast waves and to predict the permanent response of blast loaded structures [1][2][3][4]. To date, the most employed empirical model has been the model proposed by Kingery and Bulmash [5], which also laid the foundations of the CONWEP approach [6]. Although the latter approach is not valid for near-field explosions, it has been successfully implemented in many commercial finite element solvers, such as in Abaqus [7]. High fidelity numerical simulations have recently been employed to boost the performance of predictive tools in the field of blast protection engineering. Specifically, such methods have been proved to provide accurate predictions even in case complex environments and structures are considered.

Content from this work may be used under the terms of the Creative Commons Attribution 3.0 licence. Any further distribution of this work must maintain attribution to the author(s) and the title of the work, journal citation and DOI.
Published under licence by IOP Publishing Ltd

For instance, the description of blast waves propagating in confined spaces has progressed thanks to numerical frameworks, even though the response of blast loaded structures in such complex loading conditions still has to be fully characterized and understood.

Hence, this work presents an uncoupled numerical framework to compare the response of blast loaded plates in confined scenarios to that of structures in unconfined conditions. A dedicated methodology is developed to this purpose using Abaqus, since the capabilities of built-in modules in commercial finite element (FE) solvers are strongly limited when dealing with blast loads. The pressure loading acting on the structure is mapped through computational fluid dynamics (CFD) simulations. The mapped pressure is then applied in a pure Lagrangian finite element (FE) simulation to characterize the transient structural response. Results of simulations in unconfined scenarios are compared to those obtained using the built-in CONWEP module in Abaqus to validate the proposed methodology.

This work is organized as follows. The Section *Methodology* introduces the framework used to simulate blast loaded structures. The framework is then applied to describe a case study in Section *Case study*, and results are described in Section *Results*. Conclusions are drawn out from this work in the Section *Conclusions*.

2. Methodology

Over the last decades, the constantly increasing computational power led to fast development and employment of software packages to numerically simulate scenarios described by the interaction of blast waves and deformable structures. To date, the uncoupled Euleria Lagrangian (UEL) approach represents the most efficient numerical method in the blast protection engineering field, even though it cannot describe fluid-structure interaction (FSI) effects. According to the uncoupled scheme, the pressure field p(x,y,t) describing the blast pressure exerted on the structure is obtained either by empirical models or by CFD simulations including the fluid domain only. Then, the pressure distribution along the fluid boundary is applied to the structure in a Lagrangian analysis. The procedure described above is schematically represented in Figure 1 and Figure 2.

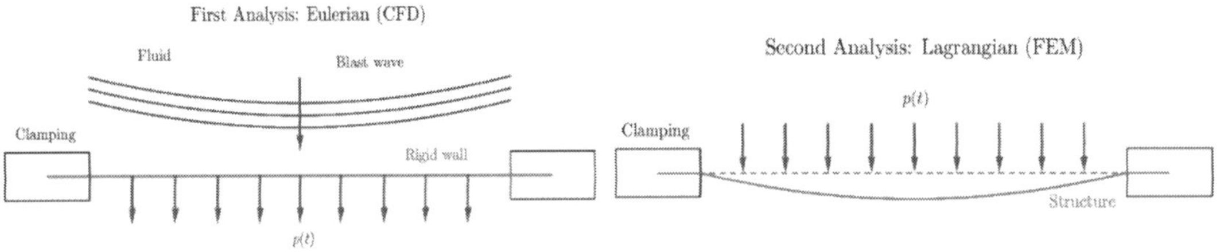

Figure 1. Uncoupled approach steps.

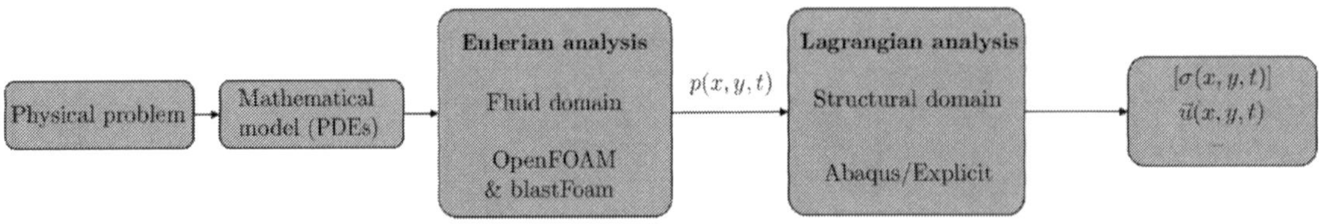

Figure 2. Required procedures to develop an Uncoupled Euler Lagrangian framework.

In this work, CFD simulations are performed in blastFoam [8], a new library dedicated to the computation of compressible flow with application to high-explosive detonation phenomena. Specifically, blastFoam is an entirely open-source toolbox based on OpenFOAM. The selected CFD framework comes with several built-in equations of state for detonation dynamics and implements numerical explicit schemes dedicated to explosive phenomena. Using this CFD software package allows accurately mapping the load history exerted by blast waves on structures. That is, CFD simulations are used to overcome the strong limitations of built-in modules of commercial FE solvers, which fail to properly characterize blast wave loading in complex scenarios.

FE simulations are performed in Abaqus. This choice is mainly made because Abaqus allows implementing user Fortran subroutines to customize simulations. This makes it possible to define user-defined material models and to bridge CFD and FE analyses. Lagrangian simulations are carried out through the Abaqus\Explicit solver, due to the dynamic nature of the problem. Confined and unconfined blast waves are differentiated directly in the CFD environment by imposing proper boundary conditions for the fluid volume. As a result, two different pressure fields p(x,y,t) are then applied as distributed loads in the FE simulations to the same steel plate.

3. Case study

Numerical simulations following the Uncoupled Eulerian Lagrangian scheme are set up to assess the effect of confinement on the structural response of steel plates. The same framework is applied to described both confined and unconfined blast events, which are differentiated through different boundary conditions. A spherical TNT charge is placed in the origin of a Cartesian reference frame centered in a cubic domain made of air (red sphere in Figure 3). In case of unconfined simulations, one rigid wall is modeled with a *slip* boundary condition, i.e., the component of velocity orthogonal to the wall is set to $v_y = \boldsymbol{v} \cdot \boldsymbol{j} = 0$ (grey faces in Figure 3). This assumption allows describing the reflection of blast waves. Other walls are set to *zeroGradient* boundary condition, that extrapolates field quantities to the patch from the nearest cell value, i.e., it allows the free propagation of blast wave out of the domain when boundaries are reached. Instead, in case of entirely confined analyses, all faces of the cube are assigned *slip* boundary conditions on the velocity component perpendicular to the wall. The symmetries of the problem are exploited to make numerical simulations run faster by modeling one eighth of the volume, as highlighted in orange in Figure 3.

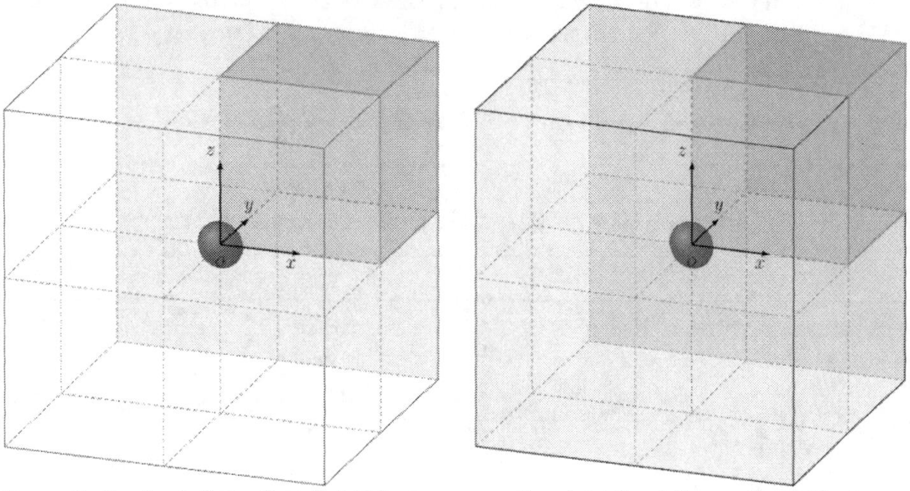

Figure 3. On the left the free-field blast case, on the right the fully confined blast case

The mathematical model employed to describe the detonation phenomenon of the explosive material is represented by the set of compressible Euler equations with reaction evolution [9]. Conservation of mass, momentum and energy equations are reported in Eq. 1

$$\begin{cases} \frac{\partial \rho}{\partial t} + \nabla(\rho \boldsymbol{u}) = 0 \\ \frac{\partial(\rho \boldsymbol{u})}{\partial t} + \nabla(\rho \boldsymbol{u} \otimes \boldsymbol{u} + p\boldsymbol{I}) = 0 \\ \frac{\partial(\rho E)}{\partial t} + \nabla((\rho E + p)\boldsymbol{u}) = 0 \end{cases} \tag{1}$$

Euler equations are not a complete set of partial differential equations. Indeed, there are six unknowns p, ρ, v_x, v_y, v_z and E, but only five equations. One additional relationship should be provided to close the system. Hence, an equation of state (EOS) is added for each phase present in the domain. In the simulations carried out in this work, two phases are present in the system: air and explosive.

Underacted energetic materials can be modelled with several relationships that describe the behavior of the material subjected to high pressure, before the occurrence of the detonation reaction. The third order Birch-Murnaghan equation [10] is adopted to model the explosive charge sphere as reported in Eq.2. Data for TNT explosive are retrieved from [8][10] and reported in Table 1.

$$p = p_{ref} + \frac{3K_0}{2K_0'}\left[\left(\frac{\rho}{\rho_0}\right)^{7/3} - \left(\frac{\rho}{\rho_0}\right)^{5/3}\right] \cdot \left\{1 + 0.75(K_0' - 4)\left[\left(\frac{\rho}{\rho_0}\right)^{2/3} - 1\right]\right\} \tag{2}$$

Table 1. Numerical coefficient of TNT for third order Birch-Murnaghan EOS [8][11].

	ρ_0 (kg/m^3)	K_0 (GPa)	K_0'(-)	p_{ref} (kPa)	γ (-)
TNT	1630	9.6	6.6	101.325	0.35

Where K_0 is the bulk modulus of the TNT explosive, K_0' is the derivative of the bulk modulus w.r.t pressure variable p and γ is the Gruneisen coefficient.

Instead, detonation products of energetic materials are modelled with the Jones-Wilkins-Lee (JWL) equation [11] reported in Eq. 3. Parameters for TNT are reported in Table 2[8][12].

$$p = A\left(1 - \frac{\omega}{R_1 V}\right)\exp(-R_1 V) + B\left(1 - \frac{\omega}{R_2 V}\right)\exp(-R_2 V) + \frac{e\omega}{V} \tag{3}$$

Table 2. Numerical coefficient of TNT for JWL EOS [8][12].

	ρ_0 (kg/m^3)	A (GPa)	B(GPa)	R_1 (-)	R_2 (-)	ω (-)	e (MJ/kg)
TNT	1630	371.21	3.231	4.15	0.95	0.30	4.294

Where A, B, R_1, R_2 ω are material constant obtained from experimental data and e is defined as the internal energy per unit mass.

Air is modelled as a perfect gas that fills the computational domain in all the volume except for the part where the explosive charge is located. Equation of state is reported in Eq.4.

$$p = (\gamma - 1)\rho e \tag{4}$$

Table 3. Numerical coefficient of air modelled as a perfect gas [8].

	ρ (kg/m³)	γ (−)	e (MJ/kg)
Air	1.225	1.4	0.224

Parameter γ is the specific heat ratio in Eq. 4.

A blast wave is characterized by sharp pressure gradient caused by the travelling of the discontinuity. Thus, discretization in both time and space plays a crucial role for the accuracy of the numerical solution. Ideally, all numerical simulations should be mesh-independent to describe the nature of the modelled phenomenon. According to the mesh sensitivity analysis performed in [13], hexahedral finite volumes with characteristic length L = 1.25 mm are adopted in the CFD analyses. That is, the mesh size determined in [13] is used in this case study, since that value already guaranteed convergence in more severe scenarios.

Instead, in regions of the domain where no sharp discontinuities are present there is no need to keep such a refined mesh. Hence, to improve computational efficiency, an adaptive mesh refinement (AMR) technique is introduced in the framework. A *scaledDelta* technique is adopted [8]. When the *scaledDelta* difference between pressures of two adjacent cells is lower than 5%, then the two cells are merged into a single cell and fluxes are recomputed.

In order to further simplify the analysis and reduce the computational burden, a built-in mapping algorithm is exploited. Such algorithm allows simulating the first part of the detonation of the explosive in a 1D mesh and mapping the pressure field into a 3D mesh before any interactions that could break the spherical symmetry occur. A graphical representation of the mapping technique is reported in Figure 4.

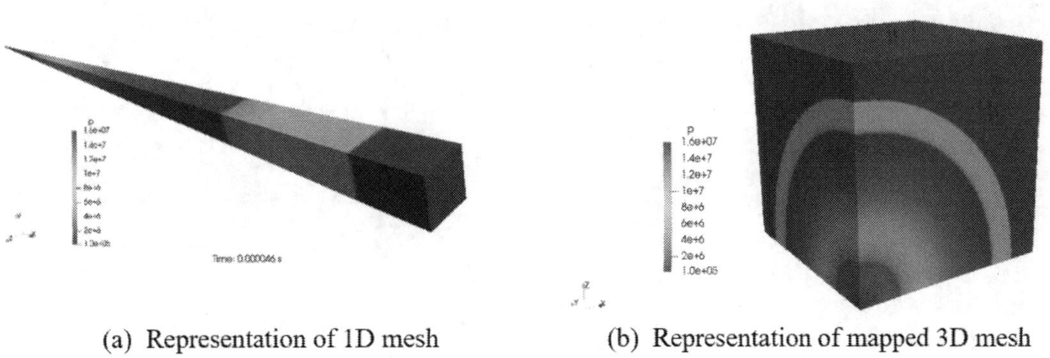

(a) Representation of 1D mesh (b) Representation of mapped 3D mesh

Figure 4. Graphical representation of the mapping technique.

A charge of 100 g of TNT at a stand-off distance of L = 371 mm is considered in the analyses. The length of the cube edge is set twice the stand-off distance to obtain a domain of the simulation with planar symmetries with respect to axes x, y and z. The choice of these parameters allows modeling one eighth of the volume as reported in Figure 3. Hence, the considered scaled distance is Z = 0.4 m/kg$^{1/3}$. Scaled distance is defined as the ratio between the stand-off distance and the cubic root of the explosive mass. According to this scaling law, blast phenomena and explosions characterized by the same scaled distance Z are described by the same properties.

A grid of virtual pressure gauges is located at the boundary wall where slip condition is applied. The pressure histories recorded from the pressure gauge located at (0, L, 0) in the unconfined and in the confined simulations are reported in Figure 5 and Figure 6, respectively.

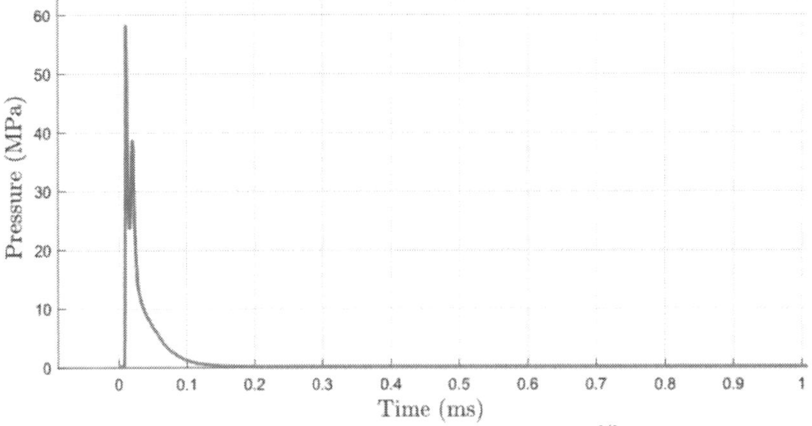

Figure 5. Unconfined blast at Z = 0.4 m/kg$^{1/3}$

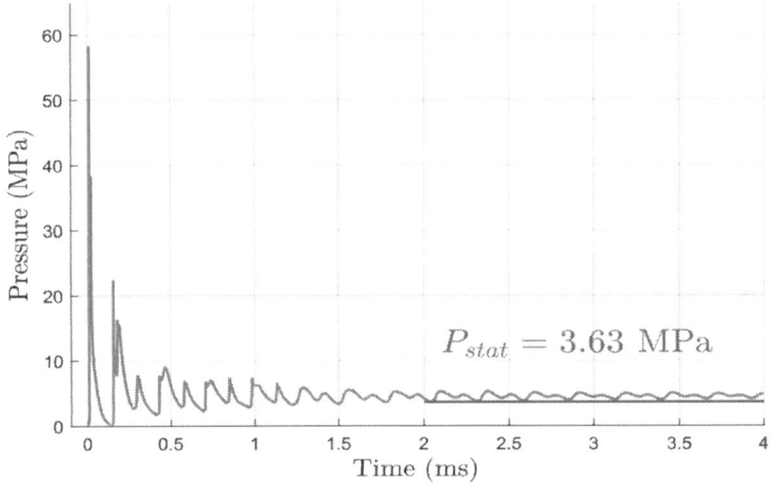

Figure 6. Fully confined blast at Z = 0.4 m/kg$^{1/3}$

The two curves of Figures 5 and 6 have the same peak overpressure, but the confined case presents multiple peaks due to the reflections that occur in the closed domain. The quasi-static pressure in the confined scenario is also estimated by means of simple thermodynamic considerations. Approximating the air and detonation products to ideal gases, the quasi-static pressure inside the confined space is only influenced by the energy introduced by the detonation and is given by Eq. 5 [14]:

$$P_{stat} = \left(\frac{p_0}{\gamma_0 - 1} \cdot \left(V_{tot} - V_{expl}\right) + e_{expl}\rho_{expl}V_{expl}\right) \cdot \frac{\gamma - 1}{V_{tot}} \tag{5}$$

Where V_{tot} is the total volume of the cubic domain; V_{expl} is the volume of the explosive charge; γ and γ_0 are the specific heat ratio of detonation products and air respectively; p_0 is the ambient pressure; e_{expl} is the energy per unit volume of TNT and ρ_{expl} is the initial density of TNT.

Three different steel materials are considered to model the plates, thus resulting in three unconfined simulation and three confined simulations. The selected steels are DOCOL 600DL, DOCOL 1000DP and DOCOL 1400M produced by Swedish Steel AB (SSAB) [15]. Steel physical parameters are shared by the three alloys and are shown in Table 4. [16] [17].

Table 4. Physical parameters common for all steels.

	ρ_0 (kg/m3)	E (MPa)	ν (-)
Steel	7850	210000	0.33

The modified Johnson Cook constitutive law, reported in Eq. 6, is employed to describe the material behavior under extreme loading conditions.

$$\sigma_y(\varepsilon_p, \dot{\varepsilon}_p, T) = \left(\sigma_0 + \sum_{i=1}^{2} Q_i \cdot \left(1 - \exp\left(-C_i \cdot \varepsilon_p\right)\right)\right) \cdot \left(1 + \frac{\dot{\varepsilon}_p}{\dot{\varepsilon}_{po}}\right)^c \cdot \left(1 - \left(\frac{T - T_r}{T_m - T_r}\right)^m\right) \quad (6)$$

The factor that considers the thermal contribution in Eq. 6 is not taken into consideration due to the adiabaticity of blast wave interaction. Once the elastic limit of the steel is reached, the material undergoes plasticization according to limit function reported in Eq.7.

$$f(\sigma_{ij}) = \Phi(\sigma_{ij}) - \sigma_y \quad (7)$$

Where Φ is the yield function according to Von Mises and σ_{ij} is the generic component of the stress tensor.

Table 5. Data for the three different steels used in the FEM simulations [17]

DOCOL	σ_0 (MPa)	Q_1(MPa)	C_1(-)	Q_2 (MPa)	C_2 (-)	c (-)	$\dot{\varepsilon}_{po}$ (s^{-1})
600DL	325.7	234.8	56.2	445.7	4.7	0.010	0.0005
1000DP	468.8	517.1	185.2	258.3	12.3	0.008	0.0005
1400M	1114.8	341.8	1111.5	143.9	20.7	0.004	0.0005

A quarter of the plate is modelled exploiting symmetries. Fixed boundary conditions are applied along all edges of the plate. Shell elements S4R are used and shell thickness t=2 mm are chosen for all the plates, according to commercially available thicknesses for the selected materials. Time duration of the FE simulations is set to 4 ms considering the end of transient effect as reported in Figures 5 and 6. The acquired pressure field p(x,y,t) is applied to a square plate with the same dimensions of a face of the Eulerian domain. The mesh size is equal to the spacing between virtual pressure gauges, so that each element in the FE simulation is assigned with the complete pressure history recorded by the corresponding probe in the CFD simulation. Thus, the mesh size of FE analyses is controlled by the number of probes in the CFD analyses. It was selected to set a 24x24 grid of evenly distanced pressure gauges over a quarter plate. Such value guarantees convergence of displacements and reduced FE simulation duration. Since no experimental data are available in the literature to validate the proposed methodology, the individual steps and their integration is validated through separate simulations. The convergence of CFD simulations is guaranteed by selecting the mesh size that was already identified in [13] for more sever case loading conditions. That mesh size was shown in [14] to accurately reproduce experimental observations. The link between the CFD and the FE solvers is validated by comparing the results in the uncoupled scenario with those generated through the built-in CONWEP module, which does not require passing data from external sources. Such a comparison is shown in Figure 7. Finally, the FE model includes material parameters already calibrated in other works, and the mesh size is selected through a convergence analysis, which is not reported here for the sake of brevity.

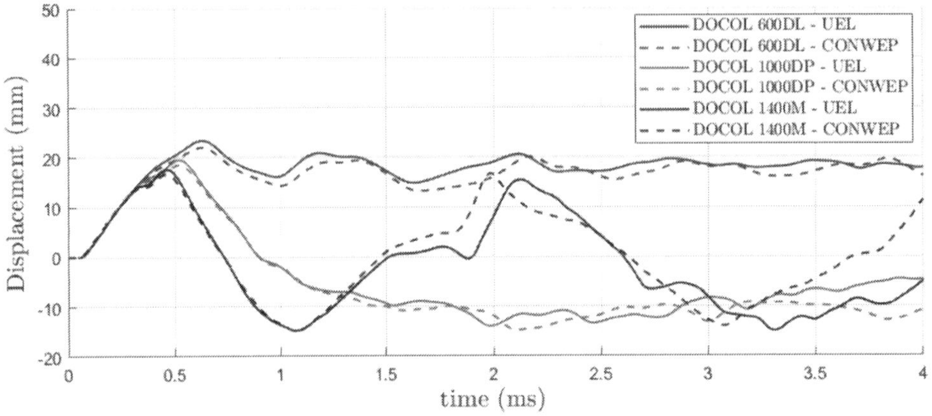

Figure 7. Comparison of midpoint displacement for different steels between CONWEP and UEL approaches.

4. Results

The results of the simulations with plates made of DOCOL 600DL, DOCOL 1000DP and DOCOL 1400M are reported in Figure 8, 9 and 10 respectively. Displacement time history of plates made of DOCOL 600DL, DOCOL 1000DP and DOCOL 1400M are reported in Figure 11, 12 and 13 respectively.

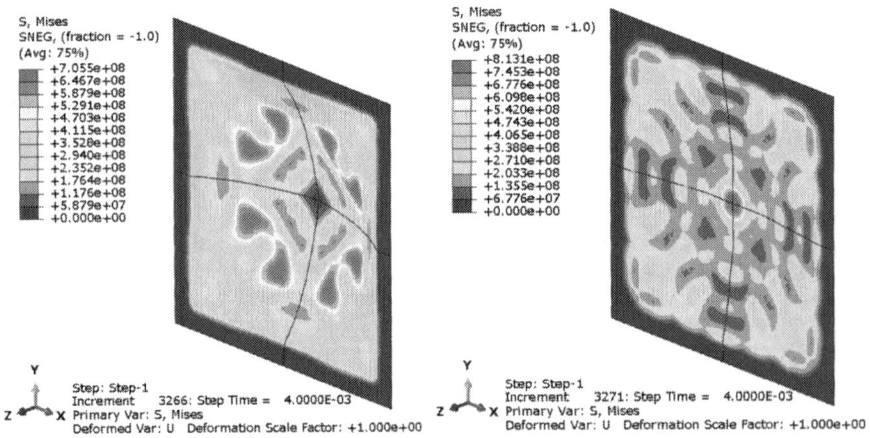

Figure 8. Von Mises stress field at the end of the simulation for the two plates made of DOCOL 600DL. Confined case (left) and unconfined case (right).

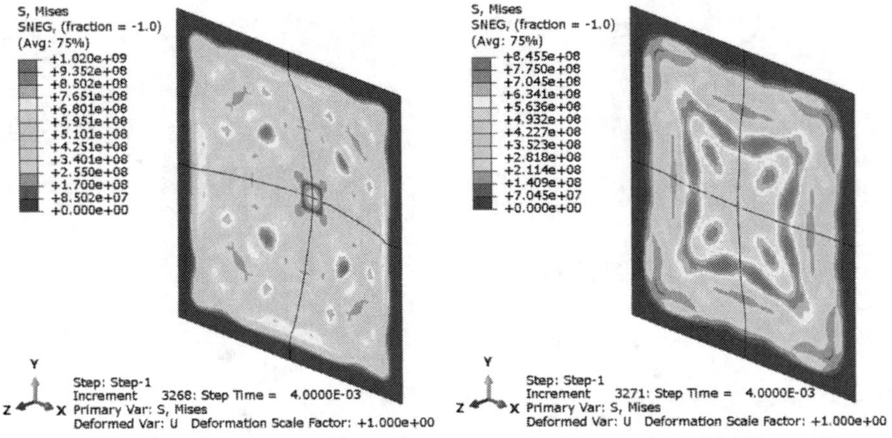

Figure 9. Von Mises stress field at the end of the simulation for the two plates made of DOCOL 1000DP. Confined case (left) and unconfined case (right).

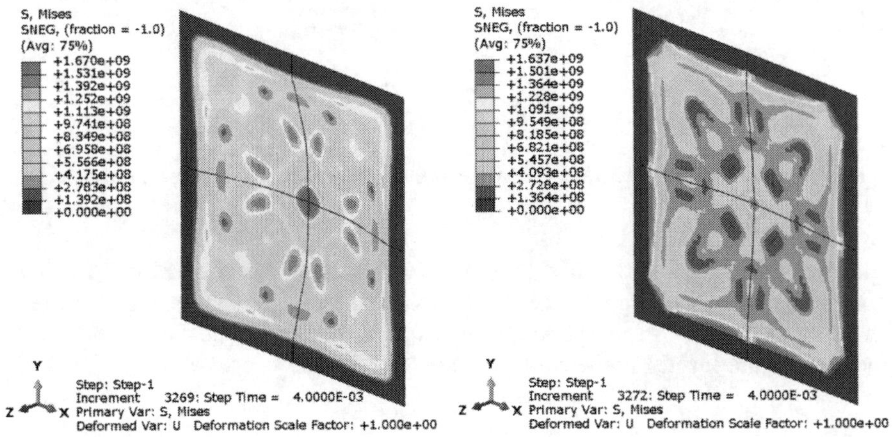

Figure 10. Von Mises stress field at the end of the simulation for the two plates made of DOCOL 1400M. Confined case (left) and unconfined case (right).

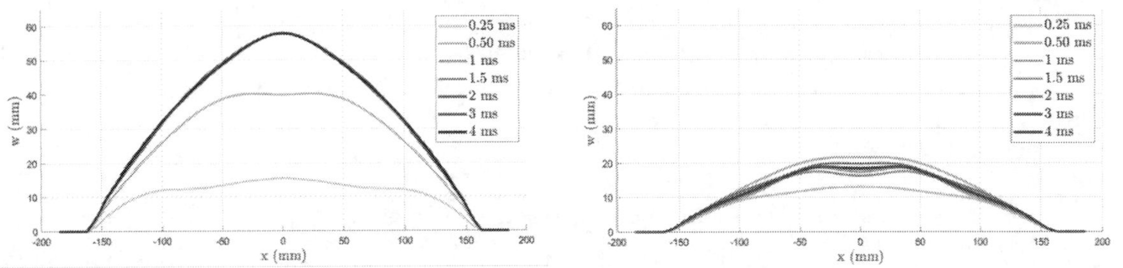

Figure 11. Displacement time history for DOCOL 600DL for the confined (left) and the unconfined (right) cases.

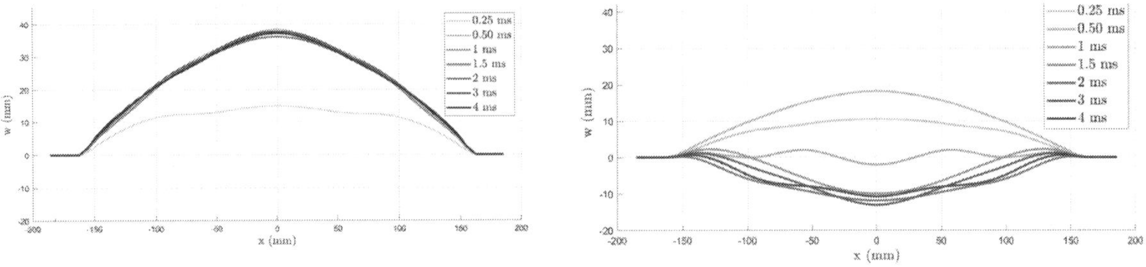

Figure 12. Displacement time history for DOCOL 1000DP for the confined (left) and the unconfined (right) cases.

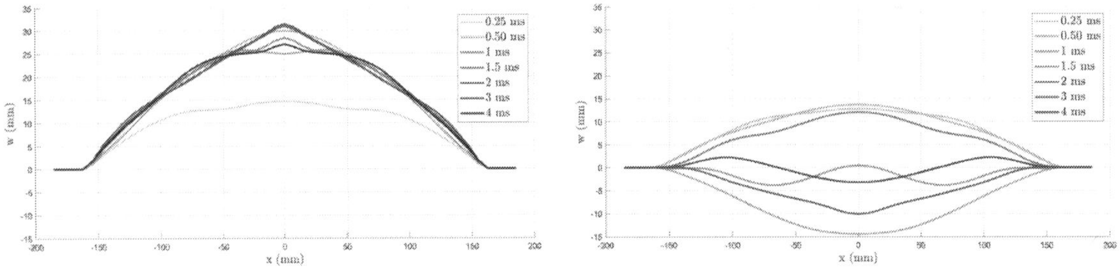

Figure 13. Displacement time history for DOCOL 1400M for the confined (left) and the unconfined (right) cases.

The most ductile of the three plates, DOCOL 600DL, experiences permanent plastic deformation in the direction of blast wave propagation in both confined and unconfined cases. What changes is the amount of displacement that is higher in the confined case. Compared to the free-field scenario, oscillations around the mean value are limited in the confined case.

The stiffest of the three plates, DOCOL 1400M, experiences the lowest permanent plastic deformation in the direction of blast wave propagation in the confined case. On the other hand, it behaves like an elastic membrane that oscillates around the equilibrium position in the unconfined case. However, plastic strain is present in the latter, as suggested by the high stress values. The overall behavior is similar to a plate that works in an elastic regime.

The behavior of the plate made of DOCOL 1000DP is unexpected. In fact, differently from what observed in the confined case in the unconfined scenario the plate shows a counter intuitive behavior, also known as reverse buckling [18], where the final permanent deflection of the plate is directed towards the blast-source.

Overall results for midpoint displacement are reported in Figure 14.

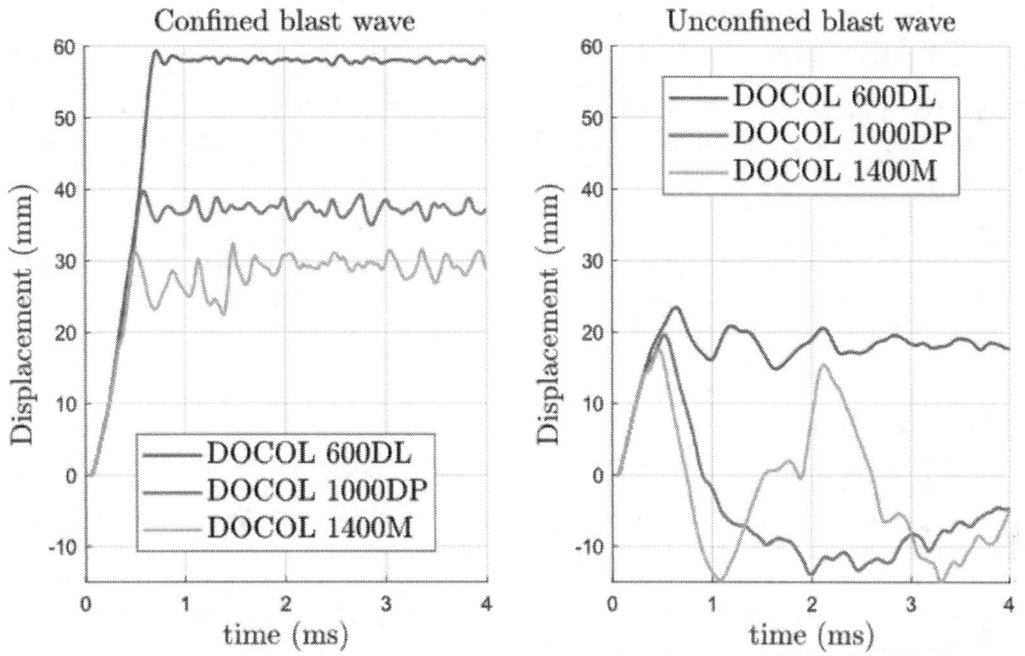

Figure 14. Representation of midpoint displacement for all the performed simulations.

5. Conclusions

In this work a high-fidelity numerical framework for uncoupled Eulerian-Lagrangian simulations of blast loaded plates is presented. The proposed approach is used to perform simulations of confined and unconfined blast scenarios. Results indicated that the presence of residual quasi-static pressure in confined blast events may prevent reverse buckling and oscillations around initial equilibrium configuration in steel plates Moreover, plates made of different steels show different qualitative behaviors in unconfined cases: DOCOL 600DL undergoes plastic deformation, DOCOL 1000DP experiences reverse buckling and DOCOL 1400M vibrates as an elastic membrane.

Little is known in the literature about the main causes of reverse buckling, which is sometimes seen in case of short duration blasts and may be determined by the combination of inertia forces, elastic unloading forces and negative phase pressure of free-field blast waves. Note that the reverse buckling is not caused by geometric imperfections, which are not accounted for in this work. Moreover, the counter intuitive behavior is not influenced by curvature effects since the pressure loading is not influenced by the structural deformation in uncoupled simulations. In addition, von Mises stress values at final time of the simulations are considerably higher that yielding stresses σ_Y before the increase due to hardening. Consequently, the current approach may result to be insufficient for fully model the behavior of steel plates for the case Z = 0.4 m/kg$^{1/3}$. A damage model capable of describing the perforation caused by ductile failure should be introduced to consider this possible event.

The introduction of a fracture model for the steel plate is expected to dramatically change the results for near-field explosions and it leads to the need for a new framework. This scenario highlights the limitation of Uncoupled Eulerian Lagrangian (UEL) approach because all considerations done in the current work rely on the presence of blast waves defined as distributed loads on each finite element of the plate. UEL simulations can be considered a fast and reliable numerical tool only when fracture does not occur. Pressure leakage through the fractures generated in the plate can be modelled only by a fully Coupled Eulerian Lagrangian (CEL), where air is modelled on both sides of the plate.

Extensive numerical and experimental campaigns are required to characterize the behavior of steel plates subjected to blast loads. The authors suggest investigating in future works the influence of different parameters to understand which are the dominant variables that control the transition between different qualitative structural responses.

References

[1] Nurick G N and Martin J B 1989 Deformation of thin plates subjected to impulsive loading—A review: Part I: Theoretical considerations *Int. J. Impact Eng.* **8** 159-170
[2] Nurick G N and Martin J B 1989 Deformation of thin plates subjected to impulsive loading—A review part II: experimental studies *Int. J. Impact Eng.* **8** 171-186
[3] Chung Kim Yuen S, Nurick G N, Langdon G S, Iyer Y 2017 Deformation of thin plates subjected to impulsive load: Part III – an update 25 years on *Int. J. Impact Eng.* **107**, 108-117
[4] Lomazzi L, Giglio M and Manes A 2021 Analytical and empirical methods for the characterisation of the permanent transverse displacement of quadrangular metal plates subjected to blast load: Comparison of existing methods and development of a novel methodological approach *Int. J. Impact Eng.* **154**, 103890.
[5] Kingery C N and Bulmash G 1984 *Air Blast Parameters from TNT Spherical Air Burst and Hemispherical Surface Burst* (Aberdeen Proving Ground, MD: U.S. Army Ballistic Research Laboratory)
[6] Hyde D 1988 *Fundamentals of Protective Design for Conventional Weapons Microcomputer Programs CONWEP and FUNPRO, Applications of TM 5-855-1 (User's Guide).* (Vicksburg, MS: Department of the Army, Waterways Experiment Station, Corps of Engineers)
[7] Smith M 2009 *ABAQUS/Standard User's Manual*, Version 6.9.
[8] Heylmun J, Vonk P and Brewer T 2021 *blastFoam 5.0 User Guide*, Synthetik Applied Technologies LLC
[9] Stewart D S 1993 *Lectures on Detonation Physics: Introduction to the Theory of Detonation Shock Dynamics* (University of Illinois, Urbana, Illinois 61801)
[10] Zheng Z and Zhao J 2016 Unreacted equation of states of typical energetic materials under static compression: A review *Chinese Physics B* **25** 076202
[11] Lee E L, Hornig H C and Kury J W 1968 *Adiabatic expansion of high explosive detonation products* (Lawrence Radiation Laboratory Report UCRL-50422)
[12] Dobratz B M 1985 *LLNL Explosives Handbook, properties of chemical explosives and explosive simulants* (Lawrence Livermore National Laboratory Report UCRL-51319)
[13] Pannell J J, Panoutsos G, Cooke S B, Pope D J and Rigby S 2021 Predicting specific impulse distributions for spherical explosives in the extreme near field using a gaussian function *Int. J. Protective Structures*, **12**, 437–459
[14] Feldgun V R, Karinski Y S and Yankelevsky D Z 2011 Some characteristics of an interior explosion within aroom without venting *Structural Engineering and Mechanics* **38**, 633-649
[15] https://www.ssab.com/en/brands-and-products/docol [Access: 2022-09-03]
[16] Aune V 2017 *Behaviour and Modelling of Flexible Structures Subjected to Blast Loading* (Doctoral theses at NTNU 2017:101)
[17] Aune V, Valsamos G, Casadei F, Langseth M and Børvik T 2017 On the dynamic response of blast-loaded steel plates with and without pre-formed holes *Int. J. Impact Eng.* **108**, 27-46
[18] Nechitailo N 2013 *Reverse buckling and post-chaotic self-organization* (Amazon Kindle Direct Publishing 476)

IOP Conf. Series: Materials Science and Engineering 1275 (2023) 012029 doi:10.1088/1757-899X/1275/1/012029

Influence of the wear rings on the buckling behaviour of hydraulic actuators

E V Arcieri[1] and S Baragetti[1]

[1]Department of Management, Information and Production Engineering, University of Bergamo, Viale Marconi 5, Dalmine 24044, Italy

sergio.baragetti@unibg.it

Abstract. During their use, hydraulic actuators are generally subjected to compressive loads which can induce buckling. In this work the critical load of a hydraulic actuator was determined for various boundary conditions and types of wear rings by theoretical and numerical models. In the theoretical model, the actuator was modelled as two connected beams, with different sections. The contribution of the wear rings on the critical load of the actuator was determined through numerical simulation based on the finite element method. The results of the mathematical model are similar to the results provided by the finite element analyses. The contribution of the investigated wear rings stiffness is very low for the materials considered for these components in this paper.

1. Introduction

Hydraulic actuators are commonly used in civil and industrial applications to apply forces. They have low construction and operation costs, reduced size and great strength. A hydraulic actuator mainly consists of two parts, a cylinder and a rod. Wear rings are typically mounted at the rod-cylinder connection in the gland and in some cases also between the external surface of the piston and the internal surface of the cylinder. The purpose of the wear rings is to guide the rod and piston, with reduced friction and improved sealing quality.

Due to their configuration, hydraulic actuators are commonly subjected to compression during their use with a consequent risk of buckling, which is one of the most common causes of failure in mechanical systems, in addition to the presence of stress concentrations [1], corrosion [2] and fatigue [3].

The bending and buckling behaviour of hydraulic actuators has only been studied in more detail since the second half of the 1900s, starting from Euler's method. Early mathematical models considered the buckling behaviour of a single beam. In [4] the device was modelled as a single beam with rod inertial properties over the entire length of the hydraulic actuator. This approximation leads to underestimate the real strength of the system. In [5] the system was modelled as a two-beam structure, with a rod and an infinite stiffness hollow cylinder, with a consequent overestimation of the actual load capacity of the actuator. Timoshenko and Gere [6] adopted a model with two different sections for rod and cylinder to consider their different moment of inertia. In [7, 8] it was shown that the presence of the fluid inside the actuator does not influence the critical load. In [9] the geometrical imperfections of the system that can cause premature deformations were examined. In fact, rectilinear imperfections induce an eccentricity of the load and a further bending moment. Ravinshankar [10] applied the finite element method to investigate the stress state in a hydraulic cylinder by modelling the system as an assembly of space frame elements. The contribution of the bearings and seals at the rod-cylinder interface was modelled with a

Content from this work may be used under the terms of the Creative Commons Attribution 3.0 licence. Any further distribution of this work must maintain attribution to the author(s) and the title of the work, journal citation and DOI.
Published under licence by IOP Publishing Ltd

rotary spring. Yoo and Siegel [11] estimated the critical column loads on telescopic power cylinders. They also considered the influence of support conditions and lateral load. In [12, 13] the influence of rectilinear imperfections was investigated analytically. Gamez-Montero and coauthors [14] considered the influence of the own weight of the actuator and fluid pressure on the buckling behaviour. The results confirmed the negligible effects of pressure, as found in [7, 8], and the validity of the analytical models in [12, 13]. Friction moments in the joint of the end supports were identified as critical for the load capacity of the actuators in [15]. In [16], the finite element method was used to conduct a sensitivity analysis in order to identify the effects of rectilinear imperfections and friction in the supports on the limit buckling load of the actuators. Zhou and coauthors [17] stress the need to consider large deflection and shear effect and bending stiffness of cylinder-rod joint in the models which describe the buckling behaviour of a horizontal hydraulic cylinder articulated at both supports. Taking into consideration the results above, the methods for evaluating the buckling load of hydraulic cylinders presented in the ISO/TS 13725:2021 standard [18] are based on the elastic buckling theory and consider the possible off-axis loading and the weight of the system.

In this work the critical buckling load of a hydraulic actuator was calculated considering different boundary conditions and types of wear rings [19]. The buckling load of the actuator was calculated using theoretical and numerical models. The presented theoretical model considers the actuator as two connected beams. Finite element analysis was used to determine the influence of the wear rings on the critical load of the whole system. The results of the mathematical and numerical models are similar and show that the influence of the investigated wear ring materials on the critical load of the actuator is limited.

2. Data for the analyses

For the study, the actuator of Figure 1 [12, 13, 16] was considered. It consists of a rod, a cylinder, a gland, a piston and four wear rings. Two wear rings are mounted at the rod-cylinder connection in the gland and two are mounted between the external surface of the piston and the internal surface of the cylinder. In the models presented in this work, the presence of the piston and the gland was not modelled. The rod and the cylinder were considered made of steel S235. Four types of wear rings were analysed, WR1, WR2, WR3 and WR4. WR1 and WR2 were commercial while WR3 and WR4 were fabricated using fused deposition modelling additive manufacturing process. Table 1 [12, 13, 16] summarizes the geometrical and mechanical properties of the hydraulic actuator components.

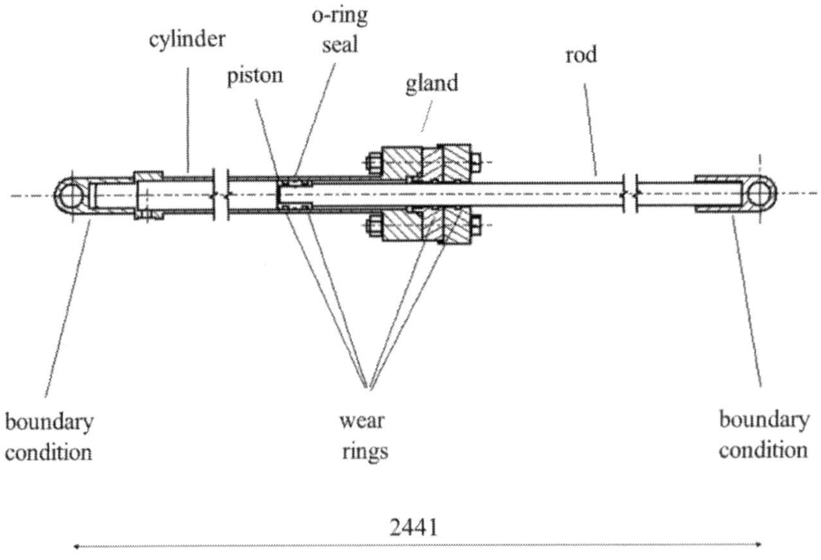

Figure 1. Actuator [12, 13, 16].

Table 1. Data [12, 13, 16].

Property	Value
Rod diameter (mm)	20
Cylinder external diameter (mm)	30
Cylinder internal diameter (mm)	25
Rod length (mm)	1165
Cylinder length (mm)	1276
Wear ring external diameter (mm)	25
Wear ring internal diameter (mm)	20
Wear ring height (mm)	5
Rod and cylinder Young's modulus – steel S235 (MPa)	206000
Rod and cylinder Poisson's ratio – steel S235	0.3
Rod and cylinder density – steel S235 (kg/m^3)	7850
WR1 Young's modulus – polyamide (MPa)	3000
WR2 Young's modulus – composite (MPa)	600
WR3 Young's modulus – polylactide (MPa)	2347
WR4 Young's modulus – nylon (MPa)	579
Wear ring Poisson's ratio	0.3

3. Mathematical model

The research method applied in this paper is the same one adopted in [20, 21]. Considering the equilibrium of an infinitesimal element of a beam, it is possible to obtain Equation 1:

$$\frac{d^4u}{dz^4} + \alpha^2 \frac{d^2u}{dz^2} = 0, \alpha^2 = N/EI \tag{1}$$

where u is the displacement in the direction normal to the axial direction z, N the axial load, E the Young's modulus of the material and I the section moment of inertia.

Equation 1 is a fourth-order differential equation whose associated homogeneous solution is defined by Equation 2:

$$u(z) = A\cos(\alpha z) + B\sin(\alpha z) + Cz + D \tag{2}$$

In the mathematical model, the actuator was modelled as two connected beams, one with a solid section (rod) and the other with a hollow section (cylinder) based on the geometry of the hydraulic actuator described in Section 2. Equation 2 was written for both the rod and the cylinder. Ground boundary conditions were modelled, namely pinned-pinned, fixed-pinned and fixed-fixed. The continuity of transversal displacement, rotation, bending moment and shear was set to model the fixed support between the two beams. The parameter to be obtained through the solution of the problem was the critical load N_{cr}, which is included in the terms α_{rod} and $\alpha_{cylinder}$ that appear when Equation 2 is written for the rod and the cylinder respectively. By assuming $\alpha_{cylinder} = \beta\alpha_{rod}$ ($\beta = (I_{rod}/I_{cylinder})^{0.5}$, where I_{rod} is the moment of inertia of the rod and $I_{cylinder}$ the moment of inertia of the cylinder), it is possible to solve the system of Equation 3 with respect to the variable α_{rod}:

$$[X]\,[Y] = [0] \tag{3}$$

where [X] represents the matrix of the coefficients, [Y] the vector of unknowns and [0] the vector of the known terms (null vector). The first solution with physical meaning of the system in Equation 3 is the critical load of the actuator and can be found by imposing det[X] = 0.

4. Finite element analyses

Under normal conditions of use, the deformation of the wear rings could influence the actuator critical load. If worn or overloaded, the wear rings may no longer be able to keep the rod aligned with the cylinder. In this case, the applied load would become eccentric with respect to the axis of the actuator, with consequent possible reduction of buckling load. Finite element analyses were conducted to calculate the radial stiffness of the wear rings and the critical load of the actuator for the different types of wear rings investigated.

4.1. Model for determining the radial stiffness of the wear rings

Given the geometry of each wear ring and the material with which it was made, a 3D finite element analysis was conducted on the model of a half of the wear ring with the Abaqus/Standard code [22] to determine its radial stiffness. The model is shown in Figure 2. The wear ring was modelled with solid elements. The load on the wear ring was applied by a semi-cylindrical surface, concentric with the wear ring. Each node of this surface was rigidly connected to a reference point (RP) positioned on the axis of the wear ring. The nodes on the outer surface of the wear ring were prevented from moving in the radial direction. A concentrated load was applied at the RP. The load direction was radial with respect to the wear ring and on the symmetry plane. Symmetry boundary conditions were set. The maximum displacement in the load direction was measured to determine the radial stiffness of the wear ring.

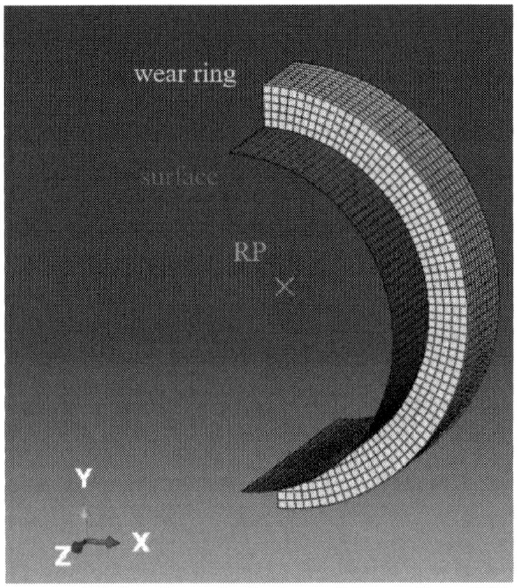

Figure 2. Finite element model for the calculation of the radial stiffness of the wear rings.

4.2. Model for determining the critical load of the actuator

To calculate the critical load of the actuator with different boundary conditions and types of wear rings, a 3D buckling analysis was conducted in Abaqus [22]. For the analysis, the cylinder, the rod and the wear rings were modelled. Shell elements were used to model the cylinder and beam elements were used to model the rod. The mechanical properties of steel S235 were assigned to the cylinder and rod. Two perpendicular linear elastic elements were used to model each wear ring in the actuator (Figure 3). The stiffness calculated with the procedure described in the previous section was assigned to the elastic elements.

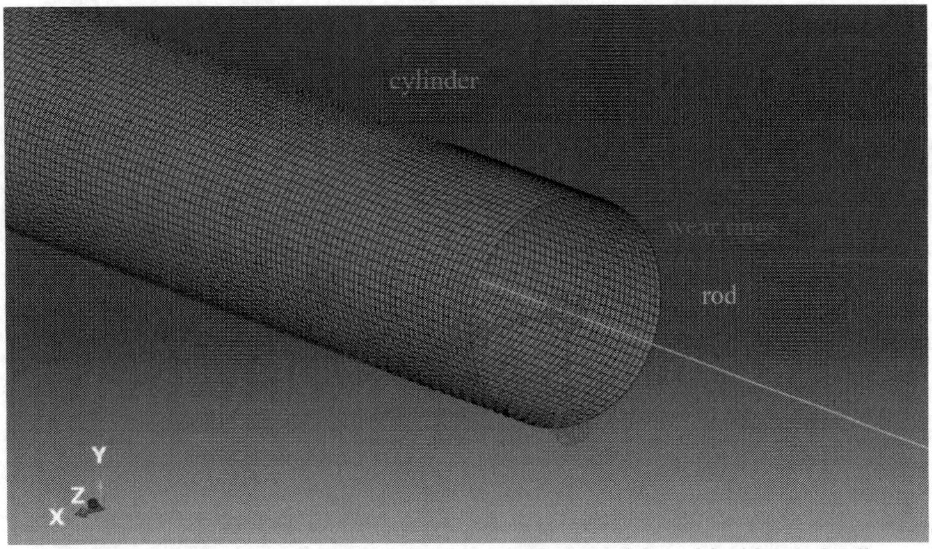

Figure 3. Finite element model for the calculation of the critical load, detail.

The fluid in the actuator was considered incompressible. To simulate this condition, the rod and cylinder were coupled with a coupling constraint and all the degrees of freedom of the coupling nodes were constrained except for the translation on the plane perpendicular to the axis of the actuator. The nodes at the end of the cylinder were rigidly connected to a RP placed in the centre of the section. Different boundary conditions were applied to this RP and to the node placed at the end of the rod. In the latter node the movement in the axial direction of the actuator was always left free. In this direction an arbitrary concentrated load was assigned. The arbitrary load, multiplied by the first critical multiplier obtained with the finite element method, is the buckling critical load of the actuator.

5. Results and discussion

Table 2 shows the values of critical load obtained with the mathematical and numerical models described in the previous sections.

Table 2. Results of the mathematical and numerical models.

Critical load [kN]	Pinned-pinned	Fixed-pinned	Fixed-fixed
Mathematical model	3.9	7.5	15.6
Numerical model, WR1	4.0	7.7	16.0
Numerical model, WR2	3.9	7.7	15.9
Numerical model, WR3	4.0	7.7	16.0
Numerical model, WR4	3.9	7.7	15.9

According to the results of the numerical models, the four types of wear ring investigated provide very similar values of critical load. The mounting of WR1 globally induces the maximum value of critical load for the actuator, while the mounting of WR4 provides the minimum value. WR1 is the stiffest wear ring of the four types investigated and WR4 is the least stiff (Table 1).

For all the three boundary conditions analysed, the theoretical results are similar to the numerical ones. The discrepancies between the results provided by the theoretical model and the finite element analyses are a few percentage points.

As expected, the critical load increases as the boundary conditions become stricter (the effective length factor becomes smaller), with the highest values of buckling critical load obtained for the fixed-fixed boundary condition.

IOP Conf. Series: Materials Science and Engineering 1275 (2023) 012029 doi:10.1088/1757-899X/1275/1/012029

6. Conclusions and future developments

In this work the critical buckling load of a hydraulic actuator was calculated for different boundary conditions and types of wear rings using theoretical and finite element models. In the theoretical model, the system was modelled as two connected beams. Finite element analyses were conducted on a model of the system containing the rod, the cylinder and the wear rings. The results of the mathematical model and finite element analyses are similar. The contribution of the wear rings stiffness is very low for the materials considered in this paper. An experimental test campaign will be planned to validate the obtained results. Some of the wear rings considered in this study were produced by additive manufacturing. Since the dimensional and geometrical tolerances obtained with this technology are poor, it would be interesting to determine how geometrical tolerances of wear rings influence the buckling behaviour of the actuator.

References

[1] Baragetti S, Borzini E and Arcieri E V 2018 *Procedia Struct. Integr.* **12** 173-182
[2] Srinivasan N, Bhaskar L K, Kumar R and Baragetti S 2018 *Mater. Des.* **160** 303-312
[3] Baragetti S and Baryshnikov A 2001 *J. Mech. Design* **123** 456-463
[4] Bleich F 1952 *Buckling Strength of Metal Structures* (New York: McGraw-Hill)
[5] Belluzzi O 1961 *Scienza delle Costruzioni* (Bologna: Zanichelli)
[6] Timoshenko S P and Gere J M 1961 *Theory of Elastic Stability* (Tokyo: McGraw-Hill)
[7] Flügge W 1973 *Stresses in Shells* (Berlin: Springer)
[8] Hoblit F 1950 *Prod. Eng.* **21** 108–112
[9] Seshasai K, Dawkins W and Iyengar S 1975 *Proc. Natl. Conf. Fluid Power*
[10] Ravishankar N 1981 *J. Mech. Des.* **103** 239-243
[11] Yoo C H and Siegel C R 1986 *Comput. Struct.* **22** 245-251
[12] Baragetti S and Terranova A 1999 *Int. J. Mater. Prod. Technol.* **14** 50–73
[13] Baragetti S and Terranova A 2001 *Proc. Inst. Mech. Eng., Part C* **215** 607–619
[14] Gamez-Montero P J, Salazar E, Castilla R, Freire J, Khamashta M and Codina E 2009 *Int. J. Mech. Sci.* **51** 105–113
[15] Gamez-Montero P J, Salazar E, Castilla R, Freire J, Khamashta M and Codina E 2009 *Int. J. Mech. Sci.* **51** 145–151
[16] Baragetti S and Villa F 2016 *Math. Probl. Eng.* **2016** 3781397
[17] Zhou J, Shi D, Di C, Zhang Y and Cheng X 2020 *Int. J. Struct. Stab. Dyn.* **20** 2050033
[18] ISO/TS 13725:2021 Hydraulic fluid power — Method for evaluating the buckling load of a hydraulic cylinder
[19] Costanzo S 2021 Instabilità progressiva di attuatori idraulici: modelli teorici, numerici e risultati sperimentali (master thesis)
[20] Baragetti S 2006 *Meccanica* **41** 443-458
[21] Baragetti S and Arcieri E V 2019 *Procedia Struct. Integr.* **24** 91-100
[22] ABAQUS, ABAQUS Documentation, Dassault Systèmes, Providence, RI, USA, 2017

IOP Conf. Series: Materials Science and Engineering 1275 (2023) 012030 doi:10.1088/1757-899X/1275/1/012030

Shape optimization of 2D beam-structures to obtain the uniform strength

C Iandiorio[1], D Milani[1] and P Salvini[1]

[1] University of Rome "Tor Vergata", Department of Enterprise Engineering, Via del Politecnico 1 – 00133 Rome, Italy, EU

Corresponding author: christian.iandiorio@uniroma2.it

Abstract. This paper provides a new method to design the optimal shape of 2D frame structures even (or especially) if not statically determined. The optimization here addressed involves obtaining the uniform strength of the whole structure, but the proposed procedure can be followed for other optimization goals. Two types of beam cross-sections are studied, in which the solution in terms of balance, kinematic and beam-shape is found analytically and involves six unknowns for every beam element. For frame structures, some internal and external constraint conditions must be satisfied; using the analytical solution within the boundary conditions, the optimization problem boils down to finding the roots of a non-linear system. This way allows to work inside the optimization workspace, avoiding the use for each iteration of other solvers (e.g. F.E. software) and obtaining the solution with a high- computation speed. Two tests are shown, which result in a uniform-strength behaviour of the maximal stress at every cross section, and a lightness gain higher than 70% by respect to the reference initial structure. This technique can be extended to other types of cross-sections or 3D frame structures.

1. Introduction

In structural design, shape optimization is the procedure that aims to improve one or more characteristics of a mechanical systems, respecting some conditions such as operating loads, imposed kinematics, and maximum or minimums dimensioning etc. [1,2].

Among the characteristics to optimize, the minimization of the weight is recurrent [3], a feature of considerable interest in aeronautical and automotive design. This is accompanied by design constraints, such as limits on maximum deflections or stresses [4,5] to control in critical regions.

The seek of optimal shape involves non-linear programming, recurring to elaborate algorithms, including the classical gradient methods or others based on heuristic approaches that simulate natural inspired phenomena (e.g. genetic algorithms, biological growth algorithms, etc.) [6-10]. Within structural analyses these algorithms often recur to Finite Element Method (F.E.M.) that typically involves solid element discretization; consequently, this implies managing a huge number of degrees of freedom (d.o.f.) and equations. Hence, the necessity to repeat the F.E. solution at every attempt implies consistent simulation times, especially for large-size structures.

For this reason, many authors have proposed algorithms based on simplified structural 1 or 2D models, significantly saving the computation time required for the optimal search. Truss structures are among the most studied [2,11-14], as they are the easiest workspace to test new algorithms and validate

Content from this work may be used under the terms of the Creative Commons Attribution 3.0 licence. Any further distribution of this work must maintain attribution to the author(s) and the title of the work, journal citation and DOI.
Published under licence by IOP Publishing Ltd

the solutions with other numerical or analytical methods. The searches for an optimum involving Frame structures (generally beams that englobe truss structures) differ in terms of purpose, weights [15,16], minimal bending moments [17,18], stress minimizations [5,19], maximal stiffnesses. Analytical solutions may be involved for statically determined beams [19-23], on the other hand, numerical approaches are recurrent for statically undetermined structures [24-25].

This work presents a new method to obtain the optimal shape that guarantees the uniform strength of a frame structure. As already known, the uniform strength is an ideal condition that involves maximum structural efficiency, hence the best utilization of the structural material; this implies the minimization of the material volume (i.e. the weight of the structure) and a suitable distribution of internal stresses.

The optimization is applied on a structure modelled through Euler-Bernoulli beams and involves a trivial numerical implementation and a small number of variables to optimize.

An exact analytical procedure for the optimal beam-shape search is presented (depending on cross-section shape), that drives the building up a non-linear system of equations to solve.

The novelty of the analytical formulation allows solving the statically undetermined frames overcoming F.E.M. analysis at every iteration. This feature is the key point of the presented method since it allows to never leaving the optimization workspace, significantly reducing the optimization time. It is worth to point out that the presented procedure works on the entire beam element (not on its sub-parts as commonly performed), involving the minimal number of equations, hence the degrees of freedom (d.o.f.) to manage.

The following optimization procedure considers only the optimal beam's shape/sizing, but does not affect the layout and topology of the structure, or node locations.

The optimized geometries would not be easily achieved using traditional manufacturing. However, composite manufacturing and the remarkable developments in the field of additive manufacturing allow the realization through 3D printing of complex and precise geometrical shapes [26].

The proposed optimization method is applied here to the case of planar straight frames with the hypothesis of linear behavior (concerning geometry and material), but the basic idea can be extended to 3D frames, distributed loads or geometrical non-linear systems, such as compliant mechanisms.

2. Analytical Formulation

In what follows the governing equations of an elastic beam with variable shape are derived, where the cross-section is assumed symmetrical along the *y-axis*. The displacement field for a planar bending excluding warping (engineering type 2D-beam model) shows up as [27,28]:

$$U(x,y) = u(x) - y\,\psi(x) \tag{1}$$

$$V(x,y) = v(x) \tag{2}$$

The positive directions of the displacement components are shown in figure 1.
The assumed beam model follows the Bernoulli-Euler hypotheses; therefore, the strain-displacements relations are:

$$\varepsilon_x(x,y) = u' - y\,\psi' \tag{3}$$

$$\gamma_{xy}(x,y) = 0 \quad or \quad \psi = v' \tag{4}$$

The beam element, of length L, is subjected only to concentrated forces at the end nodes I, J (figure 1); the equilibrium of the normal force, shear and the bending moment is governed by:

$$N(x) = -F_x^I \tag{5}$$

$$T(x) = -F_y^I \tag{6}$$

$$M(x) = F_y^I\, x - M_z^I \tag{7}$$

The nodal forces are related as follows:

$$F_x^J = -F_x^I \tag{8}$$

$$F_y^J = -F_y^I \tag{9}$$

$$M_z^J = F_y^I\, L - M_z^I \tag{10}$$

However, it is trivial to update the previous relations when involving distributed loads.
The normal force and bending moment are related with stress distribution along the cross-section as:

$$N(x) = \int_{A(x)} \sigma_x\, dA \tag{11}$$

$$M(x) = -\int_{A(x)} y \cdot \sigma_x\, dA \tag{12}$$

According to the Bernoulli-Euler model, at the section level the shear stress is null.
Using the stress-strain relationship $\sigma_x = E\varepsilon_x$ of an isotropic material in the equation (11,12) and considering the (3-7), the governing equations result:

$$u'(x) = \frac{-F_x^I}{EA(x)} \tag{13}$$

$$\psi'(x) = \frac{F_y^I\, x - M_z^I}{EI(x)} \tag{14}$$

$$v'(x) = \psi(x) \tag{15}$$

where A and I are the area and inertia moment of the cross-section.
Now, the idea is to impose the uniform strength i.e. the same stress at the top and bottom sides for the entire beam length and to obtain the shape that guarantees this assumption. Using the equations (13,14) in (3) the stress distribution as function of the nodal forces turns out:

$$\sigma_x(x,y) = \frac{-F_x^I}{A(x)} - \frac{F_y^I\, x - M_z^I}{I(x)}\, y \cong -\frac{F_y^I\, x - M_z^I}{I(x)}\, y \tag{16}$$

where in the latter, the right end comes from having assumed null the stress component due to normal force. This assumption, which involves a considerable simplification of the solution that is shown hereinafter, involves a small error inasmuch, as known, the stress due to bending moment is largely preponderant in non-wholly hinged frames. The iso-stress condition implies that all the cross-sections have the same stress magnitude σ_{iso} at the top and bottom sides $y = \pm h(x)/2$

$$\sigma_{iso} = \frac{\left|F_y^I\, x - M_z^I\right|}{I(x)} \frac{h(x)}{2} \tag{17}$$

Hence, choosing the cross-section shape it is possible to express the inertia moment $I(x)$ as a function of one-dimensional characteristics, e.g. $h(x)$ and found its trend to obtain the uniform strength.

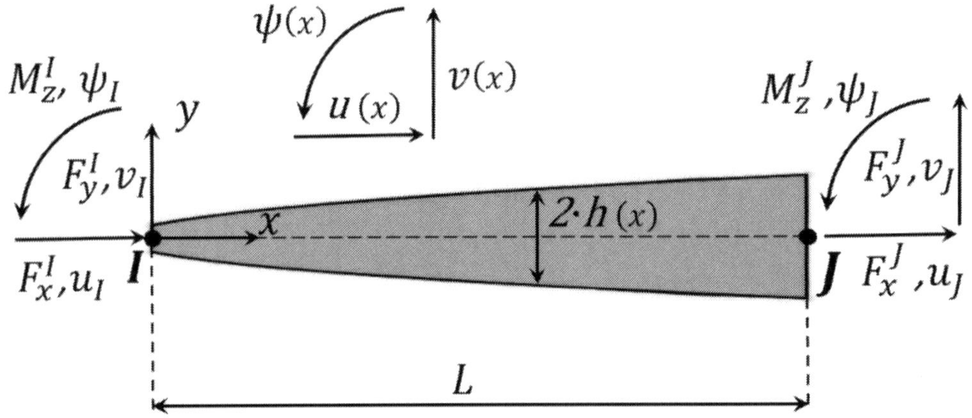

Figure 1. Force and displacement notations on the beam element

In the following two cases are considered: circular and square cross sections (figure 2) in which $A(x) = (c_A/E)\, h^2(F_y^I, M_z^I, x)$, $I(x) = (c_I/E)\, h^4(F_y^I, M_z^I, x)$ where:

$$c_A = \begin{cases} \frac{E\pi}{4} & \textit{for circular cross-section} \\ E & \textit{for square cross-section} \end{cases} \tag{18}$$

$$c_I = \begin{cases} \frac{E\pi}{64} & \textit{for circular cross-section} \\ \frac{E}{12} & \textit{for square cross-section} \end{cases} \tag{19}$$

For these cases the equations (17) turn out:

$$h(F_y^I, M_z^I, x) = \left(c_M \cdot \left|F_y^I\, x - M_z^I\right|\right)^{\frac{1}{3}} \tag{20}$$

where:

$$c_M = \begin{cases} \frac{64}{\pi\, \sigma_{iso}} & \textit{for circular cross-section} \\ \frac{6}{\sigma_{iso}} & \textit{for square cross-section} \end{cases} \tag{21}$$

Others type of cross-sections can be analysed, as the case of rectangular one with fixed height or width, or a shape in which height and width may vary.
Equation (20) presents a module function; therefore, to make easier the integration of equations (13-15), an equivalent equation is introduced:

$$h(F_y^I, M_z^I, x) = \left[c_M \cdot s(x) \cdot (F_y^I\, x - M_z^I)\right]^{\frac{1}{3}} \tag{22}$$

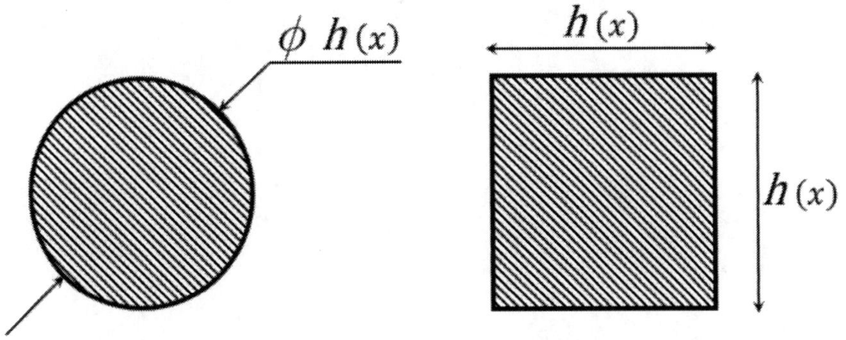

Figure 2. Considered cross-sections: Circular and Square

where $s(x) = sgn(F_y^I\, x - M_z^I)$ in which $sgn(x)$ is a function assuming the values of 1 and -1.
Using equation (22) in the ODE system (13-15) and solving it with the boundary conditions $\big(u(0), \psi(0), v(0)\big) = (u_I, \psi_I, v_I)$, it results:

$$u(F_x^I, F_y^I, M_z^I, x) = u_I + \frac{3\, s(x)\, F_x^I}{c_A\, c_M\, F_y^I}\left[h(F_y^I, M_z^I, 0) - h(F_y^I, M_z^I, x)\right] \tag{23}$$

$$\psi(F_y^I, M_z^I, x) = \psi_I + \frac{3}{2\, c_I {c_M}^2} \frac{h^2(F_y^I, M_z^I, x) - h^2(F_y^I, M_z^I, 0)}{F_y^I} \tag{24}$$

$$\begin{aligned} v\,(F_y^I, M_z^I, x) = v_I + \psi_I \cdot x - \frac{3}{2\, c_I\, {c_M}^2} \frac{x \cdot h^2(F_y^I, M_z^I, 0)}{F_y^I} + \\ + \frac{9\, s(x)}{10\, c_I\, {c_M}^3\, {F_y^I}^2}\left[h^5(F_y^I, M_z^I, x) - h^5(F_y^I, M_z^I, 0)\right] \end{aligned} \tag{25}$$

The displacements at the node J are simply $(u_J, \psi_J, v_J) = \big(u(L), \psi(L), v(L)\big)$; therefore, the forces and displacements quantities at the node J can be expressed as function of those given a node I by the equations (8-10,23-25).
It is worth pointing out that the equations (23-25), due to the law of h given in equation (20), involve a non-linear relationship among forces and displacements.
The equations (23-25) hold on the local (element) reference system. In a frame, in general, the local and global reference system are non-collinear; let ϑ be the counter-clockwise angle between the two-reference systems, the quantities referred to the global reference (identified with the superscript letter "g") can be computed from the local one by:

$$\begin{Bmatrix} {}^g u(x) \\ {}^g v(x) \\ {}^g \psi(x) \end{Bmatrix} = [\boldsymbol{R}]^T \begin{Bmatrix} u(x) \\ v(x) \\ \psi(x) \end{Bmatrix} \tag{26}$$

$$\begin{Bmatrix} {}^g F_x^I \\ {}^g F_y^I \\ {}^g M_z^I \\ {}^g F_y^J \\ {}^g F_y^J \\ {}^g M_z^J \end{Bmatrix} = \begin{bmatrix} [\boldsymbol{R}]^T & [\boldsymbol{0}] \\ [\boldsymbol{0}] & [\boldsymbol{R}]^T \end{bmatrix} \begin{Bmatrix} F_x^I \\ F_y^I \\ M_z^I \\ F_x^J \\ F_y^J \\ M_z^J \end{Bmatrix} \tag{27}$$

$$[\boldsymbol{R}] = \begin{bmatrix} \cos\vartheta & \sin\vartheta & 0 \\ -\sin\vartheta & \cos\vartheta & 0 \\ 0 & 0 & 1 \end{bmatrix} \tag{28}$$

For a statically determinate structure the nodal forces are known only by the balance law and thus solved in a one-shot; hence, the optimal shape is already known without any extra efforts by the equations (20,23-25). The resulting shape, namely the weight of the structure, remains a function of the chosen iso-stress value σ_{iso}.

For a more general frame structure, statically undetermined, the nodal forces are not given by simple equilibrium and the shape h holding uniform strength for every element is a function of these unknown forces. The problem is, therefore, to find the shape of all beam elements that imply the uniform strength of the structure. In other terms, the constraint loads are not known and are depending on structure deformations. For each beam element the unknows quantities are 6 forces $\left(F_x^I, F_y^I, M_z^I, F_x^J, F_y^J, M_z^J\right)$ and 6 nodal displacements $\left(u_I, \psi_I, v_I, u_J, \psi_J, v_J\right)$; the forces are linked each other by equilibrium equations (8-10), the nodal displacements are linked through the kinematic equations (23-25). This implies that for each beam element there are only 6 total unknowns among nodal forces and nodal displacements. As notation choice, in the following the unknows for each beam element coincide with the six quantities related to node *I*; once the solution is found, it is possible to obtain the remaining six unknowns at node *J* by the six equations (8-10,23-25).

Hence, in a frame structure composed of n_{el} elements to solve the problem $n_{B.C.} = 6 \times n_{el}$ boundary conditions are necessary. These boundary conditions produce a non-linear system of $n_{B.C.}$ equations in $n_{B.C.}$ unknows:

$$\begin{Bmatrix} f_1\left(F_{x,1}^I, \dots, F_{x,n_{el}}^I; F_{y,1}^I, \dots, F_{y,n_{el}}^I; M_{z,1}^I, \dots, M_{z,n_{el}}^I; u_{I,1}, \dots, u_{I,n_{el}}; v_{I,1}, \dots, v_{I,n_{el}}; \psi_{I,1}, \dots, \psi_{I,n_{el}}\right) \\ \vdots \\ f_{n_{B.C.}}\left(F_{x,1}^I, \dots, F_{x,n_{el}}^I; F_{y,1}^I, \dots, F_{y,n_{el}}^I; M_{z,1}^I, \dots, M_{z,n_{el}}^I; u_{I,1}, \dots, u_{I,n_{el}}; v_{I,1}, \dots, v_{I,n_{el}}; \psi_{I,1}, \dots, \psi_{I,n_{el}}\right) \end{Bmatrix}_{n_{B.C.} \times 1} = \{\boldsymbol{0}\} \tag{29}$$

Imposing the boundary condition of the structure due to the external and internal constraints and using the balance and kinematic equations derived, the optimization problem becomes simply the search the roots of the non-linear system (29).

Table 1 shows the kinematic and force boundary conditions that solve the straight case in figure 3, which involves three elements, therefore eighteen relations. In this case, the global and local reference systems coincide. A more complex case of frame structure is shown in figure 4, which involves five elements (i.e. 20 conditions) with the presence of two nodes with multiple connections. The set of boundary conditions that form the system (29) is summarized in table 2, referring to the global reference system.

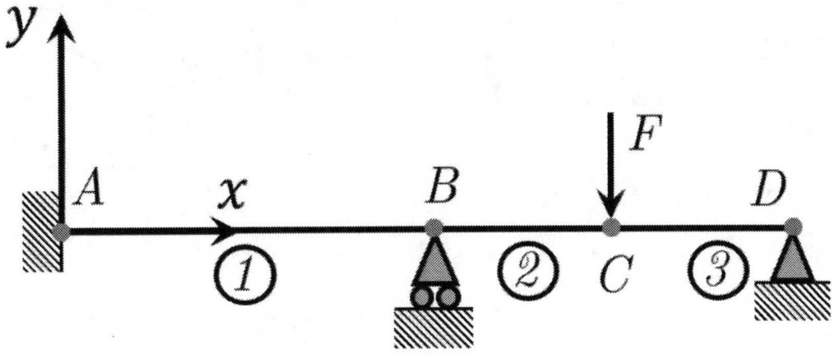

Figure 3. Multi-supported structure

3. Examples of optimized structures

This section shows two numerical applications, which involve a straight multi-supported structure (figure 5) and a framed structure (figure 6), both having circular cross-sections.

To find the optimized shape in both cases, the system (29) is built by the imposition of the boundary conditions. There are some methods to find the roots of the system (29), but most of them require an initial attempt solution, not too far from optimal. To satisfy this, we propose to carry out a preventive Finite Element analysis on a reference structure with the same topology and layout as the one to be analyzed, but formed by elements with constant cross-sections (i.e. non-optimized). From the results of the F.E.A. it is possible to extract the attempted solution, i.e. the $n_{B.C.}$ unknows in terms of nodal forces and displacements. This simple way to proceed returns a sturdy starting point for several cases examined in this research and not reported here for brevity. However, the optimal solution search is a strong non-linear problem that provides, in general, more than one solution.

To solve the system (29) the trust-region method is used, with a tolerance value 10^{-3} on the residual of (29) and 10^{-6} for the direction of descent. For the cases shown in figure 5 and 6 the reference structure has a circular cross-section with a diameter of 0.02 m.

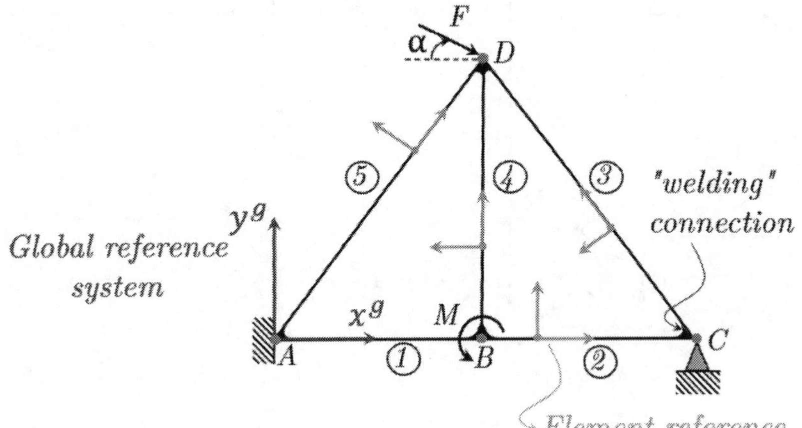

Figure 4. Frame structure with “welded” elements

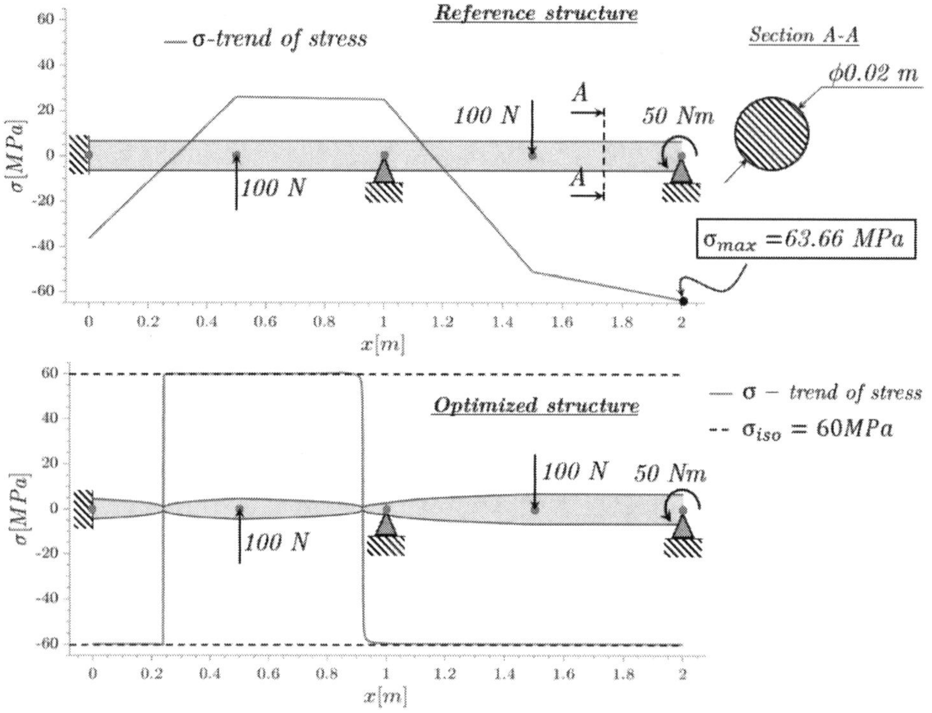

Figure 5. Optimized solution for a straight multi-constrained structure

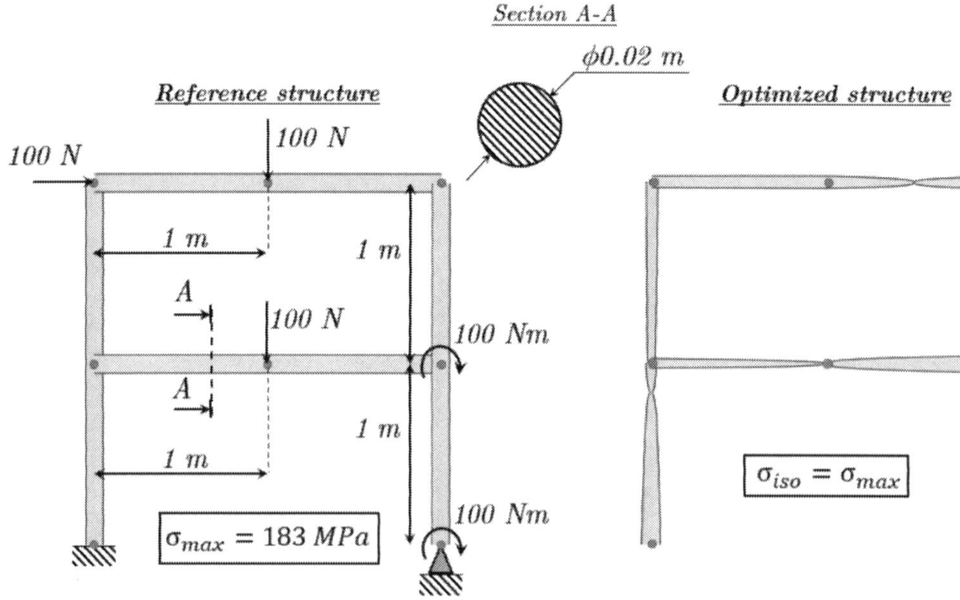

Figure 6. Optimized solution for a frame structure

The iso-stress σ_{iso} of the optimized structure is an input free-parameter; in the two test cases, it is chosen very close to the maximum stress in the initial non-optimized structure. Figures 5 and 6 show the transition between the initial structure and the optimized one. Furthermore, the linear layout of the multi-supported structure (figure 5) allows an easy representation of the stress trend of the top side along the optimized and non-optimized structures, performed by the F.E. analysis with the section shapes as returning by the optimization process.

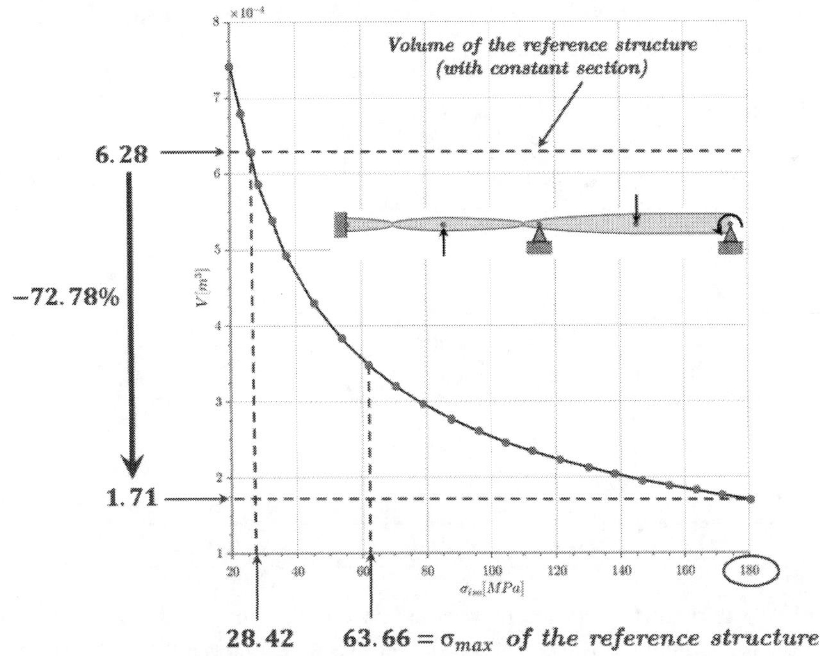

Figure 7. Overall volume vs stress of uniform strength for the case of straight multi-constrained structure

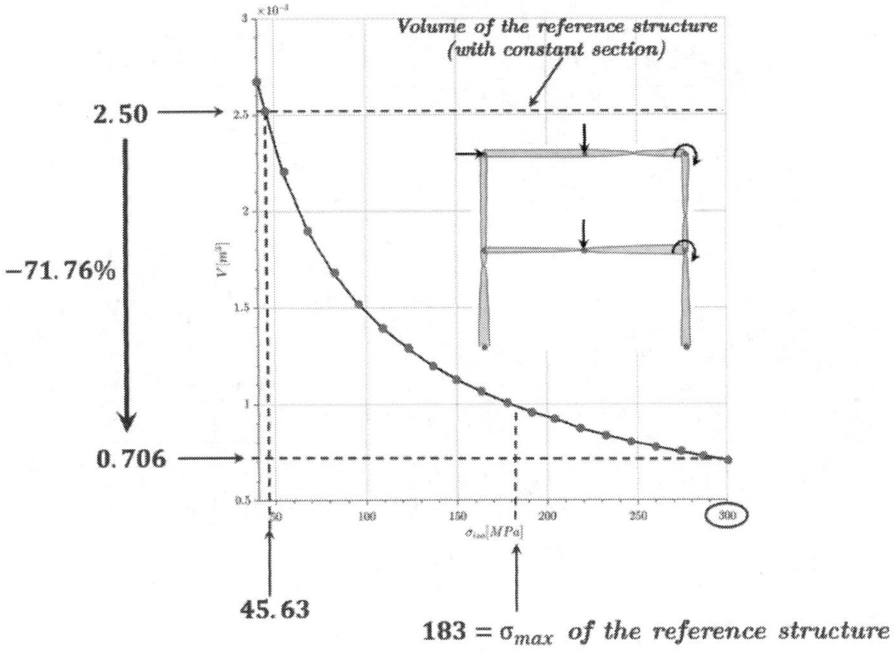

Figure 8. Overall volume vs stress of uniform strength for the case of the frame structure

In the optimized structure, the stress change in sign, as expected, after an inflection point, i.e. a point in which the curvature and the bending moment are null. Following the hypothesis of equation (16) to consider only the stresses caused by bending, the cross-sections that correspond to the inflection points are characterized by ideal zero thickness. This unrealistic situation can be avoided if a more complex stress state, considering also the normal and shear stresses, is assumed. These stresses will drive the minimal section sizes at the inflection points.

Table 1. The 18 boundary conditions of the case in figure 3

	Kinematic B.C.	*Forces B.C.*
A	$u_{I,1} = 0$ $v_{I,1} = 0$ $\psi_{I,1} = 0$	*None*
B	$u_{J,1} - u_{I,2} = 0$; $v_{J,1} = 0$ $v_{I,2} = 0$; $\psi_{J,1} - \psi_{I,2} = 0$	$F^J_{x,1} + F^I_{x,2} = 0$ $M^J_{z,1} + M^I_{z,2} = 0$
C	$u_{J,2} - u_{I,3} = 0$; $v_{J,2} - v_{I,3} = 0$ $\psi_{J,2} - \psi_{I,3} = 0$	$F^J_{x,2} + F^I_{x,3} = 0$ $F^J_{y,2} + F^I_{y,3} - F = 0$ $M^J_{z,2} + M^I_{z,3} = 0$
D	$u_{J,3} = 0$; $v_{J,3} = 0$	$M^J_{z,3} = 0$

Furthermore, it is worth pointing out that this method does not require a F.E.M. computation for every iteration search of the optimum, but only an initial analysis to give the initial condition from which to start the optimization. This is important because this way allows to not leave the optimization workspace and therefore obtain the solution in a much faster way.

Table 2. The 30 boundary conditions of the case in figure 4

	Kinematic B.C.	*Forces B.C.*
A	${}^g u_{I,1} = 0$; ${}^g u_{I,5} = 0$ ${}^g v_{I,1} = 0$; ${}^g v_{I,5} = 0$ ${}^g \psi_{I,1} = 0$; ${}^g \psi_{I,5} = 0$	*None*
B	${}^g u_{J,1} - {}^g u_{I,2} = 0$; ${}^g u_{J,1} - {}^g u_{I,4} = 0$ ${}^g v_{J,1} - {}^g v_{I,2} = 0$; ${}^g v_{J,1} - {}^g v_{I,4} = 0$ ${}^g \psi_{J,1} - {}^g \psi_{I,2} = 0$; ${}^g \psi_{J,1} - {}^g \psi_{I,4} = 0$	${}^g F^J_{x,1} + {}^g F^I_{x,2} + {}^g F^I_{x,4} = 0$ ${}^g F^J_{y,1} + {}^g F^I_{y,2} + {}^g F^I_{y,4} = 0$ ${}^g M^J_{z,1} + {}^g M^I_{z,2} + {}^g M^I_{z,4} + M = 0$
C	${}^g u_{J,2} = 0$; ${}^g u_{I,3} = 0$ ${}^g v_{J,2} = 0$; ${}^g v_{I,3} = 0$ ${}^g \psi_{J,2} - {}^g \psi_{I,3} = 0$	${}^g M^J_{z,2} + {}^g M^I_{z,3} = 0$
D	${}^g u_{J,3} - {}^g u_{J,4} = 0$; ${}^g u_{J,3} - {}^g u_{J,5} = 0$ ${}^g v_{J,3} - {}^g v_{J,4} = 0$; ${}^g v_{J,3} - {}^g v_{J,5} = 0$ ${}^g \psi_{J,3} - {}^g \psi_{J,4} = 0$; ${}^g \psi_{J,3} - {}^g \psi_{J,5} = 0$	${}^g F^J_{x,3} + {}^g F^J_{x,4} + {}^g F^J_{x,5} + F \cos\alpha = 0$ ${}^g F^J_{y,3} + {}^g F^J_{y,4} + {}^g F^J_{y,5} - F \sin\alpha = 0$ ${}^g M^J_{z,3} + {}^g M^J_{z,4} + {}^g M^J_{z,5} = 0$

Figure 7 and 8 are show the trend of the structural volume as a function of the stress of uniform strength for the two test cases. It is interesting to discuss two points: what is the σ_{iso} value for which the optimized

structure has the same volume of the reference structure, and the weight decrease of the optimized structure for high σ_{iso} by respect to the reference structure.
The multi-supported structure (figure 7) achieves the same volume of the reference one with a uniform stress $\sigma_{iso} = 28.42\ MPa$, that it is less than half of the maximum stress of the non-optimized structure.
The framed structure (figure 8) reaches the same volume of the reference one with $\sigma_{iso} = 45.63\ MPa$, which is a quarter of the maximum stress achieved in the reference structure.
Furthermore, in figures 7 and 8 is represented the trend of the structural volume up to reach a volume decrease just over 70% by respect to the volume of the reference structure, achieving $\sigma_{iso} = 180\ MPa$ for the linear structure and $\sigma_{iso} = 300\ MPa$ for the frame structure.

4. Conclusion

A new methodology to address the shape optimization of 2D frame structures is discussed in the paper. The objective is to find the shape of beam elements that form the frame structure such that it exhibits uniform strength, i.e. that has the same stress for top or bottom points along every beam.
The shape of the beam to guarantee the uniform strength is found analytically for circular and square cross-sections with the hypothesis to consider only the stress due to bending, predominant in this type of structures, and neglecting the stress due to the normal and shear forces.
According to the exposed methodology, the optimization problem becomes the search of the roots of a non-linear system of equations.
This proceeding way does not follow the standard methodologies of structural optimization; in fact, it does not involve coupling between the optimization environment and an external solver (e.g. F.E. software) for every iteration. This allows obtaining the solution by remaining only in the optimization workspace, which allows a high computation speed.
Two test cases are presented, demonstrating the high weight gain with the proposed solution when uniform-strength stress is reached.
The analytical shape and the relative kinematic equations could also be found for other type of sections or for a more refined stress field, including the normal and shear stresses.
Furthermore, this method can be extended to the case of 3D frame structures, taking into account of torsional loadings and recurring to an equivalent stress assumption.

References

[1] Arora J and Wang Q 2005 Review of formulations for structural and mechanical system Optimization *Struct Multidisc Optim* **30** pp 251–272
[2] Degertekin S, Lamberti L and Ugur I 2018 Sizing, layout and topology design optimization of truss structures using the Jaya algorithm *Applied Soft Computing 70* pp 903–928
[3] Lamberti L and Pappalettere C 2004 Improved sequential linear programming formulation for structural weight minimization *Comput. Methods Appl. Mech. Engrg.* **193** pp 3493–3521
[4] Lian H, Christiansen A, Tortorelli D, Sigmund O and Aage N 2017 Combined shape and topology optimization for minimization of maximal von Mises stress *Structural and Multidisciplinary Optimization* **55** pp 1541–1557
[5] Chai S, Chen B, Ji M, Yang Z, Lau M, Fu X and Liu L 2018 Stress-oriented structural optimization for frame structures *Graphical Models* **97** pp 80-88
[6] Porziani S, De Crescenzo F, Lombardi E, Iandiorio C, Salvini P and Biancolini M 2021 Automatic Optimization Method Based on Mesh Morphing Surface Sculpting Driven by Biological Growth Method: An Application to the Coiled Spring Section Shape : *International Conference on Computational Science ICCS 2021*, pp 479–491
[7] Steven G, Querin O and Xie M 2000 Evolutionary structural optimisation (ESO) for combined topology and size optimisation of discrete structures Comput. Methods Appl. Mech. Engrg. **188** pp 743-754

[8] Neshat M, Sepidnam G and Sargolzaei M 2013 Swallow swarm optimization algorithm: a new method to optimization *Neural Computing and Applications* **23** pp 429–454

[9] Kaveh A and Bakhshpoori T 2016 Water Evaporation Optimization: A novel physically inspired optimization algorithm *Computer & Structures* **167** *pp 69-85*

[10] Karaboga D 1995 An Idea Based on Honey Bee Swarm for Numerical Optimization *Technical report-TR06 Erciyes University Engineering Faculty Computer Engineering Department, Kayseri (Turkey)*

[11] Wang D, Zhang W and Jiang J 2002 Truss shape optimization with multiple displacement constraints *Comput. Methods Appl. Mech. Engrg.* **191** pp 3597–3612

[12] Wu J, Sigmund O and Groen 2021 Topology optimization of multi-scale structures: a review *Structural and Multidisciplinary Optimization* **63** pp 1455–1480

[13] Lamberti L and Pappalettere C 2002 Design optimization of large-scale structures with sequential linear programming *Proc Instn Mech Engrs Part C: J Mechanical Engineering Science* **216** pp 799-811

[14] Larsen S, Sigmund O and Groen J 2018 Optimal truss and frame design from projected homogenization-based topology optimization *Structural and Multidisciplinary Optimization* **57(4)**, pp 1461-1474

[15] Erbatur F and Al-Hussainy M 1992 Optimum design of frames *Computers & Structures* **45** pp 887-891

[16] Pezeshk S 1998 Design of framed structures: an integrated non-linear analysis and optimal minimum weight design *International Journal for Numerical Methods in Engineering* **41** pp 459—471

[17] An H and Huang H 2017 Topology and Sizing Optimization for Frame Structures with a Two-Level Approximation Method *AIAA Journal* **55(3)** pp 1044-1057

[18] Wang D 2007 Optimal shape design of a frame structure for minimization of maximum bending moment *Engineering Structures* **29** pp 1824–1832

[19] Rozvany G 1978 Optimal elastic design for stress constraints *Computers & Structures* **8(3-4)** pp 455-463

[20] Kozikowska A 2016 Geometry and Topology Optimization of Statically Determinate Beams under Fixed and Most Unfavorably Distributed Load *Lat. Am. j. solids struct.* **13(4)** pp 775-795

[21] Pedersen P and Pedersen N 2009 Analytical optimal designs for long and short statically determinate beam structures Struct Multidisc Optim **39** pp 343–357

[22] Abdalla H and Casagrande D 2021 Optimal area variation for maximum stiffness isostatic beams under parametric linear distributed loads *Mechanics Research Communications* **111**

[23] Nagy A, Abdalla M and Gürdal Z 2010 Isogeometric sizing and shape optimisation of beam structures Comput. Methods Appl. Mech. Engrg. **199** pp 1216–1230

[24] Tyburec M, Zeman J, Kružík M and Henrion D 2020 On optimum design of frame structures *Acta Polytechnica CTU Proceedings* 26 pp 117-125

[25] Shimoda M, Liu Y and Morimoto T 2014 Non-parametric free-form optimization method for frame structures *Struct Multidisc Optim* **50** pp 129–146

[26] Pan C, Han Y and Lu J 2020 Design and Optimization of Lattice Structures: A Review Appl. Sci. **10**

[27] Iandiorio C and Salvini P 2022 On the Formulation of an Elastic-Plastic Beam Model: the Pre-Integration Idea *IOP Conf. Ser.: Mater. Sci. Eng.* **1214 012026**

[28] Iandiorio C and Salvini P 2022 An Engineering Theory of thick Curved Beams loaded in-plane and out-of-plane: 3D Stress Analysis *European Journal of Mechanics - A/Solids* **92**

IOP Conf. Series: Materials Science and Engineering 1275 (2023) 012031 doi:10.1088/1757-899X/1275/1/012031

Correlation between Preload and Geometric Characterisation of metal meshes for space antenna reflectors

P Salvini and E Marotta

Department of Enterprise Engineering, University of Rome "Tor Vergata", Via del Politecnico 1, 00133 Rome, Italy

Email: salvini@uniroma2.it

Abstract. The paper deals with digital image processing techniques to analyse the shape of the main voids in the metallic mesh of large deployable antennas. The shape of the holes in the mesh influences the electromagnetic performances as the working frequencies and the polarisation effects. The analysis compares two types of mesh subjected to a series of equal loads in two directions; for this task, a dedicated testing machine is used. The paper presents an approach to defining an effective and averaged repetitive pattern. Thus, a statistical point of view of the entire mesh is given, applicable to studying the behaviour of a full reflector, to avoid being affected by local texture defects.

1. Introduction

For several decades, in satellite space communications, the need to replace the rigid paraboloid antenna with deployable systems equipped with flexible metal mesh reflectors has arisen [1]. The replacement is dictated by the need to improve the quality of the signal and the need to reduce launch costs related to the weight of the payload. The size of the reflector influences the quality of the transmitted signal, and for this purpose, reflectors larger than 15 m are currently being developed [2, 3]. Their design must consider numerous aspects: electromagnetic, structural, kinematic, dynamic, and thermal, without neglecting the technological features.

Large Deployable Reflectors are composed of a light opening mechanism; in particular, the reflector is formed by an intertwined metal mesh and a cable system (net) that provides tensioning of the mesh itself to preserve the concave shape of the doubly curved geometry. The task of reflecting the electromagnetic signal towards the focal point of the antenna and vice versa is entrusted to the metal mesh, generally knitted. Numerous entangled loops characterise this mesh to form a repetitive structure. Each loop is connected to contiguous ones through wires' contacts. The visual aspect of the knitted fabric presents a high ratio between voids and solids, where the size of the most significant holes affects the electromagnetic reflection capability, defining the design frequency and avoiding the onset of unwanted intermodulation products. The knitting process introduces many variables that make the final product not conform to the original design; the feature of the mesh can be predicted through models that cannot always be reliable [4]. Photographic/optical methods are often used to carry out continuous measurements and to analyse weaving processes to economise and improve the efficiency of product development and quality control [5]. During their operational life, astromeshes are subjected to a state of tension in the plane that keeps them in shape. The biaxial loading adds mesh tensioning that modifies the shape and the size of the loops, affecting the reflectivity and RF transmissibility properties [6]. The

Content from this work may be used under the terms of the Creative Commons Attribution 3.0 licence. Any further distribution of this work must maintain attribution to the author(s) and the title of the work, journal citation and DOI.
Published under licence by IOP Publishing Ltd

strain level must be kept within a defined range to avoid intermodulation of the RF signal and permanent deformation of the mesh itself in the wide thermal range encountered in flight conditions [7] (i.e. from -190 to + 140 ° C).

This work aims to measure the variations of some geometric properties of the largest voids as the applied loads vary. A specially designed tensile machine was used to load the metal meshes biaxially. Optical measurements were made for each pair of loads and subsequently processed. For the recognition of the main pattern, cross-correlation was used, and the geometric characteristics of the voids were deduced from image processing after proper identification and averaging. The main geometric features measured are related to the area, moments of inertia, principal axes, and equivalent diameter. The experimentation shows significant variations in the aforementioned geometric properties as a function of the extent of the applied load, with different behaviour for any type of mesh. The method, here aimed at checking the preload, can provide a qualitative index of the weaving process, should one have to choose between different options.

2. Methods

Astromeshes are reflective structures mounted on deployable antennas and made with metal mesh fabrics used in space communication. The production process involves hooks and needles to weave very thin threads; the final weaving consists of numerous loops knotted together. The loops can be more or less compact, with different shapes and geometries, and orientally disposed to obtain a predictable mechanical behaviour within certain limits. Astromeshes generally provide different stiffnesses in the two orthogonal directions (warp and weft) [8]. The production aspects, the knit's geometry, and the single yarn's size are essential elements to consider for an optimal design of the reflective system.

The constitutive models of astromeshes are not easily traceable; the uniaxial mechanical tests do not provide sufficient information to describe the complete behaviour. The mesh response is orthotropic; therefore, a two-dimensional analysis is mandatory. Furthermore, astromeshes often exhibit a non-linear elastic behaviour caused by local plasticisation of the yarns, friction due to the sliding of the loops, and activation of uncertain contacts. They show large deformations with relatively low loads, making measurements challenging to perform. Given the nature of the phenomenon and the objective difficulties of implementation, to obtain coherent measurements, contactless measurement techniques often employ photography and subsequent image analysis through digital correlation [9].

In the present work, a specially designed biaxial traction machine was used to characterise flat geometry meshes consisting of knitted metal structures [10]. The device is equipped with a rigid frame, two-stepping motors needed to impress the movement in the two main directions of the mesh, four load cells, and electronic control. The tests are conducted automatically through a list of commands that define the test protocol. The test management is driven by a PC with dedicated software and a graphic control interface. The same computer simultaneously makes control and data acquisition.

During the test, the mesh is conducted following the biaxial load, and a PID-type logic governs the control of the imposed loads. Reaching the prescribed load, the mesh is photographed with a resolution of 1920x1080 pixels using a telecentric lens thus avoiding perspective distortion on the picture. The test data and image analysis is carried out in post-processing in an autonomous and independent manner. Figure 1 shows a photo of the machine used in the tests and the diagram of the measurement chain referred to a single axis.

Image matching is a discipline of image processing of fundamental importance for many practical applications: to solve problems in industrial process control, automatic recognition of number plates in parking lots, biological growth phenomena, geological mapping, stereovision, video compression, and autonomous robots for space exploration. Some specific tasks are the study of liquid flows and the deformation state of materials [11].

The pictures taken during the test provide images that can be processed using digital techniques. Cross-correlation between images is a tool widely used in image analysis, especially for image filtering and matching algorithms [12]. The method creates a normalised convolution product between the

IOP Conf. Series: Materials Science and Engineering 1275 (2023) 012031 doi:10.1088/1757-899X/1275/1/012031

original image and a sub-image to be searched. The search is completed when the cross-correlation parameter defined in Eq.1 is maximised.

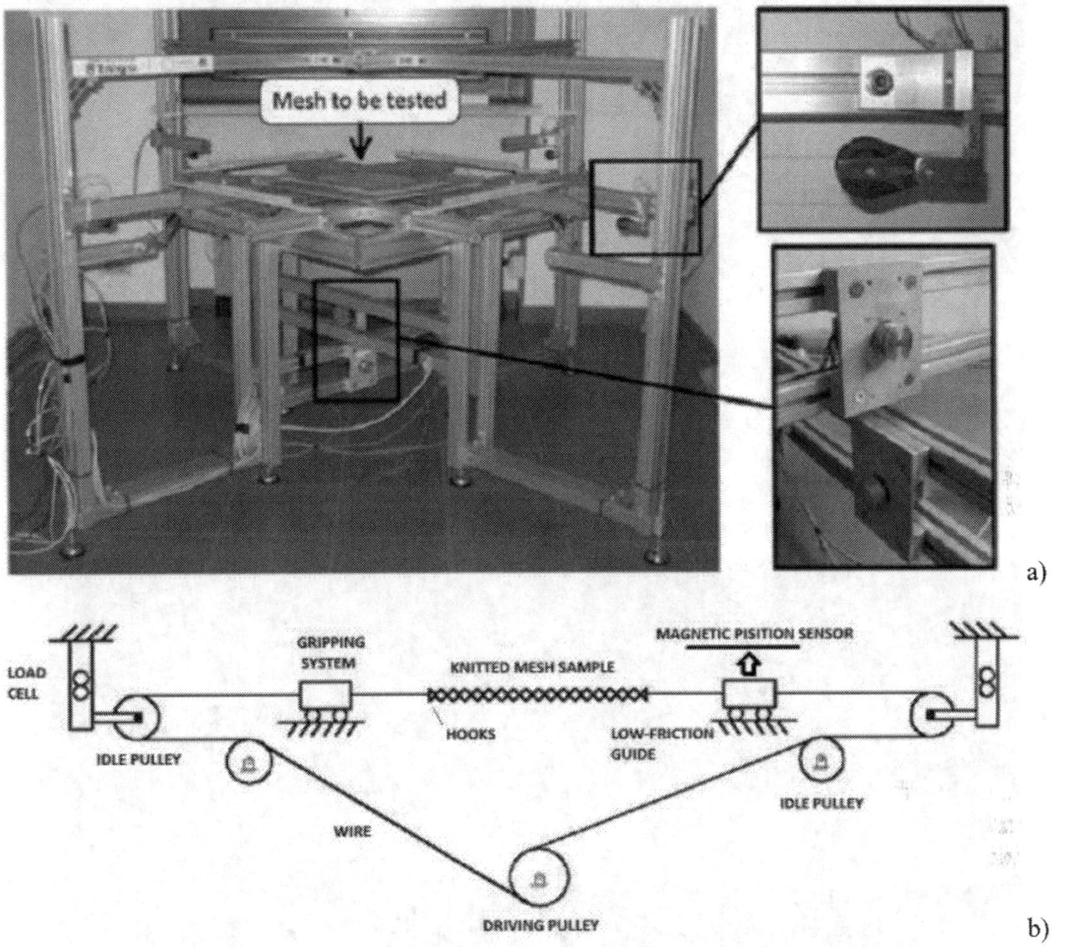

Figure 1. Picture of the biaxial traction machine a); measurement chain diagram b).

$$C(k,l) = \sum_{x=0}^{M-1} \sum_{y=0}^{N-1} X(x,y) \cdot H(x+k, y+l) \tag{1}$$

In this work, the cross-correlation technique is used to quickly and automatically determine the reference repetitive pattern, which is the minor portion of the mesh that repeats itself to form the whole mesh. A simple visual identification based on a single piece - without an appropriate averaging process - involves a too specific reference that conducts to unreliable deductions. In fact, the technique of knitting metal mesh is challenging to tune appropriately; therefore, the several loops are not precisely the same in the fabrics that are repetitive only in an average sense but not exactly.

Figure 2 a) shows the image of a mesh specially converted to black/white, the sub-image to be searched is highlighted in the red box. The latter is an arbitrary zone of any size and helps determine the prominent cross-correlation peaks shown in Figure 2 b).

IOP Conf. Series: Materials Science and Engineering 1275 (2023) 012031 doi:10.1088/1757-899X/1275/1/012031

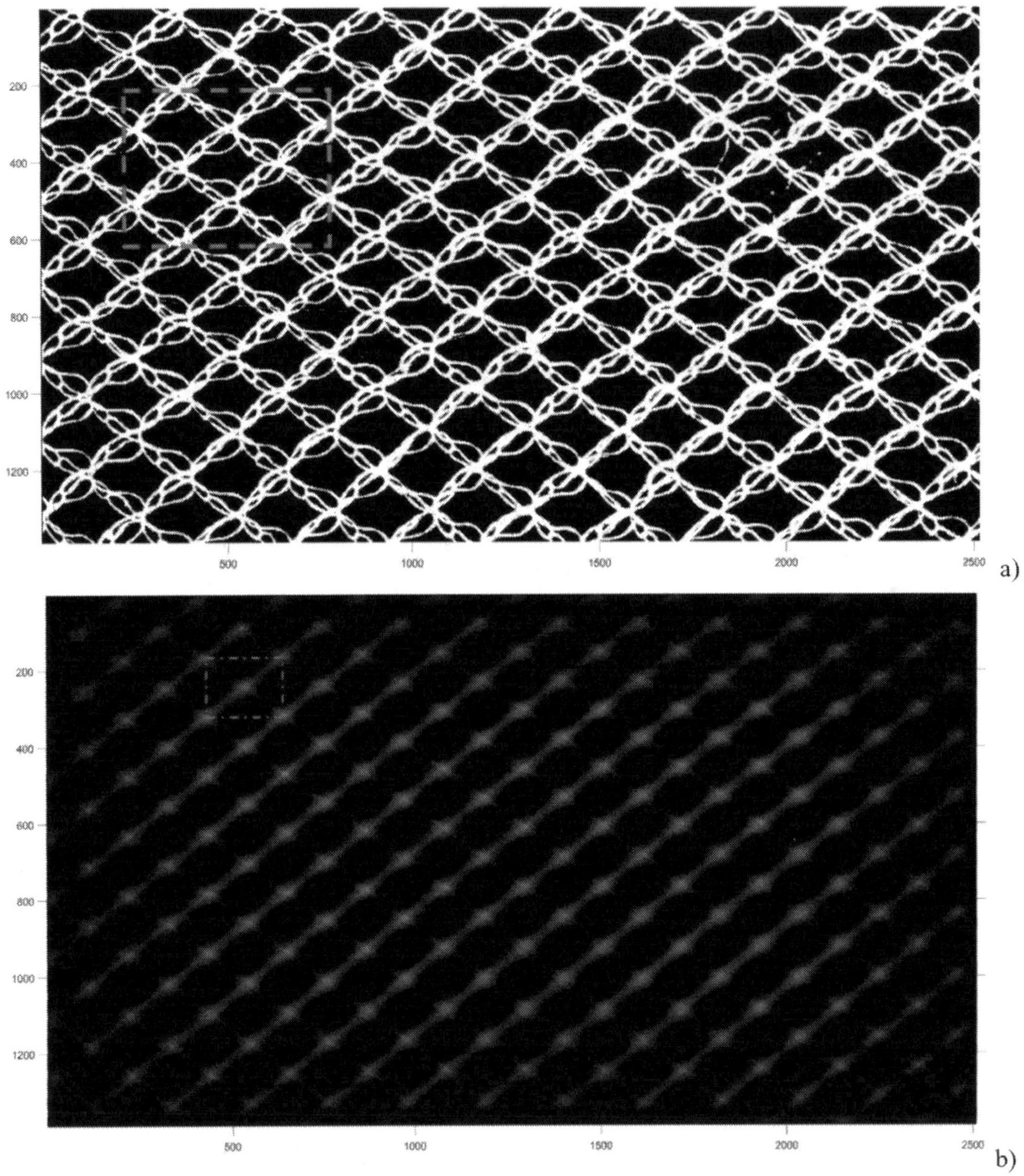

Figure 2. Cross-Correlation of images, in red the sub-image to search a); Peaks of Cross-Corr. b).

The cross-correlation peaks identify the position of the sub-image sought; furthermore, it provides the size of the elementary mesh, whatever the size of the searched repetitive pattern.

The last two pieces of information allow homogenising the geometric characteristics. Homogenisation is now possible, and it identifies a model mesh (repetitive pattern) representative of the entire mesh that is slightly affected by the variability always present in the manufacturing process. An appropriate statistical analysis of every pattern is necessary and independently performed for each picture. In the present work, after identification of the centre position of the repetitive patterns, the averaging is carried out by superimposing previously obtained grayscale images of every patterns. In this way, a normalised and homogeneous grayscale image is obtained, which characterises the average pattern of the mesh. In Figure 3, the overlapping of several sub-images with different percentages of overlaps are shown. The higher the rate, the better the model results in representative reliability. Once the total overlap is achieved, the resulting image is converted to black and white and analysed with segmentation and representation image processing techniques. Since the greater area's voids identify one of the principal mesh's electromagnetic behaviour, we focused on measuring the greatest area inside the averaged pattern.

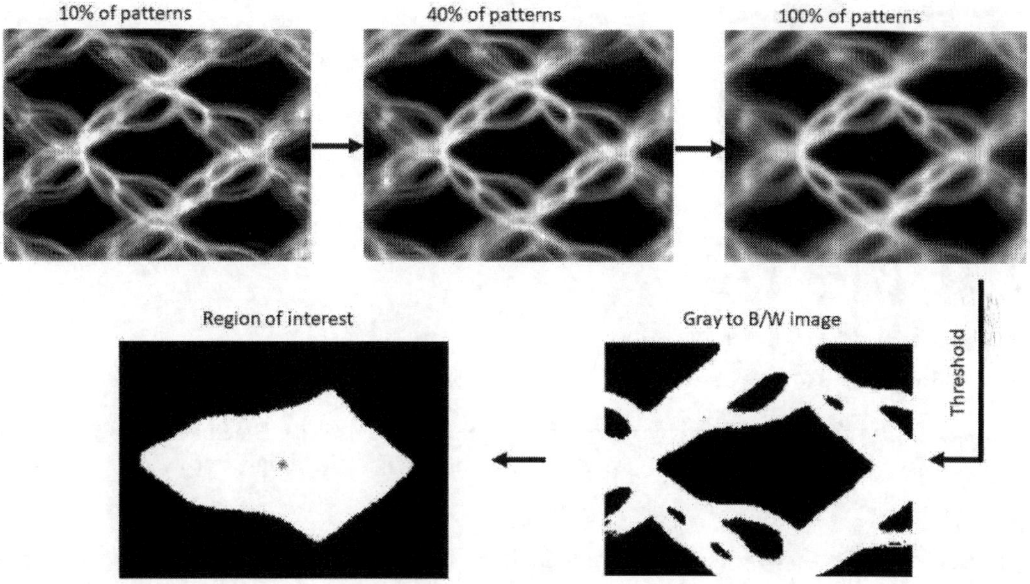

Figure 3. Overlapping of basic pattern to obtain the average one.

Switching grayscale images in black and white is the most delicate and essential operation. This transformation is called thresholding and consists in choosing a threshold value to binarise the grayscale image. A grayscale image comprises n x m pixels whose value ranges from 0 to 255. In binarisation, one moves to an image of n x m pixels that can only have two values, 0 or 1. It is necessary to establish the discriminating limit that associates the grayscale values below a specific value to 0, or above to 1.

In the paper, a basic global thresholding is used (some more sophisticated versions use more threshold values related to the illumination of part of the images) [13]. In the present application, the optimal choice is carried out by increasing the threshold value of a fixed step; at each value, the number of pixels of the largest detected area is calculated; when the value suddenly increases, the last threshold value is accepted. The sudden growth is connected with the fact that the major area of interest, due to the black and white conversion, merges with the other regions causing erroneous evaluations; for this reason, the previous value is accepted.

Choosing an incorrect thresholding value means losing information, such as the loss of some of the loops present in the mesh. Figure 4 shows the effect on the measurement of the greatest area as the threshold varies: in a) the coloured picture of the mesh considered, in b) the effect that the threshold value has on the detection of the main area; in c) it is shown how the pixel count of the main area varies according to the threshold value. A sudden increase represents the glueing of minor areas of no interest in the measurement. The optimal threshold must be acquired on every picture, since it is strongly influenced by environmental light, wire reflections and many other conditions.

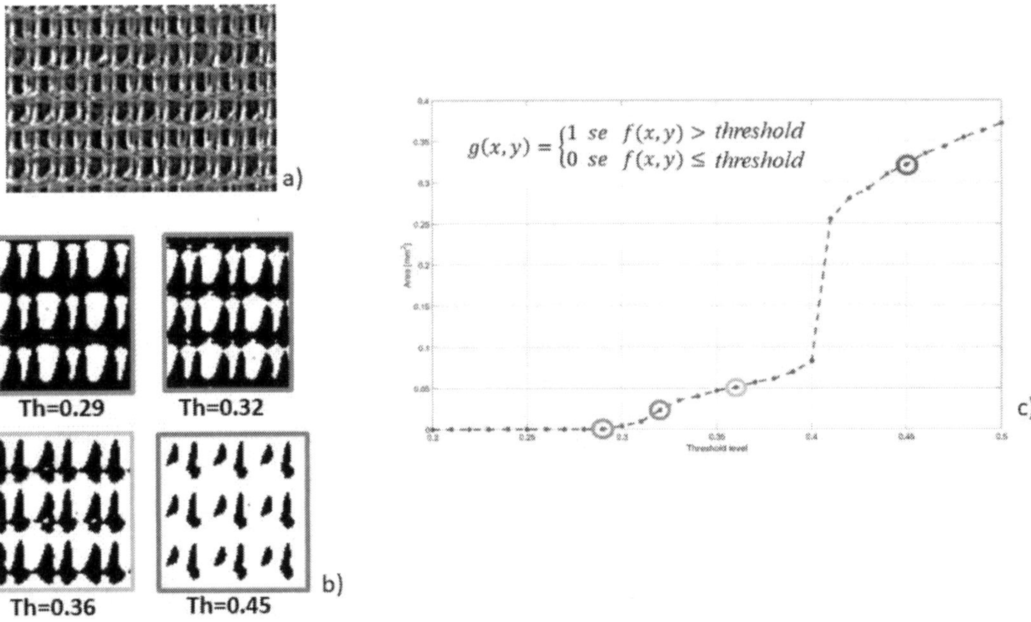

Figure 4. Effect of Thresholding on area measurement. Picture used a); Black and white result with different thresholding b); Area variation vs Thresholding value c).

3. Results

This paragraph shows the results of the analyses of the images with the previously proposed methods. The study focuses on optical measurement using imaging techniques of the principal and largest void, which has the most significant effect on the efficiency and polarisation of the electromagnetic signal.

We point out that the images are taken through a telecentric lens, so the scale conversion between mm and pixel is the same for every picture.

The meshes analysed in this work are of two types, generally used in the textile field. In fact, we aimed to verify if a general behaviour may be considered for differently knitted meshes, presenting a dissimilar shapes of the voids. Some geometric characteristics of the central void (ROI Region Of Interest) were measured after binarising the images.

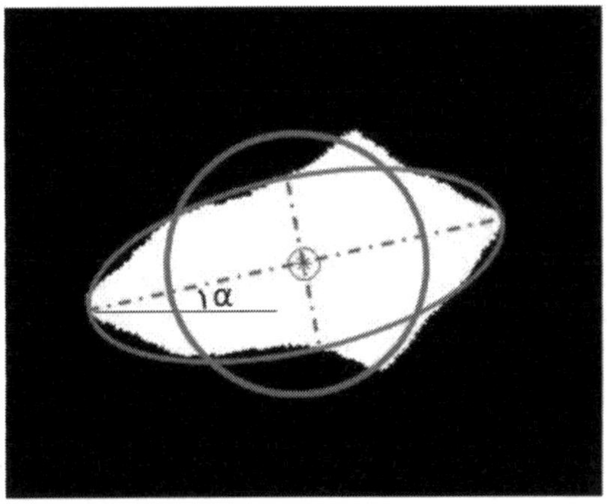

Figure 5. Some features of the area (white pixel): the red line is the major and minor axis; blu circle is the equivalent circumference; green ellipse is the inertia moment.

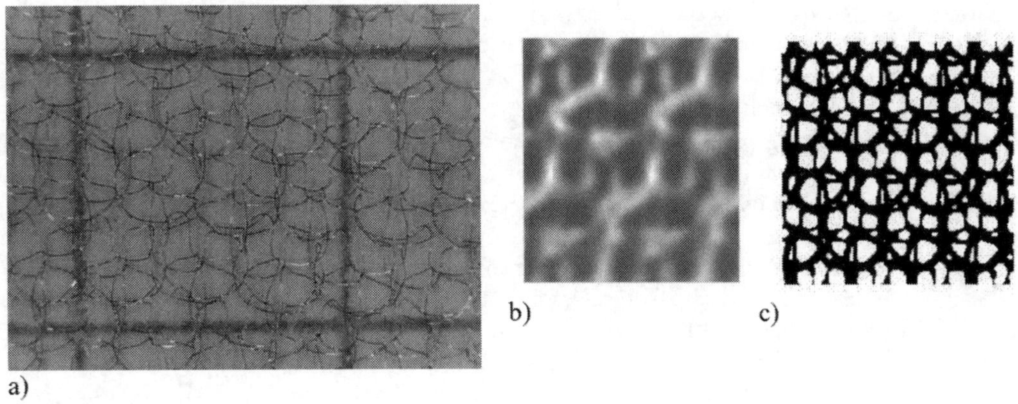

Figure 6. Type 1 mesh: a) picture of the mesh; b) average mesh in grey scale; c) effect of binarisation on average mesh.

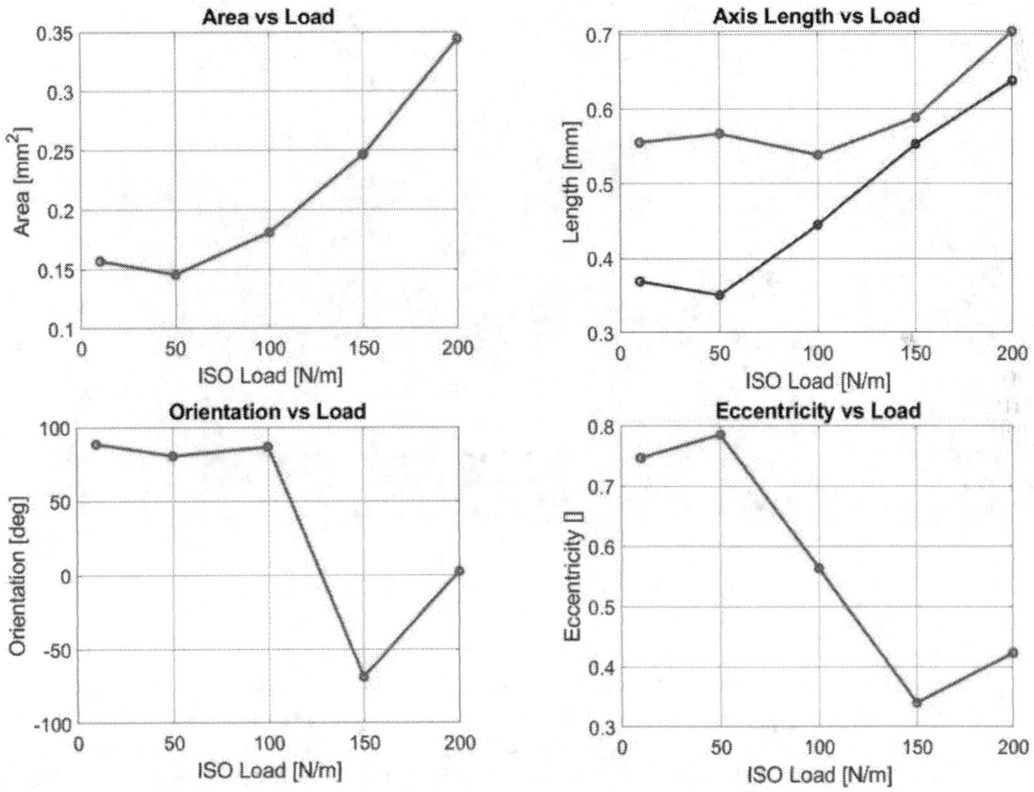

Figure 7. Measurements of some ROI features of Type 1 Mesh vs Iso-Load.

The geometric features are area, principal axes, inclination, equivalent diameter, and circularity. In Figure 5, these characteristics are sketched:

• Area: actual number of pixels in the region, returned as a scalar;

• Principal axis: length (in pixels) of the major and minor axis of the ellipse that has the same normalised second central moments as the region, returned as a scalar;

IOP Conf. Series: Materials Science and Engineering 1275 (2023) 012031 doi:10.1088/1757-899X/1275/1/012031

• Orientation: the angle between the x-axis and the major axis of the ellipse that has the same second-moments as the region, returned as a scalar. The value is in degrees, ranging from -90 degrees to 90 degrees;

• Eccentricity: eccentricity of the ellipse with the same second-moments as the region, returned as a scalar. The eccentricity is the ratio of the distance between the foci of the ellipse and its major axis length. The value is between 0 and 1. (0 and 1 are degenerate cases. An ellipse whose eccentricity is 0 is actually a circle, while an ellipse whose eccentricity is 1 is a line segment.).

The characteristics of the ROIs were measured for two different types of fabrics here: Type 1 and Type 2. The aim is to determine how the features of the ROI when the load varies for some biaxial state. The loads range between 11 N / m - 200 N/m and 11 N/m - 250 N/m, respectively, for the Type 1 and 2 meshes. Figure 6 shows the images of the first mesh a), the average image in grayscale b), and its binarisation for evaluating the ROIs c).

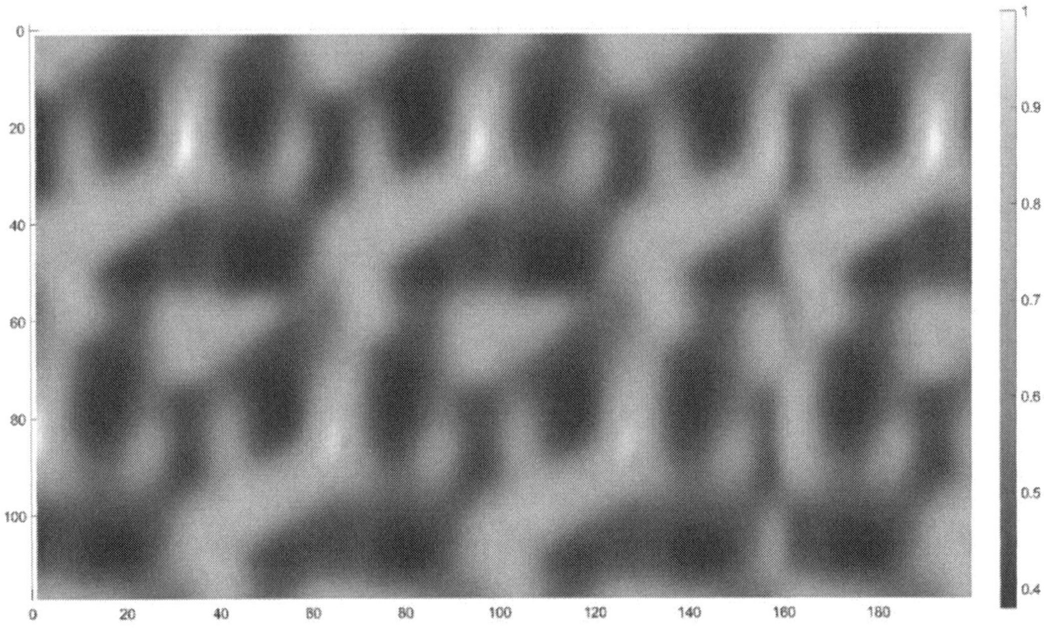

Figure 8. Mass distribution Type 1 Mesh at ISO-Load on 11 N/m.

Figure 7 shows the results of the measurements carried out on the greater void of the Type 1 mesh as the iso-load varies. In this case, there is a significant increase in the greater area with a non-linear trend according to load. There is also an increase in the axes of inertia and a reduction in eccentricity, which corresponds to a shape closer to a circle. The orientation of the major axis undergoes a sudden and non-relevant jump (180° shift is irrelevant when measuring directions); furthermore, the more the circle shape is gained the more the principal directions are non-significant.

Using the average mesh, it is possible to show the mesh's average mass distribution. Figure 8 shows the mass distribution for the Type 1 mesh; the blue colour identifies areas where the wire is never located in a statistical sense, and light yellow indicates that yarn is statistically always present. This image is particularly interesting: in fact, it gives information on both the suitability of the adopted knitting technique and the quality of the manufacturing process. This image provides the average mass distribution, and it is also an essential input for the study and modelling of the electromagnetic performance of the reflector.

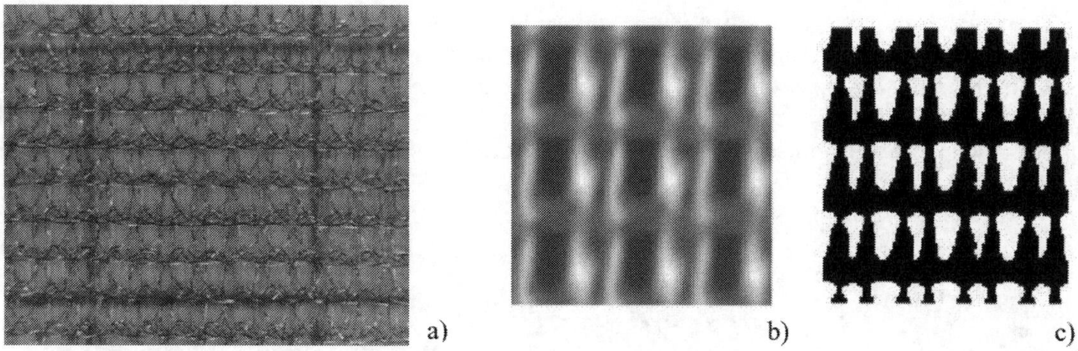

Figure 9. Type 2 mesh: a) picture of the mesh; b) average mesh in grey scale; c) effect of binarisation on average mesh.

Figure 9 shows the image of the Type 2 mesh a), the average image in grayscale b), and its binarisation for evaluating the ROIs c).

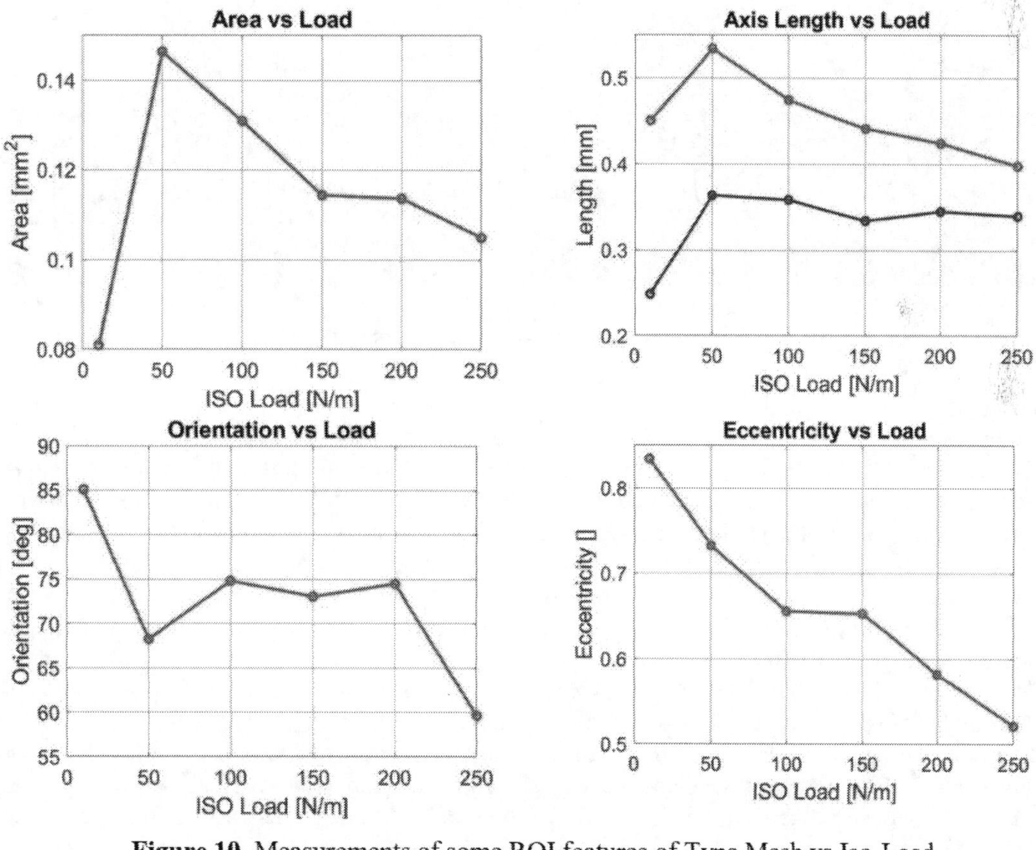

Figure 10. Measurements of some ROI features of Type Mesh vs Iso-Load.

Figure 10 shows the results of the measurements relating to the Type 2 mesh. In this case, there is a sudden increase in the greater area with a subsequent linear decreasing trend following the load. There is an almost constant length of the inertia axes towards loading accompanied by a substantial reduction of the eccentricity of the void shape. The orientation of the major axis remains relatively stable, which means that the void always keeps oriented in the same direction.

The comparison between the two meshes suggests that the void behaviour towards applied load is not independent of the knitting technique. Each mesh can present different void features with applied loads. In this work, only iso-loading was considered, but non iso-loading would open very interesting disposal of the mesh to optimise the RF characteristics.

In Figure 11, the mass distribution for the Type 2 mesh is shown; note the greater regularity of the shapes compared to the Atlas-tuch mesh.

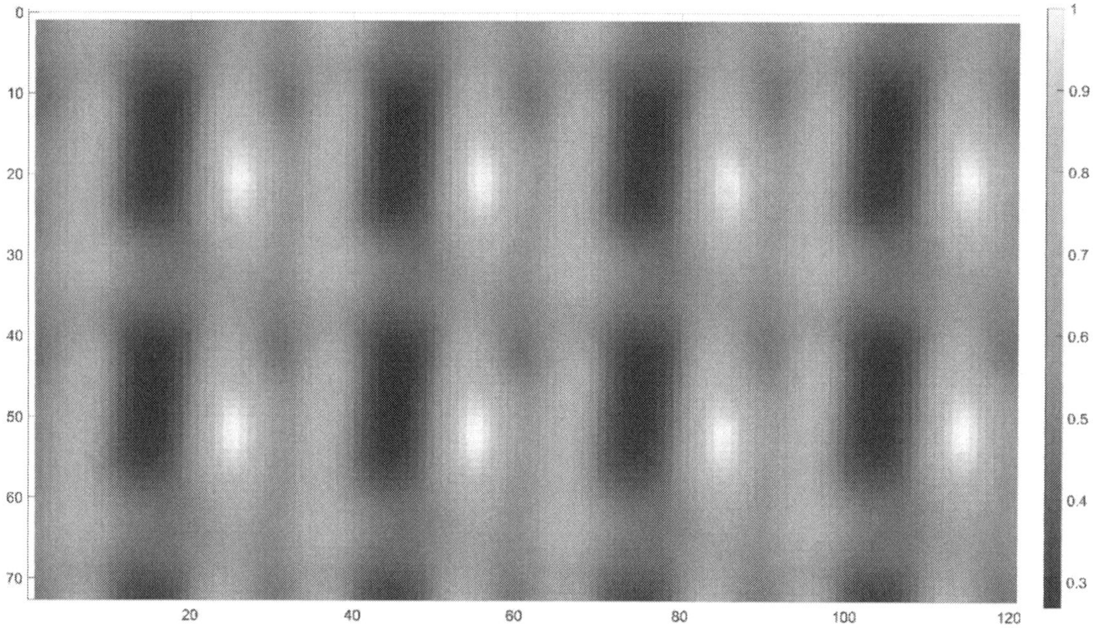

Figure 11. Mass distribution of Type 2 Mesh at ISO-Load on 11 N/m.

4. Conclusions

In the present work, some image analysis techniques have been proposed that allow direct and contactless measurements of repetitive patterns in astromeshes. Their texture shows numerous voids that influence the electromagnetic response. In the paper, the shape variation of the main voids is analyzed when the mesh is subjected to two-dimensional loading. The loads considered are the same in both directions, but the method is also suitable for analysing different combinations to study ad hoc configurations. The analysis involves image processing technique as cross-correlation and thresholding to compute some geometric features.

Experience has shown that the shape and orientation of the larger mesh area can change significantly with the applied load, generating effects on the polarisation to be kept under control for optimal design. Furthermore, the paper presents a new approach to obtaining a sort of average pattern that allows homogenization of the shape of the elementary mesh. This allows to avoid of local measurements that would introduce local measures non representative of the mesh. All measurements should be conducted through statistical analysis, if used to characterise a reflector, made of an enormous number of repetitive patterns.

With the purpose to keep under control any polarization effect of the reflectors, non-equal loading of the mesh while mounting it on the net could be considered. However, these pre-loadings should be analysed specifically for every mesh, since no general indications can be a-priori formulated.

References

[1] Thomson MW 2000 The astromesh deployable reflector, *Iutam-Iass Symposium on Deployable structures: Theory and Applications* **80** p 435

[2] Ozawa S 2014 Design Concept of Large Deployable Reflector for Next Generation L-Band S.A.R. Satellite, *Proceedings of the 2nd International Conference on Advanced Light weight Structures and Reflector Antennas* Georgia

[3] Datashvili L [et al.] 2014 Advanced architectures of large deployable mesh reflectors: from medium to very large size. *Proceedings of the 2nd International Conference on Advanced Lightweight Structures and Reflector Antennas* Georgia

[4] Eltahan E, Sultan M ans Mito A 2016 Determination of loop length, tightness factor and porosity of single jersey knitted fabric. *Alexandria Engineering Journal* **55** 851

[5] Chavkin V.P., Ilin E.P. and Molcanov A.S. 1975: Automatic control and regulation of textile materials, Moscow.

[6] Pontoppidan K and Cappellin, C 2010 Latest Achievements on Reconfigurable Reflector Antennas Modelling, *32nd ESA Antenna Workshop on Antennas for Space Applications Digest*

[7] Scialino L, Salvini P, Migliorelli, M, Pennestrì E, Valentini P P, Van't Klooster K, Santiago Prowald J and Gloy Y., 2014 Structural Characterization and Modeling of Metallic Mesh Material for Large Deployable Reflectors, *Proceedings of the 2nd Inter. Conf. on Advanced Lightweight Structures and Reflector Antennas* Georgia

[8] Valentini P P, Falcone M, Marotta E, Pennestrì E and Salvini P 2016. Theoretical and experimental characterization of a FEM element assembly for the simulation of very compliant knitted mesh, *International Journal For Numerical Methods in Engineering* **107** p 419

[9] De Salvador W, Marotta E and Salvini P 2018 Strain Measurements on Compliant Knitted Mesh Used in Space Antennas, by means of 2D Fourier Analysis *Strain*

[10] Marotta E, De Salvador W, Pennestrì E, Salvini P, Scialino L and Valentini P P 2016. Some problems arising during the experimental characterization of compliant knitted mesh. *37th Esa Antenna workshop, Esa/Estec,*

[11] Sutton M A, Orteu J J and Shreier H W 2009 Image Correlation for Shape, Motion and Deformation Measurements. Basic Concept ,Theory and Applications, *Springer New York*

[12] Gonzalez C R and Woods R E 2002 Digital Image processing, second edition *Prentice Hall* New Jersey

[13] Roy P, Dutta S, Dey N, Dey G, Chakraborty S and Ray R 2014 Adaptive thresholding: A comparative study *International Conference on Control, Instrumentation, Communication and Computational Technologies* p 1182

IOP Conf. Series: Materials Science and Engineering 1275 (2023) 012032 doi:10.1088/1757-899X/1275/1/012032

Design and implementation of an innovative wind power system for beach erosion protection and nourishment

G P Demelio[1,2], A Castellano[1] and G Bergamini[2]

[1] Polytechnic University of Bari - Department of Mechanical Engineering, Mathematics and Management (DMMM), Via Orabona 4, 70125 Bari, Italy
[2] PoliMech srl Strutture Meccaniche Innovative, Strada Scizze 7, 70126 Bari, Italy

giuseppepompeo.demelio@poliba.it

Abstract. Beach erosion phenomena have become increasingly frequent and devastating with serious economic damage on tourism. A relatively recent technique of beach protection and nourishment is based on the creation along the shoreline of an unsaturated sandy zone, in which the level of the phreatic surface is lowered by means of buried drains connected to electric suction pumps. The main problem encountered in this solution is the need to guarantee the drainage when the erosive activity is maximum in the presence of intense thunderstorms that often produce malfunction and outage problems of the power supply. This paper shows the realization of a "green" solution based on a totally mechanical design and construction approach, achieved by connecting a 3.5-meter single-bladed wind turbine to a submerged axial pump with optimized volute by means of a vertical shaft placed inside the tower. The system is managed by a mechanical pitch controller studied for starting and limiting power in exceptional winds.

1. Introduction

Beach erosion phenomena have become increasingly frequent and devastating, with considerable economic damage recorded in many areas of great tourist interest [1,2]

A relatively recent technique to counter coastal erosion and to promote the nourishment of beaches is based on the creation of an unsaturated sandy zone along the shoreline [3], in which the level of the phreatic surface is lowered by means of buried drains connected to electric suction pumps. The water carried by the wave, percolates into the drained ground and the solid material present in the water is deposited on the beach without being removed again. The process continues until obtain a new equilibrium of the beach in a more advanced position towards the sea. Such system, called "dewatering," devised and patented in 1983 by the Danish Geotechnical Institute [4,5], is neither invasive nor visible and has a relatively limited cost compared to the use of emerged or semi-submerged structures, which disfigure the coastal environment and worsen the water quality from a bathing perspective.

The application of this technique remained for many years at the level of research and experimentation on a limited number of cases, with results, moreover, sometimes even spectacular [6,7,8], when some beaches grew back spontaneously, after the installation of the systems, up to 25 to 30 meters, but with a limited life span of only a few years due to the required continuity in power supply, maintenance and management entrusted to public agencies, which often do not respond readily to the required operational needs.

Content from this work may be used under the terms of the Creative Commons Attribution 3.0 licence. Any further distribution of this work must maintain attribution to the author(s) and the title of the work, journal citation and DOI.
Published under licence by IOP Publishing Ltd

The main problem encountered in the realized applications is the need to ensure the operation of the drainage pump when erosive activity is at its highest, thus in the presence of intense thunderstorms which frequently produce problems of malfunction and interruption of the electrical power supply.

Critical analysis of a series of failures that occurred after an encouraging start sees as concomitant causes the unreliability of the pumps' power supply and the unavailability of pumps suitable for the application. These inconveniences, which were not diagnosed in a timely manner, have held back the spread of this protection and beach nourishment system, referred to as "soft", because it has no visible fixed structures or substantially alters the conformation of the beach.

The analysis of the physical mechanisms underlying the dewatering protection and nourishment system is sufficiently well established, and related mathematical models are reported in various research papers available in the literature [9,10].

2. Wind power for beach protection and nourishment

The innovative system proposed consists in a high-performance wind turbine mechanically regulated and connected directly to the suction pump for the drains [11], which also has a high content of technological innovation. It takes into account the excellent correlation between the erosive activity of the sea and the intensity of the wind that feeds the wind turbine [12] and on design assumptions that can minimize maintenance work and ensure an operational life of 20 years. This solution could be a viable alternative to the "heavy" beach nourishment techniques, carried out with conventional works based on periodic removing large amounts of sand from the seabed and transferring it to the beaches or building protective works that significantly alter the landscape.

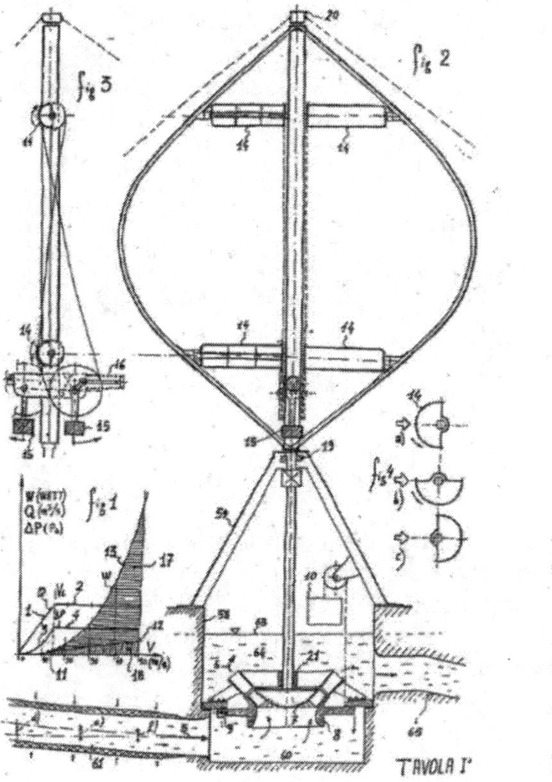

Figure 1. Vertical axis turbine configuration

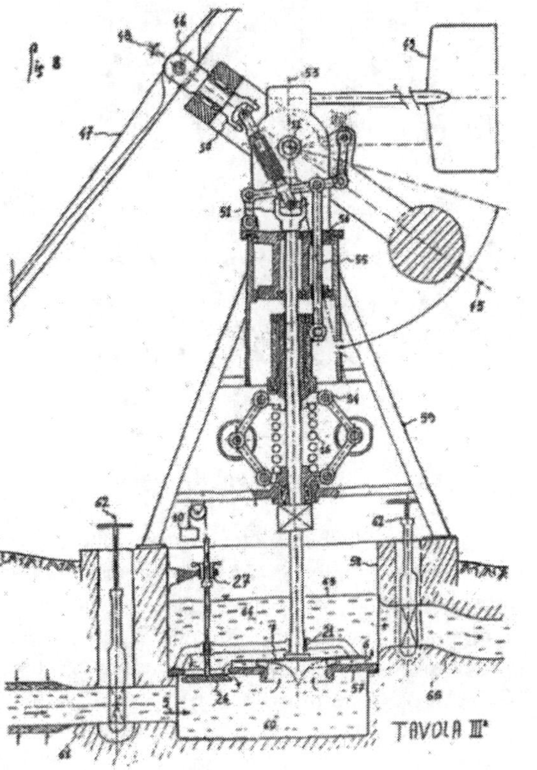

Figure 2. Fixed-pitch two-blade configuration

Different types of horizontal-axis and vertical-axis wind turbines were considered.

Figure 1 shows the technical solution adopting a vertical-axis turbine of the Darrieus type, with its starting and control system. Figure 2 shows a fixed-pitch two-blade type solution with control obtained by varying the angle that the axis of the turbine forms with the horizontal plane.

Figure 3 shows the first study carried out on a variable-pitch single-blade configuration with a horizontal axis, later adopted in the scheme of the final solution shown in Figure 4, based on a variable-pitch single-blade architecture, with the axis of rotation slightly inclined with respect to the horizontal plane and with a mechanical pitch controller placed in a bulb that allows for effective flow even at the root of the blade.

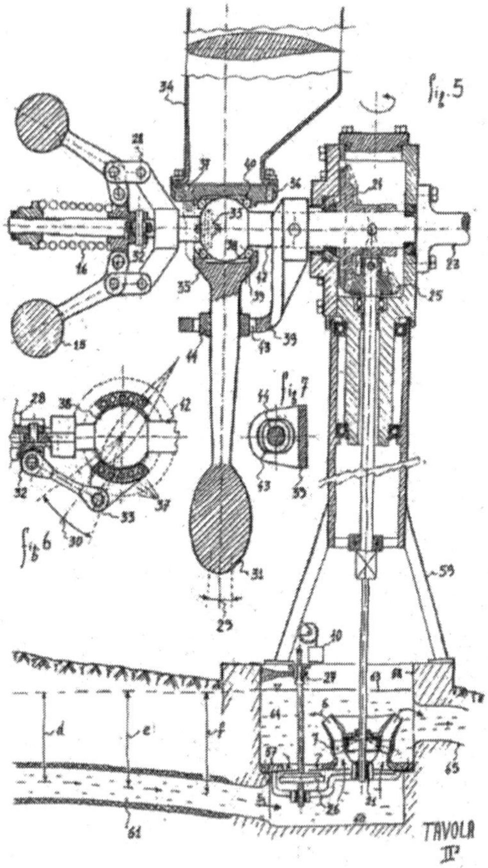

Figure 3. Variable-pitch single-blade solution

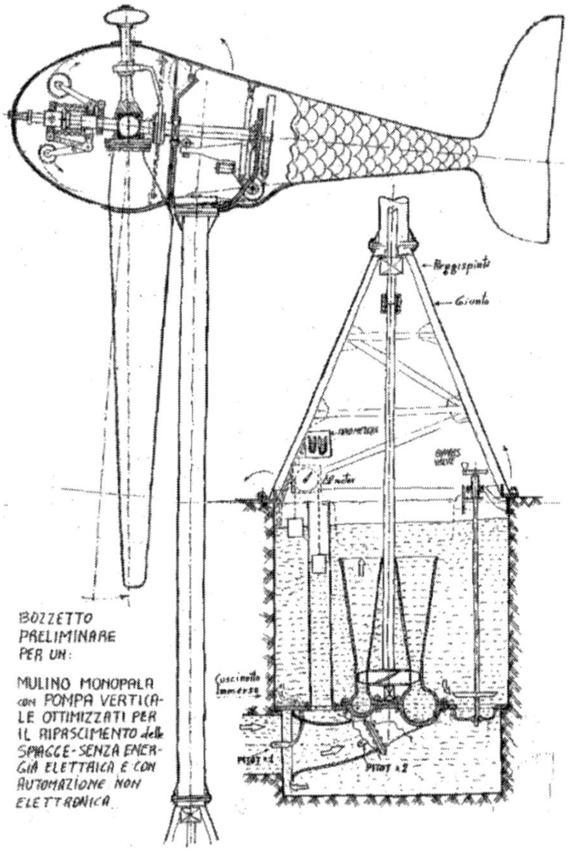

Figure 4. Final configuration

The single-blade architecture was chosen in order to reduce costs and to be able to simplify the introduction of an optimized mechanical pitch management controller both in the case of winds with limited speed, since the beach nourishment activity is more effective in the case of slightly rough sea, and in the case of winds with very large speeds, since it is necessary to take advantage of the draining capacity of the high-powered axial pump to ensure protection in the case of sustained swells.

No wind machine meeting the expected characteristics was available or even known, a special wind turbine, designed to work properly with gusts up to 200 km/h, was studied and developed.

This is high-efficiency, was made with completely mechanical components in order to ensure a life span of more than 20 years and an inherent invulnerability to lightning, the real enemies of all electrical equipment, including electric pumps used to date in "dewatering."

2.1 The wind turbine

In the study, it was taken into account that this is a very special wind application because of the coupling characteristics between the wind turbine and the pump, in that flow rates and heads increase with increasing wind energy in correspondence with greater beach erosion action, which, negligible at low wave, is maximum at large swells, which are accompanied, in turn, by maximum values of wind energy. A single-bladed aeromotor with a radius of 3.5 m was adopted, with NACA 63012 laminar profiles, suitable to ensure adequate efficiency and minimum noise (less than 35 dB(A) at 20 m).

The wind turbine is suitable for proper operation up to maximum wind speed unlike conventional wind turbines, which are blocked beyond a predetermined wind speed limit, a not permissible situation for this application.

Expecting to protect up to 1,000 to 1,200 meters of beach, the pump is designed to operate at high efficiency at variable revolutions and with hydraulic characteristics controlled by the special regulator of the blade pitch and thus of the rpm. The power performance as a function of wind speed and the pump characteristic curve are shown in Figure 5. The lay-up of the blade and the resulting realization are shown in Figure 6.

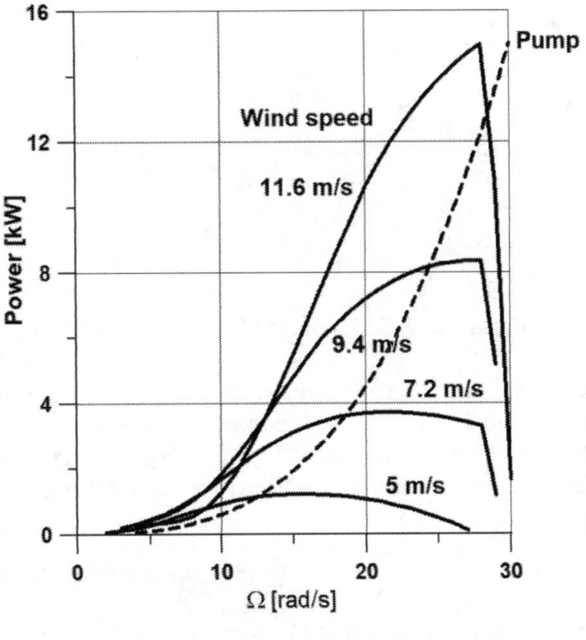

Figure 5. Power curve of turbine and pump

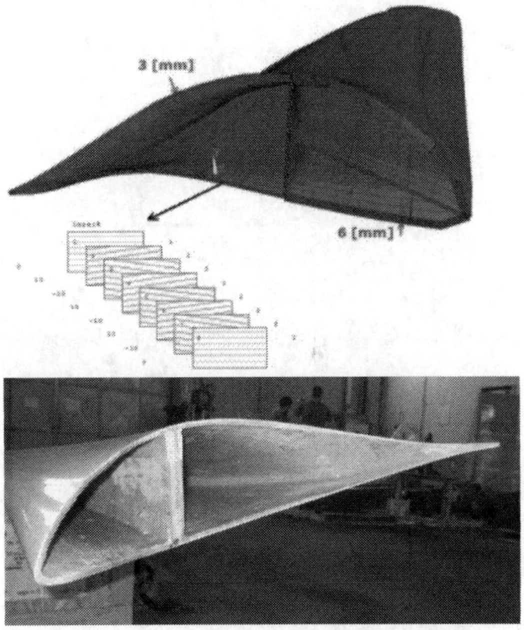

Figure 6. Blade lay-up and prototype

2.2 The mechanical pitch control

The mechanical pitch controller constitutes a particularly challenging subsystem having to ensure reliability and durability of more than 20 years.

Analysis of the causes of wind system outages has shown that one of the weak points for reliability is electronics failures, often caused by lightning, which also irreparably damage the chain of ball and roller bearings that electrically connect the blades to the tower mass. It is only in recent years, therefore, that a great deal of attention to these issues has arisen.

From these considerations came the decision to design a fully mechanical pitch controller, capable among other things of properly managing the pitch in relation to the wind speed of harnessing the associated wind power.

In this design, the bearings have been isolated and the lightning current is routed to special graphite sliding brushes.

The pitch controller consists of two distinct mechanisms driven by the centrifugal forces of masses having appropriate value, the first designed to ensure optimal start-up even in winds with limited speed so as to promote beach nourishment activity, and the second to operate the system even in very high winds by delivering maximum power to the pump but limiting the stresses acting on the structural elements. The optimal pitch in the start-up phase as a function of turbine angular speed was obtained from the analysis of the fluid dynamic behavior of the blade. The mechanism was studied with parametric geometric and mechanical modeling carried out with the Mathcad program until an almost perfect overlap was obtained between the optimal and actual curves as shown in Figure 7.

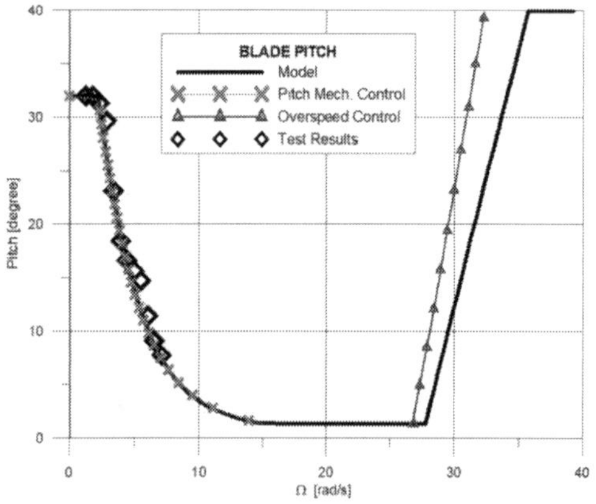

Figure 7. Blade pitch versus turbine speed

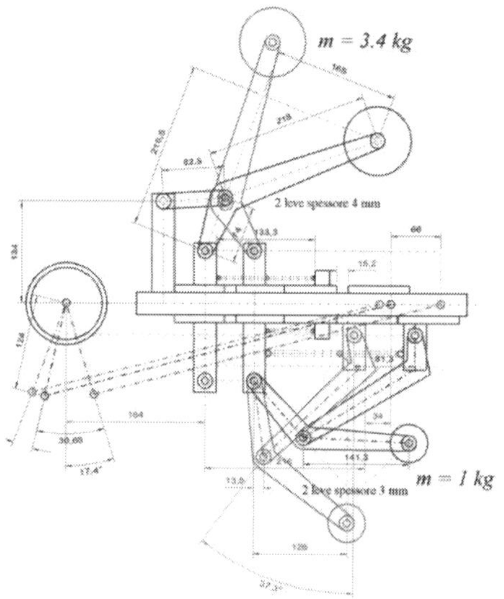

Figure 8. Functional drawing of the controller

To ensure balance during rotation, the pitch and the over speed control mechanisms were made in opposing pairs and were arranged at 90°. The drawing in Figure 8 shows the geometry of the two mechanisms. The result of the realization is shown in Figure 9. The controller was mounted on a CNC lathe to verify compliance with the design specifications. Figures 10a and 10b show the pitch and over speed controller during testing.

Figure 9. Prototype of the controller

Figure 10a. Test of the over speed controller

Figure 10b. Test of the pitch controller

2.3 The axial pump

In this design, the pump is axial type submerged in seawater, with no electrical parts and connected with a vertical axis to the wind rotor by a pair of hypoid gears.

The airfoils adopted are from the NACA 63XXX series with thicknesses varying from 6 percent at the tip to 8.6 percent at the root, with curvature (mean-line) also varying with radius and proportioned to ensure, at the design flow rate, the full power required. In this way, the profiles work in the center of the laminar behavior with minimal frictional resistance.

To ensure the long life of the pump, it was planned to be constructed entirely of fiberglass composite, using resins formulated, in the chemical plant industry, to make pumps and valves for nitric acid and sulfuric acid, agents thus far more aggressive than seawater.

In the first third of the profile on the molds a special epoxy gel-coat loaded with ceramic powder was placed, in order to have maximum resistance to abrasion due to any traces of sand.

The traditionally most delicate component, the thrust bearing, has been placed at the top, in the bevel gearbox, in an area therefore well protected and lubricated, while in the water there is only a centering bushing of the impeller, made with a rubber coating, the same device used in the axis guidance of the propellers of vessels and fishing boats and widely tested.

Even for this critical component, an operational life of twenty years has been predicted, while still being easy and inexpensive to replace. Downstream of the fluid dynamic design of the pump, the option of a bronze casting with from a marine propeller manufacturer was considered for the realization of the impeller, but in this case both long-term durability, with respect to seawater aggression, and perfect conformity of the prescribed laminar profiles, achievable only by die casting or fiberglass mold, with relative loss of hydrodynamic efficiency, would have been lost.

The final solution adopted was to produce in segments of one-sixth circumference the model of the blade of the blade support surface by means of rapid prototyping techniques, and to make the single mold of the rotor interior after putting together and bonded the six segments obtained as shown in Figure 11

This made it possible to have an impeller whose wing profiles are perfectly smooth and geometrically in accordance with the design.

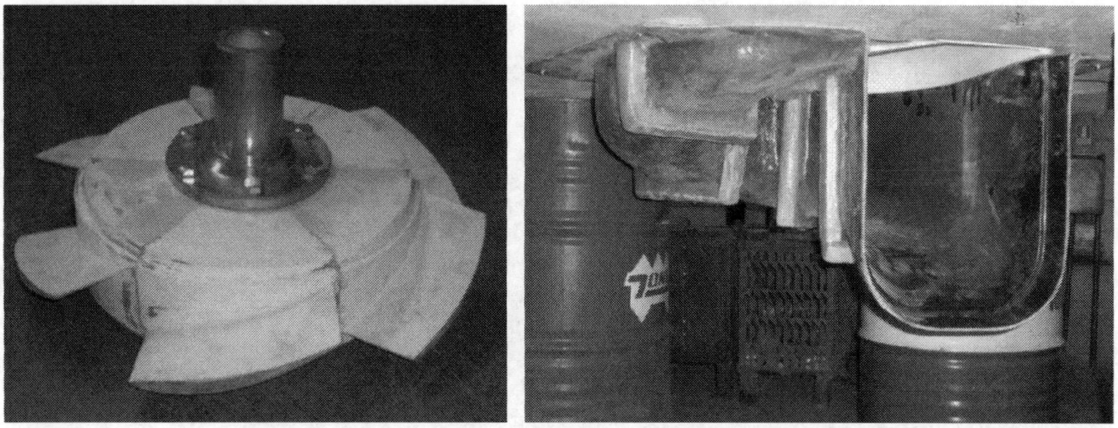

Figure 11. Prototypes of the axial pump impeller and volute

2.4 The testing system for simulating the drainage

The prototype of the complete system was assembled and tested near the factory that made it and is shown in Figure 12.While waiting for the possibility of realizing a test plant, in order to verify the effectiveness of the system and speed up the time, a drainage simulation system based on flow circulation on a specially designed valve was realized. The system consists of a cylindrical vessel with an inner diameter of 2 meters and a height of 4 meters. The vessel, filled with water, has total weight of 15 tons and the wind turbine can be mounted above it.

Figure 12. Prototype of the complete system

The wind turbine is 11 meters high at the rotor axis and the total height from the ground is 15 meters. A pressure difference can be measured above and below the pump, which is generated by the suction action of the pump itself.

Drainage resistance was obtained with a valve of considerable diameter, Ø 320 mm, characterized by a cylindrical plug whose lateral surface is fitted with a series of holes of variable diameter so as to ensure a range between a maximum cross-section at full opening of 45100 mm^2 and a minimum cross-section of 1000 mm^2.

The useful stroke simulating a given drain range is 80 mm within which the pressure at differential varies in the ratio of 10 to 1 (i.e., from 32500 to 3250 Pa) while the flow rate varies in the ratio of 3.15 to 1.

The top of the test system is shown in Figure 12. A set of steel cables have been adopted in order to ensure the stability of the tower, mounted on a relatively small base.

3. Conclusion

Standard dewatering installations, implemented using electric pumps, require systematic monitoring and maintenance that are often unreliable and costly. The analyses conducted on implementations to date do not agree that dewatering technology is 100 percent effective. A critical analysis of the causes of failures showed that the best results were obtained on systematically monitored plants where perfect efficiency was guaranteed during the monitoring period.

The proposed plant, based on the good correlation between the erosive activity of the sea and the intensity of the wind powering the wind turbine, and on design assumptions that can minimize maintenance interventions and guarantee a 20-year operational life could be a viable alternative to "heavy" beach nourishment interventions with classical techniques.

Only the wind turbine is visible, which can be made aesthetically pleasing and maintenance-free. The system has limited installation costs and virtually zero operating costs. Paradoxically, this item is also a limitation in that beach nourishment interventions based on standardized techniques, such as dredging, sand handling, and the construction of expensive fixed works are more attractive to the companies involved, thanks to spending forecasts in the tenders by a factor of 5 times higher. Large material movements and the use of large vessels offer more political visibility to administrators and are more understandable to citizens.

In any case, the possibility of providing protection without bringing power lines near the beach and with renewable energy sources should be emphasized.

4. References

[1] Silvestri F 2018 The impact of coastal erosion on tourism -A theoretical model. *Economics Letters* **8** pp 806-13

[2] Jennings S 2004 Coastal tourism and shoreline management. *Annals of Tourism Research* **31** pp 899-922

[3] Vesterby H 1994 Beach face dewatering – The European experience *Proceedings of the 1994 National Conference on Beach Preservation Technology* pp 53-68.

[4] Ovesen N K and Schuldt J C 1992 Beach management system Documentation. Job No. 300 0141. *Danish Geotechnical Institute Lyngby Denmark*

[5] Richards T 2002 Beach management system Technology Introduction. Footprints – *Beach Drainage Newsletter* **5**

[6] Fischione P, Pasquali D, Celli D, Di Nucci C and Di Risio M 2022 Beach drainage system: a comprehensive review of a controversial soft-engineering method *J. Mar. Sci. Eng.* **10**(2) p 145

[7] Turner I L and Leatherman S P 1997 Beach dewatering as a 'soft' engineering solution to coastal erosion: A history and critical review. *J.Coast. Res.* **13**, pp 1050 - 63.

[8] Ogden M R and Weisman R N 1991. Beach Stabilization using Drains – An experimental model study. Coastal Sediments ASCE **2**, pp 1955-69.

[9] Li L, Barry D A and Pattiaratchi C B, 1996 Modelling coastal ground-water response to beach dewatering *J. of Waterway, Port, Coastal and Ocean Eng.* **122** No.6 pp 273-80

[10] Baird J A and Horn D P. 1996 Monitoring and modelling groundwater behaviour in sandy beaches *J. of Coastal Research* **12** No. 3 pp 630-40

[11] Azzellino A, Lanfredi C, Riefolo L, De Santis V, Contestabile P and Vicinanza D 2019 Combined Exploitation of OffshoreWind and Wave Energy in the Italian Seas: A Spatial Planning Approach*Front. Energy Res.* **7** p 42.

[12] Bergamini G and Demelio G 2004 Sistema perfezionato di ripascimento delle spiagge basato sull'energia eolica. *Ufficio italiano brevetti e marchi* 0001322270

IOP Conf. Series: Materials Science and Engineering 1275 (2023) 012033 doi:10.1088/1757-899X/1275/1/012033

Design and simulation of a wireline core drilling device for near-shore applications

M Cocci[1], S Barletta[1], L Berzi[1*], R Capitani[1], M Delogu[1], E Mariottoni[2], M Sacchetto[3] and L Pugi[1]

[1] Università degli studi di Firenze, Department of Industrial Engineering (DIEF), Via di S.Marta 3, 50139 Firenze (FI) - Italy
[2] MESA Mariottoni Group, Via Frassina 65, 54033 Carrara (MS) - Italy
[3] Massimo Sacchetto Geotechnical and geognostic consulting, Via Bomba 13, 35030 Cinto Euganeo (PD) - Italy

*email: lorenzo.berzi@unifi.it

Abstract. The PROTEUS project involves the construction of a self-propelled and floating amphibious machine for near shore geological surveys. In this context, the development of a wireline core drilling device was tackled, evolving with respect to existing models and responding to the specificities of the new application. In the preliminary phase, comparative research was carried out both through market research and through reverse engineering activities on core drill models deemed suitable for the purpose, thus identifying the functional priorities and criticalities of the existing material through the ex-post analysis of the subject elements to wear and tear. Once the requirements of the new core drilling device were formalized, the constructive layout was defined, giving priority to functions such as sample management, interlocking and rapid disengagement from the drilling actuation system. The construction details of the mechanical coupling profiles were then optimized by simulation in a multibody software environment, in order to ensure full functionality even in harsh environments, typical of the application, making an advancement with respect to the previously identified reference solutions. Finally, the executive design of the component was carried out by introducing the corrections deriving from the simulation results and considering different solutions in relation to the material adopted and its production needs through machining processes.

1. Introduction

The current activity takes place in the context of the PROTEUS project, which is focused on the design and construction of an amphibious platform for near shore geological prospecting activities [1,2]. Such platform is aimed at facilitating and reducing the costs of drilling and exploration activities due to its characteristics, such as flexibility, compactness and ability to move on different surfaces. Such machines are expected to gain interest in the future due the increasing relevance of seaside installations, especially near shore ones, for applications frequently related to energy generation and use such as underwater data centers [3,4], wind farms whose growth is undoubted worldwide [5], wave energy extraction [6] and in general for site characterization [7].

In line with modern machine design criteria, modeling and simulation activities have been extensively performed to virtually prototype all machine subsystems and control logic [8], thus reducing development time, aiming to provide a near-finished prototype while reducing calibration and correction

Content from this work may be used under the terms of the Creative Commons Attribution 3.0 licence. Any further distribution of this work must maintain attribution to the author(s) and the title of the work, journal citation and DOI.
Published under licence by IOP Publishing Ltd

activities, which can occur in case of the first exemplary of a new machine; the semi-final design of the machine is shown in Figure 1.

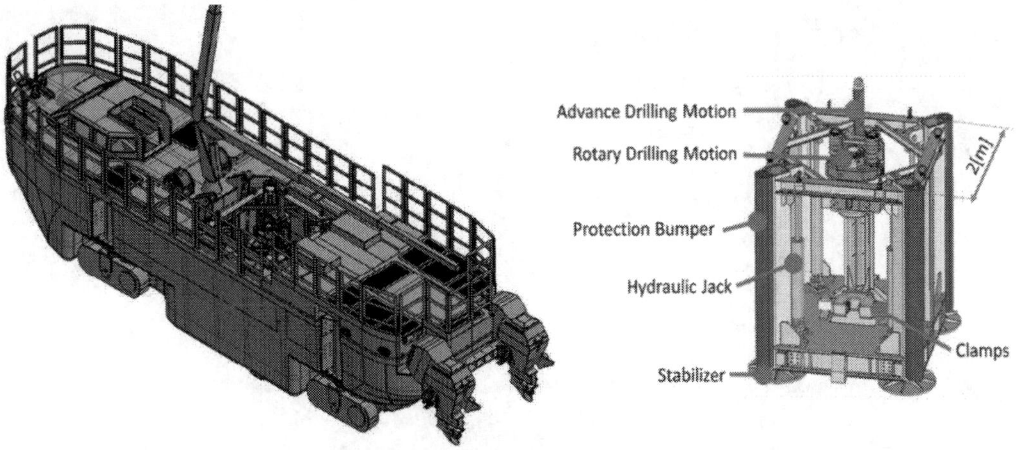

Figure 1. Overview or PROTEUS prototype: amphibious platform (left) and underwater drilling unit (right). The drilling unit is hosted at the centre of the platform and can be lowered to the seabed on occurrence. Note that the vehicle is equipped with both tracks for ground movements and outboard motors for marine propelling.

The characteristics of PROTEUS machine suggests that optimizing its ability to perform site characterization through the extraction of core samples requires the adoption of the so-called wireline technology; as known in literature [9] such technique presents a number of advantages in comparison with other surface-driven rotary drilling technologies, main one being the ability to reduce tripping pipe time.

In the Wireline drilling technique, pipes with an external diameter from 90 to 150 mm and thicknesses between 8 and 12 mm are used instead of drilling rods. Each pipe typically 1500 mm long has a conical thread at the ends (female at one end, male at the other as can be seen in Figure 2/a); in this way the pipes can be assembled in series in order to achieve arbitrarily large drilling depths such as visible in Figure 2/b . The tube acts as a drilling rod by dragging the excavation tools placed at one end into rotation, but also arms and supports the walls of the hole made. A fluid, pressurized water, is injected into the pipes, which performs the role of lubricant and coolant during excavation. The flow rate of pressurized water is typically at least three to four times higher than the volume of excavated material and ensures its removal in the form of sludge emerging from the hole on the surface. As can be seen in Figure 2/c and in Figure 2/d, tool change operations, sampling and more generally interventions on the bottom of the hole are carried out by lowering a bayonet quick coupling or other similar organ using a winch, which allows manipulation and transfer of tools and components along the tube.

Among the equipment that can be installed on the system, there are different types of tools and instruments such as the CPTWD (Cone Penetration Test While Drilling) equipment visible in Figure 3 or other equipment that allows to create cores of the soil. So, one of the requirements of the any wireline and-effector adopted is to enable the possibility to host different tools and to provide easy insertion and recovery of them.

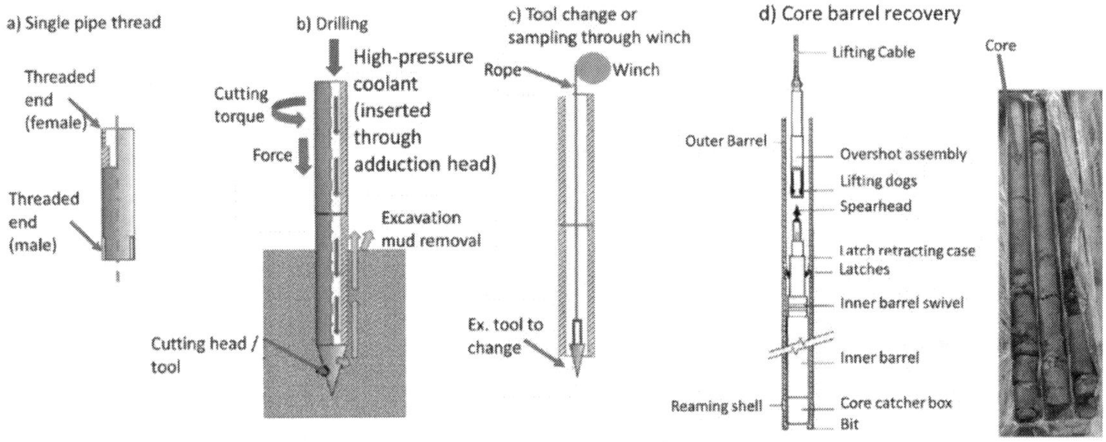

Figure 2. Single threaded tube (a), tubes assembled during drilling (b), tool change or sampling through wireline winch d) core barrel recovery [1].

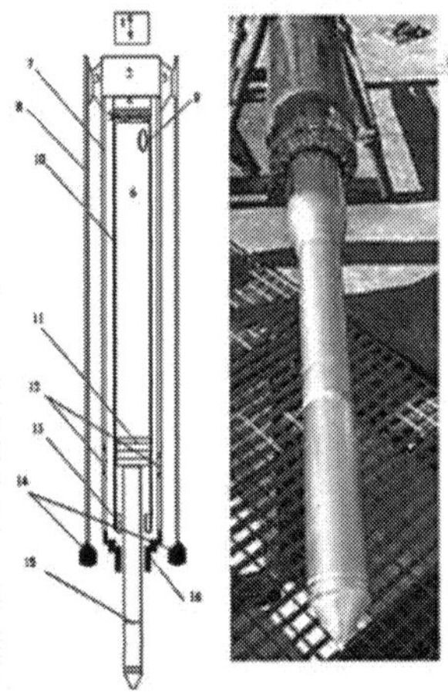

Figure 3. Cone Penetration Test While Drilling equipment to be installed on the PROTEUS platform.

Wireline core drilling is used to carry out excavations, mining research, polluting research, soil coring; this system allows to quickly lower a core barrel to the bottom of the hole and recover it without removing all the others driving rods. The recovery of the core barrel takes place through a catcher device, lowered by means of a steel cable connected to a winch, which is hooked in depth to the core barrel. This technique is usually suitable for any soil. Once that the device is outside of the drilled hole, it is possible to dismantle the tool, or extract the core and in general to perform needed operations. Such assembly takes the name of overshot connector; various designs are available, most of them being proprietary; certain of them also enable other type of operations while in the drilling hole depending on the setting of hydraulic mechanisms and valves installed on it.

There are two categories of core barrel heads as we can see in Figure 4 one is the one where the core barrel has a pointed shape (Figure 4/a) while the other is the one where the catcher has a pointed shape and the hollow core barrel as per Figure 4/b. The core barrel can be fixed or idle, it basically depends on the soil to be analyzed, if it is rocky or crumbly. Generally, as far as the thrusts are concerned, they do not exceed 5 tons and for the shots they do not exceed 12 tons. The rods remain in the hole until the tip needs to be replaced. This has great cost advantages when drilling deep holes for mining exploration.

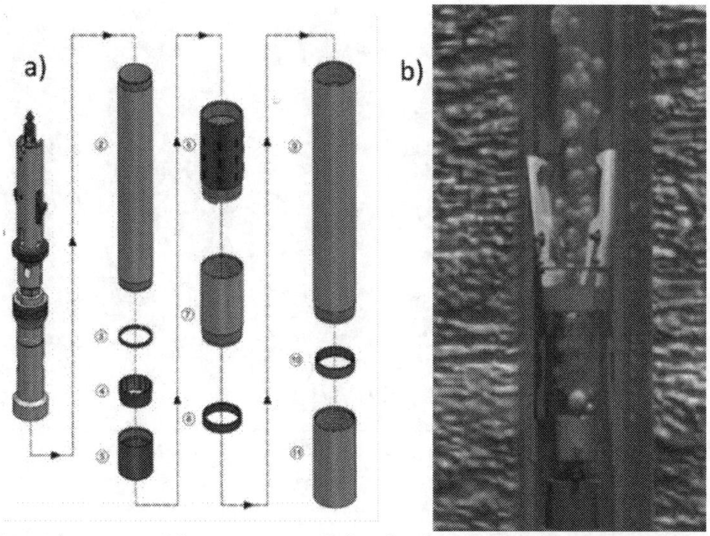

Figure 4. Overshot assembly – commercial units (Tecso [10] and Sandvik [11]).

2. Examination and re-engineering of an existing wireline tooling system

One of the main priority of the PROTEUS project was to search for technologies suitable for the purpose to create efficient and suitable cores. According to the know-how on the sector of the authors of the present text, it has been decided to use a previously tested and widely used (see Figure 5) core barrel to make Wireline cores. Such device dates back to the early 2000s and was produced in a reduced number of units by a local manufacturer, the original design being property of a small enterprise; 2D original drawings were also available. A first examination (see Figure 6) showed not only that is was partially degraded by wear and corrosion due to intense use, but also that certain parts were upgraded in comparison with existing 2D drawings, probably due to in-use modifications and improvements. Also, certain details (undercut profiles) suggested that the production activity had to include complex operations, such as manufacturing of multiple subparts, welding and additional machining. Due to the undoubted value of the device overall layout and design, it was decided to operate an operation of reverse engineering including measurements, CAD creation, resistance and holeshot connection verification, and proposal of alternative solutions.

Figure 5. Wireline assembly in use (courtesy Dott. Ing. Massimo Sacchetto [12]).

IOP Conf. Series: Materials Science and Engineering 1275 (2023) 012033 doi:10.1088/1757-899X/1275/1/012033

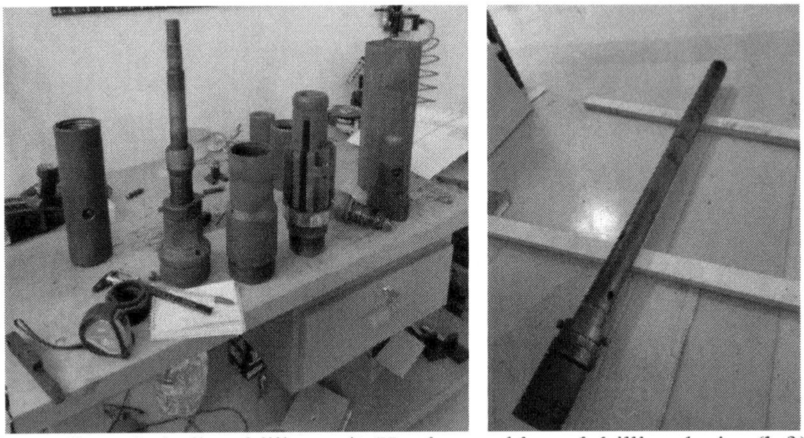

Figure 6. Dismantling of wireline drilling unit. Head assembly and drilling device (left) and catcher device (right).

Reverse engineering was carried out to reconstruct the core barrel and the various worn and damaged equipment. The reproduced core barrel can be seen in Figure 7, in this figure the two types of core barrel can be seen, the fixed one in Figure 7/a and the so called "idler one" in Figure 7/b used for different types of soil. The fixed one is directly coupled to the tube in which the core drilling is carried out; therefore, while the external tube turns, the internal core barrel also turns. In the idle core barrel, the head of the core drill is always set in rotation by the external tube, but the tube in which the core drilling is carried out is supported by a series of bearings; this enables to isolate the inner pipe – idler - from rotation, while the core is produced by simple axial force. Both solutions are needed for the PROTEUS machine, and the choice usually depends on the type of soil on which the drilling occurs.

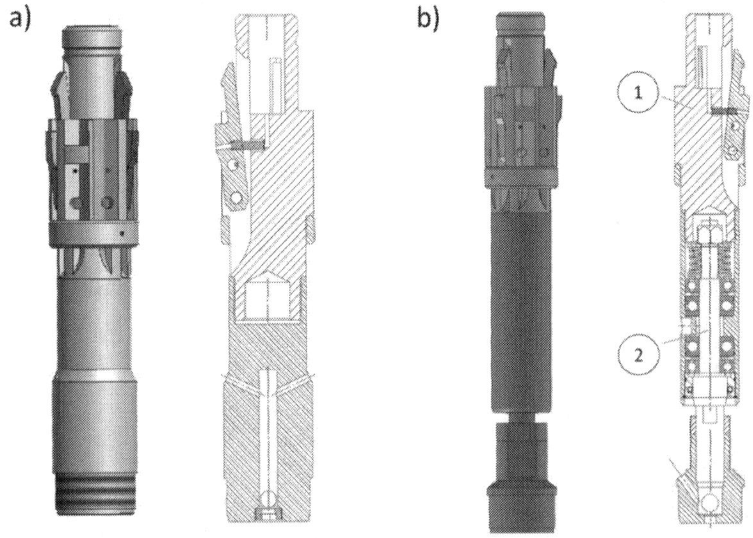

Figure 7. View of CAD reconstructed fixed (a) idler (b) devices.

Figure 7 shows a section of the idler core barrel, which is composed of two parts, the first is marked with the number 1 which is the head of the core barrel and the second marked with the number 2 which is the section to make the sampler idler. The head of the core barrel is machined in such a way as to house three teeth, which have two main functions:

IOP Conf. Series: Materials Science and Engineering 1275 (2023) 012033 doi:10.1088/1757-899X/1275/1/012033

- the first is to engage on the external tube in order to transmit the motion to the sampler thanks to the aid of some springs
- the second is to grab the fisherman during the re-leveling phase of the core.

When the teeth are in contact with the outer tube, they are slightly extracted thanks to the action of the springs, which act on each tooth; in this case they transmit torque, performing the same purpose as a key. Spring stiffness and preload is selected in order to achieve full opening with disengaged tools (see Figure 8), while opposing moderate resistance to compression during the connection phase (see Figure 11)

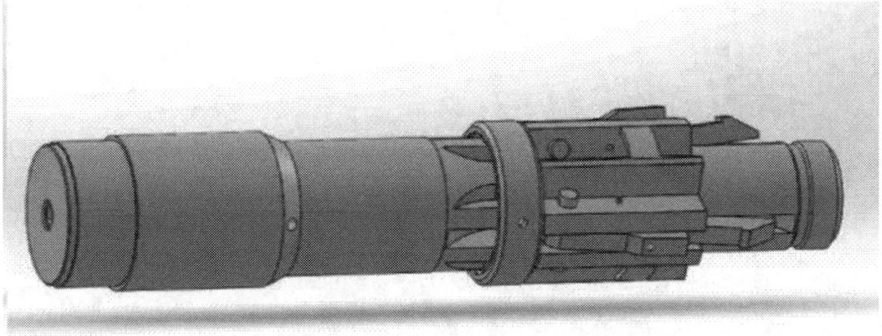

Figure 8. View of the disengaged tool: the teeth are kept fully opened by the springs.

Figure 9/c shows the section where motion is transmitted from the external tube to the internal core barrel. The position of the core barrel in the axial direction is ensured by a stop so that the teeth engage in the appropriate grooves on the external tube. The rotation torque is applied to the external tube set by a rotary motor assembly visible in Figure 9/d. Due to the design of the system, the core barrel engages on the external tube only through two of the three teeth available to the core barrel through the grooves that can be seen in Figure 9/a; also, the section view highlights the complexity of construction of a similar device, since turning is simply not possible. The results from former simulation activities carried out by the same authors of the present activity [8] have been adopted to quantify the maximum solicitations on the device, especially those related to torque transmission (see Table 1) which has been estimated at the maximum value of 16000 Nm.

Table 1. Effort on drilling end-effector depending on selected speed.

Scenario	Unit	High drilling speed	High drilling torque
Drilling speed	Nm	4000	16000
Rotation speed	RPM	140	35
Axial speed	m/s	0.02	0.02
Axial force	N	10000	10000

The external tube used in the PROTEUS project has a diameter of 127 mm, presents a thread section which is needed in order to create a column of pipes during deep drilling. For both core drills, the idle and fixed cores have sealing gaskets in order to prevent the core from leaking out during the sample re-leveling phase.

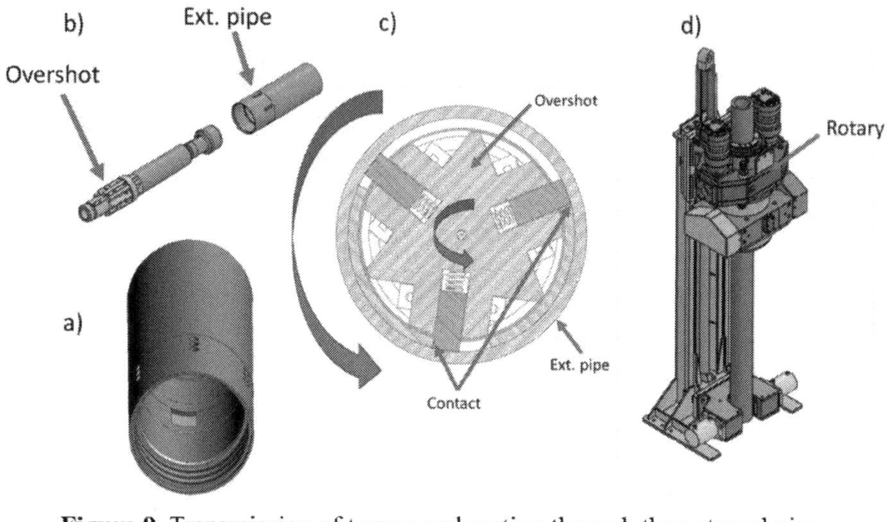

Figure 9. Transmission of torque and motion through the external pipe.

The overshot recovery unit (catcher) has also been reproduced, as shown in Figure 10. The overshot recovery has the purpose of picking up the core barrel when he has finished sampling the soil. It is lowered by a winch from the top of the pipes.

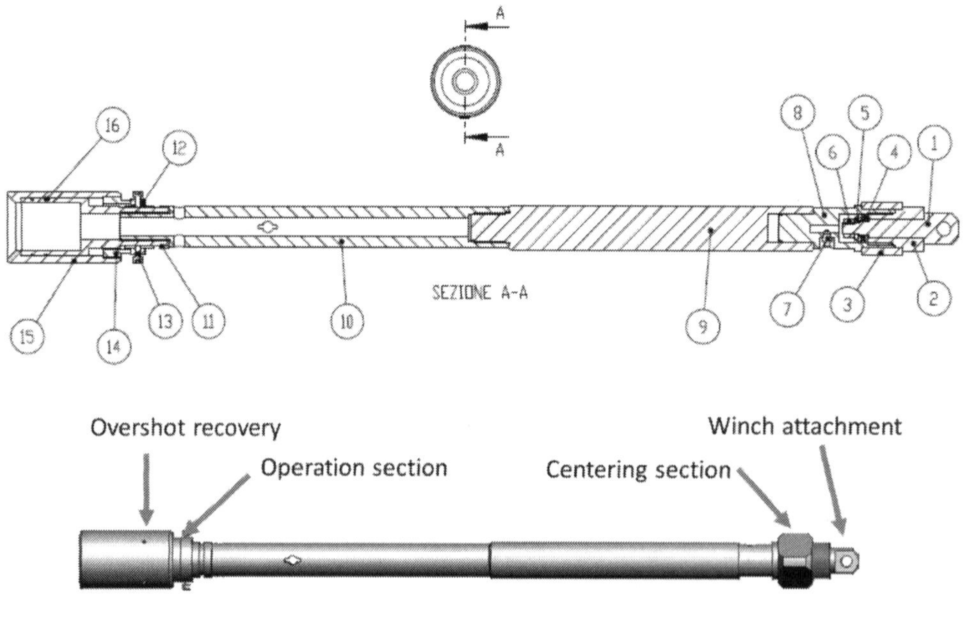

Figure 10. Overshot recovery unit (catcher).

The overshot recovery assembly is composed of a recovery bell which has the purpose of grasping the teeth of the core barrel during the re-leveling phase, a maneuvering section to release the teeth engaged for the release phase from the core barrel, another centering area to avoid undercuts of the angler and to prevent it from getting stuck in the pipes; upper parts hosts the hook for the winch.

When recovery of tool or sample is necessary, the overshot unit is lowered from the top of the pipe column by means of a winch, when it arrives on the core barrel the teeth are released from the external tube and hook onto the angler allowing the recovery of the core barrel as shown in Figure 11.

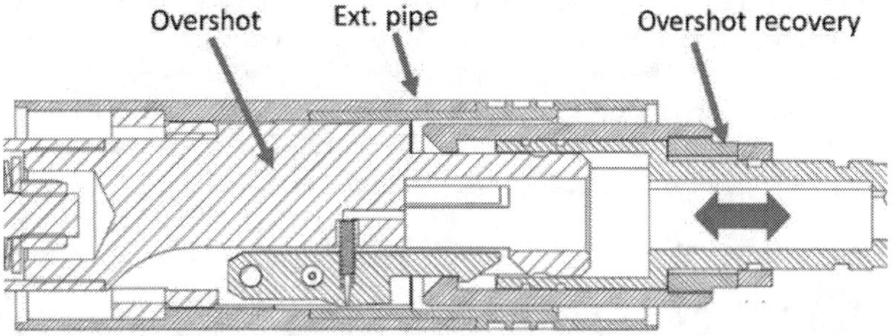

Figure 11. Tool – catcher assembly: section view of connection phase.

Figure 12 shows a section of the assembly, in particular the contact area of the three teeth, highlighted in green, with the recovery bell. The contact is ensured by the springs.

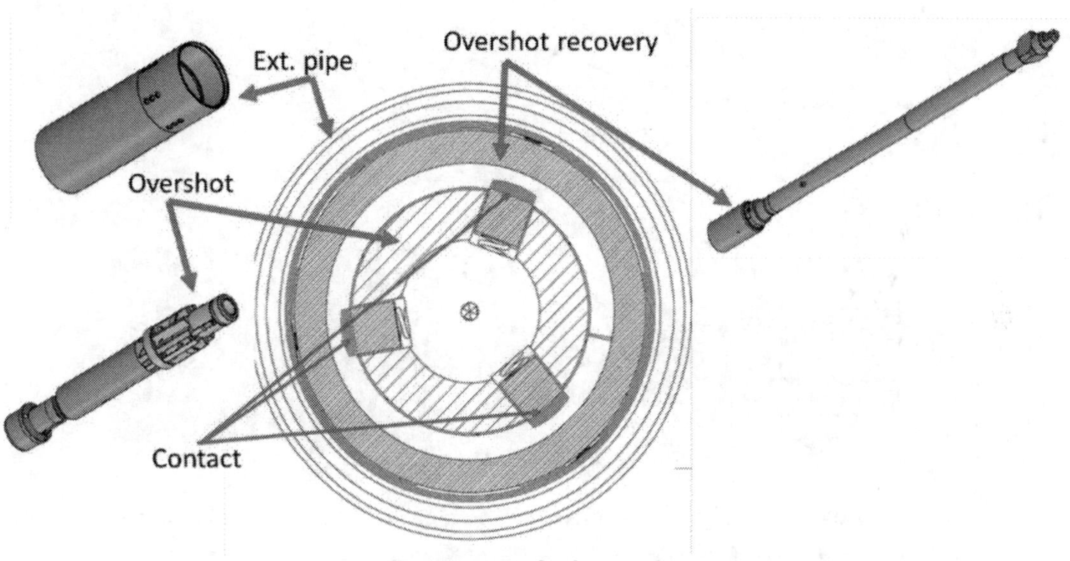

Figure 12. Gripping section.

The overshot connection needs the sliding of the teeth inside the bell in a very harsh environment, since the drilling occurs under pressure of water contaminated by drilling debris, it is necessary to ensure that the connection is not hindered by the increase of friction in comparison with steel/steel clean contact. For this reason, a simulation has been performed. Figure 13 shows the trend of the contact forces of a tooth in the various gripping positions and the directions of these. To carry out this analysis we used the Solidwork Motion software, this software allows us to animate the CAD model, and then extrapolate the information we need, for example the contact forces. After having inserted all the constraints between the various components, the constraints that the software had to simulate were left free, for example the movement of the grips, the descent of the overshot recovery. To simulate the descent of the overshot recovery, a dummy linear motor, with a speed of 20 mm/s in order to simulate the descent of the overshot recovery with a winch. In addition, the springs were inserted to simulate the movement of the grips; according to the dismantled devices, it has a stiffness of about 8 N/mm, which appeared a suitable compromise between the need to keep teeth open and the possibility to have a smooth connection during catch phase.

A friction contact model between the various components was adopted and two scenarios were made, the first with a friction coefficient of 0.25 between the two steel surfaces while the other simulation was performed by increasing the friction coefficient to 0.5, this latter to take into account degraded conditions such as mud present during the use phase. Finally, the analysis was performed by finding the contact force module and their direction for each point of contact between overshot recovery and overshot. In both simulations, the results show the occurrence of a moderate force, which is expected to be suitable for smooth connection, since its value is significantly lower than the force due to weight of the catcher assembly itself (approx. 30kgs), even considering water flotation.

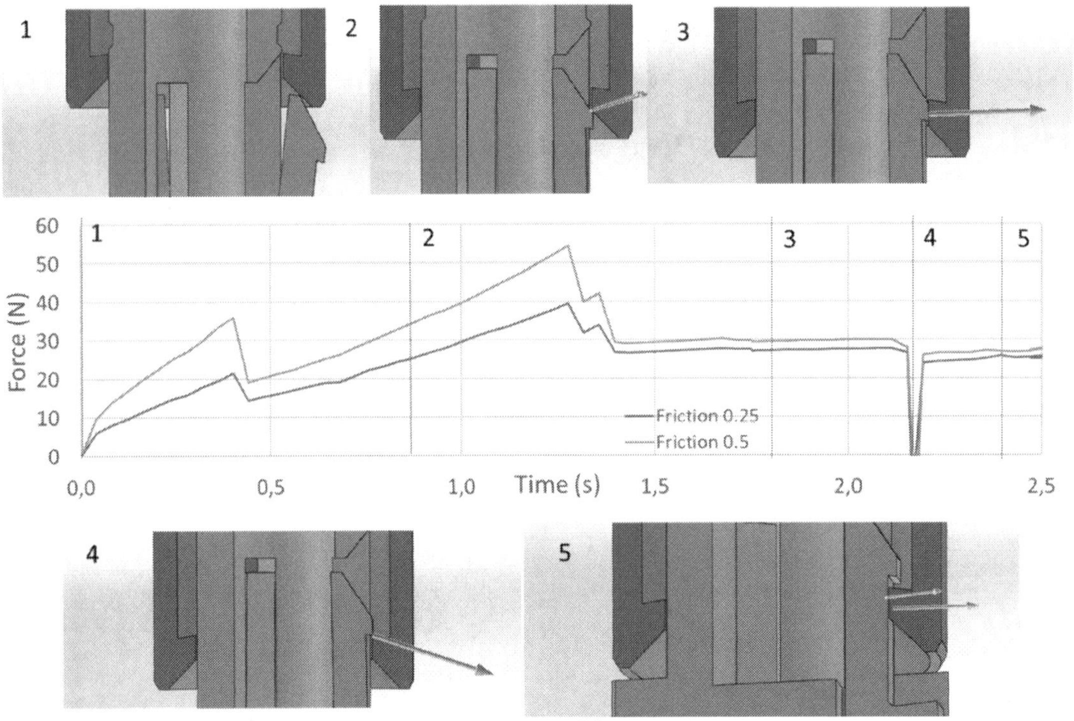

Figure 13. Trend of forces.

In addition, a finite element analysis was performed, both of the core barrel and of the recovery bell. Figure 14 shows the mesh created for the various elements.

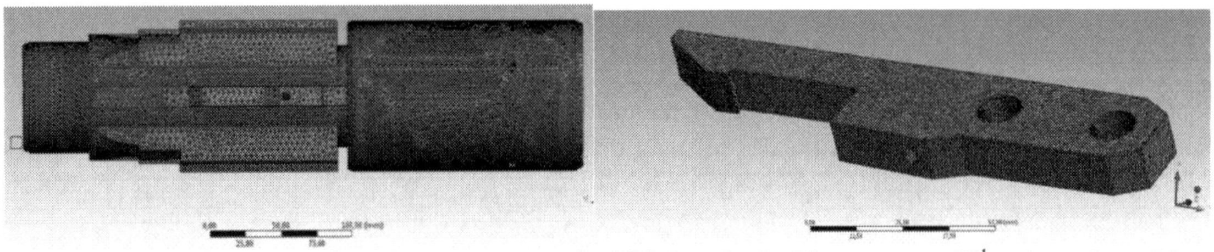
Figure 14. Mesh generated on the components for FEM analysis. Elements are 2nd order tethraedric.

Assuming that the core detachment is typically the most demanding phase of the whole catching process, a 1500 N axial force was applied to the recovery bell, a value than that which would occur in working conditions, the value of which has been estimated to be around 700 N. Figure 15 shows the analysis of the results calculated with the Ansys software, the maximum stress is approximately 85 MPa, significantly lower than the yield strength of any steel.

Finally, Table 2 shows a preliminary selection of steel alloys suitable for device construction; such list has been built through consultation with suppliers, considering the minimum required mechanical properties listed in ISO10097 [13].

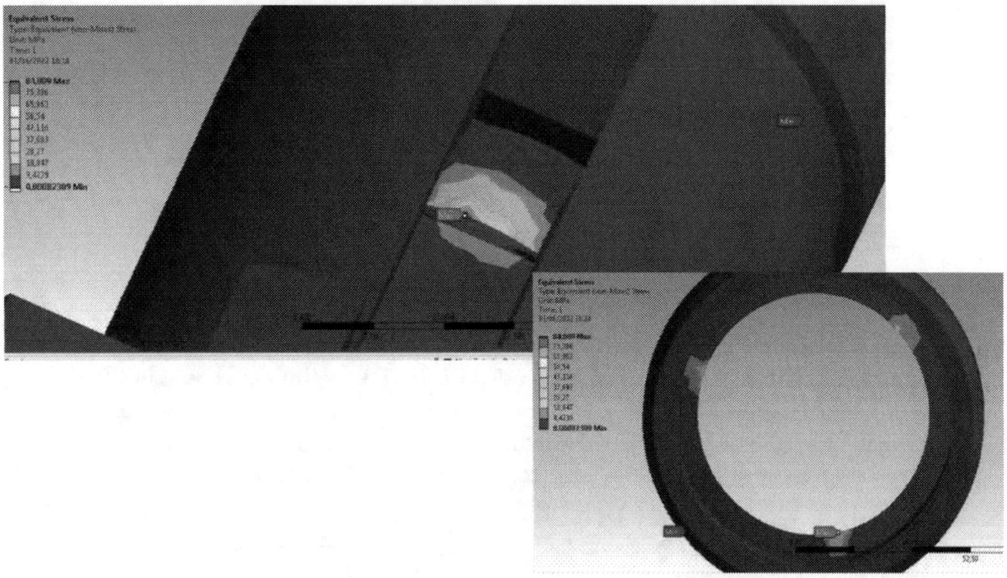

Figure 15. Results of FEM analysis on the catching device.

Table 2. Preliminary analysis of candidate materials for device construction.

Material (steel)	Yield stress (MPa)	Tensile strenght (MPa)
XJY950 (42CrMo)	950	1050
30 CrMnSiA	815	945
Si135	1080	1110
N80 type1	550	690

3. A proposal for revamping of the assembly

To make the system less expensive and reduce mechanical processing, we have to modify the core barrel head. In the current system, only two teeth are engaged on the angler to transmit motion, so we decided to reduce the teeth to two and avoid numerous milling operations. This solution is expected not to affect the overall torque transmission – since, as seen, it occurred on 2 teeth also if 3 were installed, but of course centering qualities during catch may be affected. It has been decided to experiment such design, considering the potential for a series production; a prototype will be therefore needed for on-field testing and assessment.

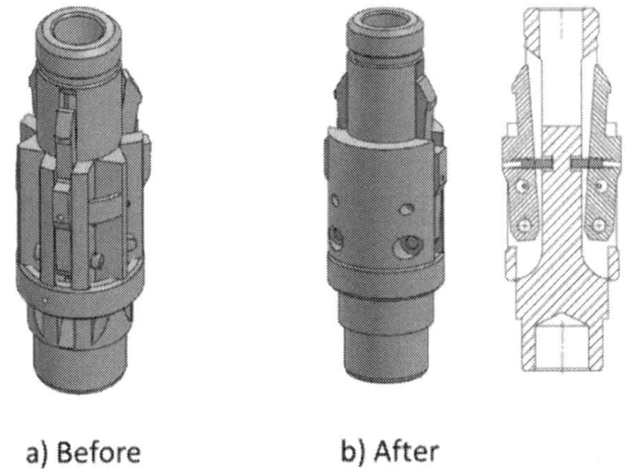

Figure 16. Head assembly group before and after revamping activity.

Figure 16 shows the core barrel head before and after the modification. Figure 16 b shows the core barrel after modification, almost all milling operations have been eliminated, making the component easy to make and reducing the costs associated with mechanical machining. The gripping teeth thickness has been increased so that they have greater grip on the recovery bell; the criteria has been to maintain the contact area at the same value between original and new solution (see Figure 17).

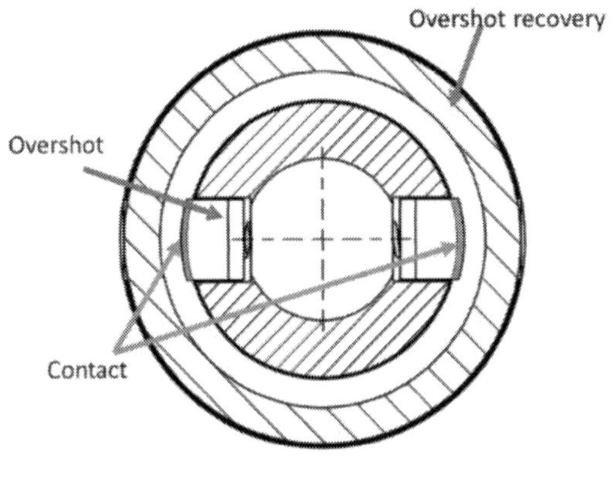

Figure 17. New gripping section.

For this modified overshot the same spring as the original overshot was used to give the grips the same reaction force. Also in this case the analysis was performed with two coefficients of friction to see how the contact forces change; again, we can see that the module of the contact forces increases as the friction coefficient increases, maximum value of the axial force being comparable.

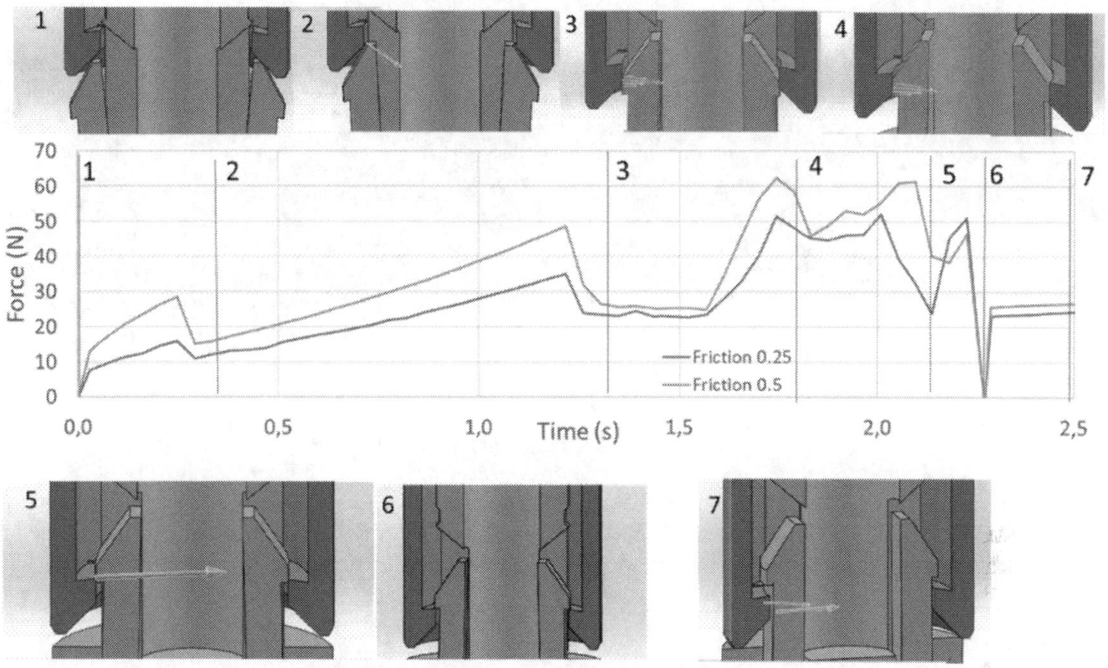

Figure 18. Contact forces trend.

Figure 18 shows the trend of the contact forces of a single tooth in the various gripping positions. Also for this modification a finite element analysis was performed, Figure 19 shows the mesh created for the tooth.

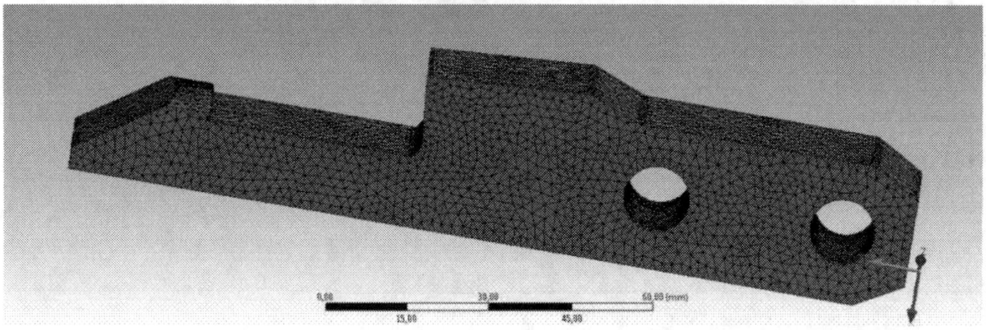

Figure 19. Tooth mesh.

Figure 20 shows the tension present on the tooth, the maximum tension in this case is about 28 MPa, this is because the contact with the bell is expected to be almost perfect.

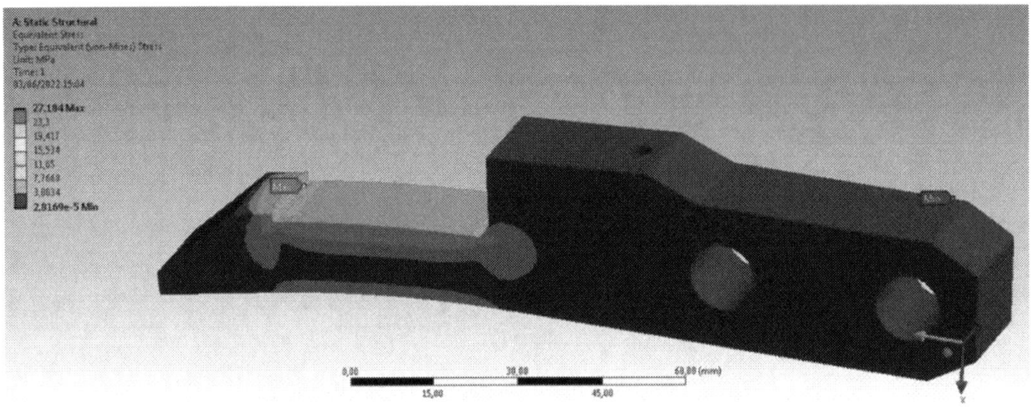

Figure 20. FEM analysis results.

Moreover, to improve grasping, we have thought of creating a tooth that follows the shape of the core barrel, so that the contact does not occur precisely on the edges but occurs uniformly over the entire surface. Figure 21 shows the optimized tooth with rounded outer surfaces.

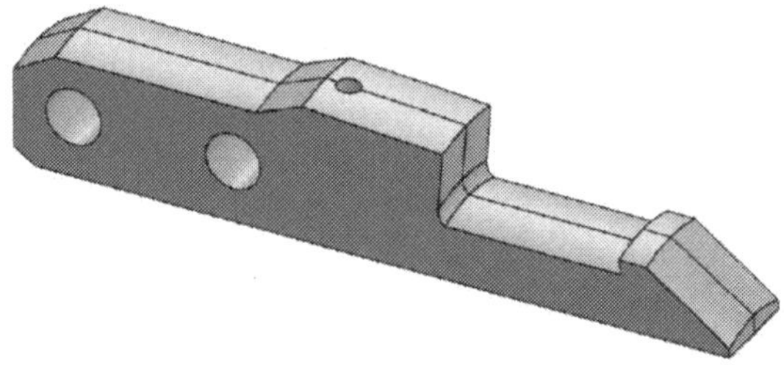

Figure 21. Tooth with optimized surfaces.

4. Conclusions and final remarks

The definition of the PROTEUS prototype needed to take into account a number of relevant requirements in order to ensure economic, flexible and efficient operation in any application field. The wireline tool is the key factor to ensure reduced time waste in case of sample coring and in case of cutting head substitution. A brief market analyses highlighted that there are different types of core drills, each of them characterized by some peculiarities. We have decided to redesign an existing core barrel by doing reverse engineering activities in order to valorize a device which was known for efficacy and easy operability, while not being covered by existing patents. A CAD model of the core barrel was reconstructed; as usual, the 3D model enabled the possibility of stress analysis and connection simulation, thus permitting to comprehend the role of the details of the existing part and to verify the expected forces. Simulations were conducted to optimize contact and understand the trend of forces in different catching scenarios. Finally, significant modifications were proposed to design a new core barrel to make it simpler and less expensive, increasing the reliability of the system as the number of components has been reduced. This core barrel will be reproduced in a mechanical workshop and tested during use.

Acknowledgements
Activities described in this work have been performed within the PROTEUS project that was financed by Regione Toscana, as a part of the POR-FESR 2014- 2020 Research Program. Also, authors wish to thank all unmentioned colleagues and partners that have supported these research activities.

References

[1] Pugi L, Mariottoni E, Sacchetto M, Cocci M, Frillici F S, Berzi L, Delogu M, Rotini F, Savino G, Pierini M and Boni E 2022 Development of a Multifunctional Amphibious Platform for Near Shore Geognostic Activities *Chemical Engineering Transactions* **91** pp 163–8

[2] Pugi L, Frillici F S, Cocci M, Rotini F, Delogu M, Berzi L, Fiorineschi L, Mariottoni E and Sacchetto M 2022 A Sustainable Anphibious Platform for Geological Prospecting of Near Shore Marine Sites *2022 IEEE International Conference on Environment and Electrical Engineering and 2022 IEEE Industrial and Commercial Power Systems Europe (EEEIC / I&CPS Europe)* pp 1–6

[3] Cutler B, Fowers S, Kramer J and Peterson E 2017 Want an Energy-Efficient Data Center? Build It Underwater *IEEE Spectrum*, Accessed: Sep. 07, 2022. [Online]. Available: https://spectrum.ieee.org/want-an-energyefficient-data-center-build-it-underwater

[4] Hu Z-H, Zheng Y-X and Wang Y-G 2022 Packing computing servers into the vessel of an underwater data center considering cooling efficiency *Applied Energy* **314** p 118986

[5] Wen Y, Kamranzad B and Lin P 2021 Assessment of long-term offshore wind energy potential in the south and southeast coasts of China based on a 55-year dataset *Energy* **224** p 120225

[6] Mouakkir L, El hou M, Mordane S and Chagdali M 2022 Wave Energy Potential Analysis in the Casablanca-Mohammedia Coastal Area (Morocco) *J. Marine. Sci. Appl.* **21** pp 92–101

[7] Tsai C-C and Lin C-H 2022 Review and Future Perspective of Geophysical Methods Applied in Nearshore Site Characterization *Journal of Marine Science and Engineering* **10** pp 344

[8] Pugi L, Cocci M, Frillici F S, Berzi L, Mariottoni E and Sacchetto M 2022 Modeling of an underwater drilling platform for very near shore applications *Proceedings of the Institution of Mechanical Engineers, Part M: Journal of Engineering for the Maritime Environment* **0** pp 1-10

[9] Wang J, Cao P, Liu C and Talalay P G 2015 Comparison and analysis of subglacial bedrock core drilling technology in Polar Regions *Polar Science* **9** pp 208–20

[10] TECSO Home *TECSO S.A. Homepage* https://www.tecso-sa.com/ (accessed Sep. 25, 2022)

[11] Sandvik Mining and Rock Solutions *Sandvik Group* https://www.home.sandvik/en/about-us/business-areas/sandvik-mining-and-rock-solutions/ (accessed Sep. 25, 2022).

[12] WEBSITE OF MASSIMO SACCHETTO *CPTWD AND OTHER GEO-INNOVATIONS* https://maxsacchetto.wixsite.com/cptwd (accessed Sep. 25, 2022)

[13] ISO *ISO 10097-1:1999 Wireline diamond core drilling equipment — System A — Part 1: Metric units*

Analysis of Wind Turbine Pitch 4-Point Contact Bearing

S Porziani[1], M E Biancolini[1] and C Brutti[1]

[1] Dept. of Engineering Enterprise "Mario Lucertini" – Univ. of Rome Tor Vergata, Italy

E-Mail: porziani@ing.uniroma2.it

Abstract. In the field of renewable energy source, the wind power is certainly one of the most important and therefore the industrial and research community are devoting increasing attention to the design, construction, management and maintenance of wind turbines. Due to the relevance, technical, economic and social of this type of installation the correct maintenance is actually a crucial challenge. As in all mechanical engineering applications, numeric models, FEM based, are widely used during both the design step and monitoring on site. In fact, in the frame of Structural Health Monitoring (SHM), the availability of a Digital Twin is fundamental to interpret the signals coming from the sensors and to check the state of wear and the progression of any damage. In general, this numerical tool must have a good accuracy and a great rapidity of response. For this reason it is essential to develop condensed and reliable models of the main elements able to represent the real behaviour of the component without excessively increasing of computational costs. One of the most important element is the pitch variation slewing bearing, which allows the optimal wing orientation. Unfortunately, a complete numeric model of the bearing requires very large dimension and so the real time reply, necessary to SHM, is difficult. In this paper, to overcome this problem, a new simplified model is proposed to represent the array of spheres and its contact condition with the rings of the bearing. The new approach is based on the adoption of a toroid-shaped element replacing the rolling elements. To study this solution, a full-detailed model of a typical slewing bearing for pitch variation has been used to properly set up the contact model and to tune the properties, geometrical and elastic, of the equivalent toroidal body. The main feature of this approach is that, with a relatively low calculation effort, the deformation behaviour and the stress state of the elements connected to the bearing is available in an easy and accurate way. Then, the comparison in real time of the numerical results with the corresponding experimental ones from the site allows to deduce initiation and growth of a damaging process.

1. Introduction

In the upcoming years, green and renewable energy is going to represent the main energy source and wind power is one of the most important ones, due to its general diffusion. In Italy, in particular, there are numerous geographical sites suitable for wind power generation. Therefore, the industrial and research community are devoting increasing attention to the design, construction, management and maintenance of wind turbines. As in all mechanical engineering applications, numeric models, FEM based, are widely used during both the design step and monitoring on site. In fact, in the frame of Structural Health Monitoring (SHM), the availability of a Digital Twin is crucial to interpret the signals coming from the sensors and to check the state of wear and the progression of any damage. Unfortunately, a complete numeric model of a wind turbine, similar to that employed in the design process, cannot be used due to its large dimension that makes impossible a real time response. For this,

Content from this work may be used under the terms of the Creative Commons Attribution 3.0 licence. Any further distribution of this work must maintain attribution to the author(s) and the title of the work, journal citation and DOI.
Published under licence by IOP Publishing Ltd

many research efforts are now given to define and validate condensed models able to represent the real behavior of the component without excessively increasing computational requirements [1].

One of the most important components in the wind turbine is the pitch variation bearing (PVB), that allows the optimal wing orientation even when the wind flow changes in direction or intensity and, in addition, to put in safety configuration both the mechanical and the electrical systems, in case wind velocity exceeds the maximum allowable. For this, PVB is an essential element during the monitoring and its behavior strongly influences the deformation and stress state of the elements connected. The design process is a known activity founded on well stated equations [2] and well recognized standards (see for example ISO standards N°76:2007 and N° 281:2007) also in the particular field of wind turbine [3]. In the frame of SHM, a digital twin of the PVB must represent correctly the transmission of loads to the elements connected taking into account the load distribution on the spheres [4], the actual contact angle [5] and the stiffness of the assembly of rolling elements and rings [6]. A detailed FEM model of this assembly requires a very large number of nodes and elements, and this can impede a quick and accurate response in real time. Many researchers proposed simplified approaches to feature this component with different levels of completeness and accuracy as simple not linear spring [7] or more complex mechanism based on beams and springs [8].

In this work, the case of a 4-Point Contact slewing bearing for the pitch variation system is analyzed with the aim to define a new simplified tool able to model the structural behavior of the rolling bearing. The new approach is based on the adoption of a toroid-shaped element in place of the rolling elements. This solution allows to replace the discrete point contact condition with more simple and regular line condition. Because the two arrangements, the real and the simplified one, are equivalent from a structural point of view, it is necessary to ensure that the stiffness of the two models is equal. Therefore, a full-detailed model of the bearing balls and rings has been used to properly set up the contact model and to tune the properties, geometrical and elastic, of the equivalent body. With this approach, it is possible, in an easy and accurate way and with a low calculation effort, obtaining the deformation behavior of the bearing in order to study the stress state of the elements connected. Furthermore, thanks to the lower requirements of the simplified model, additional optimization actions can be performed for the wind turbine components, even including advanced automatic approaches [10][11].

2. Sphere-path contact modelling

From a theoretical point of view, the application of Hertz's theory to the contact of a sphere compressed between two locally deformable rings gives the following equation to calculate from the load in the ball (Q) its deformation (δ) [5],[8], [9]:

$$\delta=\frac{8.97\cdot 10^{-4}(1-s)^{0.1946}}{D_W^{\frac{1}{3}}}Q^{\frac{2}{3}} \tag{1}$$

Where D_W is the diameter of the spheres and s is the so-called conformity factor, defined as the ratio between the ball diameter and the raceway diameter in the radial plane. But this equation does not solve completely the problem of load distribution because the rings are not flexible only locally and the balls can rotate the angle of the contact force related to the bearing axis. The first rough approximation is considering the rings rigid and to calculate the angle of rotation only basing on a linear load distribution. Unfortunately this very simple procedure leads to relevant errors especially when the bearing is subjected by an axial force and a tilting moment.

To overcome this difficulty He et al. [7] proposed to replace the balls with a not linear unidimensional springs (Figure 1). In this way, the mesh size is strongly reduced but the results are in good agreement with those obtained with FEM in full detail, only if the orientation angle of the springs are near to the real one with the further difficulty that, in the real operating conditions, this angle varies with the load intensity.

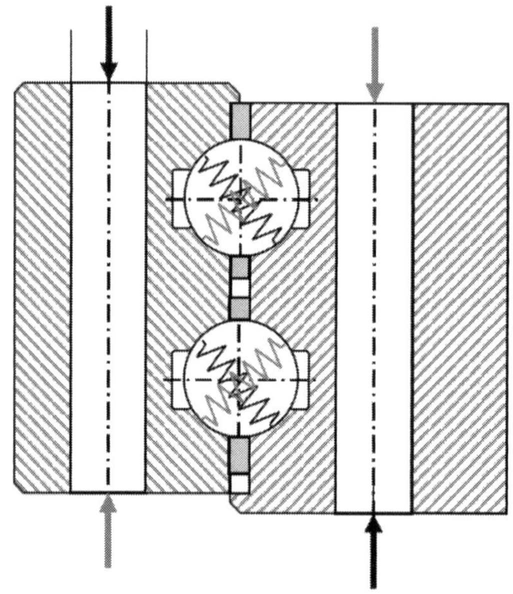

Figure 1. Not linear springs replacing the balls [7]

Another solution is that proposed by Aguirrebeitia et al. [8]: as shown in Figure 2, it is formed by two rigid shells of spherical shape (*a*) and a planar rigid mechanism (*b*) at the center of which a spring (*c*) is placed. In this way the plates allows a distributed contact condition and the mechanism can adapt the angle of the reaction. The approximation of this arrangement is that the contact area cannot vary during load application.

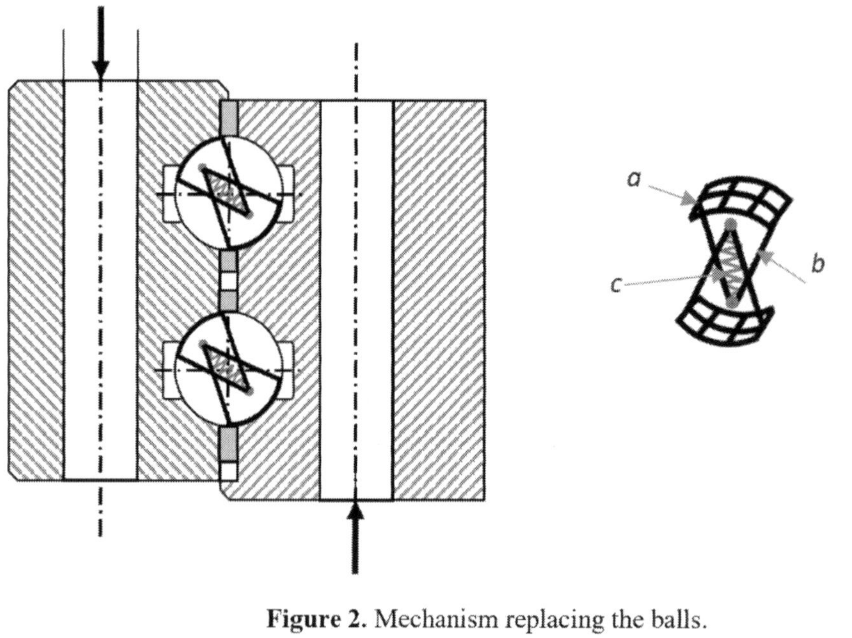

Figure 2. Mechanism replacing the balls.

In order to improve the simplified model of the rolling elements in a slewing bearing, we propose the use of a regular solid able to rotate in the space between the rings, within the limit of the conformity factor *s* and to transform the point contact in a line contact. The shape of the solid is toroidal and the section is circular with a diameter equal to the balls (Figure 3). The properties of the material forming the two toroids must be stated basing on the real structural behavior of the slewing bearing.

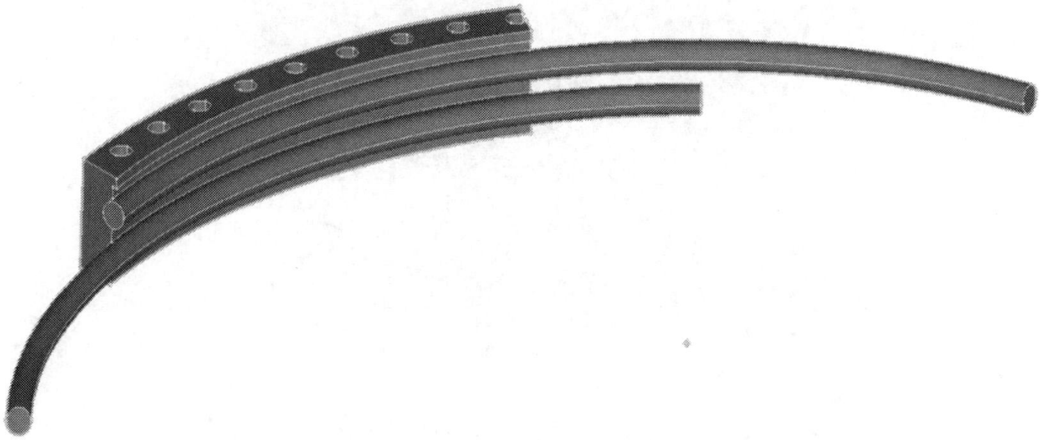

Figure 3. Shape of the toroidal bodies

3. FEM model

3.1. Full-detail model

The first step of our investigation was implementing a full detail FEM model of a typical slewing bearing for wind turbine application using ANSYS® Mechanical 2021 R2. In Figure 4, the section of the slewing bearing analyzed with its dimensions is shown.

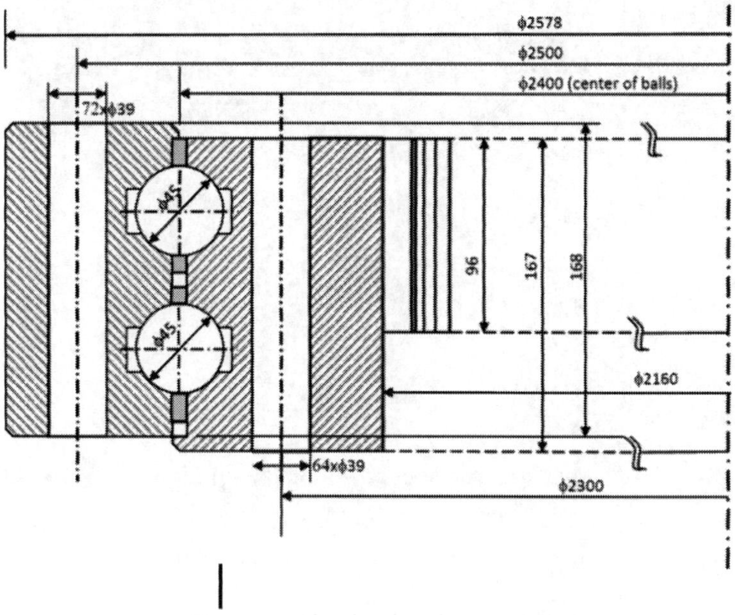

Figure 4. Slewing bearing modelled

The FEM model has the following main features: the rolling arrays have 144 spheres each; the spheres have a diameter of 44 mm, while the races in the rings have 45 mm diameter. In addition, the cyclic symmetry has been used to complete the mesh in contact zones. Moreover, the 'Frictionless' contact type was adopted between the rings and the ball.

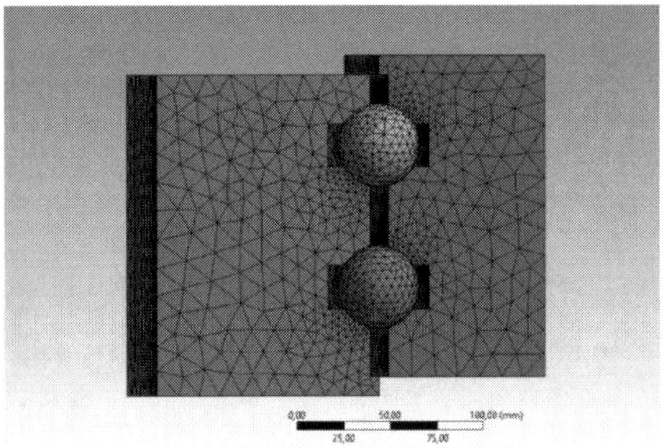

Figure 5. Section of the full detail FEM model

In Figure 5, there is the section of the model corresponding to a vertical axial plane. The mesh around the contact was optimized in order to have a good accuracy with an acceptable number of elements. The mesh was tuned examining the case of a sphere on a plane and comparing the results with the classical Hertz's results. The accuracy was about 1% for stresses.

The restraint condition was "fixed" on the base surface of the external ring, while the axial load, equal to 36000 N, was applied to the internal ring as shown in Figure 6.

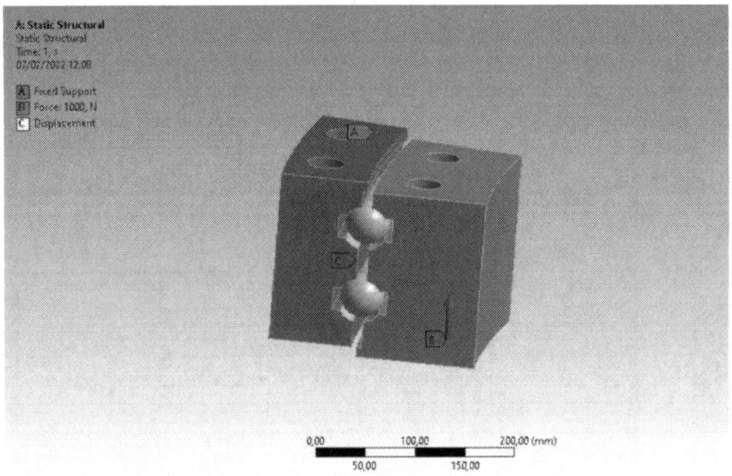

Figure 6. Restraint and load application on the full detail FEM model

The results obtained are reported in Figure 7 and Figure 8. The results so obtained were used to set the material properties of the body equivalent to the body array referencing to the stiffness of the bearing assembled.

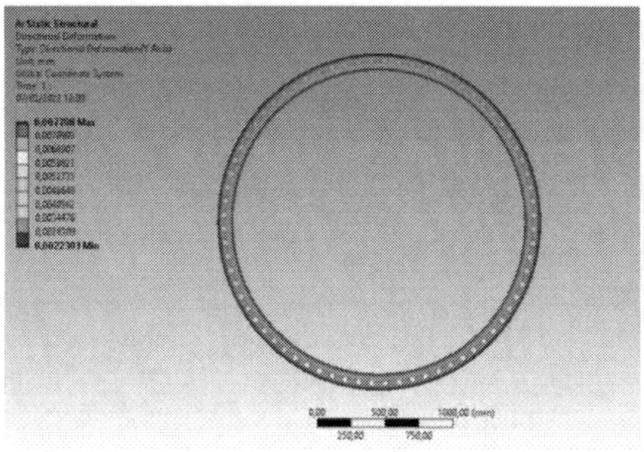

Figure 7. Deformation pattern on the ring

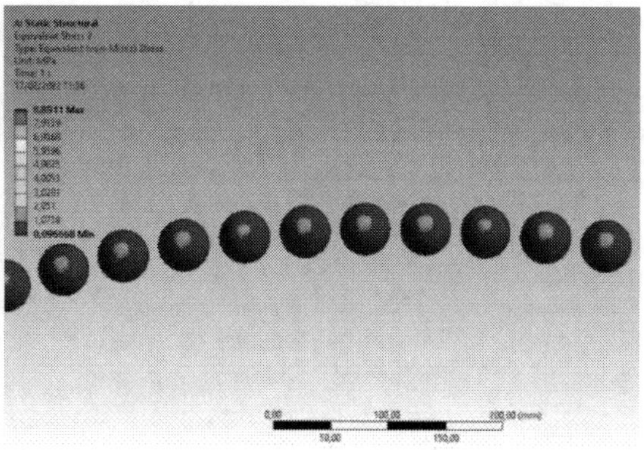

Figure 8. Contact stresses on the balls

3.2. New simplified model

As shown in Figure 3, the shape of the equivalent body is toroidal with circular section having the same diameter of the balls of the real slewing bearing. In Figure 9, there is the section of the FEM model of the equivalent arrangement.

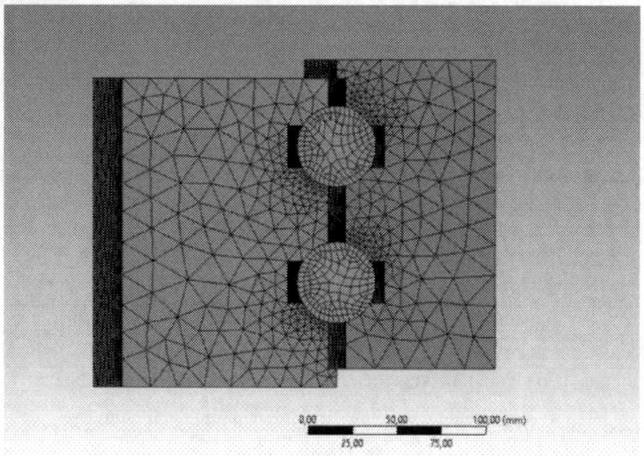

Figure 9. FEM model of the equivalent bearing with the toroidal body

The properties of the material of the toroid are depending on two main effect: the variation of material volume between the two models and the different stiffness of a line contact compared with a point contact. While the first is depending only on the dimensions of the problem the second is affected also by the load intensity. With an iterative procedure was found the Young modulus of the material of the toroid that assures the same axial stiffness of the real slewing bearing. Two solutions were found: the first, with the same Poisson coefficient equal to 0.3, resulted E = 76.13 GPa; the second, with a modified Poisson coefficient equal to 0.285, resulted E=76 GPa

The differences between the two models are resumed in the following table.

IOP Conf. Series: Materials Science and Engineering 1275 (2023) 012034 doi:10.1088/1757-899X/1275/1/012034

Table 1. Comparison between the real and equivalent material

Model	Body volume (mm^3)	N. of contacts	Total Volume (mm^3)	E (GPa)
Full-detail	4.46 e4	288	1.2844 e7	200.00
Toroid	1.15 e7	2	2.292 e7	76.13
Toroid	1.15 e7	2	2.292 e7	76.00

The axial displacement results obtained with the full detail and simplified model are shown in Figure 10. It is important to note that the use of finite models both for the full detail and the simplified one is indispensable. In fact, considering the axial displacement, if we apply the classical equations of the Hertz theory modified for bearing applications, as for example the equation (1), the value calculated differs for about 29% from the numerical corresponding one. This is due to several effects that the hand calculations cannot take into account as the distribution of load on the two arrays, the effective angle of the contact reaction.

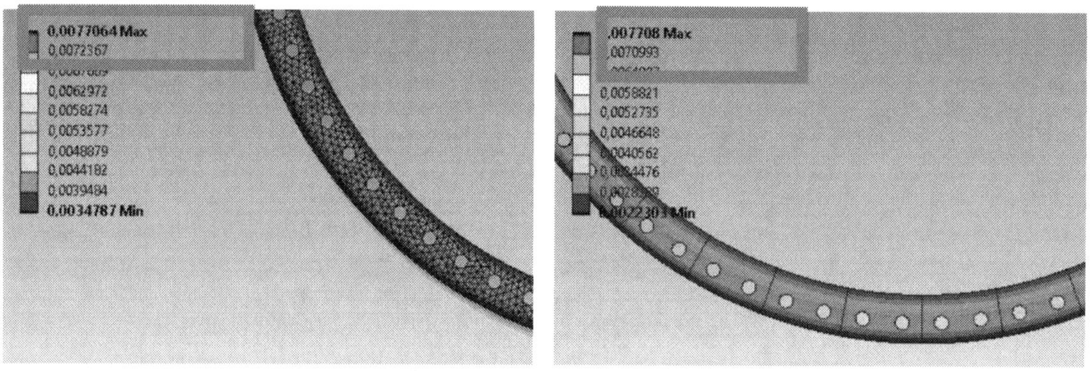

Figure 10. Comparison of the axial displacement between the full detail and the simplified model

In addition as the stiffness of contact varies with the load intensity, this effect was investigated doubling and halving the load. The results so obtained are shown in Table 2.

Table 2. Effects of load intensity variation

Load	Full-detail (mm)	Simplified contact (mm)	Variation
50%	0.00402	0.004466	-11%
100%	0.007708	0.007703	0%
200%	0.014663	0.012943	11%

4. Conclusions

In the frame of SHM of Wind Turbine, a fundamental tool is the availability of a digital twin, FEM based, able to give with acceptable computation efforts and time, results that can be compared with the output of the sensors on field. This need makes necessary to simplify all the elements that require a large number of nodes and elements. One of these is the slewing bearing used in the pitch variation system, especially regarding to the contact of the rolling elements with the races. In this paper a new simplified body, toroid shaped, is proposed to solve this problem. To check this solution a numerical investigation has been performed, developing first a full detail FEM model of a typical slewing bearing and then substituting the two balls arrays with two toroids with a circumferential section, having the same diameter of the balls. Basing on the deformation of the full detail model, the properties of the material of the toroid has been stated. In addition, as the contact stiffness depends on the load intensity, a variation of the load has been investigated.

The equivalent body here proposed, seems to be a good solution to simplify the model of the slewing bearing and to reduce consequently the computing time. Though the properties of the equivalent material depend on the load intensity, it was recognized that doubling or halving the load, the accuracy is still acceptable. Before to consider the proposed toroid a reliable solution, further investigations are necessary especially for the case of a tilting load acting.

References

[1] Mirasoli G, Brutti C, Groth C, Mancini L, Porziani S. and Biancolini ME 2022 Structural Health Monitoring of civil structures through FEM high-fidelity modelling, *In IOP Conference Series: Materials Science and Engineering* **1214** 1 012019. IOP Publishing.

[2] Harris TA and Kotzalas MN 2006 *Rolling Bearing Analysis* **2** Boca Raton FL: CRC Press.

[3] Harris T, Rumbarger JH and Butterfield CP, 2009. *Wind turbine design guideline DG03: yaw and pitch rolling bearing life*, NREL/TP-500-42362, Golden, CO National Renewable Energy Laboratory.

[4] Amasorrain JI, Sagartzazu X and Damian J 2003 Load distribution in a four contact-point slewing bearing, *Mechanism and Machine Theory* **38** 6 479–496.

[5] Daidié A, Chaib Z and Ghosn A 2008 3D simplified finite elements analysis of load and contact angle in a slewing ball bearing, *J. Mechanical Design* **130** 082601-1-8.

[6] Skyba R, Hrček S, Smetanka L and Majchrák M 2018 Stiffness analysis of slewing bearings. *IOP Conference Series: Materials Science and Engineering* **393** 1 012060. IOP Publishing.

[7] He P, Qian Q, Wang Y, Liu H, Guo E and Wang H 2021. Influence of finite element mesh size on the carrying capacity analysis of single-row ball slewing bearing, *Advances in Mechanical Engineering* **13** 4.

[8] Aguirrebeitia J , Abasolo M , Avilés R. and de Bustos I F 2012 General static load-carrying capacity for the design and selection of four contact point slewing bearings: finite element calculations and theoretical model validation, *Finite Elements in Analysis and Design* **55** 23-30.

[9] Houpert L 1997 A Uniform Analytical Approach for Ball Bearing and Roller Bearing Calculations, *ASME J. Tribol.* **119** 4 851–58.

[10] Porziani S, Groth C, Biancolini M E 2018 Automatic shape optimization of structural components with manufacturing constraints, *Procedia Structural Integrity* **12** 416–28.

[11] Porziani S, Groth C, Waldman W, Biancolini M E 2021 Automatic shape optimization of structural parts driven by BGM and RBF mesh morphing, *Int. J. of Mech. Sciences* **189**.

IOP Conf. Series: Materials Science and Engineering 1275 (2023) 012035 doi:10.1088/1757-899X/1275/1/012035

The preliminary thermo–mechanical design, simulation and optimization, of LAD onboard the eXTP space mission

G Lombardi, A Baldelli, R Fiacco, E Lombardi, M E Biancolini and R Piazzolla

INAF/IAPS-Roma, Via del Fosso del Cavaliere 100, I-00133, Rome, Italy, University of Rome "Tor Vergata", Via Cracovia 50, I-00133, Rome, Italy and ASI, Italian Space Agency, Via del Politecnico, I-00133, Rome, Italy

E-mail: giovanni.lombardi@inaf.it

Abstract. The eXTP (enhanced X-ray Timing and Polarimetry) mission is a major project of the Chinese Academy of Sciences (CAS) and China National Space Administration (CNSA) currently performing a phase B study and proposed for a launch in 2027. The eXTP scientific payload envisages a suite of instruments offering unprecedented simultaneous wide-band X-ray timing and polarimetry sensitivity. A large European consortium is contributing to the eXTP study and it is expected to provide key hardware elements, including the Large Area Detector (LAD) composed of 40 modules for a total effective area of 3.0 m^2 at 6.0 keV. In this paper, we describe the preliminary study that led to the design solutions adopted for the most important thermo-mechanical drivers of the LAD Module, which have been elaborated and used for the demonstration of compliance to the system requirements at spacecraft level. We report in particular the mechanical design for the Module and its components, the thermal analysis for each range temperature of the worst cases (coldest case) with the evaluation of the thermo-elastic deformation of a reduced module and the results of a campaign of static and dynamic finite element analysis of a simplified parametric model with an accurate mesh processing and the preliminary optimization of the critical components.

1. Introduction

The enhanced X-ray Timing and Polarimetry mission[1] (eXTP) is a flagship mission of China, with a large contribution by a European consortium, including institutions in Italy, Spain, Austria, Czech Republic, Denmark, France, Germany, the Netherlands, Poland, Switzerland and Turkey.

The eXTP mission is being designed to address the science themes of [2]understanding the behavior of matter under extreme conditions of gravity, density and magnetism. eXTP aims at determining the equation of state of ultra-dense matter in the interior of neutron stars, study the dynamics of matter in the vicinity of neutron stars and near the event horizon of black holes, where the General Relativity theory predicts large distorsions of the space-time with respect to a Newton's laws, and study the effects on the propagation of photons of the ultra–critical magnetic fields hosted in magnetar sources.

The scientific payload of the eXTP satellite consists of four scientific instruments: the Spectroscopic Focusing Array (SFA), the Polarimetry Focusing Array (PFA), the Wide Field Monitor[3] (WFM) and the Large Area Detector (LAD), the subject of this paper.
In figure 1 shown its current configuration, with the allocation of the four scientific instruments.

Content from this work may be used under the terms of the Creative Commons Attribution 3.0 licence. Any further distribution of this work must maintain attribution to the author(s) and the title of the work, journal citation and DOI.
Published under licence by IOP Publishing Ltd

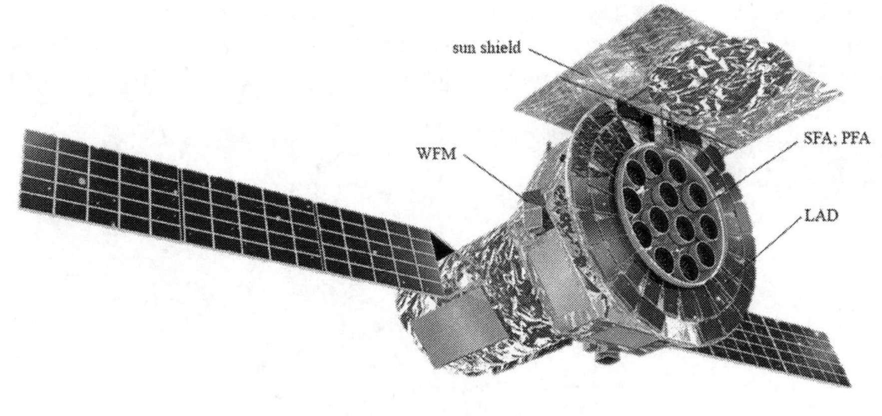

Figure 1: current configuration of eXTP satellite

The Large Area Detector[4] (LAD) is designed to be the most sensitive spectral-timing instrument for bright Galactic and extra galactic sources to date, an innovative and highly efficient technology and design allow to deploy in space an effective area as large as 3.2 m^2-class at 8 keV, adopting a modular configuration.
The LAD is organized as a modular instrument, composed of 40 coaligned Modules, each one hosting a set of 16 large area Silicon Drift Detectors (SDDs) and 16 corresponding capillary plate collimators. The 40 Modules are hierarchically interfaced by a set of 4 Panel Back End Electronics (PBEE), each one supporting 10 Modules, that are in turn interfaced to an Instrument Control Unit (ICU), in cold redundancy.

The LAD is designed as a classical collimated experiment, the requirement to acquire X-ray events with high time and spectral resolution and high statistical significance (to study astrophysical sources in the time/frequency/energy domain) has led to the LAD instrument described here, made possible by developments in large-area silicon detectors and micro-collimator structures.

The effective area requirement leads to a geometric area of ~ 6 m^2 or an active area of ~ 5 m^2. The field of view of the LAD will be limited to <1 degree by X-ray collimators. This alignment requirement must survive even thermal cycling and the ascent-to-orbit phase.

Also, everything mounted on the satellite is subject to thermal cycles related to the satellite's orbit around the earth (equatorial orbit), every hour and a half the satellite revolves around the earth and a heating and cooling thermal cycle takes place.

The first objective of this study is to conduct thermo−elastic analyzes of the LAD module by analyzing, in the most extreme thermal conditions[5] (cold thermal case), how the tensions, deformations and displacements of the structure change. This objective is made possible thanks to a preliminary examination of the models, a correct CAD import and a mesh sensitivity study. The second objective concerns the parametric optimization of the structure design. In particular, some geometric characteristics of the Collimator Frame will be parameterized[6] and an optimized design proposed.

The purpose of the optimization[7] is to understand, as the possible combinations of geometric characteristics vary, what is the trend of the main parameters of merit, all taking into account the constraints of geometry imposed on the volume and mass of the structure.
The investigation is aimed at optimizing the design of the Collimator Frame, where the parameter of merit to be minimized will be the angle that describes the misalignment of the collimator with respect to the frontal direction.

The present study aims at first to present the preliminary mechanical design of the LAD module in its entirety and then to focus on the collimator frame, object of optimization and parameterization. Then, having explained the method and the approach used, the tools used and the results obtained will be shown. The latter are encouraging from the point of view of the method but need more iterations and computational power, future objectives of this project.

2. Mechanical preliminary design

2.1. Design Overview

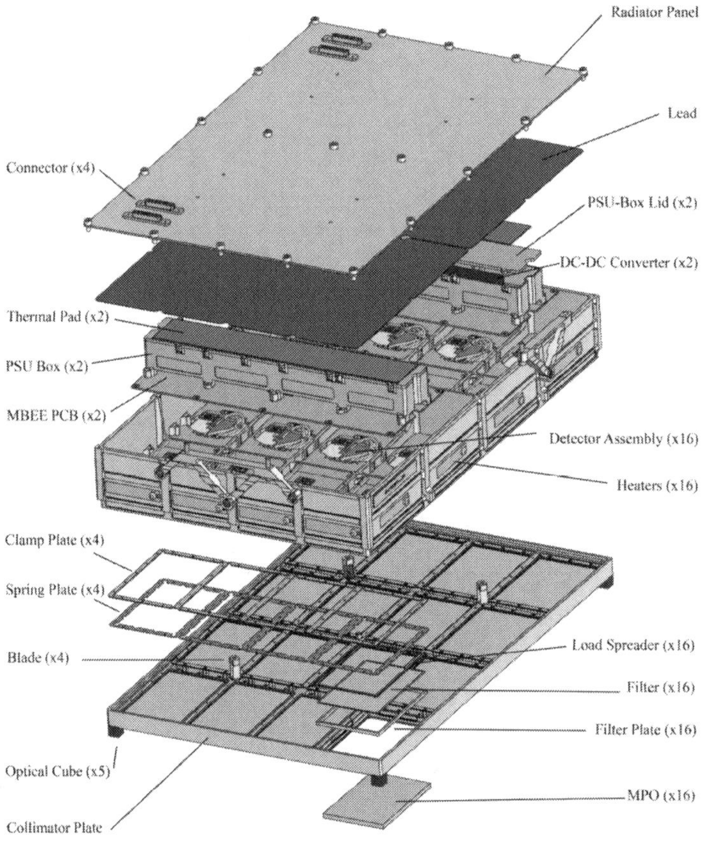

Figure 2: Module exploded view, with specification of the main components and subsystems

The module contains:

- An aluminium alloy **Collimator Frame** with 16 co−aligned **MPO** tiles (in positions corresponding to the 16 SDDs), clamped to the frame;
- Underneath each collimator tile (MPO) and the detector is a thin ($\sim$ 1 µm) Kapton film, coated with $\sim$100 nm aluminium on each side, is hosted. The filter blocks IR, visible and UV light, allowing through X−rays of $> \sim$2 keV;
- A **detector frame** containing 16 **SDD+FEE**. Each SDD has 224 anodes. Each FEE has 30 ASICs (28 analogue and 2 ADC), to read out the signals from the anodes. The ASICs amplify and digitize the anode charge pulses resulting from X-ray events;

IOP Conf. Series: Materials Science and Engineering 1275 (2023) 012035 doi:10.1088/1757-899X/1275/1/012035

- **Module Back End Electronics** (MBEE). Controls the ASICs and HV PSU and reads out the digitized events. It formats and time-stamps each event and transmits it to the PBEE; the MBEE is hosted in the same mechanical and electrical framework of the Detector Frame;
- **HV/MV/LV PSU**: This is the power supply for the SDDs; the PSU is hosted in the same mechanical and electrical framework of the Detector Frame;
- A 300 µm lead back–shield, to reduce the background events in the SDDs;
- A **radiator** to dissipate heat from the module (lower SDD temperature improves the energy resolution);

An **heater belt** is placed around the module. The heating process during annealing, which heat up the model up to 50°Calters the silicon crystal structure, recovering the damages induced by particle irradiation.

2.1.1. Collimator Tray In the upper part of the collimator frame, six optical cubes will be positioned and mounted to ensure that the stability and repeatability of the alignment are $\leq$10 *arcseconds*.

The collimator tray hosts 16 individual MPO tiles, sheets of lead glass (Pb content about 40 %) machined through the same process used to produce microchannel plates, in positions corresponding to each individual SDD detector. The MPO has a size of 111.0 $\times$ 72.5 $\times$ 5 mm^3 and fits into a recess in the aluminium alloy collimator frame.

A narrow portal frame pressure plate (load spreader) is placed on top of each MPO; also four clamp plates are positioned on top of the four spring plates. The spring guarantees sufficient preload, adequately spread on the MPO tile to prevent it from moving out of plane during vibration testing. The lateral movement of the MPO tiles is limited by the springs as well by four viton tubes, that should keep mechanical and chemical resistant properties at higher heat applications, placed at the collimator edges.

The collimator assembly is attached to the lower detector module assembly through four blades, slat fixing system, that are screwed into the lower detector module frame and pass through the collimator frame as it is lowered in place, this allows some movements between the frame assemblies.

The purpose of these blades is to reduce the thermos-elastic deformation of the collimator frame, rigidity is instead guaranteed by a central stud that, also, pass through the collimator frame.

An optical filter is used to complete the thermal and optical design, mounted on the collimator frame, above each detector, between the detector and the MPO at 3.5 mm distance to allow for any deformation at launch.

2.1.2. Detector Tray The fundamental LAD detection element is the Detector, composed of SDD+FEE (dubbed "Detector Assembly" – DA) and Collimator. The eXTP detector module assembly comprises sixteen Silicon Drift Detectors (SDD's) arranged in a four-by-four pattern.

The detectors are mounted in a monolithic aluminium alloy frame that supports each detector unit on four points with sufficient compliance to cater for CTE variation between the frame and the detector FEE PCB. For each detector there is a conductive link via the structure to the radiator.

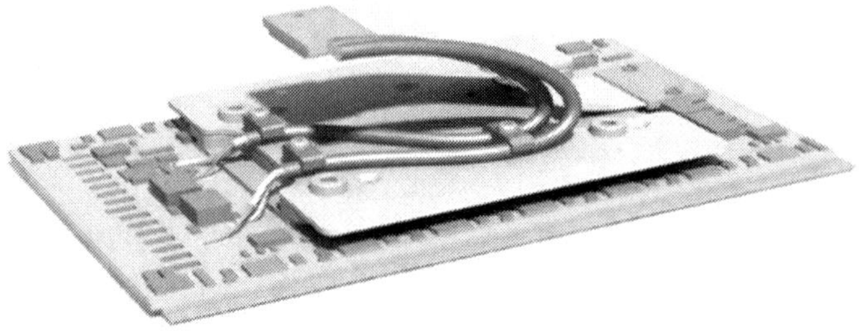

Figure 3: Silicon drift detector and front-end electronic

The FEE size is as close as possible to the SDD (120x72mm) in order to maintain a sensitive area of the module close to 100% when all FEE are mounted into the module.

Two module back-end electronics (MBEE) PCB assemblies with HV supplies attached are mounted behind the bank of detectors, each BEE/HV assembly interfacing and serving eight detectors.

A lead shield and an aluminum alloy radiator panel screwed directly to the back of the detector module body then completes the build of the lower detector module assembly.

The radiator is a single aluminum sheet of 2 mm thick, this thickness is adopted to minimize the thermal gradient across this radiator and therefore optimizes the detector temperature over each orbit.
Each module will be mounted kinematically in the payload structure to mechanically isolate each module from any deformation of the interface.

In the end the interface between the modules and the spacecraft is guaranteed by applying three bipods, two of which are standard in the long-side and one W−bipod on the short side of the module.

3. Background and Methods

For the purpose of this study, one of the modules that will be housed in the LAD instrument positioned on the Sino−European eXTP satellite was taken into consideration, giving particular importance to the collimator frame and the mission requirements related to it (e.g. alignment). The software used for this project are the following:

- **SolidWorks 2022**
  SolidWorks[13] is a three−dimensional parametric design and design software, produced and marketed by Dassault Systèmes. Useful for the design of mechanical devices, even complex ones, it involves the creation of 2D and 3D drawings of solids and surfaces, through a parametric geometric system. Parametric design allows to set up numerous types of relationships (parallelism, concentricity, perpendicularity, collinearity, equality, coaxiality, symmetry and many others).
- **ANSYS SpaceClaim**
  ANSYS Space Claim is a parametric solid modeling software available in the ANSYS package that operates mainly according to the direct modeling logic, thanks to which it is possible to define the construction and modification operations of the desired geometry, without the need to define interdependent features.
- **ANSYS Meshing**
  ANSYS Meshing is the software responsible for generating the calculation grid. This process is automated in its simplest application although it is possible to add control settings to act directly on those parameters that define the mesh in terms of morphology, topology and dimensions of the constituent cells.

IOP Conf. Series: Materials Science and Engineering 1275 (2023) 012035 doi:10.1088/1757-899X/1275/1/012035

- **RBF Morphing**
  Radial basis functions (RBFs) based mesh morphing allows to adapt the shape[8][9] of a computational grid onto a new one by updating the position of all its nodes. Usually nodes on surfaces are used as sources to define the interpolation field that is propagated into the volume mesh by the RBF. The method[10] comes with two distinctive advantages that makes it very flexible: it is mesh independent and it allows a node wise precision. There are however two major drawbacks: large data set management and excessive distortion of the morphed mesh that may occur.
- **ANSYS Mechanical**
  ANSYS Mechanical is the finite element calculation software that allows to set up and carry out structural analysis.
- **ANSYS DegignExplorer**
  ANSYS DesignXplorer is the software that allows you to generate, explore and query a certain parametric domain. In this sense it was used for the choice of the mesh in the preliminary phase of reconstruction of the simplified FEM model of the LAD module and subsequently in the various structural optimizations. The algorithms implemented in DesignXplorer, once defined the frontier of a multidimensional domain of which each dimension is represented by an input parameter previously defined in the model and calculates the desired output parameters establishing a functional relationship between them and the entire input domain. The tool available in DesignXplorer used in the project for the mesh convergence and optimization is, more precisely, the Response Surface Optimization.
- **OptisLang**
  optiSLang[12] is a software platform for CAE−based sensitivity analysis, multi−disciplinary optimization (MDO) and robustness evaluation. It provides a framework for numerical Robust Design Optimization (RDO) and stochastic analysis by identifying variables which contribute most to a predefined optimization goal. This includes also the evaluation of robustness, i.e. the sensitivity towards scatter of design variables or random fluctuations of parameters. This procedure can also be used for a pre−optimization before global optimizer (evolutionary algorithms, Adaptive Response Surface Methods, Gradient-based methods, biological-based methods) are used for a direct single-objective optimization.
- **ANSYS Workbench**
  The Workbench platform is the tool designed to put in communication the various modeling and numerical analysis software that fall within the ANSYS landscape, facilitating the management of the project that is articulated there through various work steps. Its key feature is to allow the transfer and sharing of input and output data not only between the different tools used in pre and post processing, but also between multiple simulation models (which allows for the easy implementation of sophisticated multi−physical simulations).

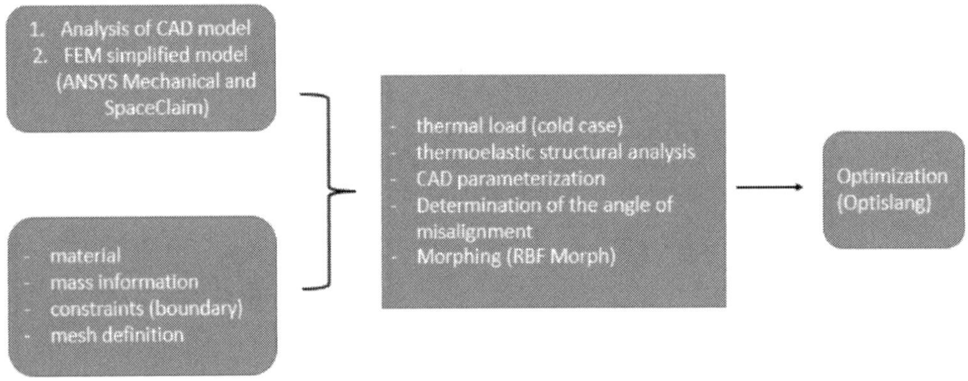

Figure 4: adopted workflow

The CAD (Computer Aided Design) model of the module was generated using Solidworks 2022 software, but the shapes were reconstructed and simplified using the standard features of the ANSYS SpaceClaim software. Furthermore, taking advantage of the advanced features of this software, a procedure of symmetry and conservation of structural constraints has been implemented even if the model is lightened. The CAD model of the module obtained with the different components highlighted is shown in figure 5:

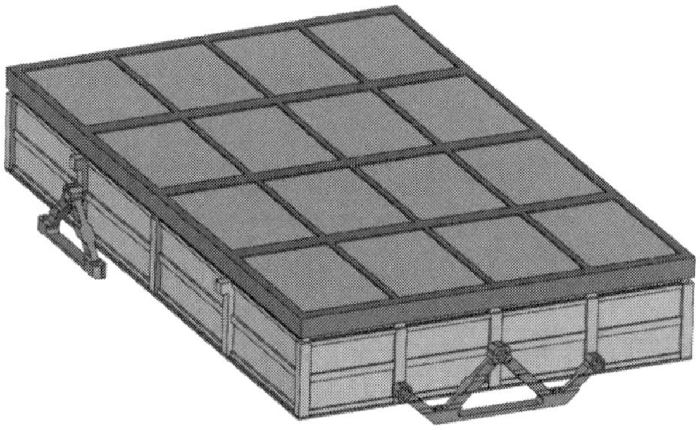

Figure 5: CAD model

In figure 6 its shown, in detail, the collimator frame of the module, object of the optimization:

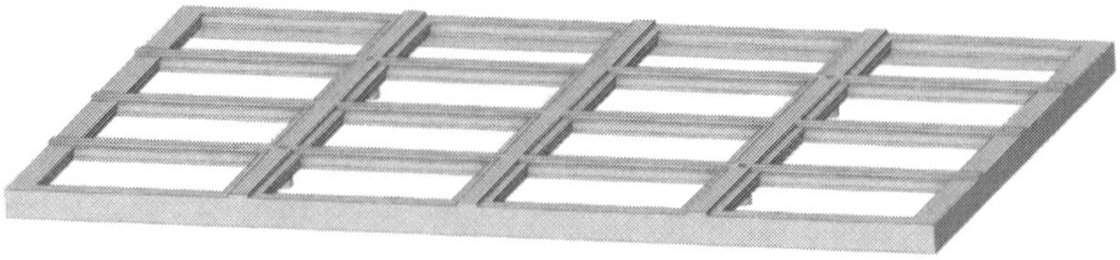

Figure 6: Collimator Frame

The LAD modules have to operate in variable thermal conditions linked to the orbital cycle of the satellite and therefore, at fluctuating temperatures that induce more or less significant deformations in the various components.

The objective of the optimization proposed in this study is to analyze the thermal deformations that occur on the Collimator Frame, in relation to the functional requirement of housing and mutually orienting the 16 detectors (Collimator), and subsequently, to improve the design of the Collimator Frame, working on the ribs, in such a way as to have better pointing requirements.

Therefore the study was carried out in the following steps:

- preliminary design of the module in a complete CAD model meeting the requirements of the mission and export of the model in ANSYS SpaceClaim obtaining a simplified defeatured[10] model but compliant in terms of mass, volume and critical part of collimator frame object of this study
- import into the simplified FEM model of the thermal load on components and interfaces associated with the most severe case (Cold Case).
- Definition of the parameters necessary to control the misalignment of the collimator with respect to the undeformed configuration of the Collimator Frame.
- Re-processing and parameterization of the CAD geometry of the Collimator Frame through the ANSYS RBF Morph tool.
- Structural optimization of the Collimator Frame through Response Surface Optimization and OptisLang.

4. Simulation and Optimization

4.1. Model Overview

A simplification of the CAD model, implemented with all its details, it is necessary in order to avoid complicated geometries (sharp edges, holes...) that implicate calculation difficulties on ANSYS. Due to the symmetry of the module, it is decided to divide the module into two identical parts and to perform the simulation campaign on only one half.

Through SpaceClaim has been added some reinforcement ribs on the Collimator's top, which it has resulted in an increase in volume of the same, the upper face of the rib has been partitioned in order to create the surfaces corresponding to the various sources of the optimization method (mesh morphing). This choice was necessary to make the improvement and the optimization of the Collimator frame. Finally, all non–functional elements that supported the Collimator were eliminated from the model where it is connected to the Detector Frame.

- **Material properties**
  A specific material is attributed to each component of the model setting its mechanical properties; the material is taken from the ANSYS library or created by inserting the values of its mechanical properties.
  Each element that makes up the module is subject to different physical and thermal loads and at the same time will have its own function within the experiment.
  Table 1 shows the elements and the material assigned to them.

Table 1: Material properties

Description	Density [kg m-3]	Young modulus [N m-2]	Poisson Ratio
7075 T6 aluminum alloy	2796	7,1E+09	0,33
Titanium alloy	4620	9,6E+09	0,36
PCB laminate FR4	1900	2,46E+09	0,136

IOP Conf. Series: Materials Science and Engineering 1275 (2023) 012035 doi:10.1088/1757-899X/1275/1/012035

- **Mass Information**
  The mass of the elements not represented in detail in the FEM model (fasteners, cables, electronic components) have been modelled as additional non−structural mass. The largest components have been represented as point masses rigidly linked to the PCB. Other electronic components, fasteners and cables have been represented as distributed mass divided into CF and DF for return to the mass budget of the CAD model.

4.2. Mesh Analysis

In order to generate a mesh of the module that provides adequate results, it is necessary to make some changes to the settings, defining, in addition to a global value of the mesh (thus defining a general element size), also specific values for some surfaces and bodies. The starting mesh is defined using the general settings and some controls to define the mesh on the critical surfaces, that are candidates to receive further refinement, in particular those with load hotspots.
For each of this parameters, the optimal value will be identified and then the optimal mesh will be calculated.

- **Model setup**
  A constrained modal analysis was set and performed without preload. The contacts already present in the model were used among the constraints and two cylindrical supports on each bipod to constrain the radial and axial displacements while leaving the tangential displacements free. Then, in order to verify the convergence, the trend of the first frequency resulting from the modal analysis and the total number of nodes were analyzed.
- **Results and convergence**
  The optimization of the mesh was performed through the use of "response surface", ANSYS tool that allows you to modify the values by adapting them with a parameterization.

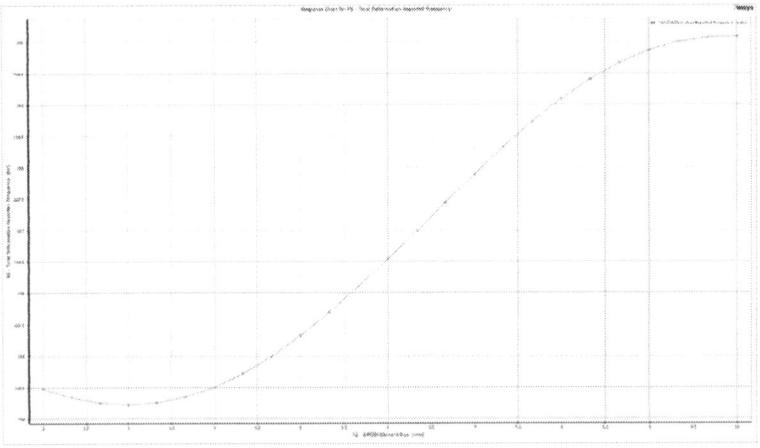

Figure 7: Response surface

The optimization constraints are:
- the total number of nodes, as a trade−off between available computing power and the need for a very dense mesh;
- constrain the first natural frequency above 120 Hz (system requirement) by a value at which the bipod frequency−size relationship is almost convergent.
At this point, once the optimization has been performed, the tool has returned some candidate points for the solution and, in accord to the graph analyzed previously, the point that has a dimension of the element of the body mesh is closest to the central part of the curve has been chosen, the one closest to the target set where the relationship is very

close to convergence.
The mesh obtained is shown in figure 8.

Figure 8: Optimized mesh of half module

4.3. Optimization

4.3.1. Thermal Load - Cold Case Due to the different thermo−physical properties of the materials, the components deform during the thermal cycles causing a misalignment of the 16 MPO and, since the structure is not free to deform due to the presence of internal and external constraints, they become tensioned.
In order to evaluate these deformations and to verify the resistance of the structure, the temperatures of the most severe thermal case, the cold one illustrated in table 2, were imported into the FEM model:

Table 2: Profile temperature of cold case

I/F bipod - truss	20 °C
I/F bipod - DF	-48 °C
short side	-67.3 °C
detector frame	-52 °C
blades	-61 °C
central bolt	-55.3 °C

From the thermoelastic analysis emerges that the temperature variation of the components with respect to the reference temperature causes a misalignment of the panels and a deformation of the internal components which causes a deformation overall.
Focusing on the Collimator Frame, the study demonstrate how it undergoes an upward inflection. Each point is subject as a whole to an almost symmetrical field of displacements along x with greater displacements on the short side not constrained by the Bipods and from a displacement field along y instead symmetrical with respect to the center of gravity of the Collimator Frame itself:

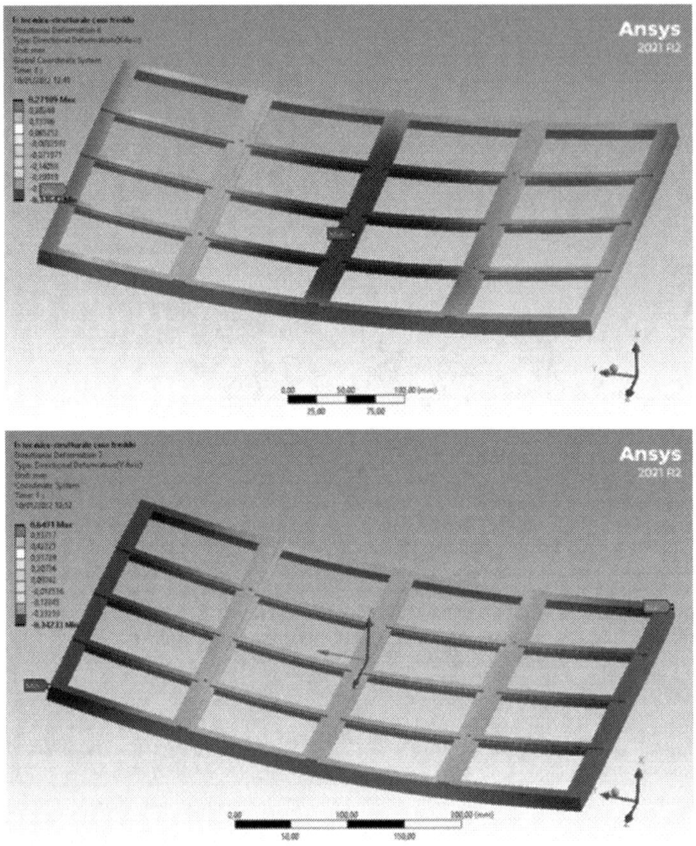

Figure 9: Deformation along x−axis and y−axis

From this displacement field it can be seen how the misalignment of the Collimator Frame is greater in the outermost windows.

4.3.2. Misalignment evaluation For the definition of the outputs the Collimator Frame has been divided into quadrants and the internal longitudinal sides of each quadrant numbered in matrix logic.

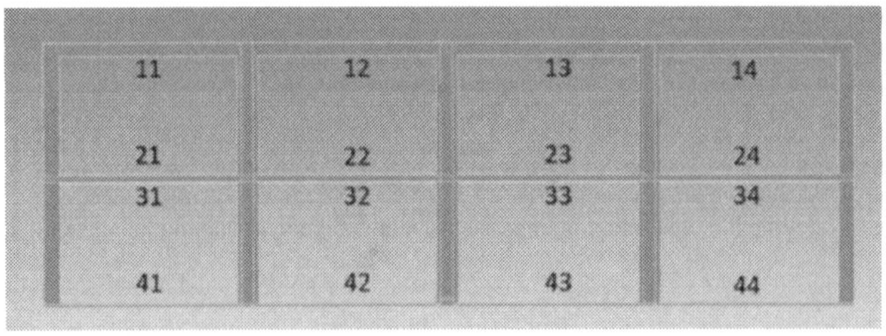

Figure 10: Subdivisions of Collimator Frame

Also, the following relation was used for the calculation of each single angle, expressed as a function of the displacements along all directions:

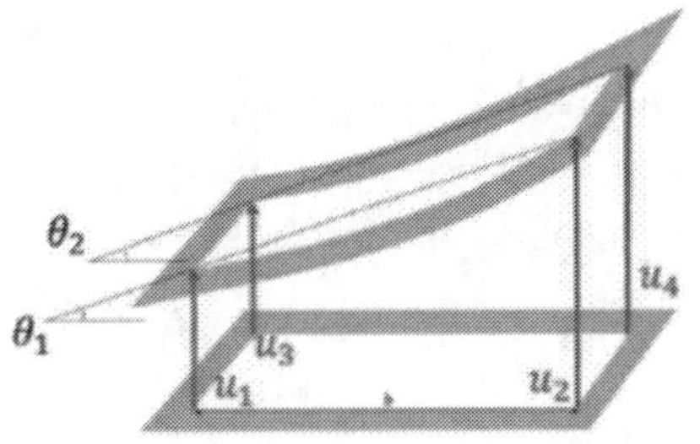

$$\theta = \left| \operatorname{atan}\left(\frac{u_{max} - u_{min}}{b} \right) \right|$$

Figure 11: Main misalignment angles of the Collimator Frame window averaged to the sides

Therefore, to represent the overall misalignment trend of all 16 panels and to have a single output merit parameter to be minimized in the optimization phase, an average of the 44 values of respective misalignment angles of the 16 panels was calculated.

$$\theta^* = \text{mean}(\theta_{11}, \theta_{12}, ., ..\theta_{44})$$

Figure 12: average of misalignment angle

Where $\Theta*$ represents the value of the misalignment calculated respectively to the original configuration, it will be a necessary condition for the success of the optimization to obtain an angle $\Theta \leq \Theta*$.

4.3.3. Geometry morphing In accordance with what has been said previously, the RBF morph tool was used to perform a geometric parameterization that implements shape modifiers directly on the computational domain. The new geometric configurations resulted from the displacement of a mesh region, through the use of algorithms, based on RBF, which smoothly propagated the displacement imposed on the surrounding volume. This tool was used to create 3 morphing regions that define the displacements:

- **Offset**: changes the height of the rib from an initial value of 3mm to an arbitrary and parametric value.
- **Inclination**: makes the height of the transverse and longitudinal rods of the rib variable variable, except for the loop in the following figure, whose height is kept constant.
- **Thinning**: makes the thickness of the internal transverse rods variable from a value of 8mm towards the center of the Collimator at an arbitrary peripheral value, makes the thickness of the rods constant external transversal and lower than the initial one by 8 mm.

Through the "Parameter Set" some dimensions have been defined, corresponding to the various instances of mesh morphing:

- **h_int**: height at the center of the rib.
- **h_est**: height adjacent to the perimeter of the rib.
- **thickness_trasv_int**: thickness of the internal transverse rods.
- **thickness_trasv_est**: thickness of the external transversal rods.

4.3.4. Optimization The limits of the optimization parametric domain have been set as:

- **h_int**: variable between 3mm and 5mm (reference value 3mm).
- **h_est**: variable between 1mm and 3mm (reference value 3mm).
- **thickness_trasv_int**: variable between 6mm and 8mm (reference value 8mm).
- **thickness_trasv_est**: variable between 6mm and 8mm (reference value 8mm).

The reference values are those respective to the original rib of the simplified model. The studied optimization corresponds to a constrained minimum problem, so it was entered as optimization objective to minimize the average misalignment and as constraint compliance with original overall volume of the Collimator Frame (ribbing included).

The chosen optimization method was selected the one suggested by the Optislang tool, called the "Adaptive Response Surface Method".

Optimizations with higher iteration numbers, produce more performing designs (reduction of misalignment greater than 3–5%), but unfortunately for this study the computational possibilities and timing have limited the simulation capacity.

4.4. Results

The optimized design was then compared, with regard to the performance of interest, with the original one.

The optimization tool was run for about five days on two different workstations and returned the following results: in order to limit the inflection of the Collimator Frame, increasing the height of the rib is more efficient than increase its thickness. In this sense, the optimization has identified the optimal design as the one characterized by:

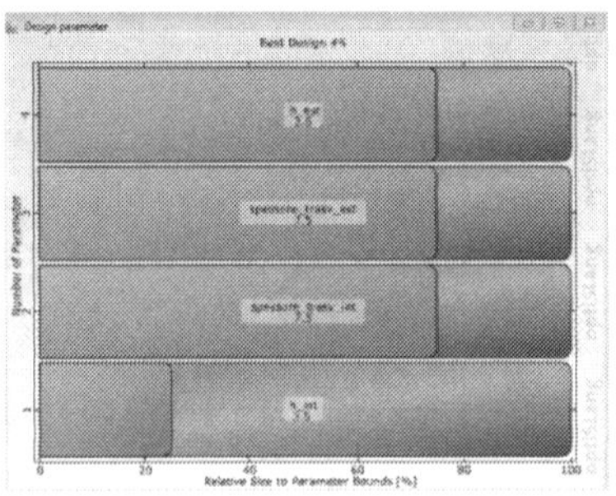

Figure 13: Optimized values

Obtaining a reduction in misalignment of approximately 5% with the shape obtained in figure 14.

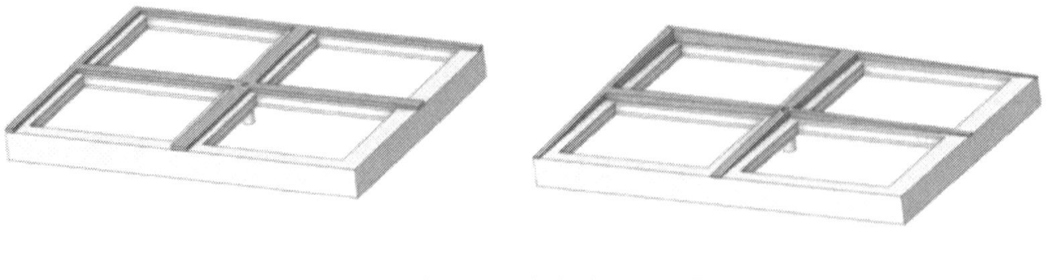

Figure 14: Optimised Collimator Frame

The optimization tool was run for about five days on two different workstations and returned the following results To keep the overall volume of the Collimator constant, note the increase in height of the rib in the area corresponding to the center of the rib is balanced by a decrease height in the peripheral area and a thinning of the transverse rods. With regard to the optimized design, taken as an example in comparison with the original one, it is possible to show the evolution of each of the angles with respect to which the output of merit has been defined. Observing the following table it can be seen how the distribution of misalignment's is better overall.

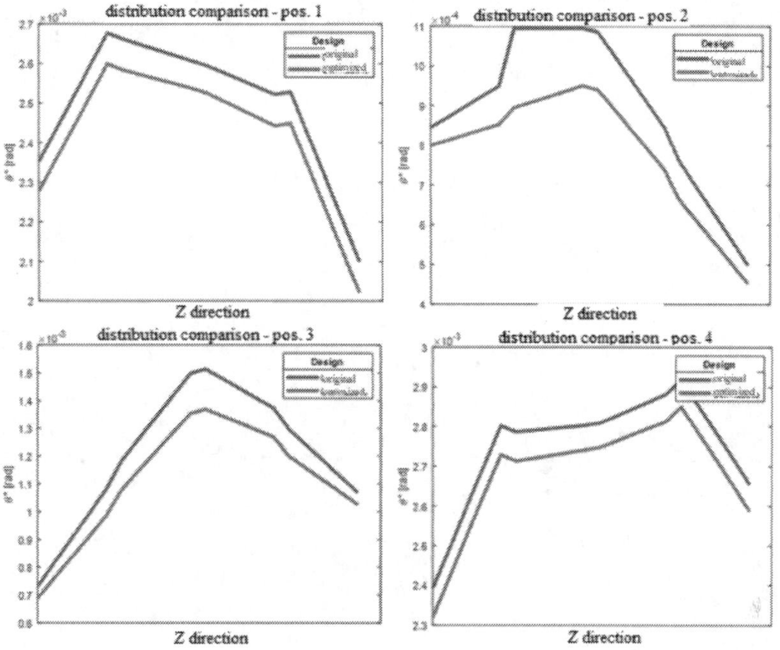

Figure 15: Comparison of the distributions of the angles of misalignment

5. Summary and Outlook

The study presented began with the analysis of the CAD model provided, which was simplified to facilitate initial simulations. Proceeded with the analysis of convergence of the mesh, from which the dimensions of the elements were defined which made it possible to obtain simulations in a short time, but with an acceptable degree of detail.

In the study of the collimator frame, a thermo–elastic structural analysis was first carried out for the cold thermal case, from which tensions and deformations of the initial geometry were determined, by analyzing the latter the possibility of carrying out a structural optimization was highlighted. Then, the CAD was modified to make the additional processing parameters parametric, object of optimization and the collimator misalignment angle was defined as a parameter of merit in order to perform a geometric optimization through the tools that ANSYS offers. The candidate point has been successfully verified in terms of misalignment, obtaining a value of about 5% less than the initial angle.

It should be emphasized that to improve computational times it has been chosen to reduce the degrees of freedom of the parameterization, by coupling some of the parameters in a symmetrical way; Therefore, the future development is to study solutions that allow the complete decoupling of the latter, calculating the angle of misalignment directly on the collimators, neglected in the study to minimize the computational burden.

6. References

[1] ShuangNan, A Santangelo, M Feroci, Y Xu et al., The enhanced X-ray Timing and Polarimetry mission—eXTP, (feb. 2019).
[2] M Feroci, M Ahangarianabhari, G Ambrosi and F Ambrosino et al., The Large Area Detector onboard the eXTP mission, Astrophysics - Instrumentation and Methods for Astrophysics (jul. 2018)
[3] M Hernanz, S Brandt, M feroci, P orleansky et al., The wide field monitor onborad the eXTP mission, Astrophysics - Instrumentation and methods for astrophysics - High Energy Phenomena (2018)
[4] V Mendes, eXTP-IANF-LAD-TN-003, Thermal Analysis of the eXTP LAD instrument (2022)
[5] M E Biancolini, Fast radial basis functions for Engineers applications, Springers (2017)
[6] G pascoletti, M Bignardi, C Conti and P Zanetti, Mandible morphing througth principal components analysis, Springer (2020)
[7] M Cali and M Olivieri, Application of and effective SIMP method with filtering for topology optimizatiuon of motorcycle tubular frame (2017)
[8] N Di Domenico, C Groth, A Wade, T Berg and M E Biancolini, Fluide structure interaction analysis, Integrity (2018)
[9] M Cali, D Speranza, U Cella and M E Biancolini, Flying shape sails by radial analysis function mesh morphing, Springer (2020)
[10] O C Zienkiewics, R L Taylor and J Z Zhu, The finite element method: Its basis and fundamentals, Butterworth-Heinmann, 6th (2005)
[11] K M Hanson, Introduction to Bayesian image analysis (1993)
[12] Dynarcdo, Optislang, www.ansys.com/product/connect/ansys-optislang
[13] Dassault Systemes, DS Soliworks 2022, www.solidworks.it

Residual stress measurements on a deep rolled aluminum specimen through X-Ray Diffraction and Hole-Drilling, validated on a calibration bench

M Beghini[1], T Grossi[1], C Santus[1], L Seralessandri[2] and S Gulisano[3]

[1] Dipartimento di Ingegneria Civile e Industriale, Università di Pisa, Pisa, Italy
[2] GNR Srl, Agrate Conturbia (NO), Italy
[3] SINT Technology Srl, Calenzano (FI), Italy

E-mail: tommaso.grossi@phd.unipi.it

Abstract. Residual stress measurements are notably affected by a high sensitivity to errors in input data. Measurements should then be presented together with an estimation of their accuracy. A common strategy is to carry out more measurements and/or to compare the results of different techniques. However, error contributions due to biases could be dangerously left unseen. In a previous work, the authors presented a calibration bench which can impose a known bending stress distribution on a specimen while simultaneously performing X-Ray Diffraction (XRD) or Hole-Drilling Method (HDM) residual stress measurements. Since the external load can freely be applied and removed, the superposition principle can be exploited to simultaneously identify either the reference bending stress distribution or the actual residual stress distribution, with the same experimental setup. A deep rolling treatment was measured and analyzed on the calibration bench with both XRD and HDM. First, residual stresses on the surface were evaluated with XRD measurements, then electrochemical material removal was performed to investigate stresses at higher depths. After that, HDM measurements were carried out and compared with the results of XRD. Both methods were also used to identify the known bending stresses, providing an additional validation of the residual stress results.

1. Introduction

Measuring residual stresses is a particularly challenging task [1]. A well-trained operator is often critical; the solution process is often grounded on a great deal of FEM calculations [2, 3, 4, 5, 6, 7] and requires the solution of an inverse problem [8, 9, 10], which is often practically ill-conditioned. The hole drilling method (HDM) [11, 12, 13, 14] is a well-known technique for measuring residual stresses, due to its low cost and its in-field applicability. It is standardized by ASTM E837 procedure [15], which can be carried out with commercially available devices, such as the MTS3000-Restan by SINT Technology. The HDM is known to be affected by a high sensitivity to noise in input data, particularly near the surface, due to the mathematical nature of the solution process. In fact, in that zone stresses depend only on the first few measurement points and are affected by errors in the identification of the zero-depth point, under the strain rosette. In addition, the ill-conditioning of the problem worsens at high depths too, where the method loses sensitivity. Due to these reasons, it is often complemented with X-Ray Diffraction (XRD) (as in [10]), which exploits Bragg's law to identify residual stresses near the specimen surface [16], with a typical penetration depth in the order of 10 μm. Stresses at higher depths

Content from this work may be used under the terms of the Creative Commons Attribution 3.0 licence. Any further distribution of this work must maintain attribution to the author(s) and the title of the work, journal citation and DOI.
Published under licence by IOP Publishing Ltd

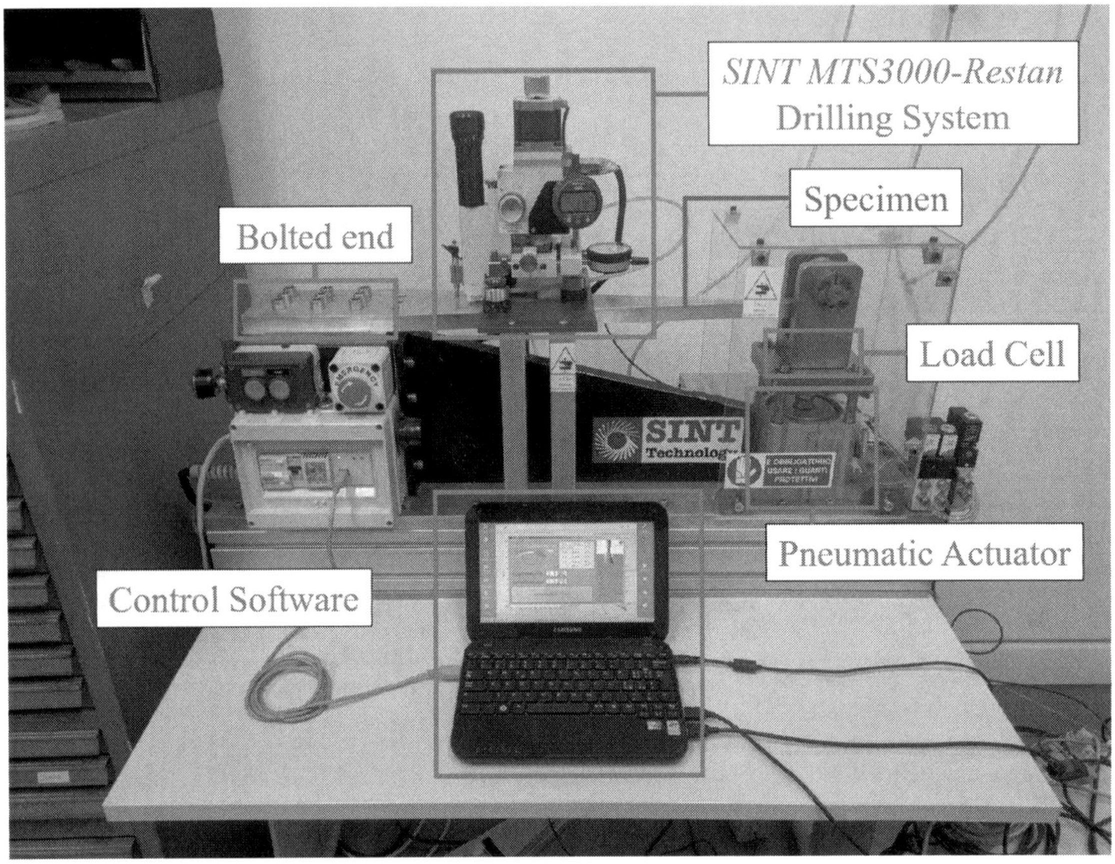

Figure 1. Description of the calibration bench.

can be profiled by removing layers of material. To this aim, electropolishing is commonly used, due to the fact that it does not introduce residual stresses in the removal process. XRD is particularly sensitive to the crystallographic properties of the material under investigation and to the strategies used to identify the diffraction peak. Therefore, its application requires a careful evaluation of the specimen and of the measurement setup, in order to gain confidence on the obtained measurements.

In a previous work [17], the authors presented a calibration bench which can impose a known bending stress distribution on a specimen while simultaneously performing X-Ray Diffraction (XRD) or Hole-Drilling Method (HDM) residual stress measurements. Most problems concerning setup, instrument calibration or material crystallographic properties generate significant differences between the reference bending distribution and its identified counterpart. Since the identification of the applied bending distribution is carried out with the same setup used to measure residual stresses, the achieved accuracy in the identification of the applied bending stresses is a direct validation of the obtained results in terms of residual stresses. For example, errors due to hole eccentricity, geometrical irregularities, zero-depth misidentification, or incorrect gauge factors are shared across the two distributions. The bench is shown in Figure 1. Details of the system are available in [17], while the decoupling procedure is described in [18]. The identification of the bending stress distribution is based on two measurements, respectively in the unloaded and in the loaded configuration. Assuming linear elasticity, the difference between the two corresponds to the effect of the applied bending

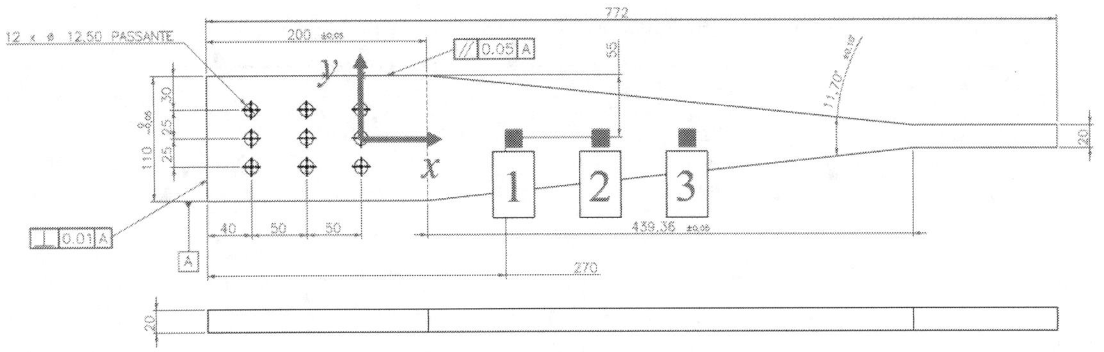

Figure 2. Technical drawing of the aluminum specimen, and definition of a reference system on its upper surface. The three measurement points are reported with red squares.

distribution alone, while measurements in the unloaded configuration corresponds to the residual stresses in the specimen.

In this work, the bench was used to analyze a deep rolling treatment on a 7075-aluminum specimen, whose technical drawing is attached in Figure 2. The corresponding residual stress distributions were analyzed through XRD and HDM measurements; both techniques were also validated on the calibration bench.

2. Methods

A deep rolling treatment was carried out on an area of $200 \times 50\,\text{mm}^2$, placed along the tapered section of the specimen (see Figure 3). A tungsten carbide roller tool (model D90-L-25-0 by DREX-TOOLS) was used. The tool was mounted on a CNC machine, and its motion was numerically controlled. The normal force was measured through a load cell placed between the tool and its support, and it was manually kept at an approximately constant value of 150 N by the CNC operator, by tuning the tool position along the direction normal to the specimen surface. The same setup is described in detail in [19]. The rolling parameters are available in Table 1.

Figure 3. Photo taken during the deep rolling process, on a CNC machine.

Table 1. Tool features and deep rolling parameters

Clearance angle	Fillet radius	Feed	Normal Force
1.6°	0.78 mm	0.18 mm	150 N

A reference system on the upper surface of the specimen is defined in Figure 2. Once the specimen was mounted on the calibration bench, the corresponding bending stress distribution applied by the pneumatic actuator was accurately characterized, as described in [17]. The vertical deflections under the external load were recorded in the tapered region with an LVDT sensor, as shown in Figure 4. Through classical beam theory, this process allowed the identification of bending strains in the specimen, which were compared with values recorded by the applied strain rosette and with a FEM model. Eventually, the achieved errors on the reference distribution are reasonably at least on order of magnitude lower than typical errors in residual stress measurements, so it can practically be considered as a benchmark.

Three measurement points were placed along the specimen x axis in the tapered region (see Figure 2), then the following tests were performed:

- Near-surface XRD residual stress measurements were carried out on all three points, with a GNR SpiderX Edge diffractometer, shown in Figure 5(a). The experimental setup was validated on the calibration bench by measuring near-surface stress values of the bending

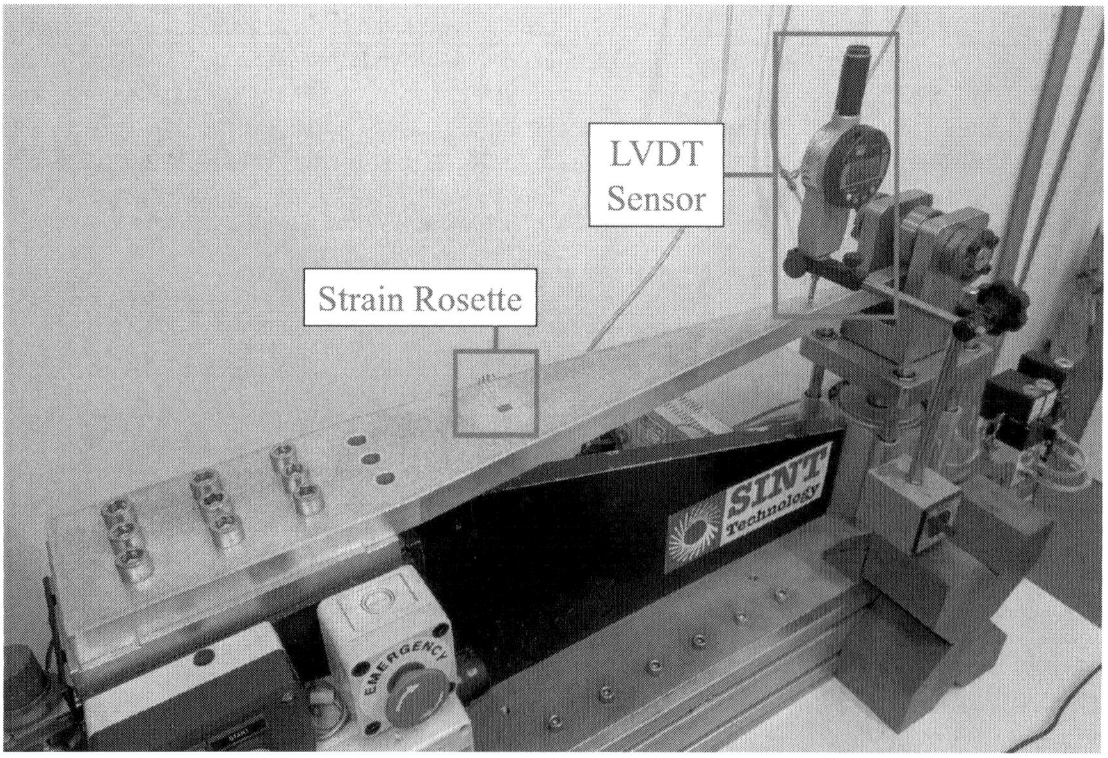

Figure 4. Specimen mounted on the calibration bench, with strain rosette applied. The vertical deflections of the specimen were measured along the whole tapered section.

(a) (b)

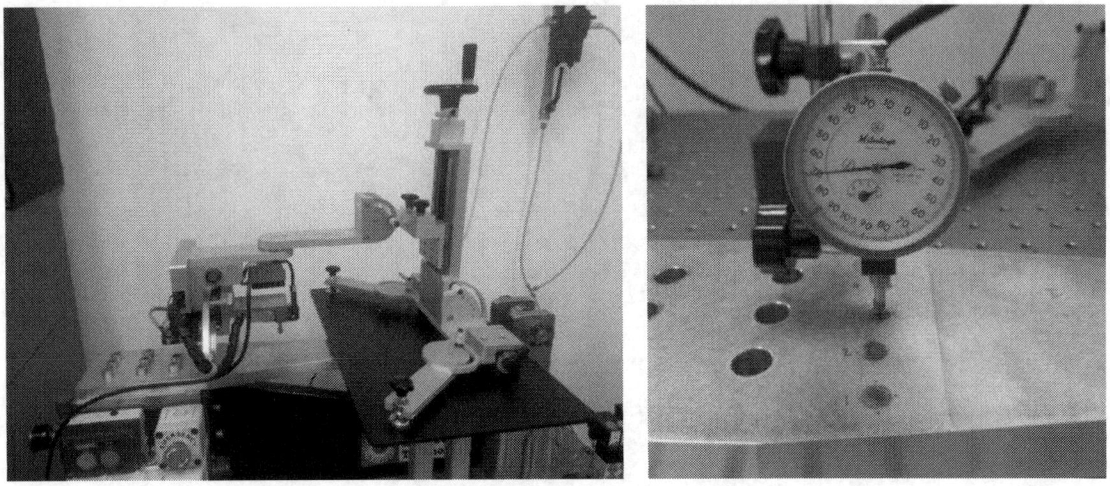

Figure 5. (a) GNR SpiderX Edge diffractometer, mounted on the calibration bench and ready to carry out a residual stress measurement. (b) Measurement of etching depth with a dial gauge.

distribution, with and without the applied load.

- HDM measurements were performed with a MTS3000-Restan system on points 1 and 2. For each drilling step, strains were sampled in both the unloaded and the loaded state, to simultaneously identify both the bending and the residual stress distribution, as mentioned before and described in [17, 18].
- XRD depth profiling through electropolishing was carried out on point 3. Diffraction measurements were performed with the same GNR SpiderX Edge diffractometer. A solution of perchloric acid was used as chemical agent. The depth was increased until clear diffraction peaks could no longer be identified.

Diffraction measurements were carried out with the $\sin 2\psi$ method, using a chrome anode

(a) (b)

Figure 6. (a) Alignment of the strain rosette (HBM RY61K) with the specimen reference system. (b) Strain rosette applied on the upper face of the specimen, after the Hole-Drilling procedure.

and analyzing the lattice spacing of {311} planes. In order to obtain the full stress tensor without any simplifying assumption, each XRD measurement was actually performed in three directions at angles of 0°, 45° and 90° with respect to the x axis. The etching depth obtained with electropolishing was measured with a micrometer dial gauge, as shown in Figure 5(b). HBM RY61K strain rosettes were used in HDM measurements and are shown in Figure 6. Holes were drilled with an inverted cone tool having a diameter of 1.8 mm, powered by an air turbine rotating at more than 300.000 rpm. A maximum depth of 1.2 mm was reached in 120 steps of 0.01 mm. The hole diameter and eccentricity were measured through an optical microscope with crosshairs, included in the MTS3000-Restan system. Eventually, the stress distributions from 0 to 1 mm depth were obtained with the *Influence Functions* method [3, 4, 5], Tikhonov regularization and the Morozov discrepancy principle [20, 21].

3. Results

In Figure 7, the validation data for XRD near-surface measurements are reported, when only the applied bending stress distribution was identified. Each graph refers to a component of the (plane) stress tensor, extracted from the three measurements at orientations of 0°, 45° and 90°. In Figure 8, the bending stress distributions evaluated through the two HDM measurements are shown. In Figure 9, the residual stress distributions obtained with HDM measurements on points 1 − 2 are compared with results from the XRD stress profiling performed on point 3.

4. Discussion

The validation process confirmed that the instrumentation setup had been properly done. In particular, XRD near-surface measurements captured the stress fields produced by the bending load with a maximum error of about 20 MPa (see Figure 7). Equivalently, HDM measurements were able to capture the bending stress distribution along the entire hole depth (see Figure 8). The achieved maximum absolute error across the whole depth range is about ±30 MPa for the largest stress component σ_{xx}, which is aligned with the state of the art [22]. Note that the error is not particularly pronounced at the surface, where HDM is instead known to be prone to higher uncertainties.

The two residual stress profiles obtained with HDM (shown in Figure 9) are consistent with the ±30 MPa maximum error achieved in the identification of the bending distribution, though at lower depths they are even more repeatable. The XRD stress profiles obtained with

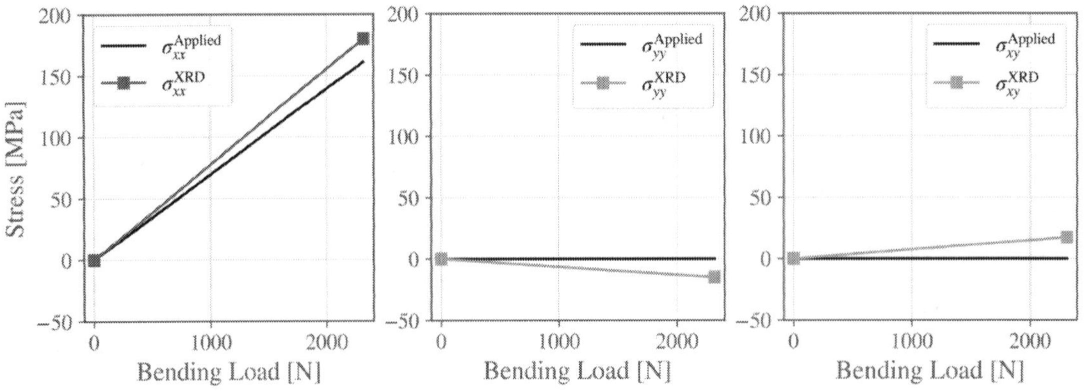

Figure 7. Validation of the XRD setup, through measurement of the reference stress fields produced by bending.

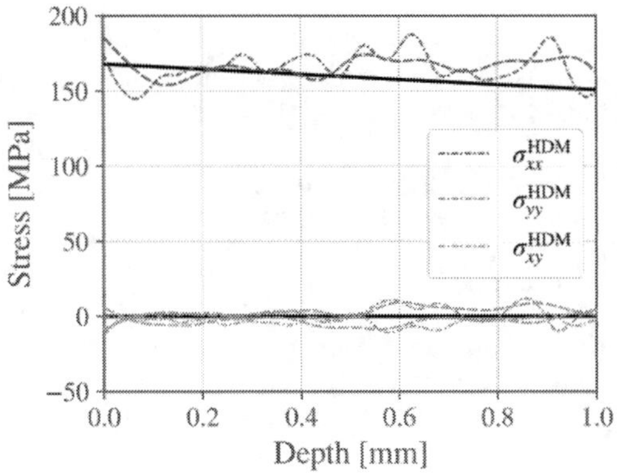

Figure 8. Bending stress fields identified through the two HDM measurements. The reference distributions are reported as black lines.

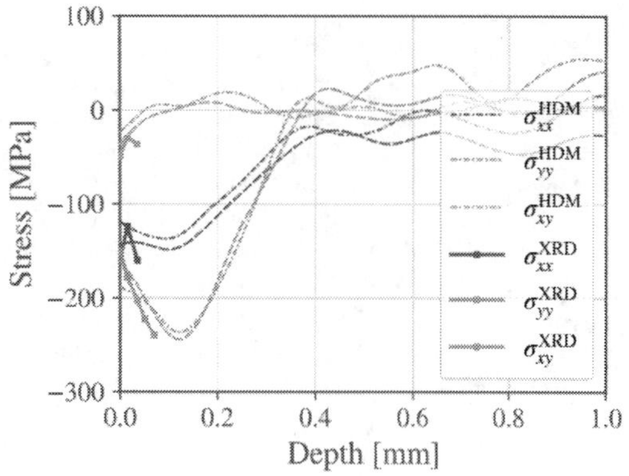

Figure 9. Obtained residual stress distributions. The HDM measurements on points $1-2$ are reported as dashed lines, while the XRD measurements on point 3 are reported as squares connected by solid lines.

electropolishing correctly mimic the trend of HDM curves, especially for the largest stresses σ_{yy}, which act in the transversal direction with respect to the rolling motion of the tool. However, the quality of the diffraction peaks quickly degraded along the depth, so the measurements turned out to be not significant after just a few removal steps. Uncertainties in the etching surface planarity probably played a role in this effect.

In addition, errors in the measurement of etching depth also affected the achieved accuracy, which was still reasonable for engineering applications. More studies will be carried out on the grain size in the specimen material and on potential texturing caused by the deep rolling process itself.

5. Conclusion

The present calibration bench allows one to validate the chosen setup before and during residual stress measurements, both with X-Ray Diffraction and the Hole-Drilling Method. For example, insufficient grain statistics, or issues in the strain gauge bonding would immediately result in huge errors in the identification of the bending distribution. Since the specimen can be designed with different materials and geometries, the calibration bench is particularly useful to investigate residual stresses produced by new surface treatments, with an increased measurement confidence provided by the validation phase.

In this work, a deep rolling treatment on a 7075-aluminum specimen was analyzed, with XRD and HDM. Both setups were validated on the calibration bench, and that allowed an estimation of the achieved accuracy on the residual stress measurements. The observed errors on the applied bending distribution are satisfactory, although the XRD depth profiling showed a rapid decrease in the quality of diffraction peaks with depth. More investigations will be carried out on the micro-structure of the specimen material.

References

[1] Schajer G S, Prime M B and Withers P J 2022 *Experimental Mechanics* ISSN 0014-4851, 1741-2765
[2] Schajer G S 1981 *Journal of Engineering Materials and Technology* **103** 157–163 ISSN 0094-4289, 1528-8889
[3] Beghini M and Bertini L 2000 *The Journal of Strain Analysis for Engineering Design* **35** 125–135 ISSN 0309-3247, 2041-3130
[4] Beghini M, Bertini L and Mori L F 2010 *Strain* **46** 324–336 ISSN 00392103, 14751305
[5] Beghini M, Bertini L and Mori L F 2010 *Strain* **46** 337–346 ISSN 00392103, 14751305
[6] Schajer G S 2020 *Experimental Mechanics* **60** 665–678 ISSN 0014-4851, 1741-2765
[7] Schajer G S 2022 *Experimental Mechanics* ISSN 0014-4851, 1741-2765
[8] Schajer G S and Prime M B 2006 *Journal of Engineering Materials and Technology* **128** 375 ISSN 00944289
[9] Smit T and Reid R 2020 *Experimental Mechanics* **60** 1301–1314 ISSN 0014-4851, 1741-2765
[10] Smit T C and Reid R G 2021 *Experimental Mechanics* **61** 1029–1043 ISSN 0014-4851, 1741-2765
[11] Lake B R, Appl F J and Bert C W 1970 *Experimental Mechanics* **10** 233–239 ISSN 0014-4851, 1741-2765
[12] Beaney E M 1976 *Strain* **12** ISSN 1475-1305
[13] Schajer G S 1988 *Journal of Engineering Materials and Technology* **110** 344–349 ISSN 0094-4289, 1528-8889
[14] Schajer G S 1992 *Strain* **28** 19–22 ISSN 00392103, 14751305
[15] American Society for Testing and Materials 2020 *West Conshohocken, PA* 10.1520/E0837-20
[16] Noyan I C and Cohen J B 2013 *Residual stress: measurement by diffraction and interpretation* (Springer) ISBN 1-4613-9570-4
[17] Beghini M, Grossi T, Santus C and Valentini E 2022 *ICRS 11–11th International Conference on Residual Stresses*
[18] Valentini E, Beghini M, Bertini L, Santus C and Benedetti M 2011 *Strain* **47** e605–e618 ISSN 00392103
[19] Beghini M, Bertini L, Monelli B, Santus C and Bandini M 2014 *Surface and Coatings Technology* **254** 175–186 ISSN 02578972
[20] Morozov V A 1984 *Methods for Solving Incorrectly Posed Problems* (New York, NY: Springer New York) ISBN 978-0-387-96059-3 978-1-4612-5280-1
[21] Schajer G S 2007 *Journal of Engineering Materials and Technology* **129** 440–445 ISSN 0094-4289, 1528-8889
[22] Schajer G S 2013 *Practical residual stress measurement methods* (John Wiley & Sons) ISBN 1-118-34237-2

IOP Conf. Series: Materials Science and Engineering 1275 (2023) 012037 doi:10.1088/1757-899X/1275/1/012037

Ductility assessment of a 17-4PH steel through simple multiaxial tests

G Cortis, G B Broggiato and L Cortese

Sapienza Università di Roma, Department of Mechanical and Aerospace Engineering, Via Eudossiana 18, 00184 Rome, Italy

gabriele.cortis@uniroma1.it

Abstract. The present work summarizes the results of an experimental campaign aimed at assessing the ductility of a wrought 17-4PH steel alloy. A simple specimen reproducing multiaxial stress states through a universal testing machine is selected. A Finite Element Model (FEM) for each test is setup to extract the local values of stress and strain in the most critical point on the onset of failure. A Digital Image Correlation (DIC) technique is employed to assess the strain field estimated via FEM. The collected data are used to analyse the material ductility, calculating the triaxiality and deviatoric parameter at the fracture strain. The proposed tests fall in the range of low triaxialities, which are less investigated in the literature. The results obtained are compared with the prediction of a damage model, previously calibrated on the material through more conventional tests. The prediction accuracy of the damage model was fully confirmed by the outcome of the new tests. Eventually, the possibility of replacing some of the conventional tests used for calibration with the proposed specimen is explored.

1. Introduction

Ductility is the ability of a material to sustain appreciable plastic strain before failure, thus its proper estimation combined with the evaluation of ultimate strength are crucial in mechanical design. In the last decades, different ductile damage models have been developed to describe how a material gets damaged with plastic strain and to predict the onset of failure. In literature, different damage models have been developed and can be classified according to their driving principle: Continuum Damage Mechanics-based models (CDM) predict collapse based on thermodynamic principles [1,2], porosity-based models analyse the failure by investigating the evolution of micro-voids, from nucleation to coalescence considering the presence of inclusion sites also [3,4], void-growth-based models link the failure to the evolution of microstructure [5,6], empirical-based models predict failure based only on experimental evidence [7,8]. In the present work only the last will be considered.

For all models, a calibration procedure involving different experimental tests on the specific material is required, which include tensile tests on round smooth or notched bar, torsion or compression tests, and often multiaxial tests, which usually require special testing setups. This means that the calibration procedure is generally non-trivial and not suitable for design companies which could not have in many cases the specific equipment at their disposal. The use of damage models as failure criteria can be applied in different scenarios: estimation of the ultimate strength of the material, evaluation of damage level of a component subjected to residual stress produced by manufacturing processes, investigation of plastic

Content from this work may be used under the terms of the Creative Commons Attribution 3.0 licence. Any further distribution of this work must maintain attribution to the author(s) and the title of the work, journal citation and DOI.
Published under licence by IOP Publishing Ltd

formability limits in forming processes, better knowledge of the behaviour of a material subjected to overloads, evaluating if components can withstand the accidental load without reaching failure.

The aim of the work is basically to assess the effectiveness of a simple specimen geometry, capable to induce multi-axial states of stress, to simplify the experimental phase necessary for the damage models calibration procedure. The specimen geometry proposed by L. Driemeier [9,10] is selected, to be tested employing a common universal testing machine; its dimensional details can be varied to reproduce several multiaxial stress states, ranging from pure shear to different combinations of tension and shear.

Along with the experimental tests of multiaxial specimens, finite element simulations of the experiments and a white light speckle digital image correlation technique are used to evaluate local values of the stress and strain states which cannot be directly measured. Using the collected data coming from the new simple tests, a ductile damage model is calibrated, and compared with the same model previously tuned using typical tests such as round bar, round notched bars, torsion and flat grooved plane strain tests [11]. The new calibration through multiaxial tests compared to the one with conventional tests proved to be equivalent, thus verifying the potential of the new geometry in replacing some of the above mentioned conventional tests. In addition, the good prediction accuracy of the ductile damage model was further confirmed.

2. Theoretical background

While first empirical damage models described damage as a function of triaxiality T only, equation (1) [12,13], recently it was found that damage accumulation also depends on a second parameter [14-17] called Lode Angle θ described by equation (2), which is often expressed in the form of a deviatoric parameter X as in equation (3). The terms I_1, J_2 and J_3 are the first invariant of the stress tensor, the second and third invariant of the deviatoric tensor respectively. In this framework, the triaxiality and the deviatoric parameter fully describe the stress state. According to the latest empirical models, ductile damage is represented by a scalar D that increases proportionally with the equivalent plastic deformation ε_p weighted on a function of the stress state $f(T, X)$, as in equation (4). The parameter D falls in the range from 0 to 1 where D equal to 1 corresponds to the onset of material failure, namely when ε_p assumes the critical value of the equivalent plastic strain to failure ε_f.

$$T = \frac{1}{3}\frac{I_1}{\sqrt{3J_2}} \tag{1}$$

$$\theta = \frac{1}{3}\arccos X \qquad 0 \leq \theta \leq \pi/3 \tag{2}$$

$$X = \frac{27}{2}\frac{J_3}{(\sqrt{3J_2})^3} \qquad -1 \leq X \leq 1 \tag{3}$$

$$D = \int_0^{\varepsilon_f} f(T, X) d\varepsilon_p \qquad 0 \leq D \leq 1 \tag{4}$$

Assuming the hypothesis of proportional loading conditions, which means T and X constant with the evolution of plastic strain, posing $D=1$ and inverting the equation (4), the following expression can be obtained (5):

$$\varepsilon_f = f^{-1}(T, X) \tag{5}$$

the above expression graphically represents a surface in the space (T, X, ε_f) usually called "Fracture Surface", which uniquely links each single stress state, namely each pair of T and X, to a single fracture strain value. Generally, T and X are not perfectly constant, exhibiting limited variations, therefore it is accepted to use their average value calculated via equations (6-7) to meet the proportional or quasi-proportional loading conditions and allowing the use of expression (5).

$$T_{avg} = \frac{1}{\varepsilon_f} \int_0^{\varepsilon_f} T(\varepsilon)\, d\varepsilon_p \quad (6)$$

$$X_{avg} = \frac{1}{\varepsilon_f} \int_0^{\varepsilon_f} X(\varepsilon)\, d\varepsilon_p \quad (7)$$

Several expressions of the function $f^{-1}(T, X)$ were devised in the literature; in the present work the formulation proposed by Coppola-Cortese-Folgarait (CCF) was adopted (8):

$$\varepsilon_f = \frac{1}{c_1} e^{-c_2 T} \left(\frac{\cos\left[\beta\frac{\pi}{6} - \frac{1}{3}arccos(\gamma)\right]}{\cos\left[\beta\frac{\pi}{6} - \frac{1}{3}arccos(X\gamma)\right]} \right)^{\frac{1}{n}} \quad (8)$$

where c_1, c_2, β e γ are material constants that define the surface shape and n represents the strain hardening exponent of the Hollomon's power law $\sigma = K\varepsilon^n$.

The calibration constants are obtained by means of the following procedure: enough experimental tests characterised by different loading conditions are conducted, typically tensile tests using round bar (RB) and round notched bar (RNB), tensile test in plane strain condition (PS) with flat grooved specimens and torsion tests with cylindrical bar (Tors) [18-20]. For each test, global quantities such as load and displacement are measured and recorded. Subsequently, a Finite Element Analysis (FEA) is performed for each test to derive the local quantities at the critical point of the specimen at the onset of failure, such as the three principal stresses and the equivalent plastic strain, from which T and X were derived through the equations (6-7). Then, a specific set of three parameters (T_{avg}, X_{avg}, ε_f) is obtained for every test at the onset of failure which represents a point in space (T, X, ε_p). Through a minimisation algorithm, the calibration constants governing the shape of the fracture surface are determined: the input data of the algorithm are the coordinates of each point, and the output are the values of the calibration constants that provide the best fit between the fracture surface and the experimental points. For the sake of brevity, please refer to the results section to see the representation of the fracture surface of the steel 17-4PH.

3. Methods and setup

A 17-4PH steel was taken as the reference material since it was already fully characterized by some of the authors of this paper [11] in order to use the results of the previous experimental campaign as a benchmark for the comparison between the fracture surface obtained with the conventional specimens (RB, RNB, PS, Tors) and that obtained employing the proposed multiaxial specimens. All 17-4PH specimens mentioned in this work were manufactured from the same wrought bars used for the previously mentioned work. The material is considered isotropic and its constitutive law is given in figure 1, in which the best fit of the Hollomon's power law is reported: the strength coefficient K and the strain hardening exponent n are equal to 1420 MPa and 0.12 respectively. For further information, please refer to [11]. Concerning the new specimens, as above mentioned in the introduction, the geometry taken into account is that proposed by L. Driemeier [9,10], whose dimensions were optimised through a sensitivity analysis carried out by some of the authors of this paper to meet specific design requirements such as: easiness of manufacturing, constant T and X in the critical area, and localization of the highest plastic strain in the centre of the specimen. Please refer to [21] for more information on the optimisation. The specimen shown in figure 2, hereafter referred to as the "Driemeier", permits to generate several stress states in low regime of triaxialities, such as pure shear or a mix of tensile-shear stress, and can be tested employing only a uniaxial testing machine.

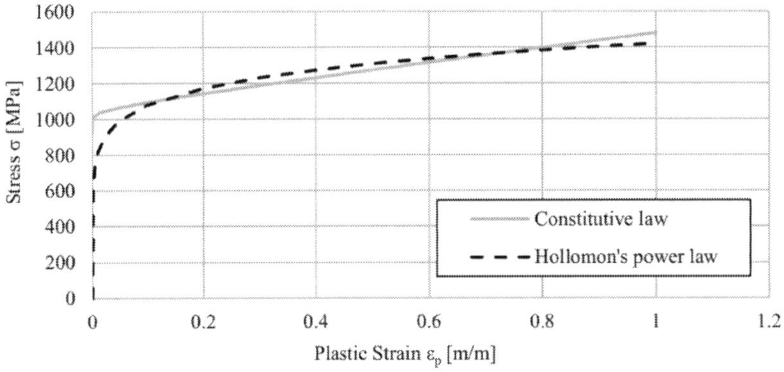

Figure 1. Experimental stress-strain constitutive behavior of 17-4PH wrought steel alloy and corresponding Hollomon's power law best fit curve.

In particular, the central holes can rotate by an alpha angle (α) along a circumference centred on the centre of the specimen, varying the proportion between shear and tensile state, figure 3a.

In the present work, two different configurations of Driemeier specimens were chosen, having the centres rotated clockwise by 10° and 30° respectively and hereinafter referred as Pos10 and Pos30, figures 3b and 3c. In detail, the Pos30 configuration generates a combination of shear-tensile stress characterized by a higher tensile component than the Pos10 configuration.

The results of the multiaxial tests will be employed to verify the prediction accuracy of the CCF model previously calibrated by conventional tests [11]. In addition, a new calibration will be performed using only the simple multiaxial tests data together with the RB and RNB test results from the previous experimental campaign. The aim is to verify the equivalence between the previously and the new fracture surface and whether it is possible to replace the PS and Tors with the new simple multiaxial tests, which are easier to perform, requiring a more common experimental setup. Finally, a further calibration containing all six tests (RB, RNB, PS, Tors, Pos10, Pos30) will be compared with the others for an additional validation.

The experimental campaign involved three repetitions for both Driemeier specimen configurations and a universal servo-hydraulic MTS with 250 kN maximum load and 150 mm stroke testing machine was used. The same equipment was used for the RB, RNB, PS and Tors tests in the previously experimental campaign [11]. Then, a finite element model was created for both tests using the commercial software ANSYS 2020 R2, to derive the equivalent plastic strain and principal stresses to evaluate the histories T and X with plastic deformation. The mesh size of the models, composed of brick elements, is fine enough to balance numerical accuracy and computational cost. The displacement was applied to the upper part of the specimen, while the lower part was kept fixed. For further validation of the finite element analyses a 2D digital image correlation was carried out to evaluate the full field distribution of the equivalent plastic strain on the specimen surface. The selected DIC software was developed by one of the authors, for more information please refer to [22,23].

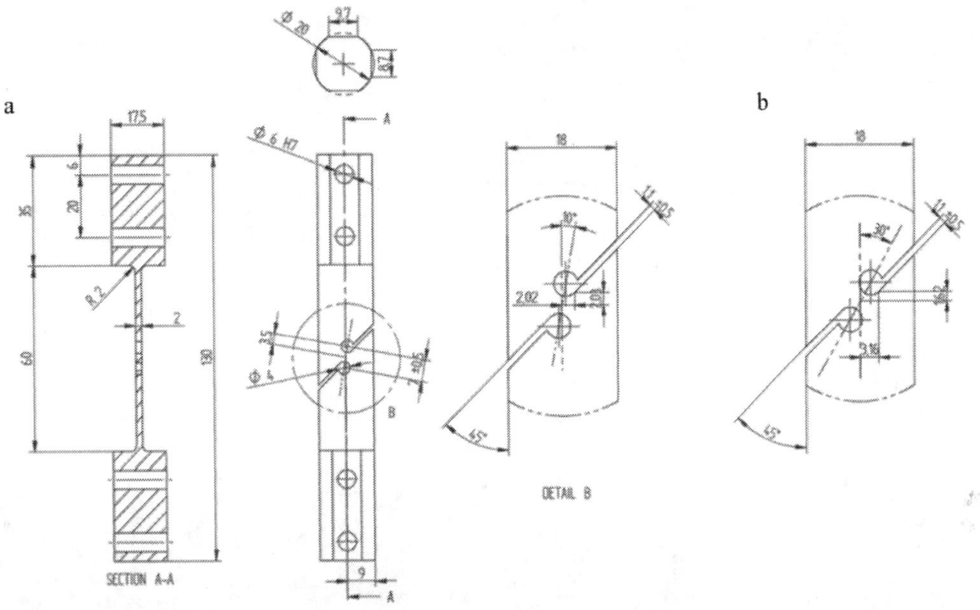

Figure 2. Technical drawings: a) Driemeier 10° (Pos10); b) Driemeier 30° (Pos30).

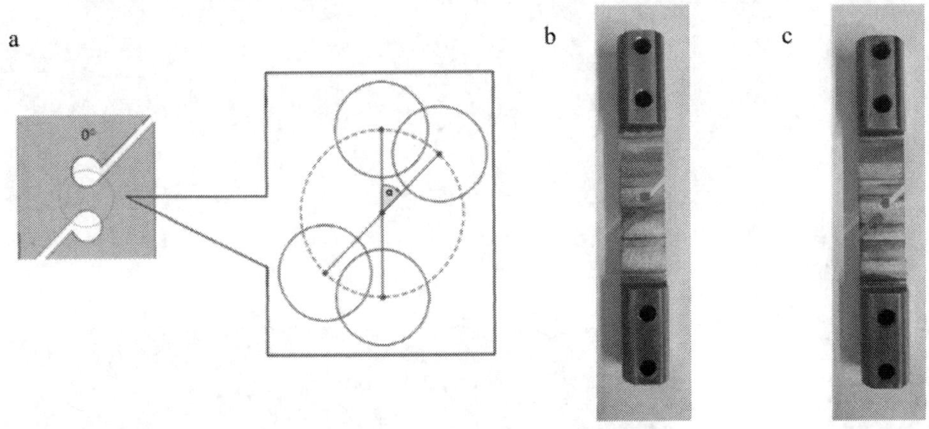

Figure 3. a) Representation of the alpha angle (α); b) Pos10 sample; c) Pos30 sample.

4. Results

The outcome of experimental campaign is reported in figure 4, along with the numerical results of the finite element analyses. Though three repetitions were performed for each specimen, only the average curve was reported. The drop in load-bearing capacity, which indicates the onset failure condition is highlighted by a red circle, and occurred for a displacement of around 0.95 mm for both specimens. The maximum load is around 3.6 kN and 4.0 kN for Pos10 and Pos30 respectively. The experimental-numerical match is very good, which confirmed the reliability of the FE analysis. The accuracy of the FEA was further assessed by comparing the numerically estimated strain field with that evaluated during the experimental test using the 2D-DIC technique, figures 5 and 6. For the sake of simplicity, only the DIC results of the first repetition are reported. As shown for both specimens, Pos10 and Pos30, the distribution of the equivalent plastic strain field is totally comparable. In terms of numerical values the strain estimated through FEA and measured using 2D-DIC technique are essentially equal, with an average value around of 0.33 and 0.46 for Pos10 and Pos30 respectively. This outcome is a further proof the confidence of local quantities retrieved via FEA.

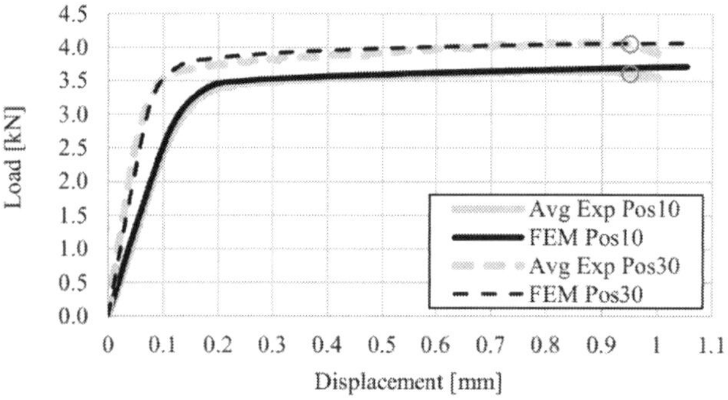

Figure 4. Numerical and experimental load-displacement curves for Pos10 and Pos30 tests. The red circles indicate the onset of failure.

Figure 5. Specimen Pos10: comparison between the strain plastic field evaluated with FEM (left) and DIC technique (right).

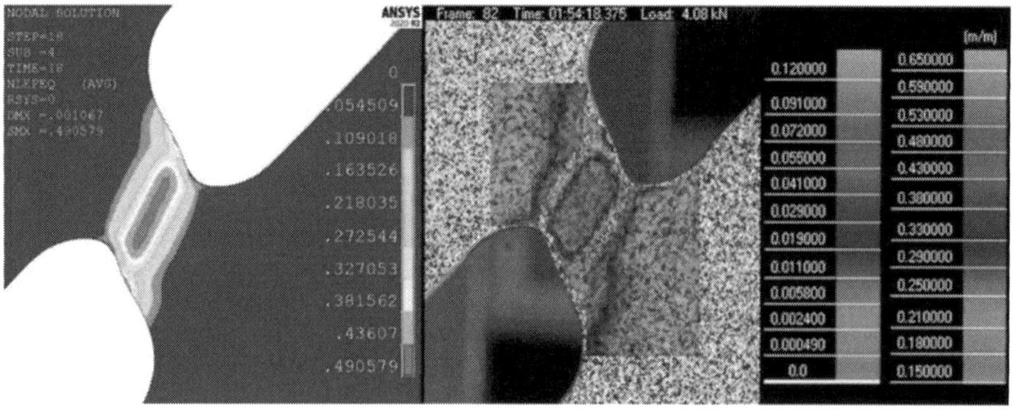

Figure 6. Specimen Pos30: comparison between the strain plastic field evaluated with FEM (left) and DIC technique (right).

Figure 7 illustrates the evolution of the instantaneous value of triaxiality and the deviatoric parameter as a function of equivalent plastic strain: since the quantities exhibit limited variations, their average value up to the onset of failure evaluated by means of equations (6-7) can be taken as a reference. In this way the hypothesis of quasi-proportional loading is met and the fracture surface can be derived according to the equation (5). The value of T_{avg}, X_{avg} and ε_f for both Driemeier specimens are provided in Table 1. For the sake of clarity, the values of the RB, RNB, PS and Tors tests evaluated in the work proposed by Nalli et al. [11] are also reported, which will be employed in the new calibrations involving the Driemeier tests.

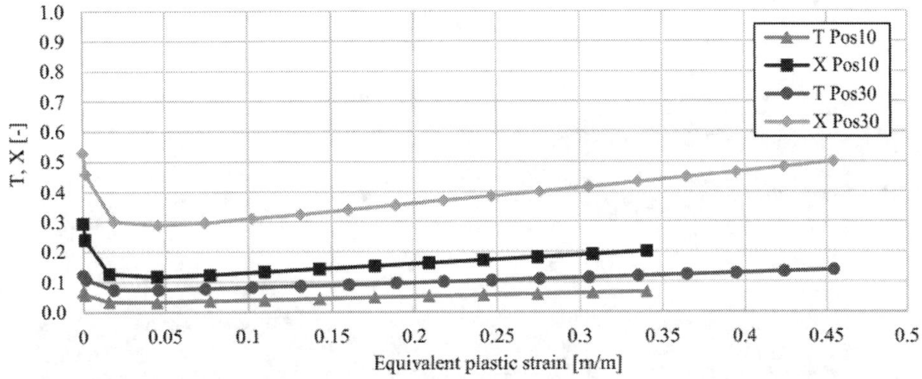

Figure 7. Evolution of the instantaneous values of the triaxiality and deviatoric parameter as function of equivalent plastic strain for Pos10 and Pos30 up to a displacement of 0.95 mm.

Table 1. Values of T_{avg}, X_{avg} and ε_f for multiaxial and conventional specimens.

Tests	T_{avg} [-]	X_{avg} [-]	ε_f [m/m]
Pos10	0.05	0.16	0.33
Pos30	0.10	0.38	0.46
RB	0.61	0.99	1.17
RNB	0.76	0.99	0.81
PS	0.70	0.00	0.66
Tors	0.00	0.00	0.47

Three different calibrations based on data of Table 1 were performed, and the results are illustrated in figures 8a-f. For the sake of clarity, in addition to the 3D graphs, a 2D view of plastic strain as function of triaxiality along X isocurves are provided in figures 8b, 8d and 8f.

In the first calibration conventional specimens are adopted, i.e. RB, RNB, PS and Tors, figures 8a and 8b. The fracture surface was calibrated ex novo. It is clear from figures 8a, 8b that the new Driemeier points, here not used for tuning, are close to the calibrated fracture surface with conventional specimens, meaning that the calibrated damage model well predicts failure under these new loading conditions. It should be noted that the Driemeier test points fall slightly lower than the fracture surface.

A second calibration is based on the new multiaxial specimens Pos10 and Pos30, together with the RB and RNB tests. In this case the two multiaxial specimens replace the PS and Tors specimens, see figures 8c and 8d. The new surface follows the same behaviour as the previous one and is slightly lower because, as already mentioned, the Driemeier specimens exhibited a lower plastic strain at failure. It is worth noting that this new fracture surface presents a good match both with the calibration points and with PS and Tors tests which are excluded from the calibration procedure. Thus, the evidence suggests that the new fracture surface successfully predicts the onset of failure, namely the use of multiaxial specimens does not affect the calibration compared to that performed using conventional specimens.

In the third calibration, reported in figures 8e and 8f, the data coming from all the tests mentioned in Table 1 are involved as input. This calibration is supposed to be the most accurate as it is based on more experimental evidence than the other two. The resulting fracture surface is very close to the experimental points showing a good fit while sharing the same surface behaviour of the other two surfaces.

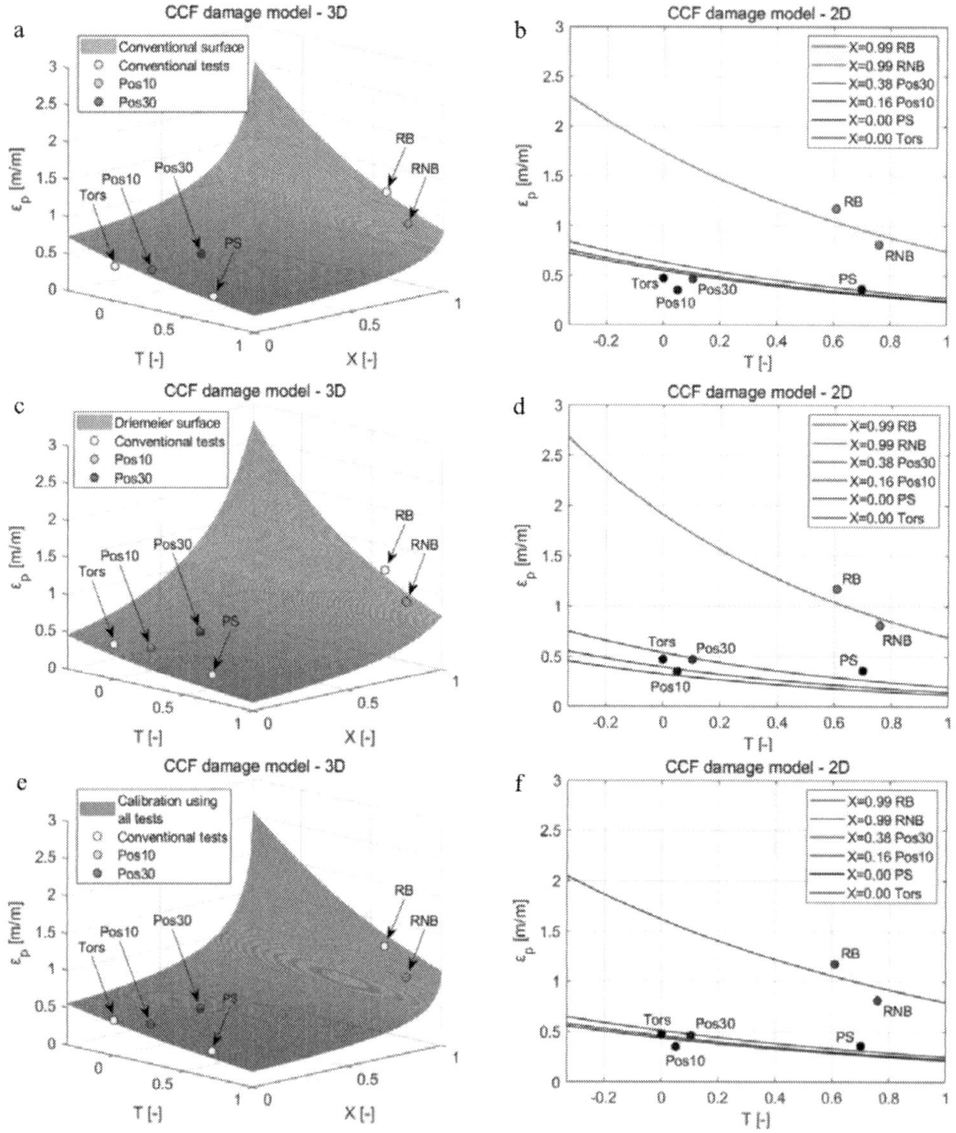

Figure 8. 3D and 2D views of fracture surface of 17-4PH obtained from different set of experimental tests: a-b) Conventional surface involving RB, RNB, PS and Tors; c-d) Driemeier surface involving Pos10, Pos30, RB and RNB; e-f) Surface involving all the tests Pos10, Pos30, RB, RNB, PS and Tors.

Lastly, a comparison of all three surfaces is shown in figure 9; they are very close one another and have the same qualitative behaviour, which means that they can be defined as almost equivalent and it is possible to replace torsion or plane strain tests, with the Driemeier multiaxial tests. Although the two fracture surfaces calibrated with the Driemeier specimens, figures 8c and 8e, are slightly lower than the conventional surface, however they are more conservative regarding the estimation of ductility and then are safety oriented.

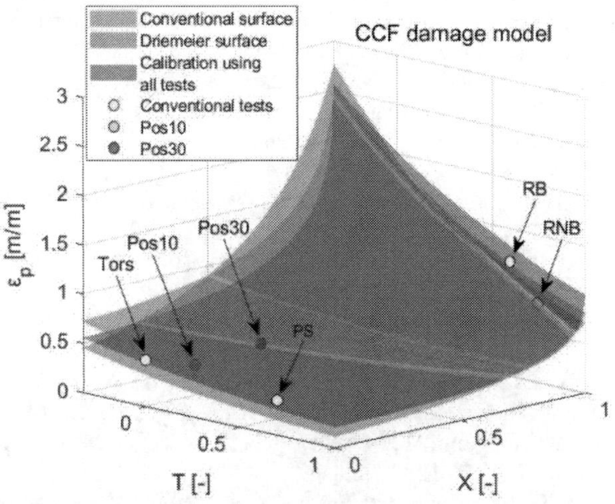

Figure 9. Comparison among the of 17-4PH fracture surfaces obtained by different set of experimental tests.

5. Conclusion

In this work an experimental campaign was carried out to successfully estimate the ductility of a wrought 17-4PH steel alloy. A damage model proposed by one of the authors was tuned to this purpose. In particular, the suitability of adopting simple multiaxial specimens for the tests, compared to more conventional ones, was investigated. The main results reported in the paper showed that together with tensile specimens such as round bars and round notched bars which are also required for elastic-plastic characterisation, it is possible to calibrate a damage model using the proposed multiaxial specimens, replacing more complex geometries, such as cylindrical bar for torsion tests and flat grooved specimens for plane strain conditions tests, thus avoiding more sophisticated testing equipment. All this should lead to a simplification of the calibration procedure, making the use of the damage models on an industrial scale more accessible and widespread. Finally, the new specimen geometry additionally permitted to investigate stress states at low triaxiality, for which few experimental evidences are available in the literature.

References

[1] Lemaitre J 1985 A continuous damage mechanics model for ductile fracture *J. Eng. Mater. Technol.* **107** 83–9

[2] Bonora N 1997 A nonlinear CDM model for ductile failure *Eng. Fract. Mech.* **58** 11–28

[3] Gurson A L 1976 Continuum theory of ductile rupture by void nucleation and growth: Part 1 – Yield criteria and flow rules for porous ductile media *J. Eng. Mater. Technol.* **99** 2-15

[4] Tvergaard V and Needleman A 1984 Analysis of the cup-cone fracture in a round tensile bar *Acta Metall.* **32** 157–69

[5] McClintock F A 1968 A criterion for ductile fracture by growth of holes *J. Appl. Mech.* **35** 363–71

[6] Hancock J W and Mackenzie A C 1976 On the mechanism of ductile failure in high-strength steel subjected to multi-axial stresses *J. Mech. Phys. Solids* **24** 147–69
[7] Cockcroft M G and Latham D J 1968 Ductility and the workability of metals *J. Inst. Metal* **96**
[8] Norris D M Jr, Reaugh J E, Moran B and Quinnones D F 1978 A plastic-strain mean-stress criterion for ductile fracture *J. Eng. Mater. Technol.* **100** 279-86
[9] Driemeier L, Brünig M, Micheli G and Alves M 2010 Experiments on stress-triaxiality dependence of material behavior of aluminum alloys *Mech. Mater.* **42** 207-17
[10] Brünig M, Chyra O, Albrecht D, Driemeier L and Alves M 2008 A ductile damage criterion at various stress triaxialities *Int. J. Plast.* **24** 1731-55
[11] Nalli F, Cortese L and Concli F 2021 Ductile damage assessment of Ti6Al4V, 17-4PH and AlSi10Mg for additive manufacturing *Eng. Fract. Mech.* **241** 107395
[12] Bao Y and Wierzbicki T 2004 On fracture locus in the equivalent strain and stress triaxiality space *Int. J. Mech. Sci.* **46** 81-98
[13] Rice J R and Tracey D M 1969 On the ductile enlargement of voids in triaxial stress fields *J. Mech. Phys. Solids* **17** 201–17
[14] Bai Y and Wierzbicki T 2008 A new model of metal plasticity and fracture with pressure and Lode dependence *Int. J. Plast.* **24** 1071-96
[15] Bai Y and Wierzbicki T 2010 Application of extended Mohr–Coulomb criterion to ductile fracture *Int. J. Fract.* **161**
[16] Dunand M and Mohr D 2014 Effect of Lode parameter on plastic flow localization after proportional loading at low stress triaxialities *J. Mech. Phys. Solids* **66** 133-53
[17] Coppola T, Cortese L and Folgarait P 2009 The effect of stress invariants on ductile fracture limit in steels *Eng. Fract. Mech.* **76** 1288-302
[18] Wierzbicki T, Bao Y, Lee Y W and Bai Y 2005 Calibration and evaluation of seven fracture models *Int. J. Mech. Sci.* **47** 719-43
[19] Kubík P, Šebek F, Hůlka J and Petruška J 2016 Calibration of ductile fracture criteria at negative stress triaxiality *Int. J. Mech. Sci.* **108-9** 90-103
[20] Cortese L, Coppola T, Campanelli F, Campana F and Sasso M 2014 Prediction of ductile failure in materials for onshore and offshore pipeline applications *Int. J. Damage Mech.* **23** 104–23
[21] Cortis G, Nalli F and Piacenti M 2022 A direct methodology for the calibration of ductile damage models from a simple multiaxial test *IOP Conf. Ser.: Mater. Sci. Eng.* **1214** 012016 (online)
[22] Broggiato G B and Cortese L 2009 White-light speckle image correlation applied to large-strain material characterization *Eur. J. Comput. Mech.* /Revue Européenne de Mécanique Numérique **18** 377-92
[23] Broggiato G B, Casarotto L, Del Prete Z and Maccarrone D 2008 Full-field strain rate measurement by white-light speckle image correlation *Eur. J. Comput. Mech.* **18**

DIC analysis of gears in bending condition

F Curà[1], A Mura[1], L Corsaro[1] and A Revilla[1]

[1] Dimeas Politecnico di Torino, corso Duca degli Abruzzi 24, 10129 Turin - Italy

E-mail: francesca.cura@polito.it

Abstract. Thin rim gears present a complex strain and stress field due to its particular geometry, above all in bending problems. From this point of view, the position of the most stressed point and the corresponding equivalent stress value are useful for an accurate design of thin rim gears. More in detail, in case of bending failure, both crack initiation point and corresponding propagation path are strictly related to the gear's geometry.

In this work, an experimental analysis was performed to evaluate how light weight gears geometry may influence the strain field close to the tooth root.

To this aim, an experimental activity was carried on by a dedicated equipment for bending tests and the gear deformation was monitored by using the 3D Digital Image Correlation (DIC) technique.

The local strain field in both tooth and web portions was measured for two types of gears (*standard* gear and *thin-rim* gear) in order to identify the stress condition due to the bending loading. A particular attention was devoted to point out the most stressed point for both gears. Results were also compared with XFEM model available in literature.

1. Introduction

Lightweight gears are generally characterized by quite thin geometries and they are realized in different shapes, according to the application requirements. These components need accurate design in order to obtain an useful compromise between high loads and high power density transmission that necessities to minimize mass.

From the bending failure point of view, crack nucleation and propagation are strongly influenced by many factors that may be classified in geometrical factors (i.e. rim thickness, web thickness, gear configuration) and load factors (i.e. centrifugal load effect) [1-8].

More in detail, if bending loading condition is considered, lightweight gears present a complex strain and stress field due to their particular geometry. So, both position of the most stressed point and corresponding equivalent stress value are useful, already in a design phase, to reliably predict a possible crack propagation path.

In particular, the definition of the strain field at the gear tooth root may provide useful information to correctly understand the role of geometry parameters and mass distribution in predict safe (through the tooth) or catastrophic (through the rim) failures [5-6] and this peculiarity is very important above all in aerospace industry.

This aspect was widely investigated from the numerical point of view, traditional 2D and 3D finite elements (FEM) and extended finite elements (XFEM) simulations were run, focusing on both crack initiation point and corresponding propagation direction. An experimental activity was recently carried

Content from this work may be used under the terms of the Creative Commons Attribution 3.0 licence. Any further distribution of this work must maintain attribution to the author(s) and the title of the work, journal citation and DOI.
Published under licence by IOP Publishing Ltd

on by the authors in order to verify the onset of the bending crack and its propagation path in spur gears with different geometries by an IR thermocamera [9].

In this work, an experimental analysis was performed in order to evaluate how light weight gears geometry may influence the strain field in lightweight gears.

To this aim, an experimental activity was carried on basing on a 3D Digital Image Correlation (DIC) technique. DIC was chosen due to its peculiarity of full-field measurement capability without contact, respect to the classical strain gauge or extensometer techniques.

Many studies are available in literature about DIC for mechanical applications, where stress and strain characterization is required to study material responses or to tune numerical models.

As an example, DIC was utilized in composite materials for determining constitutive properties [10] or to analyse non-linear shear stress-strain relations [11], in a solid solution alloy (Al-Mg) to investigate the Portevin–Le Chatelier effect [12] and to study [13] the process of deformation and failure in coarse and ultrafine grain states of Zr-1Nb alloys. Strain measurements were also carried on by DIC in welded materials, as widely described in [14-16].

For as concerns crack monitoring, DIC may be considered as a promising technique in adhesive joints [17] and in coated laminates [18].

Also regarding mechanical components, some DIC applications are present in literature to study durability, wear and failure modes in nanocomposites wheels [19-21]. In [22], DIC was also used to evaluate residual stresses generated by an induction hardened on aeronautical gears.

The experimental activity reported in this paper aims to study the strain maps generated onto two types of gears (*standard* and *thin-rim* gears) subjected to bending; the load was applied at the so called single contact point.

To this goal, a dedicated equipment for bending tests was utilized and, during different loading entities, the gear deformation was monitored by using a 3D Digital Image Correlation (DIC) approach.

The local strain field in both tooth and web portions was determined and a particular attention was devoted to point out the most stressed point for both gear geometries, responsible to the crack onset. Results were also compared with XFEM model available in literature [6].

2. Materials and Experimental Setup

Two types of spur gears (one *standard* and one *thin-rim*) were adopted for DIC analysis under bending static conditions, whose geometrical parameters are resumed in Table 1. In particular, Table 1 reports: pitch diameter (*dp*), face width (*b*), number of teeth (*z*), modulus (*m*), backup ratio (rim dimension respect to the tooth height) (*mb*), web ratio (web thickness respect to the face width) (*mw*) and gear blank factor (*CR*) [23]. Gears are made of C45 steel without thermal treatments. Technical drawings of both the tested gears (*standard* and *thin-rim*), as depicted in [9], are shown in Figure 1.

Table 1. Gears geometry.

	dp [mm]	*b* [mm]	*z*	*m* [mm]	*mb*	*mw*	C_R
Standard	96	20	32	3	1	1	1
Thin-rim	96	20	32	3	0,3	0,1	0.63

A 3D DIC equipment (*Correlated Solution*) was adopted to measure the tooth deformation under static bending loading conditions. A dedicated specimen surface preparation was required for strain measurement, so each gear was initially covered with a layer of white paint and then, by using an airbrush, with small black speckles.

Figure 2 shows on the left side the final specimen and on the right one the speckle deposition (as an example, for *standard* gear).

IOP Conf. Series: Materials Science and Engineering 1275 (2023) 012038 doi:10.1088/1757-899X/1275/1/012038

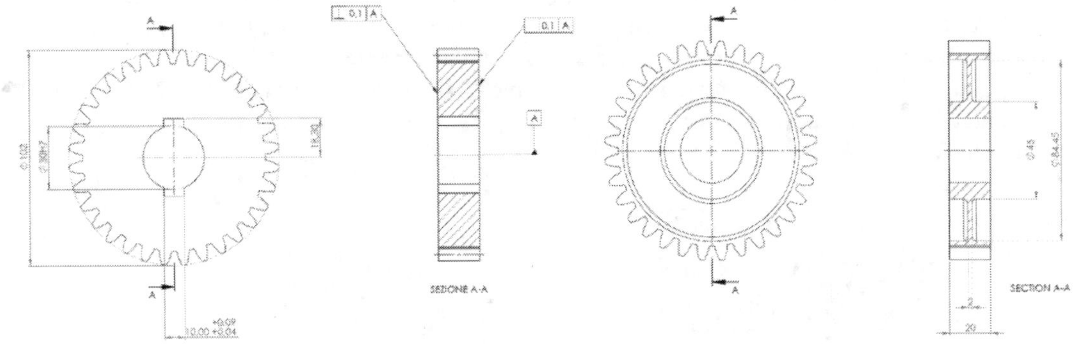

Figure 1. Technical drawings of *standard* and *thin-rim* gears [9].

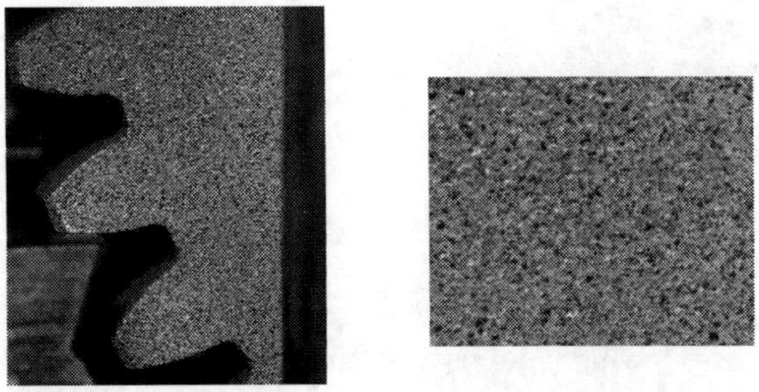

Figure 2. Speckle deposition for *standard* gear.

DIC setup was composed of two Basler ace acA4096-30um cameras (4096x2168 pixel and sensor dimensions of 14.1x7.5 mm), one blue light source and a data acquisition unit (DAQ system 6211). Each camera was equipped with a smart focus, two extension tubes (120mm and 24mm) and Rodagon 135 lens to increase the spatial resolution. Cameras and blue light source were mounted on a supporting structure located at 490 mm distance from the tested gear. Focus and brightness were adjusted to reduce the images noise level. The frame rate acquisition was 20 fps. Table 2 resumes the experimental DIC settings and performance adopted during the analysis.

Table 2. Experimental DIC settings and performance.

Technique used	3D Digital Image Correlation
Subset size	about 61x61 pixel^2
Subset weight	Gaussian weight
Step	11 pixel
Investigated area	about 9x9 mm^2
Shape function	Affine
Interpolation function	Optimized 8-tap
Correlation criterion	Normalized squared differences
Smoothing applied to the image	Low-pass filter image
Displacement accuracy rule-of-thumb	1/100 of the field of view in-plane
Strain resolution	100 με
Differentiation method	Lagrange
Smoothing method	Lagrange - dimension 25

IOP Conf. Series: Materials Science and Engineering 1275 (2023) 012038 doi:10.1088/1757-899X/1275/1/012038

A special device connected to a standard testing machine was adopted to perform bending tests for both *standard* and *thin-rim* gears. A punch integrated into the device was able to apply the loading to the gear tooth along its involute profile in both position and direction referred to the so called single contact point [24]. Each gear was tested at four different static loads, approximately: 2000 N, 4000 N, 6000 N and 8000 N. The post processing analysis was performed with VIC3D software.

Figure 3 shows a photo of the loading device (on the left) and a scheme of the DIC setup (on the right).

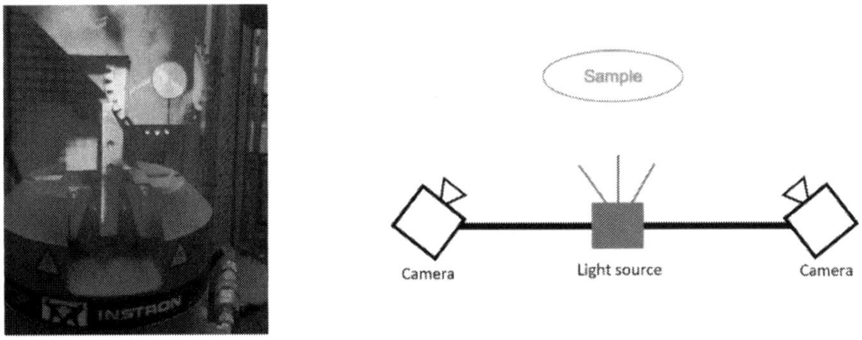

Figure 3. Digital Image Correlation Setup.

3. Results

All results are reported in the following in terms of strain maps (local strain ε_y and ε_x) of gears (*standard* and *thin-rim*), related to the maximum applied load.

The post processing analysis was referred to signals related to an Area Of Interest (*AOI*), chosen in this work as the loaded tooth. *AOI* zone was divided in 61 subset to reduce both noise level and uncertainty interval of each pixel.

A local reference system for both gears was located by considering the y-axis aligned with the centreline of the tooth and the x-axis tangential at the initial tooth root (see Figure 4). A line, named *Line 0* (see Figure 4), was defined to extract the measured local strain ε_y values, corresponding to the centre of the local reference system (y = 0).

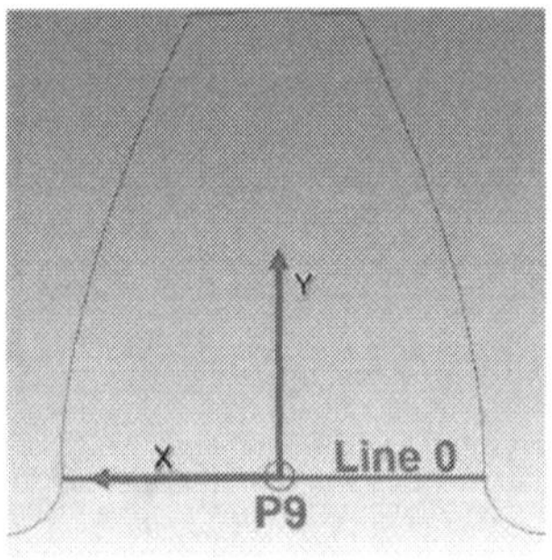

Figure 4. Line 0 and local reference system.

Local strain maps ε_y and ε_x for the *standard* gear are illustrated respectively in Figures 5 and 6 and refer to the maximum load (8000 N). Similar experimental results are illustrated in Figures 7 and 8 for the *thin-rim* gear, referring to the maximum load too (8000 N).

P9 point (see Figures 5-6-7-8) defines the local reference system centre, as in Figure 4.

IOP Conf. Series: Materials Science and Engineering 1275 (2023) 012038 doi:10.1088/1757-899X/1275/1/012038

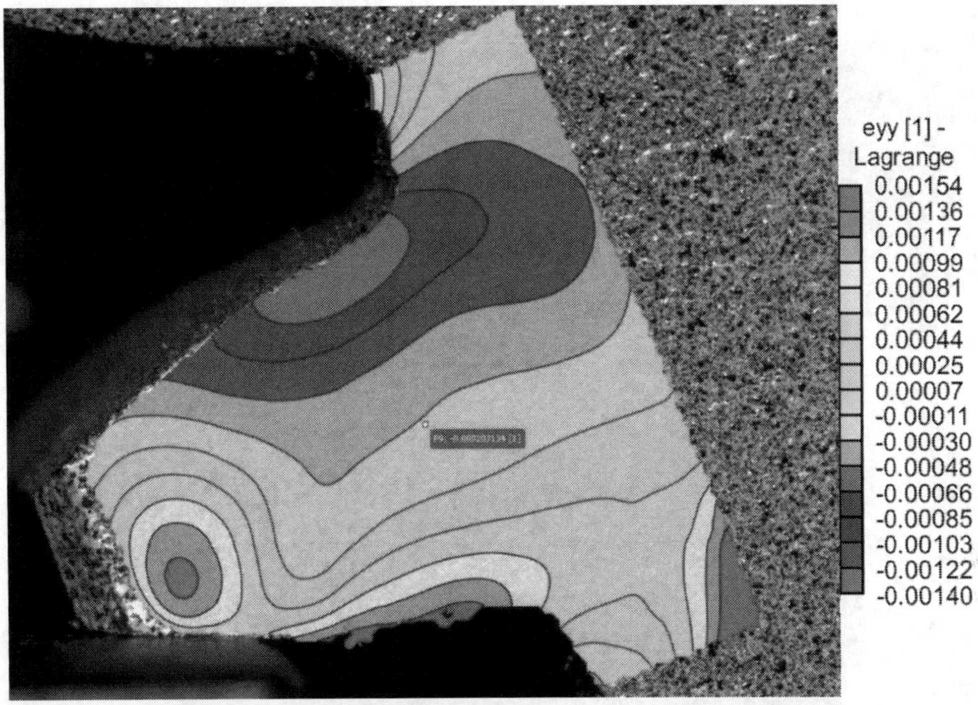

Figure 5. *Standard* gear: experimental local strain map ε_y.

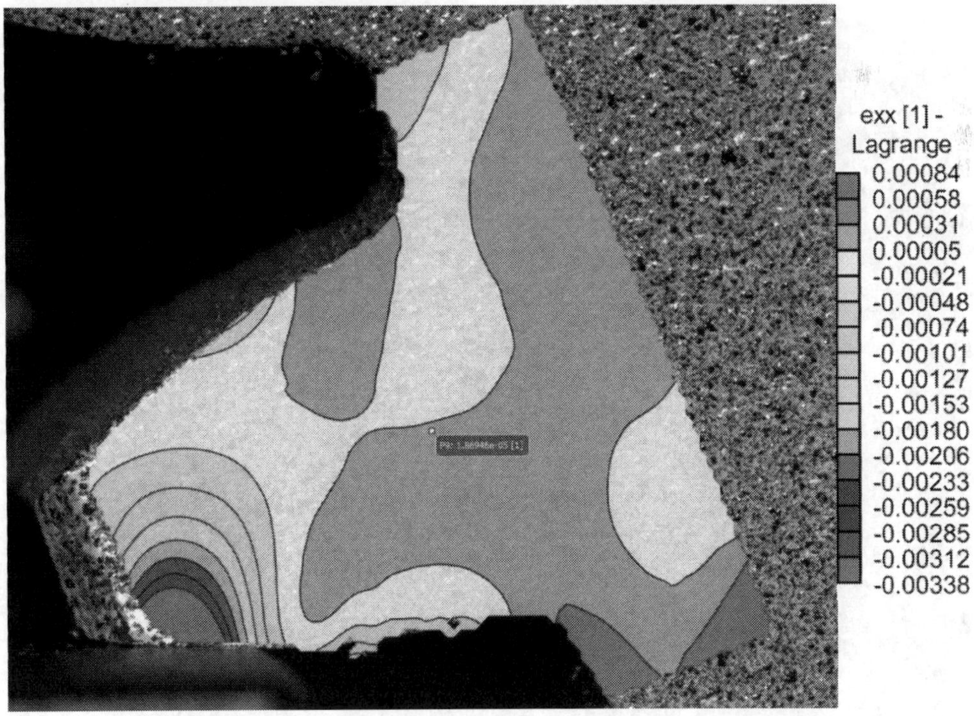

Figure 6. *Standard* gear: experimental local strain map ε_x.

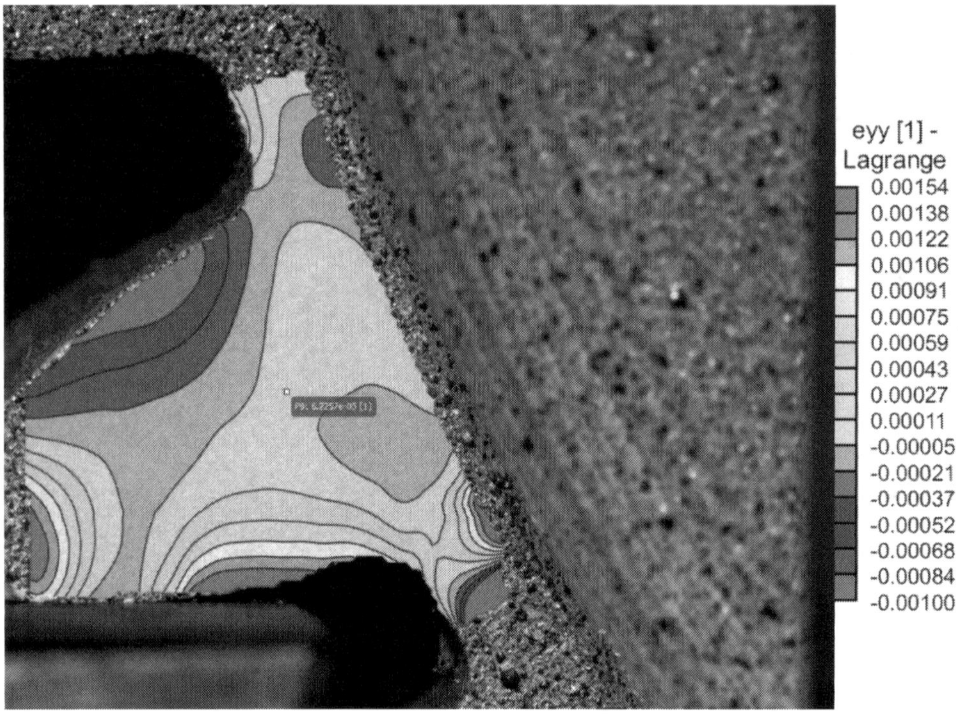

Figure 7. *Thin-rim* gear: experimental local strain map ε_y.

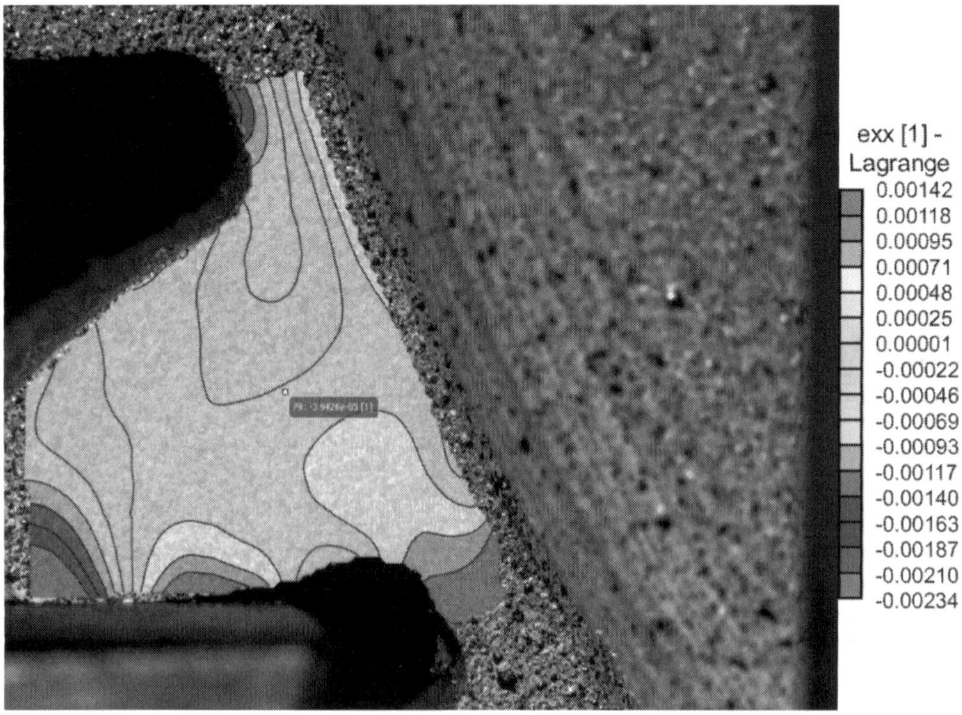

Figure 8. *Thin-rim* gear: experimental local strain map ε_x.

Figure 9 shows the experimental local strain profiles ε_y extracted from *Line 0*, on the left for *standard* gear and on the right for *thin-rim* one.

Figure 10 illustrates the comparison between normalized experimental local strains ε_y at the *Line 0* for the two gears. The normalization was done with respect to the maximum positive value of component ε_y.

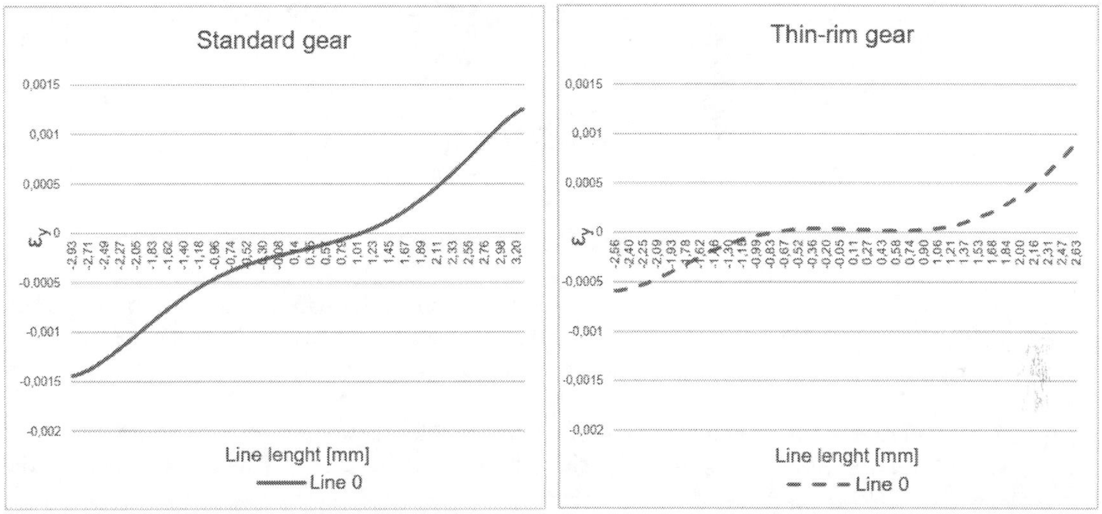

Figure 9. Experimental local strain ε_y.

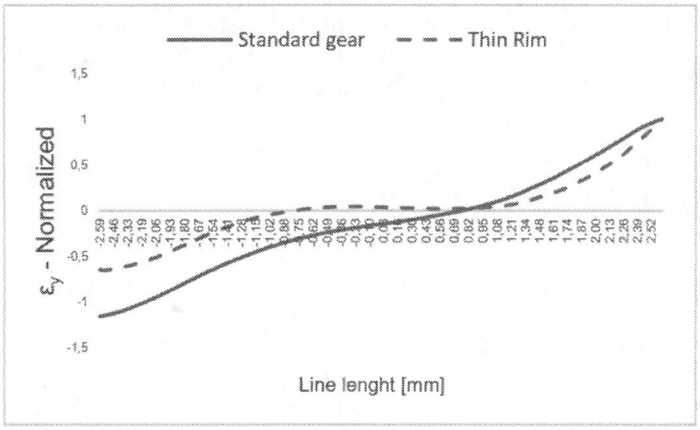

Figure 10. Normalized experimental local strain ε_y comparison (*Line 0*).

It can be observed that results of Figures 9 and 10, referring to maximum load, correspond to the ε_y strain mean value obtained from the last images (stabilized phase after the load application).

From the analysis of Figures 9 and 10, it may be observed that the *standard* gear shows a local strain ε_y comparable to the classical Lewis bending profile, while the *thin-rim* gear generates a local strain ε_y quite different, due to its low stiffness under the central part of the face width and causing a tooth rotation during the bending condition.

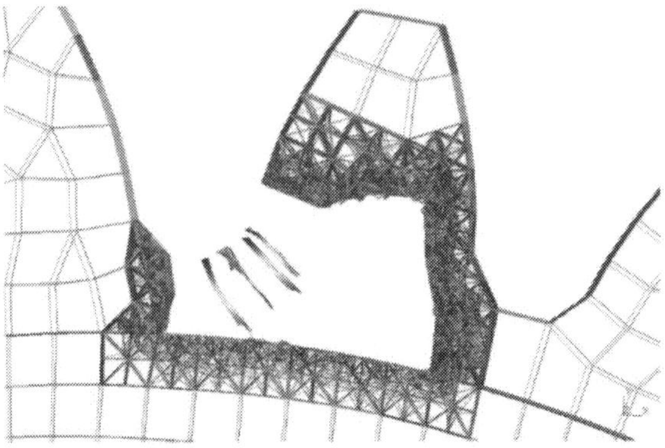

Figure 11. XFEM *Thin-rim* gear (mb = 0.3) model with different crack path [5].

From the analysis Figures 5 and 7, it can be noted that local strain maps ε_y show a different strain response under bending condition for *standard* and *thin-rim* gears respectively. In particular, the most critical zone moves from the tooth root top for *standard* gear (see Figure 5) up to the bottom for *thin-rim* gear (see Figure 7).

Furthermore, Figures 6 and 8 point out that the amount of local strain ε_x reached at the tooth root tensile zone in the *thin-rim* gear (Figure 8) is one order of magnitude higher than that related to the *standard* one (Figure 6).

These results provide an experimental verification of what obtained in literature by XFEM simulations [5], involving the presence of *catastrophic* crack propagation paths in *thin-rim* gears with a backup ratio (*mb*) lower than 0.5. In particular, Figure 11 [5] shows the presence of some crack paths starting from different points along the tooth root, all propagating in a catastrophic way (*mb* = 0.3, as considered in this paper). This behaviour is in a good agreement with ε_y and ε_x DIC maps, where the *thin-rim* gear (Figures 7 and 8) shows a bending response completely different to that generated by the *standard* one (Figures 5 and 6).

As a matter of fact, DIC maps emphasize this phenomenon for both tested gears, shifting the most stressed point (related to the maximum strain ε_y, see Figure 7) from the bottom of the tooth root zone up to the rim.

4. Conclusions

The experimental activity carried on in the present work allowed the authors to verify a wide number of numerical simulations involving the bending behaviour of lightweight gears.

To this aim, a Digital Image Correlation (DIC) technique was set up to investigate the strain maps in gears, above all to characterise how geometry parameters (rim and web) may influence the position of the most stress point at the tooth root zone.

In spite of the difficulty in setting up 3D DIC measurements to analyse complex mechanical components as standard and lightweight gears, obtained local strain maps provided a good characterisation of the stress field close to the tooth root and allowed also to interpret FEM and XFEM results available in literature.

As a matter of fact, in case of bending, both crack initiation point and corresponding propagation path are strictly related to the gear's geometry and the interpretation of the local stress and strain fields provided an effective tool to help designers to avoid catastrophic failures.

Finally, it may be concluded that DIC technique may be easily utilised to study complex mechanical components, provided that a careful calibration of the instrumentation and a good tuning of the loading system were done.

References

[1] Lewicki D G 1996 Crack propagation studies to determine benign or catastrophic failure modes for aerospace thin-rimgears *NASA Tecnical Memorandum* 107170

[2] Lewicki D G and Ballarini R 1997 Rim thickness effects on gear crack propagation life *Int. J. Fracture* **87** pp 59–86

[3] Lewicki D G 1998 Three-dimensional gear crack propagation studies U.S. Army Research Laboratory Lewis Research Center Cleveland Ohio NASA/TM-1998-208827

[4] Lewicki D G 2001 Gear crack propagation path studies *Guidelines for Ultra-Safe Design* NASA/TM-2001-211073.

[5] Curà F, Mura A and Rosso C 2014 Investigation about crack propagation paths in thin rim gears *Fracture and Structural Integrity* **8** n 30 pp 446–453

[6] Curà F, Mura A and Rosso C 2015 Effect of rim and web interaction on crack propagation paths in gears by means of XFEM technique *Fatigue Fract Eng Mater Struct* **38** n 10 pp 1237–1245

[7] Curà F, Mura A and Rosso C 2017 Influence of high speed on crack propagation path in thin rim gears *Fatigue & fracture of engineering materials & structures* **40** n 1 pp 120–129

[8] Curà F, Mura A and Rosso C 2016 Crack propagation behaviour in planet gears *Procedia Structural Integrity* **2** pp 3610–3616

[9] Curà F, Mura A and Corsaro L 2021 Experimental investigation on crack propagation paths in spur gears *IOP Conf. Series: Materials Science and Engineering* **1214** 12029

[10] He T, Liu L, Makeev A and Shonkwiler B 2016 Characterization of stress–strain behavior of composites using digital image correlation and finite element analysis *Composite structures* **140** pp 84–93

[11] Seon G, Makeev A, Cline J and Shonkwiler B 2015 Assessing 3D shear stress–strain properties of composites using Digital Image Correlation and finite element analysis based optimization *Composites science and technology* **117** pp 371–378

[12] Yamasaki S, Miike T, Mitsuhara M, Nakashima H, Akiyoshi R, Nakamura T and Kimura S 2021 Quantitative analysis of the Portevin–Le Chatelier effect by combining digital image correlation and dead-weight-type tensile test *Structural materials : properties, microstructure and processing* (*Materials science & engineering: A*) ed Elsevier **816** 141277

[13] Sharkeev Y P et al 2020 Research on the processes of deformation and failure in coarse- and ultrafine-grain states of Zr1–Nb alloys by digital image correlation and infrared thermography *Structural materials : properties, microstructure and processing* (*Materials science & engineering: A*) ed Elsevier **784** 139203

[14] Saranath K M and Ramji M 2015 Local zone wise elastic and plastic properties of electron beam welded Ti–6Al–4V alloy using digital image correlation technique: A comparative study between uniform stress and virtual fields method *Optics and lasers in engineering* **68** pp 222–234

[15] Mani C, Balasubramani S, Karthikeyan R and Kannan S 2021 Digital Image Correlation of Tensile Properties for Monel 400/SS 316L Dissimilar Metal Welding Joints *Materials* **14** n 6 1560

[16] Milosevic N, Younise B, Sedmak A, Travica M and Mitrovic A 2021 Evaluation of true stress–strain diagrams for welded joints by application of Digital Image Correlation *Engineering failure analysis* **128** 105609

[17] Sun G, Liu X, Zheng G, Gong Z and Li Q 2018 On fracture characteristics of adhesive joints with dissimilar materials – An experimental study using digital image correlation (DIC) technique *Composite structures* **201** pp 1056–1075

[18] Smith G M, Higgins O and Sampath S 2017 In-situ observation of strain and cracking in coated laminates by digital image correlation *Surface & coatings technology* **328** pp 211–218

[19] Soudmand B H and Shelesh-Nezhad K 2020 Failure and wear analysis of poly(butylene terephthalate) nanocomposite spur gears *Tribology international* **151** 106439

[20] Soudmand B H and Shelesh-Nezhad K 2021 Experimental investigation on the durability and

failure modes of polybutylene terephthalate/calcium carbonate nanocomposite gears *Engineering failure analysis* **120** p 105113
[21] Soudmand B H and Shelesh-Nezhad K 2020 Study on the gear performance of polymer-clay nanocomposites by applying step and constant loading schemes and image analysis *Wear* **458-459** p 203412
[22] Savaria V, Bridier F and Bocher P 2016 Predicting the effects of material properties gradient and residual stresses on the bending fatigue strength of induction hardened aeronautical gears *International journal of fatigue* **85** pp 70–84
[23] ISO 6336 - Calculation of load capacity of spur and helical gears, Part 1 (2019)
[24] ISO 6336 - Calculation of load capacity of spur and helical gears, Part 3 (2019)

Scooter Tire Pressure Influence on a Hard Braking Manoeuvre

M Failli, T Fontanelli, E Paoletti, M Bartolozzi, L Berzi and G Savino

Dipartimento di Ingegneria Industriale, Università degli Studi di Firenze, Via di Santa Marta 3, 50139, Italy

E-mail: giovanni.savino@unifi.it

Abstract. The mortality of road accidents quadruples when, in addition to motorists, a motorcyclist is also involved. The higher mortality is influenced by the greater exposure of the rider to injuries during the impact. This problem can be mitigated by reducing, even slightly, the impact's speed. Bench tests carried out by previous studies showed a considerable influence of the inflation pressure on the properties of motorcycle tires. It could therefore have significant effects on the deceleration achievable under braking. This study aimed to evaluate the effect of inadequate tire inflation pressure (excessive or insufficient) on the average deceleration achievable during a hard braking manoeuvre performed only with the rear wheel and with ABS intervention. Experimental tests were carried out on a Piaggio Beverly S 300 in a controlled road environment closed to traffic. Different inflation pressures of the rear braking tire were tested during multiple runs. The data were acquired through a smartphone used as an IMU. At the nominal pressure of 2.4 bar the measured average deceleration was 3.69 m/s^2 (SD = 0.12 m/s^2). When increasing the pressure to 3.0 bar and decreasing it to 1.8 bar the average deceleration was reduced respectively to 3.67 m/s^2 (-0.8%, SD = 0.10 m/s^2) and 3.59 m/s^2 (-3.0%, SD = 0.16 m/s^2). The reduction in achievable deceleration was partly mitigated by the reduced load transfer on the rear tire, which partially compensated for the reduction of the braking force coefficient. The results showed that a moderate inflation pressure variation on the rear braking tire modestly influences the achievable average deceleration.

1. Introduction

Powered Two Wheelers (PTW) riders have a much higher chance of severe body injuries than car drivers. Exposure of one's body during road accidents significantly determines the mortality rates associated with such events. Statistically, riders have a 2.7 times higher death rate than motorists (1.9 compared to 0.7 deaths every 100 accidents). [1]

More than 69% of motorists' accidents involve a frontal collision of a PTW with other vehicles or obstacles. [2] Reducing the speed at which such impacts occur would significantly reduce the severity of the injuries. One of the factors we can act upon is the entity of the deceleration achievable by the rider: a more effective braking manoeuvre will mitigate the impact effects.

Among the multiple factors that may influence the achievable deceleration while braking, the inflating pressure of the tires is one that the rider can easily control and, as suggested by previous studies, should play a role in tire behaviour.

1.1. Existing Literature on Tire Pressure Influence

Previous studies have partially examined the correlation between tire inflation pressure and its behaviour.

Content from this work may be used under the terms of the Creative Commons Attribution 3.0 licence. Any further distribution of this work must maintain attribution to the author(s) and the title of the work, journal citation and DOI.
Published under licence by IOP Publishing Ltd

The study by Mahdi Ajami et al. [3] analysed how an insufficient tire pressure setting influences a car's stopping distance: experimental tests were conducted with an ABS-equipped car on dry asphalt. Only lower than nominal inflating pressures were tested, which reduced the stopping distance from the set speeds of 90 km/h and 100 km/h.

The study by Atul Bansal et al. [4] tested, using computer simulations, the influence of various pressure settings applied to the front and rear tires of a PTW on its stopping distance from the set speed of 72 km/h. No experimental counter-proof tests were carried out.

The tests reported in the study by Vittore Cossalter et al. [5] examined the variations in the lateral response of PTW tires when changing various parameters, one of which is the inflation pressure. No investigation on the longitudinal dynamics was made.

Although multiple studies have already been conducted on this topic, none experimentally analysed the effect of tire inflation pressure on the braking performance of a PTW.

2. Methods

In order to investigate this aspect, the average deceleration of a Piaggio Beverly S 300 Scooter was evaluated during the stationary phase of the hard braking manoeuvre. An experimental activity was conducted to gather acceleration data that were post-processed and analysed in the MATLAB environment.

2.1. *Experimental Procedure*

Three rear-tire pressure levels were tested: 1.8 bar, 2.4 bar, and 3.0 bar; for each pressure setting, five runs were carried out for a total of 15 runs. During the experiment, the front tire was set to the nominal pressure recommended by the scooter manufacturer (2.4 bar).

For each run, the longitudinal acceleration (a_{long}) was recorded to derive its average value ($\bar{a}_{\mathrm{long}}$); it would be consequently averaged with the results obtained within a pressure set to calculate $\mathrm{mean}(\bar{a}_{\mathrm{long}})$ with improved accuracy.

2.2. *Test Protocol*

Before every set of runs, the rear tire was inflated to 0.2 bar above the pressure of the given run and consequently deflated to the run pressure to overcome the inaccuracy caused by any hysteresis phenomenon in the inflator pressure gauge.

The run consisted of accelerating from a standstill condition up to 50 km/h followed by a sudden hard braking manoeuvre concerning the rear wheel only that led to the complete stoppage of the scooter.

At the beginning of each run, the asphalt temperature near the expected braking point was measured. The IMU logged the longitudinal acceleration during each run and stored it on MATLAB Drive.

2.3. *Instrumentation*

The instrumentation used was an Android smartphone (Motorola Moto Z2 Play) running the MATLAB Mobile application and used as an affordable alternative to a professional-grade IMU. The device recorded the longitudinal acceleration of the vehicle with a 100 Hz sample rate. The smartphone was connected to the scooter through a 3D-printed fixture (Figure 1).

The asphalt temperature was measured through the HT 3301 infrared thermometer (Figure 2), always in the same region of the tarmac, for consistent measurement. Every measured value was rounded off to the nearest unit.

The pressure measurements were made with an analog manometer (Figure 3), meeting the ANSI 40.1 B (±2%) standard. It was also used to set the required pressure for each set of runs making use of its release valve.

Two cameras recorded the experiment: the first one was placed on the edge of the testing area (Figure 4), keeping the whole test field in the frame, and the second camera was fixed to the rider's helmet (Figure 5), focusing on the cockpit of the scooter (the brake levers and the instrument cluster were visible). The recordings of both cameras were used for qualitative data analysis and subsequent reference.

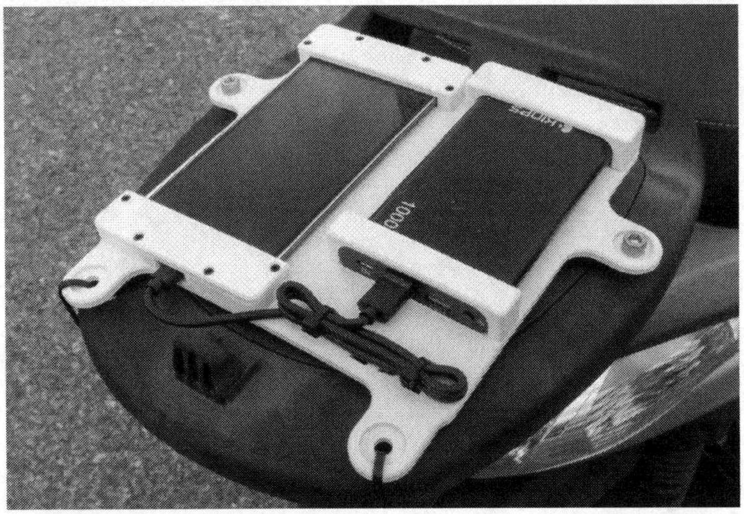

Figure 1. 3D-printed fixture

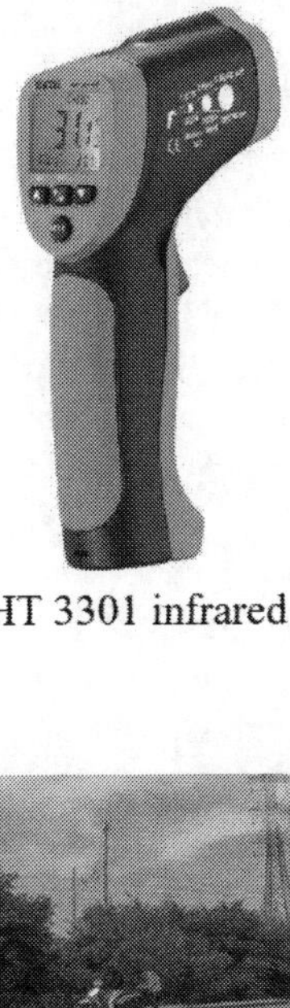

Figure 2. HT 3301 infrared thermometer

Figure 3. Manometer

Figure 4. Wide field camera

Figure 5. Helmet camera

2.4. Data Processing

Raw data were processed in MATLAB to obtain the average deceleration during each run.

Denoise. The logged longitudinal acceleration had noticeable high-frequency noise due to external causes such as engine vibrations, asphalt roughness, and electronic disturbances. A low-pass filter with a cut-off frequency of 1Hz was used to obtain a cleaner signal (Figure 6). Its effect did not negatively affect the results since the noise component, having a high frequency, had a null mean value over the intervals of interest. Therefore, removing it from the raw data did not influence the calculated average longitudinal acceleration.

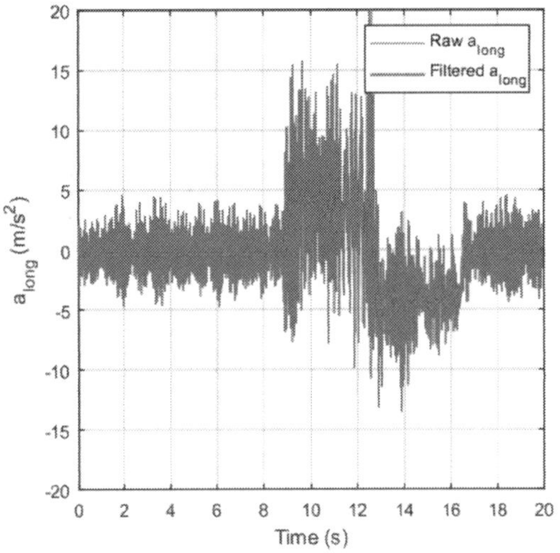

Figure 6. Signal denoise

Calibration. In steady conditions, the IMU did not lay level with the ground leading to an altered reading of the longitudinal acceleration. A component of the gravity acceleration projected along the y-axis of the IMU, resulting in an additive offset error. In order to compensate for said behaviour, a calibration was performed by calculating the average of the y-axis acceleration recorded in the samples from 50 to 500 (during the initial steady phase of each run). This quantity was subtracted from the data (Figure 7).

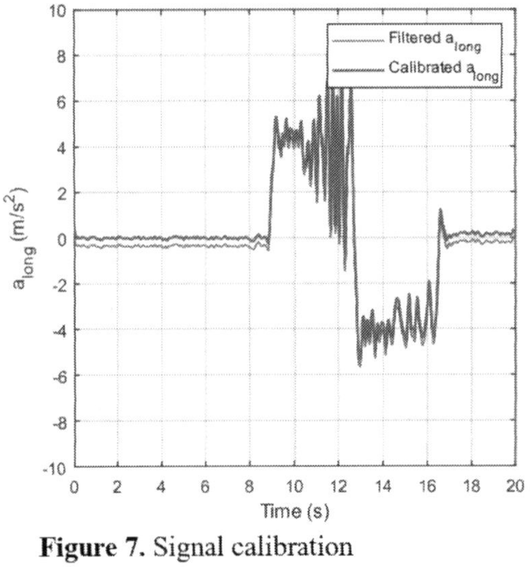

Figure 7. Signal calibration

IOP Conf. Series: Materials Science and Engineering 1275 (2023) 012039 doi:10.1088/1757-899X/1275/1/012039

Region of Interest (ROI) Identification. The stationary phase of the braking manoeuvre was considered to calculate the average longitudinal acceleration. This way, the effects of the initial and final transients were excluded. The ROI spanned from 0.2 s after the minimum peak of registered longitudinal acceleration up to 0.5 s before the positive acceleration peak generated by the rebound of the scooter suspensions (Figure 8).

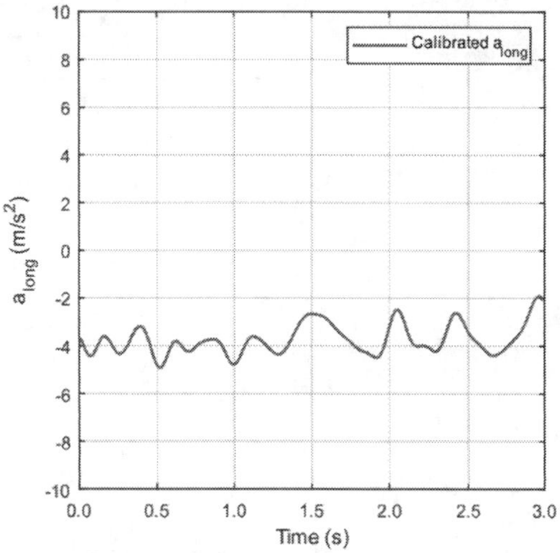

Figure 8. Region of interest

Average Longitudinal Acceleration Calculation. The mean of the acceleration values inside the ROI was calculated as the average longitudinal acceleration of each run.

Statistics of Values. The mean and standard deviation of the obtained averages were compared for each of the three pressure values.

T-test. A T-test was performed to check for the null hypothesis validity to investigate the results' meaning further. If the test was positive, the data did not provide enough information to state that the observed behaviour in the sample also existed in the population. In this case, the obtained averages might have come from normal distributions with equal means and equal but unknown variances.

Average Braking Force Coefficient Calculation. The average vertical load on the rear tire was calculated using the formula:

$$\overline{N}_{\text{rear}} = mg\frac{l_1 + h_G\frac{\bar{a}_{\text{long}}}{g}}{l}, \tag{1}$$

where $\overline{N}_{\text{rear}}$ was the average vertical load on the rear tire calculated for each run, m was the total mass of the PTW and the rider (including his safety equipment), g was the gravity acceleration, $\bar{a}_{\text{long}} < 0$ was the average longitudinal acceleration calculated for each run, l_1 was the front wheelbase of the scooter, l was the wheelbase of the scooter and h_G was the height of the centre of mass. The parameters' values are shown in Figure 9 and summarised in Table 1. [6]

Figure 9. Centre of mass position

Table 1. PTW parameters

Parameter	Value
PTW mass	187 kg
Pilot mass	90 kg
Wheelbase (l)	1.53 m
Front wheelbase (l_1)	1.02 m
Rear wheelbase (l_2)	**0.52 m**
Center of mass height (h_G)	**0.66 m**

The average braking force coefficient was calculated for each run using the formula (2), and the obtained values were averaged among every pressure set.

$$\bar{\mu} = \frac{\bar{F}_{\text{long}}}{\bar{N}_{\text{rear}}} = \frac{m\bar{a}_{\text{long}}}{mg\dfrac{l_1 + h_G\dfrac{\bar{a}_{\text{long}}}{g}}{l}} \tag{2}$$

Where $\bar{\mu} < 0$ is each run's average braking force coefficient, and $\bar{F}_{\text{long}} < 0$ is the longitudinal braking force calculated for each run.

3. Results

Figure 10 shows the longitudinal acceleration recorded during Run no.3 at low inflating pressure. The run can be divided into four phases:

1. *Initial steady phase.* The scooter was not moving; the acceleration measured in this phase was used for the calibration process.
2. *Acceleration phase.* The scooter accelerated at an approximately constant rate, up to 50 km/h.
3. *Braking phase.* The scooter was decelerating due to the hard braking manoeuvre; the recorded longitudinal acceleration was approximately constant except for the peaks caused by the ABS intervention (marked by red dots).
4. *Final steady phase.* The scooter was not moving; the run was complete.

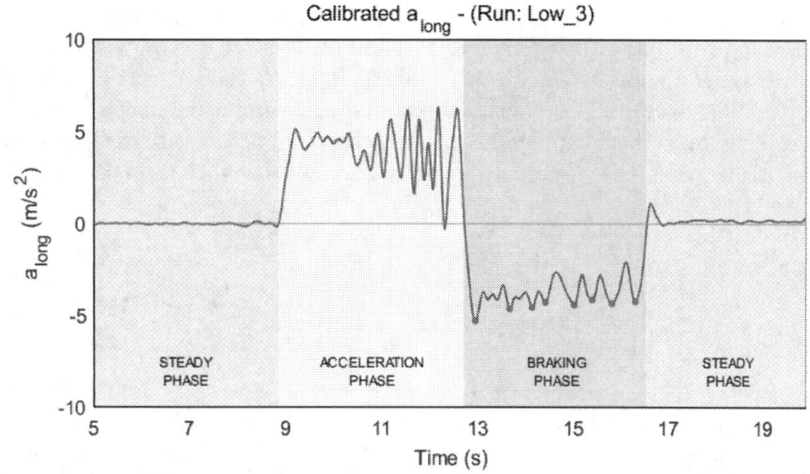

Figure 10. Acceleration logged during Run no.3 at low pressure (1.8 bar)

The results (Table 2) showed that the highest mean deceleration was achieved with the nominal tire pressure (2.4 bar), with a -3.69 m/s^2 mean and a 0.12 m/s^2 standard deviation. On the other hand, for the lower (1.8 bar) and higher (2.3 bar) pressure, we obtained means of respectively -3.58 m/s^2 and -3.66 m/s^2 with a standard deviation of 0.16 m/s^2 and 0.10 m/s^2. Underinflation reduced the magnitude of the average longitudinal acceleration by 3.0% compared to the nominal condition, while overinflation resulted in a 0.8% reduction.

The average braking force coefficient followed a similar behaviour to the average longitudinal acceleration. At nominal pressure, the mean was -0.75; in the underinflation and overinflation cases, it was -0.72 (-3.8% magnitude) and -0.74 (-1.0% magnitude).

Table 2. Results and statistics

Pressure	$\text{mean}(\bar{a}_{\text{long}})$	$\sigma\bar{a}_{\text{long}}$	$\Delta(\text{mean}\,\lvert\bar{a}_{\text{long}}\rvert)$ From Nominal	$\text{mean}(\bar{\mu})$	$\Delta(\text{mean}\,\lvert\bar{\mu}\rvert)$ From Nominal
(bar)	(m/s^2)	(m/s^2)	(%)	(-)	(%)
1.8	-3.58	0.16	-3.0	-0.72	-3.8
2.4	-3.69	0.12	-	-0.75	-
3.0	-3.66	0.10	-0.8	-0.74	-1.0

4. Discussion

The results showed a modest, albeit measurable, magnitude reduction of the achievable average deceleration. However, comparing the moderate difference among the obtained $\text{mean}(\bar{a}_{\text{long}})$ values to the calculated standard deviations, a student's T-test was performed before concluding. With the significance level set to 5%, the test did not reject the null hypothesis, assuming that the obtained averages came from normal distributions with equal means and equal but unknown variances; therefore, the observed behaviour might not exist in the population. Previous bench tests conducted to analyse the lateral dynamics of scooter tires showed modest variations in tire behaviour when changing the inflation pressure in the range we tested (1.8÷3.0 bar). It is, therefore, plausible that similar, approximately invariant behaviour was observed in the results we obtained for the longitudinal dynamics tests. The same study also reveals that, with more significant pressure differentials, the lateral response of the tire worsened in a non-linear manner, both with insufficient and excessive inflation pressure.

A trend was visible in the standard deviation, which decreased while increasing the inflating pressure. This phenomenon could be linked to the reduction of the contact area between the tire and the asphalt.

The percentage variation of the average braking force coefficient was higher than the variation in average longitudinal acceleration. This result was expected and is explained by the fact that the former is also affected by the vertical load on the tire, which in turn varies with the load transfer caused by the longitudinal acceleration. The variation, albeit more notable than the one in the longitudinal acceleration, was still modest.

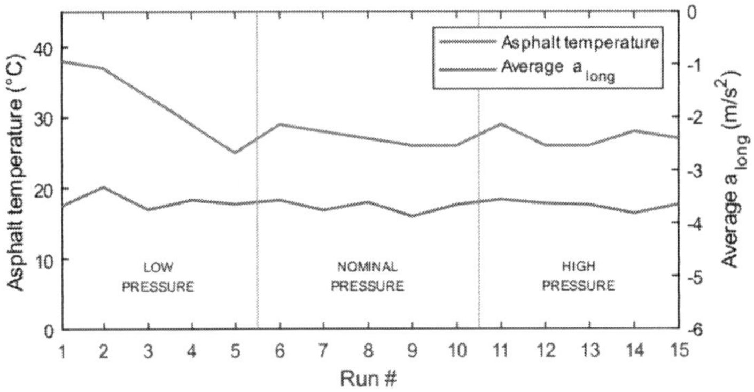

Figure 11. Asphalt temperature measurements of each run

During the experiment, the asphalt temperature changed due to the Sun irradiation, leading to a 13°C difference between the highest and lowest measured temperatures (+38°C and +25°C, respectively). A qualitative analysis showed no evident influence of the asphalt temperature on the achieved average deceleration (Figure 11).

5. Conclusion and Perspectives

Due to the small entity of the measured acceleration differences, considering the calculated standard deviation obtained for each of the three pressure values and with the support of the performed T-test, we cannot assert that the evaluated pressure differentials (±0.6 bar) measurably affected the braking performance of the performed straight-line hard braking manoeuvre.

As the tire pressure increases, the descending trend appreciable in the standard deviation values (Table 2) suggests that the achieved average deceleration value is more consistent among runs, resulting in a more predictable braking performance.

Future studies may be conducted in different conditions or with other objectives. When braking with the front wheel only, said wheel is subjected to an increase in the vertical force due to the load transfer. Such a situation is the opposite of what was analysed in this experiment. With a more significant pressure difference between the nominal and the excessive/insufficient values, as discussed above, the non-linear relation between tire pressure and its lateral dynamics should also imply a more appreciable variation in the results regarding the longitudinal dynamics. When altering the pressure of the rear tire, during the straight-line hard braking manoeuvre, the riding feel was not affected, probably because the change effects remained confined to the rear frame. Changing the inflating pressure of the front tire would likely impact the riding feel, particularly the steering torque and responsiveness, due to its influence on the steering assembly. These changes might influence the rider's behaviour during a hard braking manoeuvre, especially in an emergency.

References

[1] ISTAT (The National Institute for Statistics in Italy) 2020 Available at: https://www.istat.it/en/
[2] Penumaka A P, Savino G, Baldanzini N and Pierini M 2014 In-depth investigations of PTW-car accidents caused by human errors, *Saf. Sci.* **68** 212–21.

[3] Ajami M, Jannat H and Masih-Tehrani M 2020 The effect of tire pressure changes on braking efficiency and necessity of adjusting tire pressure before braking test at vehicle technical inspection centers. 10. 3446-56. 10.22068/ijae.10.4.3446.

[4] Bansal A, Jain A, Srivastava P, Kumar Tiwary A and Kumar Dear R 2016 Significance of tire pressure monitoring system in motorcycle, *SAE Tech. Pap.* 2016-01-1634.

[5] Cossalter V, Doria A, Giolo E, Taraborrelli L and Massaro M 2014 Identification of the characteristics of motorcycle and scooter tyres in the presence of large variations in inflation pressure, *Veh. Syst. Dyn.*, **52**:10, 1333-1354.

[6] Huertas Leyva P, Study of motorcyclist's behaviour during emergency braking in the perspective of training for safety, Doctoral Thesis, School of Engineering, University of Florence, 2017.

IOP Conf. Series: Materials Science and Engineering 1275 (2023) 012040 doi:10.1088/1757-899X/1275/1/012040

Hardware in the Loop Methodology for AEB system development

F Alfatti[1,2], C Annicchiarico[2] and R Capitani[1]

[1] DIEF, Università degli Studi di Firenze, Via Santa Marta 3, 50139 Firenze (FI), Italy
[2] Meccanica 42 srl, via Ezio Tarantelli 15, 50019 Sesto Fiorentino (FI), Italy

E-mail: federico.alfatti@unifi.it

Abstract. Thanks to ADAS systems, vehicle and occupants safety has increased. These systems have however a great impact on design complexity, increasing the time necessary for the launch of the functionality on the market. In this work, a Hardware in the Loop tool and methodology that integrates the Autonomous Emergency System (AEB) is presented, with the objective to speed up the development process. A commercial brake-by-wire braking system has been installed on a static simulator. The unit stock AEB can be activated during the real-time simulation, thanks to a custom strategy reproduced via Software in the Loop approach. In order to assess the capabilities of the proposed methodology, two different controllers have been proposed: the first reproduces the AEB functionality of a commercial car, while the second represent a possible design alternative that must be evaluated during the design phase. The two controllers have been tested in EuroNCAP relevant scenarios, and the performances have been compared in terms of Time To Collision necessary to avoid the impact, deceleration request and final distance from the obstacle. The proposed methodology highlights the role of the static simulator as a development tool that can vastly reduce the time necessary for the development of the AEB system, thanks to an extensive testing of both hardware and control strategy.

1. Introduction

The automotive industry is undergoing a rapid change. To reduce accidents and their consequences, all new production car sold in EU must be equipped with some type of Advanced Driver Assistance Systems (ADAS) [1]. Between the technologies currently available and sufficiently mature on the market, the Automatic Emergency Braking (AEB) system has the biggest potential to decrease up to the 50% [2,3] the number of Car to Car Rear collision (CCR) scenarios, applying an autonomous deceleration on the car when an unavoidable impact with an obstacle on the vehicle path is identified. For this reason, this system should be developed rapidly and efficiently, to potentially increase the number of vehicles equipped with this functionality.

The AEB system can be considered as composed by three major subsystems [4]:

- Environment Sensing
- Controls
- Brake actuators

Firstly, the environment sensing subsystem comprises both physical sensors (usually a mid or low range radar, and a camera) and software that identifies the possible obstacle and measures its relative distance and speed in respect to the vehicle [5]. The information on the obstacle presence, distance, and relative speed are sent to the high level controls of the system, which defines if the intervention

Content from this work may be used under the terms of the Creative Commons Attribution 3.0 licence. Any further distribution of this work must maintain attribution to the author(s) and the title of the work, journal citation and DOI.
Published under licence by IOP Publishing Ltd

is necessary. In most systems, the Time To Collision (TTC) is calculated, and this parameter defines the necessary intervention of the system [5], by comparing it with fixed thresholds: however, other controllers are available, such as distance based controller [5]. Lastly, the deceleration request is sent to the brake-by-wire actuator in form of brake pressure requests. This last system is responsible to slow or stop the vehicle. Given the presence of multi-domain specifications, on hardware and software reliability and efficiency, the development of this last system is very complex too. The development of the complete AEB system poses inevitable challenges that increase the time necessary to have the unit ready for the market. It is in fact necessary to test and validate the design of each subsystem independently, and only as a last step the complete integration will be tested on the real vehicle. The main problem in the development process is the high number of test needed to evaluate all the possible design variation that are present during the initial phases. Moreover, each design must be tested in a higher and higher number of intervention scenarios, that lastly takes the number of test to an high number. Moreover, the only way to this day to test the complete test is via track testing, which is reportedly very expensive in terms of costs and time.

Thanks to hardware in the loop methodologies [6,7], this work proposes a possible methodology to test the complete AEB system, with all its components, in a mixed environment, with virtual scenarios but with actual production hardware. In this way, and via a modular definition of the simulation tools, it is possible to test both controller design variations and their influence on the force actuation on the vehicle. It is important to note that the use of real time driving simulators is a must, both static and dynamic, for the development of these systems. These tools can include the driver earlier in the development process of both virtual and physical components [6].

The installation of the real time Hardware in the Loop test bench on a real-time static driving simulator is described. The entire brake system, comprising the brake actuator, brake lines, calipers and discs has been interfaced with the real time driving simulator supplied by the company Meccanica 42 S.r.l. .The environment sensing layer has been implemented virtually to reproduce valid CCR scenarios. Firstly, the description of the AEB system components and the two controllers under test is given. Then the driving simulator setup is presented. Lastly, the results of the test campaign are presented. The proposed AEB bench and the testing methodology show the potential to compare different controllers on the actual physical component, allowing for a complete test of both controller and physical actuator.

2. AEB System Description

The three available measurements produced from the environment sensing layer are the obstacle presence and distance, and its relative speed. With these, the control layer computes the Time To Collision (TTC), which is:

$$TTC = D/V \tag{1}$$

Where D is the relative distance between the ego vehicle and the obstacle, and V is their relative velocity.

Considering a system able to instantaneously impose a constant deceleration of $-8m/s^2$, the time necessary to completely stop the vehicle is dependent on its speed at the brake request instant [6]. It is thereby evident that the measured TTC must be compared with a different threshold depending on the vehicle speed. An offline computation of these thresholds however is incapable of considering the inevitable delay in signal communication and, most importantly, brake plant actual pressurization. Various methods have been proposed for the offline calculation of the intervention thresholds [4]. However, a tuning of these parameters with the physical unit is always necessary, given the importance of this delay in defining the stopping distance of the vehicle. The test scenario considered follows the EuroNCAP 2021 regulations [8]. The vehicle whose AEB system is in test is called Ego vehicle, and the possible object in the trajectory which must be avoided is referred as Target vehicle. In particular, the Car-to-Car Rear moving scenarios (CCRm) are analyzed, in order to advance the testing capabilities of the methodology and the tool from [6]. However, there are virtually no limitations on the type and number of scenarios that can be approached in virtual environment, and the methodology that is proposed in this paper can also apply to other structured test scenarios. Firstly, a simplified environment sensing

IOP Conf. Series: Materials Science and Engineering 1275 (2023) 012040 doi:10.1088/1757-899X/1275/1/012040

layer has been developed, which produces the same signals as the real sensors, i.e. distance and relative speed, via calculation of the distance between the two vehicles, considered as material points, with a global and fixed reference system X, Y.

$$Distance = (X_{target} - X_{ego})^2 + (Y_{target} - Y_{ego})^2 \tag{2}$$

The controller that reproduces a typical commercial AEB intervention during CCRs scenario is described; lastly, a different intervention strategy, based on a PID controller, has been implemented.

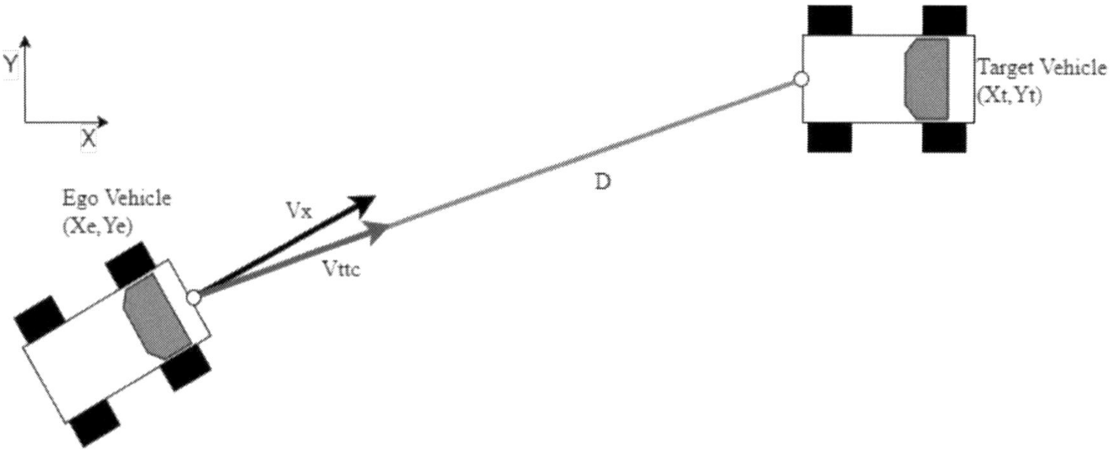

Figure 1. Environment sensing schematics

2.1. Commercial Controller

As a baseline, a controller based on a plausible intervention strategy has been implemented. Unfortunately, high level controller for AEB functionality is usually an application of the vehicle manufacturer specification and is non-disclosable material. However, a database with AEB car tests carried out on the specific CCRs scenario is available thanks to the Institute of Insurance of Highway Safety (IIHS) [9]. In a previous work, the system has been validated on CCRs scenarios taken from this database [6,14]. The controller compares the measured TTC to a pre-defined threshold, which must be speed dependent. If the measured TTC is lower than the threshold, a full brake intervention is advised, requesting a deceleration of - 8m/s2. The controller has been implemented in a virtual Simulink model. Its behavior replicates the tests taken out from the IIHS database.

2.2. PID controller

We are now interested in evaluating the performances of the system with a different intervention strategy. A possible controller has been investigated in [10]. The aim of this controller is to apply a partial brake action first, to alert with an initial lower deceleration the driver. This initial deceleration can also result in a better acceptance of the system intervention. As in [6] Two different TTC thresholds have been implemented, one corresponding to the initial deceleration, called Partial Braking Threshold (PBT), that starts the Partial Braking Phase (PBP). The second threshold, called Full Braking Threshold (FBT) corresponds to the full actuation of the brake action, or Full Braking Phase (FBP). The deceleration for the FBP is the same as the commercial controller previously described, -8m/s2. During the PBP the deceleration is imposed by a PI controller which aim is to maintain the measured TTC to the same level as the PBT, as in figure 2 and in 3.

$$\epsilon = TTC_{measured} - TTC_{pbtreshold} \tag{3}$$

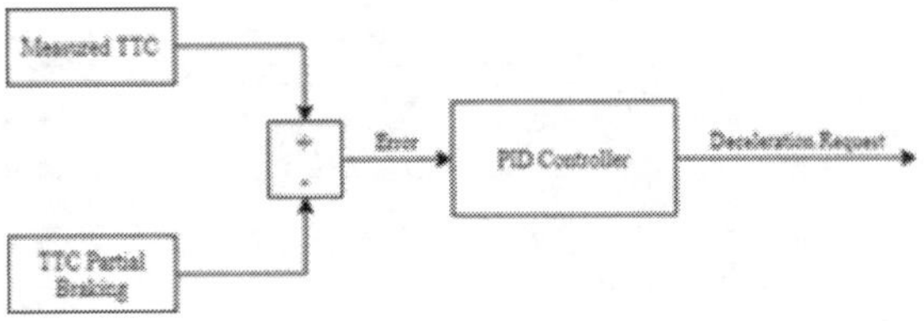

Figure 2. PID controller schematics.

Figure 3. Static simulator setup.

3. Real-time driving simulator setup

The brake bench has been installed on the driving simulator thanks to the collaboration with the company Meccanica 42. The simulator setup is the same as reported in [6,11,12,13]: for the scope of this work, the simulator can be described as composed by four main parts: the driver cockpit, the real time simulator, the EPSiL steering bench, and the Braking bench. With this specific setup, the driver can then test both steering and brake unit hardware, equipped with stock ADAS functionalities. This mixed approach, in which both hardware and possible design variation are tested into the same testing session, can be of great use in reducing the time necessary to develop the AEB system. The schematics of the simulator are reported in figure ??.

	Minimum distance to target [m]	
EGO Speed [km/h]	Commercial	PID
30	1,5	1,6
50	0,2	0,8
70	0,2	0,9

Table 1. Controller Result

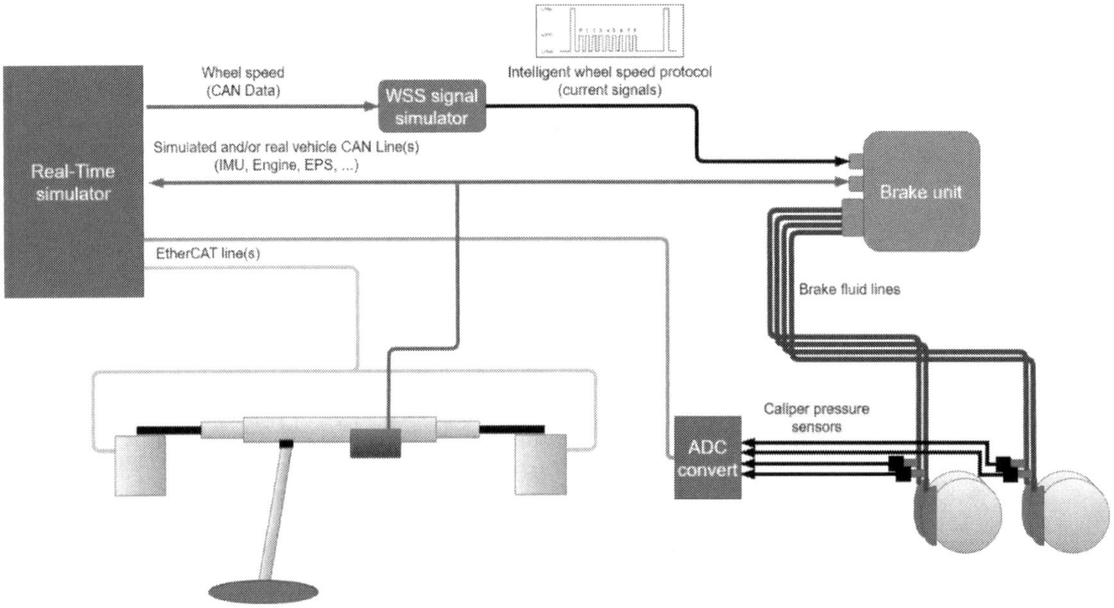

Figure 4. Simulator scheme.

4. Methodology

The testing methodology presented in [6] has been extended to evaluate the CCRm scenarios under investigation. Collecting data necessary for the complete assessment of the system in CCRm scenario, the two controllers have been compared in terms of TTC thresholds, deceleration, jerk, and minimum distance from the obstacle. To evaluate the actuator performance, the actuation delay between deceleration request and actual change in speed has been computed.

5. Results

5.1. TTC Tuning

The aim of this step was to achieve a constant and safe minimum distance from the target, to avoid unnecessary interventions caused by environment sensing false positives. The same TTC threshold values reported in [6] have been used in this scenarios, in order to avoid the definition of different thresholds in function of the vehicle movement, maintaining a simple control architecture.

5.2. Controller Comparison

In Figures 5 to 10 we can find a the performances of the two controllers during the CCRm 30, 50, 70 scenario. The target vehicle has always a constant speed of 20 km/h. The measured and requested deceleration has been compared, and the vehicle speed has been measured. The minimum distance between target and ego vehicle is reported in Table 1. In figure the result for the commercial controller are reported, while in figures the PID controller results are presented. We can see a substantial difference in the actuation: the commercial controller applies always the same deceleration, implying a very short braking event duration, especially at lower speeds. In general, the PID controller determines a longer braking intervention, but with lower intensity: however, this behaviour can alert the driver in these situations, avoiding the necessity for a complete autonomous stopping of the vehicle. In general, the PID controller yields better results in both avoiding the impact and maximising the distance from the target vehicle, as is seen in Table 1.

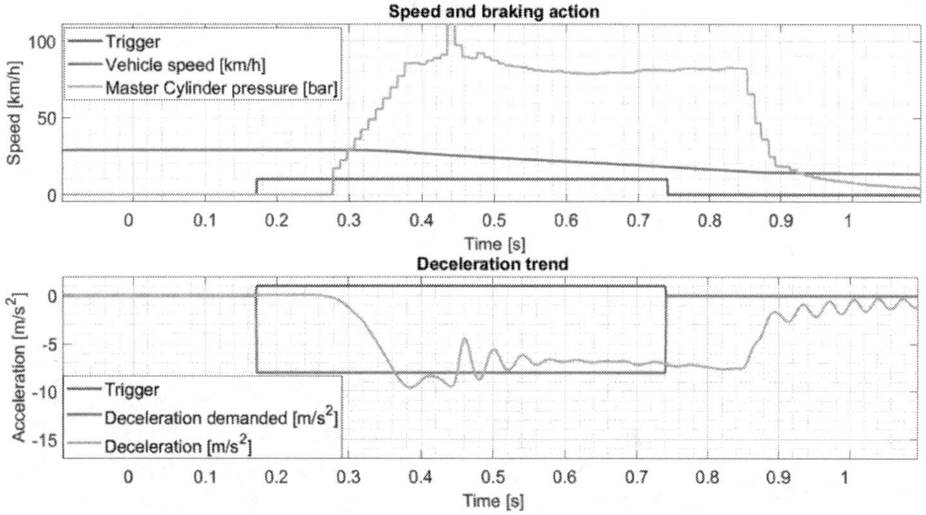

Figure 5. Commercial controller test Result: Ego speed 30 km/h

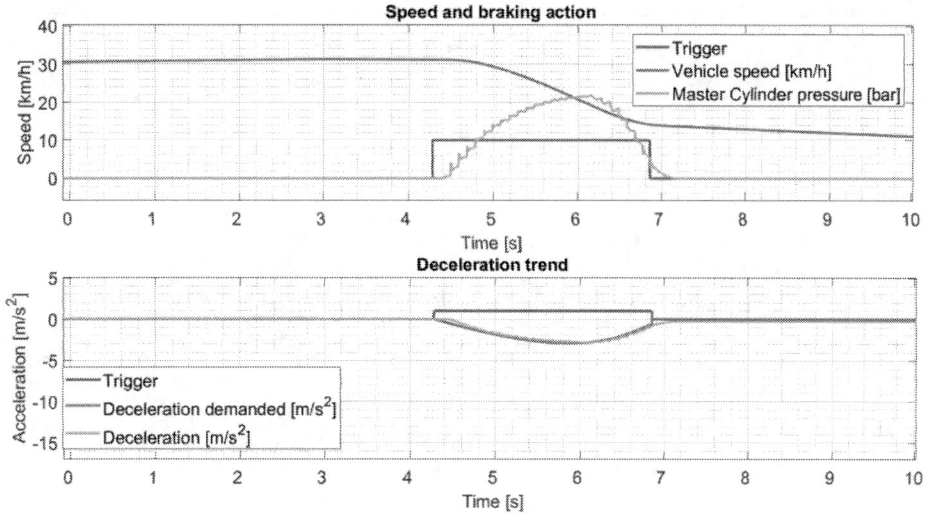

Figure 6. PID controller test Result: Ego speed 30 km/h

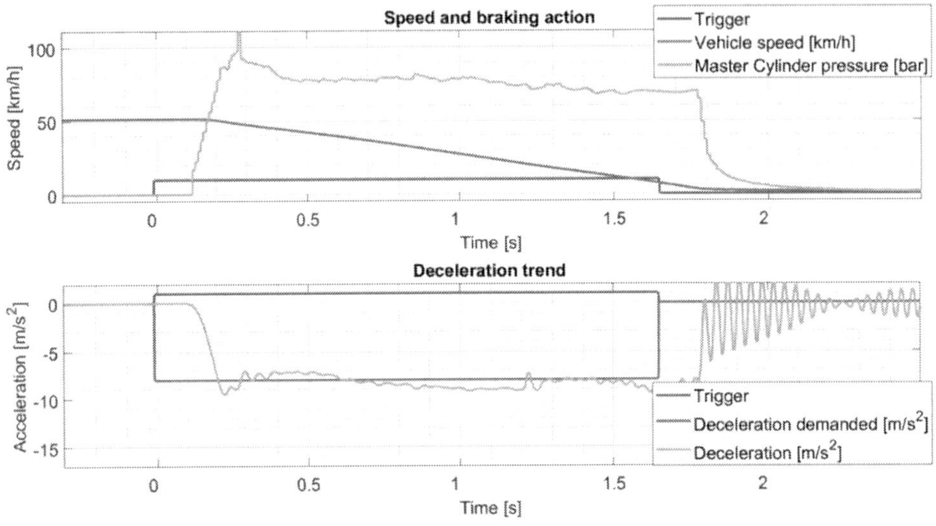

Figure 7. Commercial controller test Result: Ego speed 50 km/h

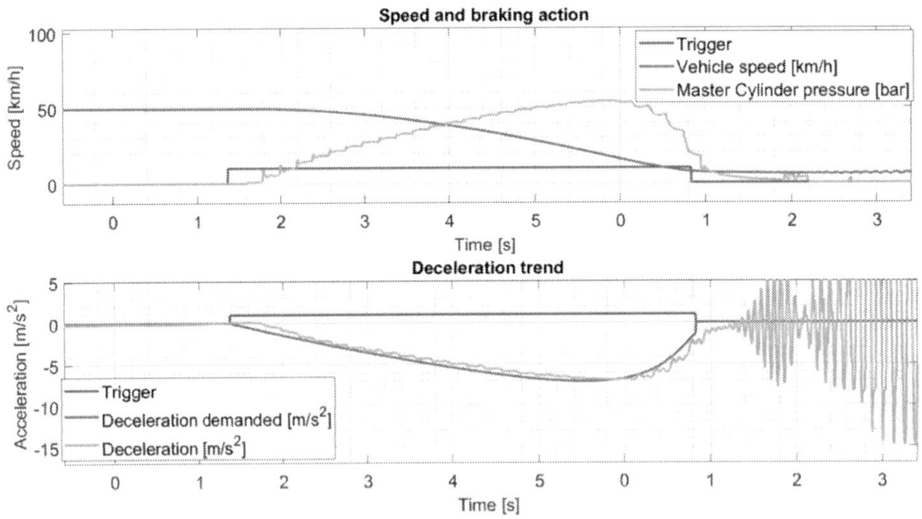

Figure 8. PID controller test Result: Ego speed 50 km/h

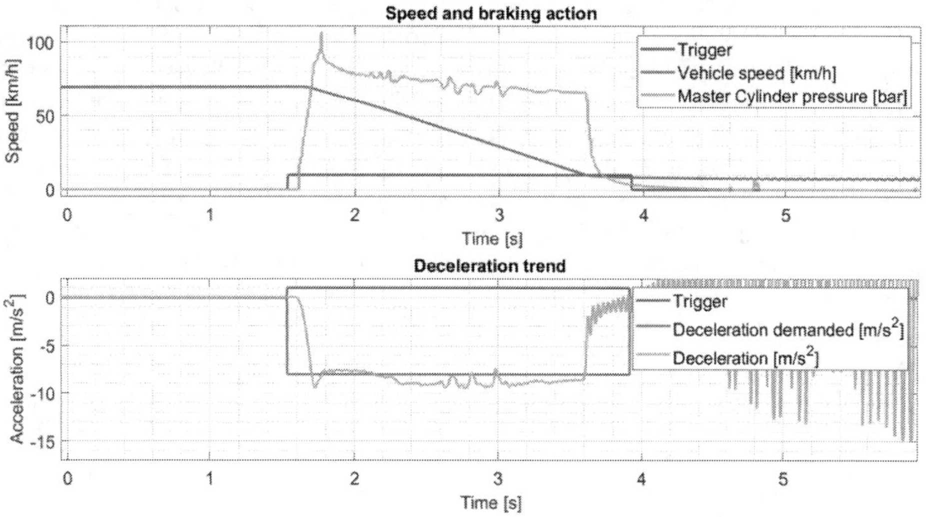

Figure 9. Commercial controller test Result: Ego speed 70 km/h

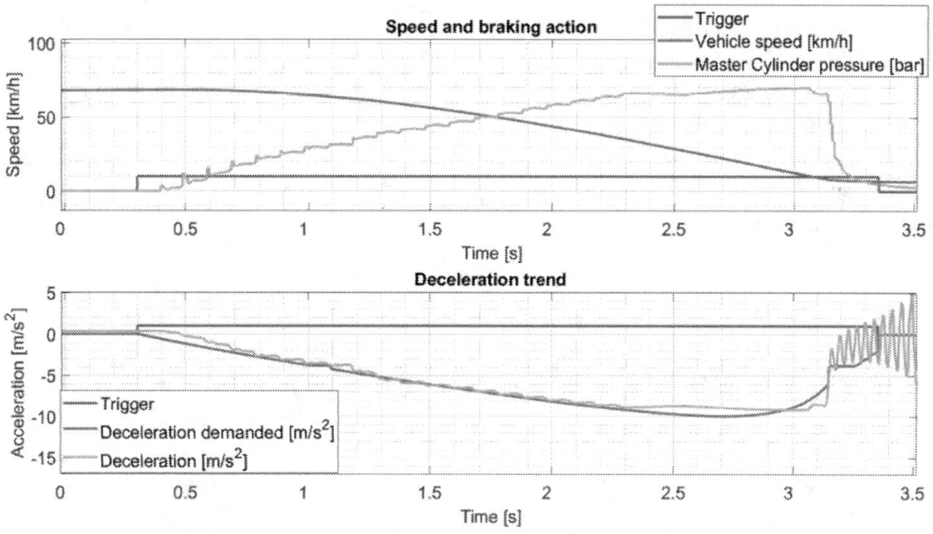

Figure 10. PID controller test Result: Ego speed 70 km/h

6. Conclusions

To reduce development time of complex active safety systems such as the AEB system, and to potentially increase the diffusion of such systems on a larger number of road vehicles, this work proposes an HiL bench architecture for real time testing of controllers and physical components of the brake-by-wire system on a static driving simulator. The installed unit, which is commercially available, has advanced AEB functions, which have been integrated on a driving simulator. A simplified environment sensing layer has been implemented, as well as two different controller for the system intervention. These two controllers, with the information made available from the previous layer, define the system intervention using two different strategies; the first one is the same as the unit stock functionality, while the second one proposes the application of two distinct levels of deceleration, with a smooth transition between the two levels. The proposed methodology starts from creating the virtual scenarios, taken from a structured testing normative. The methodology can then produce comparable results between the two controllers. Thanks to this, the HiL bench has been capable of producing solid results, allowing for a faster validation of the two controllers and for a complete assessment of the integration of the physical brake actuator. This work demonstrate the capabilities of both the simulator and the HiL bench to further explore the limits of both software and hardware of the AEB system, in a different number of reproducible scenario, in a time-effective way. Future steps will include the implementation of a radar sensor model in the bench to overcome the simplified environment sensing limitations, as well as the exploration of system performances in different scenario from the Euro-NCAP rules, with a braking or with a pedestrian-crossing scenario.

Abbreviations

The following abbreviations are used in this manuscript:

ABS	Anti-lock Braking System
ADAS	Advanced Driver Assistance System
AEB	Autonomous Emergency Braking
ASIL	Automotive Safety Integrity Level
BW	By-Wire
BBW	Brake-By-Wire
CAN	Controller Area Network
ESP	Electronic Stability Program
EPS	Electric Power Steering
EPSiL	EPS in the Loop
ESC	Electronic Stability Control
HiL	Hardware in the Loop
IMU	Inertial Measurement Unit
KPI	Key Performance Index
RT	Real Time

IOP Conf. Series: Materials Science and Engineering 1275 (2023) 012040 doi:10.1088/1757-899X/1275/1/012040

References

[1] KHAN, Imran. 2018. "Observatory." Text. . Mobility and Transport - European Commission. . June 20, 2018. $https : //ec.europa.eu/transport/road_safety/specialist/observatory_en$.

[2] Cicchino, Jessica B. 2017. "Effectiveness of Forward Collision Warning and Autonomous Emergency Braking Systems in Reducing Front-to-Rear Crash Rates." Accident Analysis and Prevention. 99 (February). https://doi.org/10.1016/j.aap.2016.11.009.

[3] Seacrist, Thomas, Ridhi Sahani, Gregory Chingas, Ethan C. Douglas, et al. 2020. "Efficacy of Automatic Emergency Braking among Risky Drivers Using Counterfactual Simulations from the SHRP 2 Naturalistic Driving Study." Safety Science. 128 (August). : 104746. . https://doi.org/10.1016/j.ssci.2020.104746.

[4] Winner, Hermann, Stephan Hakuli, Felix Lotz, and Christina Singer, eds. 2016. Handbook of Driver Assistance Systems. Cham: Springer International Publishing. . https://doi.org/10.1007/978-3-319-12352-3.

[5] Fischer, Dara. n.d. "Time to Collision and Collision Avoidance System," 13.

[6] Alfatti, F., Garinei, M., Annicchiarico, C., and Capitani, R., "Design and Hardware in the Loop Testing of AEB Controllers," SAE Technical Paper 2022-01-0099, 2022, https://doi.org/10.4271/2022-01-0099.

[7] Ye, Yifan, Jian Zhao, Jian Wu, Bing Zhu, et al. 2018. "Real-Time Automatic Test of AEB with Brake System in the Loop." In . , 2018-01–1450. . https://doi.org/10.4271/2018-01-1450.

[8] "Safety Assist — Euro NCAP." n.d. Accessed October 22, 2021. https://www.euroncap.com:443/en/for-engineers/protocols/safety-assist/.

[9] "IIHS-HLDI." n.d. IIHS-HLDI Crash Testing and Highway Safety. . Accessed October 22, 2021. https://www.iihs.org/.

[10] Shin, Seong-Geun, Dae-Ryong Ahn, Yun-Seok Baek, and Hyuck-Kee Lee. 2019. "Adaptive AEB Control Strategy for Collision Avoidance Including Rear Vehicles *." In 2019 IEEE Intelligent Transportation Systems Conference (ITSC). , 2872–78. . Auckland, New Zealand: IEEE. . https://doi.org/10.1109/ITSC.2019.8916988.

[11] Vinattieri, Francesco, Tim Wright, Renzo Capitani, Claudio Annicchiarico, et al. 2016. "Target Setting and Structural Design of an EPS-in-the-Loop Test Bench for Steering Feeling Simulation." SAE Technical Paper 2016-01–1559. . Warrendale, PA: SAE International. . https://doi.org/10.4271/2016-01-1559.

[12] Talarico, Enrico Maria, Giuseppe Raimondi, Federico Alfatti, Daniele Vitaliti, et al. n.d. "A Virtual Development Approach Using Advanced HiL Steering Bench." In . , 16. . Springer Vieweg.

[13] Alfatti, F., C. Annicchiarico, and R. Capitani. 2022. "Vehicle Stability Controller HiL Validation on Static Simulator." IOP Conference Series: Materials Science and Engineering. 1214 (1). : 012044. . https://doi.org/10.1088/1757-899X/1214/1/012044.

IOP Conf. Series: Materials Science and Engineering 1275 (2023) 012041 doi:10.1088/1757-899X/1275/1/012041

Modelling, testing and validation of an innovative AEB control logic on a Hardware-in-the-loop test bench

L Veneroso[a], F Alfatti[a], C Annicchiarico[b] and R Capitani[a]
[a] DIEF, Università degli Studi di Firenze, Via Santa Marta 3, 50139 Firenze (FI), Italy
[b] Meccanica 42 srl, Via Ezio Tarantelli 15, 50019 Sesto Fiorentino (FI), Italy
E-mail: luca.veneroso@unifi.it

Abstract. AEB, autonomous emergency braking, is an active safety system designed to prevent vehicle frontal collision. The most diffused AEB systems are based on simple *Bang Bang* control logic, which could often avoid, or at least mitigate collision effects, but their effectiveness can still be improved by increasing system repeatability. The aim of this study is to model and test an innovative AEB control logic that will increase system reliability by compensating for the non-immediate response of the braking system to braking requests. Using a hardware-in-the-loop test bench with two different braking systems implemented, the new controller was tested simulating the $CCRs$ and $CCRm$ scenarios, used by Euro NCAP for AEB system assessment. By compensating for the delay introduced by the response of two different braking systems, the innovative control logic stops the VUT, Vehicle Under Test, at the desired safety distance from the GVT, Global Vehicle Target.

1. Introduction

The main task of the AEB, Autonomous Emergency Braking, system is to reduce the severity of front collision by lowering the speed of the striking vehicle if it can't completely avoid the incident. This system is usually paired with the FCW, Forward Collision Warning, which task is to warn the driver of the imminent danger through auditory warning, blinking optical display and sometimes using haptic devices as tactors. It is demonstrated by many studies that the Autonomous Emergency Braking System offers high reductions in crash avoidance and mitigation of impact's consequences. Fildes et al. [6] analyses data from 6 different country in which vehicles equipped with AEB system and vehicle without impact the rear of another vehicle. The conclusion shows a 38% overall reduction in rear-end crash both for urban ($\leq$ 60 km/h) and rural ($>$ 60 km/h) speed zones. J.Cicchino [16] Studied the different effect of the FCW, the EB and their combination. Rear-end striking crash result reduced by 27%, 43% and 56% respectively for FCW only, EB only and their combination.

Typical AEB control logics base their decisional process on the data collected from the camera and multi-target Radar [18] mounted on the front of the vehicle. Once critical conditions are detected, from the Radar sensor can be extrapolated two important data that are: relative distance d_{rel}, and relative velocity v_{rel} between the VUT and the GVT. From these two quantities many parameters, that aid to understand the severity of the situation, can be calculated.

There are essentially two ways to determine the start of the FCW and AEB intervention and their behaviour during the braking phase.

Content from this work may be used under the terms of the Creative Commons Attribution 3.0 licence. Any further distribution of this work must maintain attribution to the author(s) and the title of the work, journal citation and DOI.
Published under licence by IOP Publishing Ltd

- Time based principle: Control logic choices are based on: HWT, TTC, ETTC, Driver estimated reaction time and system response time.
- Acceleration based principle: Control logic choices are based on the calculation of longitudinal acceleration required to avoid collision.

The use of classical *Bang Bang* control logic results in low repetitiveness of stopping distances between the two vehicles. This is due to the difficult characterization of the response of the braking system, which depends on many uncontrollable parameters such as ambient temperature or the time of use of brake fluid due to his hygroscopic properties. Therefore, in this paper is proposed a deceleration based controller that does not need precise characterization of the braking system but corrects the level of deceleration by monitoring the relative distance trend. AEB controller has been tested in two rear-end collision test scenarios: *Car-to-Car Rear stationary CCRs* and *Car-to-Car Rear moving CCRm* which are used by Euro NCAP as part of the Safety Assist assessment [11]. The validation of AEB controller has been carried out on a commercial available brake-by-wire system unit (BSU) and the Corner brake actuator (CBA) developed by Meccanica 42. Both braking systems has been integrated in a hardware-in-the-loop test bench. The two different scenarios has been simulated using a static simulator provided by Meccanica 42 in which the two different braking system were integrated with Hardware in the loop technique.

2. AEB control logic

The AEB controller is modelled with a state machine in Stateflow. The state machine allows for quick and easy modelling as it is sufficient to specify actions and transition conditions for each state. It is also very easy to make changes and add new features such as brake pre-fill or brake jerk in the future, or pass from two to three deceleration thresholds. Two types of intervention are supported by the controller, a classical one in which the requested deceleration remains constant throughout the simulation and a "*Corrected*" one in which the deceleration is updated every simulation time step. The state machine is composed by four states:

- *OFF*: This is the default state, in which no action is determined by the AEB. Each of the next states can return to the OFF state when $v_{vut} < v_{gvt}$ or $v_{vut} = 0$.

- *FCW - Forward Collision Warning*: Actuates driver warning strategies and pre-condition the vehicle. To these actions can be added the Jerk function if activated.

- *PB - Partial Braking state*: This is the first state that determines a braking intervention to avoid the collision. In the *Bang Bang* mode PB requires a constant deceleration equal to a_{pb}, while in *Corrected* mode the DR_{out} is updated at each time-step allowing to compensate brake system delay.

- *FB - Full Braking*: When the intervention of the Partial Braking state is not sufficient to avoid the collision the control logic switches to the last state with the full braking request. As in the PB state the requested deceleration remains constant in *Bang Bang* mode and is updated in *Corrected* mode.

.

The controller continuously monitors the relative distance and the relative velocity between the two vehicles which are measured by the radar sensor. During the event the controller computes three different stopping distances, these are calculated considering three different deceleration thresholds which are set to $a_{fcw} = -2m/s^2$, $a_{pb} = -4m/s^2$, $a_{fb=-8}m/s^2$. The stopping distances are then compared to a safety margin SM_d, when the following condition is true

$$DR(t) = d_{rel} - SD_x < SM_d \tag{1}$$

the corresponding state is activated. A scheme of the process is represented in figure 1, in which the FCW status should be activated.

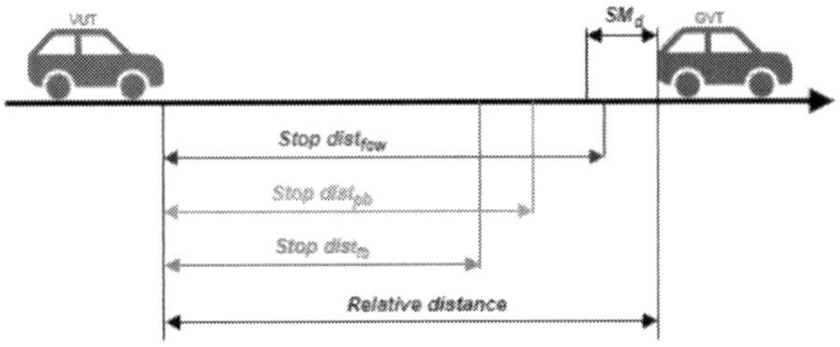

Figure 1. Stopping distances scheme

The deceleration requested to stop the vehicle at the distance SM_d from the target vehicle for both the $CCRs$ and the $CCRm$ scenario is calculated with the following formula:

$$DR_{ccrsm}(t) = \frac{v_{vut}(t)^2}{2d_{rel}(t) - SM_d} \tag{2}$$

In figure 2 is reported the required deceleration to stop the VUT at the SM_d distance from the target during a $CCRs$ or $CCRm$ scenario, splitting the plot zones corresponding to states of the state machine.

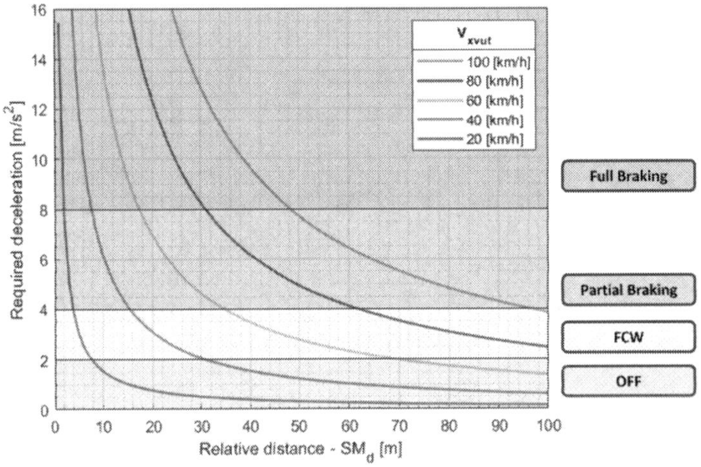

Figure 2. Deceleration required to stop the VUT vehicle at the SM_d distance

In table 1 the transition conditions and the action of each state of the state machine are shown.

3. Real-time simulator

Figure 3 shows the Hardware-in-the-loop static simulator provided by Meccanica 42 s.r.l used to test the AEB control logic with the two different braking systems. The simulator is composed by a Real-time computer which executes the simulation and manages the communication lines providing via CAN the signals needed to feed the EPS and braking units ECU, and via EtherCAT the signals necessary to control

Table 1. AEB logic steps

Transition Conditions	AEB Status
$v_{vut} > v_{gvt}$; $SD_{fcw} > d_{rel} - SM_d$	FCW
$v_{vut} > v_{gvt}$; $SD_{pb} > d_{rel} - SM_d$	Partial braking
$v_{vut} > v_{gvt}$; $SD_{fb} > d_{rel} - SM_d$	Full braking

the EPS test bench actuators, as schematized in Figure 4. The entire steering system is integrated into the simulator and the dynamic forces on the tie rods are applied by two torque motors by means of a rocker, to accurately reproduce the steering feel. The rockers are necessary to properly replicate tie-rods motion during wheel travel. Also, the complete braking system is integrated in the simulator. In fact, the commercial brake-by-wire unit is connected to brake lines with real length and diameter, that are connected to four brake callipers acting on a non-rotating brake disk. To include the CBA system in the simulation loop, as shown in Figure 5, a Micro Auto Box is added to feed the CBA ECU with the pressure request at each vehicle corner. The CBAs included in the simulation loop are the two already integrated into the experimental vehicle provided by Meccanica 42 S.r.l, hence the brake actuators pressurizes the brake line and the caliper of the vehicle acting on his brake disks.

Figure 3. Static simulator provided by Meccanica 42 S.r.l

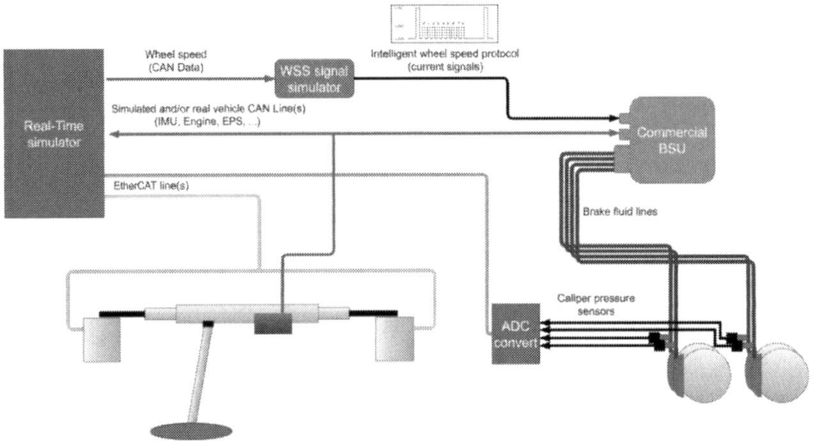

Figure 4. Static simulator layout

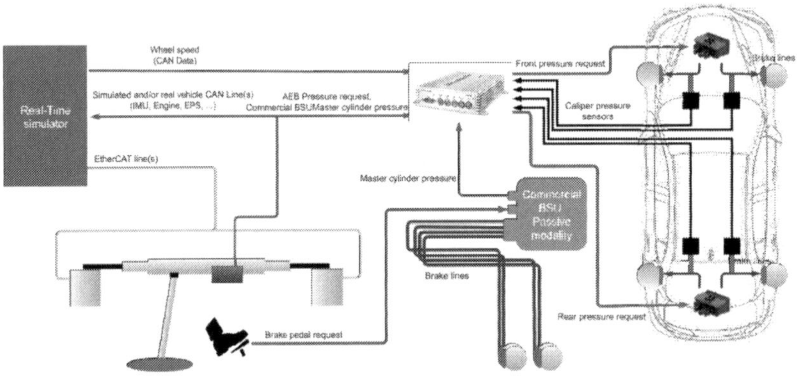

Figure 5. CBA in the loop simulator layout

4. Methodology

Once the AEB controller functioning has been debugged on offline simulation, the AEB assessment scenario has been reproduced online with the use of the driving simulator. As indicated by Euro NCAP in *Test protocol - AEB Car-to-Car systems* [11], the AEB controller has been tested on two different scenarios:

4.1. CCRs Car-to-Car rear stationary

CCRs is the simplest scenario in which the Global Vehicle Target (GVT) is stationary and the Vehicle Under Test (VUT) is moving toward the GVT at constant velocity. In table 2 can be found the test VUT velocities.

Table 2. CCRs scenario

Scenario	AEB + FCW combined AEB	 FCW	AEB only	FCW only
AEB CCRs	10-50 km/h	30-80 km/h	10-80 km/h	30-80 km/h

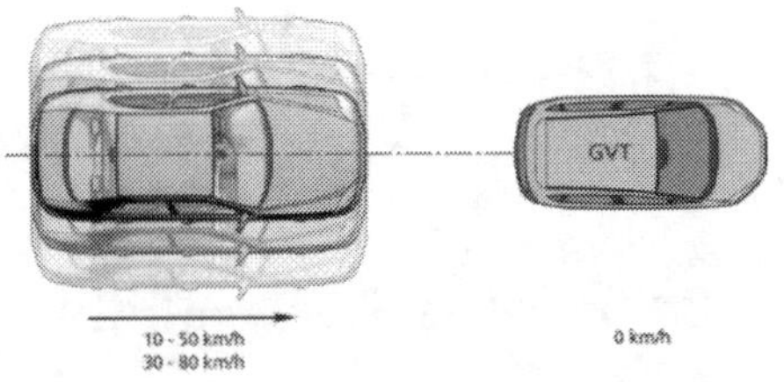

Figure 6. CCRs scenario [11]

4.2. *Car-to-Car Rear moving*

In this scenario the GVT proceed at constant velocity and the VUT is moving toward the GVT at a higher constant velocity. In table 3 can be found the test VUT velocities. In this case the controller aims to turn a rear-end collision scenario into a car-following scenario. The VUT will be decelerated so that it follows the GVT at a SM_d distance with $v_{vut} = v_{gvt}$.

Table 3. CCRm scenario

Scenario	AEB + FCW combined AEB	 FCW	AEB only	FCW only
AEB CCRs	30-80 km/h	50-80 km/h	30-80 km/h	50-80 km/h

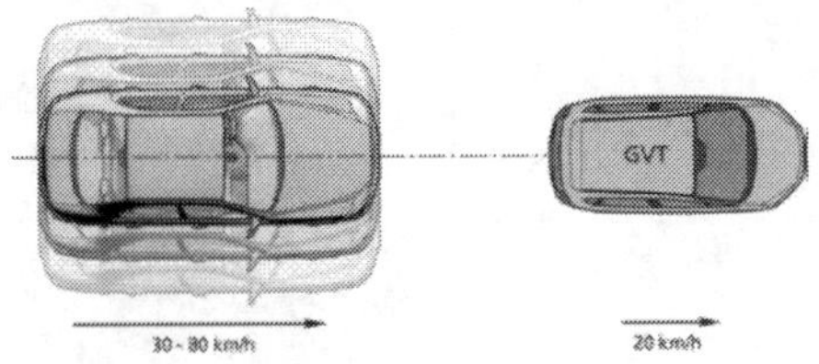

Figure 7. CCRm scenario [11]

5. Results

The AEB controller avoids car collision both in CCRs and CCRm scenarios using *Bang Bang* and *Corrected* control logic. In figure 8, and figure 9 the intervention of the two controller is compared with both braking systems in a CCRs scenario with a VUT constant velocity of 40 km/h. In the pressure curve, it can be noticed how the response of the two braking systems is different. In fact, the CBA presents a much shorter rise time to reach the target pressure compared to the commercial BSU. In addition, it can be observed that the CBA exhibits more oscillatory behaviour after reaching the target pressure value unlike the commercial BSU, which therefore exhibits a shorter stabilization time. The same kind of behaviour can be seen for both the first and second pressure demand steps. The image shows how the intervention of a classic AEB control logic requires two level of constant deceleration ignoring the system response, while the *Corrected* control logic after requesting the first deceleration threshold, increases its request to compensate for the delay in pressure build up which is evident in the first plot. Once the detachment from the ideal condition has been recovered, the required deceleration returns to values equal to those of the first threshold. As a result the stopping distance of the VUT from the vehicle GVT in the *Corrected* case is 0.98 m while in the *Bang Bang* case it is 0.76 m.

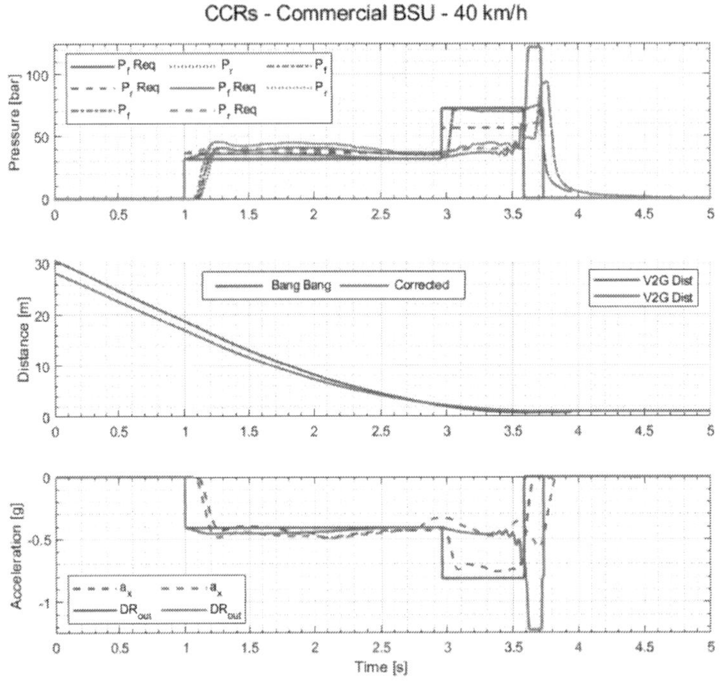

Figure 8. Comparison between *Bang Bang* and *Corrected* controller on commercial BSU

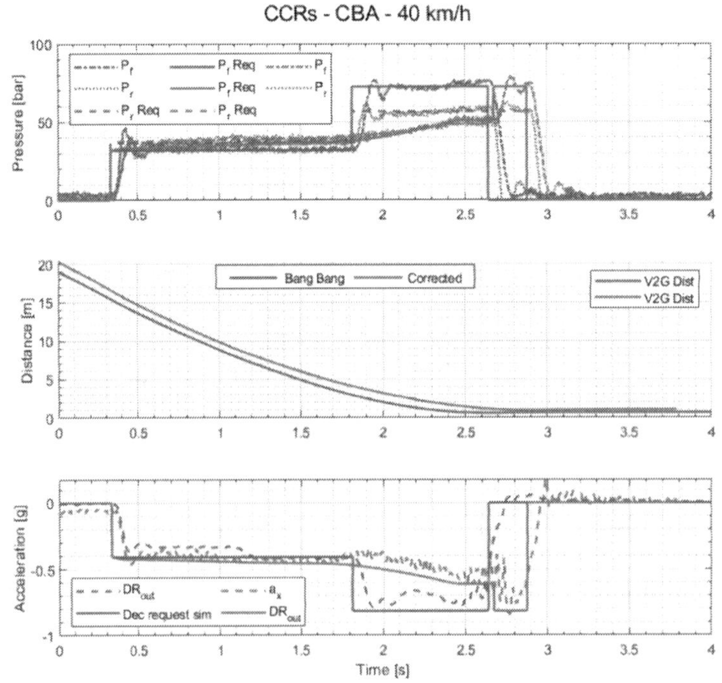

Figure 9. Comparison between *Bang Bang* and *Corrected* controller on CBA

6. Conclusion

The control logic proposed in this study made it possible to avoid crashes in all 20 simulated cases whether implemented with delay correction or not. Despite this, the correction of the system response allows to obtain more repeatable results and to stop the vehicle more safely by requiring lower average accelerations and keeping the vehicle in a more stable condition. The dispersion of stopping distances with the classic *Bang Bang* logic is very high, while the *Corrected* control logic allows very similar results to be obtained even though it is implemented on different hardware; in fact, its purpose is to cancel the effects of system delay in an adaptive manner. Figure 10 and figure 11 show the stop distance for the CCRs and CCRm scenarios highlighting the improved accuracy for the *Corrected* controller. Another advantage demonstrated by this type of algorithm is that it does not require any time-consuming calibration work. In fact, there are no tuning parameters, only parameters which affect the operating mode and the aggressiveness of the intervention. Since there is no anticipation of the intervention which takes into account the response of the braking system and vehicle and the driver's reaction time, the interventions controlled by this type of control logic are much less invasive.

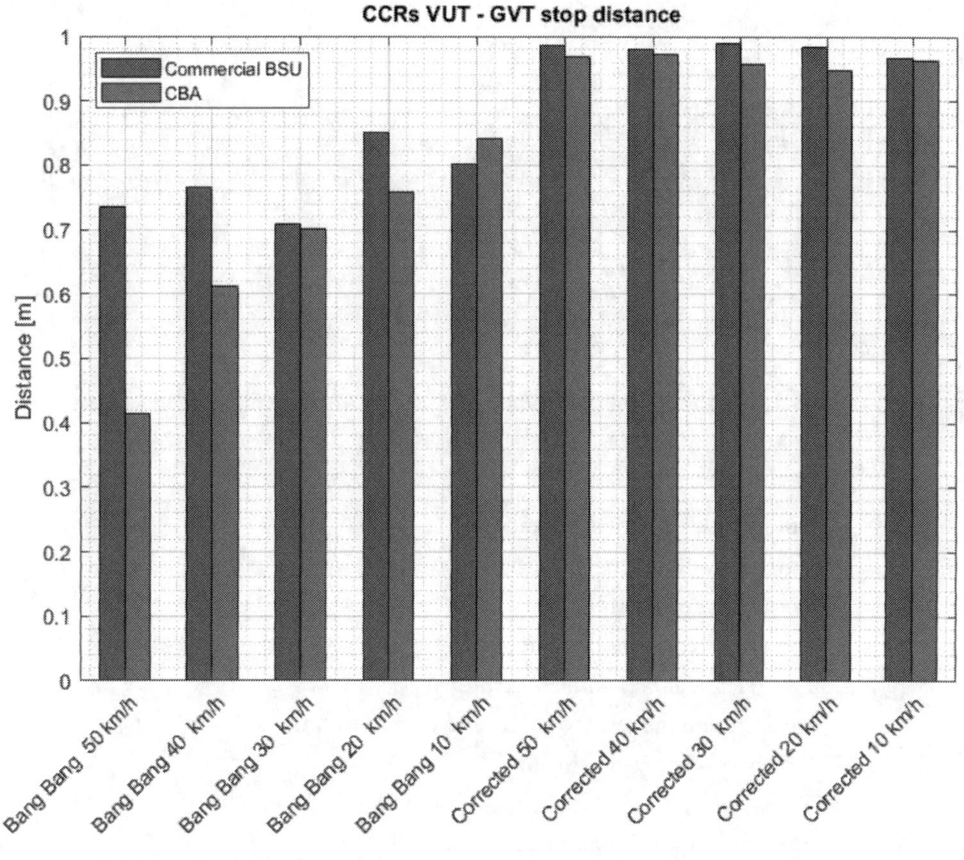

Figure 10. CCRs final distances using *Bang Bang* and *Corrected* control logic

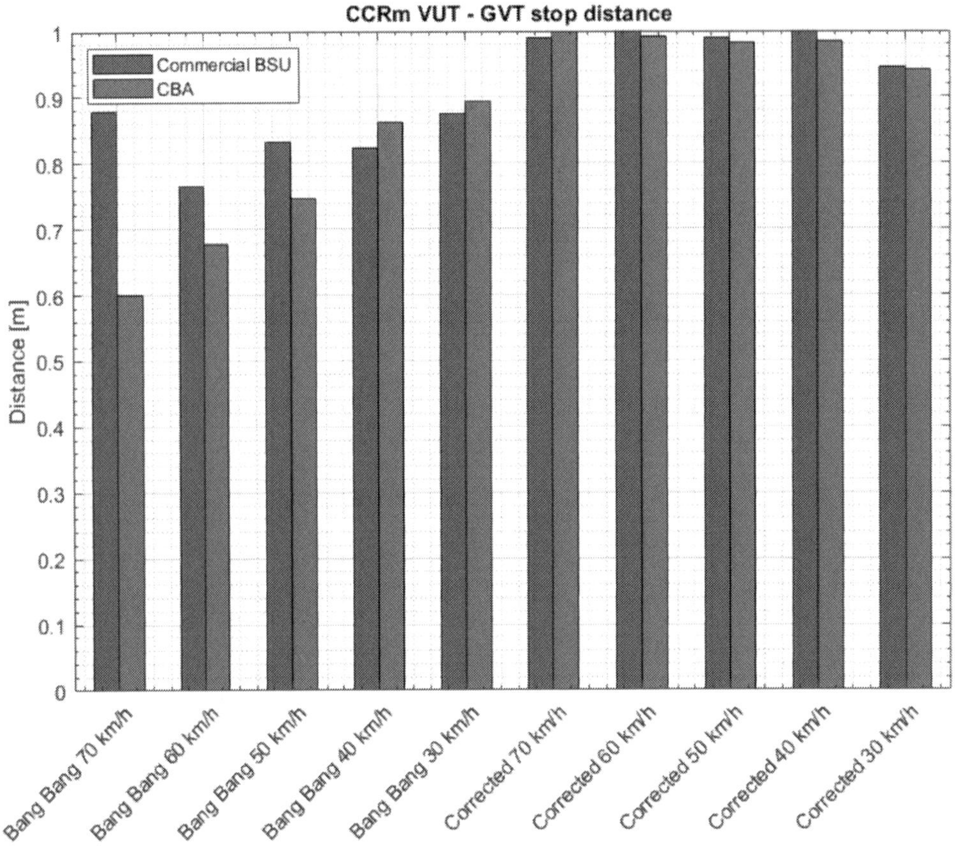

Figure 11. CCRm final distances using *Bang Bang* and *Corrected* control logic

Nomenclature

AEB	Autonomous emergency braking
BSU	Braking system unit
CBA	Corner brake actuator
$CCRm$	Car to car rear end collision moving
$CCRs$	Car to car rear end collision stationary
DR	Deceleration request for CCRs and CCRm event
$ETTC$	Enhanced Time To Collision
$Euro\ NCAP$	European New Car Assessment Programme
FCW	Forward Collision Warning
GVT	Global Vehicle target
HWT	Headway Time
SD	Stopping distance
SM_d	Safety margin distance
TTC	Time To Collision
VUT	Vehicle under test

IOP Conf. Series: Materials Science and Engineering 1275 (2023) 012041 doi:10.1088/1757-899X/1275/1/012041

7. References

[1] Bosch Professional Automotive Information 2015 *Automotive Mechatronics* (Wiesbaden: Springer Fachmedien Wiesbaden)

[2] Certosini C, Vinattieri F, Capitani R and Annicchiarico C 2019 Development of a real-time steering system model for driving simulators *Proceedings of the Institution of Mechanical Engineers, Part D: Journal of Automobile Engineering* vol 233 pp 2701–2713

[3] Zhang S, Lei B, Chen S and Sha L 2021 *E3S Web of Conferences* **268** 01037

[4] Kobiela F and Engeln A 2010 *ATZautotechnology* **10** 38–43

[5] Montani M, Capitani R and Annicchiarico C 2019 Development of a brake by wire system design for car stability controls *Procedia Structural Integrity* vol 24 (Elsevier B.V.) pp 137–154

[6] Fildes B, Keall M, Bos N, Lie A, Page Y, Pastor C, Pennisi L, Rizzi M, Thomas P and Tingvall C 2015 *Accident Analysis & Prevention* **81** 24–29

[7] van Zanten A T, Erhardt R and Pfaff G 1995 VDC, The Vehicle Dynamics Control System of Bosch *SAE Technical Papers*

[8] Vinattieri F, Morrocchi L, Capitani R and Annicchiarico C 2017 Sviluppo di un' apparecchiatura sperimentale per l'integrazione del sistema di sterzo in un simulatore di guida statico *AIAS 45° Convegno nazionale* (Associazione Italiana per l'Analisi delle Sollecitazioni) pp 6–9

[9] Alfatti F, Garinei M, Annicchiarico C and Capitani R 2022 Design and Hardware in the Loop Testing of AEB Controllers *SAE Technical Papers* 2022

[10] Alfatti F, Annicchiarico C and Capitani R 2022 Vehicle stability controller HiL validation on static simulator *IOP Conference Series: Materials Science and Engineering* vol 1214 p 012044

[11] Euro NCAP 2021 Test protocol - AEB Car-to-Car systems Tech. Rep. April Euro NCAP

[12] Vinattieri F, Wright T, Capitani R, Annicchiarico C and Danisi G 2016 *SAE Technical Papers*

[13] Gonter M and Seiffert U W 2013 *Integrated Automotive Safety Handbook* (Warrendale, PA: SAE International)

[14] Galko C, Rossi R and Savatier X 2014 Vehicle-Hardware-In-The-Loop system for ADAS prototyping and validation *2014 International Conference on Embedded Computer Systems: Architectures, Modeling, and Simulation (SAMOS XIV)* pp 329–334

[15] De Winter J C F, Van Leeuwen P M and Happee R 2012 Advantages and Disadvantages of Driving Simulators: A Discussion *Proceedings of Measuring Behavior*

[16] Cicchino J B 2017 *Accident Analysis & Prevention* **99** 142–152

[17] Montani M, Capitani R, Fainello M and Annicchiarico C 2020 Use of a driving simulator to develop a brake-by-wire system designed for electric vehicles and car stability controls *10th International Munich Chassis Symposium 2019 pp 663–684* (Springer Fachmedien Wiesbaden) pp 663–684

[18] Winner H, Hakuli S, Lotz F and Singer C 2016 *Handbook of Driver Assistance Systems* (Cham: Springer International Publishing)

[19] European Commission 2018 Advanced driver assistance systems Tech. rep. Directorate General for Transport

Design methodology of a two-degrees-of-freedom gravitational energy harvester

M Lo Monaco[a] **and C Russo**[a]

[a] Politecnico di Torino - Department of Mechanical and Aerospace Engineering, C.so Duca degli Abruzzi, Torino,10129, Italy

E-mail: mirco.lomonaco@polito.it

Abstract. The growing spread of IoT (Internet of Things) and monitoring system based on micro electro-mechanical system (MEMS) concerns also the railway systems. The recent developments have made it possible to study and realize innovative integrated systems using connected devices. The technologies allow to collect real-time data from assets, thus providing fundamental information regarding the operation conditions and offering an evaluation of safety and durability of the monitored device. The major innovation of these devices is the use of energy harvesters to support or replace their power supply, making the monitoring device completely autonomous and drastically lowering the maintenance costs. The work presented in this paper is the product of the research group knowledge in vibrational energy harvesting. In particular, we studied a two-degrees-of-freedom (2DOF) gravitational vibration-based energy harvester (GVEH), characterized by the absence of the magnet on the top end of the tube, exploiting gravity as a restoring force. Different masses configuration were tested in order to find the best configuration optimizing power output, frequency bandwidth and overall performances. The time-domain simulations realized in Matlab/Simulink environment are supported with Multiboduy simulations for a better understanding of the system dynamics.

Keywords: energy harvesting, vibrations, electro-mechanical modeling, Autonomous Internet of Things

1. Introduction

The IoT devices and miniaturized monitoring systems has reached also the railway systems. Through these connected devices it is possible to have real-time information and data regarding the monitored vehicle or component. These new data offer the opportunity to have a new vision regarding the status of the monitored device, providing information now accessible thanks to the rising of these technologies. These devices and data provide the basis for the algorithms used to create Predictive Maintenance and Digital Twins, two tools that are acquiring visibility due to their key-role in the optimization of maintenance and safety procedures. The availability of a continuous stream of data allows to have knowledge on the health of the monitored system that can be crucial for the maintenance, safety, and design field [1],[2],[3]. Wireless monitoring networks are the tools for the continuous stream of data but must comply with some strict characteristics to be effective. These systems should have low power consumption, small size, long life cycles, and reliable wireless connections. It is necessary for the device to be small scaled for safe installations in hazardous or difficult-to-reach locations and to avoid interference with

Content from this work may be used under the terms of the Creative Commons Attribution 3.0 licence. Any further distribution of this work must maintain attribution to the author(s) and the title of the work, journal citation and DOI.
Published under licence by IOP Publishing Ltd

the monitored system normal operations. Furthermore, for the purpose of having a reliable network and an affordable set of data, a strong wireless connection to the server and between the network nodes is essential. Finally, a long lifespan is necessary to lower the price of battery replacement and maintenance, following the low energy requirements. The batteries used in these networks are not sustainable in terms of cost, environmental disposal and safety. The major drawback of these systems, which should be created using the "install and forget" design principle, is the necessity of finding a reliable power supply in replacement of batteries. The absence of onboard power in freight vehicles enhances the design difficulties of these monitoring systems. Finding a solution to this issue is raising research interest in sustainable power supply options.

In literature, there are many research works regarding the design and study of energy harvesting devices in the railway field. The authors started their research with the study of one-degree-of-freedom (1DOF) vibrational energy harvester in the last decade [4]. The continuation of this work has led to recent publications [5],[6] and they are now proposing the design of two-degrees-of-freedom (2DOF) energy harvesters. These devices are studied in literature both with traditional and magnetic springs. The main advantage of 2DOF systems is the wider frequency bandwidth at high power with respect to a 1DOF system. Furthermore, the velocity amplification of the top magnet due to momentum transfer, results in higher output power. The mass ratio of the two moving magnets is a relevant design variable for these systems. Researchers have come to the conclusion that higher DOFs configurations are capable of harvesting energy over a wider frequency range, accepting lower maximum output [7]. These higher DOFs systems may be valid in applications with vibrational energy distributed over a wider frequency band, supplying devices with low power requirements. From the same research group of O'Donoghue, the work of Rodriguez et al. [8] is proposed, presenting a 2DOF velocity-amplified EM-VEH device. The novelty of this work is the possibility of varying the length of the tube, making the energy harvester frequencies tunable with the application. Always using a magnetic spring, in the work of Fan K. et al, [9] they deal with the generation of electricity exploiting ultra-low frequency excitations from human motion. This study introduces a 2DOF electromagnetic energy harvester (EMEH) as a solution to this issue. It is realized by the motion of a magnetically levitated 1DOF EMEH inside a cylindrical enclosure. The benefits of the suggested design, such as the tunable operating frequencies, increased power output, and expanded working bandwidth, are demonstrated both by experiments and modeling. The experimental results show that the proposed harvester achieves a rough 40 % improvement in power magnitude and a 152 % increase in operational bandwidth under a sinusoidal stimulation with an amplitude of 0.5 g. In railway field Chandarin U. et al [10] studies show the design and construction of a 2DOF vibration energy harvester for kinetic energy conversion via electromagnetic transduction. To successfully capture energy at two different frequencies, a coil is suspended as a second oscillating mass. The protoype resonance frequencies are set to coincide with the host structure prominent frequencies (i.e. heavy haul railcars). Peak output powers of 212 mW and 218 mW were produced via sinusoidal vibration at frequencies of 6.5 Hz and 14.5 Hz, respectively, with a peak acceleration of 0.4 g. A work from Feng Z. et al. [11] proposed a non-magnetic spring 2DOF energy harevster, using a serpentine planar spring. Compared with the conventional dual resonance harvester, the proposed system realizes an enhanced "band-pass" harvesting characteristic by increasing the relative displacement between magnet and coil among two resonance frequencies, pursuing a significant improvement in the average harvested power. The device produces an output power of 11 mW at the first resonant frequency of 58 Hz, 14.9 mW at the second resonant frequency of 74.5 Hz, and 0.52 mW at 65 Hz, which is in the middle of the two resonance frequencies. The frequency range of output power above 0.5 mW is from 55.8 Hz to 79.1 Hz.

This paper describes the design phases of a 2DOF vibrational energy harvester. Firstly, the structure of the energy harvester is described justifying the choice of the two masses. Secondly, the analytical and numerical models in Matlab/Simulink environment are proposed with the support of the Ansys Maxwell FEM and Simpack Multibody simulations. The last section reports the model results and the conclusions.

2. Materials and methods

The devices analyzed in this work are cylindrical two-degrees-of-freedom gravitational vibrational energy harvesters (2DOF GVEHs). The purpose of these systems is to supply wireless sensor nodes along with a back-up rechargeable battery for condition-based monitoring of un-electrified freight wagons. The excitation frequencies on the coach coming from wheel-railroad contact are usually low due to suspensions filtering [4], depending on the type of boogie, train speed, railroad conditions and carried weight. GVEHs are well-suited for this application since the use of magnetic suspension results in a non-linear spring having low resonance frequency (1-50 Hz, [12]). Moreover, magnetic suspensions are low-cost and simple designs that can provide stable and long-lasting performances [13]. The main challenges of designing GVEHs for vehicle applications are finding the right trade-off between small device dimensions and low resonance frequency and tuning the resonant frequency with the fundamental frequency of the application. As mentioned before, for rail-freight cars it is necessary to design an energy harvester with a large bandwidth frequency in order to obtain power both in the tare and load condition.

2.1. Harvester configuration

The main components of a 2DOF GVEH include two suspended moving magnets, a fixed magnet at the bottom and coils wrapped around a tube. The moving magnets generate variable magnetic fluxes, inducing an alternating current in the coils, and return to their equilibrium positions under gravitational force. Maximum power is generated when the external excitation is in resonance with the masses of harvester. Consequently, the dimensioning of these generators is strictly dependent of their applications. Using a non-linear stiffness broadens the frequency range around the resonance frequency, thus increasing the system efficiency. The purpose of introducing a second degree of freedom with an additional non-linear stiffness is to further increase the high-power frequency range and overall efficiency. The 2DOF GVEH has a configuration that resembles a dynamic vibration absorber. Figure 1 reports a schematic representation of the system, along with further details on the components and their main dimensions. A radial clearance between the moving magnets and internal tube diameter is necessary to allow the longitudinal oscillation. However, undesired transversal motions arise that can be reduced by adding and internal guide for the moving magnets, preventing the performances decrease. The device has several coils made of enameled copper wires with same dimensions, number of turns, wire diameter and resistance. They are placed along the tube at different heights to find the axial location of the coil that optimizes the generated power. This phenomenon is linked to the variation of the magnetic flux linkage of the coils during the motion of the suspended mass. Number of turns of each configuration is optimized through load tests on the dynamic workbench. The two moving magnets oscillate around each equilibrium position which are far enough to decouple the electro-mechanical interaction of each mass. m_1 oscillates around the lower part of the tube, inducing a voltage in the lower coils C.1 to C.4, whereas m_2 oscillates around the upper part of the tube and induces a voltage in the upper coils C.7 to C.10 (see Figure 1). Consequently, the optimum coil for each moving magnet can be evaluated considering the decoupled induction effects. These two optimum coils can then be connected in series to couple the induced voltages and maximize the harvester performances. Table 2 reports the optimized electrical variables of the system. All the magnets in the device are made of NdFeB-N42.

IOP Conf. Series: Materials Science and Engineering 1275 (2023) 012042 doi:10.1088/1757-899X/1275/1/012042

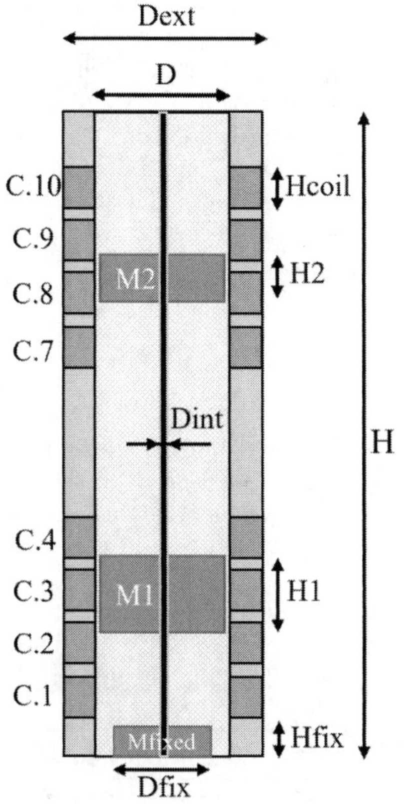

D (mm)	20
Dfix (mm)	10
H1 (mm)	20
H2 (mm)	10
Hfix (mm)	5
Dext (mm)	22
Dint (mm)	4
H (mm)	200
Hcoil (mm)	10
m1 (g)	45
m2 (g)	22.5
Dwire (mm)	0.1
Nturns	500
Rcoil (Ω)	90

Figure 1: Schematic representation of the 2DOF GVEH.

2.2. Linear system analysis

The correct tuning of the two stiffness is crucial in defining the performances of the system. The stiffness of a magnetic suspension depends on the magnetic moments of the two opposing magnets. Consequently, one of the main design variables is the mass ratio of the two moving magnets. A first analysis on a linear 2DOF GVEH model needs to be carried out in order to evaluate the relation between output power, frequency band and the mass ratio. Starting from a 1DOF configuration having optimum design variables (magnets dimensions, tube length, coil position and number of turns, resistive load), different top masses are tested and the corresponding Frequency Response Functions (FRFs) in terms of displacement amplitude and output power are computed for comparison. Figure 2 shows the results of this analysis for mass ratios m_2/m_1 of 0.25, 0.5, 0.75, 1. FRFs are computed solving the linearized version of equations (2)-(3) described in Section 2.3. Changing the mass ratios and the magnetic moments of the moving magnets effects not only the stiffness, but also damping and electromagnetic coupling coefficient values of both DOFs (see Sections 2.3.1-2.3.2). As a result, in the FRFs the resonance frequencies and peak amplitudes change with the mass ratios. As the mass of m_2 increases, also the stiffness k_1 increases even though m_1 mass has a fixed value, shifting the resonance frequenct of DOF 1 to higher values. Stiffness k_2 increases with the mass of m_2 proportionally, leading to an almost constant resonance frequency for DOF 2. As for a dynamic absorber, with increasing value of m_2 mass the displacement amplitude of m_1 decreases. The displacement amplitude of DOF 2 decreases similarly to DOF 1 because higher values of m_2 mass lead to higher electromagnetic damping values. As mass ratios tend to 1, the electromagnetic coupling coefficient of m_2 grows, leading to an higher output power peak of DOF 2. However, the undesired effect of m_2 is to damp the oscillation of m_1, reducing the output power for resonance of DOF 1. Moreover, with increasing mass ratios, the distance between the two resonance peaks grows far from the frequency value of the 1DOF configuration, resulting in the detuning of the device. The mass ratio of 0.5 shows the best trade-off between overall power enhancement and narrow frequency range between the two resonance peaks. This configuration is used for the complete non-linear analysis described in the following sections. The two eigenfrequencies are evaluated solving the following matrix equation (1) of the linearized free undamped analytical model described in Section 2.2. The linearized stiffness values are evaluated in the equilibrium positions of the two magnets as described in Section 2.3.1.

$$\begin{bmatrix} m_1 & 0 \\ 0 & m_2 \end{bmatrix} \begin{pmatrix} \ddot{x_1} \\ \ddot{x_2} \end{pmatrix} + \begin{bmatrix} k_1 + k_2 & -k_2 \\ -k_1 & k_2 \end{bmatrix} \begin{pmatrix} x_1 \\ x_2 \end{pmatrix} = \begin{pmatrix} 0 \\ 0 \end{pmatrix} \tag{1}$$

Solving the determinant of the equation in the frequency domain allows to evaluate the two eigenfrequencies. Magnetic springs have stiffness values that depend on the magnetic moments and consequently the mass of the moving magnets. Consequently, changing the masses of the moving magnets also changes the linear stiffness values. The eigenfrequencies dependence of both of this variables explains their unusual values during the mass ratio analysis. Table 1 summarizes the eigenfrequencies of the system with the tested mass ratios.

Table 1: Eigenfrequencies of the system with different mass ratios.

Mass ratio m_2/m_1	Eigenfrequency DOF 2 (Hz)	Eigenfrequency DOF 1 (Hz)
0.25	3.30	6.14
0.5	3.33	7.14
0.75	3.28	8.36
1	3.35	9.02

2.3. Analytical model

The analytical model of the generator consists in a 2DOF non-linear mass-spring-damper system. The seismic mass m_1 of the system is a moving magnet suspended thanks to the repulsive force with the fixed magnet at the bottom. The seismic mass m_2 is suspended using the repulsive force with m_1. The stiffness k_1 is related to the first magnetic suspension connecting the fixed magnet and m_1, while k_2 is related to the second one connecting m_1 and m_2, thus depending on their relative motion. The magnetic force between m_2 and the fixed magnet is neglected considering their distance. The damping characteristics derive from two components: mechanical due to viscous action of the air and friction (c_{vis}), electromagnetic due to the induction of parasitic currents in the coils (c_{em}). The mechanical damping (c_{vis2}) is placed between the two masses as its effect depends on their relative motion. Electromagnetic damping (c_{em}) effects depend on the motion of the masses respect to the coils, therefore they are placed between each mass and the fixed base. An external sinusoidal input is applied to the system, generating oscillation motions of the moving magnets. The transducing mechanism consists in the relative motion of the masses inside the tube respect to the coils, creating a time variation of the magnetic flux linkage of the coils and consequently an induced electromotive force. A representation of the system analytical mechanical model is shown in Figure 3a. In order to evaluate the transducing mechanism and the generated power, the electromagnetic subsystem needs to be studied and to be coupled with the mechanical one.

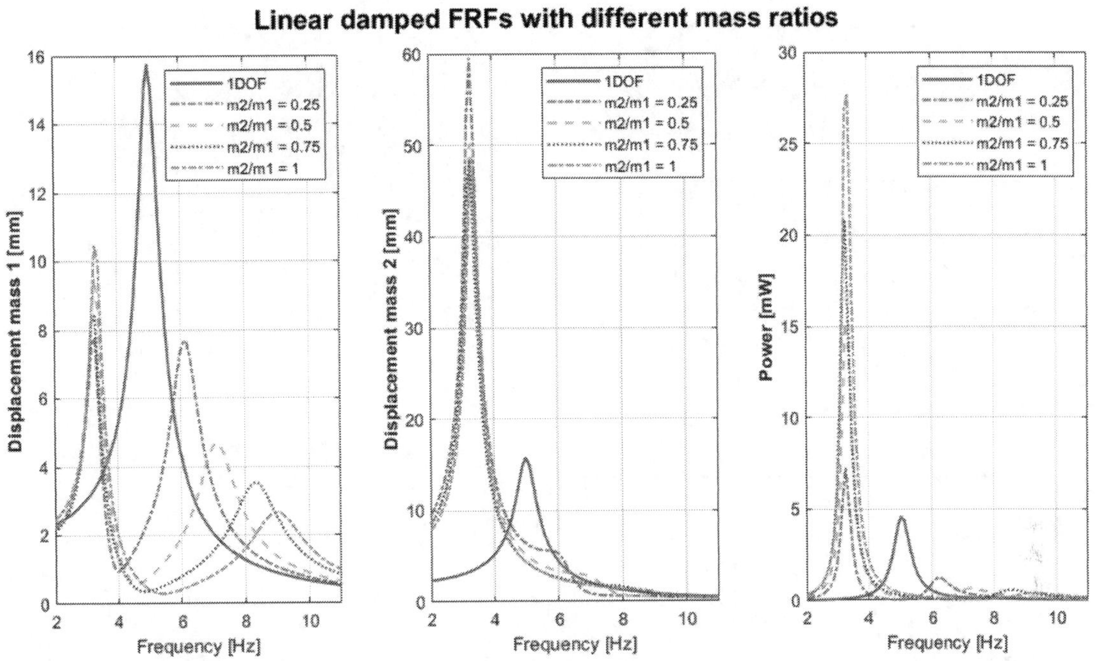

Figure 2: Comparison between FRFs of 1DOF and 2DOF linear damped systems with different mass ratios.

The main electromagnetic characteristics of the harvester are the electromagnetic damping and coupling coefficient. These non-linear physical quantities depend on the time variation of the magnetic flux linkage across the coils. The induced electromotive force and the generated power are directly dependent on the moving magnets speed. The electrical circuit of the system can be described as shown in the Figure 9. The harvester can be considered as an AC generator with its internal resistor R_{coil}, connected to the external resistive load R_{load} that needs to be properly designed through load tests to maximize power. A schematic representation of the electromagnetic circuit of the harvester is reported in Figure 3b. The equation of motion of each degree of freedom (2)-(3) is a time-dependent second order differential equation.

$$m_1\ddot{x_1} + c_{em1}(\dot{x_1} - \dot{y}) + c_{vis1}(\dot{x_1} - \dot{y}) - c_{vis2}(\dot{x_2} - \dot{x_1}) + k_1(x_1 - y) - k_2(x_2 - x_1) = 0 \quad (2)$$

$$m_2\ddot{x_2} + c_{em2}(\dot{x_2} - \dot{y}) + c_{vis2}(\dot{x_2} - \dot{x_1}) + k_2(x_2 - x_1) = 0 \quad (3)$$

Where:

- m_1, m_2 are the seismic mass of the moving magnets
- c_{vis1}, c_{vis2} are the viscous damping of DOF 1 and 2
- c_{em1}, c_{em2} are the electromagnetic damping of DOF 1 and 2
- k_1, k_2 are the non-linear stiffness of the two magnetic springs
- y is the external sinusoidal movement applied to the base, having acceleration amplitude Y_0 and frequency ω

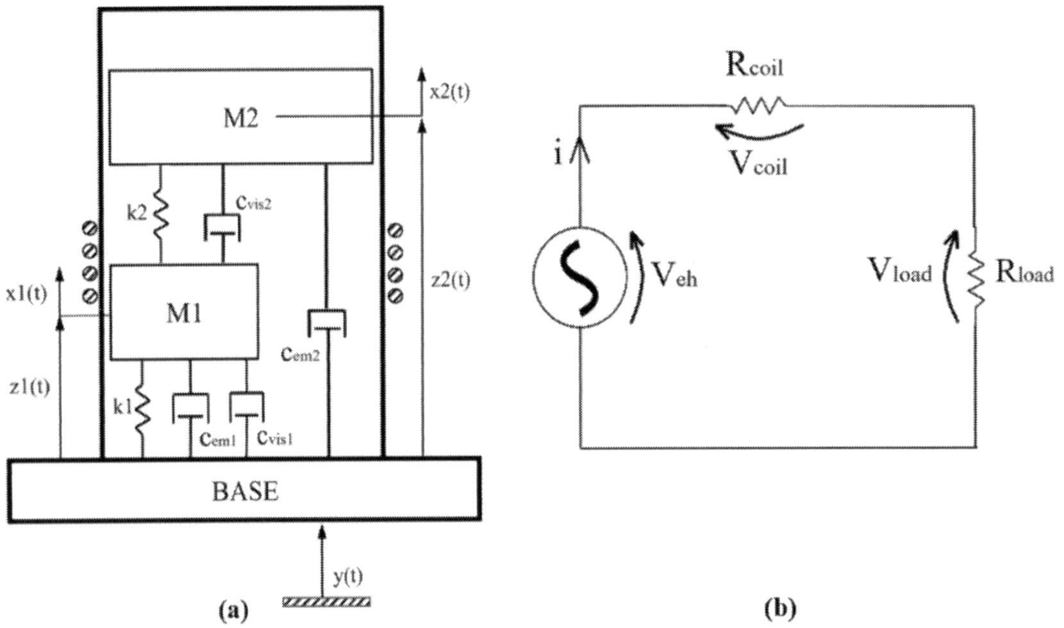

Figure 3: Analytical model of the mechanical (a) and electric (b) subsystem.

Equations (4)-(5) are derived through the implementation of Kirchoff's laws to the circuit of the electromagnetic subsystem.

$$i = \frac{V_{eh}}{R_{coil} + R_{load}} \tag{4}$$

$$P_{load} = V_{load} i \tag{5}$$

The electromagnetic coupling coefficient $k_{em}(z)$ derives from the variation of the magnetic flux during the moving magnet oscillation (6). The coefficient allows to link the voltage to the moving magnet speed (7) and define the electromagnetic damping force acting on the seismic mass (8)-(9).

$$k_{em}(z) = -\frac{d\Phi}{dz} \tag{6}$$

$$V_{eh} = -\frac{d\Phi}{dt} = -\frac{d\Phi}{dz}\frac{dz}{dt} = k_{em}\dot{z}(t) \tag{7}$$

$$F_{em} = k_{em}(z) * i(t) = k_{em}^2(z) * \frac{\dot{z}(t)}{R_{coil} + R_{load}} = c_{em}(z) * \dot{z}(t) \tag{8}$$

$$c_{em}(z) = \frac{k_{em}^2(z)}{R_{coil} + R_{load}} \tag{9}$$

Equations (1)-(8) are implemented in a MATLAB/Simulink model for dynamic simulations of the system to obtain, as solutions, the moving magnets position, speed, acceleration and the induced voltage, current and generated power on the electric load for each time step. The Simulink block scheme is shown in Figure 4. Simulink code is divided into three subsystems, two solving the equations of motion of each degree of freedom and one solving the electromagnetic subsystem. The electromechanical coupling of the three subsystems allows to evaluate the output electric variables by applying the external mechanical sinusoidal input. Equation (7) describes the computation of the induced electromotive force of each magnet, following the decoupling consideration in Section 2.1. Equation of the two masses are expressed in the variable z, which is the motion coordinate of one of the moving magnet relative to the base motion y with origin in each equilibrium position. The stiffness and damping characteristics need to be defined and inserted in the Simulink model as 1D Look-up Table blocks. Ansys Maxwell software for FEM analyses is essential for this study as it allows to numerically evaluate the characteristics mentioned above. Numerical model of the harvester is built in Ansys Maxwell environment for magnetostatic and transient electromagnetic simulations. Given the axisimmetry of the system a 2D model is sufficient.

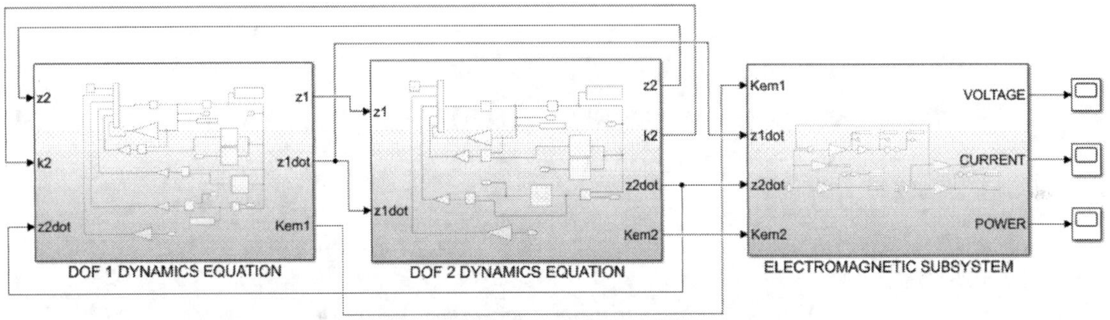

Figure 4: Simulink model.

2.3.1. Mechanical model

The main mechanical characteristics of the harvester are the stiffness of the magnetic springs $k1$, $k2$ and the viscous dampings c_{vis1}, c_{vis2}. Magnetostatic analyses in Ansys Maxwell need to be performed for the definition of the stiffness characteristics. The moving magnets' positions are swept with a fine step along a certain stroke to visualize the distribution of magnetic field vectors, as can be seen in Figure 5.

For every position of the moving magnets, Ansys Maxwell computes the corresponding coenergy $W(x)$ of the system. Coenergy is a physical quantity that is necessary for the evaluation of the harvester stiffnesess. The derivation of the coenergy respect to the suspended mass position results in the magnetic repulsive force acting between two magnets [14] that needs to be interpolated by the best fitting function.

$$F_{mag}(x) = \frac{dW(x)}{dx} \tag{10}$$

As stated in [15], the magnetic force acting between two magnetic dipoles having same magnetic moment and their axes aligned is proportional to $1/x^4$. Even though the three magnets have different magnetic moments, the proposed function fits the simulation data quite precisely. The expression of the function is the following:

$$F_{mag,fit}(x) = \frac{p_1}{x^4 + q_1x^3 + q_2x^2 + q_3x + q_4} \tag{11}$$

The coefficients are determined in MATLAB environment. The stiffness characteristic can be easily obtained by deriving again the magnetic force fitting function respect to x.

$$k(x) = \frac{dF_{mag,fit}(x)}{dx} \tag{12}$$

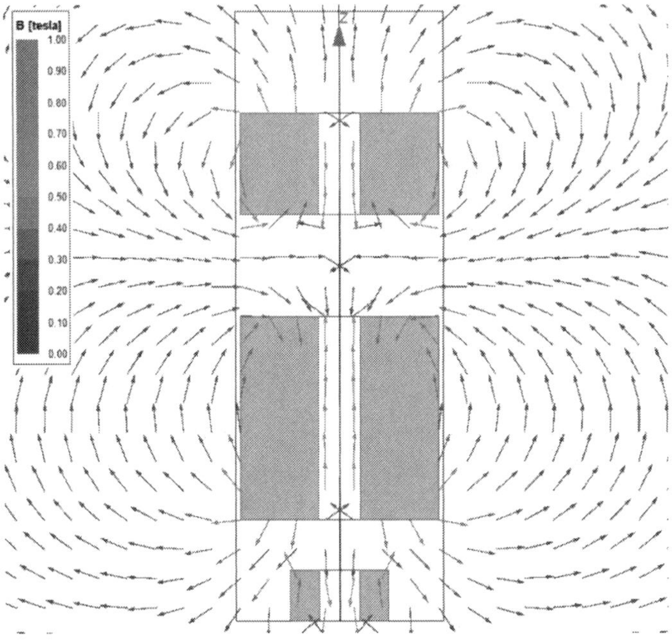

Figure 5: Magnetic field vector of the model.

The two stiffnesses of the magnetic springs are computed one at a time using different approaches. Firstly, stiffness k_2 is evaluated by sweeping m_2 relatively to m_1, starting from its upper face when the two magnets ideally touch and ending to the upper face of the tube. m_1 is positioned at a sufficient distance from the fixed magnet at the base, following the assumption of negligible interaction between m_2 and the fixed magnet. The result is a stiffness characteristic that depends on the relative position between the two moving magnets. Secondly, stiffness k_1 is evaluated by sweeping m_1 from the upper face of the fixed magnet at the base, keeping m_2 fixed in its equilibrium position respect to m_1. Figure 6 shows the plot of the stiffness curves evaluated for m_1 and m_2 in terms of absolute coordinate of the suspended masses x_1 and x_2. In all the following plots of the paper, the displacement value at $x_1 = 0mm$ is ideally considered where $m1$ and fixed magnet are touching, which is also the lower limit of the moving magnet stroke. Whereas, the displacement value at $x_2 = 0mm$ is ideally considered where m_1 and m_2 are touching. The equilibrium positions are evaluated solving the static equations of the two DOFs to obtain the linearized stiffness in the equilibrium positions used in the eigenfrequencies computation in Section 2.2.

The second mechanical characteristic is viscous damping due to air action and surface friction of the moving magnets. It can be considered as a constant and evaluated through experimental tests on the device. Harvester receives a step input and the harvester time response of output voltage signal is processed to extract the damping value. The chosen viscous damping factor equal to 0.04 is the one evaluated by experimental tests on the 1DOF optimized device.

2.3.2. Electromagnetic model

The electromagnetic variables of the circuit introduced in equations (4)-(9) are crucial in determining the output power. Increasing number of turns results in higher coupling coefficient but also greater damping, thus modifying the moving magnet speed and the variation of flux linkage in time.

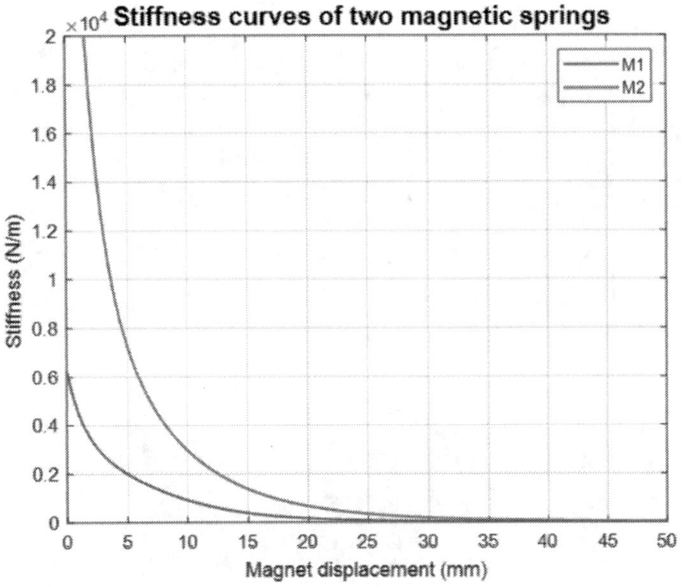

Figure 6: Stiffness curves of the two magnetic springs.

Coils dimensions and location along the tube need to be properly designed in order to extract the maximum amount of power from the system. The coil location along the tube affects the flux linkage variation during oscillation and the dimensions affect the resistance. moreover, optimum value of load resistance needs to be evaluated through load tests for maximum power generation. Transient simulations in Ansys Maxwell enable to evaluate the magnetic flux linkage across the harvester windings. The software allows to plot the magnetic flux lines distribution during the motion of the suspended masses. The plot is reported in Figure 7 with the two moving magnets in the equilibrium position. The plot is useful to visualize where the flux linkage is stronger and how fast it decreases moving away radially from the suspended magnets. As can be noticed, the magnetic flux is maximum around the mass center of the moving magnet. The tube width is a parameter that needs to be designed performing a trade-off between structural rigidity and amount of magnetic flux linkage. The output of the transient analyses is a curve that links the magnetic flux to the absolute displacement of the moving magnet for each coil. Considering that all the coils in each configuration are the same, the curves are identical but shifted along the x-axis due to the different axial locations of the coils. Afterwards, the electromagnetic coupling coefficient and damping curves can be easily determined following equations (5) and (8). These are reported in Figures 8 for two coils and for each moving magnet.

3. Results

The mechanical and electromagnetic characteristics of the harvesters obtained through Ansys Maxwell numerical simulations are used in MATLAB/Simulink environment to perform dynamic simulations of the systems. The most relevant results are the Frequency Response Functions (FRFs) curves that show the resonance peaks and the non-linear behavior of the system. FRFs are computed in terms of load power root mean square values. The sinusoidal excitation input is sweeped in terms of frequency (2-10 Hz) and amplitude (0.2 g-0.5 g). Results are computed using the optimum resistive load $R_{load} = 150\ \Omega$ and connecting one coil at a time in order to evaluate the two optimum coils, one for each moving magnet, that maximize output power.

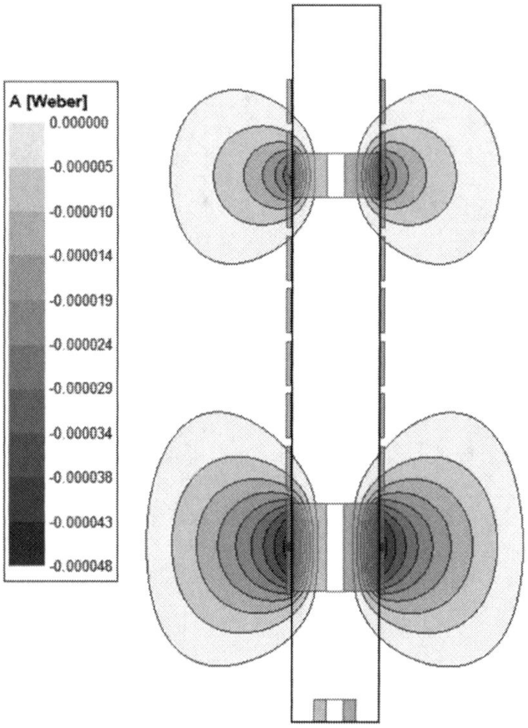

Figure 7: Magnetic flux lines of the model.

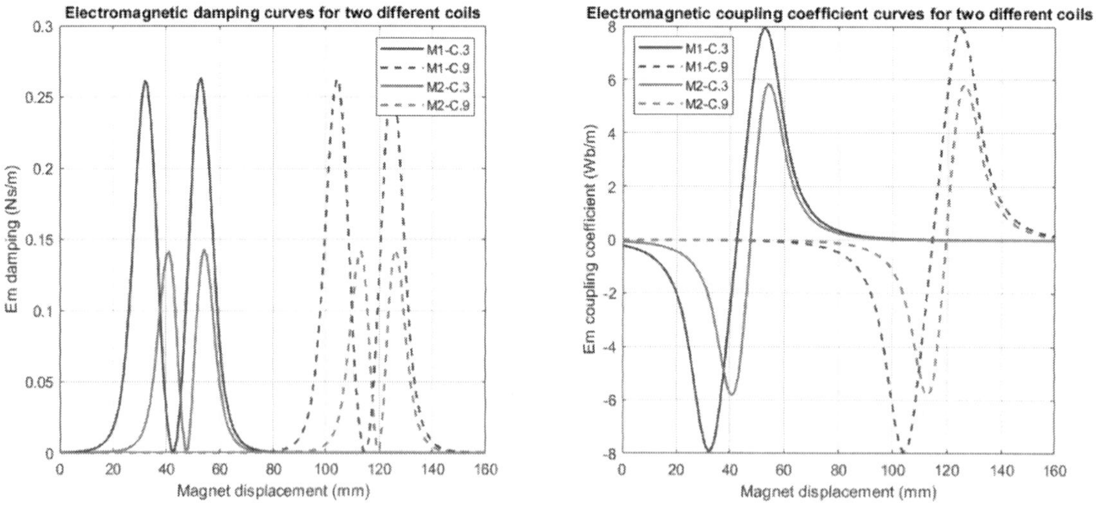

Figure 8: Electromagnetic damping and coupling coefficient curves of the system.

Table 2 shows the power resonance peaks for external excitation amplitude of 0.5 g and connecting one coil at a time. C.3 and C.9 are the optimum coils that maximize the power respectively for m_1 and m_2. Figure 9a-9b show the complete computation of the FRFs for C.3 and C.9 with excitation amplitudes from 0.2 g to 0.5 g in the frequency range of 2-10 Hz. FRF for C.3 has the highest power peaks but has low efficiency in the frequency range of 4-6 Hz.

Table 2: Power resonance peaks for different coils and external excitation amplitude of 0.5 g.

Coil number	Resonance peak 1 (mW)	Resonance peak 2 (mW)
C.1	9.42	4.79
C.2	11.67	6.74
C.3	16.28	6.87
C.4	15.57	6.30
C.7	5.52	2.00
C.8	10.26	3.34
C.9	11.9	3.68
C.10	11.58	3.58

FRF for C.9 has still high power peak and shows a wider frequency band, up to 6 Hz, where the output generation is considerable. Following the consideration in Section 2.1 of decoupled electromotive force induction of the two magnets, the series connection of C.3 and C.9 should give as output an intermediate solution that combines the benefits of the single coils. Figure 10b reports the complete FRFs for the series connection of C.3 and C.9, while Figure 10a shows the comparison between FRFs of the series connection and the single coils for 0.4 g excitation amplitude. As can be clearly noticed, the series connection shows the best results in terms of power generation and frequency broadband, thus maximizing the efficiency of the device.

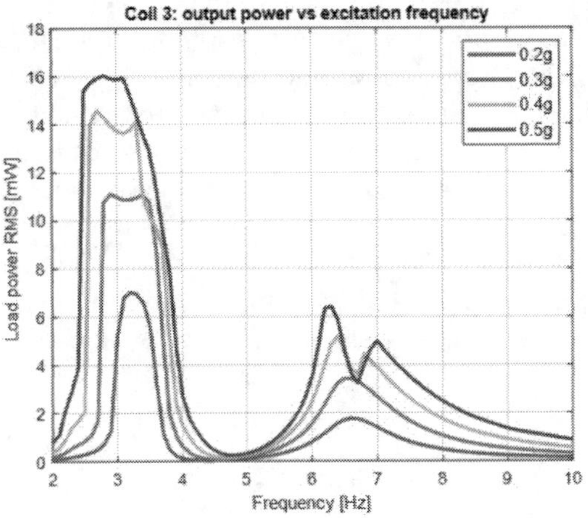

(a)

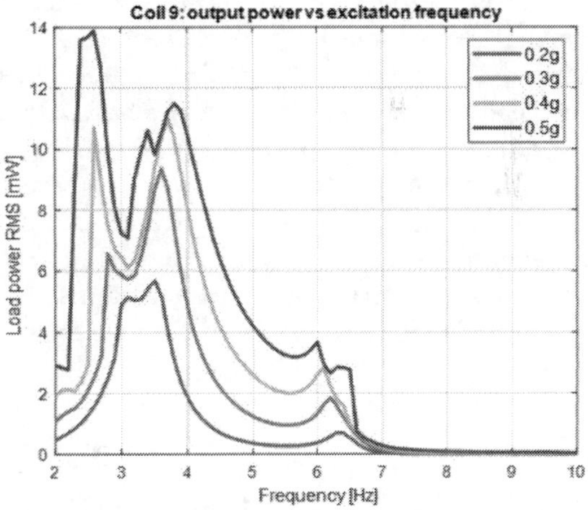

(b)

Figure 9: FRFs for C.3 (a) and C.9 (b).

4. Discussions

FRFs plots convey crucial information about generated power intensity and resonance frequencies. As can be seen in Figures 9-10, the resonance peak of DOF 1 around 6 Hz is lower than the peak of DOF 2 around 3 Hz. This behavior is connected to the damping effect that m_2 has on the vibration amplitude of m_1 due to the dynamic absorber configuration. Equation (6) shows that the induced voltage is the result of the product between the electromagnetic coupling coefficient and the magnet speed.

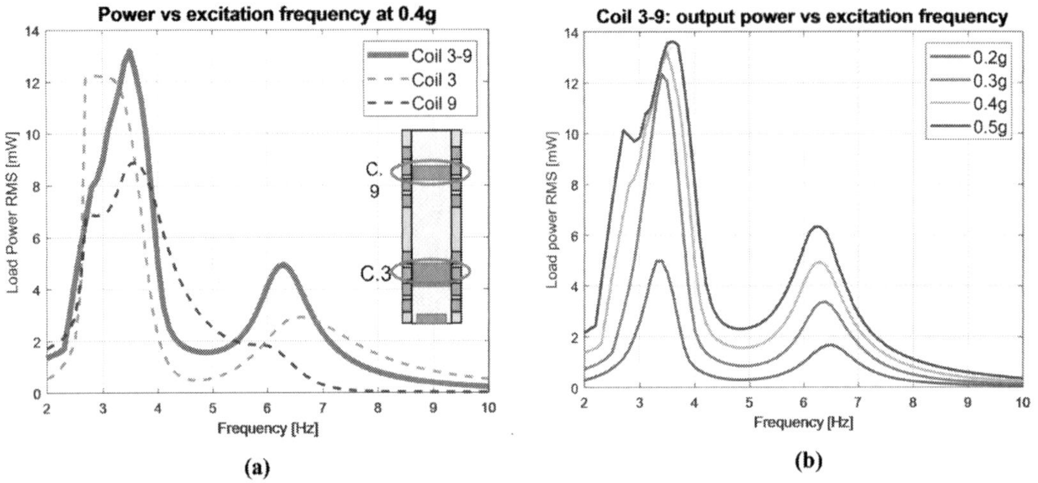

Figure 10: Comparison between FRFs with series connection and single coils (a) FRFs for series connection C.3 and C.9 (b).

Even though m_1 has a bigger magnetic moment and consequently higher coupling coefficient than m_2, the damping effect on the speed prevails. In order to have a deeper insight and better visualization of the system dynamics, a Simpack model is built for Multibody simulations and compared to the Simulink results. Resonance peak of DOF 1 has an eigenmode of vibration with anti phase motion of the two masses, while eigenmode of DOF 2 is in phase. As can be noticed in Figures 11-12, antiphase condition has lower vibration amplitudes than the in phase one, leading to the lower power peak of DOF 2. Moreover, FRFs can be analyzed to understand and quantify the benefits of a 2DOF system compared to a 1DOF one. The area under the curve for RMS power FRFs can be used as a term of comparison between the two configurations, as it considers both the peak amplitudes and the frequency band. Figure 13 reports this analysis, considering the RMS power of 2 mW as the low threshold of useful output power quantity. 2DOF system carries an increase in terms of power area of 24% and in terms of useful frequency band of 13% respect to 1DOF configuration, thus validating the efficiency benefits of a 2DOF configuration.

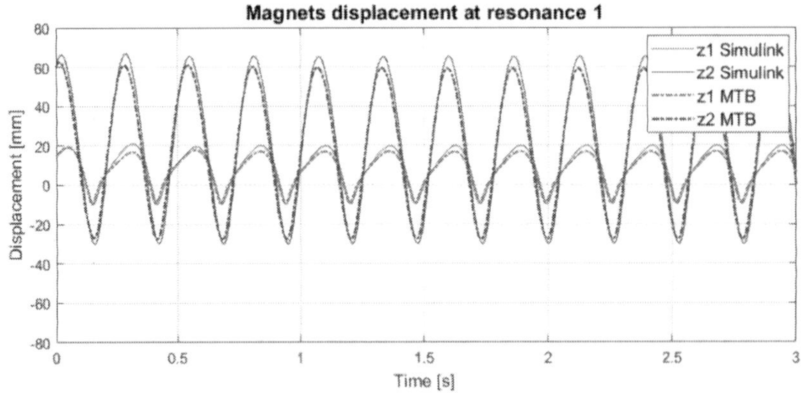

Figure 11: Multibody and Simulink simulations of in phase magnets displacement at resonance of DOF 1.

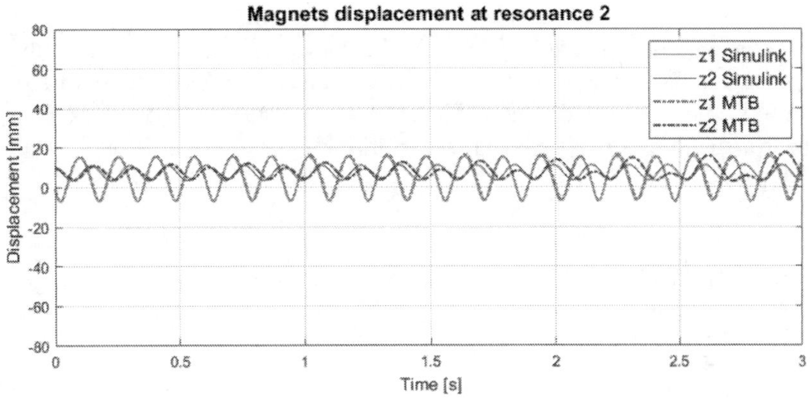

Figure 12: Multibody and Simulink simulations of antiphase magnets displacement at resonance of DOF 2.

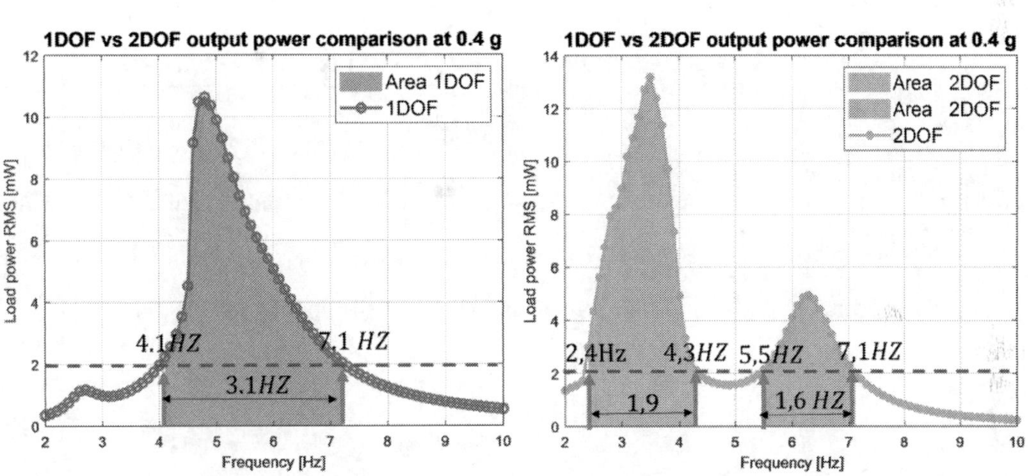

Figure 13: Area under the curve comparison of 1DOF-2DOF areas under the curve of RMS power FRFs.

$$Area\ efficiency\ index\ (\%) = \frac{Area_{2DOF} - Area_{1DOF}}{Area_{1DOF}} * 100 \tag{13}$$

$$Bandwidth\ efficiency\ index\ (\%) = \frac{Bandwidth_{2DOF} - Bandwidth_{1DOF}}{Bandwidth_{1DOF}} * 100 \tag{14}$$

4.1. Conclusions

This work presents the design methodologies of a 2DOF vibrational energy harvester. A preliminary study of this device was carried out to evaluate the power and the frequency bandwidth enhancements compared to a 1DOF device. A model in Simulink/Matlab was realized starting from the coupled mechanical dynamic equations and electromagnetic ones. For the computation of the magnetic suspension stiffness characteristic and the time-variation of the flux linkage, a model on ANSYS Maxwell was developed for FEM simulations. Moreover, Multibody simulation were executed to support the authors in the prototyping phase for further understanding of the system dynamics. The results of these simulations are promising and solid.

An increase of 26 % in terms of the area under the curve of RMS power above a 2 mW threshold was obtained, along with an increase of 16 % in high-power frequency bandwidth. These first results demonstrate how a 2DOF device can be beneficial in terms of efficiency for the supply of IoT devices, compared to a 1DOF system. The next steps of this work will regard the realization of a modular prototype in order to perform the required experimental tests to validate the model proposed in this paper.

References

[1] Dewan A, Ay S U, Karim M N and Beyenal H 2014 Alternative power sources for remote sensors: A review
[2] Singh J, Kaur R and Singh D 2021 Energy harvesting in wireless sensor networks: A taxonomic survey
[3] Majid M, Habib S, Javed A R, Rizwan M, Srivastava G, Gadekallu T R and Lin J C W 2022 *Sensors* **22**(6) ISSN 14248220
[4] Pasquale G D, Somà A and Zampieri N 2012 *Journal of Computational and Nonlinear Dynamics* **7**(4) ISSN 15551423
[5] Russo C, Monaco M L, Fraccarollo F and Somà A 2021 *Energies* **14**(15) 4622 ISSN 1996-1073
[6] Russo C, Monaco M L and Somà A 2022 *IOP Conference Series: Materials Science and Engineering* **1214**(1) 012046 ISSN 1757-8981
[7] O'Donoghue D, Frizzell R, Kelly G, Nolan K and Punch J 2016 *Smart Materials and Structures* **25**(5) ISSN 1361665X
[8] Rodriguez J C, Nico V and Punch J 2019 *Sensors (Switzerland)* **19**(17) ISSN 14248220
[9] Fan K, Zhang Y, Liu H, Cai M and Tan Q 2019 *Renewable Energy* **138** 292–302 ISSN 18790682
[10] Ung C, Moss S D and Chiu W K 2015 vol 9431 (SPIE) p 94312C ISBN 9781628415346 ISSN 1996756X
[11] Feng Z, Peng H and Chen Y 2021 *Energies* **14**(22) ISSN 19961073
[12] Carneiro P, dos Santos M P S, Rodrigues A, Ferreira J A, Simões J A, Marques A T and Kholkin A L 2020 *Applied Energy* **260** 114191 ISSN 03062619
[13] Santos M P S D, Ferreira J A, Simões J A, Pascoal R, Torrão J, Xue X and Furlani E P 2016 *Scientific Reports* **6** ISSN 20452322
[14] Fu W, Zhou P, Lin D, Stanton S and Cendes Z 2004 *IEEE Transactions on Magnetics* **40**(2) 683–686 ISSN 0018-9464
[15] Kraftmakher Y 2007 *European Journal of Physics* **28**(3) 409–414 ISSN 0143-0807

Crack propagation in viscoelastic finite-sized solids: theory and experiments

G Violano*, S De Carolis, M E Palmieri, G Carbone, L Tricarico, G P Demelio and L Afferrante

Department of Mechanics, Mathematics and Management, Polytechnic University of Bari, Via E. Orabona, 4, 70125, Bari, Italy

E-mail: *guido.violano@poliba.it

Abstract. The problem of crack propagation in viscoelastic materials is of great interest given the numerous engineering applications of such materials. Due to viscoelasticity, even the study of the basic Mode I opening represents a tricky theoretical challenge. Indeed, existing theories adopt important approximations such as i) simplistic constitutive behaviour, ii) steady-state crack propagation, iii) infinite domain of the system. In this work, we revise the theory of Persson & Brener for systems of infinite domain; specifically, we propose a solution to take into account size effects in a viscoelastic plate. The theory allows to consider the realistic constitutive behaviour of viscoelastic materials and to predict the dependence of the energy release rate with the crack tip speed. Comprehensive experimental investigations are performed to corroborate our theoretical predictions. First, dynamic mechanical analysis (DMA) is performed to characterize the complex viscoelastic modulus of PolyTetraFluoroEthylene (PTFE). Second, tensile tests are carried out on cracked PTFE samples, and pictures are recorded with an image acquisition system. Moreover, a point tracking algorithm is developed to measure the crack length and opening displacement. Moving from small to high crack tip speeds, the fracture process becomes less ductile and an increase in the maximum load is observed. In addition, experimental data show that the inclusion of finite-size effects in the theory is crucial for accurately estimating the energy release rate.

1. Introduction

Viscoelastic fracture mechanics is a topic of great engineering interest, as viscoelastic solids are largely used in soft robotics [1], tires manufacturing [2], biomechanics [3], adhesives and tapes industries [4]. For this reason, the scientific interest in crack propagation in viscoelastic materials is very topical, as evidenced by numerous recent experimental and theoretical works on the subject (see, for example, Refs. [5, 6, 7, 8, 9]).

Classical Linear Elastic Fracture Mechanics (LEFM) fails in explaining the crack propagation in viscoelastic materials, as they show a time-dependent constitutive behaviour. In fact, the energy release rate G, which is required to advance the fracture plane by one unit area, depends on the crack tip speed [9].

Mode I opening is a common fracture mechanism, where propagation occurs under the action of a tensile stress normal to the plane of the crack. Practical examples of Mode I failures are cracks in radial spokes of non-pneumatic tires [10], peeling of pressure-sensitive adhesives [11], detachment of an indenter from an elastomeric stamp [12].

Content from this work may be used under the terms of the Creative Commons Attribution 3.0 licence. Any further distribution of this work must maintain attribution to the author(s) and the title of the work, journal citation and DOI.
Published under licence by IOP Publishing Ltd

In the literature, different theoretical approaches have been derived for studying the Mode I crack propagation in linear viscoelastic solids, and we refer the reader to recent comparative studies [13, 14]. All existing models can be classified in two main schools. The first one exploits a cohesive-zone approximation to remove the stress singularity at the crack tip [15, 16, 17]. In this case, assumptions are required on the interfacial stress-separation law acting in the cohesive zone. The second school follows the energetic approach introduced qualitatively by de Gennes [18], and formally derived by Persson & Brener (PB) [19]. In PB theory, the energy dissipation involved in the viscoelastic crack propagation is divided in two contributions. The first one is associated with the non-linear process at the crack tip and is described phenomenologically; the second contribution is formally calculated from the viscoelastic dissipation in the material, in the linear viscoelastic region in front of the crack tip.

Independently of the theoretical approach, three main approximations can be found in current models: i) simplistic description of the viscoelastic constitutive behaviour; ii) steady-state propagation of the crack tip, which is supposed to move at constant speed; iii) infinite domain of the viscoelastic solid. Relaxation of such assumptions have been proposed in the last years; for example, PB theory [19] allows to consider a more realistic constitutive behaviour, providing that the frequency-dependent viscoelastic modulus of the material is known. Moreover, recent studies [8, 9] showed a good agreement between PB theory and full numerical simulations despite PB assume steady-state crack propagation.

There is an open debate on which could be the best strategy to include *finite-size effects* in current theories (see, for example, Refs. [13, 20, 21]). Under the assumption of infinite domain, the traction boundary condition is given by the linear elastic K-field for Mode I [22]. This condition is realistic only if the gross of the dissipation is confined at the crack tip [9]. However, real solids present finite dimensions that lead to non-negligible effects on the K-field. In LEFM, size effects are usually included by means of corrective shape factors [22, 23]. A similar approach has been proposed by Violano & Afferrante [21] for viscoelastic solids. In this regard, they recently revised PB theory to consider size-effects in adhesive viscoelastic Hertzian contacts. Interestingly, the solution developed in Ref. [21] could be generalized to different geometries.

In this work, we study the case of a Mode I edge crack in a viscoelastic *finite-sized* plate. Following the approach proposed in Ref. [21], we revise the theory of Persson & Brener [19] to consider size effects due to the initial length of the crack and width of the plate. The proposed model allows to predict the energy release rate in samples with finite dimensions, capturing the rate-dependent nature of the viscoelastic crack propagation.

In addition, we perform comprehensive experimental investigations of crack propagation in PolyTetraFluoroEthylene (PTFE) samples. Moreover, Dynamic mechanical analysis (DMA) is performed to measure the viscoelastic modulus of PTFE. Cracked PTFE samples are subjected to tensile tests by means of a universal testing machine. Different crosshead speeds are considered to capture rate-dependent effects. During tensile tests, pictures are recorded by using an image acquisition system and a point tracking algorithm for pictures postprocessing is developed to follow the evolution of cracks.

2. Theory for crack propagation in a finite-sized viscoelastic solid

Figure 1 shows the problem under investigation: a plate of width L, height H and thickness b, has an unconstrained crack of length a on the left edge. The bottom edge of the plate is fully constrained, while the top edge moves at constant speed V in the direction of positive z, causing a Mode I crack opening.

We assume the plate is made of linear viscoelastic material, which is initially fully relaxed. The viscoelastic modulus $E(\omega)$ of the material is a function of the frequency ω at which the material is excited (see Figure 2); at low frequencies, in the so-called rubbery region, the material

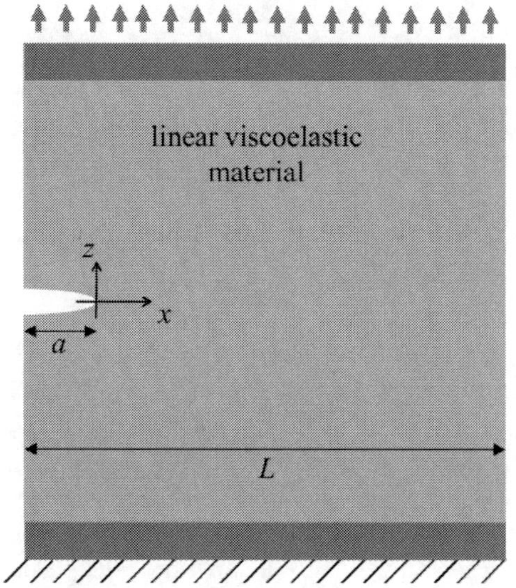

Figure 1. The problem under investigation: Mode I edge crack in a finite-sized viscoelastic plate.

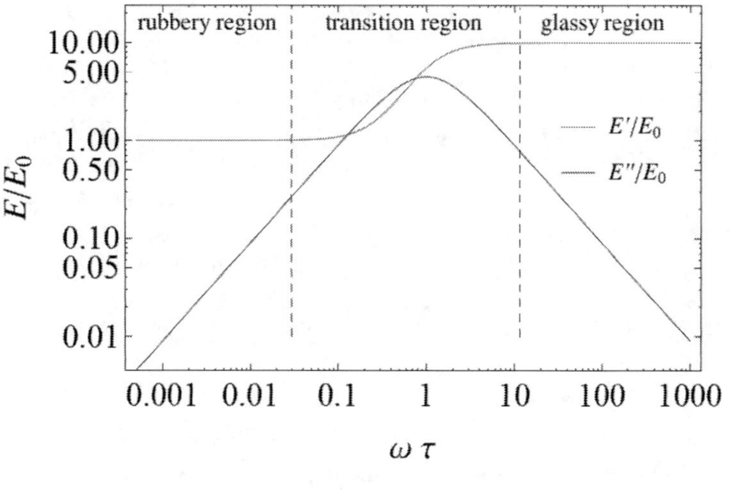

Figure 2. The viscoelastic modulus E/E_0 of a linear standard solid with relaxation time τ. Red and blue lines show the real and imaginary parts of $E(\omega)$, respectively.

behaves as an elastic solid with modulus E_0; at high frequencies, i.e. in the glassy region, the material response is again elastic, with a modulus E_∞ higher than E_0; finally, for intermediate ω (transition region), viscoelastic behaviour is expected and viscous dissipation occurs.

2.1. Theory for infinite solids

In the original theory of Persson & Brener (PB) [19], the energy release rate G that is required to advance the crack tip by one unit area is related to the crack tip radius $s(V_c)$ through

$$\frac{G(V_c)}{G_0} = \frac{s(V_c)}{s_0}, \tag{1}$$

being V_c the crack tip speed and G_0 and s_0 the energy release rate and the crack tip radius at $V_c \sim 0$. G_0 is a material property usually assumed to be constant, although its value can be affected by the temperature. However, we shall assume that the crack propagation occurs at a fixed temperature. The static crack tip radius s_0 is related to the stress σ_c required to break atomic bonds through

$$\sigma_c = \frac{K_I}{(2\pi s_0)^{1/2}} = \left(\frac{E_0 G_0}{2\pi s_0}\right)^{1/2}, \tag{2}$$

being K_I the stress intensity factor for Mode I. The value of s_0 could be estimated from experimental or numerical observations [9], but it is usually adjusted as a fitting parameter [14].

From the energy conservation balance, PB derived a solution to calculate the effective energy release rate G in terms of the crack tip speed

$$G = G_0 \left[1 - \frac{\frac{2E_\infty}{\pi}\int_0^{2\pi V_c/s} \frac{\Phi(\omega)}{\omega} Im\left(\frac{1}{E(\omega)}\right) d\omega}{1 + \frac{2E_\infty}{\pi}\int_0^{2\pi V_c/s} \frac{1}{\omega} Im\left(\frac{1}{E(\omega)}\right) d\omega}\right]^{-1} \tag{3}$$

being $\Phi(\omega) = [1 - (\omega s/(2\pi V_c))^2]^{1/2}$. In eq. (3) the viscoelastic modulus $E(\omega)$ of the material is an input quantity. Moreover, eq. (3) is based on the assumption of infinite domain of the system. In such case, G is a monotonic increasing function of V_c and reaches a plateau at high crack tip speeds, proportional to the ratio between the instantaneous and relaxed moduli

$$\frac{G_{max}}{G_0} = \frac{E_\infty}{E_0}. \tag{4}$$

2.2. *Modified theory for finite-sized solids*

The theoretical limit of eq. (4) is never reached due to the finite size of real solids [20, 21]. To include size effects, Persson [20] suggested to substitute the lower bound of the integrals of eq. (3) with a cut-off frequency $\omega_L \sim 2\pi V_c/L$ related to a characteristic dimension L of the system. However, Violano & Afferrante [21] clarified that, for viscoelastic Hertzian contacts, when dissipation is no longer confined at the contact edge but involves bulk material, Persson solution [20] is not appropriate. Therefore, an alternative approach was proposed in Ref. [21] to correct eq. (3).

Eq. (3) is directly obtained from the energy balance written around the crack tip (see Ref. [19])

$$GV_c = G_0 V_c + W \tag{5}$$

where GV_c is the flow of elastic energy into the crack, $G_0 V_c$ is the flow of fracture energy and W is the viscoelastic dissipation term. One can speculate that W can be calculated for the general case of a crack propagating with velocity V_c as

$$W = \int d^2x \dot{\varepsilon}_{ij} \sigma_{ij}, \tag{6}$$

being $\dot{\varepsilon}_{ij}$ the strain rate components and σ_{ij} the stress components. The stress in the vicinity of the crack tip is of the form

$$\sigma = K_I / (2\pi r)^{1/2}. \tag{7}$$

For a plate of width L and with an unconstrained crack of length a on the edge, (7) can be corrected by

$$\sigma = K_I / \left[(2\pi r)^{1/2} f(\beta)\right] \tag{8}$$

where $f(\beta)$ is a shape factor given by $f(\beta) = 1.12 - 0.231\beta - 10.55\beta^2 - 21.71\beta^3 + 30.382\beta^4$ and $\beta = a/L$ (see Ref. [23]). If we use (8) in (6) to calculate the dissipation energy W, we can show that eq. (3) can be rewritten in the form

IOP Conf. Series: Materials Science and Engineering 1275 (2023) 012043 doi:10.1088/1757-899X/1275/1/012043

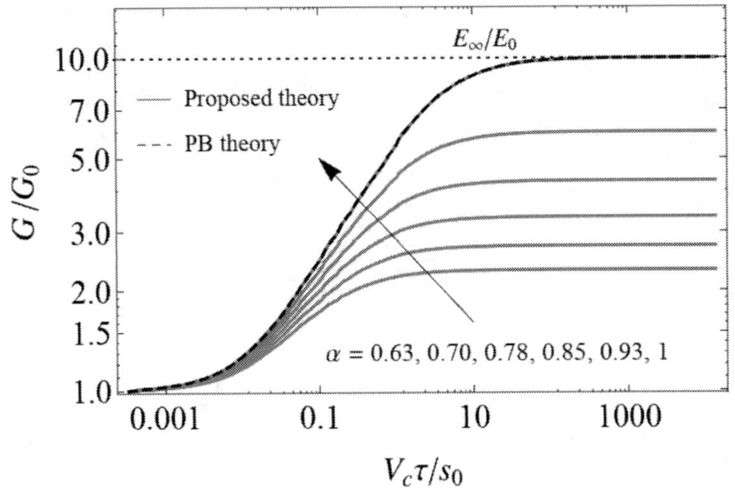

Figure 3. The energy release rate G/G_0 as a function of the adimensional crack tip speed $V_c\tau/s_0$. Red lines refer to the proposed eq. 9, while black dashed line refers to PB solution for infinite domain [19]. Calculations are performed for a standard linear solid with $E_\infty/E_0 = 10$ and relaxation time $\tau = 10^{-4}$ s. The static crack tip radius is $s_0 = 1.5$ nm.

$$G = G_0 \left[1 - \alpha \frac{\frac{2E_\infty}{\pi} \int_0^{2\pi V_c/s} \frac{\Phi(\omega)}{\omega} Im\left(\frac{1}{E(\omega)}\right) d\omega}{1 + \frac{2E_\infty}{\pi} \int_0^{2\pi V_c/s} \frac{1}{\omega} Im\left(\frac{1}{E(\omega)}\right) d\omega} \right]^{-1}, \tag{9}$$

being

$$\alpha = \left[\frac{f(\beta)}{1.12}\right]^{-2}, \tag{10}$$

where the constant 1.12 is introduced to recover the result of the original eq. (3) in the case of vanishing a/L.

Figure 2 shows a comparison between the predictions of eq. (9) and eq. (3). Calculations are performed for a linear standard solid, whose relaxation modulus is given by $E(t) = E_0 + (E_\infty - E_0)\exp(-t/\tau)$, being τ the relaxation time of the viscoelastic material. Specifically, the normalized energy release rate G/G_0 is shown in terms of the adimensional crack tip speed $V_c\tau/s_0$. Results of the proposed theory are shown in red solid lines. The original PB curve (black dashed line) is recovered only for $\alpha \to 1$, i.e., for infinite domain (vanishing a/L). In addition, we notice that finite size effects entail a significant deviation from the theoretical limit E_∞/E_0 given in eq. (4).

3. Experimental investigations

An experimental campaign has been carried out to validate the proposed model and eventually define its range of application.

3.1. Material characterization with Dynamic Mechanical Analysis

The material used in this study is the commercial PolyTetraFluoroEthylene (PTFE), a high-molecular-weight polymer characterized by viscoelastic properties. The viscoelastic frequency response of PTFE has been characterized by means of a *MCR 702 MultiDrive Dynamic Mechanical Analyzer* (DMA) manufactured by *Anton Paar*. The test, carried out in tensile conditions, has been performed on rubber strips, having a cross-section of 1.1 mm × 3.1 mm and a length of 24 mm, over a frequency interval between 0.1 and 10 Hz, and temperature varying from −50 °C to 120 °C with a 5 °C step. About the data post-processing, aimed at building up the material master curve, we calculated the shifting factors as functions of the

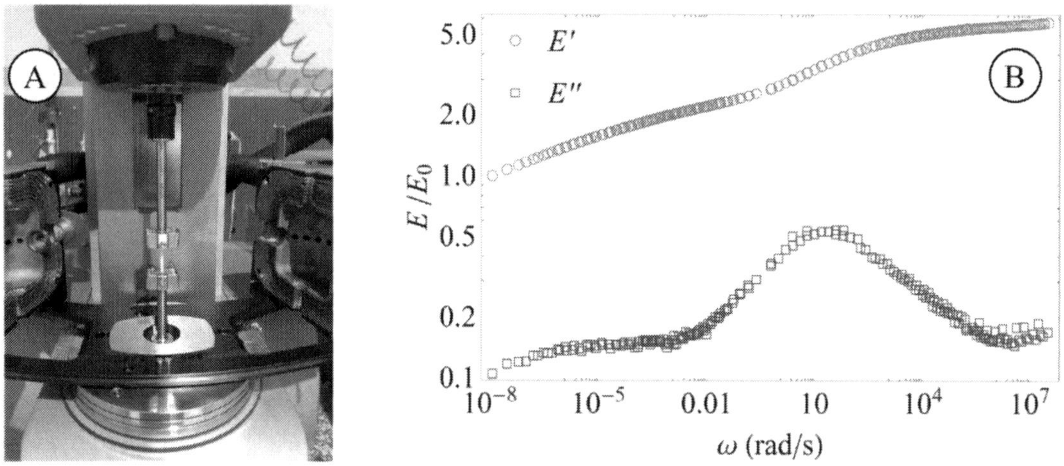

Figure 4. A) PTFE sample in the multidrive dynamic mechanical analyzer MCR 702. B) The master curve of PTFE. Experimental data are given for $T = 30$ °C. Blue squares and red circles refer to the imaginary and the real part of the viscoelastic modulus, respectively.

temperature T, by exploiting the frequency-temperature superposition effect [24]. In Figure 4, PTFE's master curve is shown for a temperature of 30 °C, over a frequency range extended for more than 15 orders of magnitude. Blue squares and red circles refer to the imaginary and real parts of the viscoelastic modulus, respectively. The three characteristic zones, typical of a viscoelastic material, can be identified with the transition region covering a very large range of frequencies.

3.2. Tensile tests and image acquisition

Tensile tests have been carried out by means of the universal tensile machine *INSTRON 4480* (Figure 5A), on single-edge cracked samples with the following geometry (Figure 5B): width 75 mm, height 160 mm, thickness 1.08 mm. Samples have been gripped across their entire width, while gripping lengths of 30 mm have been used on top and bottom of the specimen, resulting in a free height of 100 mm. Each sample had an edge crack with height 0.1 mm and initial length $a = 10$ mm.

Tests have been performed under displacement controlled conditions by fixing the crosshead speed V. To capture the rate-dependent behaviour of PTFE, we have considered different values of $V = 0.6, 1, 3, 10, 100, 200, 300$ mm/min. Moreover, three replications for each value of the crosshead speed have been carried out.

ARAMIS image acquisition system has been exploited to record pictures of the crack

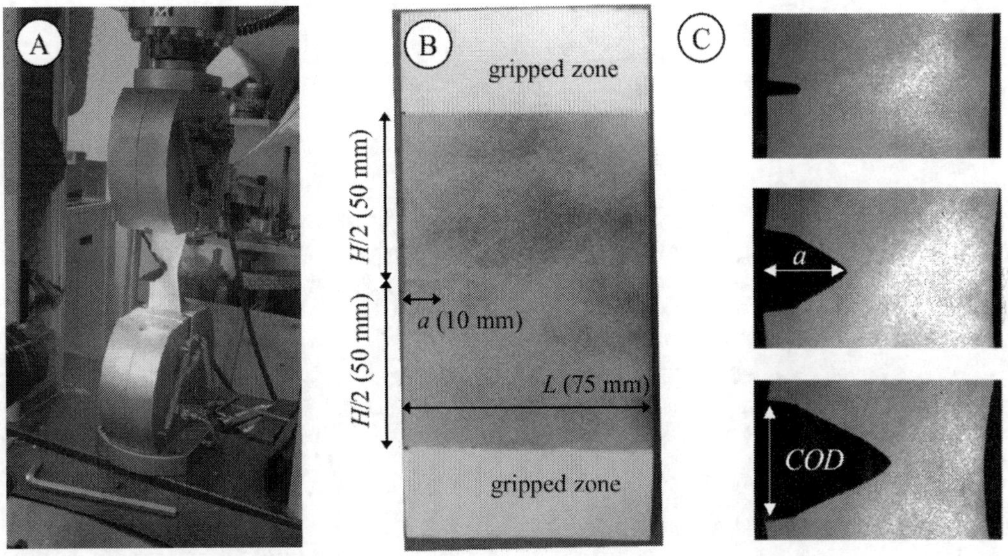

Figure 5. A) A cracked specimen in the tensile machine INSTRON 4480. B) Geometry of the cracked specimen. C) Evolution of the fracture process for $V = 1$ mm/min.

opening and propagation. Finally, we have developed a point tracking algorithm for pictures postprocessing with the aim of obtaining the crack length, monitoring the horizontal displacement of the crack tip and the crack opening displacement (COD) (here calculated as the vertical distance between the two vertices of the crack mouth). By way of example, Figure 5C shows the evolution of the fracture process for $V = 1$ mm/min.

4. Results

Figure 6 shows the tensile load F (N) as a function of the crack opening displacement COD (mm). Results are shown for different crosshead speeds $V = 0.6, 1, 3, 10, 100, 200, 300$ mm/min. Three samples have been tested for each value of V. Moving from small to high V, there is a significant reduction in the maximum COD that is reached before the occurrence of the sample rupture. At higher V the fracture process is hence less ductile, as we expect the PTFE response falls in its glassy region, where the material is stiffer. At $V = 0.6$ and 1 mm/min the curves are very similar and this suggests that the material response is close to the rubbery region, where an elastic soft behaviour is expected.

As shown in Figure 7, the maximum tensile load $F_{\max}$ increases with the speed V. Interestingly, this behaviour is similar to the detachment process of a rigid indenter from a viscoelastic substrate, where the pull-off force is a monotonic increasing function of the detachment speed [9]. At $V = 0.6$ mm/min, the maximum load is slightly higher than the

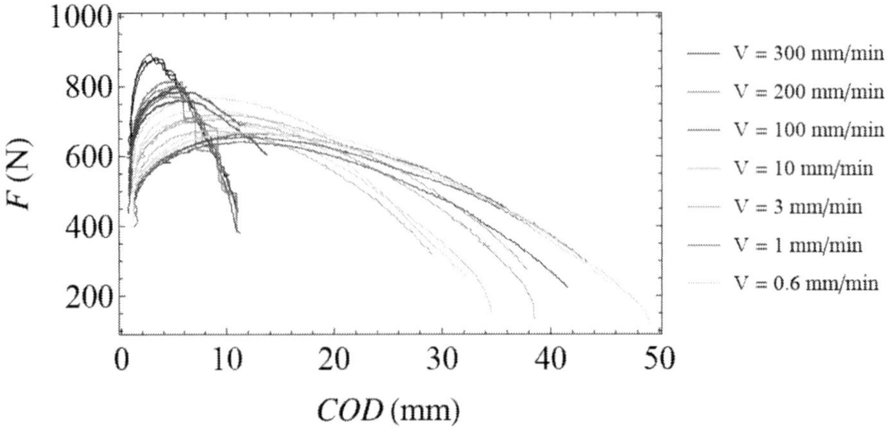

Figure 6. The tensile force F (N) as a function of the crack opening displacement COD (mm). Results are shown for different crosshead speeds $V = 0.6, 1, 3, 10, 100, 200, 300$ mm/min. Three samples have been tested for each value of V.

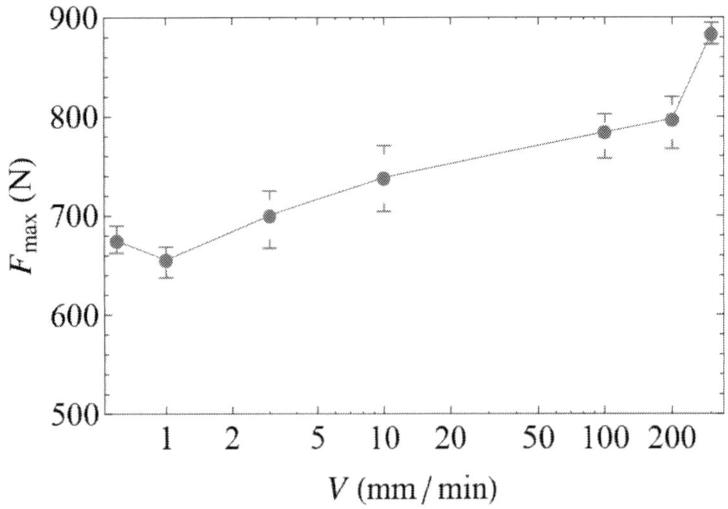

Figure 7. The maximum tensile force $F_{\max}$ (N) as a function of the crosshead speed V (mm/min). Results are shown for different crosshead speeds $V = 0.6, 1, 3, 10, 100, 200, 300$ mm/min and are averaged on three replications for each value of V. Error bars denote the maximum and minimum values of $F_{\max}$.

case at $V = 1$ mm/min; this could be related to experimental scatter or to secondary transitions in the viscoelastic behaviour of PTFE, which can present more than one relaxation time [25].

Existing theories usually assume a steady-state crack propagation [14]. In other words, the crack tip speed $V_c = da/dt$ is assumed to be constant during the entire fracture process. However, this is an ideal condition, which is very difficult to observe in real fracture phenomena.

Figure 8 shows the crack length a versus time t in a semilogarithmic representation. Results are shown again for $V = 0.6, 1, 3, 10, 100, 200, 300$ mm/min. The crack length is not a linear function of time, suggesting that the assumption of constant V_c is questionable. However, we shall observe that the steady-state assumption seems not to be crucial in the calculation of the energy release rate G, which represents a critical quantity, as it gives the energy required to advance the crack tip by one unit area and is correlated to the stress intensity factor K_{I}.

If we neglect the elastic energy stored in the viscoelastic material, the energy release rate can

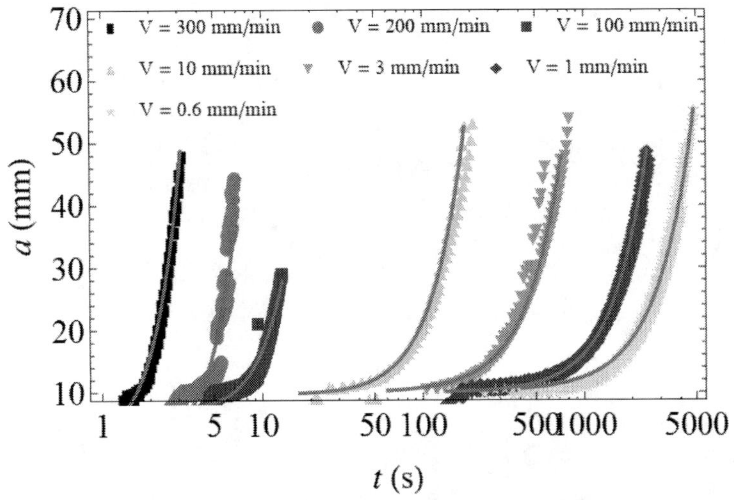

Figure 8. The crack length a versus time t in a semilogarithmic representation. Results are shown for $V = 0.6$, 1, 3, 10, 100, 200, 300 mm/min. Gray lines are a quadratic fitting of the experimental data.

be estimated from $Gab \sim Fa$ [26], where b is the specimen thickness, a is the crack length, and F is the applied tensile load.

Figure 9 shows the experimental values of G in terms of V_c in a double logarithmic chart; data are calculated at the maximum tensile force, which represents a critical situation in the fracture process. Moving towards high V_c (corresponding to high crosshead speeds V), there is a clear enhancement of G.

Theoretical calculations have been performed with the real viscoelastic modulus of PTFE (Figure 4) and using the original PB equation (3) and the new one (9) proposed in the present paper. For the adiabatic surface energy of PTFE we have considered the value $G_0 = 310$ N/mm, which is a realistic value as reported in Refs. [27, 28]. Moreover, the quasi static crack tip radius has been assumed to be equal to the initial height of the crack ($s_0 = 0.1$ mm). The coefficient $\beta = a/L \sim 0.133$ has been determined from the geometrical properties of the cracked specimens, being the initial crack length $a = 10$ mm and $L = 75$ mm. PB predictions (black solid line) quite significantly overestimate the effective energy release rate, while the proposed theory (red solid line), which takes into account finite-size effects, gives results in good agreement with the experimental data despite the crack tip propagation is assumed to occur at constant speed.

5. Conclusions

A theoretical model for the Mode I crack propagation in finite-sized viscoelastic solids has been developed and experimentally validated. We have discussed the main limitations of existing theories, which are often formulated under the assumption of crack propagation in infinite domain [14]. Specifically, we have considered the case of a viscoelastic plate with an edge crack.

Results show a significant effect due to the finite dimensions of the viscoelastic solid. In fact, the theoretical limit E_∞/E_0 of the energy release rate, which is approached at high crack tip speeds, is not reached and the maximum value $G_{\max}/G_0$ stays quite lower than that value.

A comprehensive experimental investigation carried out on cracked specimens made of PolyTetraFluoroEthylene (PTFE) has shown that, moving from small to high speeds, the fracture process becomes less ductile due to the transition of the material behaviour from the rubbery to the glassy region. For the same reason, an increase in the maximum tensile load is observed when the speed is increased.

Moreover, from experimental data, we have estimated the energy release rate G which is

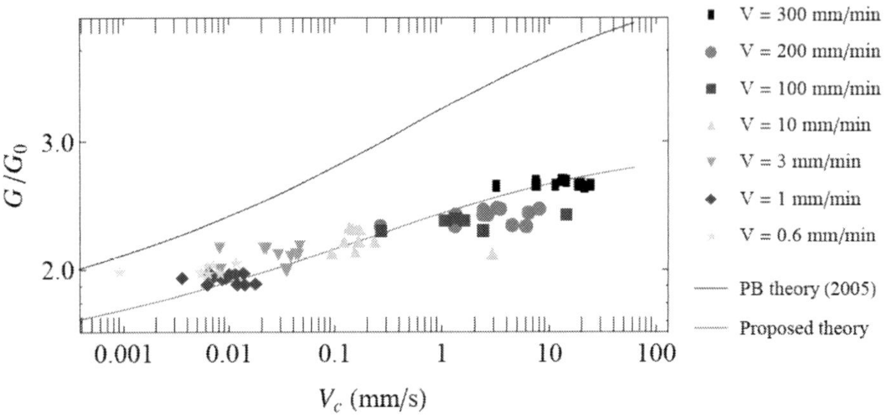

Figure 9. The energy release rate G/G_0 as a function of the crack tip speed V_c. Markers refer to the experimental data. Red line refers to the proposed eq. 9, while black line refers to PB solution for infinite domain [19]. Theoretical calculations are performed considering the viscoelastic modulus of PTFE given in Figure 4. Moreover, we fixed $s_0 = 0.1$ mm and $G_0 = 310$ N/mm.

resulted in good agreement with the theoretical predictions of the proposed theory. The quite good match between theory and experiments is a remarkable result, as the theory assumes steady-state crack propagation in a linear viscoelastic material, while in reality, the crack-tip speed is not constant and some non-linearities emerge in the fracture process at the crack tip. On the contrary, classical theories not taking into account size effects significantly deviate from experimental data.

References

[1] Elango N and Faudzi A A M 2015 *Int. J. Adv. Manuf. Technol.* **80**(5) 1027-37
[2] Zaimova D, Bayraktar E and Miskioglu I 2016 *Compos. B Eng.* **105** 203-10
[3] Sparks J L *et al* 2015 *Adv. skin wound care* **28**(2) 59-68
[4] Croll A B, Hosseini N and Bartlett M D 2019 *Adv. Mater. Technol.* **4**(8) 1900193
[5] D'Amico F. Carbone G, Foglia M M and Galietti U 2013 *Eng. Fract. Mech.* **98** 315-25
[6] Creton C and Ciccotti M 2016 *Rep. Prog. Phys.* **79**(4) 046601
[7] Rozen-Levy L, Kolinski J M, Cohen G and Fineberg J 2020. *PRL* **125**(17) 175501
[8] Müser M H and Persson B N 2022 *Europh. Let.* **137**(3) 36004
[9] Afferrante L and Violano G 2022 *JMPS* **158** 104669
[10] Fu H, Chen X, Zhao Q, Xiao Z and Liang X 2021 *Adv. Mech. Eng.* **13**(10) 16878140211052454
[11] Villey R *et al* 2015 *Soft Matter* **11**(17) 3480-91
[12] Cheng H *et al* 2013 *J. Appl. Mech.* **80**(4) 041019
[13] Persson B N J 2021 *Trib. Lett.* **69**(3) 1-8
[14] Hui C Y, Zhu B and Long R 2022 *JMPS* **159** 104748
[15] Knauss W G 2015 *Int. J. Fract.* **196**(1) 99-146
[16] Schapery R A 1975 *Int. J. Fract.* **11**(1) 141-59
[17] Greenwood J A 2004 *J. Phys. D* **37**(18) 2557
[18] de Gennes P G 1996 *Langmuir* **12**(19) 4497-500
[19] Persson B N J and Brener E A 2005 *Phys. Rev. E* **71**(3) 036123
[20] Persson B N J 2017 *EPL* **119**(1) 18002
[21] Violano G and Afferrante L 2022 *Eur. J. Mech. A* 104665
[22] Zehnder A T 2012 *Fracture mechanics* (Springer Science & Business Media)

[23] Dowling N E 1993 *Mechanical behavior of materials: engineering methods for deformation, fracture, and fatigue* (Prentice Hall international)
[24] Williams M L, Landel R F and Ferry J D 1955 *J. Am. Chem. Soc.* **77**(14) 3701-7
[25] Tan B and Stephens L S 2019 *Trib. Int.* **140** 105870
[26] Persson B N J, Albohr O, Heinrich G and Ueba H 2005 *J. Phys. Cond. Matt.* **17**(44) R1071
[27] Joyce J A 2003 *Polym. Eng. Sci.* **43**(10) 1702-14
[28] Joyce P J and Joyce J A 2004 *Int. J. Fract.* **127**(4) 361-85

Lifetime assessment of the rear axle for an off-road vehicle in a fatigue test bench by means of the Peak Stress Method

J Pelizzari[1,2], P Salvagno[2], C Dengo[2], A Campagnolo[1] and G Meneghetti[1]

[1] University of Padua, Industrial Engineering Department, Via Venezia 1, 35131, Padova, Italy
[2] Carraro S.p.a. - R & D, Virtual Validation Department, Via Olmo 37, 35010, Campodarsego, Italy

Email: giovanni.meneghetti@unipd.it

Abstract. In this study the Peak Stress Method (PSM) has been applied to the fatigue assessment of the welds of an axle equipped in an off-road vehicle. In the fatigue design of welded joints, the PSM is a FE-based method to rapidly evaluate the NSIFs relevant to the opening, sliding and tearing modes by means of the linear elastic peak stresses extracted from the FE model. First, an axle, which consists of a central body and two lateral branches composed by several plates welded to a tube, has been fatigue tested under a vertical load in a test bench designed to replicate the frame-axle interface. Then, an FE model of the axle has been calibrated to simulate the behaviour observed during the fatigue test, where the reaction forces and the strains at pre-defined locations have been considered. Once the FE model has been calibrated, the PSM has been applied to the welds of the axle, taking advantage of an automatic procedure, and the crack initiation location has been estimated. Finally, the theoretical estimations have been compared with the experimental datum.

1. Introduction

In the fatigue design of an axle for an off-road vehicle, the welds are critical zones for fatigue cracks initiation. The fatigue assessment of welded details is an issue addressed both in the literature and in the international standards. The earlier approach proposed by the international standards [1,2] is the approach based on nominal stress, which suggests to adopt the nominal stress applied to the welded detail, assuming a linear elastic material behaviour, and to compare it with the proper design curve. The approach based on nominal stress is limited by the number of welded details and the loading conditions reported in the literature, when it comes to apply the approach to complex geometries under a multiaxial stress state, the designer is in trouble to identify the proper welded detail between the ones proposed. The literature solved this issue suggesting the adoption of the local approaches [3]. The local approaches which have been proposed in the literature are mainly divided in the strain-life approach [4–6], notch stress approach [7,8], the Theory of Critical Distances (TCD) [9–12] and the critical plane approaches [5,13,14], the approach based on the Notch Stress Intensity Factors (NSIFs) [3,15,16], from which the criteria based on the averaged Strain Energy Density (SED) [17–22] and the Peak Stress Method (PSM) [23–29] are derived.

It is worth mentioning the application of the notch stress approach to evaluate the fatigue performance of an automotive axle housing [30]; in this study the PSM is proposed to evaluate the fatigue performance of the rear axle of an off-road vehicle. The PSM is a method that allows to rapidly estimate the NSIF-terms by means of the linear elastic peak stresses extracted at the V-notch tip taking

Content from this work may be used under the terms of the Creative Commons Attribution 3.0 licence. Any further distribution of this work must maintain attribution to the author(s) and the title of the work, journal citation and DOI.
Published under licence by IOP Publishing Ltd

advantage of a FE model with coarse mesh [23,24,26,27]. The PSM can be coupled with the SED approach to derive analytically the expression of an equivalent peak stress range [31,32], according to Eq. (1). Figure 1 shows the assumptions at the basis of the NSIF approach and the structural control volume, having size equal to $R_0 = 0.28$ mm for structural steels, being centred at the V-notch tip in accordance with the SED approach [17–19].

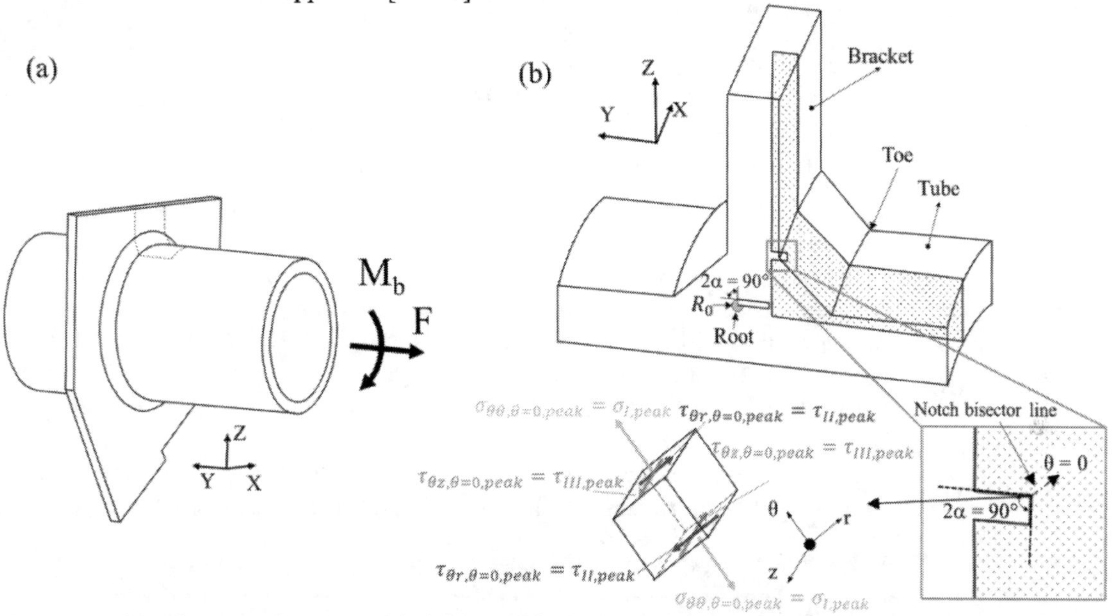

Figure 1. (a) Plate-to-tube arc welded joint with double-sided fillet welds under axial and bending loadings and (b) detail of the weld along with the assumptions of the approach based on the NSIFs, the radius R_0 adopted for the averaged SED calculation at weld root and the linear elastic peak stress to be evaluated according to the PSM, namely $\sigma_{\theta\theta,\theta=0,peak}$, $\tau_{\theta r,\theta=0,peak}$ and $\tau_{\theta z,\theta=0,peak}$.

The equivalent peak stress range can be adopted as a design parameter, according to the PSM, to estimate the fatigue life of steel as well as aluminium welded details by the adoption of the relevant PSM-based fatigue design curves [28].

$$\Delta\sigma_{eq,peak}=\sqrt{c_{w1}\cdot f_{w1}^2\cdot\Delta\sigma_{\theta\theta,\theta=0,peak}^2+c_{w2}\cdot f_{w2}^2\cdot\Delta\tau_{\theta r,\theta=0,peak}^2+c_{w3}\cdot f_{w3}^2\cdot\Delta\tau_{\theta z,\theta=0,peak}^2} \quad (1)$$

Where:

- c_{wi} (i = 1,2 or 3) is a parameter that accounts for the mean stress effects in stress-relieved welded joints. This parameter is a function of the applied nominal load ratio R_i according to the Eq. (2) [16]:

$$c_{wi}(R_i)=\begin{cases}\dfrac{1+R_i^2}{(1-R_i)^2} & \text{if stress relieved and } -1\le R_i\le 0\\ \dfrac{1-R_i^2}{(1-R_i)^2} & \text{if stress relieved and } 0\le R_i\le 1\\ 1 & \text{if as-welded for any } R \text{ value}\end{cases} \quad \text{where } i=1,\,2 \text{ or } 3 \quad (2)$$

- f_{wi} (i = 1,2 or 3) is a coefficient calculated according to Eq. (3-5):

$$f_{w1} = K_{FE}^{*} \cdot \sqrt{\frac{2e_1}{1-\nu^2}} \cdot \left(\frac{d}{R_0}\right)^{1-\lambda_1} \tag{3}$$

$$f_{w2} = K_{FE}^{**} \cdot \sqrt{\frac{2e_2}{1-\nu^2}} \cdot \left(\frac{d}{R_0}\right)^{1-\lambda_2} \tag{4}$$

$$f_{w3} = K_{FE}^{***} \cdot \sqrt{\frac{2e_3}{1-\nu^2}} \cdot \left(\frac{d}{R_0}\right)^{1-\lambda_3} \tag{5}$$

Where:

- K_{FE}^{*}, K_{FE}^{**} and K_{FE}^{***} are constants depending on the element type and formulation and on the meshing algorithm [33,34];
- d is the average mesh element dimension chosen according to the PSM requirements described more in detail in a previous contribution [28];
- R_0 is the radius of the structural volume for the averaged SED calculation [18];
- e_i (i = 1, 2 or 3) is a coefficient depending on the V-notch opening angle 2α and the Poisson's ratio ν [19], typical values of the e_i coefficient for structural steels have been reported in Table 1;
- λ_i (i = 1,2 or 3) is the stress singularity exponent function of the V-notch opening angle 2α [19], typical values of the λ_i coefficient have been reported in table 1.

Table 1. Parameters λ_i and e_i depending on the notch opening angle 2α.

2α (°)	λ_1 (a)	λ_2 (a)	λ_3 (a)	Steel ν = 0.3 (b)		
				e_1	e_2	e_3
0	0.500	0.500	0.500	0.134	0.341	0.414
90	0.545	0.909	0.667	0.146	0.168	0.310
120	0.616	-	0.750	0.130	-	0.276
135	0.674	-	0.800	0.117	-	0.259

[a] Values from [19].
[b] Values calculated under plain strain condition.

- $\sigma_{\theta\theta,\theta=0,peak}$, $\tau_{\theta r,\theta=0,peak}$ and $\tau_{\theta z,\theta=0,peak}$ are the peak stresses evaluated at the weld toe or weld root by means of linear elastic FE analyses. The peak stresses are extracted at the vertex nodes lying on the weld toe and weld root in a cylindrical coordinate system centred at the V-notch tip and with the radial direction along the V-notch bisector. In addition, if the PSM is applied to components discretized with tetra elements, the linear elastic peak stresses need to be averaged along three adjacent vertex nodes according to the following expression [35]:

$$\bar{\sigma}_{ij,peak,n=k} = \left.\frac{\sigma_{ij,peak,n=k-1}+\sigma_{ij,peak,n=k}+\sigma_{ij,peak,n=k+1}}{3}\right|_{n=node} \quad \text{where } \sigma_{ij} = \sigma_{\theta\theta,\theta=0}, \tau_{\theta r,\theta=0}, \tau_{\theta z,\theta=0} \tag{6}$$

Finally, the $\Delta\sigma_{eq,peak}$ is evaluated according to Eq. (1) and it is compared with the PSM-based design scatter band to estimate the fatigue life of the considered welded detail. The PSM-based design scatter band can be chosen based on the local biaxiality ratio λ evaluated according to Eq. (7), the reference values of λ are reported in table 2 [28].

$$\lambda = \frac{c_{w2} \cdot f_{w2}^2 \cdot \Delta\tau_{\theta r,\theta=0,peak}^2 + c_{w3} \cdot f_{w3}^2 \cdot \Delta\tau_{\theta z,\theta=0,peak}^2}{c_{w1} \cdot f_{w1}^2 \cdot \Delta\sigma_{\theta\theta,\theta=0,peak}^2} \tag{7}$$

Table 2. Criterion for selecting the PSM fatigue design scatter band for arc-welded joints made of structural steels [28].

T (mm)	λ Eq. (7)	N_A (cycles)	$\Delta\sigma_{eq,peak,A,50\%}$ (MPa)	$\Delta\sigma_{eq,peak,A,97.7\%}$ (MPa)	$\Delta\sigma_{eq,peak,A,2.3\%}$ (MPa)	k	T_σ
$T \geq 2$ mm	$\lambda = 0$	$2 \cdot 10^6$	214	156	296	3	1.90
$T \geq 2$ mm	$\lambda > 0$	$2 \cdot 10^6$	354	257	488	5	1.90

In the recent developments of the PSM, the application of the PSM to multiaxial variable amplitude fatigue is of particular interest for this study [29]. In the following, the major aspects of the application of the PSM to components under variable amplitude fatigue loadings will be covered. The equivalent peak stress range is defined as follows:

$$\Delta\sigma_{eq,peak} = \sqrt{\Delta\sigma^2_{eq,peak,I} + \Delta\sigma^2_{eq,peak,II} + \Delta\sigma^2_{eq,peak,III}} \tag{8}$$

The $\Delta\sigma_{eq,peak,I}$, $\Delta\sigma_{eq,peak,II}$ and $\Delta\sigma_{eq,peak,III}$ are the constant amplitude equivalent peak stress ranges relevant to the mode I, mode II and mode III that are calculated according to Eq. (9-11).

$$\Delta\sigma_{eq,peak,I} = f_{s1} \cdot f_{w1} \cdot \left|\Delta\sigma_{\theta\theta,\theta=0,peak,max}\right| \tag{9}$$
$$\Delta\sigma_{eq,peak,II} = f_{s2} \cdot f_{w2} \cdot \left|\Delta\tau_{r\theta,\theta=0,peak,max}\right| \tag{10}$$
$$\Delta\sigma_{eq,peak,III} = f_{s3} \cdot f_{w3} \cdot \left|\Delta\tau_{\theta z,\theta=0,peak,max}\right| \tag{11}$$

Where:

- f_{si} (i = 1, 2 or 3) is a coefficient that accounts for the Palmgren-Miner equivalency [36,37] and it is defined in the following expressions:

$$f_{s1} = \left[\sum_{i=1}^{q_I} \frac{n_{I,i}}{N_0} \cdot \left(\sqrt{c_{w1,i}} \cdot \left|\frac{\Delta\sigma_{\theta\theta,\theta=0,peak,i}}{\Delta\sigma_{\theta\theta,\theta=0,peak,max}}\right|\right)^{k_1}\right]^{1/k_1} \tag{12}$$

$$f_{s2} = \left[\sum_{j=1}^{q_I} \frac{n_{II,j}}{N_0} \cdot \left(\sqrt{c_{w2,j}} \cdot \left|\frac{\Delta\tau_{r\theta,\theta=0,peak,j}}{\Delta\tau_{r\theta,\theta=0,peak,max}}\right|\right)^{k_2}\right]^{1/k_2} \tag{13}$$

$$f_{s3} = \left[\sum_{h=1}^{q_{III}} \frac{n_{III,h}}{N_0} \cdot \left(\sqrt{c_{w3,h}} \cdot \left|\frac{\Delta\tau_{\theta z,\theta=0,peak,h}}{\Delta\tau_{\theta z,\theta=0,peak,max}}\right|\right)^{k_3}\right]^{1/k_3} \tag{14}$$

The Eq. (12-14) can be applied after having performed the following calculations. The time-history of each linear elastic peak stress relevant to mode I, mode II and mode III must be evaluated at each vertex node at the weld toe and weld root, then the Rainflow cycle counting algorithm is applied to the time-histories previously derived. Consequently, it can be defined: the maximum value of linear elastic peak stress relevant to each loading mode $\Delta\sigma_{\theta\theta,\theta=0,peak,max}$, $\Delta\tau_{r\theta,\theta=0,peak,max}$ and $\Delta\tau_{\theta z,\theta=0,peak,max}$; the number of stress levels q_I, q_{II} and q_{III}; the amount of cycles corresponding to each spectrum $N_I = \sum n_{I,i}$, $N_{II} = \sum n_{II,j}$ and $N_{III} = \sum n_{III,h}$; the reference number of cycles $N_0 = min\{N_I, N_{II}, N_{III}\}$.

Once the equivalent peak stress range has been calculated according to Eq. (8) it can be compared with the relevant PSM design scatter band for steel joints, which can be chosen on the basis of the local biaxiality ratio according to the following expression:

$$\lambda = \frac{\Delta\sigma^2_{eq,peak,I} + \Delta\sigma^2_{eq,peak,II}}{\Delta\sigma^2_{eq,peak,III}} \tag{15}$$

IOP Conf. Series: Materials Science and Engineering 1275 (2023) 012044 doi:10.1088/1757-899X/1275/1/012044

2. Geometry of the axle, fatigue test and experimental results

The axle adopted in this study is the rear axle equipped in an off-road vehicle and it is mainly composed by a central body and two lateral branches which are the assembly of a tube with a group of plates through an arc welding process. The final geometry of the axle is presented in figure 2a-c along with the relevant dimensions of the tube and the plate, more in detail the tube has an external diameter equal to 84 mm and an internal diameter of 69 mm resulting in a thickness of 7.5 mm, while all the plates are characterized by a thickness equal to 6 mm. Figure 2c shows the geometrical assumptions adopted to model the welds, the V-notch opening angle at the weld toes is supposed to be equal to $2\alpha = 135°$, while the V-notch opening angle at the weld roots is equal to $2\alpha = 90°$ for a gap greater than 0 mm while it is equal to $2\alpha = 0°$ for a null value of the gap.

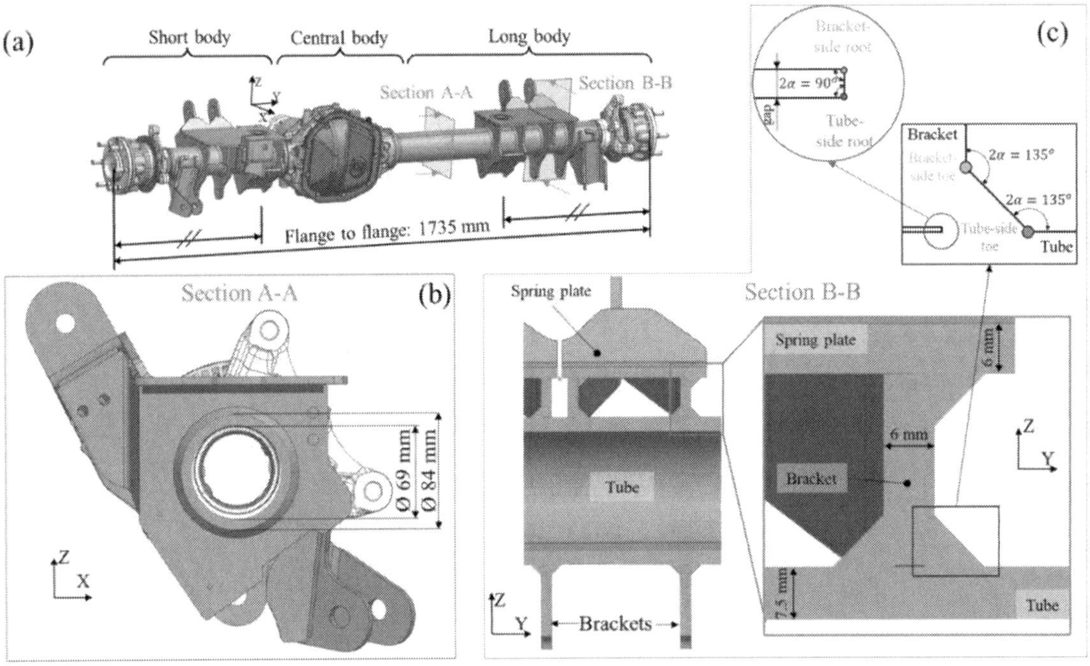

Figure 2. (a) Geometry of the axle and main geometrical parameters relevant to (b) the tube and (c) the spring plate and the bracket. The V-notch opening angle $2\alpha = 135°$ at the weld toes and the V-notch opening angle $2\alpha = 90°$ at the weld roots, for a gap greater than 0 mm, are shown. The detailed view in the welds area shows the geometrical assumptions adopted to model the welds.

The central body is made of pearlitic cast iron GJS 500-7 according to EN 1563 standard while the tubes, the spring plates and the brackets are made respectively of 25MnCr6 quenched and tempered steel and S355J2 steel, the main mechanical properties relevant to the material of the central body, the tubes, the spring plates and the brackets are summarized in table 3.

Table 3. Materials and mechanical properties of the central body, tubes, spring plates and brackets.

Component	Material	Young modulus [MPa]	Yield stress [MPa]	UTS [MPa]
Tube	25MnCr6 quenched & tempered	206 000	638	785
Spring plate & bracket	S355J2	206 000	355	510
Central body	GJS 500-7 EN 1563	169 000	320	500

The axle is mounted in a tailor-made fatigue test bench represented in figure 3, the bench resembles the axle-frame interface, more in detail the axle is anchored to the test bench by means of four longitudinal linkages, two vertical linkages, two horizontal linkages and one Panhard rod. The fatigue test bench has been equipped with four hydraulic actuators and one electric motor and it is able to control each applied load separately allowing the application of both constant amplitude and variable amplitude fatigue loads. The vertical actuator has a load capacity of 200 kN, the lateral actuator has a capacity of 100 kN, the ARB actuator has a capacity equal to 70 kN, the damper actuator has a capacity of 20 kN and the electric motor has a capacity of 2000 N m.

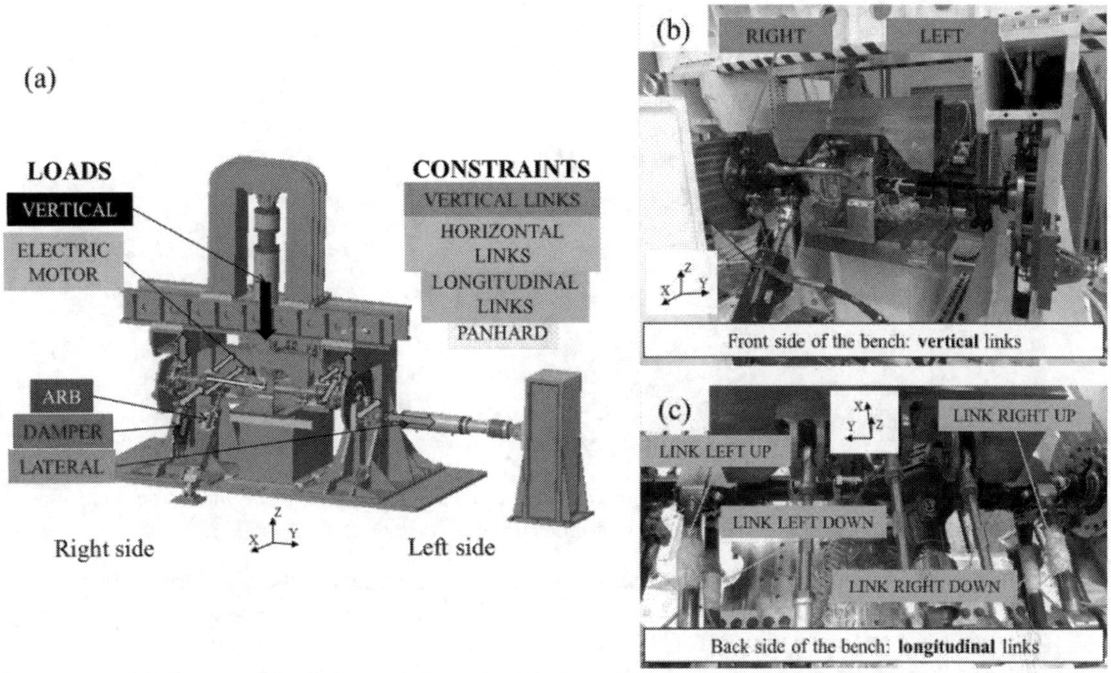

Figure 3. (a) Scheme of the fatigue test bench with the axle mounted and (b-c) front view and back view of the actual fatigue test bench with the axle mounted.

In this investigation a pure vertical load has been applied to the axle which have been constrained to the fatigue test bench by means of the two vertical linkages, the four longitudinal linkages and the panhard rod. Firstly, 500 000 cycles have been peformed with an external load range equal to $\Delta F = 35.4$ kN and a nominal load ratio $R \approx 0$, which has been defined as $R = F_{min}/F_{max}$, where F_{min} and F_{max} are respectively the minimum value and the maximum value of the applied load cycle. Secondly, 85 567 cycles have been perfomed with an external load range $\Delta F = 68.4$ kN and a nominal load ratio $R \approx 0$, table 4 summarizes the parameters relevant to the load steps applied to the axle.

Table 4. Load history applied to the axle.

Applied load	Minimum load value [kN]	Maximum load value [kN]	Load range [kN]	Nominal load ratio R	Number of cycles	Load frequency [Hz]	Joint condition
Vertical load	0.4	35.8	35.4	≈ 0	500 000	3	As-welded
Vertical load	0.4	68.8	68.4	≈ 0	85 567 (55 000*)	1	

*number of cycles to technical crack initiation

The fatigue test has been ended on the basis of the visual inspection of a fatigue crack as shown in figure 4. The fatigue crack initiated at the tube-side weld toe in correspondence of the lower side of the axle at the tube-bracket lower link weld as defined in the next figure 12.

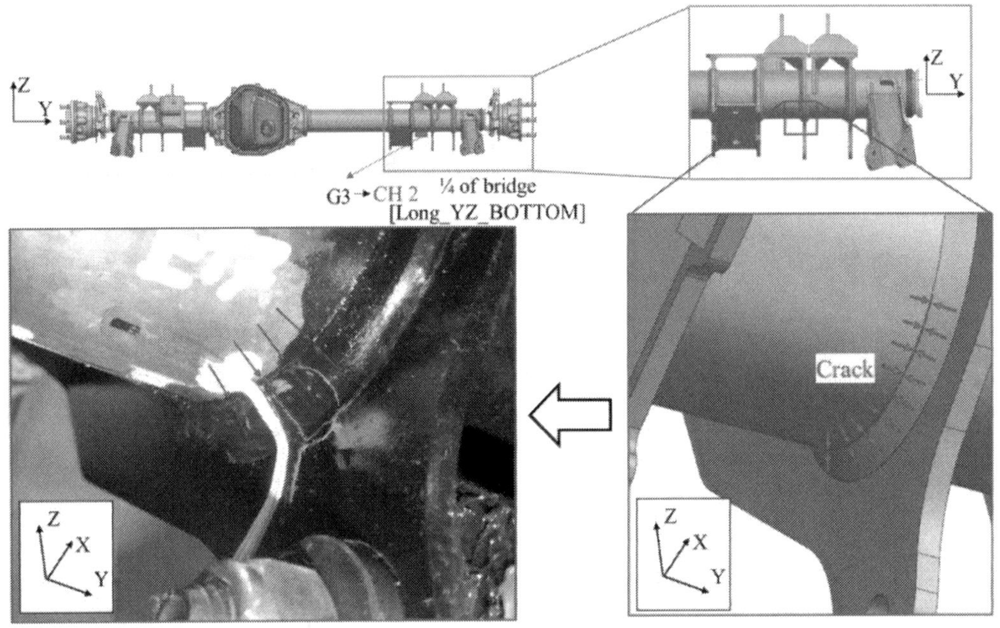

Figure 4. Fatigue crack identified after 585 567 cycles in correspondence of the tube-side weld toe at the lower side of the axle.

The axle has been equipped with KYOWA axial strain gauges model KFGS 1-120-C1-11 L3M3R characterized by a gauge length equal to 1 mm to acquire the strains at pre-defined locations. The strain gauges marked in black in figure 5 (G1, G2 and G6) were characterized by a gauge factor K = 2.06 while the strain gauges marked in red in figure 5 (G3, G4 and G5) were characterized by a gauge factor K = 2.11. Furthermore, figure 5 shows the strain gauges Gi (i = 1, 2, 3, 4, 5 and 6) configuration and the channels CHi (i = 1, 2, 3, 4 and 5) scheme, in addition the names given to the channels have been reported between square brackets. The strain gauges G1 and G2 have been connected together to obtain a Wheatstone half bridge configuration sensitive to the bending moment on the Z-Y plane, the strain gauges from G3 to G6 have been connected in a Wheatstone quarter bridge configuration to measure the strains in correspondence of chosen points.

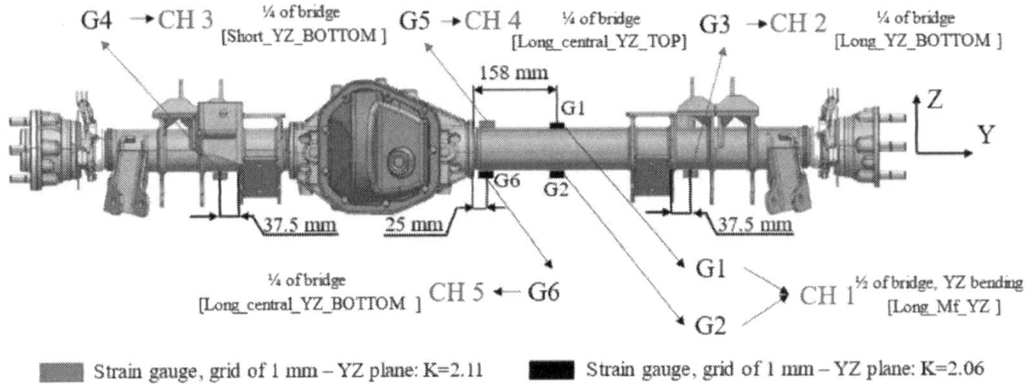

Figure 5. Strain gauges configuration and channels scheme adopted in the tested axle.

Figure 6 shows the maximum strain value, the minimum strain value and the range = maximum strain value – minimum strain value of the CH2 channel signal, the values have been measured with steps of 1200 cycles. The decrease of the channel two (CH2) reading, relevant to the strain gauge G3 in the proximity of the failure location, has been adopted as a paramater to identify the number of cycles to technical crack initiation; a 10% decrease of the signal has been observed after performing 55 000 cycles at a load range equal to $\Delta F = 68.4$ kN and consequently the overall number of cycles to technical crack initiation has been set equal to $N_i = 555\ 000$ cycles.

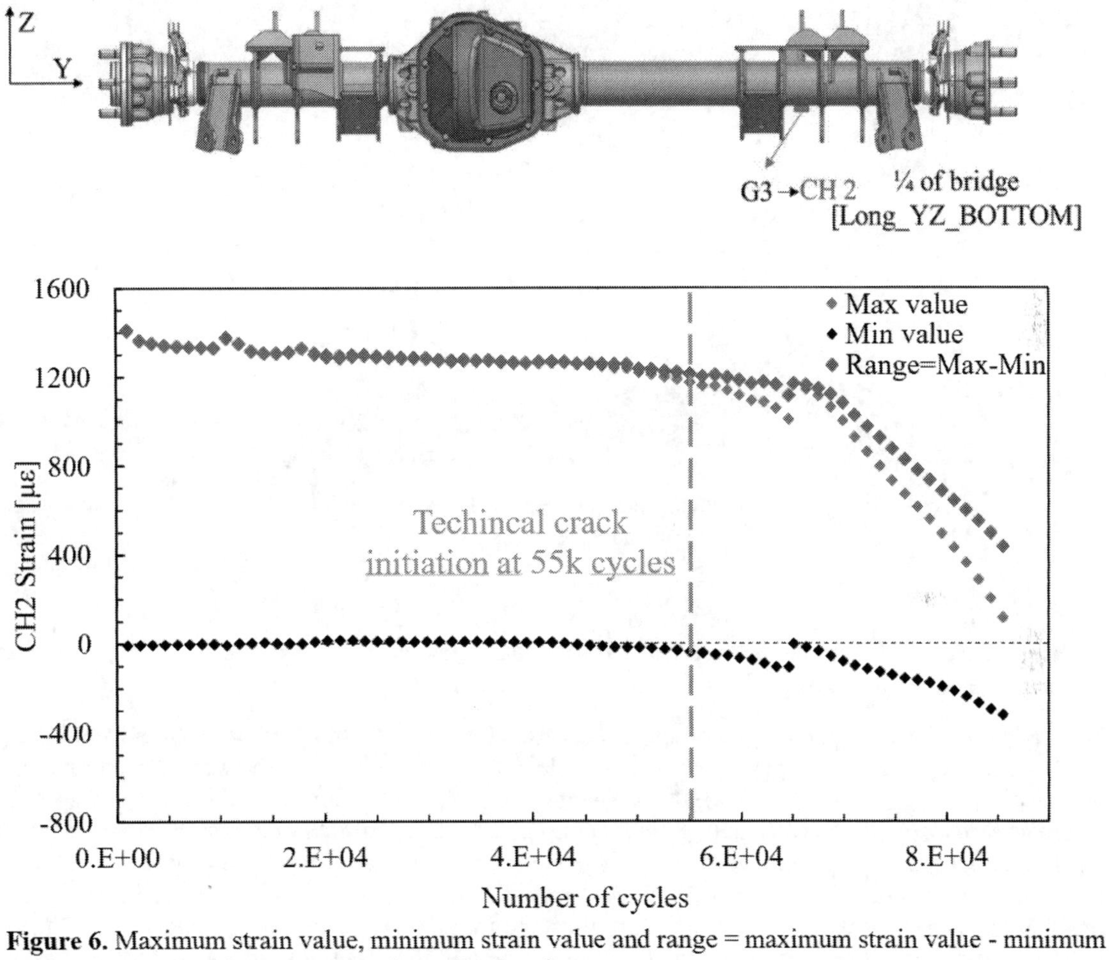

Figure 6. Maximum strain value, minimum strain value and range = maximum strain value - minimum strain value derived from the CH2 channel's signal.

3. Experimental data analysis and FEM calibration

The CAD model of the axle has been introduced in Ansys® Workbench, due to the complexity of the geometry, 10-node quadratic tetrahedral elements (SOLID 187) have been chosen to discretize the axle adopting the free meshing algorithm available in the FE software. A different value of the mesh average element size has been applied to the short body, the central body and the long body, in particular a value equal to 11 mm has been adopted for the short body, a value equal to 8 mm has been applied to the central body and a value equal to 8.5 mm has been chosen to discretize the long body. Figure 7 shows the boundary conditions applied to the model, more in detail a null displacement in the Z direction has been applied at the wheels center point, a null displacement in the Y direction has been set in correspondence of the Panhard rod and the vertical force, equal to $\Delta F = 68.4$ kN, has been evenly distributed in correspondence of the vertical springs housings. Moreover, a bonded contact has been applied to the overlapping surfaces of the central body and the tubes to simulate the adhesion between the three components which is obtained in reality by means of the press-fitting and plug welding process. Furthermore, no separation contact has been applied between the facing surfaces of the spring plate and the brackets relevant to the long body.

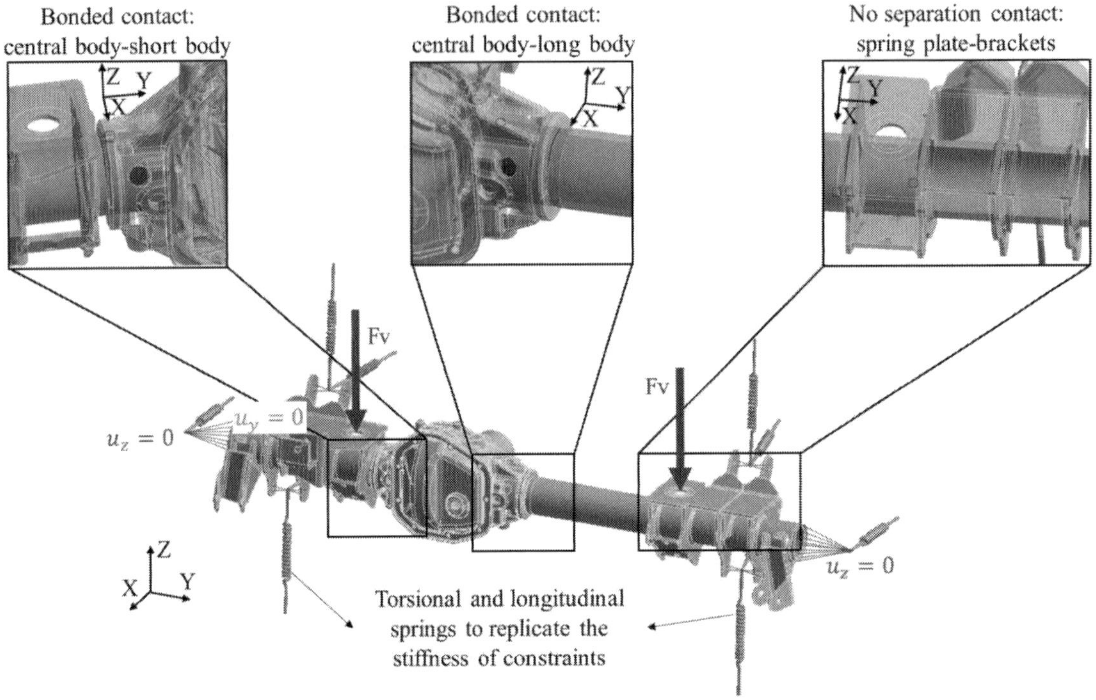

Figure 7. Scheme of the model adopted to replicate the behaviour of the axle undergoing the fatigue test under a pure vertical load.

In addition, the calibration of the preloads and of the torsional and longitudinal stiffness associated to the springs represented in figure 7, which resemble the connecting elements between the axle and the test bench, has been needed to replicate the behaviour of the axle undergoing the fatigue test with applied a pure vertical load. Through an iterative process, based on trial and error, the stiffness, both longitudinal and torsional, of the springs, and the associated preloads, have been varied until a satisfactory agreement between the experimental readings and the FE analysis output has been reached; more in detail, the reaction forces in correspondence of the constraints and the strains at the strain gauges locations have been considered as parameters to check for the quality of the FE model. Figure 8 shows the values of stiffness and the preloads, relevant to the modelled springs, obtained from the calibration procedure. It is worth nothing from figure 8a that the higher values of longitudinal stiffness k are associated to the springs that resemble the longitudinal linkages while a lower value is associated to the longitudinal stiffness introduced with the vertical springs. On the other hand, figure 8b shows that higher values of torsional stiffness k_θ are associated to the springs parallel to the Z and X directions while a lower value of torsional stiffness is associated to the springs that are parallel to the longitudinal linkages.

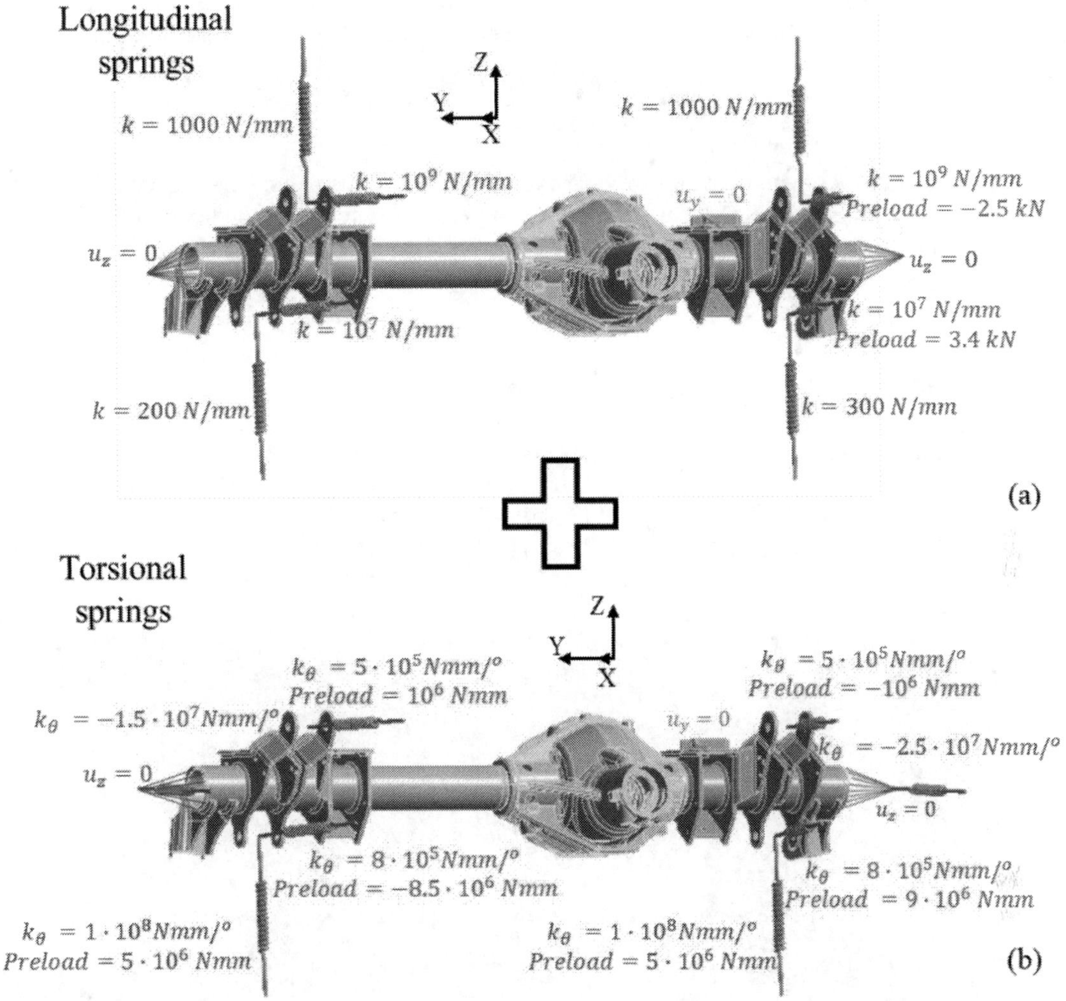

Figure 8. (a) Values of longitudinal stiffness and preloads applied to the longitudinal springs modelled in the FE model and (b) values of torsional stiffness and preloads applied to the torsional springs modelled in the FE model.

The FE model obtained from the calibration procedure has been validated on the basis of the experimental readings which have been acquired during the fatigue tests. Figure 9a shows the comparison between the actual values and FEA values of the reaction forces. The actual vertical reaction forces relevant to the vertical linkages ($F_{vertLink,BENCH}$) have been compared with the reaction forces extracted from the FE analysis ($F_{vertLink,FEM}$), the absolute percentage error falls between the 4% and the 14%. The experimental reaction forces along the longitudinal linkages directions ($F_{Link,BENCH}$) have been compared with the corresponding FEA reactions ($F_{Link,FEM}$), the absolute percentage error ranges between the 2% and the 37%. In addition, four longitudinal vertical springs, which are not associated to a geometrical entity in the fatigue test bench, have been introduced in the FE model to resemble the vertical reaction forces exerted by the longitudinal linkages, in this case only the values relevant to the FE analysis ($F_{Spring,FEM}$) have been reported in figure 9a, since such forces could not be measured directly in the experimental tests. Figure 9b shows the comparison between the strain values extracted from the axle under test and the strain values evaluated at the corresponding locations in the FE model, the absolute percentage error ranges between the 2% and the 28%. The strain values have been evaluated along the Y direction and the percentage error Er% has been calculated according to Eq. (16):

$$\mathrm{Er\%} = \frac{\mathrm{FEM - BENCH}}{\mathrm{BENCH}} \cdot 100 \tag{16}$$

Furthermore, it is worth noting that low values of absolute percentage error have been reached in the nominal area of the tube where no local effects are present, namely in correspondence of the CH1, CH4 and CH5 channels, while the absolute percentage error value raises considering areas strongly affected by local effects such as the locations of the CH2 and CH3 strain gauge channels.

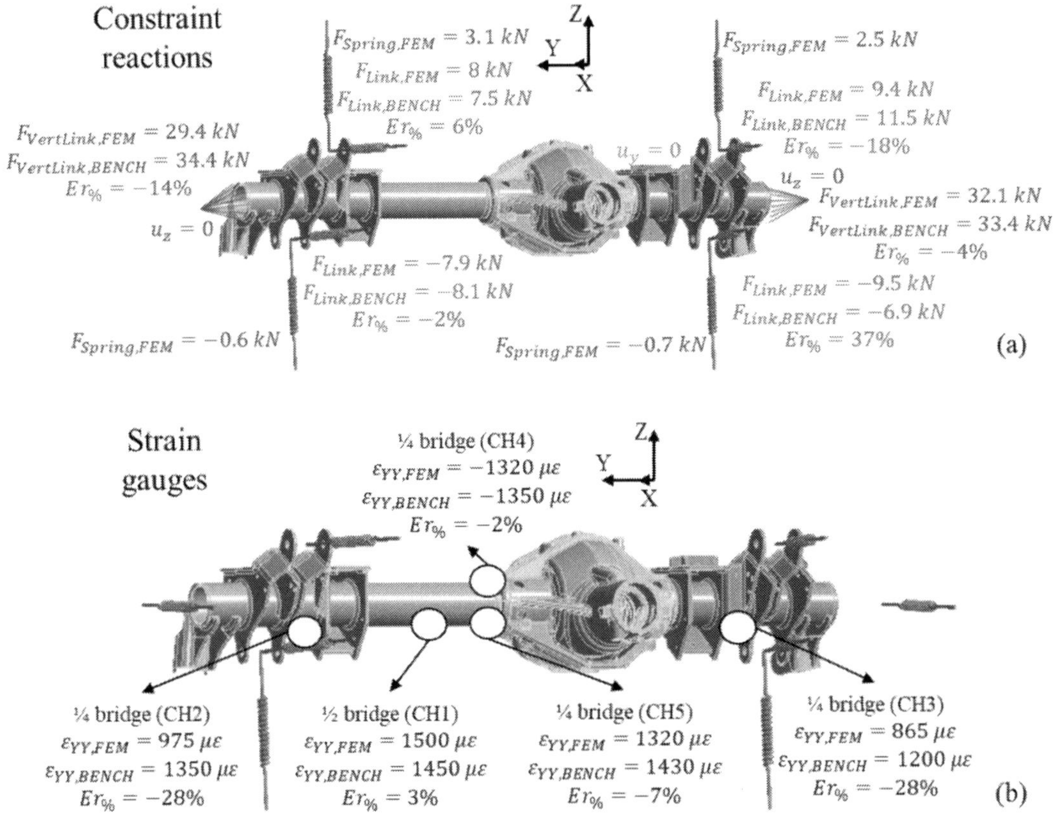

Figure 9. Comparison of the experimental readings and the FEA results relevant to (a) reaction forces and (b) strains evaluated along the Y direction.

4. Application of the PSM

Given the calibrated FE model of the axle, it has been necessary to prepare the model for the application of the PSM, more in detail the maximum average element size to apply the PSM has been evaluated for each weld. Figure 10 shows the procedure to determine the maximum average element dimension to apply the PSM with respect to two different welded details, a double-sided fillet weld between the tube and the bracket and a single-sided fillet weld between the plate and the bracket. Firstly, the values of penetration and gap have been evaluated through optical analysis of several sections made in correspondence of the welds, then the welds have been grouped in classes based on the actual values of penetration and gap, six classes have been identified. The mean values of penetration and gap associated to each class have been introduced in the CAD model of the axle as displayed in figure 10c. Finally, the parameters *h*, *l*, *t* and *z* have been evaluated to define the maximum value of the average element size

d_{max} according to the PSM [28]; in the following, the calculation of d_{max} will be cover much more in detail for a double-sided fillet weld between the tube and the bracket and for a single-sided fillet weld between the bracket and the plate. The double-sided welds between the tube and the bracket have been grouped in the first class, which is characterized by an average gap equal to 0.2 mm, an average penetration equal to 1.27 mm, a z dimension equal to 6 mm, a tube thickness equal to 7.5 mm and a bracket thickness equal to 6 mm; the a parameter introduced in the mesh density ratio calculation has been set equal to the difference between half bracket thickness and the average penetration value, consequently the following equation applies $a = 2l/2 = 1.73$ mm and finally $d_{max}=a/3 = 1.73/3 = 0.57$ *mm*. The single-sided welds between the plate and the brackets have been included in the second class which is characterized by a null value of the gap and an average penetration equal to 1.24 mm, a z dimension equal to 6 mm and the thickness of both the plate and the bracket is equal to 6 mm; having in hand these parameters the d_{max} value can be defined as $d_{max} = a/3 = 4.76/3 = 1.58$ *mm*.

Figure 11 shows a summary of the classes adopted to group the welds relevant to the long body, the maximum average element size ranges between 0.57 mm and 2 mm. In addition, a detail of the mesh, with an average element dimension equal to 0.57 mm, applied in the failure location surroundings has been reported.

The fatigue assessment based on the calibrated FE model has been carried out with reference to the welds belonging to the long body, the welds relevant to the short body have not been analysed because of the symmetry of the loading and boundary conditions and based on the results of the preliminary analyses which have not been reported here for the sake of brevity. The mesh size required for the application of the PSM has been adopted in the surrounding areas of the welds by means of the sphere of influence command of Ansys® Workbench, while the same element size applied at the calibration stage, summarized in figure 12, has been adopted to discretize the whole axle; more in detail, the mesh size applied is equal to 11 mm for the short body, 8 mm for the central body and 8.5 mm for the long body.

Due to the absolute dimensions of the axle, it has been necessary to run different FE analyses to fatigue assess the welds of the axle. The summary of the data concerning the FE analyses carried out to apply the PSM to the welds between the tube and the brackets, which belong to the long body, has been reported in table 5, it is worth noting that the welds between the tube and the brackets require the lowest value of FE size and this turns into the longest computational times. Each FE analysis has been characterized by approximately 29 million degrees of freedom and a RAM usage which ranges between 42 GB and 51 GB, while the cores usage is a function of the cores availability and it ranged between two and ten cores. The solution time is a function of the cores usage and it has been around 1 hour with all the cores available, while it has been increased to 3 hours and 42 minutes with two cores only.

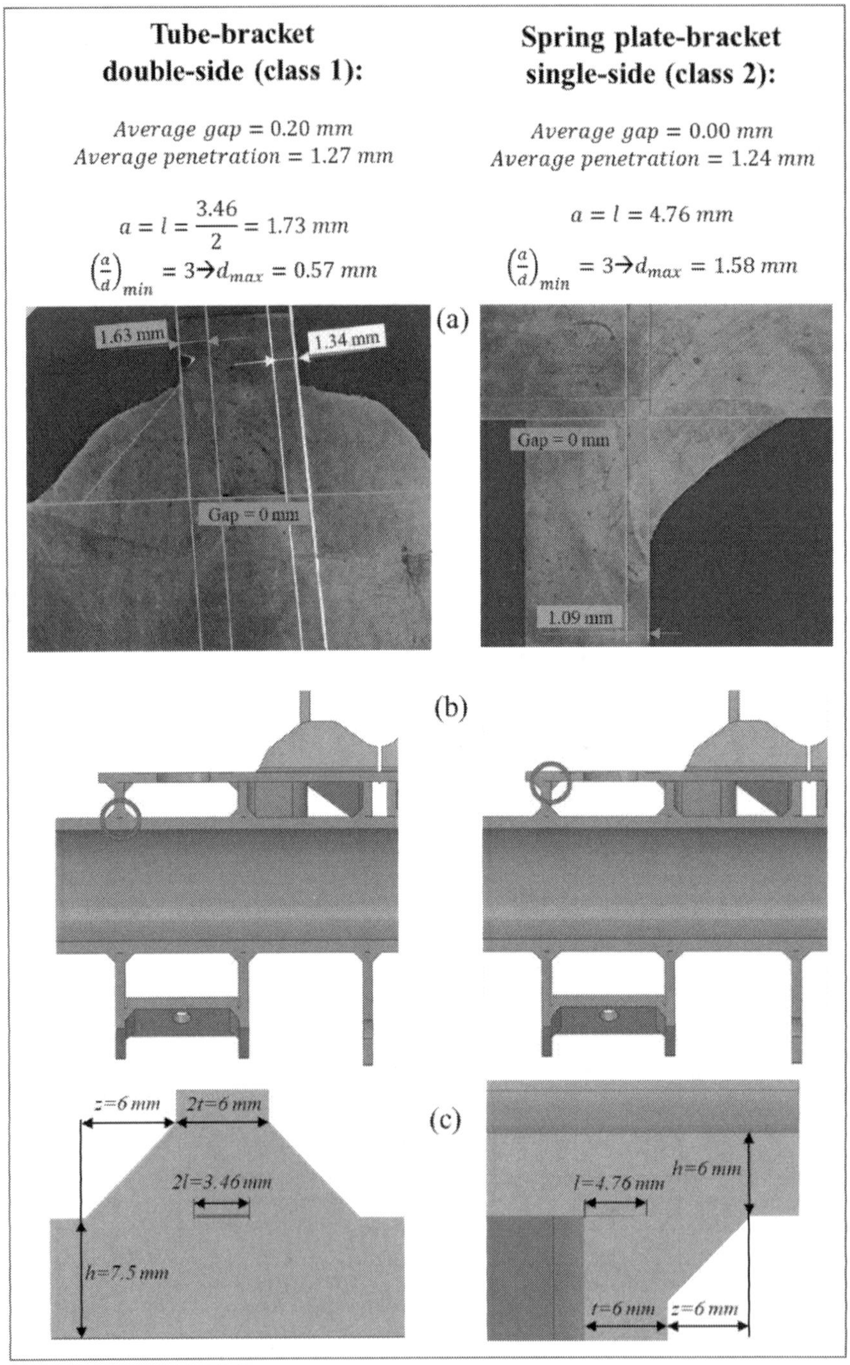

Figure 10. Procedure to evaluate the average element size according to the PSM. (a) Evaluation of the actual gap and penetration values from a welded detail cut from the axle, (b) and (c) details of the CAD geometry of the welds and definition of the weld parameters relevant to the calculation of the maximum mesh element size according to the PSM.

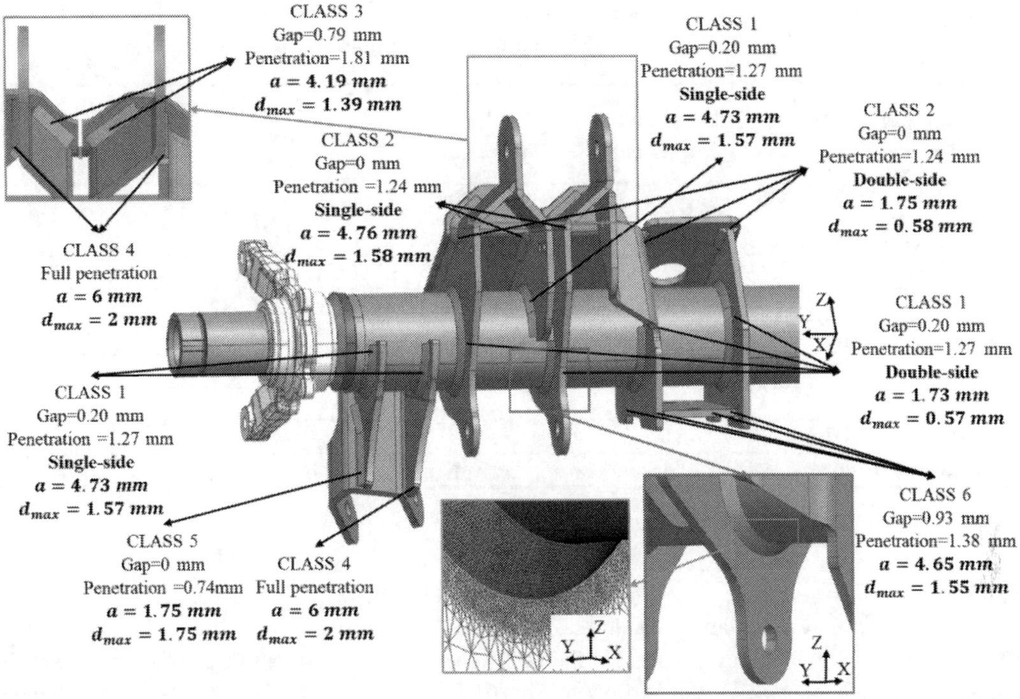

Figure 11. Summary of the classes adopted to group the welds relevant to the long body, for each class the average gap value, the average penetration value and the maximum mesh dimension d_{max} according to the PSM have been reported. Detail of the mesh generated to apply the PSM in correspondence of the experimental failure location.

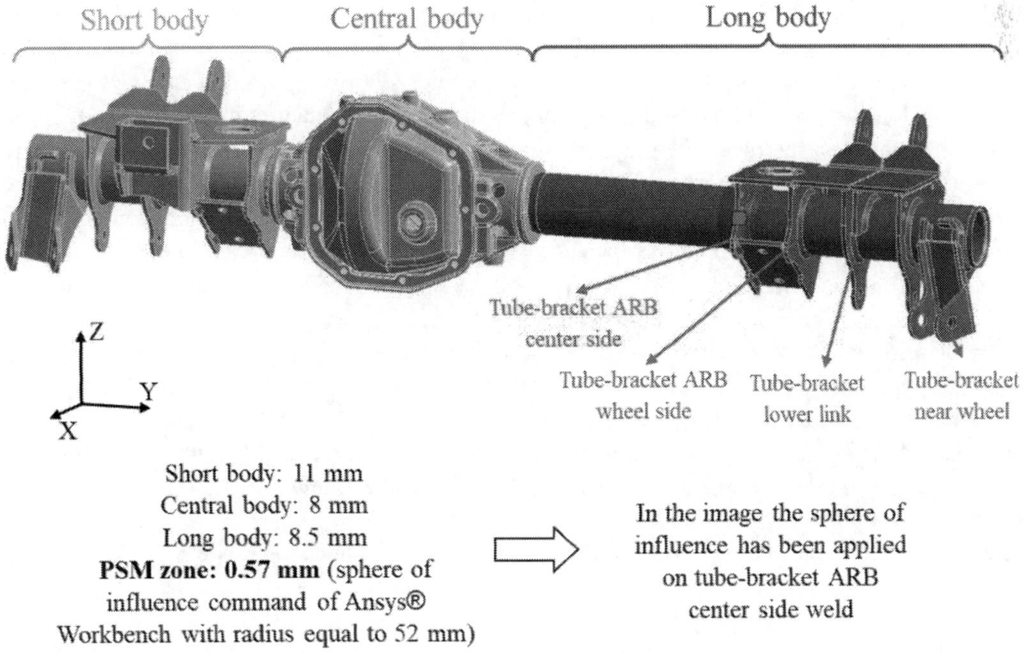

Figure 12. Average element size applied to the whole axle and mesh size defined in the surroundings area of the welds using the sphere of influence command of Ansys® Workbench.

Table 5. Summary of the data relevant to the FE analyses performed to apply the PSM to the welds between the tube and the brackets belonging to the long body.

Weld analyzed in the model	Tube-bracket near wheel	Tube-bracket lower link	Tube-ARB bracket wheel side	Tube-ARB bracket center side
FE size for PSM [mm]	0.57	0.57	0.57	0.57
DOFs	29 162 043	29 213 301	28 929 069	28 778 724
CPU	Intel Xeon W-2155 3.3 GHz			
Cores availability	10	10	10	10
Cores usage	10	10	2	5
RAM availability [GB]	64	64	64	64
RAM usage [GB]	51	51	42	45
Solution time	1 hr	56 min	3 hr 42 min	1 hr 8 min

The fatigue assessment of the welds of the long body has been performed by taking advantage of a tool which automates the application of the PSM [38]. Firstly, the linear elastic peak stresses relevant to mode I, mode II and mode III are extracted at the nodes laying on the weld toes and weld roots lines together with the V-notch opening angle 2α, then the moving average across three adjacent vertex nodes is applied to the peak stress values according to Eq. (6). Secondly, the f_{wi} (i = 1,2, or 3) coefficients are evaluated, after calculating the values of the coefficients e_i (i = 1 2, or 3) and the stress singularities exponents λ_i (i = 1 2, or 3). Finally, all the above-mentioned parameters are introduced in the Eq. (1) to evaluate the range of the equivalent peak stress $\Delta\sigma_{eq,peak}$. The local biaxiality ratio λ is computed according to Eq. (7) and it allows to choose the PSM design scatter band.

For the sake of brevity, the equivalent peak stress range distribution, which has been evaluated along the weld toes and weld root lines, relevant to the two welds which demonstrated the higher absolute values of $\Delta\sigma_{eq,peak}$ is presented in figure 13. Figure 13a shows the $\Delta\sigma_{eq,peak}$ distributions along the weld toe and weld root lines of the tube-bracket ARB center side weld, which displayed the highest value of $\Delta\sigma_{eq,peak}$ across the analysed welds; more in detail, the maximum value of $\Delta\sigma_{eq,peak} = 1023\ MPa$ is located on the tube-side toe at an angle $\beta = 0°$, which corresponds to the upper part of the axle that is on the compression side. The area surrounding the $\Delta\sigma_{eq,peak} = 1023\ MPa$ value is dominated by the mode I compression linear elastic peak stress contribution, which results in a local biaxiality ratio $\lambda = 0$. Figure 13b shows the $\Delta\sigma_{eq,peak}$ distribution along the weld toe and weld root lines of the tube-bracket lower link weld, the peak of $\Delta\sigma_{eq,peak} = 919\ MPa$ is reached at the tube-side toe at an angle equal to $\beta = 180°$ on the tension side, the relevant local biaxiality ratio $\lambda = 0$ suggest that the crack initiation is mostly driven by the tension mode I linear elastic peak stress contribution.

Even if the $\Delta\sigma_{eq,peak}$ value in correspondence of the experimental crack initiation location is 11% below the maximum value detected, it is more likely here that the crack will initiate because of the tension mode I contribution, on the contrary, the absolute maximum of $\Delta\sigma_{eq,peak} = 1023\ MPa$ comes with a compression mode I contribution which tends to close the fatigue crack; this consideration is supported also by a previous study [39].

The experimental fatigue datum has been converted from the applied load range ΔF to the equivalent peak stress $\Delta\sigma_{eq,peak}$, in addition, to account for the two load blocks applied to the axle, the Linear Damage Rule (LDR) proposed by Palmgren and Miner has been coupled with the PSM as reported in the Introduction section in agreement with the guidelines given in a previous contribution [29]. The resulting value of $\Delta\sigma_{eq,peak}$, relevant to the crack initiation point, is equal to $\Delta\sigma_{eq,peak} = 562\ MPa$ with an associated number of cycles to technical crack initiation N_i = 555 000 cycles. The experimental datum expressed in terms of $\Delta\sigma_{eq,peak}$ is presented along with the PSM design scatter band in figure 14, which shows the theoretical estimations being on the safe side with respect to the experimental datum. Furthermore, the experimental data relevant to steel welded joints under both constant amplitude [25] and variable amplitude loadings [29], being used to validate the PSM design scatter band, are presented in figure 14 for comparison purposes.

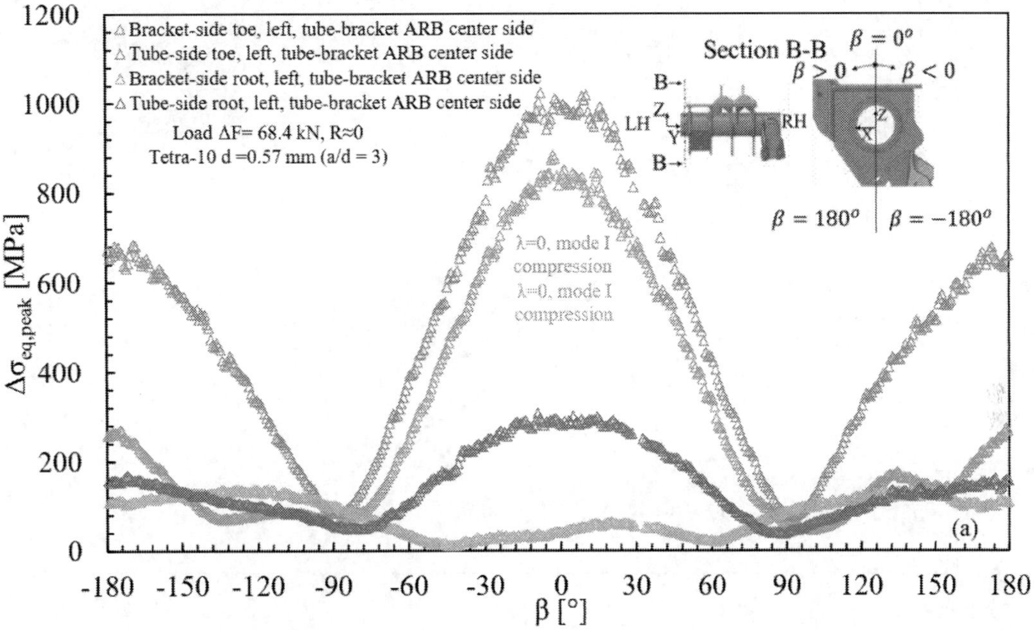

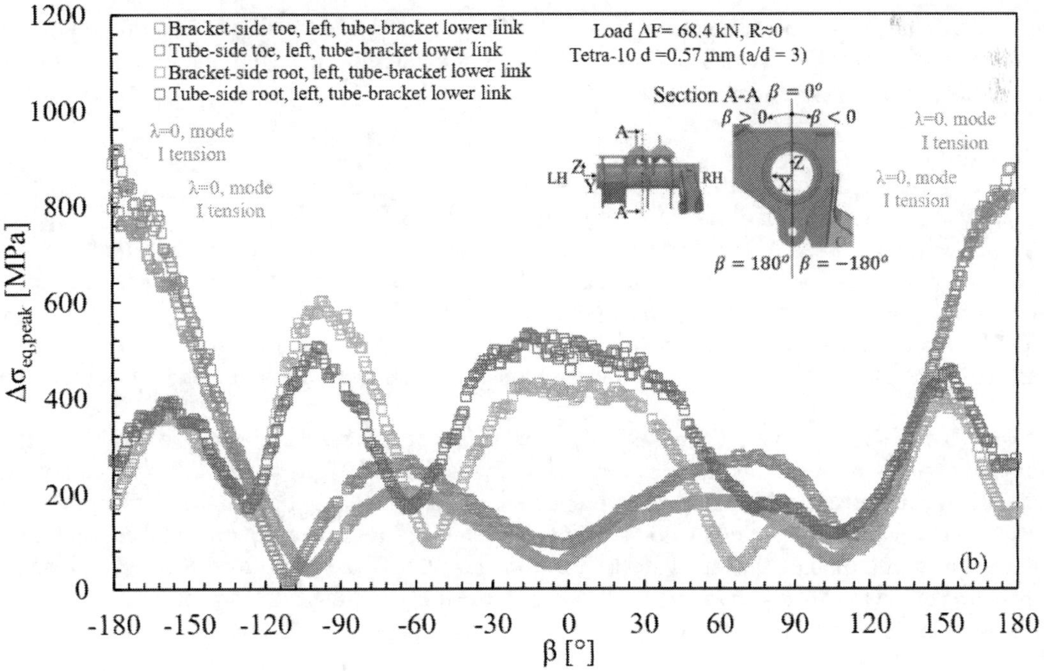

Figure 13. Distribution of the equivalent peak stress range along the weld toe and weld root lines for (a) tube-bracket ARB center side weld and (b) tube-bracket lower link weld. The stress values are related

to a nominal force range applied equal to 68.4 kN and they have been evaluated from Eq. (1). The biaxiality ratios have been calculated from Eq. (7).

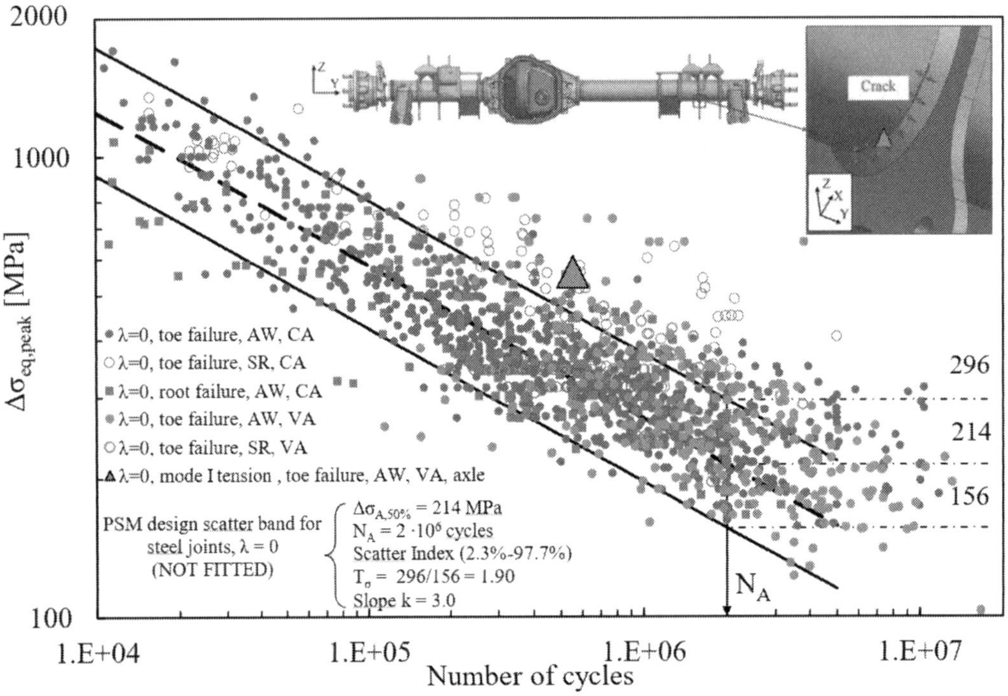

Figure 14. Fatigue assessment according to the PSM of the tube-side toe failure in correspondence of the tube-bracket lower link weld. The PSM design scatter band relevant to the fatigue assessment steel joints under prevailing mode I was previously calibrated in Ref. [25].

5. Conclusions

In this study a rear axle for an off-road vehicle has been fatigue tested under a pure vertical load, taking advantage of a tailor-made fatigue test bench. The fatigue crack in correspondence of the fatigue test end was over 20 mm long. The welds of the axle have been grouped based on the penetration and gap values, afterwards a mean value of gap and penetration has been associated to each group of welds and then they have been introduced in the CAD model of the axle.

The FE model of the axle has been calibrated to replicate the experimental reaction forces and strains measured in the fatigue test bench; as a result a good agreement between numerical and experimental results has been achieved.

The welds of the long body of the axle have been fatigue assessed according to the PSM and an automatic tool has been adopted to speed up the analyses. The PSM proved to correctly estimate the crack initiation point according to the experimental evidences which highlighted the crack initiation and propagation being driven by the prevailing tensile mode I peak stress. The experimental datum has been evaluated in terms of the equivalent peak stress range, which has been calculated considering the contribution of the two load steps applied. By entering the relevant PSM design scatter band, the fatigue life estimation resulted on the safe side with respect to the experimental datum.

Acknowledgements
Carraro S.p.a. is gratefully acknowledged for financial support to the present research.

References

[1] 2005 Eurocode 3: Design of steel structures - part 1-9: Fatigue (CEN)

[2] Hobbacher A F 2016 *Recommendations for Fatigue Design of Welded Joints and Components. International Institute of Welding - IIW Collection.* (Springer International Publishing)

[3] Radaj D, Sonsino C M and Fricke W 2006 *Fatigue Assessment of Welded Joints by Local Approaches* (Cambridge, UK: Woodhead Publishing)

[4] Jakubczak H and Glinka G 1986 Fatigue analysis of manufacturing defects in weldments *Int J Fatigue* **8** 51–7

[5] Sonsino C M 1995 Multiaxial fatigue of welded joints under in-phase and out-of-phase local strains and stresses *Int J Fatigue* **17** 55–70

[6] Fiedler M and Vormwald M 2016 Considering fatigue load sequence effects by applying the Local Strain Approach and a fracture mechanics based damage parameter *Theoretical and Applied Fracture Mechanics* **83** 31–41

[7] Fricke W 2012 *IIW Recommendations for the Fatigue Assessment of Welded Structures By Notch Stress Analysis: IIW-2006-09* (Woodhead Publishing)

[8] Radaj D and Vormwald M 2013 *Advanced methods of fatigue assessment* (Springer-Verlag Berlin Heidelberg)

[9] Susmel L 2008 Modified Wöhler curve method, theory of critical distances and Eurocode 3: A novel engineering procedure to predict the lifetime of steel welded joints subjected to both uniaxial and multiaxial fatigue loading *Int J Fatigue* **30** 888–907

[10] Susmel L 2009 Three different ways of using the Modified Wöhler Curve Method to perform the multiaxial fatigue assessment of steel and aluminium welded joints *Eng Fail Anal* **16** 1074–89

[11] Baumgartner J, Schmidt H, Ince E, Melz T and Dilger K 2015 Fatigue assessment of welded joints using stress averaging and critical distance approaches *Welding in the World* **59** 731–42

[12] Karakaş Ö, Leitner M and Tüzün N 2022 Application of critical distance approach for fatigue assessment of welded and HFMI-treated steel joints *Int J Fatigue* **154**

[13] Susmel L 2009 *Multiaxial notch fatigue* (Cambridge, UK: Woodhead Publishing)

[14] Carpinteri A, Spagnoli A and Vantadori S 2009 Multiaxial fatigue life estimation in welded joints using the critical plane approach *Int J Fatigue* **31** 188–96

[15] Lazzarin P and Tovo R 1998 A notch intensity factor approach to the stress analysis of welds *Fatigue Fract Eng Mater Struct* **21** 1089–103

[16] Lazzarin P, Sonsino C M and Zambardi R 2004 A notch stress intensity approach to assess the multiaxial fatigue strength of welded tube-to-flange joints subjected to combined loadings *Fatigue Fract Eng Mater Struct* **27** 127–40

[17] Lazzarin P and Zambardi R 2001 A finite-volume-energy based approach to predict the static and fatigue behavior of components with sharp V-shaped notches *Int J Fract* **112** 275–98

[18] Livieri P and Lazzarin P 2005 Fatigue strength of steel and aluminium welded joints based on generalised stress intensity factors and local strain energy values *Int J Fract* **133** 247–76

[19] Lazzarin P, Livieri P, Berto F and Zappalorto M 2008 Local strain energy density and fatigue strength of welded joints under uniaxial and multiaxial loading *Eng Fract Mech* **75** 1875–89

[20] Berto F and Lazzarin P 2009 A review of the volume-based strain energy density approach applied to V-notches and welded structures *Theoretical and Applied Fracture Mechanics* **52** 183–94

[21] Fischer C, Fricke W and Rizzo C M 2016 Review of the fatigue strength of welded joints based on the notch stress intensity factor and SED approaches *Int J Fatigue* **84** 59–66

[22] Fischer C, Fricke W and Rizzo C M 2016 Experiences and recommendations for numerical analyses of notch stress intensity factor and averaged strain energy density *Eng Fract Mech* **165** 98–113

[23] Meneghetti G 2008 The peak stress method applied to fatigue assessments of steel and aluminium fillet-welded joints subjected to mode I loading *Fatigue Fract Eng Mater Struct* **31** 346–69
[24] Meneghetti G and Lazzarin P 2007 Significance of the elastic peak stress evaluated by FE analyses at the point of singularity of sharp V-notched components *Fatigue Fract Eng Mater Struct* **30** 95–106
[25] Meneghetti G and Lazzarin P 2011 The peak stress method for fatigue strength assessment of welded joints with weld toe or weld root failures *Welding in the World* **55** 22–9
[26] Meneghetti G 2012 The use of peak stresses for fatigue strength assessments of welded lap joints and cover plates with toe and root failures *Eng Fract Mech* **89** 40–51
[27] Meneghetti G 2013 The peak stress method for fatigue strength assessment of tube-to-flange welded joints under torsion loading *Welding in the World* **57** 265–75
[28] Meneghetti G and Campagnolo A 2020 State-of-the-art review of peak stress method for fatigue strength assessment of welded joints *Int J Fatigue* **139** 105705
[29] Campagnolo A, Vecchiato L and Meneghetti G 2022 Multiaxial variable amplitude fatigue strength assessment of steel welded joints using the peak stress method *Int J Fatigue* **163** 107089
[30] Prabhakar M, Prasad A K, Paswan M K and Tendulkar V 2019 Fatigue life Prediction of welding in Axle Housing using Notch Stress approach *Mater Today Proc* **22** 2233–40
[31] Meneghetti G, Campagnolo A and Rigon D 2017 Multiaxial fatigue strength assessment of welded joints using the Peak Stress Method – Part I: Approach and application to aluminium joints *Int J Fatigue* **101** 328–42
[32] Meneghetti G, Campagnolo A and Rigon D 2017 Multiaxial fatigue strength assessment of welded joints using the Peak Stress Method – Part II: Application to structural steel joints *Int J Fatigue* **101** 343–62
[33] Meneghetti G, Campagnolo A, Avalle M, Castagnetti D, Colussi M, Corigliano P, de Agostinis M, Dragoni E, Fontanari V, Frendo F, Goglio L, Marannano G, Marulo G, Moroni F, Pantano A, Rebora A, Scattina A, Spaggiari A and Zuccarello B 2018 Rapid evaluation of notch stress intensity factors using the peak stress method: Comparison of commercial finite element codes for a range of mesh patterns *Fatigue Fract Eng Mater Struct* **41** 1044–63
[34] Meneghetti G, Campagnolo A, Visentin A, Avalle M, Benedetti M, Bighelli A, Castagnetti D, Chiocca A, Collini L, de Agostinis M, de Luca A, Dragoni E, Fini S, Fontanari V, Frendo F, Greco A, Marannano G, Moroni F, Pantano A, Pirondi A, Rebora A, Scattina A, Sepe R, Spaggiari A and Zuccarello B 2022 Rapid evaluation of notch stress intensity factors using the peak stress method with 3D tetrahedral finite element models: Comparison of commercial codes *Fatigue Fract Eng Mater Struct* **45** 1005–34
[35] Campagnolo A, Roveda I and Meneghetti G 2019 The Peak Stress Method combined with 3D finite element models to assess the fatigue strength of complex welded structures *Procedia Structural Integrity* **19** 617–26
[36] Palmgren A 1924 Die lebensdauer von kugellargern *Zeitschrift Des Vereines Dtsch Ingenieure* **68** 339–41
[37] Miner M A 1945 Cumulative Damage in Fatigue *J Appl Mech* **12** A159–64
[38] Visentin A, Campagnolo A, Babini V and Meneghetti G 2022 Automated implementation of the Peak Stress Method for the fatigue assessment of complex welded structures *Forces in Mechanics* **6** 100072
[39] Meneghetti G, Campagnolo A, Babini V, Riboli M and Spagnoli A 2019 Multiaxial fatigue assessment of welded steel details according to the peak stress method: Industrial case studies *Int J Fatigue* **125** 362–80

IOP Conf. Series: Materials Science and Engineering 1275 (2023) 012045 doi:10.1088/1757-899X/1275/1/012045

Simulated and real world tests to compare drivers performance in dynamic wireless technology perspective

L Berzi*, R Barbieri, D Uggiosi and M Delogu

Università degli Studi di Firenze, Department of Industrial Engineering (DIEF), Via di S. Marta, 50139, Firenze – Italy.

*e-mail: lorenzo.berzi@unifi.it

Abstract. The ongoing electrification of road vehicles needs to be supported by proper growth of charging infrastructure. In this context, dynamic wireless charging can provide a number of advantages, main being the possibility to extend vehicle range without increasing on board battery capacity, potentially reducing cost, mass, and tank-to-wheel energy consumption. The development of such solutions however poses various questions, including the acceptance and the capability of users in driving according to infrastructure characteristics; in particular, the misalignment reduction while a driver follows a straight path positively influence the charging efficiency in dynamic wireless technology. In this paper, authors describe a tests campaign to determine driving perfomances using both a simulator and a real world equipped vehicle. The research question of this paper is to assess and quantify differences between the two approaches. To reach this objective, in a first phase, data have been collected through a driving simulator (i.e. a full car body mounted on a parallel linked kinematics with a large screen, proposing a virtual city scenario), and in a second phase through a car equipped with a camera. As a post processing phase, statistical tools have been used to describe driving performance indexes and related impacts on wireless charging infrastructure by determining the secondary voltage on the vehicle. Data coming from the activity will be functional to be used by scenario analyst to develop characterization tests only with a simulative approach to decrease costs.

1. Introduction

In the past years the interest for environmental impact of vehicles has experienced a strong increase due to the approval of the directives of the Kyoto protocol, and the effects of pollution on human health have been seriously studied [1], making car industry to produce more and more efficient vehicles, according to EURO standards [2].

Pollution effects have been discovered acting both on human health and monuments and historical masterpieces causing premature corrosion effects [3], [4].

Several studies [5],[6],[7] have been conducted showing that the effects of pollution on human and mental health cannot be neglected. Considering that transportation sector is responsible of approximately 25% of the total Greenhouse Gases (GHG) emissions [8] and more than 75% of these are attributable to ICE (Internal Combustion Engine) vehicles and the vast majority of the NOx, sulfides and particulate matter [9], it is obvious why a more eco-friendly mobility model should be preferred to the classical one. These problems especially affect many large European cities; their historical city centres often have centuries-old mobility infrastructure that cause traffic congestions and local pollution increases. One important example is the city of Firenze (Italy) [10], [11] where 5 km^2 of the city centre

Content from this work may be used under the terms of the Creative Commons Attribution 3.0 licence. Any further distribution of this work must maintain attribution to the author(s) and the title of the work, journal citation and DOI.
Published under licence by IOP Publishing Ltd

was declared UNESCO site; it hosts a large number of historical buildings together with a high population density. Nowadays, significant percentage of the urban population in the EU is exposed to pollutant concentrations higher than limit/target values stated by EU air quality legislation [11],[12]. Moreover, the high concentration of tourists led to the need of mobility, where the economy is highly dependent on the tourism [13].
Vehicle ownership is high for developed countries and tends to increase in developing ones [14].

In order to reduce the number of vehicles on the road and, consequently, pollution, several vehicle – sharing systems have been developed, but these have had little effect on new vehicle registrations, reducing them by only a few percentage points [15].
This work is aimed to deal with "old" environmental and mobility problems and new European Union plans, known as "Fit for 55" [16] package which goal is to reduce greenhouse gas emissions up to 55 % (related to 2021 goals) by 2030 and to achieve carbon neutrality by 2050. Climate change is a multidimensional matter and have to be treated as a real emergency, so that every sector has been considered in Fit for 55: from energy production up to vehicles and aircrafts [17].
According to data available, cars and van are responsible, as mentioned before, for about 75% of GHG release in the past few years, increasing every year due to higher number of vehicles on the road [18]. The European Green Deal poses various ambitious goals in front of us, considering that, as stated previously, emissions are to be zero by 2050, several questions must be analysed:

- How to develop new environmental strategies?
- How to develop less pollutant fuels in the next years?
- How to design and built new charging infrastructures?

Focusing on the latest question, "sustainable and Smart Mobility Strategy" and "TEN-T" network have to be considered: this programs are aimed to install up to 600 kW each 60 km of highway path by 2035. To achieve long-terms benefits from new environmental strategies, new methods have to be followed: smart cities are, now, really needed [19], [20].
The main drivers for developing electric vehicles (EVs) are, then, clearly outlined: environmental issues, energy crisis and decarbonization [21]. Considering that by 2035 no more internal combustion vehicles can be sold, it is legitimate to ask how existing charging infrastructures can be modified and integrated to make them more usable, facing with vehicle-grid integration (VGI) issues (see [22], [23], [24]). Nowadays electric vehicles (EV) are charged using static methods: the driver has to stop and "refill" the car according to charging power and rest time. Typically, EVs are charged during the night and this can lead to grid characteristics deterioration, congestion and stress on the charging infrastructure [25].
Charging power is one of the most important values that have to be considered while developing infrastructures: it is self-evident that facilities have to co-evolve with the EV sale rate [26], even because electric vehicles users are often experiencing "range anxiety" [27], due to short mileage of their vehicle. The trend to build and upgrade new infrastructure may lead to expensive grid expansion and high power demand [28].
With a view to reducing charging times, developing more efficient electric vehicles and reducing energy density in parking slots [29], dynamic wireless charging technology has been investigated; the application of such technology needs the solution of various challenging engineering and management problems, one of these being the acceptability from drivers and their capability to use it. In this work, authors prove that a misalignment reduction while a driver follows a straight path positively influence the charging efficiency. This technology has previously been developed in the earlier 90's and used by several researchers in the world, building prototypes and simplified models [30].
Lane keeping ability is strictly connected to dynamic wireless charging systems. Average drivers tend not to assume a good lane keeping behaviour, as stated in [31], various possible reasons can be effective on poor lane keeping conditions in a traffic flow such as road surface deformations, uneducated drivers, invisible lane markings, etc. It is expected that the continuous improvements of driver assistance systems (i.e. Advanced Driver Assistance Systems, ADAS) and automated driving technologies will progressively mitigate or avoid the issues related to imprecise lane keeping [32–34]; however most circulating vehicles are still not equipped with such devices. In addition, ADAS devices are mandatory

since July 2022 worldwide but can anyway be disabled for a number of reasons, not far investigated in this paper. All these elements suggests that an investigation on unassisted driver's capability is necessary.

The aim of this paper is to carry out simulated and real-world tests, can be difficult to manage and costly to conduct [35], to compare drivers performance in dynamic wireless technology perspective.

2. Methods

In this section researching methods are presented: a tests campaign to determine driving performances using both a simulator and a real world equipped vehicle is described. In a first phase, data have been collected through a driving simulator (i.e. a full car body mounted on a parallel linked kinematics with a large screen, proposing a virtual city scenario), and in a second phase through a car equipped with a camera. Data coming from both simulation and real-world test have been used after adopting a physical model related to magnetic flux and its evaluation as offset flux and air gap flux. As a post processing phase, statistical tools have been used to describe driving performance indexes and related impacts on wireless charging infrastructure by determining the secondary voltage on the vehicle. The results coming from the activity will be functional to be used by scenario analyst to develop characterization tests only with a simulative approach to decrease costs and reduce vehicle batteries weight.

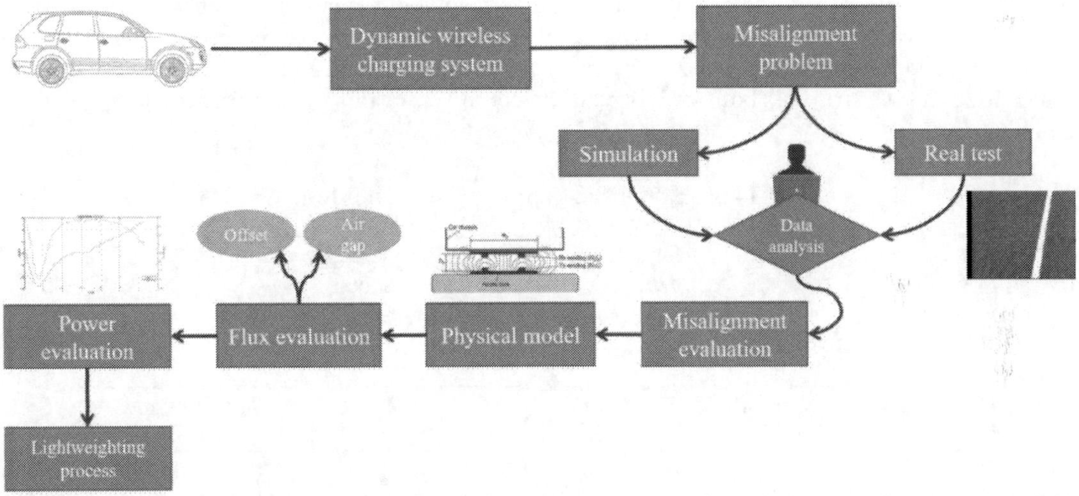

Figure 1. Method using to evaluate misalignment problem using both real and simulated tests.

2.1. Simulator test

Simulator tests have been conducted in LaSiS laboratory of University of Florence (Italy): a full car body has been mounted on a parallel linked kinematics (Figure 2, a) with a large screen, proposing a virtual city scenario (Figure 2, b) [36].

a) b)

Figure 2. a) LaSiS simulator b) A view of simulated urban path.

The participants were recruited on a voluntary basis among the students and teachers of the University of Florence. Twenty-eight subjects participated in the research, as can be noticed in Table 1. None of the subjects had previous experience with a driving simulator. The simulation has been carried out by uploading a city scenario with both straight path and corners, to evaluate performances in various driving conditions.

Table 1. Statistic sample for simulator test.

Gender		Age		Driving licence years		Driving frequency	Average speed (km/h)	Path type
M	F	min	max	min	max			
19	9	22	45	4	20	1-5	20-60	Straight and corner

The studies about gender, age and the link to risk perception have been conducted in the past (see [37], [38]). The driving frequency score was discretised on a scale of 1 to 5, with the former being a value relating to sporadic driving and the latter to very frequent driving.

The scenography, showed in Figure 3, includes traffic lights and parked cars, in order to achieve a more realistic driving experience.

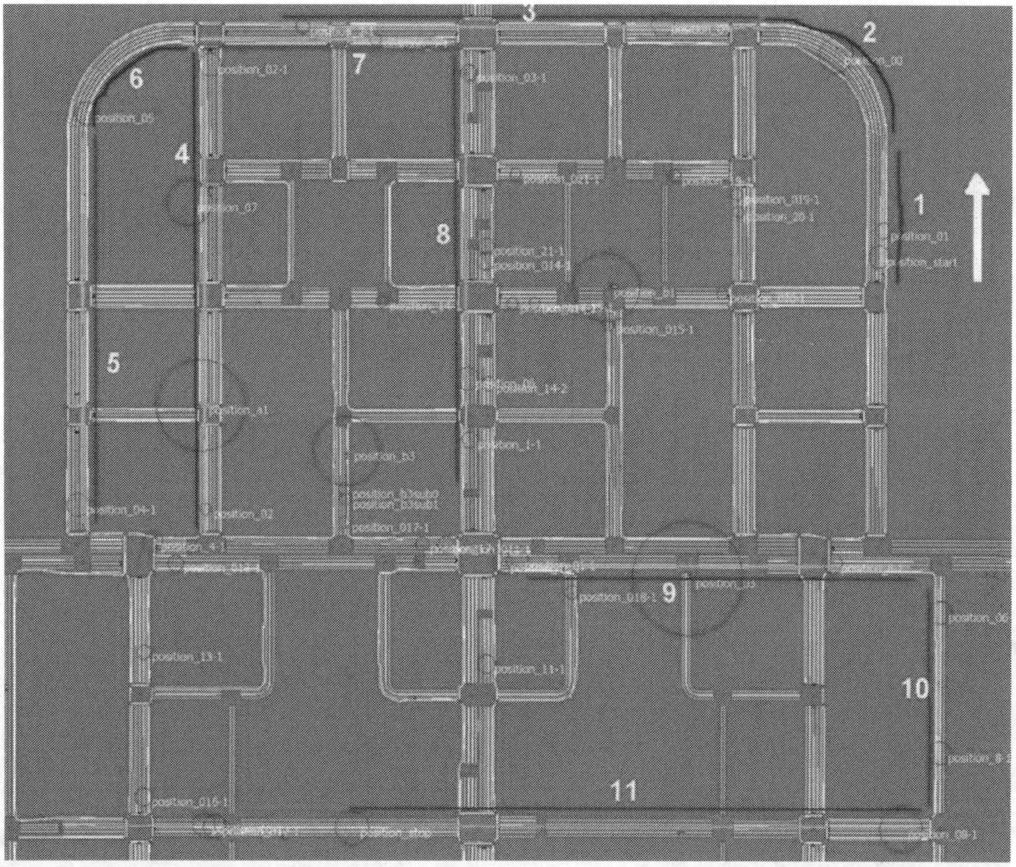

Figure 3. City scenario used during the simulated test. It includes both straight path and corners.

2.2. *Real-world test*

As for the real-world tests, they were carried out in a closed road in a safe zone, referring to Figure 4, Segment 'A' is about 50 metres long while the curved line 'B' is about 20 metres long, with a white path painted on the ground (Figure 5, a): the chosen road has all the characteristics of a normal road and in addition there were several cars parked in a fish-bone pattern during the test. The test car was equipped with a camera mounted on a specific support anchored to the vehicle bonnet (Figure 5, b). Additionally, the vehicle was not fitted with ADAS, such as lane keeping aid and adaptive cruise control [39], that are becoming more and more important in four and two-wheels vehicles [40], as the main goal of the experiment was to identify human influence in following a line while driving. Experimental sessions have been conducted in a speed range between 20 and 60 km/h . The regenerative braking systems and aspects related to performance and perception [41] are not included in this study, as a traditional ICE vehicle has been used.

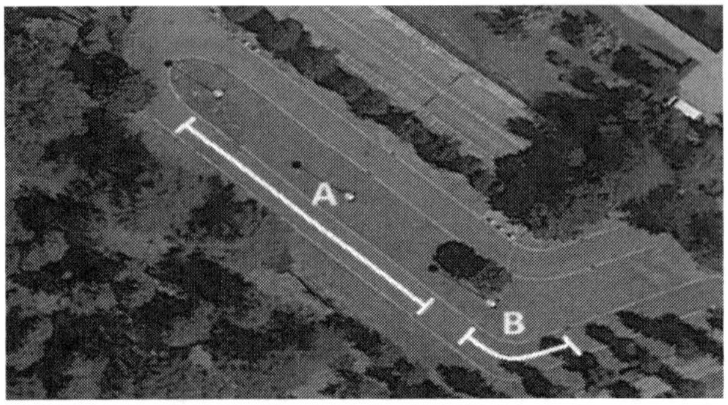

Figure 4. Test site in Florence.

a) b)

Figure 5. a) The line represents the electrified charging path. At the end of the longitudinal line there is a perpendicular one, used during image processing phase. b) Equipped test vehicle.

To carry out the test, each driver was asked to drive the route three times:

- on the first lap no support was given to the driver;
- at the beginning of the second lap simple indications were given regarding the side where the most misalignment persisted;
- on the third lap continuous indications were given to help the driver align the car.

In this way, it was possible to understand whether simple information and/or driving training could help the driver to improve his driving performance, this is a method often used during simulator tests (see [42], [43]). By means of a template, it was ascertained that the position of the camera allowed a field of view on the asphalt of 0,62 m and in case the same experiment be repeated in the future, the position of the camera could be reconstructed.

3. Post processing

In this section authors will describe how the information about misalignment have been extrapolated by footages. As a first step, frames recorder while the driver was out of the test field shown in Figure 4, where there was no white line painted, have been excluded.

Considering camera position in reference to vehicle centre of gravity, it is possible to understand how it must follow a larger path while driving along corners (see Figure 6 and Figure 7). Nevertheless, this error has not been considered during data analysis.

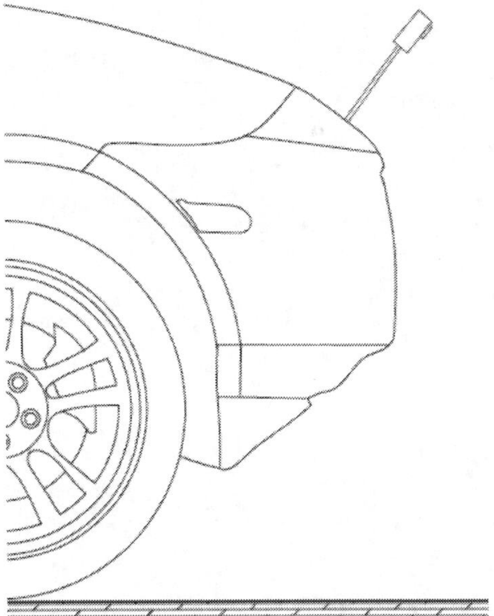

Figure 6. Lateral view of camera (in red) mounted on the vehicle using a rod (in blue).

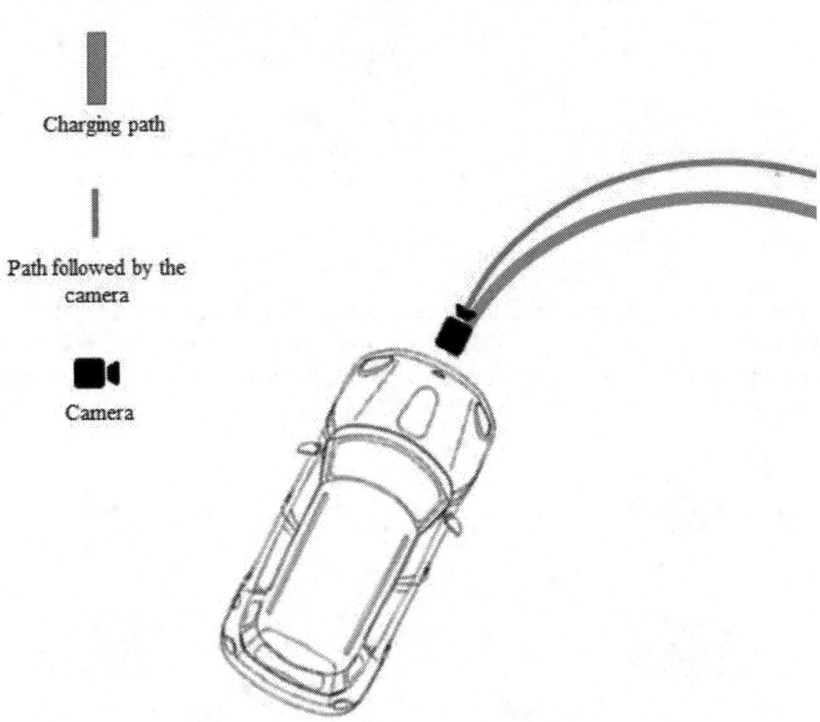

Figure 7. A schematic illustration of the path followed by the camera.

3.1. Simulink tool for footage analysis

In order to evaluate misalignment, cut videos have been analysed in Matlab-Simulink environment (see Figure 8).

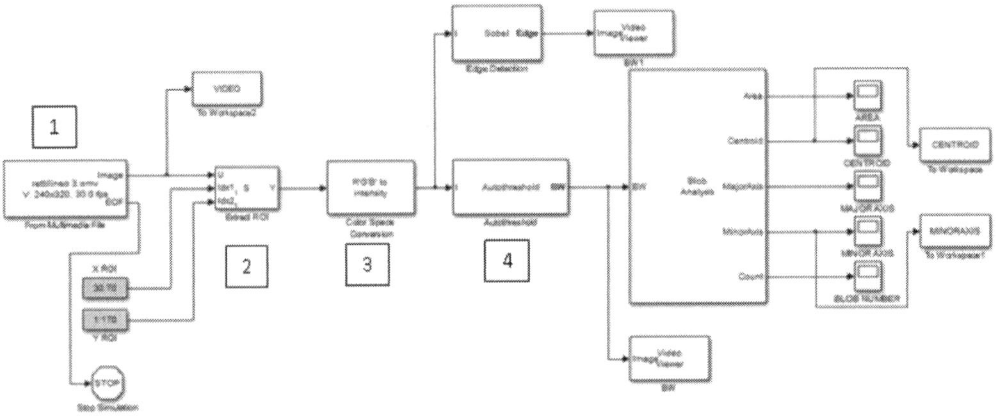

Figure 8. Matlab-Simulink code for footage analysis.

Main functions provided by the code are summarized in Table 2. This algorithm is based on "Blob analysis" function [44].

Table 2. Function involved in code and main operation provided

Execution order	Operation
1	Coloured video upload
2	Footage downsizing
3	Gray-scale transformation
4	From grey-scale to black and white
5	Blob analysis
6	Output

This process is repeated for each frame and gives a vectorial output with misalignment and time. A convention has been adopted whereby positive deviations are to the left of the main line and negative deviations are to the opposite part. Analysing data coming from real simulation, it is emerged a positive misalignment trend for every driver, this could be attributable to parallax error [45] as reported in Figure 9.

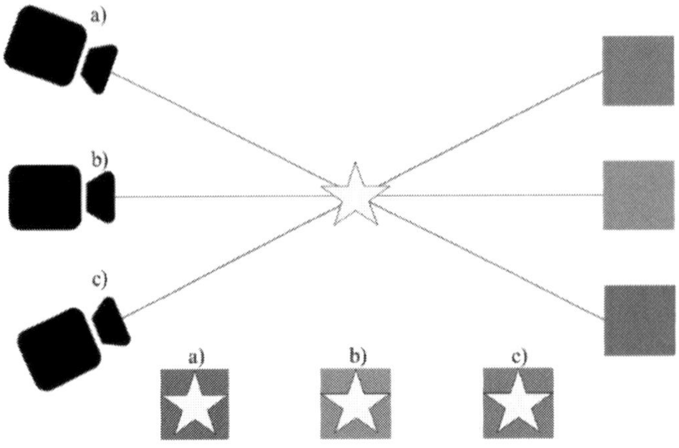

Figure 9. Parallax error.

Figure 10. A screenshot taken during real world tests.

The analysis of the films (see Figure 10) carried out with the aforementioned code also revealed some erroneous findings due to spots on the road or external elements, such as manholes and pedestrian crossings. Outliers, highlighted in red circles (see Figure 11), emerged during the analysis have not been considered, because they were not taken into account as physical data.

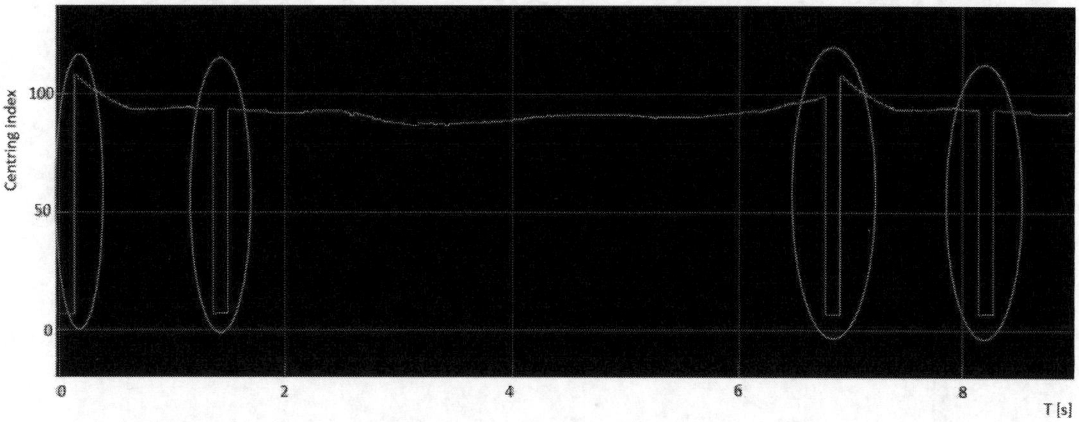

Figure 11. Centring index obtained by footage analysis vs time measured in seconds.

Centring index, provided by *blob analysis* block has been calibrated in order to link it to a physical value. Average duration of driving test analysed is between six and ten seconds.

4. Data analysis

In this section, the tools used on the data obtained from the two tests will be analysed in order to assess their statistical consistency. Data coming from simulated and real-world tests have been processed according to the procedure shown in Figure 12.

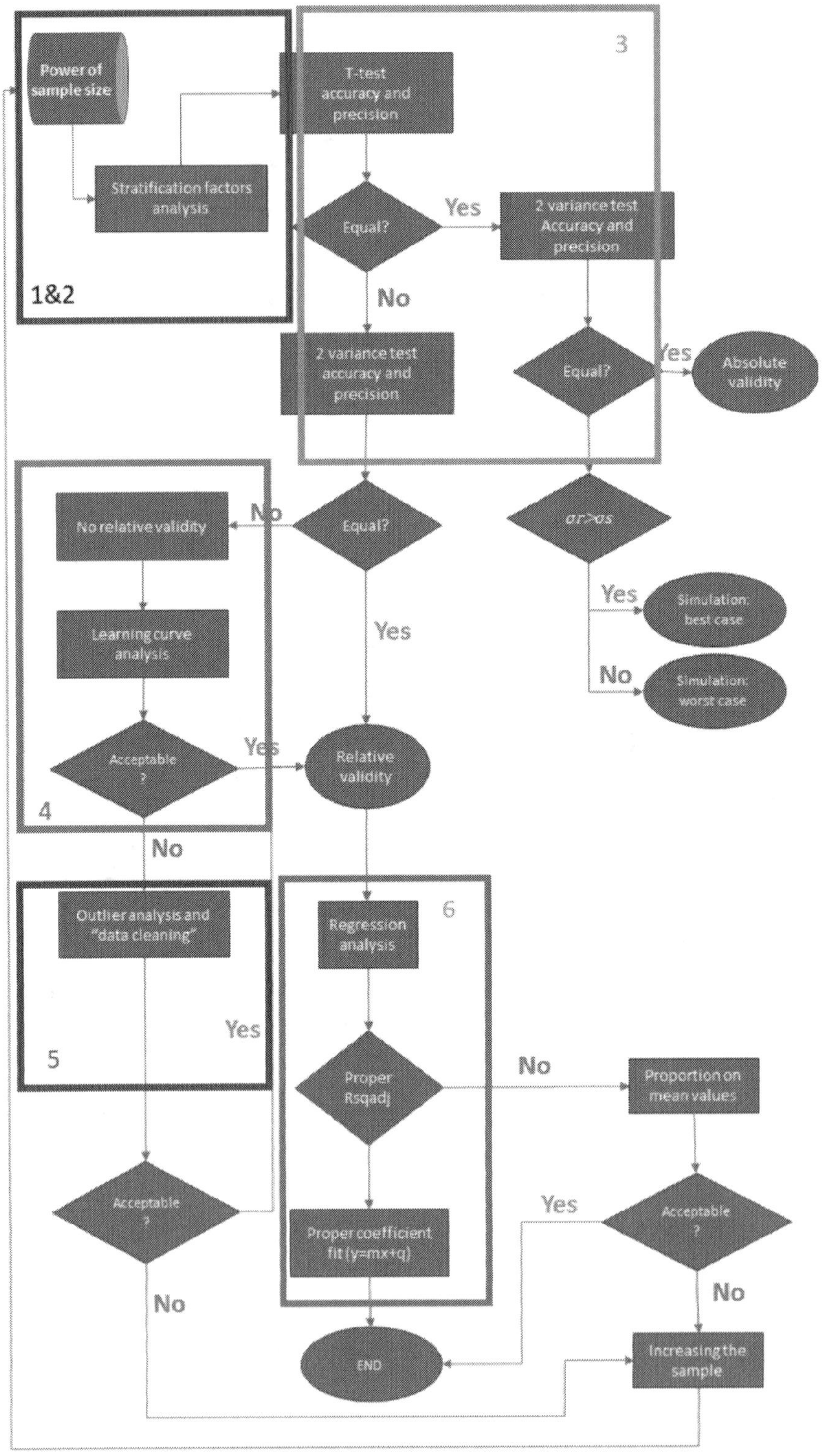

Figure 12. Flowchart used to evaluate data collected from real and simulated driving session.

The above flowchart describes the approach authors used to transform data in information: the aim of this work, in fact, is to define a model to investigate the problem of simulated vs real world tests, so it is needed to determine a repeatable procedure to compare different approaches.
Before further inspection of the statistical tools, it is necessary to distinguish between accuracy and precision in this specific test.

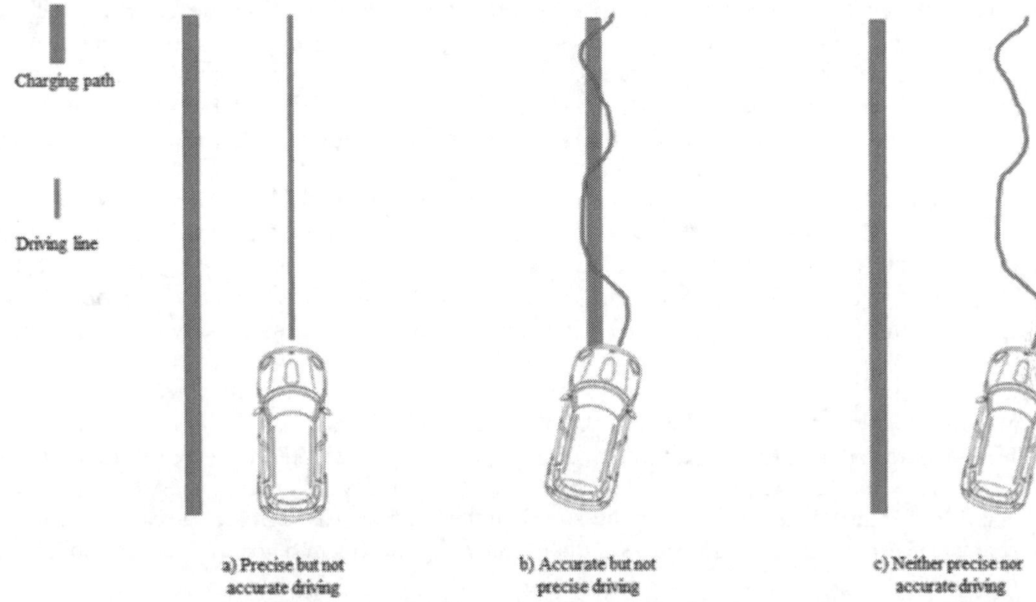

Figure 13. Precision and accuracy during driving sessions.

As visualized in Figure 13, authors define the Precision as the driver ability to keep a "straight line" path without lateral deviations from the reference line and the Accuracy as the driver ability to keep the vehicle close to the electrified path identified by the painted line on the street-floor. These parameters are inferred from the data through the study of a variable, the electric captor barycentre distance from the reference line. From here, this parameter will be called "offset". Next step is to calculate both offset average and offset standard deviation to determine a continuous statistical curve that will be used as process model. In Figure 14 it is possible to visualize an "offset" distribution with average and standard deviation so Accuracy and Precision parameter:

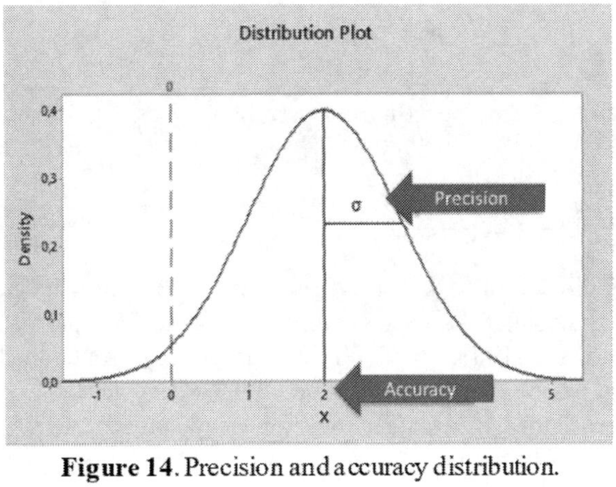

Figure 14. Precision and accuracy distribution.

Once the accuracy and precision are defined, it is possible to consider the Figure 12 algorithm. In fact, authors set up a method to infer real world behaviour starting from the simulator results, considering the actual test as a reference to compare future simulated case studies. The method main steps are described below, number of sub paragraphs are reported also in Figure 12:

1. Starting point is a Power of sample size analysis to determine if the variability in the data is adequately explained by the numerosity of the chosen sample. It is a preliminary analysis necessary to be confident of the sample significance before to start any other consideration.
2. Then it is important to choose the stratification factors. This activity will be important during the regression part that will be analysed here below. Stratification factors are factors that can be used to determine sub-population in the data to make differentiations in the sample and family analysis.
3. 2Sample t-tests and 2 variance tests: these tests compare the population average and standard deviations of the offset, respectively. The method is applied for both Accuracy and Precision. If the compared results are equal then the "absolute validity" is reached; if any of the two test average or standard deviations are different, "relative validity" is reached. "Absolute" and "relative" validity are two parameters that authors implement in this study to discriminate between two possible cases: if tests say that no difference are revealed, absolute validity means that it is possible to use simulator results for real world behaviour directly. If relative validity is reached, further studies, described in next steps, have to be carried out.
4. It is possible that learning curves behaviour can distort the driver attitude in the simulator. It is a very easy-to-use tool and so it is very easy to learn how to drive. This characteristic brings to a visible difference from the first simulator test and the last one. Through the paired t-test, authors evaluate the learning curves in the simulator between the first and the last lap searching for improvement.
5. If the results are still not acceptable authors propose an outliers analysis and their causes and proceed to eliminate the discarded values if any outlier is found. Point 4 and 5 are an iterative process to be repeated until relative validity is reached.
6. Last step is the regression analysis to determine the influencing variables in driving performance; in fact, if relative validity is reached it will be possible to determine which are the factors distorting the results. The regression analysis is a statistical method to quantify this distortion and to determine the stratification factor statistical significance. Considering Rsq-adj the correct parameter to discriminate between an acceptable and a not acceptable model, a regression coefficient proportioning in simulated test with those of the real world reference test is made.

In this algorithm, all the possibilities to go from simulated test to real world behaviour are investigated.

4.1. Physical model

According to previous studies, magnetic flux in dynamic wireless can be considered as consisting of two components:

- Air gap flux
- Displacement air flux

With regard to the first contribution, it is strictly connected with vehicle receptors' height respect to the street. In earlier applications of this charging systems (see [46],[47]), the air gap was minimized using complex mechanical systems. KAIST applied wireless charging system to an electric bus (see [48], [49]) using an average air gap of 0,25 m: this value can be retained acceptable.
The authors assumed as valid the empirical flux variation law, presented in [46] and reported in Figure 15, derived from previous experimentation.

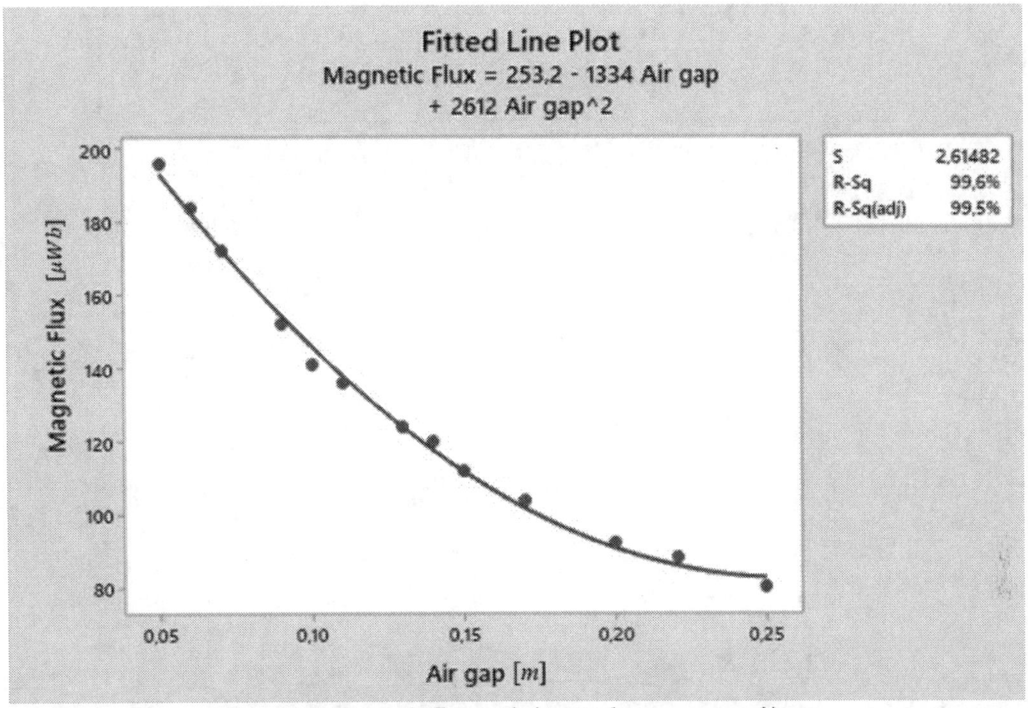

Figure 15. Magnetic flux variation vs air gap measured in m.

To assess the actual influence of this parameter, it is necessary to make a complete vehicle model, evaluating suspension displacement and tyre reaction model, developed both for automotive and motorcycle applications (see [50], [51]).

In this case, air gap, shown in Figure 16 has been considered constant during the test, assuming a smooth driving (constant speed) with no emergency brakes. Pitch angle sensible variation may affect magnetic flux and charging dynamics, in case of regenerative suspensions are considered [52].

Figure 16. Air gap evaluation on the test vehicle. During the test it is assumed to be constant due to the test circuit morphology.

Adopted relationship between magnetic flux and offset is available in Figure 17. Offset is the main parameter analysed in this study, as it is strongly dependent from drivers ability.

Previous studies [53] have shown that this factor is important in determining the magnetic flux intensity, with this in mind, vehicles with 'anti-offset abilities' were developed [54] and research activities were oriented through charging pad design [55].

The offset evaluation, provided by Matlab code previously presented, is depicted in Figure 18.

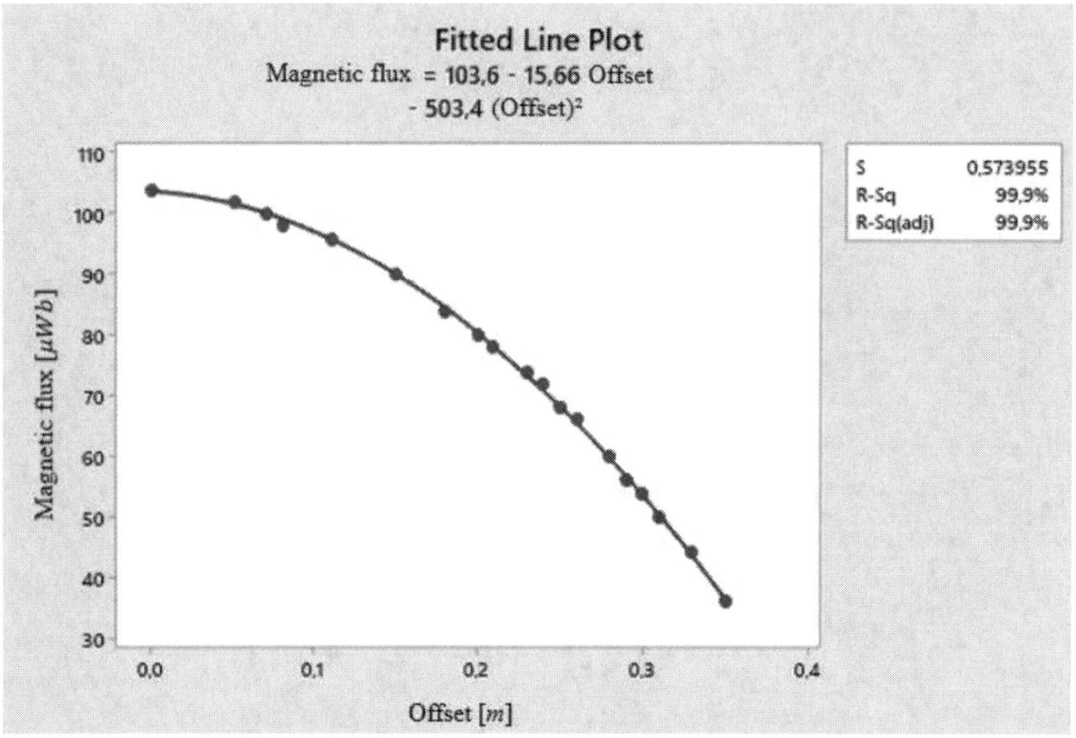

Figure 17. Magnetic flux variation vs offset measured in m.

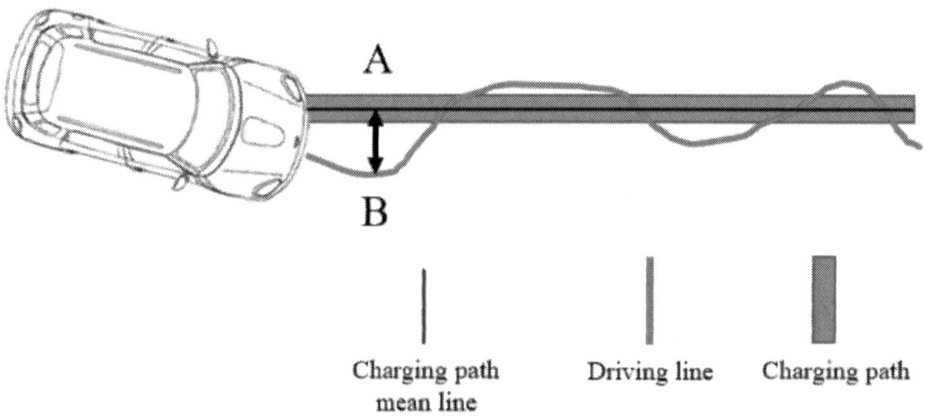

Figure 18. A schematic illustration of offset evaluation.

Authors adopted the simplified physical model presented in [53] and reported in equation (1), considering voltage on secondary coil as result of magnetic interaction between road circuit and vehicle one.

$$V = \frac{n\omega_0 \mu_0 I_s A_0}{2h_0} \cong n\omega_0 \phi_0 \quad (1)$$

Table 3. Parameters involved in **Errore. L'origine riferimento non è stata trovata.**)

Parameter	Significance	Unit	Adopted value
$\boldsymbol{V}$	Pick up voltage	[V]	
$\boldsymbol{n}$	Number of windings		45
$\boldsymbol{\omega_0}$	Electrical Angle	[rad/s]	
$\boldsymbol{\mu_0}$	Vacuum magnetic permeability		
$\boldsymbol{I_s}$	Current on primary circuit	[A]	
$\boldsymbol{A_0}$	Receptor's area	[m²]	
$\boldsymbol{h_0}$	Air gap	[m]	
$\boldsymbol{\phi_0}$	Magnetic flux	[Wb]	Depending on offset value

In wireless charging applications, the pick up voltage is considered as a secondary parameter, but it can easily be linked to power transferred from the road charging path due to its direct proportionality with respect to power.

In order to evaluate the battery State of Charge (SOC), model presented in [56] has been adopted and reported in equation (2), due to the complexity and multitude of factors involved in the model (and summarized in

Table **4**, the authors will only mention the main quantities and proportionalities.

$$SOC_{k+1}(r) = \begin{cases} SOC_k(r) - \dfrac{ECR'(r)d_k(r)}{BatCap(r)} + \dfrac{\eta_c \eta_{b,t} P\left(\frac{\tau}{60}\right)}{BatCap(r)} & if\ SOC_{k+1}(r) \leq SOC_{ub} \\ SOC_{ub}\ otherwise \end{cases} \quad (2)$$

Table 4. Parameters involved in State of Charge formula

Parameter	Significance	Unit
$\boldsymbol{k}$	The count of charging stations ina particular bus encounters during a daily operation	
$\boldsymbol{r}$	Ruote ID	
$\boldsymbol{SOC_k(r)}$	State of charge (SOC) at stop k for each route r	
$\boldsymbol{ECR'(r)}$	The adjusted energy consumption rate for each route r by considering the lightweighting effects	[kWh/mile]
$\boldsymbol{d_k(r)}$	The distance between stop k and stop k + 1 for each route r	[miles]
$\boldsymbol{\eta_c}$	The wireless charging efficiency (grid-to-battery)	
$\boldsymbol{\eta_{b,t}}$	The battery round-trip efficiency at particular year t, which fades over time from 90% to 72%	
$\boldsymbol{P}$	The charger power rate	[kW]
$\boldsymbol{\tau}$	Dwell time at a particular stop	[min]
$\boldsymbol{BatCap(r)}$	Vector of battery capacity for each route	[kWh]
$\boldsymbol{SOC_{ub}}$	The maximum SOC threshold (upper bound)	

Considering the aforementioned formula, direct proportionality between the state of charge and the dwell time, which, in turn, is inversely proportional to the speed of the vehicle, it's self-evident.

5. Conclusion and final remarks

In this chapter, authors present conclusion and final remarks, applying physical model presented above and providing dynamic wireless charging SWOT analysis.

5.1. Driving test results

As mentioned before, misalignment is an important factor in this kind of application. Authors considered several complex models presented in literature and decided to analyse voltage variation with respect to time [57].

In the next figure (from Figure 19 to Figure 21) three driving behavior have been studied:

- A low performance driving session
- A medium performance driving session
- A high-performance driving session

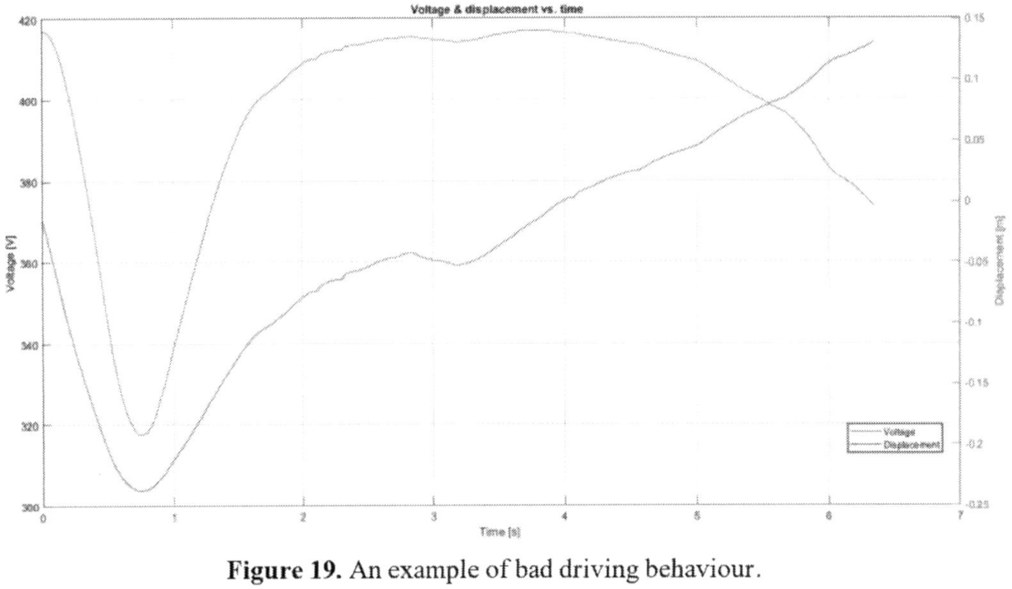

Figure 19. An example of bad driving behaviour.

In Figure 19 displacement curve, represented in red, highlights how the driver tends to derive towards positive offset after a first phase. This behaviour is reflected in the voltage which, after 2 s, drops considerably.
Considering longer driving sessions, this driving style is highly inefficient and is not suitable for a good battery charging cycle. The same considerations can also be made for driving in curved tracts, where maintaining a position consistent with that of the charging path could be challenging.

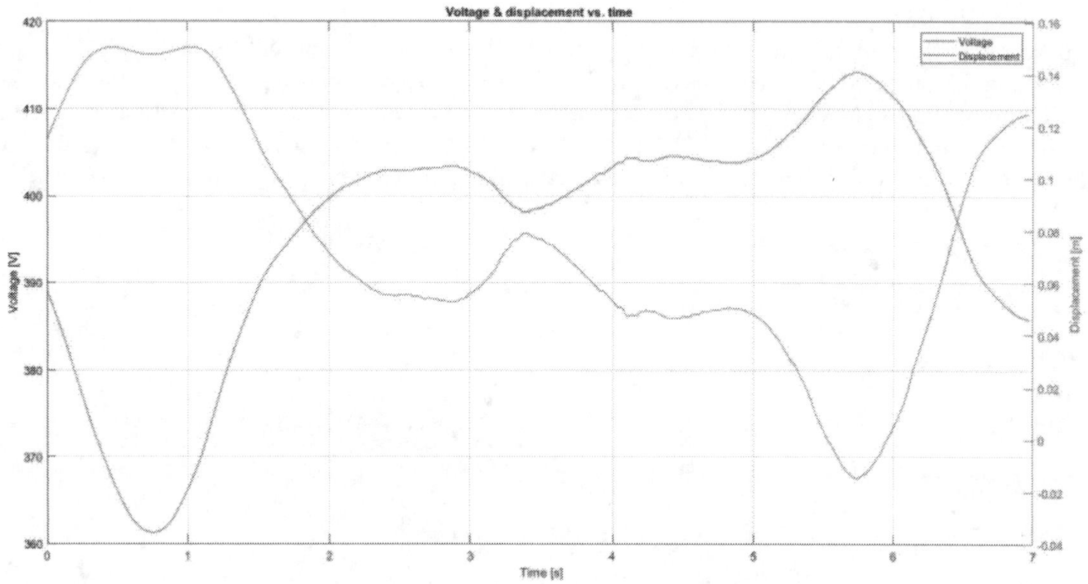

Figure 20. An example of average driving behaviour.

In Figure 20 the driver, between 2 and 5 seconds, tends to keep a constant offset, but still too far from the charging path. The average voltage value is better than the previous test-case presented.
It is important to show offset minimum and maximum: in these phases the driver tends to correct the trajectory.

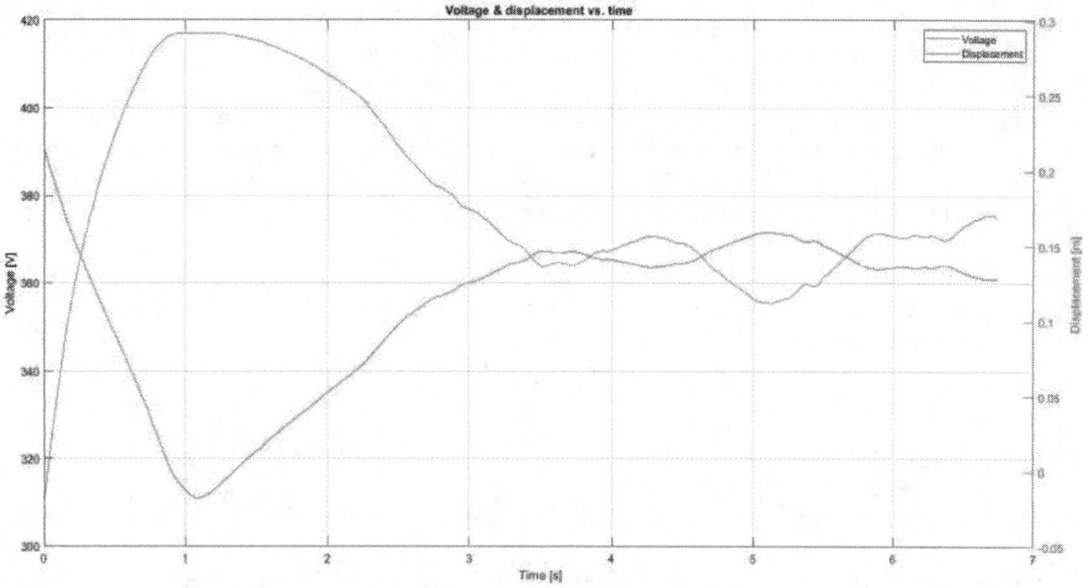

Figure 21. An example of good driving behaviour.

In Figure 22 collected data show how the driver tends to maintain a constant offset value and, thus, the voltage can be considered constant during the sampling time.

5.2. SWOT analysis
To conclude the research, authors developed a SWOT analysis to highlight technology's Strengths, Weakness, Opportunities and Threads (Figure 22). This approach is very useful in industrial applications because puts the evidence on future developments. It is interesting to underline as a final remark that technology improvements in artificial view [58] or in other field of technology will support both experimental data analysis or ADAS-like tools that will help the driver in raise up their efficiency.

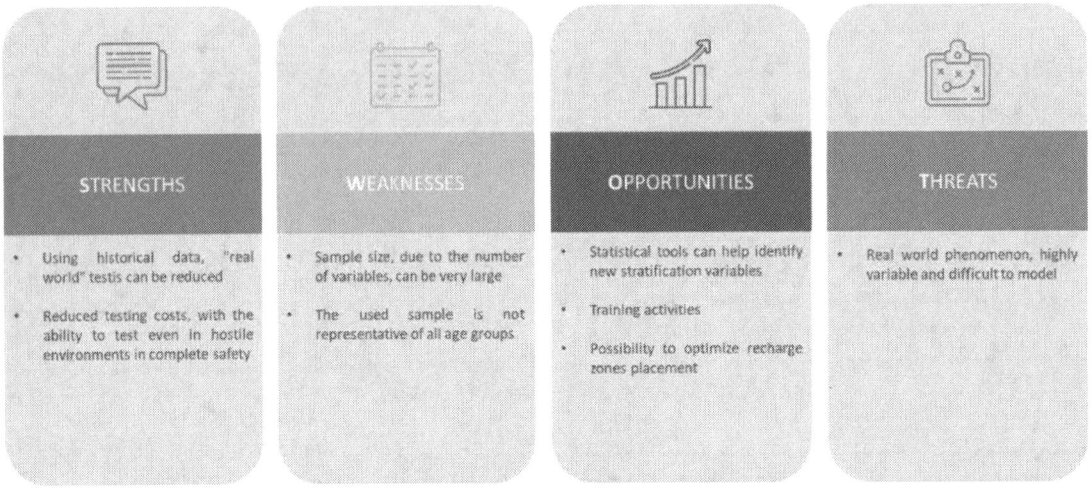

Figure 22. SWOT analysis.

5.3. Conclusions
In conclusion, dynamic wireless charging systems are suitable for use on motorway sections or on city routes where there are traffic light intersections and high traffic flows.

This work has shown how data from real-world tests can be used in order to reduce them and perform simulator tests, at the same time highlighting the importance of misalignment during the charging phase. A future development of this work will be to analyse real traffic data to create an optimisation tool for the positioning of recharging zones within a city route, training drivers with simulator driving in adverse driving conditions (snow, wet roads, cobblestones, etc.).

Acknowledgments
Authors would like to thank the researchers from the La.Si.S. laboratory of the University of Florence, Industrial Engineering school, for the support and the help.

References

[1] Kittelson D, Watts W, Johnson J, Baltensperger U and Burtscher H 2003 Gasoline Vehicle Exhaust particle sampling study *9th Diesel Engine Emissions Reduction (DEER) Workshop* (Newport, RI (US))

[2] Smith T W, Axon C J and Darton R C 2013 The impact on human health of car-related air pollution in the UK, 1995–2005 *Atmospheric Environment* **77** pp 260–6

[3] Gan W Q, Davies H W, Koehoorn M and Brauer M 2012 Association of long-term exposure to community noise and traffic-related air pollution with coronary heart disease mortality *Am J Epidemiol* **175** pp 898–906

[4] De la Fuente D, Vega J M, Viejo F, Díaz I and Morcillo M 2011 City scale assessment model for air pollution effects on the cultural heritage *Atmospheric Environment* **45** pp 1242–50

[5] Lozhkin V, Lozhkina O and Dobromirov V 2018 A study of air pollution by exhaust gases from cars in well courtyards of Saint Petersburg *Transportation Research Procedia* **36** pp 453–8

[6] Leroutier M and Quirion P 2022 Air pollution and CO2 from daily mobility: Who emits and Why? Evidence from Paris *Energy Economics* **109** p 105941

[7] Braithwaite I, Zhang S, Kirkbride J B, Osborn D P J and Hayes J F 2019 Air Pollution (Particulate Matter) Exposure and Associations with Depression, Anxiety, Bipolar, Psychosis and Suicide Risk: A Systematic Review and Meta-Analysis *Environ Health Perspect* **127** p 126002

[8] Al-Hanahi B, Ahmad I, Habibi D and Masoum M A S 2021 Charging Infrastructure for Commercial Electric Vehicles: Challenges and Future Works *IEEE Access* **9** pp 121476–92

[9] Luo Z, Wang Y, Lv Z, He T, Zhao J, Wang Y, Gao F, Zhang Z and Liu H 2022 Impacts of vehicle emission on air quality and human health in China *Science of The Total Environment* **813** p 152655

[10] Del Carmine P, Lucarelli F, Mandò P A, Valerio M, Prati P and Zucchiatti A 1999 Elemental composition of size-fractionated urban aerosol collected in Florence, Italy; preliminary results *Nuclear Instruments and Methods in Physics Research Section B: Beam Interactions with Materials and Atoms* **150** pp 450–6

[11] Gualtieri G, Crisci A, Tartaglia M, Toscano P, Vagnoli C, Andreini B P and Gioli B 2014 Analysis of 20-year air quality trends and relationship with emission data: The case of Florence (Italy) *Urban Climate* **10** pp 530–49

[12] 2008 *Directive 2008/50/EC of the European Parliament and of the Council of 21 May 2008 on ambient air quality and cleaner air for Europe* vol 152

[13] Jurkauskas A, Micevičiene D and Prunskiene J 2010 The main principles of modelling the interaction between transport infrastructure development and economy *Transport* **20** pp 117–22

[14] Dargay J, Gately D and Sommer M 2007 Vehicle ownership and income growth, worldwide: 1960-2030 *Energy Journal* **28** pp 143–70

[15] Intini M and Percoco M 2021 Car Sharing and the Impact on New Car Registration *International Encyclopedia of Transportation* ed R Vickerman (Oxford: Elsevier) pp 225–9

[16] 2019 *Regulation (EU) 2019/631 of the European Parliament and of the Council of 17 April 2019 setting CO2 emission performance standards for new passenger cars and for new light commercial vehicles, and repealing Regulations (EC) No 443/2009 and (EU) No 510/2011 (recast) (Text with EEA relevance.)* vol 111

[17] Ovaere M and Proost S 2022 Cost-effective reduction of fossil energy use in the European transport sector: An assessment of the Fit for 55 Package *Energy Policy* **168** p 113085

[18] Directorate-General for Mobility and Transport (European Commission) 2020 *EU transport in figures: statistical pocketbook 2020* (LU: Publications Office of the European Union)

[19] Gonzalez R A, Ferro R E and Liberona D 2020 Government and governance in intelligent cities, smart transportation study case in Bogotá Colombia *Ain Shams Engineering Journal* **11** pp 25–34

[20] Richter M A, Hagenmaier M, Bandte O, Parida V and Wincent J 2022 Smart cities, urban mobility and autonomous vehicles: How different cities needs different sustainable investment strategies *Technological Forecasting and Social Change* **184** p 121857

[21] Sadeghian O, Oshnoei A, Mohammadi-ivatloo B, Vahidinasab V and Anvari-Moghaddam A 2022 A comprehensive review on electric vehicles smart charging: Solutions, strategies, technologies, and challenges *Journal of Energy Storage* **54** p 105241

[22] Li B, Ma Z, Hidalgo-Gonzalez P, Lathem A, Fedorova N, He G, Zhong H, Chen M and Kammen D M 2021 Modeling the impact of EVs in the Chinese power system: Pathways for implementing emissions reduction commitments in the power and transportation sectors *Energy Policy* **149** p 111962

[23] Yu H, Niu S, Shang Y, Shao Z, Jia Y and Jian L 2022 Electric vehicles integration and vehicle-to-grid operation in active distribution grids: A comprehensive review on power architectures, grid connection standards and typical applications *Renewable and Sustainable Energy Reviews* **168** p 112812

[24] Lei X, Shang Y, Shao Z, Jia Y and Jian L 2022 Grid integration of electric vehicles for optimal marginal revenue of distribution system operator in spot market *Energy Reports* **8** pp 1061–8

[25] Brinkel N, AlSkaif T and van Sark W 2022 Grid congestion mitigation in the era of shared electric vehicles *Journal of Energy Storage* **48** p 103806

[26] Kubli M 2022 EV drivers' willingness to accept smart charging: Measuring preferences of potential adopters *Transportation Research Part D: Transport and Environment* **109** p 103396

[27] Unterluggauer T, Rich J, Andersen P B and Hashemi S 2022 Electric vehicle charging infrastructure planning for integrated transportation and power distribution networks: A review *eTransportation* **12** p 100163

[28] Muratori M 2018 Impact of uncoordinated plug-in electric vehicle charging on residential power demand *Nat Energy* **3** pp 193–201

[29] Zhang X, Yuan Z, Yang Q, Li Y, Zhu J and Li Y 2016 Coil Design and Efficiency Analysis for Dynamic Wireless Charging System for Electric Vehicles *IEEE Transactions on Magnetics* **52** pp 1–4

[30] Lukic S and Pantic Z 2013 Cutting the Cord: Static and Dynamic Inductive Wireless Charging of Electric Vehicles *IEEE Electrification Magazine* **1** pp 57–64

[31] Metin A 2020 A new evaluation method to quantify drivers' lane keeping behaviors on urban roads *Transportation Letters* **12** pp 738–49

[32] Cho R L-T, Liu J S and Ho M H-C 2021 The development of autonomous driving technology: perspectives from patent citation analysis *Transport Reviews* **41** pp 685–711

[33] Chetan N B, Gong J, Zhou H, Bi D, Lan J and Qie L 2020 An Overview of Recent Progress of Lane Detection for Autonomous Driving *2019 6th International Conference on Dependable Systems and Their Applications (DSA)* 2019 6th International Conference on Dependable Systems and Their Applications (DSA) pp 341–6

[34] Nidamanuri J, Nibhanupudi C, Assfalg R and Venkataraman H 2022 A Progressive Review: Emerging Technologies for ADAS Driven Solutions *IEEE Transactions on Intelligent Vehicles* **7** pp 326–41

[35] Carsten O, Kircher K and Jamson S 2013 Vehicle-based studies of driving in the real world: The hard truth? *Accident Analysis & Prevention* **58** pp 162–74

[36] Domenichini L, La Torre F, Vangi D, Virga A and Branzi V 2017 Influence of the lighting system on the driver's behavior in road tunnels: A driving simulator study *Journal of Transportation Safety & Security* **9** pp 216–38

[37] Rhodes N and Pivik K 2011 Age and gender differences in risky driving: The roles of positive affect and risk perception *Accident Analysis & Prevention* **43** pp 923–31

[38] Song X, Yin Y, Cao H, Zhao S, Li M and Yi B 2021 The mediating effect of driver characteristics on risky driving behaviors moderated by gender, and the classification model of driver's driving risk *Accident Analysis & Prevention* **153** pp 106038

[39] Monfort S S, Reagan I J, Cicchino J B, Hu W, Gershon P, Mehler B and Reimer B 2022 Speeding behavior while using adaptive cruise control and lane centering in free flow traffic *Traffic Injury Prevention* **23** pp 85–90

[40] Savino G, Lot R, Massaro M, Rizzi M, Symeonidis I, Will S and Brown J 2020 Active safety systems for powered two-wheelers: A systematic review *Traffic Injury Prevention* **21** pp 78–86

[41] Berzi L, Delogu M and Pierini M 2016 Development of driving cycles for electric vehicles in the context of the city of Florence *Transportation Research Part D: Transport and Environment* **47** pp 299–322

[42] Yahoodik S and Yamani Y 2021 Effectiveness of risk awareness perception training in dynamic simulator scenarios involving salient distractors *Transportation Research Part F: Traffic Psychology and Behaviour* **81** pp 295–305

[43] Prohn M J and Herbig B 2020 Evaluating the effects of a simulator-based training on knowledge, attitudes and driving profiles of German ambulance drivers *Accident Analysis & Prevention* **138** p 105466

[44] Mathworks 2022, Blob Analysis - Statistics for labeled regions - Simulink, https://it.mathworks.com/help/vision/ref/blobanalysis.html (accessed Nov. 05, 2022).

[45] Stephens J, Bostjancic M and Koskulitz T 2019 A Study on Parallax Error in Video Analysis *The Physics Teacher* **57** p 193

[46] Choi S Y, Gu B W, Jeong S Y and Rim C T 2015 Advances in Wireless Power Transfer Systems for Roadway-Powered Electric Vehicles *IEEE Journal of Emerging and Selected Topics in Power Electronics* **3** pp 18–36

[47] Lee C H, Jung G, Hosani K A, Song B, Seo D and Cho D 2020 Wireless Power Transfer System for an Autonomous Electric Vehicle *2020 IEEE Wireless Power Transfer Conference (WPTC)* 2020 IEEE Wireless Power Transfer Conference (WPTC) pp 467–70

[48] Jung G, Song B, Shin S, Lee S, Shin J, Kim Y, Lee C and Jung S 2013 Wireless charging system for On-Line Electric Bus(OLEB) with series-connected road-embedded segment *2013 12th International Conference on Environment and Electrical Engineering* 2013 12th International Conference on Environment and Electrical Engineering pp 485–8

[49] Shin J, Song B, Lee S, Shin S, Kim Y, Jung G and Jeon S 2012 Contactless power transfer systems for On-Line Electric Vehicle (OLEV) *2012 IEEE International Electric Vehicle Conference* 2012 IEEE International Electric Vehicle Conference pp 1–4

[50] Muzakkir Ahamed M and Natrayan L 2022 Structural and dynamic analysis of automotive quarter car model suspension system for different materials *Materials Today: Proceedings* **62** pp 2291–6

[51] Bartolozzi M, Savino G and Pierini M 2021 Novel high-fidelity tyre model for motorcycles to be characterised by quasi-static manoeuvres – rationale and numerical validation *Vehicle System Dynamics* **0** pp 1–27

[52] Crolla D A and Cao D 2012 The impact of hybrid and electric powertrains on vehicle dynamics, control systems and energy regeneration *Vehicle System Dynamics* **50** pp 95–109

[53] Choi Y, Lee S and Kim J 2010 The Lateral Position Estimation System for Autonomous Driving of the Online Electric Vehicle *IFAC Proceedings Volumes* **43** pp 407–12

[54] Wen F, Zhang D, Han C, Zhang G, Li G, Zhang X, Ma J, Yao Z and Yu K 2022 Anti-offset performance optimization of coupling coils in wireless power transfer system based on genetic algorithm *Energy Reports* **8** pp 1–9

[55] Ahmad A, Alam M S and Mohamed A A S 2019 Design and Interoperability Analysis of Quadruple Pad Structure for Electric Vehicle Wireless Charging Application *IEEE Transactions on Transportation Electrification* **5** pp 934–45

[56] Bi Z, Keoleian G A and Ersal T 2018 Wireless charger deployment for an electric bus network: A multi-objective life cycle optimization *Applied Energy* **225** pp 1090–101

[57] Wang Z, Shi Y, Ma X and Wei Y 2020 A coreless hybrid magnetic coupling coil structure for dynamic wireless power charging system *Energy Reports* **6** pp 843–50
[58] Galteri L, Seidenari L, Bertini M, Uricchio T and Del Bimbo A 2019 Fast Video Quality Enhancement using GANs *Proceedings of the 27th ACM International Conference on Multimedia* MM '19 (New York, NY, USA: Association for Computing Machinery) pp 1065–7

Time-of-Flight measurement on a rail by laser-ultrasound

D Vangi and M S Gulino*

Università degli Studi di Firenze, Department of Industrial Engineering of Florence, Via di Santa Marta 3, 50139, Firenze, Italy

E-mail: michelangelosanto.gulino@unifi.it

Abstract. Stress states of thermal origin on the Continuously Welded Rail (CWR) are a source of possible inconvenience due to winter breakage or dangerous summer buckling. To date, alongside procedures based on the measurement of rail displacements or rail cutting, several methods for measuring the stress state in the rail have been proposed, based on X-ray diffraction, Barkhausen noise, and the acoustoelastic effect; all these approaches have some critical issues for extensive use in the field. The acoustoelastic method is based on the dependence of the ultrasonic wave speed on the stress state of the material and involves the measurement of the Time of Flight (ToF). Because of the modest acoustoelastic effect, the speed variations caused by the stress states are very limited, which is reflected in the need for ToF measurements with an accuracy of the order of a few nanoseconds.

In the past, the research group successfully used the acoustoelastic method to monitor a 3 km portion of a rail between Florence and Pontassieve, with fixed measurement stations, where the ultrasonic probes were permanently positioned on the neutral axis of the rail. Today, the need is to have a mobile measuring system, which can measure in several points or even continuously as the system moves along the rail. This appears to be possible through the use of, for example, non-contact systems such as those based on laser and air-coupled probes or lasers alone. In the first case, however, the acoustic wave partly propagates in the air before reaching the piezoelectric probe, and this path is subjected to variations in temperature and thus ToF. In the second case, the ultrasonic waves are generated and detected by optical methods with lasers. For the generation of acoustic waves, the use of pulsed lasers has become a widespread, reliable practice, which allows researchers to obtain signals of high amplitude and wide bandwidth. For the reception of the acoustic signal, the precision required in the measurement involves careful analysis and experimentation to identify the most appropriate laser technique, depending on the quality of the surface on which the measurement is performed, the sensitivity of the method, and the possibility of taking measurements while moving.

In the present work, a technique for laser detection of ultrasonic waves, Optical Beam Deflection (OBD), is analysed. The technique is based on the deflection of an optical beam: the beam irradiates the surface of the rail and is reflected at angles that depend on the orientation of the surface at the point of measurement, which is perturbed by the passage of the acoustic wave. The technique, developed in the 1990s, has limited applications in non-destructive testing, where interferometer-based techniques are more widespread; still, they require more sophisticated and expensive equipment. Different experimental arrangements and initial experimental results are analysed and presented, to optimise their sensitivity and verify their suitability for the purpose.

Content from this work may be used under the terms of the Creative Commons Attribution 3.0 licence. Any further distribution of this work must maintain attribution to the author(s) and the title of the work, journal citation and DOI.
Published under licence by IOP Publishing Ltd

IOP Conf. Series: Materials Science and Engineering 1275 (2023) 012046 doi:10.1088/1757-899X/1275/1/012046

1. Introduction

The introduction of Continuously Welded Rails (CWRs) into railway operation from the 1960s onwards, dictated by demands for running comfort and increased train speeds, led to the need for careful control of the rail stress state. At all stages of rail manufacturing, installation and operation, stresss are produced whose superimposition determines the overall stress state. Of these, that produced by thermal variations is undoubtedly the highest and most dangerous for the rail. The rails are laid in such a way that the thermal stress state is zero at a given temperature, known as the neutral temperature. The expansion of the rail is prevented, in the ideal case, by the set of constraints constituted by the attaching organs, the sleeper, and the ballast. This prevented expansion causes an increase in stress of approximately 2.5 MPa/°C, which corresponds, for a 60 UIC type rail, to approximately 18.6 kN/°C. Variations from the neutral temperature of more than 30°C can be reached between summer and winter, resulting in very high loads and the danger of buckling in hot weather. However, the restraints that hold the rail do not guarantee the absence of displacements with respect to the ground, which manifest themselves as sliding of the rail with respect to the attachments or movement of the sleepers on the ballast. These sliding movements cause the neutral temperature to vary over time, so that it is no longer sufficient to measure the temperature of the rail to know the thermal stress state. To avoid excessively high stress states, it is therefore necessary to know the neutral temperature or stress state of the rail at various points and over time. This has led to the study and application of various measurement systems, which, however, have disadvantages in terms of both applicability and invasiveness and cost in terms of time and instrumentation. The main methods used in maintenance practice to assess the state of stress in rails to date are [1, 2]:

- the cutting of the rail in a given section and the subsequent measurement of the displacements of the two edges. In this way, it is possible to determine, to a certain degree of approximation, the value of longitudinal stress present in the rail prior to cutting;
- measurement of the relative displacement of the rail relative to the ground of several points. If the distance between the points decreases, then there is a state of compression in that section and vice versa; it is also possible to tell in which direction the rail material is moving. Although this technique is not destructive, it is time-consuming and not very accurate or reliable;
- measurement of the temperature difference from the nominal neutral temperature, assuming completely prevented expansion. This system is approximate in that it does not take into account any failure of the constraints, which cause the rail to slip;
- lifting the rail, detached from the restraints for a certain distance upstream and downstream, until a predetermined height is reached. The force required for lifting is related to the axial stress – there is a commercial instrumentation called "Vertical Rail Stiffness Equipment" (VERSE) to perform such a procedure.

Faced with these techniques, which are either too costly or approximate, other non-destructive techniques have been studied, summarised in Vnenk and Culek [1], based on X-ray diffraction, the correlation between Barkhausen magnetic noise from the material surface and the stress state, and techniques based on the acoustoelastic effect, which correlates the propagation speed of bulk (and guided) acoustic waves with the stress state [3, 4, 5]. The latter technique has developed into different methods, depending on the type of ultrasonic waves used and the wave generation and reception technique.

Despite the long period of experimentation, a satisfactory solution to the problem has not yet been reached and only in recent years some applications have emerged from the prototype or laboratory stage, which have been field-tested and made operational in a semi-engineered form. In the past [6, 7], the research group successfully used the acoustoelastic method with critically refracted longitudinal bulk waves to monitor a 3 km stretch of rail between Florence and Pontassieve, with fixed measurement stations, where the ultrasonic probes were permanently positioned on the neutral axis of the rail. Today, the need is to have a mobile measurement system, which can take measurements at several points or even continuously while the system moves along the rail. This has oriented international research towards non-contact measuring systems such as laser-ultrasound or EMAT [3]. The precision required in the measurement entails careful analysis and experimentation to identify the most appropriate technique, depending on the quality of the surface on which the measurement is made, the sensitivity of the method and the possibility of taking measurements while moving.

In the present work, techniques based on acoustoelasticity are considered. The principle behind these techniques is the variation of the speed of ultrasonic waves as a function of the stress state in the material, due to the non-perfect linearity between stress and deformation, which is also present in the macroscopically considered linear elastic field. To describe this non-linearity, the two elastic constants of Lamé are generally added to the third-order constants of Murnaghan, which allow the speed dependence of longitudinal, transverse, and surface waves to be expressed as a function of the stress state [8]. For an uniaxial stress state, a relation between Time-of-Flight (ToF) and stress of the type is obtained:

$$\frac{t_0 - t}{t} = K\sigma \tag{1}$$

Wave velocity is also affected by temperature and microscopic structure of the material; therefore, it is necessary to use temperature-dependent acoustoelastic constants and make calibrations to account for the structure of the particular material being measured.

The velocity measurement is generally made on a reference of known length and is derived by measuring ToF. Velocity variations due to the state of stress are of the order of tens of ppm, so it is generally necessary to carry out the ToF measurement with the accuracy of a few ns. This causes some difficulties for those techniques that use contact probes to obtain an absolute measurement of the state of stress, linked to the exact knowledge of the propagation path (which can vary depending on the thickness of the couplant and also with temperature).

Since the acoustoelastic effect has a different intensity depending on the type of ultrasonic wave used, techniques that allow two different types of waves to be propagated and received with a single probe are particularly useful. For example, with electromagnetic probes (EMAT), it is possible to propagate two transverse waves polarised orthogonally to each other and receive them with the same probe, so that their acoustic path is the same. In this case, the difference in the arrival times of the waves is proportional to the voltage state (acoustic birefringence):

$$t_1 - t_2 = k\sigma \tag{2}$$

A similar effect occurs using Raleigh waves, which propagate with very low attenuation on the surface of the material and in which the particles perform an elliptical motion. The components of horizontal motion (aligned to the uniaxial stress state) and vertical motion (perpendicular to the uniaxial stress state), have different speed and this generates a polarisation that also depends on the stress state [3, 9, 10].

In the present work, the ultrasonic wave detection technique Optical Beam Deflection (OBD) [11] is analysed. The technique is based on the deflection of a laser beam: the beam strikes the rail surface and is reflected at angles that depend on the interaction with the surface, which is perturbed by the passage of the acoustic wave. The technique has few applications in non-destructive inspection, where interferometer-based techniques are more widespread; still, these require more sophisticated and costly instrumentation and precise focusing of the laser, which is unsuitable for moving inspection [12]. The work presents the results of an initial approach to the OBD technique, with the aim of assessing the parameters influencing the measurement, the best experimental arrangements and the suitability for measuring ToF on the rail and the potential for making measurements in motion.

2. The OBD technique

The OBD technique, developed since the 1990s, has found its main application in the field of atomic force microscopy [11], while it is not used in non-destructive testing mainly because it requires highly-reflective surfaces. The principle is based on the deflection of a laser beam when it insists on a surface where an ultrasonic wave propagates [13]. The ultrasonic wave causes a perturbation of the surface that results in a change in the local angle of incidence of the beam with respect to the surface, consequently changing its angle of reflection. The deflection of the reflected laser beam is measured by means of a photodiode. Various experimental configurations have been applied in the literature, the most common of which are :

1 knife-edge configuration. In this configuration (see Figure 1), the laser beam is focused on the measuring point (at distance f_1 from the focusing lens), passed by through a knife-edge after reflection, and then focused once again on a photodiode by a lens (at distance L from the measuring point). Changes in the reflection angle due to the surface perturbation caused by the acoustic wave in the piece cause the intensity of the beam, not intercepted by the knife edge, to vary at the photodiode. The response V_s at the photodiode due to a variation θ is proportional to the incident light power P_s and can be written [13], referring to the symbols used in the figure, as:

$$V_s = V_0 \theta \frac{8 f_1}{\pi w} \tag{3}$$

where V_0 is the photodiode voltage in the absence of an ultrasonic wave. The photodiode response is independent of the angle of incidence of the laser beam and the distance L.

2 double photodiode (see Figure 2) or position-sensitive photodetector configuration. Conceptually this configuration is analogous to the first, only the last part on how the reflected beam is detected changes: there is no knife edge but the beam is split into two, each focused on a photodiode. This doubles the sensitivity of the system as, when the reflection angle varies, the positive light intensity variation on one photodetector is added to the negative variation on the other.

The parameters that chiefly influence the measurement are reported below:

a the quality of the surface on which the laser beam insists; the more reflective the surface, the lower the scattering, and the greater the sensitivity. High scattering produces a high aperture reflected cone, requiring large diameter lenses to collect all the reflected light.

b the size of the beam waist, i.e. the diameter of the beam at the focal point, which impinges on the surface of the piece. Since the ultrasonic wave produces a perturbation of the piece surface with a sinusoidal geometry, it is necessary for the diameter of the incident beam to be much smaller than the ultrasonic wavelength, so that all the beam is reflected in one direction. If, on the other hand, the beam has a large diameter compared to the ultrasonic wavelength, individual areas of the spot will interact with areas of the surface that have different slopes and will be reflected at different angles, reducing the sensitivity of the method. The system actually behaves like a low-pass filter, because the high frequencies (with small wavelength compared to the beam diameter) are averaged and with almost zero amplitude;

c if moving the measuring system changes the distance from the surface, the beam spot no longer irradiates the surface in the same way, varying the spot diameter, the diameter of

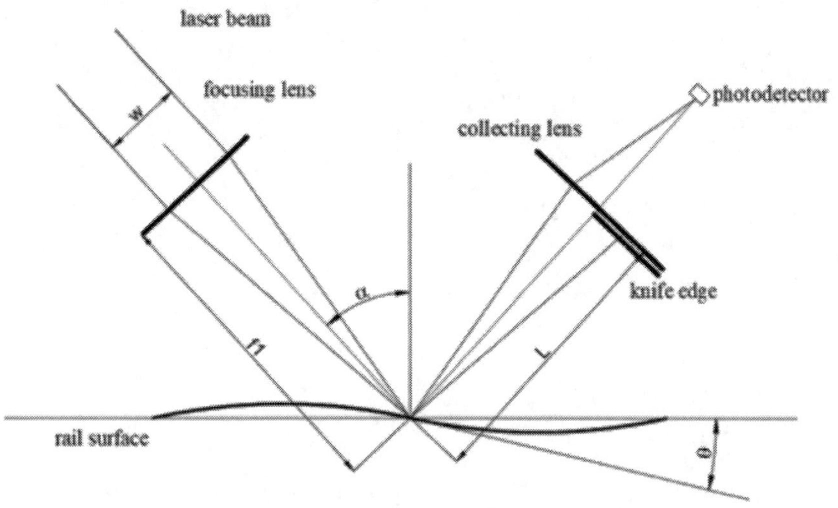

Figure 1. Configuration of the OBD technique with the knife-edge

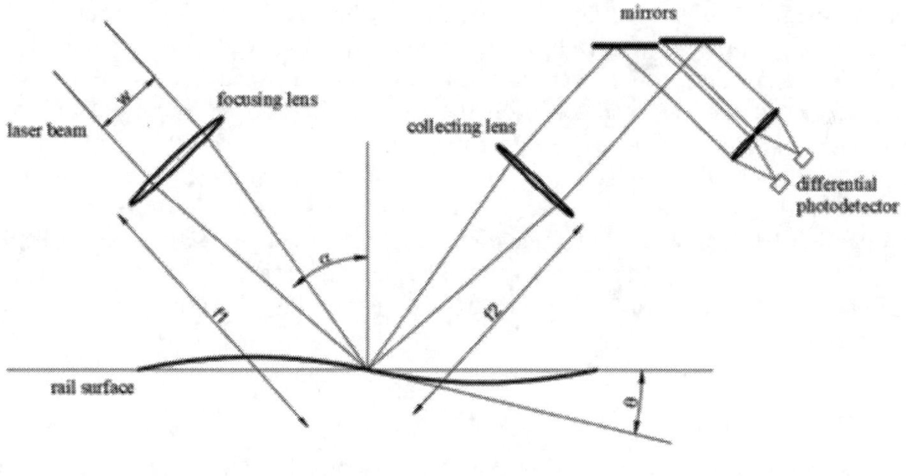

Figure 2. Configuration of the OBD technique with the differential photodiode

the reflected incident cone on the final lens and also the position of the spot on the lens itself. Some precision is therefore required in adjusting the distance from the spot to focus the beam on the surface.

d it can be seen from Equation (1) that, the smaller the beam of the initial laser beam and the greater the focal length of the converging lens, the greater will be the response to the photodiode, the θ angle being the same. However, to have a small beam waist, it is necessary to have a starting laser beam with a large diameter and a small focal length, in contrast with the previously reported points;

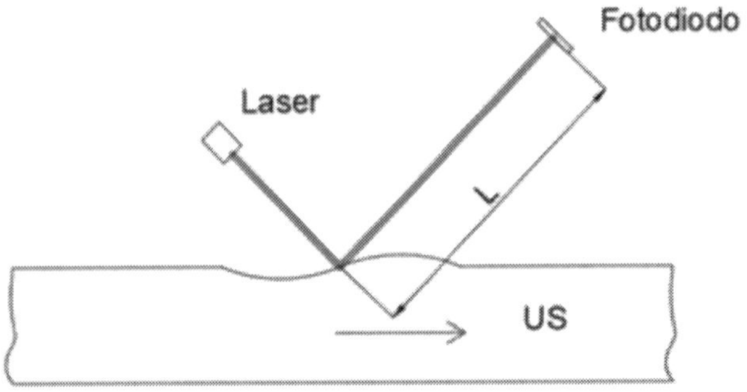

Figure 3. Employed configuration of the OBD technique

e the response does not depend on the distance L at which the lens collecting the reflected beam is placed (i.e., given the geometrical characteristics of the incident beam, the response remains constant);

f the θ angle. The greater the wavelength, the smaller the θ angle; conversely, the greater the amplitude of the wave, the greater the θ angle and thus, by Equation (1), also the response of the system.

The main advantage of using the above configurations lies in the fact that high accuracy is not required in the positioning of the photodiode and the lens that collects the reflected beam. If the latter is sufficiently larger than the laser spot insisting on it, the system is stable even at variations in the reflection angle on the surface, due to the roughness of the surface. In the present work, due to the high scattering of the signal on the rail surface and to limit the diameter of the lens that collects the reflected beam, a different configuration of the OBD technique was used, illustrated in Figure 3. The reflected beam is sent directly to a position-sensitive photodetector without a converging lens, which is able to detect the beam displacement, with twice the sensitivity of the knife-edge. This configuration, besides being simpler, has the advantage of having a distance-dependent response to the photodiode L according to the following Equation:

$$V_s = V_0 \theta \frac{8L}{\pi w} \tag{4}$$

where w is the diameter of the laser beam. Comparing Equations (1)-(2) and considering that a small beam waist requires a small f_1 focal length, it is evident a higher sensitivity can be achieved with the employed experimental layout compared to literature arrangements. In fact, by using a collimated beam, a high L length can be used and sensitivity can be improved. Generally, the laser beam has a divergence with angle 2, and the response to the photodiode becomes:

$$V_s = V_0 \theta \frac{8L}{\pi(2\beta L + w)} \tag{5}$$

The beam divergence limits the distance L at which the photodiode can be positioned, as the spot diameter must always be smaller than the diameter of the photodiode. In this configuration, the incident beam, not being focused at the spot where it insists on the piece, has a non-negligible diameter, in our case 0.7 mm. The considered ultrasonic wave has a frequency of 1 MHz, with a wavelength of approximately 3 mm. The response of the photodiode therefore depends on the

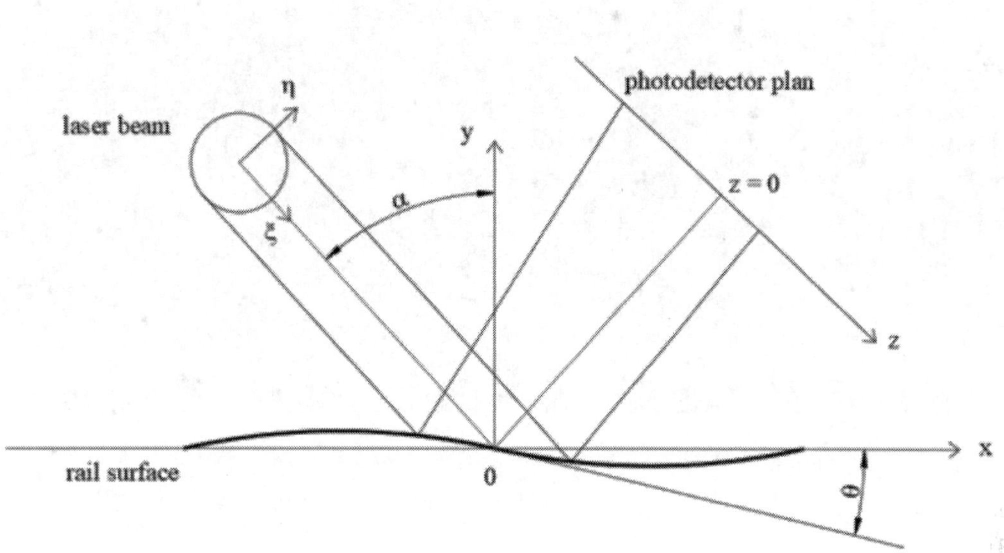

Figure 4. Reflection of a finite diameter laser beam on a surface subjected to ultrasonic propagation

ratio w/λ between the diameter of the incident beam and the acoustic wavelength. Referring to Figure 4, we consider a Gaussian beam whose intensity varies according to the following equation:

$$I(\xi, \eta) = \frac{P_0}{\pi w^2} e^{-\frac{\eta^2+\xi^2}{w^2}} \tag{6}$$

In the absence of the acoustic wave, the ray of coordinates (η=0,ξ=0), hits the surface at the x= 0 point with an angle of incidence and is reflected back to the photodiode at the z= 0 point. The surface acoustic wave perturbs the surface according to a sinusoidal profile that can be described as:

$$y(x,t) = A\sin(kx - \omega t) \tag{7}$$

where $k = 2\pi/\lambda$ is the angular wave number. At a generic time, the surface at point x= 0 is inclined at an angle θ to the undisturbed horizontal surface. Given the small value of the angle θ, it can be written:

$$\theta \approx \tan\theta = \frac{dy(x,t)}{dx} = kA\cos(kx - \omega t) = kA\cos(-\omega t) \tag{8}$$

The incident ray at the point x= 0 in the presence of the angle is reflected back to the photodiode at a distance z:

$$z(0,t) = 2L\theta = 2LkA\cos(-\omega t) \tag{9}$$

A ray of co-ordinates (η,ξ) incident at a generic point x is reflected back to the photodiode at point:

$$z(x,t) = 2LkA\cos(kx - \omega t) + x\cos\alpha \tag{10}$$

IOP Conf. Series: Materials Science and Engineering 1275 (2023) 012046 doi:10.1088/1757-899X/1275/1/012046

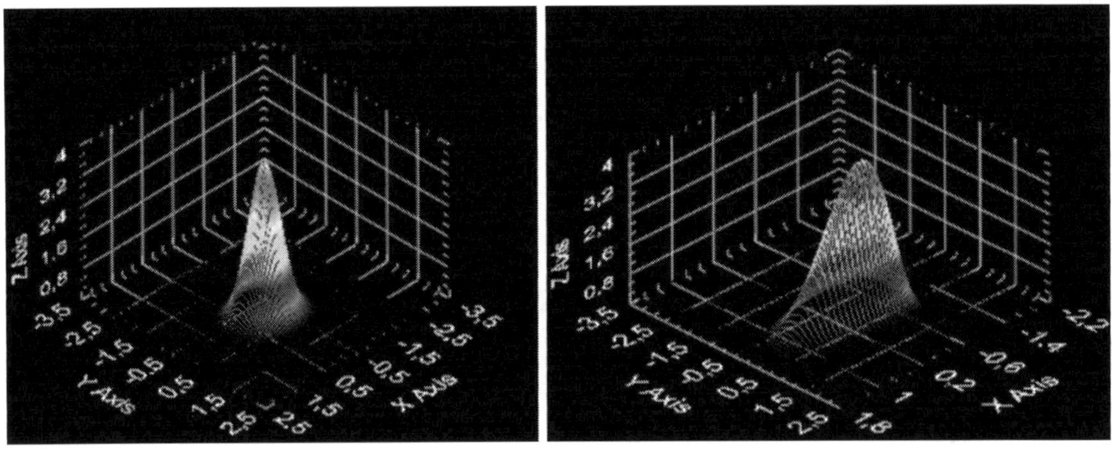

Figure 5. Distribution of light intensity in the optical beam before (left) and after (right) the reflection on the perturbed surface of the acoustic wave, obtained from numeric simulations

If one considers the position-sensitive photodetector consisting of two separate sensitive elements at the point $z = 0$, one can numerically simulate its response by assigning each ray its intensity $I(\eta,\xi)$ and the position $z(x,t)$ where it is reflected. Generally, as x varies, there are different angles and thus different z-positions where the ray impinges on the photodiode; as a consequence, one will obtain profiles of the beam incident on the photodiode no longer with a Gaussian intensity distribution. Figure 5 shows, for example, the shape of the source beam and that at the photodiode after reflection on the surface of the workpiece perturbed by an acoustic wave, at a given time, obtained from numerical simulations.

From the simulations carried out, the response of the photodiode appears to be dependent not only on the distance L and the acoustic wavelength λ, but also on the diameter of the incident beam w and, slightly, on the angle of incidence α. In Figure 6, the response to the photodetector as w and λ vary is shown. Responses of less than 1 are hardly visible, taking into account the noise on the signal. It is evident that the greatest responses are obtained with high ultrasonic frequencies (small wavelengths) and small laser beam diameters.

3. Application of the OBD technique for ToF measurement on the rail

ToF measurement tests were carried out on a piece of rail, performing the measurements on the rolling surface. The ultrasonic waves were generated with a piezoelectric contact probe with a frequency of 1 MHz, with a plexiglass supporting wedge inclined at the angle necessary to generate a surface ultrasound. The laser used is a TOPTICA iBeam Smart 640 with a power of 2 mW, a wavelength of 640 nm and a beam collimated on an elliptical spot of 1.2 x 0.7 mm^2 (measured by a beam profiler), with a divergence of 1 mrad.

The photodiode used is a Red Enhanced Quad Cell Silicon Photodiode (SD 197-23-21-014) developed by Luna Optoelectronics. The cells are connected to create a bi-cell detector with the divider aligned perpendicular to the deflection of the laser beam. The bi-cell is connected to the detection circuit consisting of two AD817 operational amplifiers and passive components: this converts the currents received by the sensor and amplifies the difference, so as to obtain a signal directly proportional to the displacement of the laser beam. The resulting signal is pre-processed with a Brüel & Kjær 2638 amplifier/filter. The filter is a bandpass filter from 300 kHz to 2 MHz.

The OBD configuration used is as in Figure 3, with a distance between laser and surface of 15 cm and a distance L between surface and photodiode of 30 cm (see Figure 7), which is the

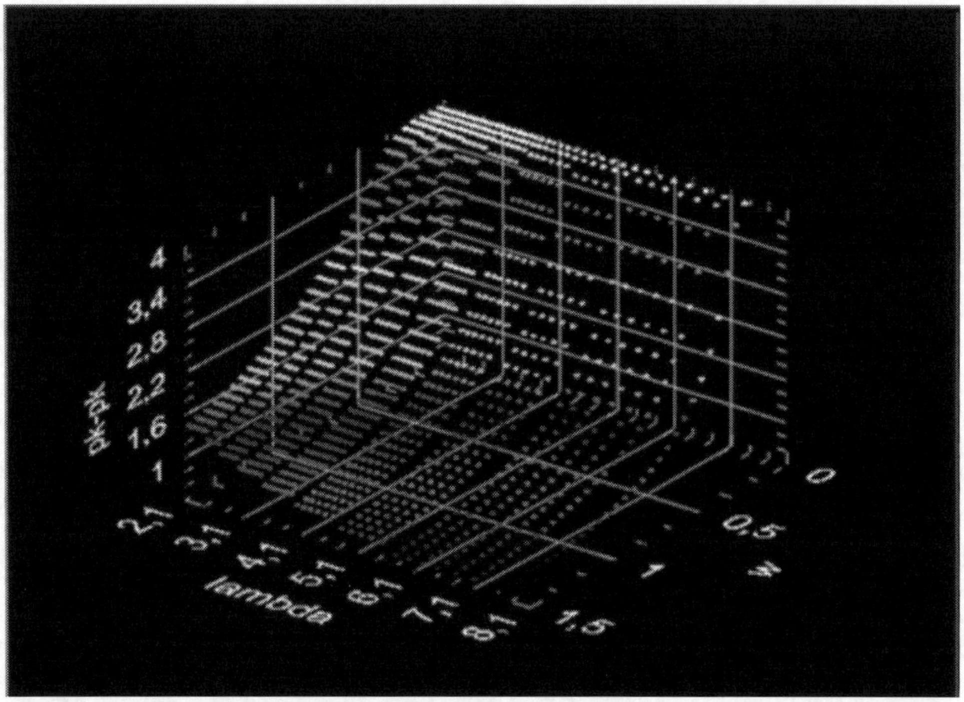

Figure 6. Response of the system in terms of peak-to-peak amplitude of the signal, as a function of the beam diameter w and the acoustic wavelength λ

maximum distance that can be used, taking into account beam divergence and scattering in relation to the diameter of the photodiode.

The surface, although apparently smooth and polished for the passage of trains, needs a light cleaning with 600-grit sandpaper to reduce the scattering of the laser beam and make it more reflective. This causes difficulties for the application of the technique on a moving measurement system, as it would require a system upstream of the measurement system to clean the surface. Figure 8 shows an example of a signal received after cleaning the surface by applying 512 ensemble averages.

Since the tests were carried out on a rail which is not subjected to external constraints, evaluations of the velocity variation due to the stress state can only be first carried out indirectly: the error on the calculation of the speed variation caused by the acoustoelastic effect is proportional to the error made on the ToF measurement. The error on the measurement of the ToF of the ultrasonic wave was assessed by considering the electronic noise at the photodiode (depending on electromagnetic interference by the laser circuit) as well as the distance between laser, photodiode, and piece:

a For electronic noise at the photodiode, it is useful to average the signal and the result depends on the type of processing performed to derive the ToF. If a threshold method is used, the higher the signal frequency, the steeper the rising edge of the signal over time and thus the less influence of noise on the calculation. The frequency of the received signal, however, depends not only on the frequency of the acoustic wave but also on the ratio of the beam diameter to the acoustic wavelength w/λ. As already indicated, the system behaves like a low-pass filter. Experimentally and from the numerical simulations performed, it can be seen that wavelengths of less than 4-5 times the beam diameter are filtered out. With

Figure 7. Experimental layout

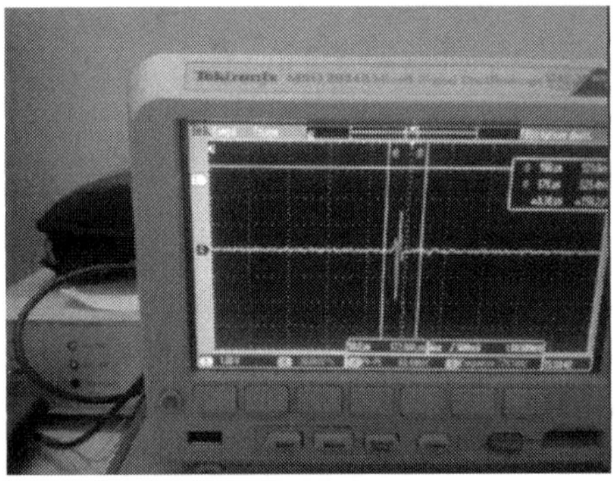

Figure 8. Example of signal received with the layout in Figure 7, performing 512 ensemble averages, with a signal-to-noise ratio of approximately 18

the considered beam, the maximum visible frequencies are around 800 kHz. Figure 9 shows the trend, as a function of the number of averages taken on the signal, of the standard deviation calculated on 10 ToF measurements. The ToF was calculated using:

- the thresholding method directly on the averaged signal;
- the thresholding method on the signal envelope obtained with the Hilbert transform;
- the thresholding method applied on the averaged signal, previously filtered with the Gabor transform;
- the cross-correlation between the averaged signal and a reference signal;
- the cross-correlation between a Gabor-filtered signal and a reference signal.

The filtering of the signal with the Gabor transform is performed with single-level hard threshold in an iterative manner, varying the threshold until the amplitude of the signal in a noise-only window is less than one-tenth of that of the original signal. As can be

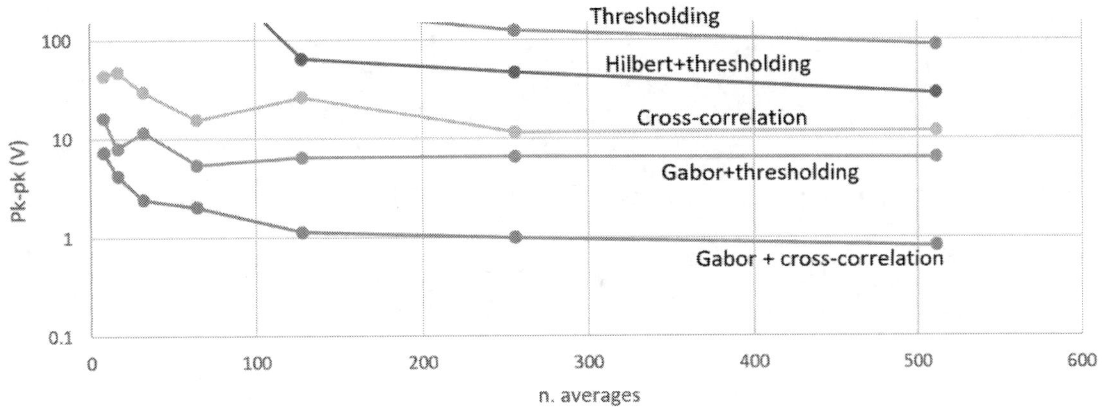

Figure 9. Trend of the standard deviation as a function of the number of averages on the signal and for different techniques of ToF calculation,, calculated based on 10 ToF measures

seen from the figure, applying cross-correlation to the filtered signal gives the best results, and even with a low number of averages, changes in ToF limited to a few ns are obtained. Methods based on simple thresholding without prior signal processing are not applicable without a high number of averages on the signal. By using crosscorrelation, the frequency of the ultrasonic signal is of less importance and even lower frequencies can be used, with the advantage of having less signal attenuation in the rail and thus a longer measurement length. As an example, on a measurement length of 1 m between the ultrasonic source and the detection point, an error of 1 ns in the time measurement results in an error on the voltage of 0.4 MPa, i.e., 0.17 °C in the neutral temperature variation. Appropriate signal processing therefore enables one to use the technique without averaging the signal, or to carry out a limited number of averages, compatible with a moving measurement system.

b For what regards the distance between laser, photodiode, and piece, it must be considered that the rolling surface of the rail has a roughness whose limit is fixed by Regulations, as the ISO 3095. Thus, by moving the measuring system along the rail, variations in the distance between the point of incidence and the laser can occur due to the roughness of the rolling surface, as illustrated in Figure 10; these variations cause z-shifts in the signal at the photodiode. A displacement of the surface d generates a displacement $d/\cos\alpha$ of the beam from the centre of the photodiode, which is read as a variation caused by a different reflection angle of the beam. Similar variations in z occur because of a change in the slope of the surface as a consequence of the roughness. However, even assuming a moving measurement system, variations in z due to roughness would be 'slow' variations, compared to the speed at which the system is moving; in the worst case, i.e., roughness wavelengths of 3 mm (below this length, with the usual laser beam diameters, the signal is filtered out), these give rise to signals with a limited frequency of a few thousand Hz. These variations would then be easily filtered out, compared to the ultrasonic signal, which has a frequency three orders of magnitude higher. Such z-variations are contained within a few tens of μm and mean that, in the absence of the ultrasonic wave, the signal from the photodiode is no longer zeroed, i.e., zero drifts occur on which the high-frequency ultrasonic signal is modulated. This aspect has no relevance for evaluating the ToF.

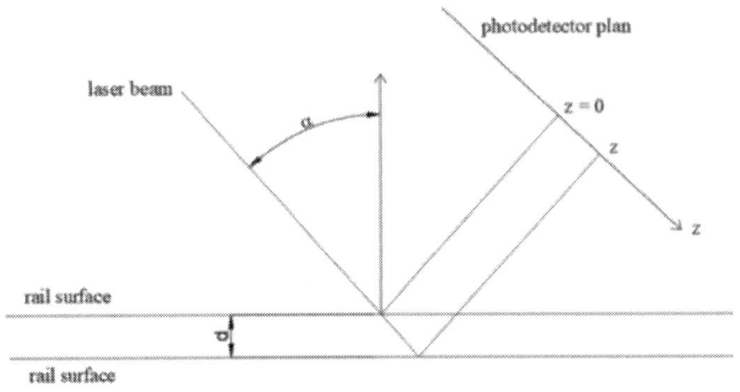

Figure 10. Variation of the z distance of reflection for the laser beam as a function of the distance between the laser and the rail surface

4. Conclusions

In the present work, the Optical Beam Deflection (OBD) technique for measuring the ToF of ultrasonic surface waves was considered. The application is aimed at the measurement of the stress state and neutral temperature on railway tracks. The technique allows ultrasonic waves to be detected without direct contact with the rail, and is therefore potentially suitable for measurements with mobile or otherwise non-fixed measuring systems. The analysis was conducted by considering different experimental arrangements and researching the most influential parameters on the system's response, to optimise its sensitivity. The quality of the signal received and the error on the ToF measurement, which propagates linearly on the measurement of the stress, was also verified to assess the applicability of the technique to moving measurement systems. The results obtained, which are still to be considered preliminary, show some limitations of the technique, mainly because of the poor reflectivity of the rolling surface of the rail. For an optimal application of the technique, at least a light cleaning of the surface with sandpaper is necessary beforehand to make it more reflective. On the other hand, the technique has several advantages, including the simplicity of the experimental apparatus, which translates in the low cost of the equipment, a sufficiently robust experimental setup with regard to the geometry of the assembly of the various components, and a good sensitivity with a high signal-to-noise ratio; as a result, the measurements can be obtained without averaging the signal or by averaging a limited number of signals. This is an indispensable quality to enable measurements with a mobile system.

References

[1] Vnenk P and Culek B 2017 *Acta Polytechnica CTU Proceedings* **11** 91
[2] Bray D E 2000 *Proceedings of the 15th World Conference on Non-Destructive Testing (Rome)* pp 1209–1215
[3] MacLauchlan D T, Burns L R and Alers GA 1988 *Review of Progress in Quantitative Nondestructive Evaluation* pp 1399–1404.
[4] Duan X, Yu Z, Zhu L and Xu X 2016 *WIT Transactions on The Built Environment* **162** pp 239–249.
[5] Szelazek J 1998 *NDTnet* **3** 6 pp 1–8
[6] Vangi D and Virga A 2007 *Experimental mechanics* **47** 5 pp 617–623
[7] Vangi D and Virga A 2009 *Railway Transportation: Policies, Technology and Perspectives*
[8] Xiangyu D, Liqiang Z, Zujun Y and Xining X 2020 *Periodica Polytechnica Transportation Engineering* **48** pp 45–51
[9] Chandrasekaran N and Salama K 1984 *Nondestructive Methods for Material Property Determination* pp 393–403
[10] Lüthi T 1990 *NDT international* **23** 6 pp 351–356

[11] Hermans R I, Dueck B, Ndieyira J W, McKendry R A and Aeppli G 2016 *Scientific reports* **6** 1 pp 1–12
[12] Oliveira V, Polushkin NI, Conde O and Vilar R 2012 *Optics & Laser Technology* **44** 7 pp 2072–2075
[13] Noui L and Dewhurst RJ 1993 *Ultrasonics* **31** 6 pp 425–432.

Effective ways to compare two families of freight trains

L Cantone
University of Rome Tor Vergata, Dept. of Enterprise Engineering "Mario Lucertini", Via del Politecnico 1, 00133, Rome, Italy

E-mail: Luciano.Cantone@uniroma2.it

Abstract. The paper shows several use cases of the "relative approach" method, envisaged by international recommendations, for evaluating the safety of two families of freight trains. This is relevant when a decision is needed about a new technology or a new operating mode of freight trains to keep the same safety as existing trains. The paper discusses the heterogeneity of the database of trains and provides examples of the classic application of the "relative approach", highlighting the effect of train operation. Furthermore, it shows the application to DAC and radio-controlled Traction Units. Finally, it introduces the "Iterative Proportional" algorithm to generate a freight train from an existing one just by changing the hauled mass of each wagon. This algorithm is helpful when the "relative approach" is applied to trainsets having articulated wagons and running with a similar braking regime.

1. Introduction

Unlike typical passenger trains, freight trains have wagons that do not brake simultaneously. The freight trains still predominantly use a braking system derived by George Westinghouse. They usually miss the electrification and the second pipe used in electro-pneumatic (EP) brake or electronically controlled pneumatic (ECP) brake. One of the main benefits of such devices is the possibility of reducing the in-train or the Longitudinal Forces (LFs) during braking [1], [2]. The reduction of LFs is critical for safety and efficiency reasons: high Longitudinal Compressive Forces (LFC) can cause train derailment; high Longitudinal Tensile Forces (LTF) can cause train disruption, i.e., the division of the train into two parts and the need for a recovery train.

The accurate determination of LFs is a difficult task since it requires the contemporary solution of the pneumatic problem (determination of air pressure in the braking system) and mechanical problem (determination of the relative movement of wagons along the track). The reference [3] extensively reviews the problem of computing the in-train forces, considering the numerical solvers, the connection models of rail vehicles, the traction and dynamic brake models and other relevant aspects of Longitudinal Train Dynamics (LTD). [4] benchmarks several worldwide codes, showing results differences depending on the simulation scenario; however, such a benchmark neglects the pneumatic issue because of the relevant differences among the LTD codes.

The LFs penalize the freight train efficiency more than the infrastructure limitation, or, in other words, they do not allow extracting all the capacity from current railway infrastructures. Several European projects have been launched in the past years to overcome such problems. The references [5]-[7] report the LFs of trains equipped with Distributed Power System, i.e. a train in which the Traction Units (TUs) communicate between them by radio and can vent the brake pipe. More recently, within Horizon Europe, the joint undertaking programme Europe's Rail has launched the Digital

Content from this work may be used under the terms of the Creative Commons Attribution 3.0 licence. Any further distribution of this work must maintain attribution to the author(s) and the title of the work, journal citation and DOI.
Published under licence by IOP Publishing Ltd

Automatic Coupler (DAC) to overcome all the barriers that prevent the freight trains from using all the network capacity [8][9]

The UIC software TrainDy solves both pneumatic and mechanic problems and satisfactorily matches experimental data as reported in [10]. This software is suggested by the IRS 40421 [11] to compute the LTD of freight trains. The IRS 40421 describes the "relative approach" method, which provides an effective way to compare two families of freight trains and establish relative safety; this method is also recalled by the European Railway Agency (ERA) in [12], which provides acceptable means of compliance for freight trains.

After describing the "relative approach" method in §2, the paper shows several use cases of the "relative approach" method: the classic application, the application to DAC and DPS system. Then it introduces the Iterative Proportional ("IP") algorithm in 3.6. as a way to handle the articulated wagons in new trainsets. Articulated wagons are not explicitly handled by [11], and the paper describes a way to handle them appropriately; §3.6.1. and 3.6.2. provide ways for their usage applying the "relative approach" method.

2. The "relative approach" method

This paper applies the "relative approach" described in the IRS 40421 [11]; the recent document [12], which provides acceptable means of compliance for freight trains, recalls this method. The "relative approach" compares a new train family with an accepted train family in terms of LCF and compares the number of potential derailments of these two families. If this number is lower for the new train family, it can circulate safely. Trainsets respecting the IRS 40421 are examples of accepted train families, e.g., G-braked trains (all vehicles brake in position "G") having a mass up to 4000 T and length up to 700 m (traction unit(s) excluded). An example of a new train family is the one in which a radio-controlled traction unit works at the trainset end. The comparison requires the checking of a series of parameters, noting the value in reference and new train family (or system), describing the (statistical) changes from reference to the new system and commenting on them (optionally). An example of parameters to check is:

- Train composition (arrangement of vehicles) in terms of mass and length.
- Brake equipment of vehicles, e.g., considering empty/load devices and auto-continuous devices.
- The means the braking is controlled: e.g., pneumatically, via radio, by wire.
- Braking regime, P-braked or G-braked trains
- Type of wagons, e.g., the wagons are the same, but the permissible LCF is different, as it is for DAC
- Initial braking speed
- Braking mode: emergency braking or full-service braking, from coasting or acceleration (traction). The most common braking mode is electro-dynamic braking, followed by service braking, full-service braking, emergency braking from coasting (cruising), and emergency braking from acceleration. The train derailment and disruption risks follow
- Wagon load, e.g., positioning of empty wagons at the train end.
- Coupling feature, e.g., the new trainsets have the same wagons but different couplings.

As much as possible, the changes must be limited to the parameters that want to be investigated. In the example of a radio-controlled TU, the reference trains must be generated with a TU at the train end behaving as a wagon; on the contrary, this traction unit in the new system is radio-controlled. In this way, the behaviour of the traction unit is the only difference; see §3.5. for another example of applying the "relative approach" in radio-controlled TUs.

In this paper, the algorithm used to generate a family of trains is that described in [13], except when the algorithm described in §3.6. is employed.

IOP Conf. Series: Materials Science and Engineering 1275 (2023) 012047 doi:10.1088/1757-899X/1275/1/012047

3. Use cases of the "relative approach" method

The following subsections report several use cases of the "relative approach" method. Primary wagon data are listed in Table 1 to ease the recalculation; cast iron is the friction material for block brakes. Figure 1 reports buffers and draw gears data; buffers are "side buffers" if not specified differently.

Table 1. Main wagon data

Name	Length [m]	Tare [T]	Mass [T] vs % of braked weight [%]	Buffer	Draw gear
Bogie1	16.94	18.98	$\begin{bmatrix} 18 & 57 & 69.5 \\ 100 & 100 & 82 \end{bmatrix}$	Buff_L	Draw_L
Bogie2	19.9	24	$\begin{bmatrix} 24 & 58 & 80 \\ 100 & 100 & 72.5 \end{bmatrix}$	Buff_NL	Draw_NL
Art1	34	34.7	$\begin{bmatrix} 34 & 88 & 135 \\ 100 & 100 & 65.1 \end{bmatrix}$	Buff_L	Draw_L
Art2	26.6	28.7	$\begin{bmatrix} 28.7 & 109 & 135 \\ 100 & 100 & 80.7 \end{bmatrix}$	Buff_NL	Draw_NL

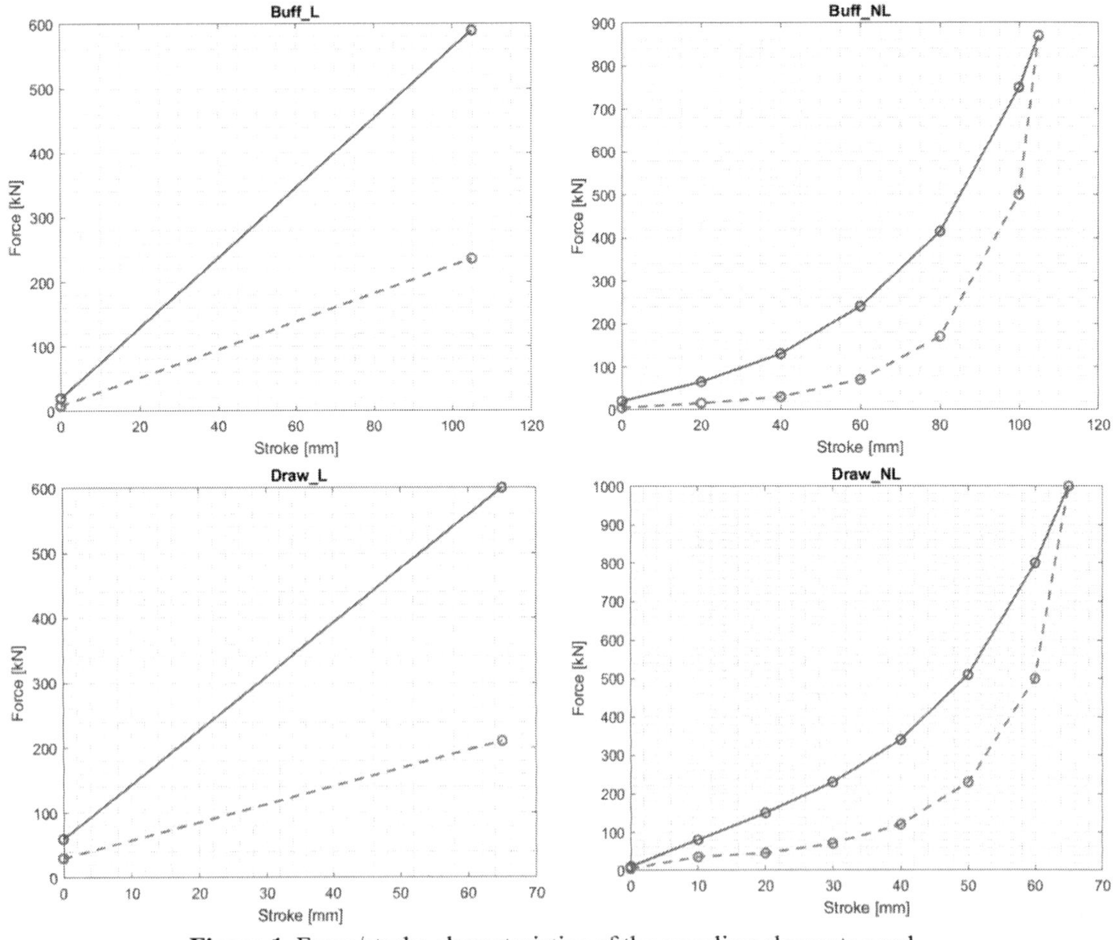

Figure 1. Force/stroke characteristics of the coupling elements used.

The results shown in this paper employ the permissible values for the LCF as in [11]. In contrast, the paper assumes for permissible LTF (PLTF) a fixed value of 550 kN since this value is the threshold for the activation of the fatigue process in typical draw gears.

3.1. The payload for articulated wagons

Before explicitly addressing different use cases for the "relative approach" method, it is worthwhile to mention that articulated wagons can be unevenly loaded, affecting the payload used when the PLCF is of concern.

The payload of articulated wagons is divided into two parts, loading at the maximum possible (according to a maximum mass per axle of 22.5 t) one of the two parts and considering as payload, for the IRS 40421 extrapolation, the remaining payload; in this way, the calculation of PLCF is conservative. In formulas:

- $m_{pa} = 22.5$
- $L = L_H + L_L$. The total wagon payload (L) is the sum of the load of the heavy part (L_H) and the load of the light part (L_L).
- n_a is the number of axles of the wagon
- T is the tare of the wagon (wagon mass is $M = L + T$, of course)

$$L_H = \min\left(L,\ \frac{m_{pa} \cdot n_a - T}{2}\right) \tag{1}$$

$$L_L = L - L_H \tag{2}$$

L_L is used as the wagon payload for the extrapolation rules of IRS 40421. The PLCF for articulated wagons is computed parametrically, considering them as axle wagons (conservatively), as bogie wagons (as an assumption) and with an intermediate value. §3.6.1. and §3.6.2. show the application of this parametrization.

3.2. Heterogeneity issue of the trains database

The generation of virtual trains requires a database of trains from which to extract the probability distribution of employed wagons and, for each wagon, the payload distribution. If the original train database is homogeneous (it contains similar trains in terms of mass and length), the generated trains match the real trains in terms of percentages of the train derailment and disruption (breaking of the train in two parts because of a draw gear breakage). Figure 2 reports (in (a) and in blue) the cumulative frequency of the LFs (compressive forces have negative values) for 42407 real trains running in GP mode (i.e., the TU brakes in G and the wagons in P). The trains have a mass in the interval 800-1200 T and perform an emergency braking from 30 km/h after a full acceleration (TEB): each dot represents the maximum LF (in modulus) for each train. This figure also reports the maximum ratio with the permissible force for each train in green: negative values for LCF/PLCF and positive for LTF/PLTF. The label reports the percentages of train derailment and disruption between brackets: they are linked to the number of trains having a ratio with the permissible value bigger than 1 (in an absolute sense). Part (b) refers to 1000 virtual trains generated from the trains database: the probabilities of train derailment and disruption are pretty similar. Moreover, the relatively high probability of potential derailment for the actual trains is the result of a conservative calculation: i) the PLCF values in IRS 40421 are conservative; ii) the LTF/PLTF ratio assumes that the wagon subjected to the highest LCF is also running on a track section with a low radius of curvature; iii) the likelihood of TEB is low (around 10^{-4}).

Part (c) refers to 993 real trains having a length between 10 and 620 m and a mass in the interval [10-2000] T; by repeating the generation of virtual trains, as in (d), the results differ more than before. The reason is the heterogeneity of the actual train database. Extracting a small number of trains from the trains database, as in (e), and generating virtual trains from this reduced but more homogeneous database improves the results regarding train derailment and disruption probability (f).

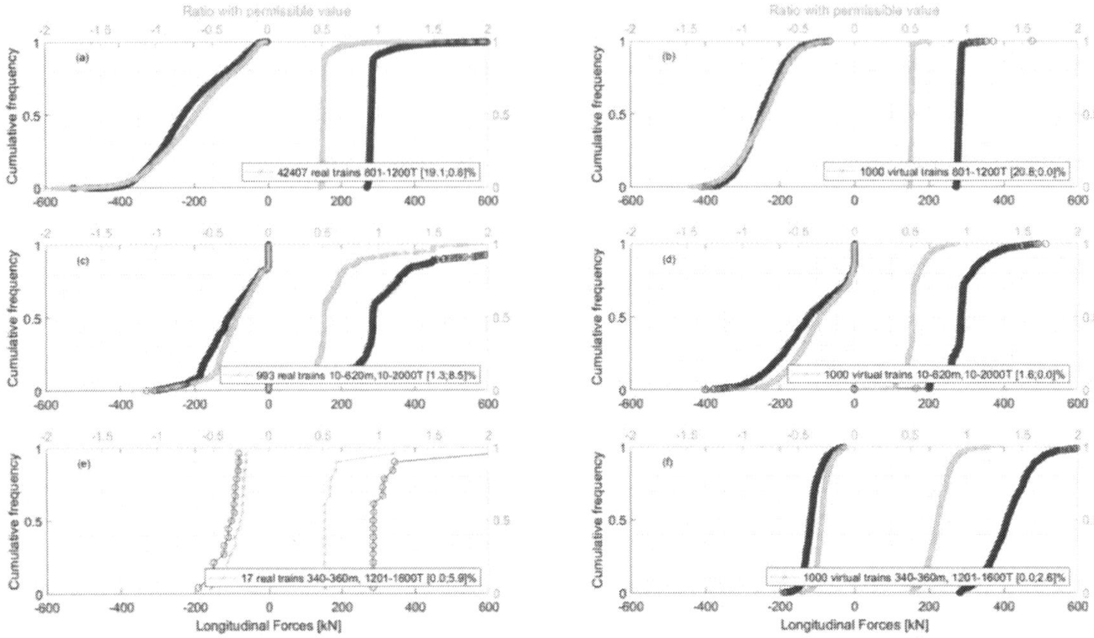

Figure 2. Longitudinal forces and ratio with permissible value for real and generated trains are on the left and right of the figure, respectively.

Consequently, the trains database used to generate the virtual trains must be homogeneous to obtain virtual trains representative of real trains. It is impossible to give general guidance on the heterogeneity of trains; however, trains with the same service (connecting the same places and carrying similar goods) are examples of homogeneous trains.

3.3. A classic application of "relative approach"

This paper considers as reference train families those explicitly allowed by UIC IRS 40421: trains with hauled mass in the range [1200-1600] T with Long Locomotive regime (LL) and trains with hauled mass in the ranges [1600-2500] T and [2500-4000] T in G; all with train lengths in [640-740] m (TU included), but without DAC or EP brake. The first two train families are likely to have empty wagons, which does not happen for the third train family because of the higher hauled mass. Remember that in the LL regime, the first five wagons and the leading TU are in freight position G (G braked); in contrast, the remaining wagons are in passenger position P (P braked). As the results of the following sections show, the higher minimum hauled mass of the last train family is beneficial in terms of safety. Figure 3 part (a) shows the cumulative frequency of LFs, for an emergency braking from 30 km/h in coasting conditions. The number of virtual trains is 1000, which provides a satisfactory evaluation of the mean and standard deviation of LCF (as required by [11]), and the trains contain only bogie wagons. Part (b) shows the ratios of LCF/PLCF and LTF/PLTF, as in Figure 2. Looking at (a) and (b), [2500-4000] T trains have LCF like [1200-1600] T trains in UIC IRS 40421 but have a lower risk of derailment because of higher wagon mass, as also confirmed by the histograms in (c) and (e).

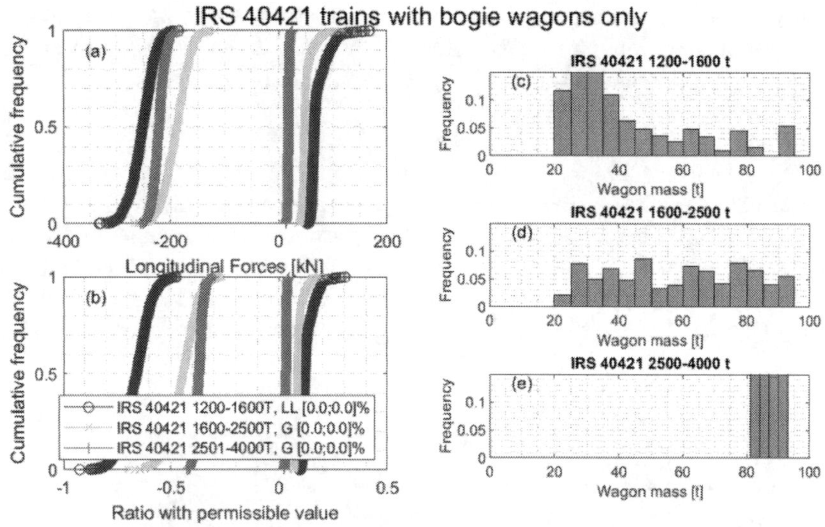

Figure 3. Trains explicitly allowed by IRS 40421.

Figure 3 shows that the trains allowed by the IRS 40421 have a different degree of safety and the safest trains are those in G with a hauled mass between 2500 and 4000 T. These results have been obtained by statistically generating three families of virtual trains, hence applying the "relative approach" classically, i.e., as foreseen by the IRS 40421. The following sections show different ways to apply the "relative approach" method.

3.3.1. Effect of train operation. Before showing different ways to apply the "relative approach" method, this subsection shows that this method's results do not usually depend on the train operation. The train operation (highly) affects the probability of train derailment, but usually, it does not affect the comparison between different train families. If this is not the case, it is better to consider a new train family safer than the reference train family when this statement is true for different train operations, e.g., for emergency braking or full-service braking from coasting and acceleration.

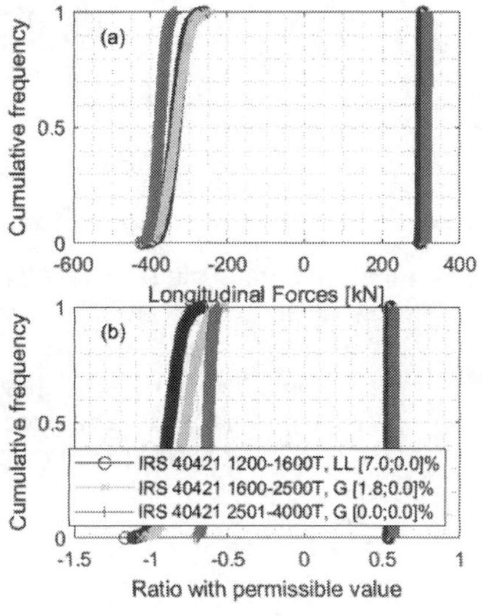

Figure 4. Trains explicitly allowed by IRS 40421, performing an emergency braking from 30 km/h after an acceleration.

Figure 4 shows the LF in (a) and the ratio with the permissible value in (b) for the same trains as Figure 3, but performing an emergency braking from 30 km/h after an acceleration. As anticipated, the safest train family is the one running in G and hauling a mass between 2500 and 4000 T. The main effect of this new train operation is to increase the LFs and the probabilities of a train derailment but not to change the result of the relative approach application, in this case.

3.4. Benefits of DAC

The European DAC Delivery Program (EDDP) from Shift2Rail (S2R) is currently developing the Digital Automatic Coupler (DAC), intending to increase productivity in freight operations in a safe way. Depending on the DAC level (see Figure 5), DAC may reduce LCF generated in operated trains. In the figure, "AC 1" and "AC 2" refer to automatic couplers (AC) that are just "mechanical DAC", and the label "DAC2" is used in this paper. Any level of DAC increases and even doubles the PLCF of the wagons. "AC 1", "AC 2", and "DAC 3" are the same in terms of mechanical characteristics and PLCF.

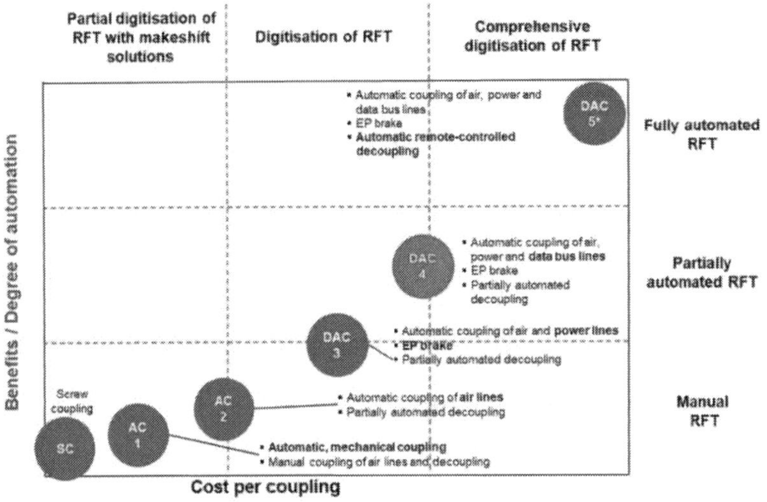

Figure 5. Different levels of train automation depending on DAC type

As DAC is not yet in production, this paper presents results in which the PLCF of wagons equipped with DAC equals 1.5 times the PLCF of corresponding non-DAC wagons. This figure must be confirmed by subsequent studies, even if it is realistic.

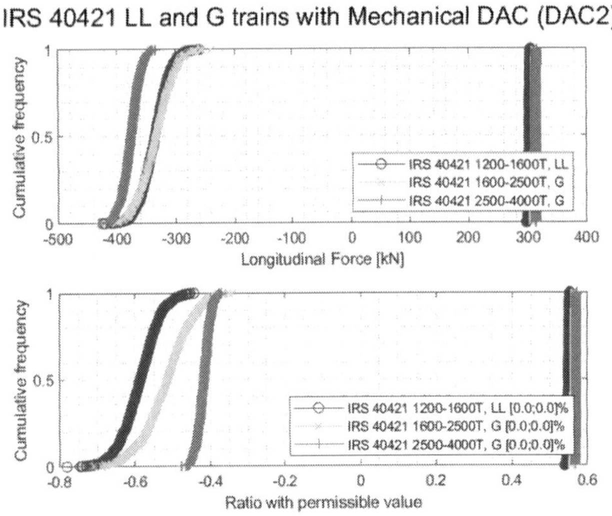

Figure 6. Trains, explicitly allowed by IRS 40421, and performing an emergency braking from 30 km/h after acceleration and equipped with mechanical DAC (DAC2).

Figure 6 shows on top LFs and below the corresponding ratios with permissible values. The trains and the operating conditions are identical to Figure 4, but the PLCF has been increased by DAC2, resulting in a lower probability of train derailment for all train families. Consequently, it is possible to find an upper value bigger than 1600 T for trains with DAC2, keeping the same level of safety and proving the benefits of DAC2.

It is important to note that the results of Figure 6 assume that the mechanical characteristic of DAC is identical to the set "side buffers plus draw gear" and that only the wagon PLCF changes. DAC should have substituted the current couplers for a more realistic comparison, but this was not the case since official mechanical DAC characteristics are not yet available. This way, the benefits of just PLCF increment can be promptly and correctly measured.

3.5. Implementation of radio communication in Traction Units

Reference [7] describes the set of simulations performed to study the LFs of an experimental train equipped with three radio-controlled TUs with a length of 640 m and a train mass of 1735 T. To perform this study by applying the "relative approach", the train is simulated considering the 2nd and 3rd TU as wagons during the braking (i.e., they do not discharge the brake pipe); hence as a train allowed to traffic (REF). The trains with radio-controlled TUs (DPS) have three different layouts:

- LWL indicates a trainset where the active TUs are at the beginning and the end. The TU in the middle is not active.
- LWLW indicates a trainset where the active TUs are at the beginning and middle. The TU at the end is not active.
- LWLWL indicates a trainset in which all TUs are active.

Randomly changing the wagon's position and performing the simulation of LF for an emergency braking from 30 km/h and after an acceleration, it is possible to compare the two systems (with and without radio-controlled TU) comparing the same trains but having different technologies for TUs. Figure 7 reports the LF for the REF train on the x-axis for each train and the corresponding LF for the DPS train on y axis: first row refers to LCF, the second to LTF. If the point, representing a train with a specific wagon order, is below the bisectrix, it means that the REF train is worst than the corresponding DPS. Depending on wagon ordering, the DPS train is usually safer than the corresponding REF train, even if, for some wagon arrangements (LWLW and LTF), it is vice-versa.

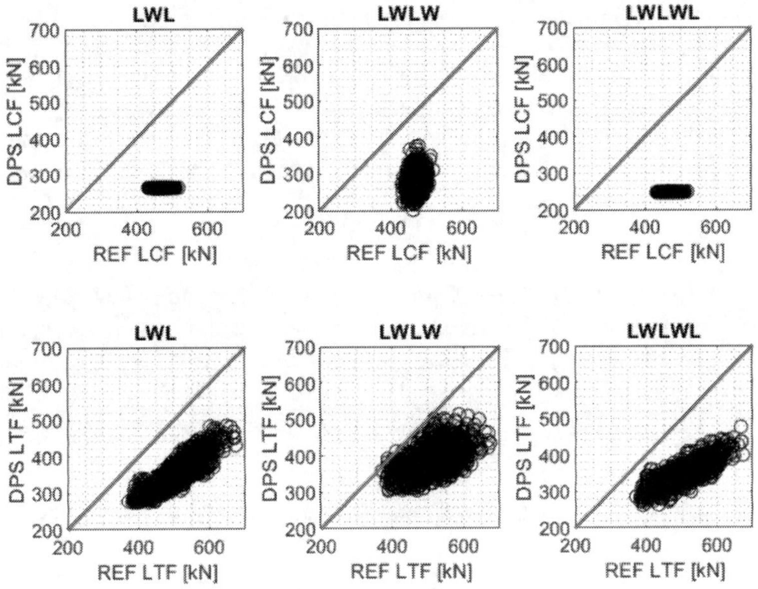

Figure 7. Comparison between classic (REF) trains and trains with radio-controlled traction units (DPS), for different train layouts. 1st row refers to LCF, 2nd to LTF.

It is essential to note that the "relative approach" application differs from the standard way: only one family of trains is generated just by permuting the wagons' position from an initial train configuration. The comparison is achieved by activating or not the radio-controlled communication for DPS and REF, respectively.

3.5.1. Effect of radio technology. The following results compare the LFs of trains formed by coupling two long trains with a total length of up to 1500 m, with a TU placed at the end of the train (the label "SW" applies). The TUs are radio-controlled. The radio link uses GSMR, the current standard, or LTE, as bridge technology for the 5G network (FRMCS). Figure 8 illustrates the probability of potential derailment for IRS 40421 [1200-1600] T trains in LL as blue and red vertical broken lines for TEB and EB, respectively. It also shows trains in which the TUs communicate via GSMR (a) or LTE (b). It is worth noting that the delay in radio communication between the TUs is considered constant for each train: 1.6 s for GSMR and 0.8 s for LTE (see [14]). This time is required to discharge the brake pipe where the guided TU is located and is not just the communication time between the radios. Therefore, the radio technology gives the only difference between the trains in (a) and (b).

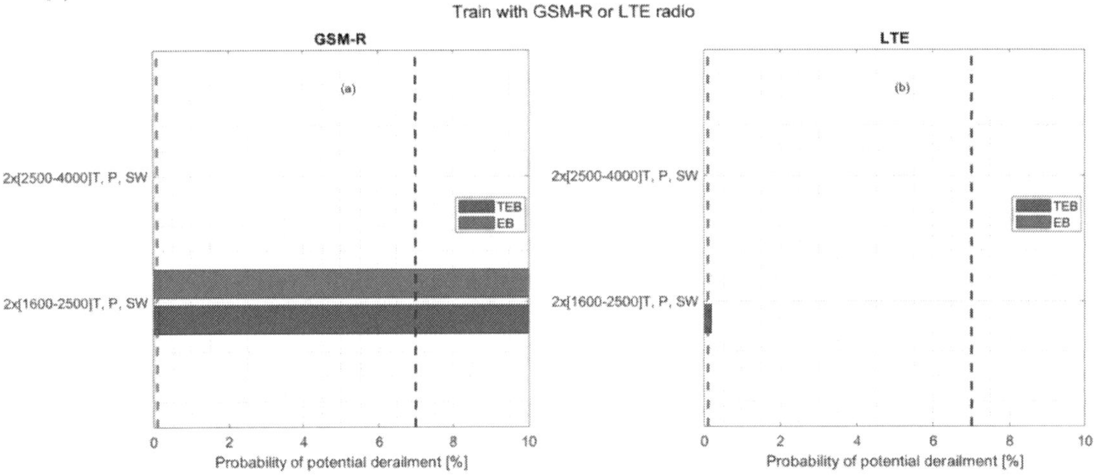

Figure 8. Different probabilities of train derailment according to radio link technology.

As before, there is just one generation of virtual trains; only the radio link technology changes to improve the consistency of the comparison. All families of trains listed in the figure are safer than the reference train family (IRS 40421 LL) independently from the radio technology used, except the trains family formed by coupling two trains running in P, each with a hauled mass of [1600-2500]T and with a 3rd TU at the train end.

3.6. The Iterative Proportional ("IP") Algorithm

This paper proposes an algorithm to change the train mass from an initial value, *m*, to a final value *M*. This algorithm is called Iterative Proportional (IP). The initial mass *m* is the hauled mass of a general train within the reference train family; the new mass *M* is the corresponding mass of the train in the new train family, computed by linear interpolation as:

$$M = M_1 + \frac{M_2 - M_1}{m_2 - m_1}(m - m_1) \tag{3}$$

Where m_1, m_2 and M_1, M_2 are the mass boundaries for reference and new train families, respectively.

Through the "IP" algorithm, it is possible to modify the train mass from *m* to *M*, iteratively. Figure 9 reports the flowchart of this algorithm. If $M > m$, the ratio β between the target mass M and the mass of the new train is computed. If this ratio is 1, the new train already has the target mass, and the algorithm ends. If the ratio is bigger than 1, the ratio between the maximum wagon mass and the current mass is computed: β_{wi}. Then, the minimum of these ratios is computed: β_i. If this ratio is lower than β, it means that there is a few difference in terms of the maximum loaded wagon and its maximum capacity; if this is the case, $\beta = \beta_i$ and the mass of each wagon is incremented as: $m_{wi} = \beta \cdot m_{wi}$; therefore, the new hauled mass is $m^* = \sum m_{wi}$. The process can restart computing $\beta = \frac{M}{m^*}$. With this algorithm, wagons having the higher mass have a mass increment up to their maximum value. In this process, all wagons having a mass lower than their maximum value have a mass increment proportional to their value ($m_{wi} = \beta \cdot m_{wi}$), until the target mass is reached.

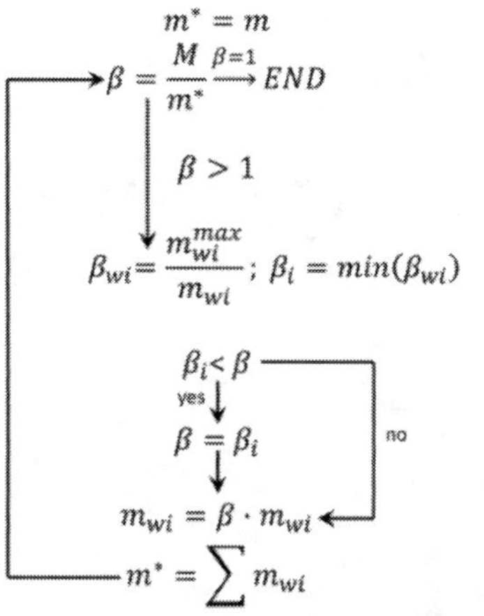

Figure 9. Flowchart of Iterative Proportional (IP) method.

3.6.1. Articulated wagons in the trainset. The IRS 40421 provides explicit values for PLCF only for axle and bogie wagons but not articulated wagons. The results shown in previous sections have proved that the LFs of a train are affected by the wagon type and position. If articulated wagons are in the trainset, the proposed way to apply the "relative approach" is to consider in the reference system the articulated wagons, e.g., trains in the LL regime having a mass between 1200 and 1600 T.

This system is compared in Figure 10 with another train system having a higher mass (1600-1800 T) and obtained applying or not the IP algorithm. The new system has a different braking regime: the Extended Long Locomotive (ELL), with 7 wagons in G (instead of 5) at the beginning of the trainset. The PLCF of articulated wagons are computed following §3.1. The probability of derailment considerably increases for IRS 40421 LL trains, and it is like that of new trains, depending on the assumption made on the PLCF of articulated wagons. If the PLCF of articulated wagons is like that of axle wagons, the new system is equivalent to the reference; otherwise, this statement is not valid, even if this conclusion depends on how the new trains are generated. In Figure 10, the label "noIP" means that the train family has been generated using the classic IRS 40421 algorithm without any guarantee that the number of articulated wagons per train is the same and that they are placed in the same positions along the trainset. Therefore, the results labelled with "noIP" should be discarded.

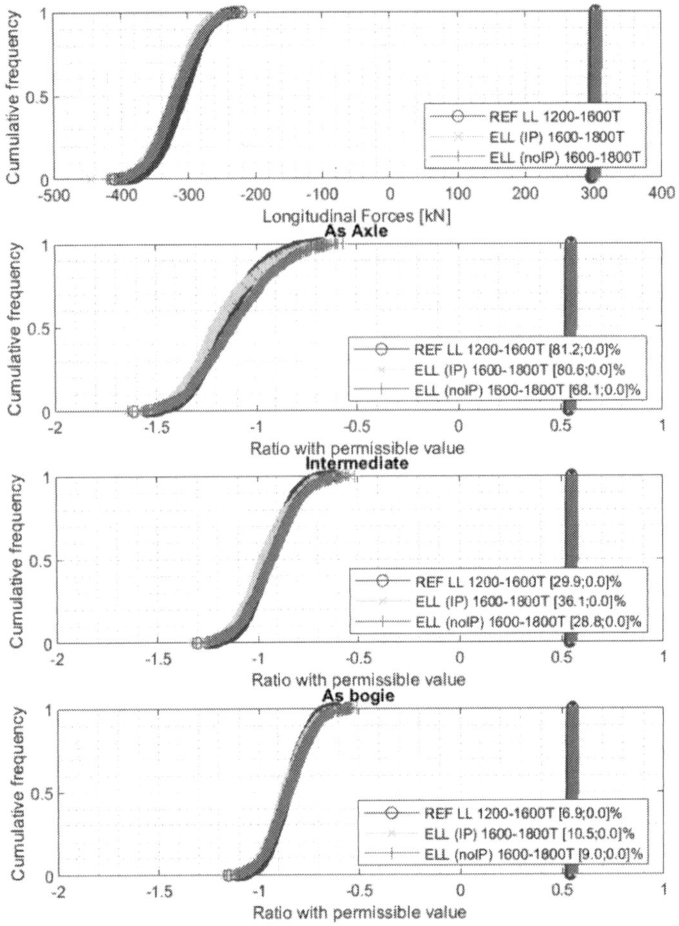

Figure 10. Longitudinal forces for IRS 40421 trains in LL and 1600-1800 T trains in ELL with and without IP algorithm.

3.6.2. Trainset with just articulated wagons. Suppose the trainset is made of only articulated wagons. In that case, the comparison with IRS 40421 LL and ELL (1600-1800T) is given in Figure 11, for emergency braking from coasting (cruising) conditions commanded at 30 km/h.

The figure shows that changing the assumption on the PLCF of articulated wagons, in this case, does not affect the comparison considerably: the mass interval of 1600-1800 T in ELL has the same level of safety as 1200-1600 T in LL. This statement is also true for emergency braking following an acceleration, even if the results are not reported here.

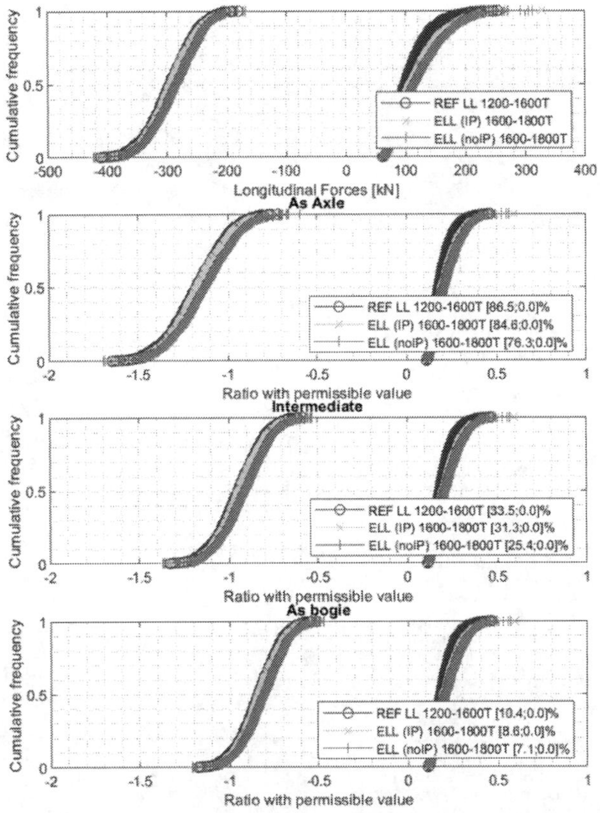

Figure 11. Longitudinal forces for IRS 40421 LL trains, having articulated wagons, applying or not the IP algorithm, compared to 1600-1800 T trains with ELL, performing an emergency braking from 30 km/h.

3.6.3. Application of IP to trains listed by IRS 40421. In this last section, Figure 12 reports the LFs and the ratios with permissible values for IRS 40421 trains, obtained by 1200-1600 LL trains applying ("IP" is used in the label) or not the IP algorithm ("noIP" label is useed). Unlike previous results, the assumption made on the PLCF of articulated wagons does not affect the comparison of train families, and the application of the IP algorithm does not affect the derailment probabilities significantly; therefore is useless. The main reason for this behaviour is that the train families are too different in hauled mass and brake regime; therefore, the number and position of articulated wagons are masked by those other and more relevant differences.

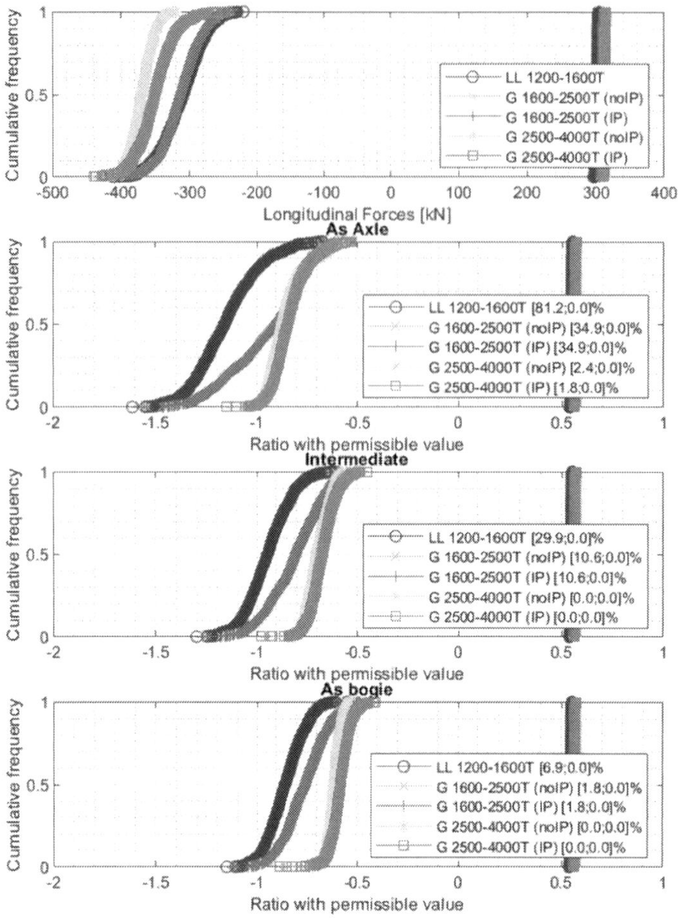

Figure 12. Longitudinal forces for IRS 40421 trains, having articulated wagons, applying or not the IP algorithm.

4. Conclusions

This paper recalls the “relative approach” method envisaged by [11] and [12], which are essential reference documents for Railway Undertakings when a new trainset has to go into service. It provides several use cases of this method application, including the Digital Automatic Couplers (DAC) and radio-controlled traction units. It shows that it is critical to minimize the differences when comparing two families of trains to make the comparison consistent. Moreover, for articulated wagons, which are not explicitly considered by [11], the paper introduces the Iterative Proportional (“IP”) algorithm to generate new trains from a given family of trains, keeping the position and type of wagons but changing the hauled mass. After the introduction of the Extended Long Locomotive (ELL) regime, considering trains with only articulated wagons, the paper finds the equivalence between the IRS 40421 LL trains and [1600-1800] T trains in ELL, using the “IP” algorithm. Finally, the paper shows that the “IP” algorithm is helpful when the two families of trains are similar, e.g., it is not necessary when the comparison happens between LL trains and G braked trains.

References

[1] Mandelbaum R (2008) Brake-by-wire comes to freight trains: electronically controlled pneumatic brakes could help stop accidents and speed cargo. IEEE spectrum 45(2):18.
[2] Aboubakr AK, Volpi M, Shabana AA, et al (2016) Implementation of electronically controlled pneumatic brake formulation in longitudinal train dynamics algorithms. Proc IMechE Part K: J Multi-body Dynamics 230(4):505-526.

[3] Cole C, Spiryagin M, Wu Q and Sun Y Q 2017 Modelling, simulation and applications of longitudinal train dynamics *Veh. Syst. Dyn.* **55** 1498-1571.

[4] Wu Q *et al* 2018 International benchmarking of longitudinal train dynamics simulators: results, *Veh. Syst. Dyn.* **56** 343-65.

[5] Cantone L and Toubol A 2019 Marathon2Operation, a Shift2Rail project for radio communication and simulation of train dynamics for distributed power within long trains *Procedia Structural Integrity* **24** 437-447, https://doi.org/10.1016/j.prostr.2020.02.040.

[6] Cantone L, Laghi D and La Rovere S 2021 A statistical study on long freight trains equipped with radio communication within Shift2Rail *IOP Conf. Ser.: Mater. Sci. Eng.* **1038** 012034

[7] Cantone L 2022 Simulation of freight trains with up to three traction units in radio communication *IOP Conf. Ser.: Mater. Sci. Eng.* **1214** 012039.

[8] EUROPE'S RAIL MASTER PLAN, available at https://shift2rail.org/wp-content/uploads/2021/12/20211216-Master-Plan_agreed-in-princ_clean.pdf

[9] Europe's Rail Joint Undertaking Multi-Annual Work Programme, available at https://shift2rail.org/wp-content/uploads/2021/12/20211222_mawp_v1_agreed-in-principle_clean.pdf

[10] Cantone L 2011 *TrainDy*: the new Union Internationale des Chemins de Fer software for freight train interoperability *Proc. Inst. Mech. Eng. F* **225** 57-70.

[11] IRS 40421, "Rules for the consist and braking of international freight trains", 1st edition, Dec. 2021.

[12] ERA - Operations and traffic management system TSI, Acceptable means of compliance on checks and tests before departure, including brakes and checks during operation, July 2022 https://www.era.europa.eu/sites/default/files/library/docs/opinion-advice/tsi-ope-amoc-tests_checks_braking-v2_en.pdf

[13] Cantone L New Random Trains Generation According to UIC Code 421. *International Journal of Mechanical Engineering and Technology*, **9**(10), 2018, pp. 890-899

[14] Toubol A., 2020 *D2.1 - GSM-R or LTE radio, Design solution*, https://www.marathon2operation.eu/web/pdf/D2p1%20GSM-R%20or%20LTE%20radio,%20Design%20solution.pdf

AUTHOR INDEX